M. Waldschmidt P. Moussa J.-M. Luck C. Itzykson (Eds.)

From Number Theory to Physics

Springer
Berlin
Heidelberg
New York
Barcelona
Budapest
Hong Kong
London
Milan
Paris
Tokyo

M. Waldschmidt P. Moussa
J.-M. Luck C. Itzkyson (Eds.)

From Number Theory to Physics

With Contributions by
P. Cartier J.-B. Bost H. Cohen D. Zagier
R. Gergondey H. M. Stark E. Reyssat
F. Beukers G. Christol M. Senechal A. Katz
J. Bellissard P. Cvitanović J.-C. Yoccoz

With 93 Figures

Springer

Editors

Michel Waldschmidt
Université Pierre et Marie Curie (Paris VI)
Département de Mathématiques
Tour 45-46, 4 Place Jussieu
F-75252 Paris Cedex 05, France

Pierre Moussa
Jean-Marc Luck
Claude Itzykson ✝

Centre d'Etudes de Saclay
Service de Physique Théorique
F-91191 Gif-sur-Yvette Cedex, France

Second Corrected Printing 1995

Lectures given at the meeting 'Number Theory and Physics', held at the 'Centre de Physique', Les Houches, France, March 7-16, 1989.

Photograph on page XII courtesy of Georges Ripka

Cover illustration:
© 1960 M. C. Escher/Cordon Art–Baarn–Holland

Mathematics Subject Classification (1991): 11F, 11G, 11H, 11M, 11R, 11S, 12F, 12H, 12J, 14H, 14K, 19D, 30F, 30G, 33E, 34A, 39B, 42A, 46L, 47B, 51P, 52B, 52C, 55R, 58A, 58F, 70K, 81Q, 82D

ISBN 3-540-53342-7 Springer-Verlag Berlin Heidelberg New York
ISBN 0-387-53342-7 Springer-Verlag New York Berlin Heidelberg

Library of Congress Cataloging-in-Publication Data

From number theory to physics / M. Waldschmidt ... [et al.] (eds.) ;
with contributions by P. Cartier ... [et al.].
p. cm.
"Second corrected printing"--T.p. verso.
"Lectures given at the meeting 'Number Theory and Physics' held at the Centre de physique, Les Houches, France, March 7-16, 1989"--T.p. verso.
Includes bibliographical references and index.
ISBN 3-540-53342-7 (hardcover)
1. Number theory--Congresses. 2. Mathematical physics--Congresses. I. Waldschmidt, Michel, 1946- .
QA251.F76 1992b
512'.7--dc20

Typesetting: Camera-ready output prepared by authors/editors using a Springer TEX macro package
SPIN 10507834 41/3143-5 4 3 2 1 0 Printed on acid-free paper

Preface

The present book contains fourteen expository contributions on various topics connected to Number Theory, or Arithmetics, and its relationships to Theoretical Physics. The first part is mathematically oriented; it deals mostly with elliptic curves, modular forms, zeta functions, Galois theory, Riemann surfaces, and p-adic analysis. The second part reports on matters with more direct physical interest, such as periodic and quasiperiodic lattices, or classical and quantum dynamical systems.

The contribution of each author represents a short self-contained course on a specific subject. With very few prerequisites, the reader is offered a didactic exposition, which follows the author's original viewpoints, and often incorporates the most recent developments. As we shall explain below, there are strong relationships between the different chapters, even though every single contribution can be read independently of the others.

This volume originates in a meeting entitled *Number Theory and Physics*, which took place at the Centre de Physique, Les Houches (Haute-Savoie, France), on March 7 – 16, 1989. The aim of this interdisciplinary meeting was to gather physicists and mathematicians, and to give to members of both communities the opportunity of exchanging ideas, and to benefit from each other's specific knowledge, in the area of Number Theory, and of its applications to the physical sciences. Physicists have been given, mostly through the program of lectures, an exposition of some of the basic methods and results of Number Theory which are the most actively used in their branch. Mathematicians have discovered in the seminars novel domains of Physics, where methods and results related to Arithmetics have been useful in the recent years.

The variety and abundance of the material presented during lectures and seminars led to the decision of editing two separate volumes, both published by Springer Verlag. The first book, entitled *Number Theory and Physics*, edited by J.M. Luck, P. Moussa, and M. Waldschmidt (Springer Proceedings in Physics, vol. **47**, 1990), contained the proceedings of the seminars, gathered into five parts: (I) Conformally Invariant Field Theories, Integrability, Quantum Groups; (II) Quasicrystals and Related Geometrical Structures; (III) Spectral Problems, Automata and Substitutions; (IV) Dynamical and Stochastic Systems; (V) Further Arithmetical Problems, and Their Relationship to Physics.

The present volume contains a completed and extended version of the lectures given at the meeting.

The central subject of Arithmetics is the study of the properties of rational integers. Deep results on this subject require the introduction of other sets. A first example is the ring of Gaussian integers

$$\mathbb{Z}[i] = \{a + ib; (a, b) \in \mathbb{Z}^2\},$$

related to the representation of integers as sums of two squares of integers. This ring has a rich arithmetic and analytic structure; it arises in this volume in many different guises: in chapter 1 in connection with quadratic forms, in chapters 2 and 3 as the group of periods of an elliptic function, in chapter 6 as a ring of integers of an algebraic number field, in chapter 7 it gives rise to a complex torus, in chapter 10 it is used as the main example of a lattice.

A second example is the field $\mathbb{C}$ of complex numbers which enables one to use methods from complex analysis. Analytic means have proved to be efficient in number theory; the most celebrated example is Riemann's zeta function, which provides information on the distribution of prime numbers.

There is a zeta function associated with the ring $\mathbb{Z}[i]$; it is constructed in chapter 1 by Cartier, in connection with quadratic forms, and also defined in chapter 6 by Stark, as the simplest case of the (Dedekind) zeta function of a number field. Other examples of Dirichlet series show up in this book: in chapter 1, Hurwitz zeta functions, in chapter 3, the Hasse-Weil zeta function of an elliptic curve, in chapter 4, the Hecke L-series attached to a modular form, in chapter 6, the Artin L-functions attached to a character; there are even p-adic L-functions in chapter 9. The mode-locking problem in chapter 13 involves another type of zeta function, which in some cases reduces to a ratio of two Riemann zeta functions.

Lattices, tori, and theta functions are also met in several chapters. The simplest lattice is $\mathbb{Z}$ in $\mathbb{R}$. The quotient is the circle (one-dimensional torus), which is studied in chapter 14. Lattices are intimately connected with elliptic curves and Abelian varieties (chapters 1, 2, 3, 5); they play an important role in Minkowski's geometry of numbers (chapter 10) and in Dirichlet's unit theorem (chapter 6). They arise naturally from the study of periodic problems, but their role extends to the study of quasiperiodic phenomena, especially in quasicrystallography (chapter 11). They deserve a chapter for their own (chapter 10). Theta functions were used by Jacobi to study sums of four squares of rational integers. They can be found in chapters 1, 2, 3, 5 and 10.

Let us now give a brief description of the content of each chapter.

In chapter 1 Cartier investigates properties of the Riemann's zeta function, with emphasis on its functional equation, by means of Fourier transformation, Poisson summation formula and Mellin transform. He also decomposes the zeta function attached to $\mathbb{Z}[i]$ into a product of the Riemann zeta function and a Dirichlet L-series with a character. This chapter includes exercises, which refer to more advanced results.

The set $\mathbb{Z}[i]$ is the simplest example of a lattice in the complex plane. When L is a general lattice in $\mathbb{C}$, the quotient group $\mathbb{C}/L$ can be given the structure of an algebraic Abelian group, which means that it is an algebraic variety, and that the group law is defined by algebraic equations.

If L is a lattice in higher dimension (a discrete subgroup of $\mathbb{C}^n$) the quotient $\mathbb{C}^n/L$ is not always an algebraic variety. Riemann gave necessary and sufficient conditions for the existence of a projective embedding of this torus as an Abelian variety. In this case theta functions give a complex parametrization. The correspondence between Riemann surfaces (see chapter 7), algebraic curves, and Jacobian varieties is explained by Bost in chapter 2. He surveys various definitions of Riemann surfaces, characterizes those defined over the field of algebraic numbers, discusses the notion of divisors and holomorphic bundles, including a detailed proof of the Riemann-Roch theorem. Various constructions of the Jacobian are presented, leading to the general theory of Abelian varieties. This thorough presentation can be viewed as an introduction to arithmetic varieties and Diophantine geometry.

Coming back to the one-dimensional case, a possible definition of an elliptic curve (chapter 3) is the quotient of $\mathbb{C}$ by a lattice L. The case $L = \mathbb{Z}[i]$ is rather special: the elliptic curve has non trivial endomorphisms. It is called of complex multiplication (CM) type. We are not so far from down-to-earth arithmetic questions; Cohen mentions a connection between the curve $y^2 = x^3 - 36x$ (which is 'an equation' of our elliptic curve) and the congruent number problem of finding right angle triangles with rational sides and given area. One is interested in the rational (or integral) solutions of such an equation. One method is to compute the number of solutions 'modulo p' for all prime numbers p. The collection of this data is recorded in an analytic function, which is another type of Dirichlet series, namely the L-function of the elliptic curve. According to Birch and Swinnerton-Dyer, this function contains (at least conjecturally) a large amount of arithmetic information.

Once the situation for a single elliptic curve is understood, one may wonder what happens if the lattice is varied. One thus comes across modular problems. A change of basis of a lattice involves an element of the modular group $\mathrm{SL}_2(\mathbb{Z})$, acting on the upper half plane. Once more analytic methods are relevant: one introduces holomorphic forms in the upper half plane, which satisfy a functional equation relating $f(\tau)$ to $f\big((a\tau + b)/(c\tau + d)\big)$. The modular invariant $j(\tau)$, the discriminant function Δ, Eisenstein series satisfy such a property. Taking sublattices induces transformations on these modular forms which are called Hecke operators. These operators act linearly on a vector space of modular forms; they have eigenvectors, and the collection of eigenvalues is included in a Dirichlet series, which is Hecke's L-series. An interesting special case is connected with the Δ function: the coefficient of q^n in the Fourier expansion is Ramanujan's τ function. In chapter 4 Zagier completes this introduction to modular forms by explaining the Eichler-Selberg trace formula which relates the trace of Hecke operators with the Kronecker-Hurwitz class number (which counts equivalence classes of binary quadratic forms with given discriminant).

In chapter 5 Gergondey also considers families of elliptic curves. He starts with the function

$$\vartheta_3(z \mid \tau) = \sum_{n \in \mathbb{Z}} e^{2i\pi\left(nz+(n^2/2)\tau\right)}$$

which is a solution of the heat equation. For fixed τ, this is an example of a theta function with respect to the lattice $\mathbb{Z} + \mathbb{Z}\tau$; quotients of such theta functions give a parametrization of points on an elliptic curve. For fixed z, the variable τ parametrizes lattices; but the so-called 'theta-constant' $\vartheta_3(\tau) = \vartheta_3(0 \mid \tau)$ is not a modular function: it is not invariant under isomorphisms of elliptic curves. The solution which is proposed is to change the notion of isomorphism by adding extra structures (it will be harder for two objects to be isomorphic: the situation will be rigidified). Moduli spaces thus obtained are nicer than without decoration.

Stark discusses in chapter 6 classical algebraic number theory. With almost no prerequisite one is taught almost the whole subject, including class field theory! He starts with Galois theory of algebraic extensions (with explicit examples: all subfields of $\mathbb{Q}(i, 2^{1/4})$ are displayed; another example involves a Jacobian of a curve of genus 2). He studies the ring of integers of an algebraic number field by means of divisor theory, avoiding abstract algebraic considerations. The Dedekind zeta function of a number field is introduced, with its functional equation, and its decomposition into a product of L-functions. Dirichlet class number formula proves once again the efficiency of analytic methods. The Cebotarev density theorem (which generalizes Dirichlet's theorem on primes in an arithmetic progression) is also included.

We mentioned that the quotient $\mathbb{C}/\mathbb{Z}[i]$ has the structure of an algebraic variety. This is a Riemann surface, and a meromorphic function on this surface is just an elliptic function. More generally, to a plane algebraic curve is associated such a surface, and coverings of curves give rise to extensions of function fields. Therefore Galois theory applies, as described by Reyssat in chapter 7. A useful way of computing the genus of a curve is the Riemann-Hurwitz formula. An application is mentioned to the inverse Galois problem: is it true that each finite group is the Galois group of an algebraic extension of $\mathbb{Q}$?

The quotient of the upper half plane by the modular group $\Gamma(7)$ is a curve of genus 3 with a group of 168 automorphisms; this curve is connected with a tessellation of the unit disc by hyperbolic triangles. By comparing Figure 12 of chapter 2, or Figure 14 of chapter 7, with the illustration of the front cover, the reader will realize easily that M.C. Escher's 'Angels and Demons' has been chosen for scientific reasons[1], and not because the meeting gathered Physicists and Mathematicians.

There is still a third type of Galois correspondence, in the theory of linear differential equations. This is explained in chapter 8 by Beukers (who could

[1] If one forgets about the difference between angels and demons, the group of hyperbolic isometries preserving the picture is generated by the (hyperbolic) mirror symmetries in the sides of a triangle with angles $\pi/2$, $\pi/4$, $\pi/6$. If one really wants to distinguish between angels and demons, one has to take a basic triangle which is twice as big, with angles $\pi/2$, $\pi/6$, $\pi/6$. Such groups are examples of Fuchsian groups associated with ternary quadratic forms, or quaternion algebras.

not attend the meeting). While the classical Galois theory deals with relations between the roots of an algebraic equations, differential Galois theory deals with algebraic relations between solutions of differential equations. Algebraic extensions of fields are now replaced by Picard-Vessiot extensions of differential fields. Kolchin's theorem provides a solution to the analytic problem which corresponds to solving algebraic equations with radicals. Another type of algebraic groups occurs here: the linear ones. Beukers also gives examples related to hypergeometric functions.

Analytic methods usually involve the field of complex numbers: $\mathbb{C}$ is the algebraic closure of $\mathbb{R}$, and $\mathbb{R}$ is the completion of $\mathbb{Q}$ for the usual absolute value. But $\mathbb{Q}$ has other absolute values, and ultrametric analysis is also a powerful tool. This is the object of chapter 9 by Christol. There is a large family of p-adic functions: exponential, logarithm, zeta and gamma functions, etc. Connections between the ring of adelic numbers $\widehat{\mathbb{Z}}$ and the Parisi matrices are pointed out.

Chapter 10, by Marjorie Senechal, deals with lattice geometry, a vast subject at the border between mathematics and physics, with applications ranging from integer quadratic forms to crystallography. The topics of Voronoï polytopes, root lattices and their Coxeter diagrams, and sphere packings, are covered in a more detailed fashion.

The next chapter is devoted to quasiperiodic lattices and tilings, which model the quasicrystalline phases, discovered experimentally in 1984. Katz uses the description of quasiperiodic sets of points as cuts of periodic objects in a higher-dimensional space. These objects are periodic arrays of 'atomic surfaces', which are placed at the vertices of a regular lattice. Several aspects of quasicrystallography are considered within this framework, including the Fourier transform and Patterson analysis, considerations about symmetry (point groups, self-similarity), and the possibility of growing a perfect quasiperiodic lattice from local 'matching rules'.

The last three chapters involve concepts and results related to the theory of dynamical systems, in a broad sense, namely, the study of temporal evolution, according to the laws of either classical or quantum mechanics.

In chapter 12, Bellissard presents an overview of the consequences of algebraic topology, and especially K-theory, on the spectra of Hamiltonian or evolution operators in quantum mechanics. The main topic is the gap labelling problem. Several applications are discussed, including the propagation of electrons on a lattice in a strong magnetic field, the excitation spectra of quasicrystals, and various one-dimensional spectral problems, in connection with sequences generated by automata or substitutions.

Cvitanović deals with circle maps in chapter 13. These provide examples of classical dynamical systems which are both simple enough to allow for a detailed and comprehensive study, and complex enough to exhibit the many features referred to as 'chaos'. The mathematical framework of this field involves approximation theory for irrational numbers (continued fraction expansions, Farey series).

The last chapter is devoted to yet a different aspect of dynamical systems, known as 'small divisor problems'. This name originates in the occurrence of small divisors in the calculations of the stability of a periodic orbit of a Hamiltonian dynamical system under small perturbations. Yoccoz reviews the progress made in the understanding of the behavior of periodic orbits throughout this century, starting with the pioneering works by Poincaré and Denjoy.

Let us finally emphasize that the title of this book reveals our conviction that number-theoretical concepts are becoming more and more fruitful in many areas of the natural sciences, as witnessed by the success of the meeting during which part of the material of this book has been presented.

*Saclay and Paris, May 1992.**

M. Waldschmidt, P. Moussa, J.M. Luck, C. Itzykson

We take advantage of this opportunity to thank again Prof. N. Boccara, the Director of the Centre de Physique, for having welcomed the meeting on the premises of the Les Houches School, with its unique atmosphere, in a charming mountainous setting, amongst ski slopes (see 'Quasicrystals: The View from Les Houches', by M. Senechal and J. Taylor, The Mathematical Intelligencer, vol. 12, pp. 54–64, 1990).

The Scientific Committee which organized *Number Theory and Physics* was composed of: J. Bellissard (Theoretical Physics, Toulouse), C. Godrèche (Solid State Physics, Saclay), C. Itzykson (Theoretical Physics, Saclay), J.M. Luck (Theoretical Physics, Saclay), M. Mendès France (Mathematics, Bordeaux), P. Moussa (Theoretical Physics, Saclay), E. Reyssat (Mathematics, Caen), and M. Waldschmidt (Mathematics, Paris). The organizers have been assisted by an International Advisory Committee, composed of Profs. M. Berry (Physics, Bristol, Great-Britain), P. Cvitanović (Physics, Copenhagen, Denmark), M. Dekking (Mathematics, Delft, The Netherlands), and G. Turchetti (Physics, Bologna, Italy).

The following institutions are most gratefully acknowledged for their generous financial support to the meeting: the Département Mathématiques et Physique de Base of the Centre National de la Recherche Scientifique; the Institut de Recherche Fondamentale of the Commissariat à l'Énergie Atomique; the Direction des Recherches, Etudes et Techniques de la Délégation Générale pour l'Armement (under contract number 88/1474); the French Ministère de l'Éducation Nationale; the French Ministère des Affaires Étrangères; and the Commission of the European Communities. The regretted absence of support from NATO finally turned out to allow a more flexible organization.

---

* For the second printing, which we prepared in May 1995, we have corrected a few misprints, and added some references to Chapters 3 and 12. The only significant modification of the text takes place in Chapter 3, page 234, and reports on the recent progress made by Wiles on the Taniyama-Weil conjecture, which provides a proof of Fermat's celebrated "Last Theorem".

Claude Itzykson

While the second printing of this volume was going to press, we learned of the untimely death of Claude Itzykson on May 22, 1995. Born in 1938, Claude graduated from the École Polytechnique in Paris, and spent his scientific career in the Theoretical Physics Department of Saclay. His vision of bringing together physicists and mathematicians inspired the Les Houches meeting and the present book. His great expertise in Quantum Field Theory and his vast culture both in Mathematics and in Physics were essential to creating a common ground for workers in the two fields. His colleagues and friends will also remember him as a tremendously warm and enthusiastic person, always ready to encourage and to give advice, while claiming only a modest role for himself.

The contributors to the present book and Claude's fellow editors would like to express their deepest regrets over the loss of an esteemed friend and colleague.

List of Contributors

Authors

Chapter 1: Prof. P. Cartier, École Normale Supérieure, Centre de Mathématiques, 45 rue d'Ulm, 75230 Paris cedex 05, France.

Chapter 2: Prof. J.-B. Bost, Institut des Hautes Études Scientifiques, 35 route de Chartres, 91440 Bures-sur-Yvette, France.

Chapter 3: Prof. H. Cohen, CeReMaB, U. F. R. de Mathématiques et Informatique, 351 Cours de la Libération, 33405 Talence cedex, France.

Chapter 4: Prof. D. Zagier, Max Planck Institut für Mathematik, Gottfried-Claren-Str. 26, 5300 Bonn 3, Germany.

Chapter 5: Prof. R. Gergondey, U.S.T.L.F.A., U.F.R. de Mathématiques de Lille I, B.P. 36, 59655 Villeneuve-d'Ascq cedex, France.

Chapter 6: Prof. H. M. Stark, University of California, Dept. of Mathematics, La Jolla CA 92093, U.S.A.

Chapter 7: Prof. E. Reyssat, Université de Caen, Dépt. de Mathématiques et Mécanique, 14032 Caen cedex, France.

Chapter 8: Prof. F. Beukers, University of Utrecht, Dept. of Mathematics, Budapestlaan 6, P.O. Box 80.010, 3508 TA Utrecht, The Netherlands.

Chapter 9: Prof. G. Christol, Université P. et M. Curie (Paris VI), Dépt. de Mathématiques, Tour 45-46, 4 Place Jussieu, 75252 Paris cedex 05, France.

Chapter 10: Prof. M. Senechal, Smith College, Dept. of Mathematics, Northampton MA 01063, U.S.A.

Chapter 11: Prof. A. Katz, École Polytechnique, Centre de Physique Théorique, 91128 Palaiseau cedex, France.

Chapter 12: Prof. J. Bellissard, Centre de Physique Théorique, Campus de Luminy, Case 907, 13288 Marseille cedex, France. Address since 1991: Laboratoire de Physique Quantique, Université Paul Sabatier, 118 Route de Narbonne, 31062 Toulouse cedex, France.

Chapter 13: Prof. P. Cvitanović, Niels Bohr Institute, Blegdamsvej 17, 2100 Copenhagen Ø, Denmark.

Chapter 14: Prof. J.-C. Yoccoz, Université de Paris-Sud, Dépt. de Mathématiques, 91405 Orsay cedex, France.

Editors

1. Prof. M. Waldschmidt, Université P. et M. Curie (Paris VI), Dépt. de Mathématiques, Tour 45-46, 4 Place Jussieu, 75252 Paris cedex 05, France.

2. Dr. P. Moussa, Centre d'Études de Saclay, Service de Physique Théorique, 91191 Gif-sur-Yvette cedex, France.

3. Dr. J.-M. Luck, Centre d'Études de Saclay, Service de Physique Théorique, 91191 Gif-sur-Yvette cedex, France.

4. Dr. C. Itzykson, Centre d'Études de Saclay, Service de Physique Théorique, 91191 Gif-sur-Yvette cedex, France.

Contents

Chapter 1

An Introduction to Zeta Functions

by Pierre Cartier

Table of Contents

Notations : $\mathbb{N}$ set of integers $0, 1, 2, \cdots$, $\mathbb{Z}$ set of rational integers $0, \pm 1, \pm 2, \pm 3, \cdots$, $\mathbb{R}$ set of real numbers, $\mathbb{C}$ set of complex numbers, $Re\ z$ and $Im\ z$ real and imaginary part of a complex number z. 0 is counted as a positive number. We say *strictly positive* for a positive number different from 0. We use the standard notation $f(x) = O\big(g(x)\big)$ for x near x_0 to mean that there exists a constant $C > 0$ such that $|f(x)| \leq C|g(x)|$ for all x in a neighborhood of x_0.

Introduction

In this Chapter, we aim at giving an elementary introduction to some functions which were found useful in number theory. The most famous is *Riemann's zeta function* defined as follows

$$\zeta(s) = \sum_{n=1}^{\infty} n^{-s} = \prod_{p} (1 - p^{-s})^{-1} \tag{1}$$

(where p runs over all prime numbers). This function provided the key towards a proof of the *prime number distribution* : as it was conjectured by Gauss and Legendre before 1830 and proved by Hadamard and de La Vallée Poussin in 1898, the number $\pi(x)$ of primes p such that $p \leq x$ is asymptotic to $x/\log x$.

Around 1740 Euler, in an amazing achievement, was able to calculate the sums of the series $\zeta(2r) = \sum_{n=1}^{\infty} n^{-2r}$ for $r = 1, 2, \ldots$. In particular he found the following results

$$\sum_{n=1}^{\infty} n^{-2} = \frac{\pi^2}{6} \quad , \quad \sum_{n=1}^{\infty} n^{-4} = \frac{\pi^4}{90}, \quad \ldots .$$

In general $\zeta(2r)/\pi^{2r}$ is a rational number, closely connected to the Bernoulli numbers. These numbers B_m are defined by their generating series

$$\frac{z}{2} \coth \frac{z}{2} = 1 + \sum_{m=2}^{\infty} B_m z^m / m! . \tag{2}$$

Euler knew both the series expansion and the product formula for $\zeta(s)$ given in (1). These definitions make sense in the half-plane $Re\ s > 1$ but at the time of Euler there was little justification for considering $\zeta(s)$ beyond this natural domain of existence. This fact didn't prevent Euler from considering $\zeta(0), \zeta(-1), \ldots$ and one of his most striking results may be expressed as follows

$$B_m = -m\, \zeta(1 - m) \qquad \text{for} \qquad m = 2, 3, \ldots . \tag{3}$$

Around 1850, Riemann clarified the meaning of the analytic continuation and used immediately this new tool in the case of $\zeta(s)$, thus vindicating the previous formula.

In Section 1, we give a very elementary exposition of the theory of $\zeta(s)$ which should be accessible to any reader with a working knowledge of infinitesimal calculus and of the basic facts connected with Fourier series. It is truly Eulerian mathematics. We perform the analytic continuation of $\zeta(s)$ by using Euler-MacLaurin summation formula. The method is elementary and very direct and extends to various generalizations of $\zeta(s)$, namely Hurwitz zeta function $\zeta(s, v)$ and the Dirichlet L-series $L(\chi, s)$ associated to the Dirichlet characters. We prove all the above mentioned results. While mentioning the functional equation for $\zeta(s)$, we refrain from proving it at this stage, and we postpone such a proof to Section 3.

In Section 2, we explain the geometric and number theoretical facts connected with *Gaussian integers.* The theory of these numbers, of great intrinsic beauty, is also important in the plane crystallography; it provides one of the test examples for the methods of Geometry of Numbers, and similar considerations play a very important rôle in the theory of elliptic functions.

We are primarily interested in the corresponding zeta function

$$Z_4(s) = \Sigma'(m^2 + n^2)^{-s} , \tag{4}$$

the summation being extended over all pairs (m, n) of integers (of both signs). The dash over Σ indicates that the pair $m = n = 0$ has to be excluded. Of particular importance is the factorization

$$Z_4(s) = 4\zeta(s)L(s) \tag{5}$$

where $\zeta(s)$ is Riemann's zeta function, the other factor being defined by

$$L(s) = 1^{-s} - 3^{-s} + 5^{-s} - \cdots . \tag{6}$$

This very compact formula contains the main results obtained by Fermat and Jacobi about the representation of an integer as a sum of two squares.

The quadratic form $m^2 + n^2$ is intimately connected with the Gaussian integers of the form $m + ni$, where $i^2 = -1$ and m, n are ordinary integers. We give all necessary details in this case and just give hints about a very parallel case, namely the quadratic form $m^2 + mn + n^2$ and the corresponding numbers $m + nj$ where j is a cubic root of unity ($j^3 = 1$). It was left to Dirichlet (around 1840) to extend these results to the general binary quadratic form $am^2 + bmn + cn^2$, after arithmetical investigations by Gauss. The corresponding zeta functions are of the form $\sum'_{m,n}(am^2 + bmn + cn^2)^{-s}$ and can be generalized in two different directions:

a) Epstein's zeta functions are associated with quadratic forms in any number of variables $\sum_{i,j} a_{ij}m_i m_j$.

b) Dedekind's zeta functions are associated with fields of algebraic numbers and form the subject of Stark lectures in this volume.

In Section 3, we introduce two powerful analytical tools : *Poisson summation formula* and *Mellin transformation.* Both are classical and have been extensively used in analytic number theory. In such a subject, it's difficult to innovate, but we believe that the method used for the analytic continuation of Mellin transforms is not completely orthodox. It relies on Hadamard finite parts and one of its main advantages is the ease with which one uncovers the structure of poles of the extended functions. We prove the functional equation by one of the classical methods, namely the one connected with *theta series.* These series provide one of the most efficient tools in all sectors of analytic number theory. In the oral lectures we gave more emphasis to them, especially in the multidimensional case. They would deserve a more thorough treatment than the one offered in this Chapter. Our regret is somewhat alleviated by the existence of many wonderful textbooks, among which we give a special mention to Bellman (Bellman 1961) and Mumford (Mumford 1983).

Warning : The so-called 'exercises' are an integral part of this Chapter. Much information has been given there is a more sketchy form than in the main text, and is necessary for a complete understanding of the theory.

1. Riemann's zeta function

1.1. Definition

We consider at first the values of $\zeta(s)$ for complex numbers in the half-plane defined by $Re\ s > 1$. We can use the convergent series, where n runs over all strictly positive integers

$$\zeta(s) = \sum_{n=1}^{\infty} n^{-s} , \tag{1.1}$$

or an infinite product extended over all prime numbers

$$\zeta(s) = \prod_{p} \frac{1}{1 - p^{-s}} . \tag{1.2}$$

To prove the convergence of the series (1.1), one can compare it with the integral $\int_1^\infty x^{-s} dx$, which converges exactly when $Re\ s > 1$. According to classical results, the convergence of the product (1.2) is equivalent to that of the series $\sum_p p^{-s}$, where p runs over the prime numbers. This last series can be written as $\sum_{n=1}^\infty \varepsilon_n n^{-s}$, where ε_n is 1 or 0 depending on the fact that n is a prime or not. The convergence is implied by that of the series (1.1). To show that the series $\sum_{n=1}^\infty n^{-s}$ defines the same function of s as the product $\prod_p (1 - p^{-s})^{-1}$, expand every term in the product as a geometric series

$$(1 - p^{-s})^{-1} = 1 + \sum_{n=1}^{\infty} a_{p,n} \tag{1.3}$$

with $a_{p,n} = p^{-ns}$. The product of these series, for p running over all primes, is a sum of all terms of the form

$$a_{p_1,n_1} a_{p_2,n_2} \cdots a_{p_k,n_k} \tag{1.4}$$

where $p_1, \ldots, p_k$ are prime numbers in increasing order $p_1 < p_2 < \cdots < p_k$ and where $n_1, \ldots, n_k$ are integers equal to $1, 2, \ldots$. The product (1.4) is nothing else than N^{-s} where N is equal to $p_1^{n_1} \cdots p_k^{n_k}$. According to the unique decomposition of a number as a product of primes, every power N^{-s} with N equal to $1, 2, \ldots$, occurs exactly once in the development of the product $\prod_p (1 - p^{-s})^{-1}$. This proves the equivalence of the definitions (1.1) and (1.2).

We want now to prove the following result :

The function $\zeta(s)$ extends to a meromorphic function in the complex plane $\mathbb{C}$, where the only singularity is a pole of order 1 for $s = 1$, with residue equal to 1 (in other terms $\zeta(s) - \frac{1}{s-1}$ extends to an entire function).

Our proof is an elementary one based on Bernoulli numbers, Bernoulli polynomials and Euler-MacLaurin summation formula. It's one proof among many known ones.

1.2. Bernoulli polynomials

We use the following generating series for Bernoulli polynomials

$$\sum_{n=0}^{\infty} B_n(x) z^n / n! = z e^{xz} / (e^z - 1). \tag{1.5}$$

The n-th Bernoulli number is the constant term of the n-th Bernoulli polynomial, namely

$$B_n = B_n(0). \tag{1.6}$$

These numbers are given by the following generating series

$$\sum_{n=0}^{\infty} B_n z^n / n! = z / (e^z - 1). \tag{1.7}$$

Here are some simple consequences. It is obvious that the function

$$\frac{z}{e^z - 1} + \frac{z}{2} = \frac{z}{2} \frac{e^{z/2} + e^{-z/2}}{e^{z/2} - e^{-z/2}}$$

is an even function. Hence we get $B_1 = -1/2$, while B_n is 0 for any odd number $n \geq 3$.

A variant of Cauchy's rule for multiplying power series is obtained as follows. The product of the two exponential generating series

$$\Gamma(z) = \sum_{n=0}^{\infty} c_n z^n / n! \tag{1.8}$$

and

$$\Delta(z) = \sum_{n=0}^{\infty} d_n z^n / n! \tag{1.9}$$

is given by

$$\Gamma(z)\Delta(z) = \sum_{n=0}^{\infty} e_n z^n/n! \tag{1.10}$$

where the rule for the coefficients is

$$e_n = \sum_{k=0}^{n} \binom{n}{k} c_{n-k} d_k. \tag{1.11}$$

An interesting case occurs when $\Delta(z) = e^{xz}$, that is $d_n = x^n$. We have there

$$\Gamma(z)e^{xz} = \sum_{n=0}^{\infty} C_n(x) z^n/n! \tag{1.12}$$

where the polynomials $C_0(x), C_1(x), \ldots$ are derived from the constants $c_0, c_1, \ldots$ by means of the definition

$$C_n(x) = \sum_{k=0}^{n} \binom{n}{k} c_{n-k} x^k. \tag{1.13}$$

The function $G(x,z) = \Gamma(z)e^{xz}$ is completely characterized by the following two properties

$$\frac{\partial}{\partial x} G(x,z) = z\, G(x,z) \tag{1.14}$$

$$G(0,z) = \Gamma(z). \tag{1.15}$$

These properties can be translated as the following characterization of the polynomials $C_n(x)$, where $n = 0, 1, 2, \ldots$

$$\frac{d}{dx} C_n(x) = n\, C_{n-1}(x), \tag{1.16}$$

$$C_n(0) = c_n. \tag{1.17}$$

The case $n = 0$ of formula (1.16) has to be interpreted as $dC_0/dx = 0$, that is $C_0(x)$ is the constant c_0; by induction on n, it follows that $C_n(x)$ is a polynomial of degree n in x.

Let us specialize these results. If we assume that the Bernoulli numbers are already known, the Bernoulli polynomials are characterized by the following formulas (for $n = 0, 1, 2, \ldots$)

$$\frac{d}{dx} B_n(x) = n\, B_{n-1}(x), \tag{1.18}$$

$$B_n(0) = B_n. \tag{1.19}$$

Note also the generating series

$$\sum_{n=0}^{\infty} B_n(x)z^n/n! = e^{xz} \cdot \sum_{n=0}^{\infty} B_n z^n/n! \tag{1.20}$$

and the explicit formula

$$B_n(x) = \sum_{k=0}^{n} \binom{n}{k} B_{n-k} x^k. \tag{1.21}$$

The series $\beta(z) = \sum_{n=0}^{\infty} B_n z^n/n!$ is in fact characterized by the relation $e^z\beta(z) - \beta(z) = z$. Using (1.5), this amounts to

$$\sum_{n=0}^{\infty} \{B_n(1) - B_n(0)\} z^n/n! = z \tag{1.22}$$

that is

$$B_n(1) = B_n(0) \qquad \text{for } n \geq 2, \tag{1.23}$$

$$B_1(1) = B_1(0) + 1. \tag{1.24}$$

More explicitly, we get the inductive formula

$$\sum_{k=0}^{m} \binom{m+1}{k} B_k = 0 \qquad \text{for } m \geq 1 \tag{1.25}$$

with initial condition $B_0 = 1$. This enables us to calculate easily the following table:

$B_2 = +1/6$	$B_8 = -1/30$
$B_4 = -1/30$	$B_{10} = +5/66$
$B_6 = +1/42$	$B_{12} = -691/2730.$

Recall that $B_1 = -1/2$ and $B_3 = B_5 = B_7 = \cdots = 0$. A more extended table is given by Serre (Serre 1970, page 147), but notice that our B_{2k} is denoted by $(-1)^{k+1}B_k$ in Serre's book.

Exercise 1 : (symbolic method). If a function $f(x)$ can be developed as a power series $f(x) = \sum_{n=0}^{\infty} a_n x^n$, one defines symbolically $f(B)$ as the sum $\sum_{n=0}^{\infty} a_n B_n$ (the power B^n is interpreted as the n-th Bernoulli number B_n). With this convention, express the definition of Bernoulli numbers by $e^{Bz} = z/(e^z - 1)$, or else $e^{(B+1)z} = e^{Bz} + z$. The recurrence equation (1.25) can be written as $B^n = (B+1)^n$ for $n \geq 2$, and the definition of Bernoulli polynomials can be expressed by $B_n(x) = (x+B)^n$.

Exercise 2 : Generalize formulas (1.23) and (1.24) as follows

$$B_n(x+1) = B_n(x) + nx^{n-1} \quad \text{for } n = 1, 2, \ldots.$$

Derive the summation formula

$$1^n + 2^n + \cdots + N^n = \frac{B_{n+1}(N+1) - B_{n+1}}{n+1}.$$

Exercise 3 : Plot the graphs of the functions $B_n(x)$ in the range $0 \leq x \leq 3$, $1 \leq n \leq 4$.

1.3. Euler-MacLaurin summation formula

Let us consider a real valued function $f(x)$ defined in an interval $a \leq x \leq b$ whose end points a and b are integers; assume that $f(x)$ is n times continuously differentiable for some integer $n \geq 1$. Any real number t can be written uniquely in the form $t = m + \theta$ where m is an integer and $0 \leq \theta < 1$; we call m the integer part of t and denote it by $[t]$.

We denote by $\overline{B}_n(x)$ the function of a real variable x which is periodic, with period 1, and coincides with $B_n(x)$ in the fundamental interval $[0, 1[$. Explicitly, we get

$$\overline{B}_n(x) = B_n(x - [x]). \tag{1.26}$$

It is clear that $\overline{B}_0(x) = 1$ and that

$$\overline{B}_1(x) = x - [x] - \frac{1}{2}. \tag{1.27}$$

Hence the function $\overline{B}_1(x)$ admits discontinuities at the integral values of x, with a jump equal to -1 (see fig. 1). For $n \geq 2$, the function $\overline{B}_n(x)$ is continuous.

Exercise 4 : Plot the graphs of the functions $\overline{B}_n(x)$ in the range $0 \leq x \leq 3$, $2 \leq n \leq 4$.

According to formula (1.18), the function $\overline{B}_n(x)$ admits derivatives of any order $k \leq n-2$, which are continuous functions given by

$$\overline{B}_n^{(k)}(x) = n(n-1)\cdots(n-k+1)\overline{B}_{n-k}(x). \tag{1.28}$$

The derivative of order $n-1$ of $\overline{B}_n(x)$ is equal to $n!\ \overline{B}_1(x)$ with a jump for every integral value of x. One more derivation gives

$$\frac{d^n}{dx^n}\,\overline{B}_n(x) = n!(1 - \sum_{k \in \mathbb{Z}} \delta(x-k)) \tag{1.29}$$

using Dirac δ-functions.

We state now *Euler-MacLaurin summation formula*

$$\sum_{r=a}^{b-1} f(r) = \int_a^b f(x)\,dx + \sum_{k=1}^{n} \frac{B_k}{k!}\left\{f^{(k-1)}(b) - f^{(k-1)}(a)\right\} + \frac{(-1)^{n-1}}{n!} \int_a^b \overline{B}_n(x) f^{(n)}(x)\,dx\,. \tag{1.30}$$

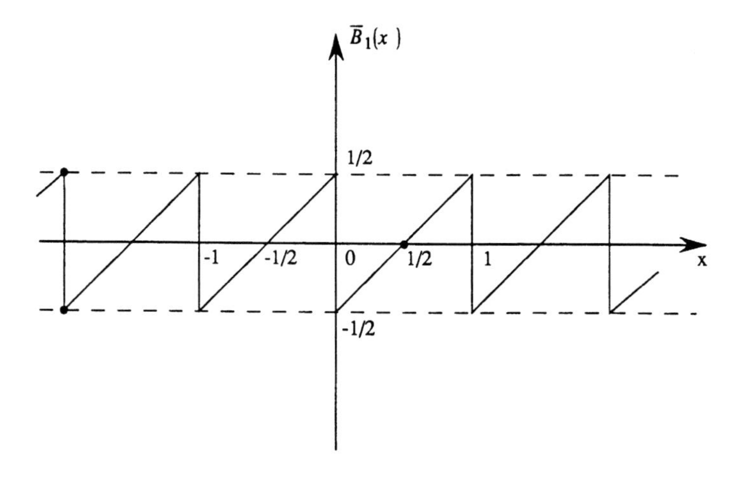

Fig. 1. Graph of function $\overline{B}_1(x)$.

Proof. A) Any integral over the interval $[a,b]$ splits as a sum of $b-a$ integrals, each one extended over an interval $[r,r+1]$ for $r=a,a+1,\ldots,b-1$. The above formula is an immediate consequence of a similar formula for each of these subintervals, and this fact enables us to reduce the proof to the case of an interval of length $b-a$ equal to 1. Replacing the integration variable x by $y+a$ where y runs over $[0,1]$ reduces therefore the proof to the case $a=0$, $b=1$.

B) In this last case, we have to prove the formula

$$\begin{aligned} f(0) = \int_0^1 f(x)\,dx &+ \sum_{k=1}^{n} \frac{B_k}{k!}\left\{f^{(k-1)}(1) - f^{(k-1)}(0)\right\} \\ &+ \frac{(-1)^{n-1}}{n!}\int_0^1 B_n(x) f^{(n)}(x)\,dx\,. \end{aligned} \tag{1.31}$$

We begin with the case $n=1$, which reads as follows

$$f(0) = \int_0^1 f(x)\,dx + B_1(f(1)-f(0)) + \int_0^1 B_1(x)\,df(x). \tag{1.32}$$

But an integration by part gives

$$\int_0^1 B_1(x)\,df(x) = B_1(x)\,f(x)\,\Big|_{x=0}^{x=1} - \int_0^1 f(x)\,dB_1(x)$$

and since $B_1(x) = x - \frac{1}{2}$, we get

$$\int_0^1 B_1(x)\ df(x) = \frac{1}{2}\Big\{f(1) + f(0)\Big\} - \int_0^1 f(x)\ dx.$$

Formula (1.32) follows since B_1 is equal to $-1/2$. To make the induction from n to $n+1$ in formula (1.31), we transform in a similar way the last integral by integration by part. Since the derivative of $B_{n+1}(x)/(n+1)!$ is equal to $B_n(x)/n!$, we get

$$\begin{aligned}\frac{(-1)^{n-1}}{n!}\int_0^1 B_n(x)\ f^{(n)}(x)\ dx &= \frac{(-1)^{n-1}}{(n+1)!}\int_0^1 dB_{n+1}(x)\ f^{(n)}(x)\\ &= \frac{(-1)^{n-1}}{(n+1)!}B_{n+1}(x)\ f^{(n)}(x)\ \Big|_{x=0}^{x=1} + \frac{(-1)^n}{(n+1)!}\int_0^1 B_{n+1}(x)\ df^{(n)}(x)\\ &= \frac{(-1)^{n+1}}{(n+1)!}B_{n+1}\Big\{f^{(n)}(1) - f^{(n)}(0)\Big\} + \frac{(-1)^n}{(n+1)!}\int_0^1 B_{n+1}(x)\ f^{(n+1)}(x)\ dx.\end{aligned}$$

But, let us remark that for $n \geq 1$, B_{n+1} is 0 if n is even, and that $(-1)^{n+1}$ is 1 if n is odd; we can drop therefore the sign $(-1)^{n+1}$ in front of B_{n+1}. This calculation establishes the equality

$$\begin{aligned}\frac{(-1)^{n-1}}{n!}\int_0^1 B_n(x)\ f^{(n)}(x)\ dx &= \frac{B_{n+1}}{(n+1)!}\Big\{f^{(n)}(1) - f^{(n)}(0)\Big\}\\ &+ \frac{(-1)^n}{(n+1)!}\int_0^1 B_{n+1}(x)\ f^{(n+1)}(x)\ dx\end{aligned}$$

and this formula provides us with the inductive step from n to $n+1$ in (1.31). □

Exercise 5 : Let $g(x)$ be any primitive function of $f(x)$. If we let n go to infinity in Euler-MacLaurin formula, we get

$$\sum_{r=a}^{b-1} f(r) = \Big\{g(x) + \sum_{k=1}^{\infty} g^{(k)}(x)\ B_k/k!\Big\}\Big|_{x=a}^{x=b};$$

according to the symbolic method of exercise 1, we can write

$$\sum_{r=a}^{b-1} f(r) = g(b+B) - g(a+B) \tag{1.33}$$

or even more boldly

$$\sum_{r=a}^{b-1} f(r) = \int_{a+B}^{b+B} f(x)\ dx. \tag{1.34}$$

The particular case $b - a = 1$ can also be written as

$$g'(x) = g(x + B + 1) - g(x + B). \tag{1.35}$$

Exercise 6 : Let us denote by D the derivation operator. In operator form, Taylor's formula is expressed as

$$g(x + 1) = (e^D g)(x). \tag{1.36}$$

If we use the generating series $\beta(z) = \sum_{n=0}^{\infty} B_n z^n / n!$ for the Bernoulli numbers, we can write

$$g(x + B) = (\beta(D)g)(x). \tag{1.37}$$

The operator form of formula (1.35) is therefore

$$D = (e^D - 1)\beta(D)$$

and this agrees with formula (1.7) which reads as $\beta(z) = z/(e^z - 1)$.

Exercise 7 : Using the fact that $B_1 = -1/2$ and that $B_k = 0$ for odd k, $k \geq 3$, transform Euler-MacLaurin formula as follows

$$\begin{aligned} \sum_{r=a+1}^{b} f(r) = \int_a^b f(x)\, dx + \sum_{k=1}^{n} (-1)^k \frac{B_k}{k!} \left\{ f^{(k-1)}(b) - f^{(k-1)}(a) \right\} \\ + \frac{(-1)^{n-1}}{n!} \int_a^b \overline{B}_n(x)\, f^{(n)}(x)\, dx\ . \end{aligned} \tag{1.38}$$

According to the symbolic method of exercise 5, derive the formulas

$$\sum_{r=a+1}^{b} f(r) = \int_{a-B}^{b-B} f(x)\, dx = \int_a^b f(x - B) dx \tag{1.39}$$

and

$$g'(x) = g(x - B) - g(x - B - 1). \tag{1.40}$$

1.4. Analytic continuation of the zeta function

We consider a complex number s and two integers $n \geq 1$, $N \geq 2$. We use Euler-MacLaurin formula in the case

$$f(x) = x^{-s} \quad , \quad a = 1 \quad , \quad b = N$$

and derive the relation

$$(1.41)\quad \begin{cases} \sum_{r=1}^{N} r^{-s} = \frac{1-N^{1-s}}{s-1} + \frac{1+N^{-s}}{2} \\ \qquad + \sum_{k=2}^{n} B_k s(s+1)\cdots(s+k-2)(1-N^{-s-k+1})/k! \\ \qquad - \frac{1}{n!} s(s+1)\cdots(s+n-1) \int_1^N \overline{B}_n(x)\, x^{-s-n} dx. \end{cases}$$

We consider first the case $Re\ s > 1$ where the series $\sum_{r=1}^{\infty} r^{-s}$ converges with sum $\zeta(s)$. If we let N go to infinity in the previous formula, we get

$$(1.42)\quad \begin{cases} \zeta(s) = \frac{1}{s-1} + \frac{1}{2} + \sum_{k=2}^{n} B_k s(s+1)\cdots(s+k-2)/k! \\ \qquad - \frac{1}{n!} s(s+1)\cdots(s+n-1) \int_1^\infty \overline{B}_n(x)\, x^{-s-n} dx. \end{cases}$$

All these formulas, for $n = 1, 2, \cdots$, are valid in the said half-plane. In particular, for $n = 1$, we get

$$(1.43)\qquad \zeta(s) = \frac{1}{s-1} + \frac{1}{2} - s \int_1^\infty (x - [x] - \frac{1}{2}) x^{-s-1}\, dx.$$

The fundamental remark is that the function $\overline{B}_n(x)$ is periodic, namely we have $\overline{B}_n(x+1) = \overline{B}_n(x)$, hence it remains bounded over the whole interval $[1, +\infty[$. Hence the integral

$$\int_1^\infty \overline{B}_n(x) x^{-s-n}\, dx$$

will converge provided that $Re\ s > 1 - n$. It follows that the right-hand side of formula (1.42) defines a function $\zeta_n(s)$ holomorphic in the half-plane defined by $Re\ s > 1 - n$. Since the derivative of $\overline{B}_n(x)$ is equal to $n\overline{B}_{n-1}(x)$, an integration by part shows that $\zeta_n(s)$ and $\zeta_{n+1}(s)$ agree on their common domain $Re\ s > 1 - n$. It follows that *there exists in the complex plane* $\mathbb{C}$ *a meromorphic function* $\zeta(s)$ *whose only singularity is a pole of order* 1 *at* $s = 1$, *which is given in the half-plane* $Re\ s > 1 - n$ *by the formula* (1.42). In particular, formula (1.43) is valid whenever $Re\ s > 0$.

Exercise 8 : Using formulas (1.41) and (1.42) show that, for fixed s and for N going to infinity, the quantity

$$(1.44)\qquad E_s(N) = 1^{-s} + 2^{-s} + \cdots + N^{-s} - \zeta(s)$$

has an asymptotic expansion

$$\frac{N^{1-s}}{1-s} + \frac{N^{-s}}{2} - \frac{B_2}{2} s N^{-1-s} - \cdots - \frac{B_{n+1}}{(n+1)!} s(s+1)\cdots(s+n-1) N^{-s-n} - \cdots.$$

Let us truncate these series after the last term which doesn't tend to 0 with $1/N$ (make this precise!). Then one gets, for every complex number $s \neq 1$,

$$\zeta(s) = \lim_{N \to \infty} \{1^{-s} + 2^{-s} + \cdots + N^{-s} - E'_s(N)\}$$

where $E'_s(N)$ is the truncated series.

Exercise 9 : Using our symbolic notations, transform the previous formulas as follows

$$E_s(N) = (N - B)^{1-s}/(1-s) \tag{1.45}$$

$$\zeta(s) = 1 - (1 - B)^{1-s}/(1-s). \tag{1.46}$$

1.5. Some special values of the zeta function

Letting $n = 2$ in formula (1.42), we get the following representation for $\zeta(s)$ in the half-plane $Re\ s > -1$

$$\zeta(s) = \frac{1}{s-1} + \frac{1}{2} + \frac{1}{12}s - \frac{1}{2}s(s+1)\int_1^\infty \overline{B}_2(x)x^{-s-2}\,dx. \tag{1.47}$$

We get immediately

$$\zeta(0) = -\frac{1}{2}. \tag{1.48}$$

More generally, let $m \geq 1$ be an integer and put $s = -m, n = m+2$ in formula (1.42). The coefficient of the integral is equal to the value for $s = -m$ of the product $s(s+1)\cdots(s+m)(s+m+1)$, hence vanishes. Similarly, in the summation over k, the term with $k = m+2$ vanishes for $s = -m$. Hence we get

$$\zeta(-m) = -\frac{1}{m+1} + \frac{1}{2} - \sum_{k=2}^{m+1}(-1)^k B_k \frac{m(m-1)\cdots(m-k+2)}{k!}. \tag{1.49}$$

We know that B_k is 0 if k is odd and $k \geq 2$. We can therefore replace $(-1)^k B_k$ by B_k in the preceding formula, hence

$$-(m+1)\zeta(-m) = \sum_{k=0}^{m+1} \binom{m+1}{k} B_k. \tag{1.50}$$

Taking into account the recurrence formula (1.25) for the Bernoulli numbers, we conclude

$$\zeta(-m) = -\frac{B_{m+1}}{m+1}. \tag{1.51}$$

Distinguishing the cases where m is even or odd, we can conclude :

$$\begin{cases} \zeta(-2r) = 0 \\ \zeta(-2r+1) = -B_{2r}/2r \end{cases} \tag{1.52}$$

for $r = 1, 2, \ldots$. We include a short table

$$\zeta(0) = -\frac{1}{2}\,,\ \zeta(-1) = -\frac{1}{12}\,,\ \zeta(-3) = \frac{1}{120}\,,\ \zeta(-5) = -\frac{1}{252}$$

$$\zeta(-2) = \zeta(-4) = \zeta(-6) = \cdots = 0.$$

Exercise 10 : Fix an integer $m \geq 1$. Using notations as in exercise 8, show that the asymptotic expansion of $E_{-m}(N)$ has finitely many terms, and deduce the equality

$$1^m + 2^m + \cdots + N^m = \zeta(-m) + E_{-m}(N).$$

Here $E_{-m}(N)$ is a polynomial in N, written symbolically

$$E_{-m}(N) = (N - B)^{m+1}/(m+1).$$

Using exercise 2, prove that the constant term of the polynomial $E_{-m}(N)$ is equal to $-\zeta(-m)$ and give a new proof for formula (1.51).

Using the Fourier series expansions for the periodic functions $\overline{B}_n(x)$, we shall compute $\zeta(2), \zeta(4), \zeta(6), \ldots$. Let us introduce the Fourier coefficients

$$c(n,m) = \int_0^1 \overline{B}_n(x) e^{-2\pi i m x}\, dx. \tag{1.53}$$

These can be computed using Euler-MacLaurin summation formula (1.30) for the case $a = 0, b = 1$, $f(x) = e^{-2\pi i m x}$. We get in this way

$$1 = \int_0^1 e^{-2\pi i m x}\, dx + (-1)^{n-1}(-2\pi i m)^n c(n,m)/n! \tag{1.54}$$

and derive easily the following values

$$\begin{cases} c(n,0) & = 0 \\ c(n,m) & = -n!(2\pi i m)^{-n} \quad \text{for} \quad m \neq 0. \end{cases} \tag{1.55}$$

For $n \geq 2$, the series $\sum_{m \neq 0} m^{-n}$ converges absolutely, hence the Fourier series for $\overline{B}_n(x)$ is absolutely convergent, and we get

$$\overline{B}_n(x) = -n!(2\pi i)^{-n} \sum_{m \neq 0} m^{-n} e^{2\pi i m x} \tag{1.56}$$

(here x is any real number, n any integer such that $n \geq 2$). The special case $x = 0$ reads as follows

(1.57)
$$B_n = -n!(2\pi i)^{-n} \sum_{m \neq 0} m^{-n}$$

for $n = 2, 3, \ldots$. The sum of the series is obviously 0 if n is *odd*, and we prove again that B_n is 0 for n odd, $n \geq 3$. In case $n = 2r$ is even we get

(1.58)
$$\zeta(2r) = (-1)^{r-1} B_{2r}\, 2^{2r-1}\, \pi^{2r}/(2r)!.$$

Since $\zeta(2r)$ is the sum of the series $1^{-2r} + 2^{-2r} + \cdots$ with positive summands, we get $\zeta(2r) > 0$, hence we prove again that the sign of B_{2r} is equal to $(-1)^{r-1}$; hence $B_2 > 0, B_4 < 0, B_6 > 0, \ldots$. We give a short table, using the values given above (see end of Section 1.2) for the Bernoulli numbers :

$$\zeta(2) = \frac{\pi^2}{6}\ , \quad \zeta(4) = \frac{\pi^4}{90}\ , \quad \zeta(6) = \frac{\pi^6}{945}$$

$$\zeta(8) = \frac{\pi^8}{9450}\ , \quad \zeta(10) = \frac{\pi^{10}}{93555}\ , \quad \zeta(12) = \frac{691\pi^{12}}{638\ 512\ 875}.$$

Comparing formulas (1.52) and (1.58), we may remark (after Euler)

(1.59)
$$\zeta(1-2r) = (-1)^r\, 2^{1-2r} \pi^{-2r} (2r-1)!\ \zeta(2r)$$

for $r = 1, 2, \ldots$. The functional equation for $\zeta(s)$, to be established later, is a generalization of this relation connecting $\zeta(s)$ and $\zeta(1-s)$ for a complex number s. Moreover, we noticed that the function $\zeta(s)$ vanishes for $s = -2, -4, -6, \ldots$; these zeroes are dubbed 'trivial zeroes'. After Riemann, everyone expects that the other zeroes of $\zeta(s)$ are on the *critical line* $Re\ s = \frac{1}{2}$. Despite overwhelming numerical evidence, no mathematical proof is in sight.

Exercise 11 : Using formula (1.43), deduce

$$\lim_{s \to 1} \{\zeta(s) - \frac{1}{s-1}\} = 1 - \int_1^\infty (x - [x]) x^{-2}\, dx$$

and by evaluating the integral conclude that $\zeta(s) - \frac{1}{s-1}$ tends, for s going to 1, to the *Euler constant*

$$\gamma = \lim_{N \to \infty} \left(\frac{1}{1} + \frac{1}{2} + \cdots + \frac{1}{N} - \log\ N\right).$$

Exercise 12 : Using formula (1.43), deduce

$$\zeta'(0) = -1 - \int_1^\infty (x - [x] - \frac{1}{2}) x^{-1}\, dx$$

and by evaluating the integral prove that $\zeta'(0)$ is equal to $-\frac{1}{2} \log\ 2\pi$ (*Hint* : use Stirling's formula).

Exercise 13 : Since $\zeta'(s)$ is given by the series

$$-\sum_{n=1}^{\infty}(\log\ n)\cdot n^{-s}$$

in the half-plane $Re\ s > 1$, one may interpret the previous result as asserting that a 'renormalized sum' for the divergent series $\sum_{n=1}^{\infty} \log\ n$ is equal to $\frac{1}{2}\log\ 2\pi$, or that $\sqrt{2\pi}$ is the renormalized value for the divergent product $\prod_{n=1}^{\infty} n$. Shorthand : $\infty! = \sqrt{2\pi}$.

Exercise 14 : From formula (1.58) infer the relation $\pi^2 = \lim_{r\to\infty} c_r$, where c_r is given by

$$c_r = r(r-1/2)|B_{2r-2}/B_{2r}|.$$

Here is a short list of values

$$c_2 = 15\ ,\ c_3 = 10.5\ ,\ c_4 = 10\ ,\ c_5 = 9.9\ ,\ c_6 = 9.877\ ,\ \ldots$$

to be completed using Serre's table (Serre 1970, page 147). Remember $\pi^2 = 9.8696\cdots$.

Exercise 15 : Calculate π with an accuracy of 10 digits using formula $\pi^{10} = 93555.\zeta(10)$.

Exercise 16 : Show how to deduce from each other the following formulas

$$(*) \qquad B_n = -n!(2\pi i)^{-n}\sum_{m\neq 0} m^{-n},$$

$$(**) \qquad \pi\ \text{cotg}\ \pi z = 1/z + \sum_{n=1}^{\infty} 2z/(z^2-n^2) = \lim_{N\to\infty}\sum_{n=-N}^{N} 1/(z-n),$$

$$(***) \qquad \sin\ \pi z = \pi z\prod_{n=1}^{\infty}(1-z^2/n^2).$$

Euler gave a direct proof of formula $(***)$ thereby providing another proof of the formula (1.58) giving $\zeta(2r)$.

Exercise 17 : a) Fix a complex number $z \neq 0$ and develop into a Fourier series the function e^{xz} for x running over the real interval $]0,1[$. The result is

$$(*) \qquad \frac{e^{xz}}{e^z-1} = \frac{1}{z} + \sum_{m\neq 0}\frac{e^{2\pi i m x}}{z-2\pi i m}$$

where the series must be summed symmetrically

$$\sum_{m\neq 0} = \lim_{N\to\infty}\sum_{\substack{|m|\leq N\\ m\neq 0}}.$$

b) By letting z go to 0, deduce the relation

$$(**) \qquad \frac{1}{2} - x = \sum_{m \neq 0} \frac{e^{2\pi i m x}}{2\pi i m} \quad \text{for} \quad 0 < x < 1$$

(symmetrical summation). This is the limiting case $n = 1$ in formula (1.56).

c) Expanding both sides of formula $(*)$ as power series in z, give a new proof of formula (1.56).

d) Using classical results about Fourier series, evaluate the half-sum of the limiting values for $x = 0$ and $x = 1$ in formula $(*)$ and deduce the formula

$$(***) \qquad \coth \frac{z}{2} = 2 \sum_{m=-\infty}^{+\infty} (z - 2\pi i m)^{-1}$$

(symmetrical summation). Compare with the formulas in exercise 16.

1.6. Hurwitz zeta function

This is the function defined by the series

$$\zeta(s, v) = \sum_{n=0}^{\infty} (n + v)^{-s} \tag{1.60}$$

for $v > 0$. Like the series for $\zeta(s)$, it converges absolutely for $Re\ s > 1$. In this half-plane, we get the obvious relations

$$\zeta(s) = \zeta(s, 1) \tag{1.61}$$

$$\zeta(s, v + 1) = \zeta(s, v) - v^{-s}. \tag{1.62}$$

We now get the following generalization of formula (1.41) by specializing the Euler-MacLaurin summation formula to the case $f(x) = (x + v)^{-s}$, $a = 0$, $b = N$:

$$\begin{aligned} \sum_{r=0}^{N-1} (r + v)^{-s} &= \frac{v^{1-s} - (v + N)^{1-s}}{s - 1} + \frac{v^{-s} - (v + N)^{-s}}{2} \\ &+ \sum_{k=2}^{n} B_k\, s(s+1) \cdots (s + k - 2)\big(v^{-s-k+1} - (v + N)^{-s-k+1}\big)/k! \\ &- \frac{1}{n!}\, s(s+1) \cdots (s + n - 1) \int_0^N \overline{B}_n(x)(x + v)^{-s-n}\, dx. \end{aligned} \tag{1.63}$$

If we let N go to $+\infty$, we get the following representation in the convergence half-plane $Re\ s > 1$

$$\zeta(s,v) = \frac{v^{1-s}}{s-1} + \frac{v^{-s}}{2} + \sum_{k=2}^{n} B_k\ s(s+1)\cdots(s+k-2)\ v^{-s-k+1}/k! \tag{1.64}$$
$$-\frac{1}{n!}\ s(s+1)\cdots(s+n-1)\int_0^\infty \overline{B}_n(x)(x+v)^{-s-n}\ dx.$$

Arguing like in Section 1.4, we conclude that *for every real number* $v > 0$, *there exists a function* $s \mapsto \zeta(s,v)$ *meromorphic in the complex plane* $\mathbb{C}$ *whose only singularity is a pole of order* 1 *at* $s = 1$, *which is given by formula* (1.64) *in the half-plane* $Re\ s > 1 - n$ (for $n = 1, 2, 3, \ldots$). In particular, for $n = 1$, we get the following representation in the half-plane $Re\ s > 0$

$$\zeta(s,v) = \frac{v^{1-s}}{s-1} + \frac{v^{-s}}{2} - s\int_0^\infty (x - [x] - \frac{1}{2})(x+v)^{-s-1}\ dx \tag{1.65}$$

and the representation

$$\zeta(s,v) = \frac{v^{1-s}}{s-1} + \frac{v^{-s}}{2} + \frac{sv^{-s-1}}{12} - \frac{1}{2}\ s(s+1)\int_0^\infty \overline{B}_2(x)(x+v)^{-s-2}\ dx \tag{1.66}$$

in the half-plane $Re\ s > -1$.

From these formulas, one derives

$$\zeta(0,v) = \frac{1}{2} - v. \tag{1.67}$$

More generally, the formula

$$\zeta(-m,v) = -\frac{B_{m+1}(v)}{m+1} \quad \text{for} \quad m = 0, 1, \ldots \tag{1.68}$$

follows easily from (1.64). We leave to the reader the derivation of the following formulas

$$\lim_{s\to 1}\left\{\zeta(s,v) - \frac{1}{s-1}\right\} = -\Gamma'(v)/\Gamma(v) \tag{1.69}$$

$$\zeta'(0,v) = \log\ \Gamma(v) - \frac{1}{2}\ \log(2\pi) \tag{1.70}$$

where $\zeta'(s,v)$ is the derivative of $\zeta(s,v)$ *w.r.t.* s. As in exercise 13, this last formula can be interpreted as follows :

the 'renormalized product' $\prod_{n=0}^{\infty}(n+v)$ *is equal to* $\dfrac{\sqrt{2\pi}}{\Gamma(v)}$.

The particular case $v = \frac{1}{2}$ is worth mentioning : since $\Gamma(\frac{1}{2}) = \sqrt{\pi}$, we see that the 'renormalized product' $\prod_{n=0}^{\infty}(n + \frac{1}{2})$ is equal to $\sqrt{2}$.

Exercise 18 : a) Deduce from formula (1.64) the asymptotic expansion

$$(*) \qquad \zeta(s,v) \sim \frac{v^{1-s}}{s-1} + \frac{v^{-s}}{2} + \sum_{k\geq 2} B_k\, s(s+1)\cdots(s+k-2)v^{-s-k+1}/k!$$

for fixed s and v a real number tending to $+\infty$.

b) Using the functional equation

$$\zeta(s,v) - \zeta(s,v+1) = v^{-s}$$

and the previous asymptotic expansion, show that $\zeta(s,v)$ can be calculated as follows

$$(*\,*) \qquad \zeta(s,v) = \lim_{N\to\infty} \left\{ \sum_{n=0}^{N} (n+v)^{-s} - E'_s(N+v) \right\}$$

where $E'_s(w)$ represents the series

$$\frac{w^{1-s}}{1-s} + \frac{w^{-s}}{2} - \frac{B_2 s}{2}\, w^{-s-1} - \cdots - \frac{B_{n+1}}{(n+1)!}\, s(s+1)\cdots(s+n-1)w^{-s-n} - \cdots$$

truncated after the last term which doesn't tend to 0 for $w \to +\infty$ [compare with exercise 8].

c) Let s be a complex number different from $0, -1, -2, \ldots$. Show that $\eta(v) = \zeta(s,v)$ is the unique function of a variable $v > 0$ satisfying the functional equation

$$(*\,*\,*) \qquad \eta(v) - \eta(v+1) = v^{-s}$$

and admitting of an asymptotic expansion

$$(*\,*\,*\,*) \qquad \eta(v) \sim \sum_{k\geq 0} c_k\, v^{-s_k}$$

for v tending to $+\infty$, with nonzero exponents s_k tending to $+\infty$ with k. [*Hint*: replace $\eta(v)$ by its asymptotic expansion in the functional equation $(***)$ and show that $\eta(v)$ and $\zeta(s,v)$ have the same asymptotic expansion for $v \to +\infty$; then repeat the reasoning in b) to show that $\eta(v)$ is given by the right-hand side of formula $(*\,*)$].

d) Prove the formulas (1.67) to (1.70) by a similar reasoning, using difference equations satisfied by the left-hand side in these formulas.

1.7. Dirichlet *L*-series

A generalization of Riemann's zeta function is obtained as follows : let $f \geq 1$ be any integer and $\{\theta(n)\}$ a sequence of numbers, periodic with period f

$$(1.71) \qquad \theta(n+f) = \theta(n) \quad \text{for } n \text{ in } \mathbb{Z}.$$

We define the series

$$L(\theta, s) = \sum_{n=1}^{\infty} \theta(n) n^{-s}. \tag{1.72}$$

Since the sequence $\{\theta(n)\}$ is periodic, it is bounded and the above series converges for $Re\ s > 1$. The relation with Hurwitz zeta function is obtained as follows

$$\begin{aligned} L(\theta, s) &= \sum_{a=1}^{f} \sum_{m=0}^{\infty} \theta(a+mf)(a+mf)^{-s} \\ &= \sum_{a=1}^{f} \theta(a) \sum_{m=0}^{\infty} (a+mf)^{-s} \\ &= \sum_{a=1}^{f} \theta(a) f^{-s} \sum_{m=0}^{\infty} (m + \frac{a}{f})^{-s} \end{aligned}$$

and finally

$$L(\theta, s) = f^{-s} \sum_{a=1}^{f} \theta(a)\ \zeta(s, \frac{a}{f}). \tag{1.73}$$

From the analytic properties of the function $\zeta(s, v)$, it follows that *the function* $L(\theta, s) - \frac{\Theta}{s-1}$, *where* Θ *is the mean value*

$$\Theta = \lim_{N \to \infty} \frac{1}{N} (\theta(1) + \cdots + \theta(N)) = \frac{1}{f} \sum_{a=1}^{f} \theta(a) \tag{1.74}$$

extends to a holomorphic function in the complex plane $\mathbb{C}$. *In particular, if the mean value* Θ *is* 0, *then* $L(\theta, s)$ *itself extends to an entire function.*

For the special values of $L(\theta, s)$ we note the following

$$L(\theta, -m) = -\frac{f^m}{m+1} \sum_{a=1}^{f} \theta(a) B_{m+1}\left(\frac{a}{f}\right) \tag{1.75}$$

for $m = 0, 1, \ldots$. If the mean value Θ is 0, we derive from (1.69) the value

$$L(\theta, 1) = -\frac{1}{f} \sum_{a=1}^{f} \theta(a) \frac{\Gamma'(a/f)}{\Gamma(a/f)}. \tag{1.76}$$

The most interesting case occurs with Dirichlet characters. Such a character is a function $\chi(n)$ of an integer n which fulfills the following assumptions

a) *periodicity* : $\chi(n + f) = \chi(n)$

b) *multiplicativity* : $\chi(mn) = \chi(m)\chi(n)$ if m and n are prime to f

c) *degeneracy* : $\chi(n) = 0$ if n and f have a common divisor $d > 1$.

The integer f is called the *conductor* of χ. The *principal character* ε_f of conductor f is given by

(1.77) $$\varepsilon_f(n) = \begin{cases} 1 & \textit{if } n \textit{ is prime to } f \\ 0 & \textit{otherwise} . \end{cases}$$

The mean value of ε_f is $\varphi(f)/f$ where $\varphi(f)$ is the number of integers n prime to f between 1 and f. A standard argument shows that the mean value of any character $\chi \neq \varepsilon_f$ is 0.

Any number N prime to f can be written as a product $p_1^{n_1} \cdots p_k^{n_k}$ where the primes $p_1, \ldots, p_k$ don't divide f. Moreover we have

$$\chi(N)N^{-s} = \chi(p_1)^{n_1}\, p_1^{-n_1 s} \cdots \chi(p_k)^{n_k}\, p_k^{-n_k s}$$

in this case for any character χ of conductor f. By repeating the argument in Section 1.1, one proves the following result :

For any character χ of conductor f, one has

(1.78) $$L(\chi, s) = \prod_{p \nmid f} \frac{1}{1 - \chi(p)p^{-s}}$$

where the product is extended over the primes p which don't divide f.

In particular, one gets

(1.79) $$L(\varepsilon_f, s) = \zeta(s) \prod_{p|f} (1 - p^{-s})$$

for the principal character ε_f. Hence $L(\varepsilon_f, s)$ *extends as a meromorphic function in the complex plane, the only singularity being a simple pole at* $s = 1$, *with residue* $\prod_{p|f}(1-p^{-1}) = \varphi(f)/f$. *For any nonprincipal character* χ *of conductor* f, *the function* $L(\chi, s)$ *defined in the half-plane Re* $s > 1$ *extends to an entire function. The special values are given by*

(1.80) $$L(\chi, -m) = -\frac{B_{m+1,\chi}}{m+1},$$

for $m = 0, 1, 2, \ldots$ *where the* χ*-Bernoulli numbers* $B_{m,\chi}$ *are defined as follows*

(1.81) $$B_{m,\chi} = f^{m-1} \sum_{a=1}^{f} \chi(a) B_m\left(\frac{a}{f}\right).$$

Exercise 19 : Derive the following generating series

(*) $$\sum_{m=0}^{\infty} B_{m,\chi}\, z^m/m! = z \frac{\sum_{a=1}^{f} \chi(a) e^{az}}{e^{fz} - 1}.$$

Exercise 20 : Prove the identity

$$(*) \qquad \sum_{a=0}^{f-1} \zeta(s, \frac{v+a}{f}) = f^s \zeta(s, v)$$

for every integer $f \geq 1$. Using formulas (1.67) and (1.68) derive the relation

$$(**) \qquad B_m(v) = f^{m-1} \sum_{a=0}^{f-1} B_m(\frac{v+a}{f}) \qquad (m \geq 0)$$

for the Bernoulli polynomials $B_m(v)$. Using formula (1.70) derive the multiplication formula of Gauss and Legendre

$$(***) \qquad \prod_{a=0}^{f-1} \Gamma\left(\frac{v+a}{f}\right) = (2\pi)^{\frac{f-1}{2}} f^{\frac{1}{2}-v} \Gamma(v).$$

2. Gaussian integers

2.1. A modicum of plane crystallography

We consider an Euclidean plane, with Cartesian coordinates x, y. The corresponding complex coordinate is $z = x + yi$, and enables one to identify complex numbers with points in the plane. The distance between two points z and z' is therefore the modulus $|z - z'|$.

Given two complex numbers ω_1, ω_2, both nonzero, whose ratio ω_2/ω_1 is not real, we denote by Λ the set of complex numbers of the form $z = m_1\omega_1 + m_2\omega_2$ where m_1, m_2 runs over the set $\mathbb{Z}^2$ of pairs of integers (of either sign). We shall say that Λ is the *lattice* with basis (ω_1, ω_2). A given lattice Λ has infinitely many bases, given by

$$(2.1) \qquad \begin{cases} \omega_1' = a\omega_1 + b\omega_2 \\ \omega_2' = c\omega_1 + d\omega_2 \end{cases}$$

where (ω_1, ω_2) is a fixed basis and a, b, c, d are integers such that $ad - bc = \pm 1$. It's a standard practice to consider only *positive* bases, that is to assume that $\omega_2/\omega_1 = \tau$ has a positive imaginary part; this condition amounts to the inequality

$$(2.2) \qquad i(\overline{\omega}_2\omega_1 - \overline{\omega}_1\omega_2) > 0.$$

This being assumed, the formula (2.1) defines a positive basis (ω_1', ω_2') if and only if $ad - bc = 1$.

In this Section, we shall study the following lattices :

a) The *Gauss lattice* Λ_4 (see fig. 2) is the lattice with basis $\omega_1 = 1, \omega_2 = i$. Its elements are the *Gaussian integers* of the form $z = m+ni$, with m, n integral. Like every lattice, Λ_4 is a commutative group *w.r.t.* the addition, namely

$$\begin{cases} (m+ni)+(p+qi) &= (m+p)+(n+q)i \\ -(m+ni) &= -m+(-n)i \ . \end{cases} \tag{2.3}$$

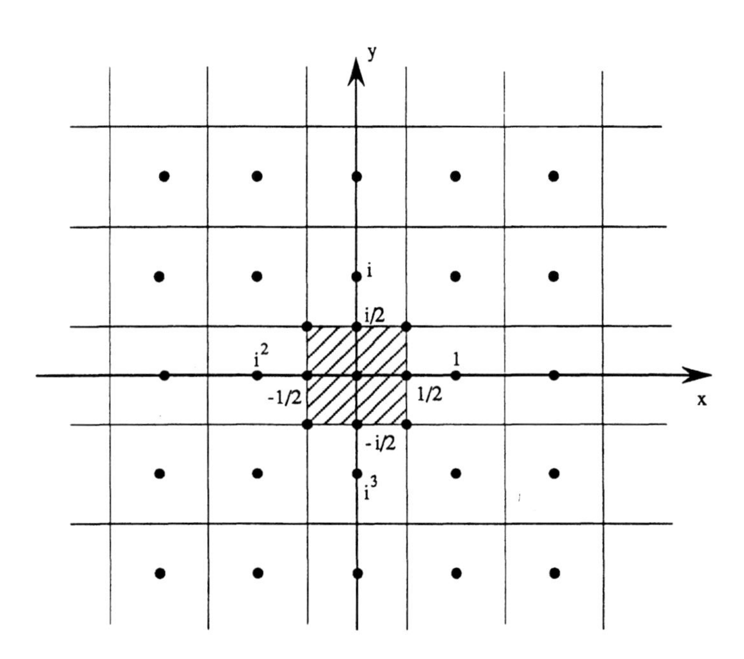

Fig. 2. The Gaussian lattice Λ_4 (the square C is hashed).

But, more specifically, Λ_4 is closed under multiplication

$$(m+ni)(p+qi) = (mp-nq)+(mq+np)i$$

and the unit 1 belongs to Λ_4. Hence Λ_4 is a subring of the ring $\mathbb{C}$ of complex numbers, indeed the smallest subring containing i. We express this property by saying that Λ_4 *is the ring obtained by adjoining* i *to the ring* $\mathbb{Z}$ *of integers*, and we denote it by $\mathbb{Z}[i]$.

b) Let us consider the cube root of unity

$$j = e^{2\pi i/3} = -\frac{1}{2} + \frac{\sqrt{3}}{2} i. \tag{2.4}$$

From the identity $j^2 + j + 1 = 0$, one deduces

$$(2.5) \qquad (m + nj)(p + qj) = (mp - nq) + (mq + np - nq)j.$$

Hence the lattice Λ_6 with basis $(1, j)$ is the ring $\mathbb{Z}[j]$ obtained by adjunction of j to $\mathbb{Z}$ (see fig.3).

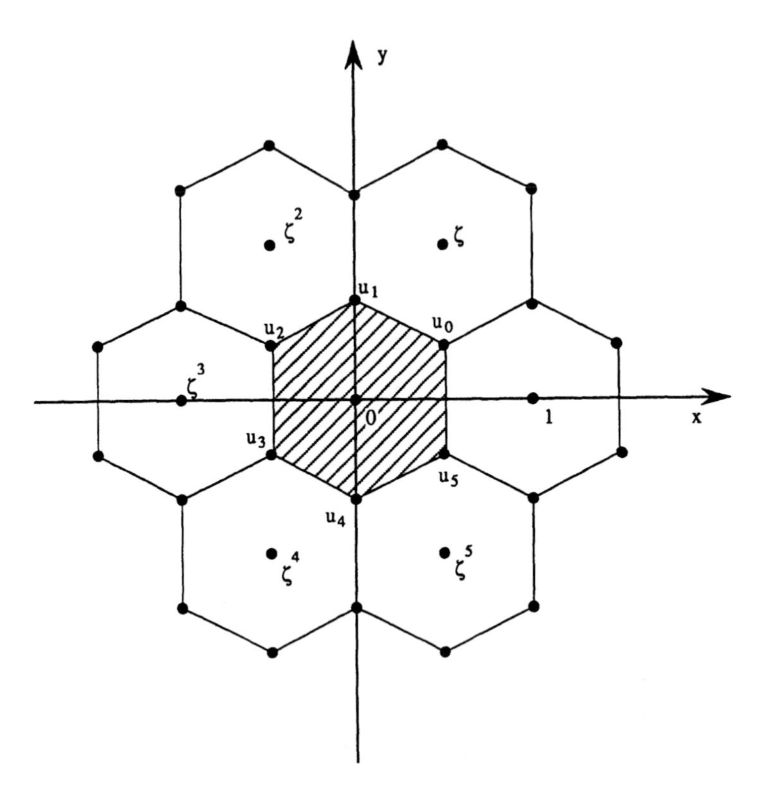

Fig. 3. The lattice Λ_6.

The lattice Λ_4 defines a tessellation of the plane into equal squares. The fundamental square C consists of the complex numbers $z = x + yi$ with the condition $|x| \leq \frac{1}{2}, |y| \leq \frac{1}{2}$. The origin 0 is the center of the square C, its vertices are the numbers $(\pm 1 \pm i)/2$, and the length of the sizes is equal to 1. For every point ω in Λ_4, denote by C_ω the square deduced from C by the translation moving 0 to the point ω : the squares C_ω form the aforementioned tessellation of the plane. Notice also the following characterization of C_ω :

z belongs to C_ω if and only if $|z - \omega| \leq |z - \omega'|$ for every point ω' in the lattice Λ_4.

Let us denote by $T(\Lambda_4)$ the group of translations of the plane of the form $t_\omega(z) = z + \omega$ for ω running over Λ_4. The square C is a *fundamental domain for the group $T(\Lambda_4)$* (or for the lattice Λ_4) namely :

- the plane is the union of the squares $C_\omega = t_\omega(C)$ obtained from C by the translations t_ω in $T(\Lambda_4)$;

- for t_ω and $t_{\omega'}$ distinct (that is for $\omega \neq \omega'$), the squares $t_\omega(C)$ and $t_{\omega'}(C)$ are disjoint or share at most a part of their boundary.

A *scaling transformation* in the plane is a transformation of the form $h_\lambda(z) = \lambda z$, where λ is a fixed nonzero complex number. Such a transformation transforms the lattice Λ_4 with basis $(1, i)$ into the lattice $\lambda\Lambda_4$ with basis $(\lambda, \lambda i)$. A fundamental domain for the lattice $\lambda\Lambda_4$ is the square λC with vertices $(\pm\lambda \pm \lambda i)/2$.

The scaling transformation h_i maps $x+yi$ into $-y+xi$; it is a rotation $r_{\pi/2}$ around 0 of angle $\pi/2$ in the positive direction (counterclockwise). The relation $i(m+ni) = -n+mi$ implies that the lattice Λ_4, hence also every scaled lattice $\Lambda = \lambda\Lambda_4$ is invariant under the rotation $r_{\pi/2}$.

We prove the converse :

Any lattice Λ in the plane which is invariant under the rotation $r_{\pi/2}$ is one of the lattices $\lambda\Lambda_4$.

Proof. a) Since Λ is a discrete subset of the plane, every disc $D(0, R)$ of center 0 and finite radius R contains only finitely many points in Λ. Hence there exists an element $\lambda \neq 0$ in Λ whose length is minimal among the nonzero elements of Λ.

b) Consider the lattice $\Lambda' = \lambda^{-1}\Lambda$. It contains 1 and is stable under the rotation $h_i = r_{\pi/2}$. Hence it contains $i = r_{\pi/2}(1)$ and the lattice Λ_4 with basis $(1, i)$ is contained in Λ'. By definition of λ, any element $\mu \neq 0$ in Λ' satisfies the relation $|\mu| \geq 1$.

c) Since C is a fundamental domain for the lattice Λ_4, any point μ in Λ' is of the form $\mu = \omega + \nu$ where ω belongs to Λ_4 (hence to Λ') and ν belongs to C. Hence ν belongs to $\Lambda' \cap C$. Assume that $\nu \neq 0$. Since ν belongs to Λ' we get $|\nu| \geq 1$ by b). Since ν belongs to C we can write $\nu = x + yi$, with $|x| \leq \frac{1}{2}, |y| \leq \frac{1}{2}$, hence

$$|\nu| = (x^2 + y^2)^{1/2} \leq 1/\sqrt{2} < 1.$$

Contradiction! Hence we get $\nu = 0$, that is $\mu = \omega$ belongs to Λ_4.

d) We have proved the equality $\Lambda' = \Lambda_4$, hence

$$\Lambda = \lambda\Lambda' = \lambda\Lambda_4.$$

□

The lattice Λ_6 shares similar properties, which are established in the same way as above.

For every ω in Λ_6 denote by H_ω the set of points in the plane which are closer to ω than to any other point in Λ_6, namely

$$z \in H_\omega \Leftrightarrow |z - \omega| \leq |z - \omega'| \quad \textit{for every } \omega' \textit{ in } \Lambda_6.$$

The sets H_ω are regular hexagons and form a tessellation of the plane. The hexagon H_ω is derived by the translation t_ω from the hexagon $H = H_0$, which is therefore a fundamental domain for the lattice Λ_6.

Put $\zeta = e^{\pi i/3} = \frac{1}{2} + \frac{\sqrt{3}}{2} i$. We get $\zeta^2 = j$, hence the 6-th roots of unity are enumerated as

$$1, \zeta = -j^2, \zeta^2 = j, \zeta^3 = -1, \zeta^4 = j^2, \zeta^5 = -j.$$

The scaling transformation h_{ζ^k} is the rotation $r_{k\pi/3}$ around the origin. Since the numbers ζ^k belong to the ring $\Lambda_6 = \mathbb{Z}[j]$, the rotation $r_{\pi/3}$ generate a cyclic group C_6 of order 6 of rotations leaving the lattice Λ_6 invariant. Similarly, the rotation $r_{\pi/2}$ generates a cyclic group C_4 of order 4 of rotations leaving the lattice Λ_4 invariant.

The hexagon H is centered at 0, and its vertices are the points

$$u_k = \frac{1}{3}(\zeta^k + \zeta^{k+1}) \quad \text{for} \quad 0 \leq k \leq 5. \tag{2.6}$$

In particular, we get $u_1 = i\sqrt{3}/3$ hence $|u_1| = 1/\sqrt{3}$, and the vertices of H lie on the circle of center 0 and radius $1/\sqrt{3}$. We conclude $|z| \leq 1/\sqrt{3} < 1$ for every point z in the fundamental domain H for the lattice Λ_6.

We can repeat the proof given above for Λ_4 and conclude :

Any lattice Λ which is invariant under the group C_6 of rotations of angle $k\pi/3$ around the origin is of the form $\Lambda = \lambda\Lambda_6$ for some nonzero complex number λ.

Exercise 1 : Let Λ be a lattice. Then any rotation around 0 leaving Λ invariant belongs to C_4 or C_6. In particular, there is no place for five-fold symmetry in crystallography, but see the Chapters on quasi-crystals in this book! [*Hint* : any rotation r_θ of angle θ is a linear transformation in the plane considered as a real vector space. In the basis $(1, i)$ the matrix of r_θ is

$$\begin{pmatrix} \cos\,\theta & -\sin\,\theta \\ \sin\,\theta & \cos\,\theta \end{pmatrix}.$$

In a basis (ω_1, ω_2) of the lattice Λ, the rotation r_θ is expressed by a matrix

$$\begin{pmatrix} a & b \\ c & d \end{pmatrix}$$

where a, b, c, d are integers. Computing the trace, we see that $2\cos\,\theta = a + d$ is an integer, ...]

As a consequence :

a) The group of rotations leaving invariant the lattice Λ_4 (and hence the scaled lattices $\lambda\Lambda_4$) is C_4.

b) Similar statement for Λ_6 and C_6.

c) If a lattice Λ is not a scaled lattice $\lambda\Lambda_4$ or $\lambda\Lambda_6$, then its group of rotations is the group C_2 consisting of the transformations $z \mapsto z$ (identity) and $z \mapsto -z$ (symmetry *w.r.t.* 0).

2.2. Divisibility of Gaussian integers

In the ring $\mathbb{Z}[i]$ of Gaussian integers, we define the notion of divisibility in the obvious way :

Let z and z' be nonzero Gaussian integers. One says that z divides z', or that z' is a multiple of z (notation $z|z'$) *if there exists an element u in $\mathbb{Z}[i]$ such that $z' = uz$, that is if z'/z is a Gaussian integer.*

It is obvious that z divides z; if z divides z' and z' divides z'', then z divides z''. An important feature is the following : it may be that z divides z' and at the same time z' divides z. Indeed call a Gaussian integer u a *unit* if $1/u$ is also a Gaussian integer. Then the previous circumstance holds if and only if z'/z is a unit.

Here is a fundamental result :

The units of the ring $\mathbb{Z}[i]$ of Gaussian integers are the 4-th roots of unity

$$1, i, i^2 = -1, i^3 = -i.$$

For the proof we use the norm $N(z) = |z|^2$ of a complex number $z = x + yi$, that is

$$N(z) = z\overline{z} = x^2 + y^2$$

where $\overline{z}$ denotes the complex conjugate of z. If $z = m+ni$ is a Gaussian integer, its norm is the positive integer m^2+n^2. If z is a unit in $\mathbb{Z}[i]$, there exists another Gaussian integer z' with $zz' = 1$, hence

$$N(z) \cdot N(z') = 1.$$

Since both $N(z)$ and $N(z')$ are positive integers, we get $N(z) = 1$, that is $m^2 + n^2 = 1$. There are obviously four possibilities

$$\left\{ \begin{array}{lll} m = 0 & n = 1 & z = i \\ m = 0 & n = -1 & z = -i \\ m = 1 & n = 0 & z = 1 \\ m = -1 & n = 0 & z = -1 \,. \end{array} \right.$$

Exercise 2 : The units of the ring $\mathbb{Z}$ of ordinary integers are $1, -1$.

Exercise 3 : The units of the ring $\mathbb{Z}[j]$ are the 6-th roots of unity 1, -1, j, $-j$, j^2, $-j^2$.

Exercise 4 : Give an *a priori* proof of the following fact : a unit in the ring $\mathbb{Z}[i]$ is any complex number u such that the scaling transformation $h_u(z) = uz$ be a rotation mapping the lattice $\Lambda_4 = \mathbb{Z}[i]$ into itself.

Exercise 5 : Same as exercise 4 for the lattice Λ_6.

Let us interpret the main geometrical result of Section 2.1 in terms of divisibility. Given any Gaussian integer z, consider the set of its multiples uz, where u runs over $\mathbb{Z}[i]$. Geometrically speaking, it is the *lattice* $z\Lambda_4$; from the number-theoretic point of view, it is an *ideal* of $\mathbb{Z}[i]$, that is a subgroup of $\mathbb{Z}[i]$ (for the addition), stable under multiplication by any element in $\mathbb{Z}[i]$. The set of multiples of z (including 0) is the principal ideal generated by z, to be denoted by (z). For instance, the ideal (0) consists of 0 only, it is called the *zero ideal.*

Let us state the main properties of ideals in $\mathbb{Z}[i]$:

a) *Every nonzero ideal in $\mathbb{Z}[i]$ is a principal ideal.* Namely, let I be such an ideal. From a geometrical point of view, it is a nonzero subgroup of Λ_4, stable under the rotation $r_{\pi/2}$ mapping z into iz. This property precludes the case where I is the set of multiples $n\omega$ of a fixed Gaussian integer ω, with n running over $\mathbb{Z}$. By the elementary divisor theorem, there exists a basis (ω_1, ω_2) for the lattice Λ_4 and integers $d_1 \geq 1, d_2 \geq 1$ such that I be a lattice with basis $(d_1\omega_1, d_2\omega_2)$. Since I is a lattice, and it is invariant under the rotation $r_{\pi/2}$, it is of the form $z\Lambda_4$ by our geometrical results, hence $I = (z)$ as stated.

b) *Let z and z' be nonzero Gaussian integers. Then z divides z' if and only if the principal ideal (z) contains the principal ideal (z')* : obvious.

c) *Two principal ideals (z) and (z') are equal if and only if z'/z is a unit in the ring $\mathbb{Z}[i]$* : follows from b).

d) *A principal ideal (z) is equal to the ideal $(1) = \mathbb{Z}[i]$ if and only if z is a unit* : follows from c).

It is possible to refine statement a). Let I be a nonzero ideal in $\mathbb{Z}[i]$; any Gaussian integer z such that $I = (z)$ is called a *generator* of I. If z is such a generator, there are exactly 4 generators for I namely $z, -z, iz$ and $-iz$ since there are 4 units $1, -1, i, -i$. Moreover, any element in I is of the form $z' = uz$ with a Gaussian integer u and we have $N(z') = N(u)N(z)$ as well as the following classification

$$\begin{cases} N(u) = 0 & \textit{if} \quad u = 0 \\ N(u) = 1 & \textit{if} \quad u \textit{ is a unit} \\ N(u) > 1 & \textit{otherwise} . \end{cases} \tag{2.7}$$

Hence the generators of I are the elements of minimal norm in the set I^* of nonzero elements of I.

Exercise 6 : Extend the previous results to the ring $\mathbb{Z}[j]$.

Exercise 7 : Let I be an ideal in $\mathbb{Z}[i]$ and z a generator of I. Prove the equality $N(z) = (\mathbb{Z}[i] : I)$. More generally, if Λ is any lattice invariant by the rotation $r_{\pi/2}$, the lattice $z\Lambda$ is contained in Λ, and the index $(\Lambda : z\Lambda)$ is equal to the norm $N(z)$ for any nonzero Gaussian integer z.

Exercise 8 : Same as exercise 7 for the ring $\mathbb{Z}[j]$.

2.3. Gaussian primes and factorization

The definition of Gaussian primes is complicated by the existence of units. In the standard arithmetic of integers, the units in the ring $\mathbb{Z}$ are $1, -1$, every nonzero ideal I is principal, with two generators $n, -n$, where n may be taken strictly positive. Let $a \neq 0$ be an integer. The following two properties are equivalent :

a) *the number a is of the form $\pm p$, where p is a prime*[1] *number;*

b) *the number a is not a unit, and for every factorization $a = bc$, either b or c is a unit in $\mathbb{Z}$.*

This suggests the following definition : *a Gaussian prime is any Gaussian integer $\varpi \neq 0$ which is not a unit* (in $\mathbb{Z}[i]$) *and such that for every factorization $\varpi = \lambda\lambda'$ into Gaussian integers, then either λ or λ' is a unit* (in $\mathbb{Z}[i]$).

The product of a Gaussian prime by a unit is again a Gaussian prime. Hence the property of a Gaussian integer ϖ to be prime depends only on the ideal $I = (\varpi)$ and the above definition can be reformulated as follows[2]

ϖ is prime if and only if the ideal $I = (\varpi)$ is different from $\mathbb{Z}[i]$, and I and $\mathbb{Z}[i]$ are the only ideals in $\mathbb{Z}[i]$ containing I.

As a preliminary step towards the factorization of Gaussian integers into Gaussian primes, we establish two lemmas :

(1.) Lemma. (Bezout identity) : *Let ϖ be a Gaussian prime and z a Gaussian integer. If z is not divisible by ϖ, there exist two Gaussian integers u, v such that*

$$uz + v\varpi = 1. \tag{2.8}$$

Remark : If z is divisible by ϖ, every combination $uz + v\varpi$ is divisible by ϖ and this forbids the relation (2.8).

Proof. The linear combinations $uz + v\varpi$, where u, v run independently over $\mathbb{Z}[i]$, form an ideal J in $\mathbb{Z}[i]$ (check it!). Since $0 \cdot z + v\varpi$ belongs to J, the ideal J contains the ideal $(\varpi) = I$. Since $z = 1 \cdot z + 0 \cdot \varpi$ belongs to J, but not to I, the above definition of Gaussian primes leaves open the possibility $J = \mathbb{Z}[i]$ only. Hence 1 belongs to J, and that means equality (2.8) holds for suitable Gaussian integers u, v. □

(2.) Lemma. (Gauss' lemma) : *Let ϖ be a Gaussian integer. Assume that ϖ is neither 0 nor a unit in $\mathbb{Z}[i]$. Then ϖ is a Gaussian prime if and only if it satisfies the following criterion :*

---

[1] In modern times, it was agreed that 1 is *not* a prime number.

[2] In the standard algebraic terminology this amounts to saying that the ideal $I = (\varpi)$ is *maximal*.

(G) *Whenever ϖ divides a product of two Gaussian integers, it divides one of them.*

Proof. Suppose first that (G) holds and consider a factorization $\varpi = \lambda\lambda'$ into Gaussian integers. Since ϖ divides $\varpi = \lambda\lambda'$, it divides λ or λ' according to (G). Assume ϖ divides λ; since λ divides $\varpi = \lambda\lambda'$, the number $\lambda' = \varpi/\lambda$ is a unit in $\mathbb{Z}[i]$. Similarly, if ϖ divides λ', then $\lambda = \varpi/\lambda'$ is a unit. Hence ϖ is a Gaussian prime.

Conversely, assume ϖ is a Gaussian prime. We have to prove that if ϖ doesn't divide the Gaussian integers z and z', it doesn't divide zz'. According to Bezout identity, there exist Gaussian integers u, v, u', v' such that

$$1 = uz + v\varpi = u'z' + v'\varpi.$$

Multiplying out, we get

$$1 = uu' \cdot zz' + v''\varpi$$

with $v'' = uzv' + u'z'v + vv'\varpi$. According to the remark after lemma 1, this precludes zz' from being a multiple of ϖ. □

[In customary parlance, an ideal $\mathfrak{p}$ in a commutative ring is called *prime* if $1 \notin \mathfrak{p}$ and the relations $a \notin \mathfrak{p}, b \notin \mathfrak{p}$ imply $ab \notin \mathfrak{p}$. An ideal $\mathfrak{m}$ is called *maximal* if $1 \notin \mathfrak{m}$ and every ideal containing $\mathfrak{m}$, but not 1, is equal to $\mathfrak{m}$. It's a general property that every maximal ideal is prime. The converse property (that every nonzero prime ideal is maximal) is true only in special rings like the rings $\mathbb{Z}[i]$ or $\mathbb{Z}[j]$, in general in rings in which every ideal is a principal ideal. The content of lemma 2 is that ϖ is a Gaussian prime if and only if the ideal (ϖ) in $\mathbb{Z}[i]$ is prime.]

If ϖ is a Gaussian prime, then $u\varpi$ is also a prime when u runs over the units $1, i, -1, -i$ of $\mathbb{Z}[i]$. We call ϖ a *normalized Gaussian prime* in case ϖ is of the form $m + ni$, with $m > 0$ and $n \geq 0$. Then every Gaussian prime can be written, in a unique way, as a product $u\varpi$ where u is a unit and ϖ a normalized Gaussian prime.

(1) Theorem. (Gauss) : *Let z be a nonzero Gaussian integer.*

a) The number z admits the factorization $u\varpi_1 \cdots \varpi_N$ where u is a unit and $\varpi_1, \ldots, \varpi_N$ are normalized Gaussian primes.

b) If $z = u'\varpi_1' \cdots \varpi_{N'}'$ is another factorization of the same kind, then $u = u', N = N'$ and the sequence $(\varpi_1', \ldots, \varpi_{N'}')$ differs from $(\varpi_1, \ldots, \varpi_N)$ by a permutation.

Proof. Existence : If z is a unit or a Gaussian prime, we're done! Otherwise, we argue by induction on the norm $N(z)$ of z, an integer $n \geq 2$. Since z is not a Gaussian prime, nor a unit, there exists a decomposition $z = z'z''$ where z' and z'' are not units. Hence we get

$$N(z) = N(z')N(z'') \ , \ N(z') > 1 \ , \ N(z'') > 1$$

and it follows

$$N(z') < N(z) \ , \ N(z'') < N(z).$$

By the induction assumption we may find decompositions

$$z' = u'\varpi_1' \cdots \varpi_{N'}' \ , \ z'' = u''\varpi_1'' \cdots \varpi_{N''}''$$

of the sought-for type. Multiplying out we get the required decomposition for $z = z'z''$ since $u = u'u''$ is a unit. □

Proof. Uniqueness : Suppose given two decompositions

$$z = u\varpi_1 \cdots \varpi_N \ , \ z = u'\varpi_1' \cdots \varpi_{N'}'.$$

Without loss of generality, assume $N \leq N'$. Since the Gaussian prime ϖ_1' divides $z = u\varpi_1 \cdots \varpi_N$ and does not divide the unit u, it divides one of the factors $\varpi_1, \ldots, \varpi_N$ (lemma 2). For instance assume ϖ_1' divides ϖ_1. Since both ϖ_1, ϖ_1' are normalized Gaussian primes, this implies $\varpi_1 = \varpi_1'$. We are done if $N' = 1$. Otherwise after simplifying we get

$$u\varpi_2 \cdots \varpi_N = u'\varpi_2' \cdots \varpi_{N'}'$$

and continuing the previous argument, we may assume that, after a permutation of factors if necessary, we have

$$\varpi_1 = \varpi_1', \ldots, \varpi_N = \varpi_N' \ , \ u = u'\varpi_{N+1}' \cdots \varpi_{N'}.$$

If $N = N'$, we derive $u = u'$ from these equalities and we are done. If $N < N'$, by simplifying we get

$$u = u'\varpi_{N+1}' \cdots \varpi_{N'}'$$

and the Gaussian prime $\varpi_{N'}'$ would divide the unit u. But a divisor of a unit is a unit and a Gaussian prime is not a unit. Hence the case $N < N'$ is impossible. □

We can express the decomposition theorem in a more invariant way. Let us denote by $\mathcal{P}$ the set of normalized Gaussian primes. If z is a nonzero Gaussian integer and ϖ a Gaussian prime, the number ϖ appears the same number of times, to be denoted $ord_\varpi(z)$, in any decomposition of the type $z = u\varpi_1 \cdots \varpi_N$. We can therefore write the decomposition into Gaussian primes as follows :

$$z = u \prod_{\varpi \in \mathcal{P}} \varpi^{ord_\varpi(z)} \quad (u \text{ is a unit}). \tag{2.9}$$

Notice that $ord_\varpi(z)$ is a positive integer for every ϖ and that these integers are 0 except for finitely many ϖ's in $\mathcal{P}$.

In terms of ideals, the map $\varpi \mapsto (\varpi)$ establishes a bijection between $\mathcal{P}$ and the set of prime ideals in $\mathbb{Z}[i]$. The number $ord_\varpi(z)$ depends only on the prime ideal $\mathfrak{p} = (\varpi)$ and can be denoted $ord_{\mathfrak{p}}(z)$. Formula (2.9) can be written as follows

$$(z) = \prod_{\mathfrak{p}} \mathfrak{p}^{ord_{\mathfrak{p}}(z)} \tag{2.10}$$

where the product extends over the *prime ideals* $\mathfrak{p}$. In this form, the factorization theorem was generalized by Kummer (around 1840) to all algebraic numbers.

Exercise 9 : Extend the results of Section 2.3 to the elements of $\mathbb{Z}[j]$.

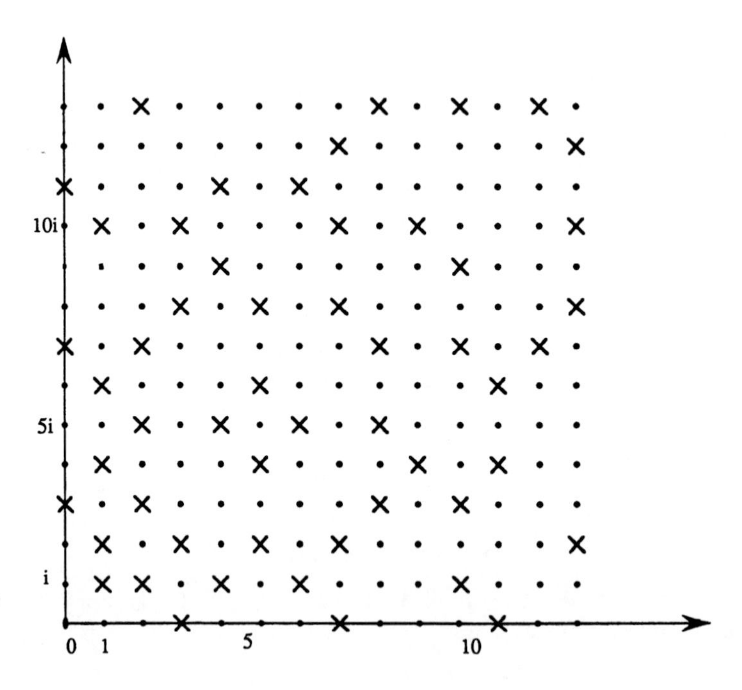

Fig. 4. Gaussian primes

2.4. Classification of Gaussian primes

Once it is proved that every ideal in the ring $\mathbb{Z}[i]$ is principal, the arguments leading to the factorization theorem parallel closely the classical ones for ordinary integers and primes. The only difference is that extra care is needed to handle units.

The problem is now to classify the Gaussian primes. The figure 4 displays the Gaussian primes of the form $\varpi = a + bi$ with $0 \leq a \leq 13, 0 \leq b \leq 13$.

(3.) Lemma. *Any Gaussian prime divides an ordinary prime.*

Proof. Let ϖ be a Gaussian prime. The number 1 is not a multiple of ϖ otherwise ϖ would be a unit. There exist ordinary integers $n \geq 1$ which are multiple of ϖ, namely $N(\varpi) = \varpi \cdot \overline{\varpi}$. Let p be the smallest among the ordinary integers $n \geq 1$ which are multiple of ϖ in $\mathbb{Z}[i]$. For any factorization $p = a \cdot b$ with $1 < a < p, 1 < b < p$, the Gaussian prime ϖ doesn't divide a and b by the minimality property of p. By Gauss' lemma this contradicts the fact that ϖ divides $p = ab$. Hence p is an ordinary prime. □

To classify the Gaussian primes, we have therefore to factorize in $\mathbb{Z}[i]$ the ordinary primes. Let p be such a prime. From a decomposition

$$p = u\varpi_1 \cdots \varpi_N$$

we get

$$\left\{ \begin{array}{l} p^2 = N(p) = N(\varpi_1) \cdots N(\varpi_N) \\ N(\varpi_j) > 1 \quad \text{for} \quad j = 1, \ldots, N. \end{array} \right. \tag{2.11}$$

There are therefore three possibilities :

1) One has $p = u\varpi^2$ where u is a unit, ϖ is a normalized Gaussian prime and $N(\varpi) = p$. We say p is *ramified* in $\mathbb{Z}[i]$.

2) One has $p = u\varpi\varpi'$ where u is a unit, ϖ and ϖ' are normalized Gaussian primes with

$$\varpi \neq \varpi' \ , \ N(\varpi) = N(\varpi') = p.$$

We say p is *split* in $\mathbb{Z}[i]$.

3) The number p is a normalized Gaussian prime : for every factorization $p = \lambda\lambda'$ into Gaussian integers, either λ or λ' is a unit in $\mathbb{Z}[i]$. We say p is *inert* in $\mathbb{Z}[i]$.

Exercise 10 : Extend the previous discussion to the ring $\mathbb{Z}[j]$.

The following theorem explains how to categorize the ordinary primes as ramified, split and inert.

(2) Theorem. (Gauss) : *a) The only ramified prime is* 2.

b) The split primes are the primes of the form $p = 4r + 1$ *(with an integer* $r \geq 1$*).*

c) The inert primes are the primes of the form $p = 4r + 3$ *(with an integer* $r \geq 1$*).*

With the standard notation $a \equiv b$ mod. m meaning $(a-b)/m$ is an integer, the split primes satisfy $p \equiv 1$ mod. 4 and the inert primes satisfy $p \equiv 3$ mod. 4.

Table 1 : Primes $p \leq 100$

Ramified	2								
Split	5,	13,	17,	29,	37,	41,	53,	61,	73,
	89,	97							
Inert	3,	7,	11,	19,	23,	31,	43,	47,	59,
	67,	71,	79,	83					

Table 2 : Decomposition of primes ≤ 100

$2 = 1^2 + 1^2$	$29 = 5^2 + 2^2$	$61 = 6^2 + 5^2$
$5 = 2^2 + 1^2$	$37 = 6^2 + 1^2$	$73 = 8^2 + 3^2$
$13 = 3^2 + 2^2$	$41 = 5^2 + 4^2$	$89 = 8^2 + 5^2$
$17 = 4^2 + 1^2$	$53 = 7^2 + 2^2$	$97 = 9^2 + 4^2$

Proof. The prime 2 is ramified according to the formulas

$$2 = -i(1+i)^2 \quad , \quad N(1+i) = 2.$$

The prime number p is ramified or split if and only if there exists a Gaussian integer ϖ with $p = N(\varpi)$. Putting $\varpi = m + ni$, this relation amounts to $p = m^2 + n^2$. According to the relations

$$(2s)^2 = 4s^2 \quad , \quad (2s+1)^2 = 4(s^2 + s) + 1$$

any square is congruent to 0 or 1 mod. 4. Hence the sum of two squares is congruent to $0 = 0 + 0, 1 = 1 + 0$ or $2 = 1 + 1$ mod. 4, hence never to 3 mod. 4. It follows that every prime $p \equiv 3$ mod. 4 is inert. We know already that 2 is ramified. Hence it remains to prove that

any prime number $p \equiv 1$ mod. 4 *is split in* $\mathbb{Z}[i]$.

According to Fermat 'small theorem', every integer a not divisible by p satisfies the congruence

$$a^{p-1} \equiv 1 \quad \text{mod. } p. \tag{2.12}$$

Moreover, there exists a *primitive root modulo* p, namely an integer α not divisible by p such that any integer a not divisible by p is congruent mod. p to some power of α. In other words, consider the powers

$$\alpha^0 = 1 \ , \ \alpha^1 = \alpha, \alpha^2, \dots, \alpha^{p-2}$$

and the remainders of the division by p

$$\beta_1, \dots, \beta_{p-2}, \beta_{p-1}$$

(β_j is the remainder of the division of α^{j-1} by p). Then the previous numbers form a permutation of the numbers

$$1, 2, \dots, p-1.$$

By assumption $r = (p-1)/4$ is an integer. Choose a primitive root α modulo p and put $a = \alpha^r$. Then $a^2 - 1 = \alpha^{(p-1)/2} - 1$ is not divisible by p, but $a^4 - 1 = \alpha^{p-1} - 1$ is divisible by p. Since $(a^2+1)(a^2-1) = a^4 - 1$, it follows from the ordinary Gauss lemma that p divides a^2+1. Hence, we found integers a, b with the following properties

$$\text{(2.13)} \qquad \begin{cases} a^2 + 1 = pb \\ a \text{ is not divisible by } p \ . \end{cases}$$

We introduce now the lattice Λ with basis $(p, a-i)$. It follows from the formulas

$$\text{(2.14)} \qquad \begin{cases} i \cdot p = a \cdot p - p \cdot (a - i) \\ i \cdot (a - i) = b \cdot p - a \cdot (a - i) \end{cases}$$

that the lattice Λ is stable by multiplication by i, that is by the rotation $r_{\pi/2}$. Otherwise stated, Λ is an ideal in the ring $\mathbb{Z}[i]$, we have

$$p \cdot \mathbb{Z}[i] \subset \Lambda \subset \mathbb{Z}[i]$$

and it is easy to check that Λ is distinct from both $p \cdot \mathbb{Z}[i]$ and $\mathbb{Z}[i]$. Since every ideal is principal, we may choose a generator ϖ of Λ. Then ϖ is a Gaussian integer and since p belongs to the ideal $\Lambda = (\varpi)$, there is another Gaussian integer ϖ' such that $p = \varpi \cdot \varpi'$. Since the ideals $(p) = p \cdot \mathbb{Z}[i], (\varpi) = \Lambda$ and $(1) = \mathbb{Z}[i]$ are distinct, ϖ is not a unit, nor $\varpi' = p/\varpi$ is. We get

$$N(\varpi) > 1 \ , \ N(\varpi') > 1 \ , \ N(\varpi)N(\varpi') = N(p) = p^2$$

and from this it follows

$$N(\varpi) = N(\varpi') = p.$$

We have found a Gaussian integer $\varpi = a+bi$, with norm p, that is $p = \varpi \cdot \overline{\varpi}$. It remains to show that ϖ is a Gaussian prime and that $\overline{\varpi}/\varpi$ is not a unit

a) ϖ is a Gaussian prime : namely, for any factorization $\varpi = \lambda\lambda'$ into Gaussian integers, we get

$$N(\lambda)N(\lambda') = N(\varpi) = p.$$

Since p is a prime, either $N(\lambda) = 1$ and λ is a unit, or $N(\lambda') = 1$ and λ' is a unit. Hence ϖ is a Gaussian prime.

b) We have $\varpi = a + bi$, $\overline{\varpi} = a - bi$, and the units are $1, -1, i, -i$. If $\overline{\varpi}/\varpi$ is a unit, then by inspection we are left with the following cases

$$a = 0 \ , \ b = 0 \ , \ a = b \ , \ a = -b.$$

Then $p = a^2 + b^2$ is a square, or twice a square which contradicts the assumption that p is a prime different from 2. □

From the previous proof, one obtains an explicit description of the factorizations :

a) The *four* numbers $\pm 1 \pm i$ are the Gaussian primes of norm 2, they generate the same ideal.

b) For every prime $p \equiv 3$ mod. 4, there are *four* Gaussian primes with norm p^2, namely $p, -p, ip, -ip$.

c) For every prime $p \equiv 1$ mod. 4, there exists a decomposition $p = a^2 + b^2$ as sum of two squares. We may assume $0 < a < b$, and there are *eight* Gaussian primes of norm p, namely

$$a + bi \ , \ i(a + bi) \ , \ -(a + bi) \ , \ -i(a + bi)$$

$$b + ai \ , \ i(b + ai) \ , \ -(b + ai) \ , \ -i(b + ai).$$

Geometrically, we have *eight* points in the square lattice Λ_4 at a distance $\sqrt{p}$ of the origin, namely $(\pm a, \pm b)$ and $(\pm b, \pm a)$.

Exercise 11 : Describe the prime numbers in the ring $\mathbb{Z}[j]$ [*Hint* : 3 is ramified, any $p \equiv 1$ mod. 3 is split, any $p \equiv 2$ mod. 3 is inert.]

2.5. Sums of squares

Fermat considered the following problem :

Represent, if possible, an integer $n \geq 1$ as a sum of two squares

$$n = a^2 + b^2.$$

It amounts to represent n as the norm of some Gaussian integer $\varpi = a + bi$. Using the factorization theorem established in Section 2.3, we can write ϖ as a product $u\varpi_1 \cdots \varpi_N$, hence

$$a^2 + b^2 = N(a + bi) = N(\varpi_1) \cdots N(\varpi_N). \tag{2.15}$$

According to the results in Section 2.4, the norm of a Gaussian prime ϖ_j is equal to 2, to a prime number $p \equiv 1$ mod. 4, or to the square of a prime number $p \equiv 3$ mod. 4. The following criterion, due to Fermat, follows immediately:

An integer $n \geq 1$ is a sum of two squares if and only if every prime divisor of n congruent to 3 mod. 4 *appears with an even exponent in the prime factor decomposition of n.*

A similar problem pertaining to the sum of *three* squares was mentioned by Bachet and solved by Gauss in the form

'Demonstravi num. $= \Delta + \Delta + \Delta$'.

Here is the meaning : *a triangular number* is a number of the form (see fig. 5.)

$$1 + 2 + \cdots + a = \frac{a(a+1)}{2}$$

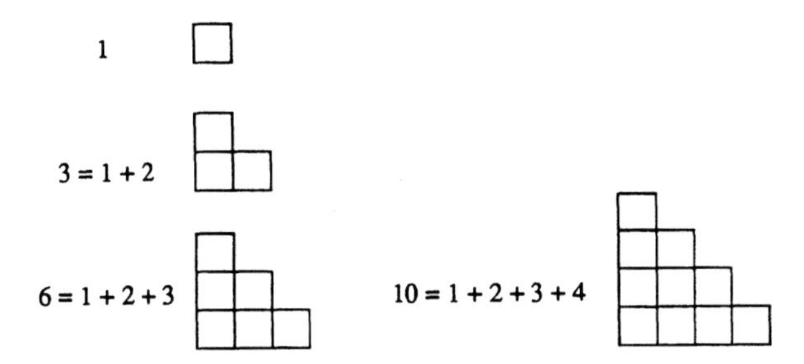

Fig. 5. The triangular numbers.

Gauss' theorem is the possibility of representing any integer $n \geq 1$ in the form

$$n = \frac{a^2 + a}{2} + \frac{b^2 + b}{2} + \frac{c^2 + c}{2}.$$

Otherwise stated, every number of the form $8n + 3$ is the sum of three odd squares.

Let us mention also Lagrange's theorem :

Any integer $n \geq 1$ is a sum of four squares.

In order to give a quantitative meaning to similar results, one introduces the following definition. Let k, n be integers with $k \geq 1, n \geq 0$. The number $r_k(n)$ denotes the number of solutions of the equation

$$x_1^2 + \cdots + x_k^2 = n \tag{2.16}$$

where $x_1, \ldots, x_k$ are integers (of either sign). Hence Lagrange's theorem states

$$r_k(n) > 0 \quad \text{for} \quad k \geq 4\ ,\ n \geq 0. \tag{2.17}$$

It is convenient to introduce generating series in the form of *theta functions.* Put

$$\theta(q) = \sum_{r \in \mathbb{Z}} q^{r^2} = 1 + 2q + 2q^4 + 2q^9 + \cdots. \tag{2.18}$$

The radius of convergence of this series is obviously 1. Moreover, we get

$$\theta(q)^k = \sum_{n=0}^{\infty} r_k(n) q^n \tag{2.19}$$

according to the following calculation

$$\begin{aligned}
\theta(q)^k &= \sum_{a_1} q^{a_1^2} \sum_{a_2} q^{a_2^2} \cdots \sum_{a_k} q^{a_k^2} \\
&= \sum_{a_1 \cdots a_k} q^{a_1^2 + \cdots + a_k^2} \\
&= \sum_{n=0}^{\infty} \sum_{a_1^2 + \cdots + a_k^2 = n} q^n \\
&= \sum_{n=0}^{\infty} r_k(n) q^n .
\end{aligned}$$

Hence the series $\sum_{n=0}^{\infty} r_k(n) q^n$ converges for $|q| < 1$. More precisely an easy geometric argument about volumes shows that $r_k(n) \leq C_k \; n^{(k/2)}$ where the constant C_k is independent of n.

Fermat's theorem about sums of two squares and Lagrange's theorem about sums of four squares were given a quantitative form by Jacobi in 1828, namely

$$\theta(q)^2 = 1 + 4 \sum_{n=0}^{\infty} \frac{(-1)^n q^{2n+1}}{1 - q^{2n+1}} \tag{2.20}$$

$$\theta(q)^4 = 1 + 8 \sum_{m} \frac{m q^m}{1 - q^m} \tag{2.21}$$

(sum extended over the integers m not divisible by 4). We shall give in the next Section a proof of formula (2.20) depending on Gaussian methods. Jacobi's proofs of the above mentioned formulas were purely analytical.

2.6. The zeta function of $\mathbb{Z}[i]$

Besides the generating series

$$\theta(q)^k = \sum_{n=0}^{\infty} r_k(n) q^n = \sum_{a_1 \cdots a_k} q^{a_1^2 + \cdots + a_k^2},$$

we shall consider the following Dirichlet series

$$\sum_{n=1}^{\infty} r_k(n) n^{-s} = \sum_{a_1 \cdots a_k}' (a_1^2 + \cdots + a_k^2)^{-s}. \tag{2.22}$$

The summation $\sum'_{a_1 \cdots a_k}$ extends over all systems $(a_1, \ldots, a_k)$ of k integers, except for $a_1 = \cdots = a_k = 0$. In particular, we get

$$Z_4(s) = \sum_{a,b}'(a^2 + b^2)^{-s} \tag{2.23}$$

the summation being extended over the pairs $(a, b) \neq (0, 0)$ in $\mathbb{Z}^2$.

Using the notions connected with Gaussian integers, we get the alternative form

$$Z_4(s) = \sum_{z \neq 0} N(z)^{-s}, \tag{2.24}$$

where the sum is extended over the nonzero Gaussian integers. According to the factorization theorem for Gaussian integers, nonzero Gaussian integers are parametrized by units u and family of positive integers $m(\varpi)$ according to the formula

$$z = u \prod_{\varpi \in \mathcal{P}} \varpi^{m(\varpi)} \tag{2.25}$$

(here $\mathcal{P}$ is the set of normalized Gaussian primes). We then have

$$N(z) = \prod_{\varpi \in \mathcal{P}} N(\varpi)^{m(\varpi)}. \tag{2.26}$$

Repeating the proof given in Section 1.1 for Riemann's zeta function we get

$$Z_4(s) = 4 \prod_{\varpi \in \mathcal{P}} (1 + N(\varpi)^{-s} + N(\varpi)^{-2s} + \cdots) \tag{2.27}$$

hence

$$Z_4(s) = 4 \prod_{\varpi \in \mathcal{P}} \frac{1}{1 - N(\varpi)^{-s}} \tag{2.28}$$

(notice that there are 4 units in $\mathbb{Z}[i]$!).

Taking into account the three categories of primes according to their decomposition law in $\mathbb{Z}[i]$, we get the following factors in the product expansion (2.28) :

a) the Gaussian prime $1 + i$ of norm 2 gives a factor $(1 - 2^{-s})^{-1}$;

b) for every prime $p \equiv 1$ mod. 4, there exists two normalized Gaussian primes of norm p, hence a factor $(1 - p^{-s})^{-2}$;

c) for every prime $p \equiv 3$ mod. 4, p itself is a Gaussian prime, of norm p^2, hence a factor

$$(1-p^{-2s})^{-1} = (1-p^{-s})^{-1}(1+p^{-s})^{-1}.$$

If we recall the definition of $\zeta(s)$ as the infinite product $\prod_p (1-p^{-s})^{-1}$ extended over all primes p, we get

$$Z_4(s) = 4\zeta(s)L(s). \tag{2.29}$$

The function $L(s)$ is given as a product

$$L(s) = \prod_{p'} \frac{1}{1-p'^{-s}} \prod_{p''} \frac{1}{1+p''^{-s}} \tag{2.30}$$

where p' (resp. p'') runs over all prime numbers congruent to 1 (resp. 3) modulo 4.

One defines a character of conductor 4 as follows

$$\chi_4(n) = \begin{cases} 0 & \text{if} \quad n \quad \text{is even} \\ (-1)^{(n-1)/2} & \text{if} \quad n \quad \text{is odd.} \end{cases} \tag{2.31}$$

We can rewrite formula (2.30) as follows

$$L(s) = \prod_p \frac{1}{1-\chi_4(p)p^{-s}}. \tag{2.32}$$

Otherwise stated, $L(s)$ is the Dirichlet L-series corresponding to the character χ_4 (*cf.* Section 1.7). We can therefore write

$$L(s) = \sum_{n=1}^{\infty} \chi_4(n)n^{-s} = 1^{-s} - 3^{-s} + 5^{-s} - 7^{-s} + \cdots \tag{2.33}$$

the series converging for $Re\ s > 1$. According to formula (2.29) we have therefore

$$Z_4(s) = 4(1^{-s}+2^{-s}+3^{-s}+\cdots)(1^{-s}-3^{-s}+5^{-s}-\cdots). \tag{2.34}$$

Multiplying out these two series, and remembering that $r_2(n)$ is the coefficient of n^{-s} in the series $Z_4(s)$, we get

$$r_2(n) = 4\sum_{n=jk} (-1)^{(k-1)/2} \qquad \text{for } n \geq 1, \tag{2.35}$$

where the summation extends over $j \geq 1$ and $k \geq 1$ odd. This formula can be transformed as follows. Firstly, if we denote by $d_+(n)$ the number of divisors k of n such that $k \equiv 1$ mod. 4 and similarly $d_-(n)$ for divisors with $k \equiv 3$ mod. 4, we get

$$r_2(n) = 4\big(d_+(n) - d_-(n)\big) \quad \text{for } n \geq 1. \tag{2.36}$$

Moreover, we get

$$\begin{aligned} \sum_{n=1}^{\infty} r_2(n) q^n &= 4 \sum_{k \text{ odd}} (-1)^{(k-1)/2} \sum_{j=1}^{\infty} q^{jk} \\ &= 4 \sum_{k \text{ odd}} (-1)^{(k-1)/2} \frac{q^k}{1-q^k}. \end{aligned} \tag{2.37}$$

Finally we get Jacobi's formula

$$\theta(q)^2 = 1 + \sum_{n=1}^{\infty} r_2(n) q^n = 1 + 4 \sum_{r=0}^{\infty} \frac{(-1)^r q^{2r+1}}{1-q^{2r+1}}. \tag{2.38}$$

To conclude, we describe the corresponding results for the ring $\mathbb{Z}[j]$. By definition we have

$$Z_3(s) = \sum_{z \neq 0} N(z)^{-s} \tag{2.39}$$

where the sum is extended over the nonzero elements in $\mathbb{Z}[j]$. Since the norm of $a - bj$ is equal to $a^2 + ab + b^2$, we get

$$Z_3(s) = \sum_{a,b}{}' (a^2 + ab + b^2)^{-s} \tag{2.40}$$

(summation over pairs of integers a, b excluding $a = b = 0$). There are 6 units in $\mathbb{Z}[j]$ and using the decomposition laws of prime numbers in $\mathbb{Z}[j]$ we get

$$Z_3(s) = 6\zeta(s) L(\chi_3, s) \tag{2.41}$$

where the character χ_3 of conductor 3 is defined as follows

$$\chi_3(n) = \begin{cases} 0 & \text{if} \quad n \equiv 0 \text{ mod. } 3 \\ 1 & \text{if} \quad n \equiv 1 \text{ mod. } 3 \\ -1 & \text{if} \quad n \equiv -1 \text{ mod. } 3. \end{cases} \tag{2.42}$$

More explicitly

$$L(\chi_3, s) = \sum_{r=1}^{\infty} \{(3r+1)^{-s} - (3r+2)^{-s}\}. \tag{2.43}$$

The number of representations of an integer $n \geq 1$ by the quadratic form $a^2 + ab + b^2$ is equal to $6\big(\delta_+(n) - \delta_-(n)\big)$ where $\delta_+(n)$ (resp. $\delta_-(n)$) is the number of divisors of n which are congruent to 1 (resp. -1) modulo 3. For the theta series we get

$$\sum_{a,b} q^{a^2+ab+b^2} = 1 + 6 \sum_{n=0}^{\infty} \frac{q^{3n+1}}{1-q^{3n+1}} - 6 \sum_{n=0}^{\infty} \frac{q^{3n+2}}{1-q^{3n+2}}. \tag{2.44}$$

3. Functional equation

3.1. A short account of Fourier transformation

We first recall Fejer's fundamental theorem about Fourier series. Let $f(x)$ be a function of a real variable, assumed to be bounded, measurable and periodic with period 1, namely

$$f(x+1) = f(x). \tag{3.1}$$

The *Fourier coefficients* are defined by

$$c_n = \int_0^1 f(x)\, e^{-2\pi i n x}\, dx \tag{3.2}$$

for any integer n in $\mathbb{Z}$. The partial sums of the Fourier series are the following

$$\sigma_N(x) = \sum_{n=-N}^{N} c_n\, e^{2\pi i n x}. \tag{3.3}$$

Let x_0 be any real number such that the function $f(x)$ admits left and right limiting values

$$f(x_0 \pm 0) = \lim_{\varepsilon \to 0_+} f(x_0 \pm \varepsilon). \tag{3.4}$$

Then Fejer's formula is the following

$$\frac{f(x_0+0)+f(x_0-0)}{2} = \lim_{N\to\infty} \frac{\sigma_0(x_0)+\cdots+\sigma_N(x_0)}{N+1}. \tag{3.5}$$

A simplification occurs when $f(x)$ is continuous at $x = x_0$, and the partial sums $\sigma_N(x_0)$ converge ; we have then

$$f(x_0) = c_0 + \sum_{n=1}^{\infty} \left\{ c_n\, e^{2\pi i n x_0} + c_{-n}\, e^{-2\pi i n x_0} \right\}. \tag{3.6}$$

The most favorable case occurs when $f(x)$ is a continuous function and the series $\sum\limits_{n=-\infty}^{\infty} |c_n|$ is finite. Then the function is represented by the absolutely convergent Fourier series

$$f(x) = \sum_{n=-\infty}^{\infty} c_n\, e^{2\pi i n x}. \tag{3.7}$$

Consider now a function $F(x)$ of a real variable which is continuous and (absolutely) integrable, namely the integral

$$\| F \|_1 = \int_{-\infty}^{+\infty} |F(x)|\, dx \tag{3.8}$$

is finite. The *Fourier transform* is normalized as follows

$$\widehat{F}(u) = \int_{-\infty}^{+\infty} F(x)\, e^{-2\pi i x u}\, dx; \tag{3.9}$$

it's a continuous function of u. Assuming that $\widehat{F}$ is integrable, namely

$$\int_{-\infty}^{+\infty} |\widehat{F}(u)|\, du < \infty,$$

Fourier inversion formula holds

$$F(x) = \int_{-\infty}^{+\infty} \widehat{F}(u)\, e^{2\pi i u x}\, du; \tag{3.10}$$

that is the Fourier transform of $\widehat{F}(u)$ is $f(-x)$. Notice that the integration kernel is $e^{-2\pi i u x}$ in (3.9) and $e^{+2\pi i u x}$ in (3.10) ; putting the 2π in the exponential gives the most symmetrical form of the inversion formula. Here is a short table of Fourier transforms :

Table 3 : Fourier transforms

Function	Transform
$F(x)$	$\widehat{F}(u)$
$F(x+a)$	$e^{2\pi i a u}\widehat{F}(u)$
$e^{2\pi i x b}F(x)$	$\widehat{F}(u-b)$
$F(t^{-1}x)$	$\|t\|\widehat{F}(tu)$
$F'(x)$	$2\pi i u\ \widehat{F}(u)$
$x\ F(x)$	$\frac{i}{2\pi}\widehat{F}'(u)$
$e^{-c\|x\|}\ (Re\ c > 0)$	$2c/(c^2+4\pi^2u^2)$
$e^{-\pi x^2}$	$e^{-\pi u^2}$

In the previous table, $F'(x)$ is the derivative of $F(x)$ and $\widehat{F}'(u)$ that of $\widehat{F}(u)$. We give now a few details about the last two examples.

a) For $F(x) = e^{-c|x|}$ we get

$$\begin{aligned}\widehat{F}(u) &= \int_{-\infty}^{+\infty} e^{-c|x|}\, e^{-2\pi i u x}\, dx \\ &= \left(\int_0^{\infty} + \int_{-\infty}^{0}\right)\cdots \\ &= \int_0^{\infty} e^{-(c+2\pi i u)x} dx + \int_0^{\infty} e^{-(c-2\pi i u)x} dx \\ &= \frac{1}{c+2\pi i u} + \frac{1}{c-2\pi i u} = \frac{2c}{c^2+4\pi^2 u^2}.\end{aligned}$$

We used the elementary identity

$$\int_0^{\infty} e^{-px} dx = \frac{1}{p}$$

where the integral converges absolutely when $Re\ p > 0$; this is why we need the assumption $Re\ c > 0$.

b) For $F(x) = e^{-\pi x^2}$, we want $\widehat{F}(u) = e^{-\pi u^2}$, that is

$$\int_{-\infty}^{+\infty} e^{-\pi x^2} e^{-2\pi i x u}\, dx = e^{-\pi u^2}. \tag{3.11}$$

This relation amounts to

$$\int_{-\infty}^{+\infty} e^{-\pi(x+iu)^2} dx = 1. \tag{3.11bis}$$

The proof is given in two steps :

- For $u = 0$ we need the relation

$$\int_{-\infty}^{+\infty} e^{-\pi x^2} dx = 1. \tag{3.12}$$

Call I the previous integral. Then calculate I^2 and go to polar coordinates ; this standard trick runs as follows :

$$\begin{aligned} I^2 &= \int_{-\infty}^{+\infty} e^{-\pi x^2} dx \int_{-\infty}^{+\infty} e^{-\pi y^2} dy \\ &= \int_{\mathbb{R}^2} e^{-\pi(x^2+y^2)} dx\, dy \\ &= \int_0^{\infty} r\, dr \int_0^{2\pi} e^{-\pi r^2}\, d\theta \qquad [x = r\ \cos\theta\ ,\ y = r\ \sin\theta] \\ &= \int_0^{\infty} 2\pi\ r\ e^{-\pi r^2} dr \\ &= \int_0^{\infty} e^{-\pi r^2} d(\pi r^2) \\ &= \int_0^{\infty} e^{-v} dv = 1 \qquad [v = \pi r^2]. \end{aligned}$$

Moreover I is the integral of a positive function, hence $I > 0$ and since $I^2 = 1$, we get $I = 1$ as required.

• Put $H(x,u) = e^{-\pi(x+iu)^2}$. Then the following differential equation holds

$$\frac{\partial}{\partial u} H(x,u) = i \frac{\partial}{\partial x} H(x,u) = -2\pi i(x+iu)\, H(x,u).$$

Integrating *w.r.t.* x, we get

$$\begin{aligned}\frac{\partial}{\partial u}\int_{-\infty}^{+\infty} H(x,u)dx &= \int_{-\infty}^{+\infty}\frac{\partial}{\partial u} H(x,u)dx\\ &= i\int_{-\infty}^{+\infty}\frac{\partial H(x,u)}{\partial x}\,dx\\ &= i\; H(x,u)\,|_{x=-\infty}^{x=+\infty} = 0.\end{aligned}$$

It follows that the integral $\int_{-\infty}^{+\infty} H(x,u)dx$ is independent of u, but for $u = 0$, it's equal to $\int_{-\infty}^{+\infty} H(x,0)dx = \int_{-\infty}^{+\infty} e^{-\pi x^2} dx$, hence to 1 by the previous calculation. Finally we get

$$\int_{-\infty}^{+\infty} H(x,u)dx = 1,$$

that is the sought-for relation (3.11*bis*).

3.2. Poisson summation formula

Let $F(x)$ and $\widehat{F}(u)$ be a pair of Fourier transforms. Poisson summation formula reads as follows :

$$\sum_{n\in\mathbb{Z}} F(n) = \sum_{m\in\mathbb{Z}} \widehat{F}(m). \tag{3.13}$$

Since $F(x+v)$ and $e^{2\pi i v u}\ \widehat{F}(u)$ form another pair of Fourier transforms for any real v, substitution of this pair into Poisson summation formula gives the identity

$$\sum_{n\in\mathbb{Z}} F(n+v) = \sum_{m\in\mathbb{Z}} \widehat{F}(m)\; e^{2\pi i m v}. \tag{3.14}$$

Conversely, putting $v = 0$ in this formula, we recover (3.13). The proof is now obvious :

a) Put $G(v) = \sum_{n\in\mathbb{Z}} F(n+v)$. If this series converges absolutely and uniformly in v, then $G(v)$ is a continuous function of v.

b) Obviously, one has $G(v) = G(v+1)$. The Fourier coefficients of $G(v)$ are given as follows

$$\begin{aligned} c_m &= \int_0^1 G(v)\, e^{-2\pi i m v}\, dv \\ &= \sum_{n\in\mathbb{Z}} \int_0^1 F(n+v)\, e^{-2\pi i m v}\, dv \\ &= \sum_{n\in\mathbb{Z}} \int_n^{n+1} F(x)\, e^{-2\pi i m x}\, dx \qquad [x = n+v] \\ &= \int_{-\infty}^{+\infty} F(x)\, e^{-2\pi i m x}\, dx = \widehat{F}(m). \end{aligned}$$

c) If the series $\sum_{m\in\mathbb{Z}} |c_m|$ converges, then we can represent $G(v)$ by its Fourier series $\sum_{m\in\mathbb{Z}} c_m\, e^{2\pi i m v} = \sum_{m\in\mathbb{Z}} \widehat{F}(m)\, e^{2\pi i m v}$, and we are done !

From the previous proof, it follows that formula (3.14) holds whenever the left-hand side converges absolutely and uniformly in v, and moreover the sum $\sum_{m\in\mathbb{Z}} |\widehat{F}(m)|$ is finite. Under these assumptions, Poisson summation formula (3.13) holds also.

A further generalization is obtained by using formula (3.14) for the pair of Fourier transforms $F(t^{-1}x)$, $|t|\widehat{F}(tu)$, namely

$$\sum_{n\in\mathbb{Z}} F(v+n/t) = \sum_{m\in\mathbb{Z}} |t|\ \widehat{F}(mt)\ e^{2\pi i (mt) v}. \tag{3.15}$$

If we let t tend to 0, every term $v + n/t$ for $n \neq 0$ tends to $\pm\infty$; hence if F is small enough at infinity, the limit of the left-hand side in (3.15) is $F(v)$. The right-hand side is a Riemann sum corresponding to the subdivision of the real line into the intervals $[mt, mt+t[$ of length t. For t going to 0, the right-hand side approximates an integral, and in the limit we get

$$F(v) = \int_{-\infty}^{+\infty} \widehat{F}(u)\ e^{2\pi i u v}\ du, \tag{3.16}$$

that is Fourier inversion formula. The various steps in this derivation are justified for instance if $F(x)$ admits two continuous derivatives and vanishes off a finite interval.

Exercise 1 : Prove the following symmetrical generalization of Poisson summation formula :

$$\sum_{n\in\mathbb{Z}} F(n+v)\ e^{-\pi i w(2n+v)} = \sum_{m\in\mathbb{Z}} \widehat{F}(m+w)\ e^{\pi i v(2m+w)} \tag{$*$}$$

for v and w real.

We give now a simple application of Poisson summation formula. Let us consider a complex constant c with $Re\ c > 0$ and the pair of Fourier transforms

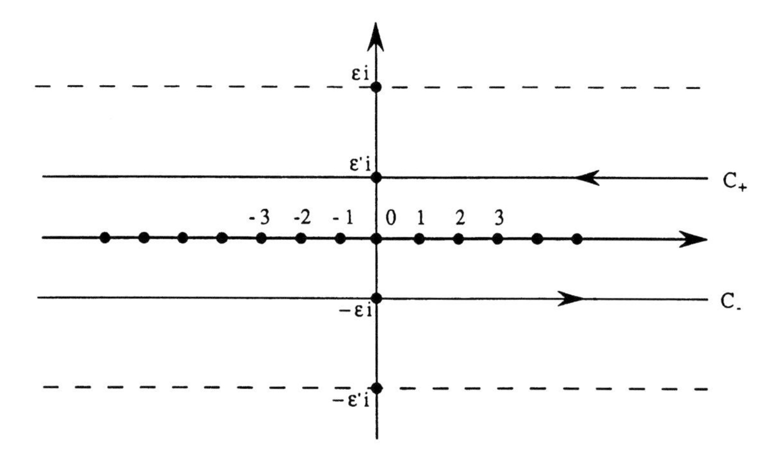

Fig. 6. The contour C of integration.

$$F(x) = e^{-c|x|} \quad , \quad \widehat{F}(u) = \frac{2c}{c^2 + 4\pi^2 u^2}.$$

One gets the identity

$$1 + 2 \sum_{n=1}^{\infty} e^{-cn} = \frac{2}{c} + 2 \sum_{m=1}^{\infty} \frac{2c}{c^2 + 4\pi^2 m^2} \tag{3.17}$$

which one easily rearranges as

$$\coth \frac{c}{2} = 2 \sum_{m=-\infty}^{\infty} \frac{1}{c - 2\pi i m} \tag{3.18}$$

(symmetrical summation). By the change of variable $c = 2\pi i z$ one gets

$$\pi \ \text{cotg} \ \pi z = \sum_{m=-\infty}^{\infty} \frac{1}{z - m} \tag{3.19}$$

(symmetrical summation). The above derivation works for z in the lower half-plane $Im\ z < 0$, but both sides in formula (3.19) representing odd functions of z, this identity remains true for every non real complex number z.

Exercise 2 : a) Suppose that the function $F(x)$ extends to a function of a complex variable holomorphic in some strip $-\varepsilon < Im\ x < \varepsilon$. Using Cauchy residue theorem, give conditions of validity for the formula

$$2\pi i \sum_{n \in \mathbb{Z}} F(n) = \int_C \pi \ \text{cotg} \ \pi z \ F(z) dz \tag{$*$}$$

where the contour of integration C is depicted in figure 6.

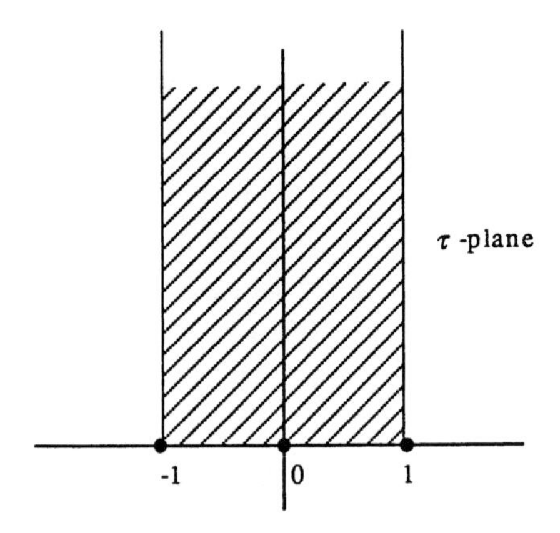

Fig. 7. Domain of values for τ.

b) Use the expansion

$$\pi \cot g \ \pi z = \begin{cases} -\pi i - 2\pi i \sum_{m=1}^{\infty} e^{2\pi i m z} & Im \ z > 0 \\ \pi i + 2\pi i \sum_{m=1}^{\infty} e^{-2\pi i m z} & Im \ z < 0 \end{cases}$$

to transform the right-hand side of formula $(*)$ as follows :

$$\int_C \pi \cot g \ \pi z \ F(z) dz = \left(\int_{-i\varepsilon' - \infty}^{-i\varepsilon' + \infty} - \int_{i\varepsilon' - \infty}^{i\varepsilon' + \infty} \right) \cdots$$

$$= 2\pi i \int_{-\infty}^{+\infty} F(x) dx + 2\pi i \sum_{m=1}^{\infty} \int_{-\infty}^{+\infty} F(x) (e^{-2\pi i m x} + e^{2\pi i m x}) dx$$

$$= 2\pi i \sum_{m \in \mathbb{Z}} \widehat{F}(m).$$

c) In conclusion, Poisson summation formula follows from Euler's identity (3.19) for functions extending as holomorphic functions in a strip $|Im \ z| < \varepsilon$.

3.3. Transformation properties of theta functions

We mentioned already (see Section 2.5) the theta series

$$\theta(q) = 1 + 2q + 2q^4 + 2q^9 + \cdots.$$

It is convergent for $|q| < 1$. We shall use instead a complex variable τ in the upper half-plane $Im \ \tau > 0$ connected to q by the relation $q = e^{\pi i \tau}$. Notice that $e^{\pi i \tau} = e^{\pi i \tau'}$ holds if and only if $\tau' - \tau$ is an even integer, hence we can if necessary normalize τ by $-1 \leq Re \ \tau < 1$ to insure uniqueness of a number τ associated to a given q (see fig. 7).

We introduce a two-variable theta function

$$\theta_3(z|\tau) = \sum_{n\in\mathbb{Z}} q^{n^2} u^n \tag{3.20}$$

with $q = e^{\pi i\tau}$, $u = e^{2\pi i z}$, that is

$$\begin{aligned}\theta_3(z|\tau) &= \sum_{n\in\mathbb{Z}} e^{\pi i(n^2\tau+2nz)} \\ &= 1 + 2q \cos 2\pi z + 2q^4 \cos 4\pi z + 2q^9 \cos 6\pi z + \cdots\end{aligned} \tag{3.21}$$

Here τ and z are both complex numbers and τ is subjected to the restriction $Im\ \tau > 0$.

The *fundamental transformation formula* is

$$\theta_3(z|\tau) = \frac{1}{\sqrt{-i\tau}}\, e^{-\pi i z^2/\tau}\, \theta_3\left(\frac{z}{\tau}\Big|-\frac{1}{\tau}\right). \tag{3.22}$$

The square root of $-i\tau$ is the branch that takes the value 1 for $\tau = i$, holomorphic in the (simply-connected) upper half-plane. To prove this, start from the integral (3.11) and make the change of variable $(x, u) \mapsto (x\sqrt{t}, u/\sqrt{t})$ for some real number $t > 0$. We get

$$\int_{-\infty}^{+\infty} e^{-\pi t x^2} e^{-2\pi i x u}\, dx = \frac{1}{\sqrt{t}}\, e^{-\pi u^2/t}, \tag{3.23}$$

that is, we have a pair of Fourier transforms

$$F(x) = e^{-\pi t x^2} \quad , \quad \widehat{F}(u) = \frac{1}{\sqrt{t}}\, e^{-\pi u^2/t}.$$

If we specialize Poisson summation formula (3.14) to this case we get

$$\sum_{n\in\mathbb{Z}} e^{-\pi t(n+v)^2} = \frac{1}{\sqrt{t}} \sum_{m\in\mathbb{Z}} e^{-\pi m^2/t}\, e^{2\pi i m v}. \tag{3.24}$$

The relation (3.22), written in full, reads as follows

$$\sum_{n\in\mathbb{Z}} e^{\pi i n^2\tau + 2\pi i n z} = \frac{1}{\sqrt{-i\tau}}\, e^{-\pi i z^2/\tau} \sum_{m\in\mathbb{Z}} e^{(-\pi i m^2 + 2\pi i m z)/\tau}. \tag{3.25}$$

Hence (3.24) is the particular case $\tau = it$, $z = itv$ of (3.25). It follows that (3.22) is true when τ is purely imaginary and the general case follows by analytic continuation.

We shall use the particular case $z = 0$ of the transformation formula (3.22). Since $\theta(q) = \theta_3(0|\tau)$ for $q = e^{\pi i\tau}$ we get

$$\theta(e^{\pi i\tau}) = \frac{1}{\sqrt{-i\tau}}\, \theta(e^{-\pi i/\tau}) \quad \text{for} \quad Im\ \tau > 0. \tag{3.26}$$

Exercise 3 : Write the previous relation as

$$(*) \qquad \sum_{n\in\mathbb{Z}} e^{-\pi n^2 t} = \frac{1}{\sqrt{t}} \sum_{m\in\mathbb{Z}} e^{-\pi m^2/t} \qquad (t>0).$$

Multiplying both sides by $e^{-c^2t/4\pi}$ and integrating over t in $]0,+\infty[$ show how to recover formula (3.18).

3.4. Mellin transforms : general theory

We consider a function $f(x)$ of a positive real variable. Its *Mellin transform* is given by

$$M(s) = \int_0^\infty f(x)\, x^{s-1}\, dx. \tag{3.27}$$

Let us assume that there exist two real constants a and b such that $a<b$ and that $f(x)=O(x^{-a})$ for x close to 0 and $f(x)=O(x^{-b})$ for x very large. Then the previous integral converges for s in the strip $a < \mathit{Re}\ s < b$ and $M(s)$ is a holomorphic function in this strip. The following *inversion formula* is known : assume that there exists a constant $C>0$ such that

$$\int_{-\infty}^{+\infty} |M(\sigma+it)|\, dt \leq C \tag{3.28}$$

for $a<\sigma<b$. Then

$$f(x) = \frac{1}{2\pi i}\int_{\sigma-i\infty}^{\sigma+i\infty} M(s)\, x^{-s}\, ds \tag{3.29}$$

for $a<\sigma<b$. By the change of variable $x=e^{-u}$ where u runs from $-\infty$ to $+\infty$, we express $M(s)$ as a Laplace transform

$$M(s) = \int_{-\infty}^{+\infty} e^{-us}\, F(u)\, du \tag{3.30}$$

where $F(u)=f(e^{-u})$ is of the order $O(e^{ua})$ for u near $+\infty$ and of the order $O(e^{ub})$ for u near $-\infty$. The inversion formula (3.29) is therefore reduced to the classical inversion formula for Laplace transforms

$$F(u) = \frac{1}{2\pi i}\int_{\sigma-i\infty}^{\sigma+i\infty} M(s)\, e^{us}\, ds. \tag{3.31}$$

In turn, this formula is a consequence of Fourier inversion formula (3.10).

We want to generalize the Mellin transform to cases where $M(s)$ extends to a meromorphic function in the complex plane $\mathbb{C}$. Assume that $f(x)$ admits of asymptotic expansions

$$f(x) \sim \sum_{k=0}^{\infty} c_k \, x^{i_k} \qquad \text{for} \quad x \to 0 \tag{3.32}$$

$$f(x) \sim \sum_{\ell=0}^{\infty} d_\ell \, x^{j_\ell} \qquad \text{for} \quad x \to +\infty \tag{3.33}$$

where the real exponents satisfy the assumptions

$$i_0 < i_1 < i_2 < \cdots \qquad \lim_{k \to \infty} i_k = +\infty, \tag{3.34}$$

$$j_0 > j_1 > j_2 > \cdots \qquad \lim_{\ell \to \infty} j_\ell = -\infty. \tag{3.35}$$

We split $M(s)$ as a sum of two integrals $M(s) = M_1(s) + M_2(s)$ where

$$M_1(s) = \int_0^1 f(x) \, x^{s-1} \, dx \ , \quad M_2(s) = \int_1^\infty f(x) \, x^{s-1} \, dx. \tag{3.36}$$

The integral $M_1(s)$ converges absolutely for $Re\ s > -i_0$. Moreover by definition of an asymptotic expansion, we can write

$$f(x) = \sum_{k=0}^{K-1} c_k \, x^{i_k} + r_K(x) \tag{3.37}$$

with a remainder $r_K(x)$ of order $O(x^{i_K})$ for x close to 0. The formula

$$M_1(s) = \sum_{k=0}^{K-1} \frac{c_k}{s + i_k} + \int_0^1 r_K(x) \, x^{s-1} \, dx \tag{3.38}$$

gives an analytic continuation in the domain $Re\ s > -i_K$. Since K is arbitrary and i_K tends to $+\infty$ with K, we conclude that $M_1(s)$ extends to a meromorphic function in $\mathbb{C}$ with simple poles at $-i_0, -i_1, -i_2 \cdots$ and a residue equal to c_k at $s = -i_k$.

We can treat the integral $M_2(s)$ in a similar way or reduce it to the previous case by a change of variable :

$$M_2(s) = \int_0^1 f\Big(\frac{1}{x}\Big) \, x^{-s-1} \, dx. \tag{3.39}$$

The conclusion is : $M_2(s)$ extends to a meromorphic function in $\mathbb{C}$, with simple poles at $-j_0, -j_1, -j_2, \cdots$ and a residue equal to $-d_\ell$ at $s = -j_\ell$.

The sum of $M_1(s)$ and $M_2(s)$ is therefore a meromorphic function $M(s)$ with two series of poles located at (see fig. 8)

$$-i_0 > -i_1 > -i_2 > \cdots$$
$$-j_0 < -j_1 < -j_2 < \cdots$$

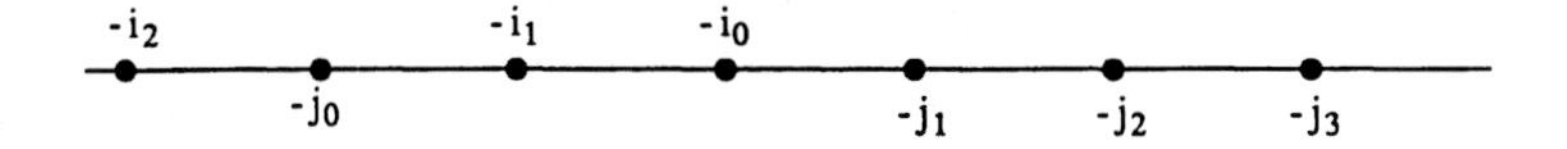

Fig. 8. Poles of $M(s)$.

We continue to write $M(s)$ as an integral $\int_0^\infty f(x)\, x^{s-1}\, dx$ which can be understood as follows : in the case of convergence, let $g(x)$ be a primitive function of $f(x)\, x^{s-1}$

$$dg(x) = f(x)\, x^{s-1}\, dx. \tag{3.40}$$

Then one gets

$$\int_0^\infty f(x)\, x^{s-1}\, dx = \lim_{x\to\infty} g(x) - \lim_{x\to 0} g(x) \tag{3.41}$$

by the fundamental theorem of calculus. In the general case (assuming s is not a pole of the function $M(s)$), the primitive function $g(x)$ admits an asymptotic expansion

$$g(x) \sim \sum_{k=0}^{\infty} c'_k\, x^{i'_k} \quad \textit{for } x \text{ near } 0. \tag{3.42}$$

By derivating term by term we should obtain the asymptotic expansion derived from (3.32)

$$f(x)\, x^{s-1} \sim \sum_{k=0}^{\infty} c_k\, x^{s+i_k-1}. \tag{3.43}$$

Hence everything is completely determined in (3.42) except for the constant term, corresponding to an exponent i'_k equal to 0, depending on the choice of the primitive $g(x)$. This constant term is called the *finite part of* $g(x)$ *at* $x = 0$, to be denoted

$$\underset{x=0}{FP}\, g(x).$$

Then the generalized integral is defined by

$$\int_0^\infty f(x)\, x^{s-1}\, dx = \underset{x=\infty}{FP}\, g(x) - \underset{x=0}{FP}\, g(x) \tag{3.44}$$

where the finite part at $x = \infty$ is defined in analogy with the finite part at $x = 0$.

The integral $\int_0^\infty f(x)\, x^{s-1}\, dx$ converges in the usual sense for s in the strip $i_0 < Re\ s < j_0$, which is non empty only if $i_0 < j_0$. But notice that the generalized integral $\int_0^\infty \Phi(x)\, x^{s-1}\, dx$ is identically 0 if $\Phi(x)$ is a finite linear combination of monomials x^α. By subtracting a suitable $\Phi(x)$ from $f(x)$ we can achieve the relation $i_0 < j_0$, hence providing a strip where the Mellin transform is defined in the usual way.

3.5. Some examples of Mellin transforms

The *gamma function* is defined by the following Mellin integral

$$\Gamma(s) = \int_0^\infty e^{-x}\, x^{s-1}\, dx. \tag{3.45}$$

We have the convergent power series expansion

$$e^{-x} = \sum_{k=0}^{\infty} (-1)^k\, x^k/k! \qquad \text{for } x \text{ near } 0 \tag{3.46}$$

and $e^{-x} \sim 0$ for x near infinity, meaning $e^{-x} = O(x^{-N})$ for every integer N. According to the general theory, $\Gamma(s)$ is a meromorphic function in $\mathbb{C}$, with simple poles at

$$0, -1, -2, \cdots$$

and a residue equal to $(-1)^k/k!$ for $s = -k$. The functional equation $\Gamma(s+1) = s\,\Gamma(s)$ is obtained from the *generalized integration by part principle*

$$\int_0^\infty u(x)\, v'(x)\, dx = \underset{x=\infty}{FP}\ u(x)\, v(x) - \underset{x=0}{FP}\ u(x)\, v(x) - \int_0^\infty u'(x)\, v(x)\, dx, \tag{3.47}$$

by putting $u(x) = x^s$, $v(x) = e^{-x}$.

We consider now a general Dirichlet series of the form

$$L(s) = \sum_{n=1}^{\infty} c_n\, n^{-s}. \tag{3.48}$$

A linear change of variable is admissible in our generalized integrals

$$\int_0^\infty \Phi(ax)\, dx = \frac{1}{a} \int_0^\infty \Phi(x)\, dx \tag{3.49}$$

(a real positive). From the definition of $\Gamma(s)$ we get

$$\int_0^\infty e^{-ax}\, x^{s-1}\, dx = a^{-s}\, \Gamma(s). \tag{3.50}$$

It follows that

$$\Gamma(s)\ L(s) = \sum_{n=0}^{\infty} c_n \int_0^\infty e^{-nx}\ x^{s-1}\ dx. \tag{3.51}$$

If integrating term by term is legitimate for s in a suitable half-plane $Re\ s > \sigma$, then we get

$$L(s) = \frac{1}{\Gamma(s)} \int_0^\infty F(x)\ x^{s-1}\ dx \tag{3.52}$$

with

$$F(x) = \sum_{n=1}^{\infty} c_n\ e^{-nx}. \tag{3.53}$$

If $F(x)$ admits of suitable asymptotic expansions for $x = 0$ and $x = \infty$, then interpreting (3.52) as a generalized Mellin transform, we get the analytic continuation of the series $\sum_{n=1}^{\infty} c_n\ n^{-s}$ to a meromorphic function in $\mathbb{C}$.

We illustrate this principle in the case of $\zeta(s)$, that is $c_n = 1$ for $n = 1, 2, \ldots$. In this case

$$F(x) = \sum_{n=1}^{\infty} e^{-nx} = \frac{e^{-x}}{1 - e^{-x}} = \frac{1}{e^x - 1}. \tag{3.54}$$

According to the definition (1.7) of Bernoulli numbers, we get an asymptotic expansion near 0

$$\frac{1}{e^x - 1} \sim \sum_{k=-1}^{\infty} \frac{B_{k+1}}{(k+1)!}\ x^k \tag{3.55}$$

while $\frac{1}{e^x - 1} \sim 0$ for x near ∞. Hence $\zeta(s)$ is a meromorphic function given by

$$\zeta(s) = \frac{1}{\Gamma(s)} \int_0^\infty \frac{x^{s-1}}{e^x - 1}\ dx. \tag{3.56}$$

The structure of poles is as follows :

- for $\Gamma(s)$ $\quad s = 0, -1, -2, \ldots$ $\quad$ residue $(-1)^k/k!$ at $s = -k$
- for $\int_0^\infty \frac{x^{s-1}}{e^x - 1} dx$ $\quad s = 1, 0, -1, -2, \ldots$ residue $\frac{B_{k+1}}{(k+1)!}$ at $s = -k$.

It follows that the poles cancel except for $s = 1$. Hence the result :

$\zeta(s)$ extends to a meromorphic function in the plane $\mathbb{C}$, the only pole is $s = 1$, with residue 1. For $k = 0, -1, \ldots$ we get

$$\zeta(-k) = (-1)^k\ \frac{B_{k+1}}{k+1}. \tag{3.57}$$

Using the properties

$$B_1 = -\frac{1}{2} \quad , \quad B_3 = B_5 = \cdots = 0$$

we easily transform the previous relation as follows

$$\zeta(0) = -\frac{1}{2} \quad , \quad \zeta(-2) = \zeta(-4) = \cdots = 0$$

$$\zeta(-1) = -\frac{B_2}{2} \quad , \quad \zeta(-3) = -\frac{B_4}{4} \quad , \quad \zeta(-5) = -\frac{B_6}{6}, \ldots$$

Suppose now that $\theta(1), \theta(2), \ldots$ is a periodic sequence of coefficients

$$\theta(n+f) = \theta(n) \qquad \text{for} \quad n \geq 1 \tag{3.58}$$

for some fixed period $f \geq 1$ (see Section 1.7). We consider the Dirichlet L-series

$$L(\theta, s) = \sum_{n=1}^{\infty} \theta(n)\; n^{-s}. \tag{3.59}$$

By specializing formula (3.52), we get an integral representation

$$L(\theta, s) = \frac{1}{\Gamma(s)} \int_0^{\infty} \Theta(x)\; x^{s-1}\; dx \tag{3.60}$$

with

$$\Theta(x) = \sum_{n=1}^{\infty} \theta(n)\; e^{-nx}. \tag{3.61}$$

Using the periodicity of the coefficients $\theta(n)$, we transform $\Theta(x)$ as follows

$$\begin{aligned} \Theta(x) &= \sum_{a=1}^{f} \sum_{m=0}^{\infty} \theta(a+mf)\; e^{-(a+mf)x} \\ &= \sum_{a=1}^{f} \theta(a)\; e^{-ax} \sum_{m=0}^{\infty} e^{-mfx} \end{aligned}$$

hence

$$\Theta(x) = \frac{\sum_{a=1}^{f} \theta(a)\; e^{(f-a)x}}{e^{fx} - 1}. \tag{3.62}$$

For $x = 0$, the numerator takes the value $\sum_{a=1}^{f} \theta(a)$, while in the denominator $e^{fx} - 1$ has a series expansion $fx + \frac{1}{2} f^2 x^2 + \cdots$. It follows that, for x near 0, $\Theta(x)$ has an asymptotic expansion with leading term Θ/x where

$$\Theta = \frac{1}{f} \sum_{a=1}^{f} \theta(a). \tag{3.63}$$

More precisely, using the definition of Bernoulli polynomials

$$\frac{e^{py}}{e^y - 1} = \sum_{k=-1}^{\infty} \frac{B_{k+1}(p)}{(k+1)!} \, y^k \tag{3.64}$$

with the substitution $y = fx$, $p = \dfrac{f-a}{f}$ together with the symmetry property for Bernoulli polynomials

$$B_k(1-p) = (-1)^k \, B_k(p), \tag{3.65}$$

we get the following asymptotic expansion

$$\Theta(x) = \frac{\Theta}{x} + \sum_{k=0}^{\infty} \frac{(-1)^{k+1} \, B_{k+1,\theta}}{(k+1)!} \, x^k \tag{3.66}$$

with the definition

$$B_{m,\theta} = f^{m-1} \sum_{a=1}^{f} \theta(a) \, B_m\left(\frac{a}{f}\right). \tag{3.67}$$

According to the general theory, the Mellin transform

$$M_\theta(s) = \int_0^\infty \Theta(x) \, x^{s-1} \, dx \tag{3.68}$$

extends to a meromorphic function in $\mathbb{C}$, with a simple pole at $s = 1$ with residue Θ as well as a sequence of simple poles at $s = 0, -1, -2, \ldots$, the residue at $s = -k$ being equal to $(-1)^{k+1} B_{k+1,\theta}/(k+1)!$. Dividing by $\Gamma(s)$ with simple poles at $s = 0, -1, -2, \ldots$ and a residue equal to $(-1)^k/k!$ at $s = -k$, we conclude that

$L(\theta, s) = M_\theta(s)/\Gamma(s)$ *extends to a meromorphic function in* $\mathbb{C}$, *with a single pole at* $s = 1$, *residue equal to* Θ *and the special values*

$$L(\theta, -k) = \frac{-B_{k+1,\theta}}{k+1}. \tag{3.69}$$

In the special case where $\Theta = 0$, that is $\sum_{a=1}^{f} \theta(a) = 0$, then $L(\theta, s)$ *is an entire function.*

Hurwitz zeta function can be treated in the same spirit. Namely, we start from

$$(n+v)^{-s} \, \Gamma(s) = \int_0^\infty e^{-x(n+v)} \, x^{s-1} \, dx. \tag{3.70}$$

The series

$$\zeta(s, v) = \sum_{n=0}^{\infty} (n+v)^{-s} \tag{3.71}$$

which converges in the half-plane $Re\ s > 1$ extends to a meromorphic function in $\mathbb{C}$ given as a Mellin transform

$$\zeta(s,v) = \frac{1}{\Gamma(s)} \int_0^\infty \frac{e^{-xv}}{1-e^{-x}}\, x^{s-1}\, dx. \tag{3.72}$$

Using formulas (3.64) and (3.65) we get the power series expansion

$$\frac{e^{-xv}}{1-e^{-x}} = \frac{e^{x(1-v)}}{e^x - 1} = \frac{1}{x} + \sum_{k=0}^{\infty} \frac{(-1)^{k+1}\, B_{k+1}(v)}{(k+1)!}\, x^k.$$

The Mellin transform

$$M(s,v) = \int_0^\infty \frac{e^{-xv}}{1-e^{-x}}\, x^{s-1}\, dx \tag{3.73}$$

is meromorphic with simple poles at $s = 1, 0, -1, \ldots$. Since $\zeta(s,v)$ is equal to $M(s,v)/\Gamma(s)$, the poles of $\Gamma(s)$ cancel those of $M(s,v)$ except the one at $s = 1$, and we get the result :

$\zeta(s,v)$ is a meromorphic function in $\mathbb{C}$, with a single pole at $s = 1$, residue equal to 1, and the special values

$$\zeta(-k,v) = -\frac{B_{k+1}(v)}{k+1}. \tag{3.74}$$

From this result, the properties of the Dirichlet series $L(\theta, s)$ can be recovered using formula (1.73), namely

$$L(\theta,s) = f^{-s} \sum_{a=1}^{f} \theta(a)\, \zeta(s, \frac{a}{f}). \tag{3.75}$$

3.6. Functional equation of Dirichlet series

To express the functional equation of $\zeta(s)$, one introduces after Riemann the following meromorphic function

$$\xi(s) = \pi^{-s/2}\, \Gamma\Big(\frac{s}{2}\Big) \zeta(s). \tag{3.76}$$

Then the functional equation reads as follows

$$\xi(s) = \xi(1-s). \tag{3.77}$$

This formula has some important consequences. For instance, we know that $\zeta(s)$ is meromorphic in the half-plane $Re\ s > 0$ with a single pole at $s = 1$ with residue 1. Since $\Gamma(s)$ is holomorphic in the same half-plane, the only singularity of $\xi(s)$ in the half-plane $Re\ s > 0$ is a simple pole at $s = 1$ with residue $\pi^{-1/2} \cdot \Gamma(\frac{1}{2}) \cdot 1 = 1$. By the functional equation, in the half-plane $Re\ s < 1$

obtained from the previous one by the symmetry exchanging s and $1-s$, the only singularity of $\xi(s)$ is a simple pole at $s=0$, with residue -1. Hence, the function $\Xi(s) = s(s-1)\,\xi(s)$ is an entire function satisfying the symmetry $\Xi(s) = \Xi(1-s)$. The function $\Gamma(\frac{s}{2})$ having poles at $s=-2,-4,-6,\ldots$, and $\xi(s)$ being regular at these points, the poles are cancelled by zeroes of $\zeta(s)$ hence, we recover the result

$$\zeta(-2) = \zeta(-4) = \cdots = 0.$$

The gamma function satisfies two classical identities

$$\textit{Complement formula}: \quad \Gamma(s)\,\Gamma(1-s) = \frac{\pi}{\sin \pi s}$$

$$\textit{Duplication formula}: \quad \Gamma\left(\frac{s}{2}\right)\,\Gamma\left(\frac{s+1}{2}\right) = \pi^{1/2}\,2^{1-s}\,\Gamma(s)\,.$$

Hence the functional equation (3.77) can be written as follows

$$\zeta(1-s) = 2^{1-s}\,\pi^{-s}\,\Gamma(s)\cos\frac{\pi s}{2}\cdot\zeta(s). \tag{3.78}$$

For the *proof* of the functional equation, we start from the transformation formula for theta functions (see exercise 3)

$$\Theta(t) = t^{-1/2}\,\Theta\left(\frac{1}{t}\right) \tag{3.79}$$

for the series

$$\Theta(t) = \sum_{n=-\infty}^{+\infty} e^{-\pi n^2 t} = 1 + 2\sum_{n=1}^{\infty} e^{-\pi n^2 t}. \tag{3.80}$$

Take the Mellin transform. From our conventions, one gets that the Mellin transform of the constant 1 is 0. Moreover one gets

$$\int_0^\infty e^{-\pi n^2 t}\,t^{s-1}\,dt = \pi^{-s}\,n^{-2s}\,\Gamma(s).$$

Summing over n, we deduce

$$\frac{1}{2}\int_0^\infty \Theta(t)\,t^{s-1}\,dt = \pi^{-s}\,\Gamma(s)\,\zeta(2s) = \xi(2s)\,.$$

Moreover by changing t into $(1/t)$ we get

$$\begin{aligned}\frac{1}{2}\int_0^\infty t^{-1/2}\,\Theta\left(\frac{1}{t}\right)\,t^{s-1}\,dt &= \frac{1}{2}\int_0^\infty t^{-1/2-s}\,\Theta(t)\,dt \\ &= \frac{1}{2}\int_0^\infty t^{(\frac{1}{2}-s)-1}\,\Theta(t)\,dt = \xi\left(2(\frac{1}{2}-s)\right)\,.\end{aligned}$$

From the functional equation (3.79) we conclude

$$\xi(2s) = \xi(1-2s).$$

There is nothing mysterious about e^{-x^2} and theta functions as shown by the following exercise, inspired by Tate's thesis (see Cassels and Fröhlich 1967).

Exercise 4 : a) Let $\mathcal{S}(\mathbb{R})$ be the class of infinitely differentiable functions $F(x)$ such that $x^p(\frac{d}{dx})^q F(x)$ be bounded in x for all integers $p \geq 0$ and $q \geq 0$. The Fourier transform $\widehat{F}(u)$ of a function $F(x)$ in $\mathcal{S}(\mathbb{R})$ is also in $\mathcal{S}(\mathbb{R})$ and the Poisson summation formula holds, namely

$$(*) \qquad \sum_{n\in\mathbb{Z}} F(tn) = |t|^{-1} \sum_{m\in\mathbb{Z}} \widehat{F}(t^{-1}\, m)$$

for any real $t \neq 0$.

b) Taking the Mellin transform of both sides of formula $(*)$, derive the relation

$$(**) \qquad \zeta(s) \int_{-\infty}^{+\infty} F(t)\, |t|^{s-1}\, dt = \zeta(1-s) \int_{-\infty}^{+\infty} \widehat{F}(u)\, |u|^{-s}\, du.$$

c) Derive the formula

$$(***) \qquad W(s) \int_{-\infty}^{+\infty} \widehat{F}(u)\, |u|^{-s}\, du = \int_{-\infty}^{+\infty} F(t)\, |t|^{s-1}\, dt$$

where $W(s) = 2(2\pi)^{-s}\, \Gamma(s) \cos\frac{\pi s}{2}$ [*Hint :* insert the convergence factor $e^{-\varepsilon|u|}$ in the first integral, replace $\widehat{F}(u)$ by its definition as an integral (3.9), interchange the integrations and use the extension of formula (3.50) to complex numbers a with $Re\ a > 0$. At the end let ε tend to 0.]

d) Derive the functional equation $\zeta(1-s) = W(s)\ \zeta(s)$ as well as the relation $W(s)\ W(1-s) = 1$. As a corollary rederive the complement formula.

We extend now the functional equation to L-series. As before, consider a sequence of numbers $\theta(n)$ (for n in $\mathbb{Z}$) with period f, that is $\theta(n+f) = \theta(n)$. We distinguish two cases :

a) *Even case :* $\theta(n) = \theta(-n)$. Because of the periodicity, it suffices to check the equality $\theta(a) = \theta(f-a)$ for $1 \leq a \leq f-1$. Then we set

$$(3.81) \qquad \xi(\theta, s) = (f/\pi)^{s/2}\, \Gamma\left(\frac{s}{2}\right)\, L(\theta, s).$$

b) *Odd case :* $\theta(n) = -\theta(-n)$ or $\theta(a) = -\theta(f-a)$ for $1 \leq a \leq f-1$ and $\theta(0) = 0$. We set

$$(3.82) \qquad \xi(\theta, s) = (f/\pi)^{\frac{s+1}{2}}\, \Gamma\left(\frac{s+1}{2}\right)\, L(\theta, s).$$

Moreover we define the finite Fourier transform $\widehat{\theta}$ of θ by

$$\widehat{\theta}(m) = f^{-1/2} \sum_{n=0}^{f-1} \theta(n)\, e^{2\pi i mn/f}. \tag{3.83}$$

The functional equation then reads as follows

$$\xi(\widehat{\theta}, 1-s) = \xi(\theta, s). \tag{3.84}$$

More specifically, assume that χ is a Dirichlet character with conductor f, *primitive* in the following sense : there cannot exist a proper divisor f_1 of f and a character χ_1 of conductor f_1 such that $\chi(n) = \chi_1(n)$ for n prime to f. By a group theoretic argument, it can be shown that $\widehat{\chi}$ is given by

$$\widehat{\chi}(n) = W(\chi)\, \overline{\chi(n)} \tag{3.85}$$

($\overline{\chi(n)}$ imaginary conjugate of $\chi(n)$). The constant $W(\chi)$ is obtained by putting $n = 1$; it is a so-called *Gaussian sum*

$$W(\chi) = f^{-1/2} \sum_{a=1}^{f-1} \chi(a)\, \zeta_f^a \tag{3.86}$$

with $\zeta_f = e^{2\pi i/f}$. Then the functional equation reads as follows

$$\xi(\overline{\chi}, 1-s) = W(\chi)^{-1}\, \xi(\chi, s). \tag{3.87}$$

By using twice this equation one derives the identity

$$W(\chi)\, W(\overline{\chi}) = 1. \tag{3.88}$$

When χ is the unit character of conductor 1, that is $\chi(n) = 1$ for all $n \geq 1$, then $\chi = \overline{\chi}$, $W(\chi) = 1$ and $\xi(\chi, s) = \xi(s)$ hence equation (3.87) reduces to the functional equation $\xi(1-s) = \xi(s)$. Otherwise, we get $f > 1$ and $\sum_{a=1}^{f} \chi(n) = 0$, hence both $\xi(\chi, s)$ and $\xi(\overline{\chi}, 1-s)$ are entire functions of s.

We shall not give the proof of the functional equation (3.84) (but see the following exercise). When θ is even, it can be deduced from the transformation law (3.22) for theta functions.

Exercise 5 : a) Consider the following series

$$\zeta(s|v, w) = \sum_{n=0}^{\infty} (n+v)^{-s}\, e^{2\pi i n w} \tag{$*$}$$

which converges absolutely for $Re\ s > 1$ and $Im\ w > 0$ (no restriction on the complex number v). For $w = 0$, this sum reduces to $\zeta(s, v)$. Define the analytic continuation by the methods of this Section.

b) Establish Lerch's transformation formula

$$
(**)\qquad \begin{aligned} \zeta(s|v,w) &= i\, e^{-2\pi i v w}\, (2\pi)^{s-1}\, \Gamma(1-s) \\ &\times \left\{e^{-\pi i s/2}\, \zeta(1-s|w,-v) - e^{i\pi s/2}\, e^{2\pi i v}\, \zeta(1-s|1-w,v)\right\}. \end{aligned}
$$

[*Hint :* use formula $(*)$ in exercise 1 for a suitable function F.]

c) By specialization, derive the functional equation for Hurwitz zeta function

$$
(***)\qquad \zeta(s,v) = 2(2\pi)^{s-1}\, \Gamma(1-s) \sum_{n=1}^{\infty} n^{s-1}\, \sin(2\pi n v + \frac{\pi s}{2})
$$

for $0 < v \leq 1$ and $Re\ s < 0$.

d) Derive the functional equation for $L(\theta, s)$ using the following representation

$$
(****)\qquad L(\theta,s) = f^{-s} \sum_{a=1}^{f} \theta(a)\, \zeta(s, \frac{a}{f}).
$$

3.7. Application to quadratic forms

We shall revisit the zeta function connected with the Gaussian integers, namely

$$
Z_4(s) = \Sigma'_{m,n}\ (m^2+n^2)^{-s}. \tag{3.89}
$$

In Section 2.6, we established the factorization

$$
Z_4(s) = 4\zeta(s)\ L(\chi_4, s) \tag{3.90}
$$

where χ_4 is the character of conductor 4 given by the table of values

$$
\chi_4(0) = 0\ ,\ \ \chi_4(1) = 1\ ,\ \ \chi_4(2) = 0\ ,\ \ \chi_4(3) = -1.
$$

We know that $\zeta(s)$ extends to a meromorphic function with a pole at $s = 1$, residue 1, and no other singularity and that $L(\chi_4, s)$ extends to an entire function. Hence $Z_4(s)$ extends as a meromorphic function in $\mathbb{C}$, whose only singularity is a pole at $s = 1$ with residue $4L(\chi_4, 1)$. But $L(\chi_4, 1)$ is given by the series $1 - \frac{1}{3} + \frac{1}{5} - \frac{1}{7} \cdots$ and Leibniz proved that it is equal to $\frac{\pi}{4}$. Hence

$$
L(\chi_4, 1) = \frac{\pi}{4} \tag{3.91}
$$

and the *residue of* $Z_4(s)$ *at* $s = 1$ *is equal to* π.

We derive now the functional equation for $Z_4(s)$. We could use the known functional equations for $\zeta(s)$ and $L(\chi_4, s)$ but it is more expedient to use theta functions. Indeed from the Mellin transform

$$
\pi^{-s}\, \Gamma(s)\, (m^2+n^2)^{-s} = \int_0^{\infty} e^{-\pi t(m^2+n^2)}\, t^{s-1}\, dt \tag{3.92}
$$

one gets

$$\pi^{-s}\,\Gamma(s)\,Z_4(s)=\int_0^\infty \Theta(t)^2\,t^{s-1}\,dt \tag{3.93}$$

since

$$\Theta(t)^2=\Big(\sum_m e^{-\pi t m^2}\Big)^2=\sum_{m,n} e^{-\pi t(m^2+n^2)}.$$

[*Reminder* : the Mellin transform of the constant term 1 corresponding to $m=n=0$ is 0]. We can now use the functional equation

$$\Theta(t)^2=t^{-1}\,\Theta\Big(\frac{1}{t}\Big)^2$$

and imitate the proof of the functional equation for $\zeta(s)$. As a result, the function $\pi^{-s}\,\Gamma(s)\,Z_4(s)$ is invariant under the symmetry $s\leftrightarrow 1-s$

$$\pi^{-s}\,\Gamma(s)\,Z_4(s)=\pi^{s-1}\,\Gamma(1-s)\,Z_4(1-s). \tag{3.94}$$

Since $\pi^{-s/2}\,\Gamma(\frac{s}{2})\,\zeta(s)$ is also invariant under the symmetry $s\leftrightarrow 1-s$, and since $Z_4(s)=4\zeta(s)\,L(\chi_4,s)$, it follows that

$$\frac{\Gamma(s)}{\Gamma(\frac{s}{2})}\,\pi^{-s/2}\,L(\chi_4,s)$$

is invariant under $s\leftrightarrow 1-s$. Using the duplication formula we get the functional equation

$$\xi(\chi_4,s)=\xi(\chi_4,1-s) \tag{3.95}$$

where

$$\xi(\chi_4,s)=\Big(\frac{4}{\pi}\Big)^{\frac{s+1}{2}}\,\Gamma\Big(\frac{s+1}{2}\Big)\,L(\chi_4,s) \tag{3.96}$$

(notice that χ_4 is an odd character!) It would be easy to calculate directly that $W(\chi_4)$ is 1, hence (3.95) follows from the general functional equation (3.87). But this direct derivation of (3.95) is typical of *the use of analytic methods to produce an arithmetical result like* $W(\chi_4)=1$.

Exercise 6 : a) Calculate the values $L(\chi_4,-m)$ for $m=0,1,2,\ldots$ and using the functional equation (3.95) derive the values $L(\chi_4,m)$ for $m\geq 3$, m *odd* (notice that $L(\chi_4,-m)=0$ for m odd, $m\geq 1$).

b) Using the functional equation (3.94) show that $Z_4(-m)=0$ for $m=1,2,3,\ldots$. No information can be obtained about the values $Z_4(2), Z_4(3),\ldots$.

Exercise 7 : a) Derive a functional equation for the zeta function $Z_3(s)=\sum'_{m,n}(m^2+mn+n^2)^{-s}$ using the factorization $Z_3(s)=6\zeta(s)\,L(\chi_3,s)$.

b) Show that $\pi^{-s}\,\Gamma(s)\,Z_3(s)$ is the Mellin transform of the theta function

$$\Theta_3(t) = \sum_{m,n} e^{-\pi t(m^2+mn+n^2)}.$$

Working backwards, derive a functional equation for $\Theta_3(t)$ from the functional equation for $Z_3(s)$.

c) Show that $Z_3(s)$ vanishes for $s = -1, -2, \ldots$. Calculate $Z_3(0)$ and the residue of $Z_3(s)$ at $s = 1$.

References

A. General textbooks about number theory and works of historical value.

Blanchard, A. (1969) Introduction à la théorie analytique des nombres premiers, Dunod, 1969
Borevich, Z. and Shafarevich, I. (1966) Theory of numbers, Academic Press, 1966
Cassels, J. and Fröhlich, A. (1967) Algebraic number theory, Academic Press 1967
Chandrasekharan, K. (1968) Introduction to analytic number theory, Springer, 1968
Eichler, M. (1963) Einführung in die Theorie der algebraischen Zahlen und Funktionen, Birkhäuser, 1963
Eisenstein, G. (1967) Mathematische Abhandlungen, Holms Verlag, 1967
Hardy, G. H. (1940) Ramanujan, Cambridge Univ. Press, 1940
Hardy, G. H. and Wright, E. M. (1945) An introduction to the theory of numbers, Oxford Univ. Press, 1945
Hecke, E. (1938) Dirichlet series, modular functions and quadratic forms, Edwards Bros., 1938
Lang, S. (1968) Algebraic number theory, Addison-Wesley, 1968
Minkowski, H. (1967) Gesammelte Abhandlungen, Chelsea, 1967
Samuel, P. (1967) Théorie algébrique des nombres, Hermann, 1967
Serre, J.-P. (1970) Cours d'arithmétique, Presses Universitaires, 1970

B. Basic references for analytical methods.

Bellman, R. (1961) A brief introduction to theta functions, Holt, Rinehart and Winston, 1961
Bochner, S. (1932) Vorlesungen über Fouriersche Integrale, Leipzig, 1932
Erdelyi, A. (editor), (1953) Higher transcendental functions, vol. 1, McGraw Hill, 1953
Igusa, J. (1972) Theta functions, Springer, 1972
Mumford, D. (1983) Tata Lectures on Theta I, Prog. Math., vol. 28, Birkhäuser, 1983
Jacobi, C. G. (1829) Fundamenta Nova Theoriae Functionum Ellipticarum, in Gesammelte Werke, tome I, Königsberg, 1829 (reprinted by Chelsea).
Titchmarsh, E. C. (1930) The zeta-function of Riemann, Cambridge Univ. Press, 1930
Titchmarsh, E. C. (1937) Introduction to the theory of Fourier integrals, Oxford Univ. Press, 1937
Whittaker, E. T. and Watson, G. N. (1935) A course of modern analysis, Cambridge Univ. Press, 1935

Chapter 2

Introduction to Compact Riemann Surfaces, Jacobians, and Abelian Varieties

by Jean-Benoît Bost

Table of Contents

My aim in these notes is to give an introduction to the theory of *Abelian varieties.*

At many places, I will not give detailed proofs, but I will try to discuss the 'concrete' origins of the theory. Therefore I will spend most of these lectures explaining basic results about *Riemann surfaces* and *algebraic curves* and about the Abelian varieties which are associated to them, the *Jacobian varieties.*

The content of these notes is quite classical: most of it is known since the beginning of the century. We follow a rather geometrical and analytical approach to the subject, in order to stay 'as concrete as possible'; the arithmetical aspects will only be mentioned at the very end of the lectures. I have tried to use a language familiar to theoretical physicists: the only prerequisites are some familiarity with the basic facts concerning holomorphic functions, differential forms and manifolds.

Introduction

0.1. In order to motivate the definition of Abelian varieties, let us recall the geometric interpretation of the theory of elliptic functions in terms of elliptic curves.

Let $\mathfrak{H} = \{\tau \in \mathbb{C} \mid \mathrm{Im}\ \tau > 0\}$ denote the Poincaré upper half-plane. For any $\tau \in \mathfrak{H}, \Gamma_\tau = \mathbb{Z} + \tau\mathbb{Z}$ is a lattice in $\mathbb{C}$. An *elliptic function* with respect to Γ_τ is a meromorphic function on $\mathbb{C}$ which is Γ_τ-periodic. The *elliptic curve* associated to Γ_τ is the quotient $E_\tau = \mathbb{C}/\Gamma_\tau$. It is a compact Riemann surface, and elliptic functions with respect to Γ_τ can be identified with meromorphic functions on E_τ.

Elliptic curves satisfy the following properties (*cf.* [Coh], [Ge], [Z]):

i) Any elliptic curve E_τ may be holomorphically embedded in the complex projective plane $\mathbb{P}^2\mathbb{C}$. Indeed, if $\wp$ denotes the Weierstrass function associated with the lattice Γ_τ, defined by

$$\wp(z) = \frac{1}{z^2} + \sum_{\gamma \in \Gamma_\tau - \{0\}} \left[\frac{1}{(z-\gamma)^2} - \frac{1}{\gamma^2}\right], \tag{0.1}$$

then the couple of elliptic functions $(\wp, \wp')$ defines an embedding of $E_\tau \backslash \{0\}$ in $\mathbb{C}^2$, which extends to an embedding $i : E_\tau \hookrightarrow \mathbb{P}^2\mathbb{C}$. Moreover, it follows from the differential equation satisfied by the $\wp$-function that $i(E_\tau)$ is an algebraic curve of equation

$$y^2 = 4x^3 - g_2(\tau)x - g_3(\tau) \tag{0.2}$$

where

$$g_2(\tau) = 60 \sum_{\gamma \in \Gamma_\tau - \{0\}} \gamma^{-4} \tag{0.3}$$

and

$$g_3(\tau) = 140 \sum_{\gamma \in \Gamma_\tau - \{0\}} \gamma^{-6}. \tag{0.4}$$

In fact the study of elliptic functions associated to Γ_τ is nothing else than function theory on the cubic curve defined by (0.2).

ii) Two elliptic curves E_τ and $E_{\tau'}$ $(\tau, \tau' \in \mathfrak{H})$ are isomorphic as Riemann surfaces, or equivalently, as algebraic curves iff there exists

$$\gamma = \begin{pmatrix} a & b \\ c & d \end{pmatrix}$$

in $\mathrm{SL}(2, \mathbb{Z})$ such that

$$\tau' = \gamma \cdot \tau := \frac{a\tau + b}{c\tau + d}\,.$$

Hence the quotient space $\mathfrak{H}/\mathrm{SL}(2,\mathbb{Z})$ can be identified with the set of isomorphism classes of elliptic curves, the so-called moduli space of elliptic curves, $\mathcal{M}_1$, and any modular function may be seen as the assignment of a complex number to any (isomorphism class of) elliptic curve(s). More generally, modular forms of weight $2k$ on $\mathfrak{H}$ (with respect to some congruence subgroup of $\mathrm{SL}(2,\mathbb{Z})$) may be interpreted as holomorphic differentials of weight k on some covering of $\mathcal{M}_1$. One of the basic recipes to construct modular forms is to form theta series. The basic example of such series is

$$\theta(z,\tau) = \sum_{n=-\infty}^{+\infty} \exp(\pi i n^2 \tau + 2\pi i n z) \tag{0.5}$$

defined for any $(z,\tau) \in \mathbb{C} \times \mathfrak{H}$.

0.2. Abelian varieties are higher dimensional generalizations of elliptic curves: a complex manifold A of dimension g is an Abelian variety[1] if it is isomorphic to a complex torus of the form $\mathbb{C}^g/(\mathbb{Z}^g + \Omega\mathbb{Z}^g)$, where Ω is a matrix in *Siegel's upper half-space* $\mathfrak{H}_g$, *i.e.*, a complex square matrix of size g which is symmetric and whose imaginary part $\mathrm{Im}\ \Omega := (\mathrm{Im}\ \Omega_{ij})_{1\leq i,j\leq g}$ is positive (*i.e.*, such that $v \in \mathbb{R}^g\backslash\{0\} \Rightarrow {}^t v \cdot \mathrm{Im}\ \Omega \cdot v > 0$).

An Abelian variety is thus a complex torus of a special kind. Indeed the condition on a lattice in $\mathbb{C}^g$ to be of the form $\mathbb{Z}^g + \Omega\mathbb{Z}^g$, after a complex linear transformation, is non-trivial (count the parameters describing the space of these lattices and the space of all lattices in $\mathbb{C}^g$!). We will see that this condition implies that the complex manifold $A = \mathbb{C}^g/(\mathbb{Z}^g + \Omega\mathbb{Z}^g)$ can be embedded as an algebraic subvariety of some projective space $\mathbb{P}^N\mathbb{C}$. However, this is not possible for an arbitrary complex torus.

Thus, Abelian varieties are higher dimensional generalizations of elliptic curves for which the analogue of i) holds. The assertions in ii) also generalize in higher dimensions: if $\Omega, \Omega' \in \mathfrak{H}_g$, the two Abelian varieties $\mathbb{C}^g/(\mathbb{Z}^g + \Omega\mathbb{Z}^g)$ and $\mathbb{C}^g/(\mathbb{Z}^g + \Omega'\mathbb{Z}^g)$ are isomorphic iff there exists $\gamma \in \mathrm{Sp}(2g,\mathbb{Z})$ such that $\Omega' = \gamma \cdot \Omega$. Here $\mathrm{Sp}(2g,\mathbb{Z})$ denotes the group of symplectic integral $2g \times 2g$ matrices, *i.e.*, matrices of the form

$$\gamma = \begin{pmatrix} A & B \\ C & D \end{pmatrix},$$

where A, B, C, D are matrices[2] in $M_g(\mathbb{Z})$, such that

[1] We are cheating a little at this point: what we are defining here are the complex manifolds underlying principally polarized Abelian varieties.

[2] For any ring A, we denote the ring of square matrices of size g with entries in A by $M_g(A)$. The unit matrix $\begin{pmatrix} 1 & & 0 \\ & \ddots & \\ 0 & & 1 \end{pmatrix} \in M_g(A)$ is denoted I_g.

$$A \cdot {}^tB = B \cdot {}^tA$$
$$C \cdot {}^tD = D \cdot {}^tC$$
$$A \cdot {}^tD - B \cdot {}^tC = I_g.$$

This group acts on $\mathfrak{H}_g$ by

$$\gamma \cdot \Omega := (A\Omega + B)(C\Omega + D)^{-1}.$$

The quotient $\mathfrak{H}_g/\mathrm{Sp}(2g, \mathbb{Z})$ may be identified with a moduli space of Abelian varieties, and there is a theory of modular forms on $\mathfrak{H}_g$, which generalizes the theory of modular forms on $\mathfrak{H} = \mathfrak{H}_1$. Such modular forms may be constructed by means of theta series in several variables, which generalize (0.5).

Modular forms on $\mathfrak{H}$ have striking arithmetic applications (see [Coh] and [Z]). Similarly, modular forms on $\mathfrak{H}_g$ have been used by Siegel to prove deep results on quadratic forms over number fields.

0.3. The purpose of these lectures is to explain how Abelian varieties were introduced in the mathematical world. The story goes back to the last century, when the work of Abel, Jacobi and their followers (Göpel, Rosenhain, Hermite, Weierstrass, Riemann ...) led to the great discovery that *to any complex algebraic curve* X (= compact Riemann surface) *of genus* g *is associated an Abelian variety* $J(X)$ *of complex dimension* g, the so-called *Jacobian variety of* X, in which X may be embedded, and that function theory on X is made much simpler if one uses this embedding. In the following pages, we try to present their results.

Since the beginning of this century, Abelian varieties have been studied for their own sake, and have been the subject of an impressive amount of work which involves analytic and algebraic geometry, arithmetics, automorphic forms, representation theory,... and which, like Siegel's theory alluded to above, often establishes deep interactions between these domains of mathematics. However, as examplified by the recent proof by Faltings of Mordell's conjecture, Abelian varieties are still a key tool to understand, not only function theory on Riemann surfaces, but also the geometric and arithmetic properties of algebraic curves, thanks to constructions which go back to Abel and Jacobi.

These notes cover slightly different topics than the oral conferences, where the content of §I.5, §B.7 to B.9, Appendix C, §II.4 and §III.6 was not discussed. On the other hand, a lecture providing an introduction to Arakelov geometry was given and is not reproduced here. I would like to thank Norbert A'Campo and the mathematicians of the 'Mathematisches Institut der Universität Basel' for discussions on the topics of §I.5 and Marc-Henri Dehon and Patrick Gérard for remarks on the content of Appendix C. Finally, I am very grateful to Melle Cécile Gourgues for the great skill and patience with which she typed these notes.

I. Compact Riemann surfaces and algebraic curves

The first part of these lectures is devoted to a discussion of basic facts concerning Riemann surfaces. It is intended mainly to provide a large supply of compact Riemann surfaces and to introduce some notations and definitions which will be needed later. More technical results, concerning the topology of compact Riemann surfaces and holomorphic line bundles are presented in three appendices to this Section.

More precisely, Sections I.1 to I.4 provide a short survey of the theory of Riemann surfaces. The reader is expected to read them without any serious difficulty, as we include (almost) no proof (see the 'bibliographical comments' at the end of this Chapter for references to the relevant literature). The content of §I.1, I.2, I.4.1 and I.4.2 is specially important for the sequel. Section I.3, concerning uniformization and classification of Riemann surfaces, is not strictly needed for understanding the rest of these notes. We included it because it contains a most remarkable classification theorem which provides an 'overview' of all Riemann surfaces. Section I.5, whose content is not used later in these notes, can be skipped at first reading; however, it presents a recent and striking result which illustrates the various concepts introduced so far. Appendix A contains some basic results concerning the topology of surfaces (without proof). Appendix B contains more technical results about holomorphic line bundles on Riemann surfaces. At first reading, the reader should browse through §B.1 to B.5 and return only later to a more detailed study of this Appendix. Appendix C contains proofs of the results in Appendix B, which avoid any explicit use of sheaf cohomology and which we think to be simpler than in the usual treatments of analysis on compact Riemann surface found in modern literature.

I.1. Basic definitions

A Riemann surface is defined as a complex manifold of complex dimension 1, that is, roughly speaking, a topological space in which a neighbourhood of any point looks like the complex plane.

More precisely, this definition means that a Riemann surface X is a Hausdorff space X such that, for any point P of X, we are given an open neighbourhood Ω of P and a homeomorphism

$$\varphi : \Omega \to \varphi(\Omega) \subset \mathbb{C}$$

from Ω to an open domain in $\mathbb{C}$. These homeomorphisms satisfy the following consistency condition: let P_1, P_2 be any two points of X such that the associated neighbourhoods Ω_1, Ω_2 overlap; then $\varphi_1 \circ \varphi_2^{-1}$ is holomorphic on $\varphi_2(\Omega_1 \cap \Omega_2)$.

Such a pair (Ω, φ) is called a *holomorphic chart* on X and the map φ is called a *local coordinate* at P. More generally, if f is any holomorphic function defined on a neighbourhood of $z = \varphi(P)$, such that $f'(z) \neq 0$, then $f \circ \varphi$ defines the same complex structure on a neighbourhood of P as φ and is also a local coordinate at P.

Clearly, any open domain in $\mathbb{C}$ defines a Riemann surface. The simplest compact Riemann surface is the ***Riemann sphere or projective line***

$$\mathbb{P}^1\mathbb{C} = \mathbb{C} \cup \{\infty\}.$$

Topologically, it is the one point compactification of $\mathbb{C}$, *i.e.*, a sphere. Its holomorphic structure is defined by the charts $(U,\varphi),(V,\psi)$ given by

$$U = \mathbb{P}^1\mathbb{C} - \{\infty\} = \mathbb{C}\ ;\ \varphi(z) = z$$

and

$$V = \mathbb{P}^1\mathbb{C} - \{0\} = \mathbb{C}^* \cup \{\infty\}\ ;\ \psi(z) = z^{-1} \text{ if } z \neq \infty\ ;\ \psi(\infty) = 0.$$

The standard notions from the theory of functions of a complex variable may be transferred from the complex plane to Riemann surfaces:
- a complex valued function ψ defined on an open set $U \subset X$ is ***holomorphic*** iff for any $P \in U$ and any local coordinate φ at P, $\psi \circ \varphi^{-1}$ is holomorphic near $z = \varphi(P)$. Moreover, ψ has a zero of order n at P iff $\psi \circ \varphi^{-1}$ has a zero of order n at z.
- in the same way, one defines *meromorphic* functions on Riemann surfaces and the order of a pole of a meromorphic function.
- a continuous map $f : Y \to X$ between two Riemann surfaces is ***holomorphic*** if, for any holomorphic chart $\varphi : \Omega \to \mathbb{C}$ in X, the map $\varphi \circ f : f^{-1}(\Omega) \to \mathbb{C}$ is holomorphic. Such a map is called an *isomorphism* of Riemann surfaces, or a *biholomorphic* map, when it is a homeomorphism and its inverse is holomorphic. Any meromorphic function on a Riemann surface X may be seen as a holomorphic function from X to $\mathbb{P}^1\mathbb{C}$.

I.2. Four constructions of Riemann surfaces

I.2.1. Conformal structures.

Let X be a C^∞ real surface[3].

Let g be any C^∞ Riemannian metric on X. A classical theorem[4] says that, *locally on X, the metric g is conformally flat*, *i.e.*, that X may be covered by charts U, such that *there are C^∞ local coordinates (x,y) on U in terms of which g takes the form*

$$g = \lambda(dx^2 + dy^2)$$

where λ is a C^∞ function from U to $\mathbb{R}_+^*$.

Furthermore, suppose that X is oriented. Permuting x and y if necessary, we can suppose that the charts $(U;x,y)$ are compatible with the given orientation. Then the complex valued charts $(U, x+iy)$ are such that the coordinate

---

[3] *i.e.*, a C^∞ manifold of dimension 2.

[4] Due to Gauss when X and g are real analytic and to Korn and Lichtenstein in the C^∞ case ; *cf.* [Ch] for a simple proof.

changes between them are oriented and conformal diffeomorphisms, *i.e., biholomorphic transformations.* Hence these charts define a structure of complex manifold on X (of complex dimension 1).

Fig. 1. A compact Riemann surface two thousand years before Euclid (Cycladic art, 2400–2200 b.c., Athens National Museum).

Moreover, for any function $\mu \in \mathcal{C}^\infty(X, \mathbb{R}_+^*)$ the Riemannian metric μg clearly defines the same holomorphic structure on X as g.

Thus we see that the data of *an orientation and of a conformal class of* C^∞ *metrics on* X *determines a structure of Riemann surface on* X. Furthermore, one easily checks that one gets in that way a one to one correspondence:

$$\text{(I.2.1)} \qquad \left\{ \begin{array}{c} \textit{holomorphic structures} \\ \textit{on } X \textit{ compatible with} \\ \textit{its } C^\infty \textit{ structure} \end{array} \right\} \longleftrightarrow \left\{ \begin{array}{c} (\textit{conformal class of metrics} \\ \textit{on } X, \textit{ orientation on } X) \end{array} \right\}$$

A nice feature of this construction is that it immediately provides a large supply of Riemann surfaces: any oriented surface embedded in a Riemannian manifold (*e.g.*, in the 'physical' Euclidean three space $\mathbb{R}^3$ as in fig. 1) gives a Riemann surface !

The correspondence (I.2.1) shows that the study of Riemann surfaces may be seen as the study of conformally invariant properties of two-dimensional Riemannian manifolds. That is the reason why Riemann surfaces occur in recent topics of theoretical physics such as string theory or conformal field theory.

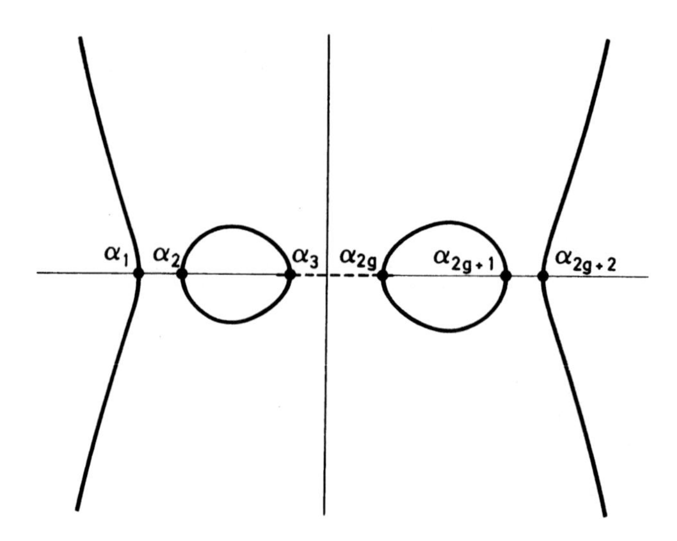

Fig. 2. The real points of the curve $y^2 = \prod_{i=1}^{2g+2}(x - \alpha_i)$.

I.2.2. Algebraic curves.

Historically, the theory of Riemann surfaces arose from the study of algebraic functions and of their integrals. The Riemann surfaces occurring in this study are Riemann surfaces attached to *algebraic curves*, which are defined by the vanishing of a family of complex polynomials.

Here is the simplest example of an algebraic curve: let $P \in \mathbb{C}[X, Y]$ be an irreducible (non-constant) polynomial; then

$$C_P = \left\{(x, y) \in \mathbb{C}^2 \mid P(x, y) = 0 \text{ and } \left(\frac{\partial P}{\partial X}(x, y), \frac{\partial P}{\partial Y}(x, y)\right) \neq (0, 0)\right\}$$

is a one-dimensional complex submanifold of $\mathbb{C}^2$; hence it is a Riemann surface.

Examples. Here are three examples of increasing order of complexity.

i) Let $P(X, Y) = X^2 - Y^3$. Then

$$C_P = \left\{(x, y) \in \mathbb{C}^2 \backslash \{(0, 0)\} \mid x^2 = y^3\right\}.$$

The parametrization $x = t^3$, $y = t^2$ shows that C_P is isomorphic (as a Riemann surface) to the pointed complex plane $\mathbb{C}^*$.

ii) Let $Q(X) \in \mathbb{C}[X]$ be a polynomial with simple roots. Then $P(X, Y) := Y^2 - Q(X)$ is irreducible and

$$C_P = \left\{(x, y) \in \mathbb{C}^2 \mid y^2 = Q(x)\right\}$$

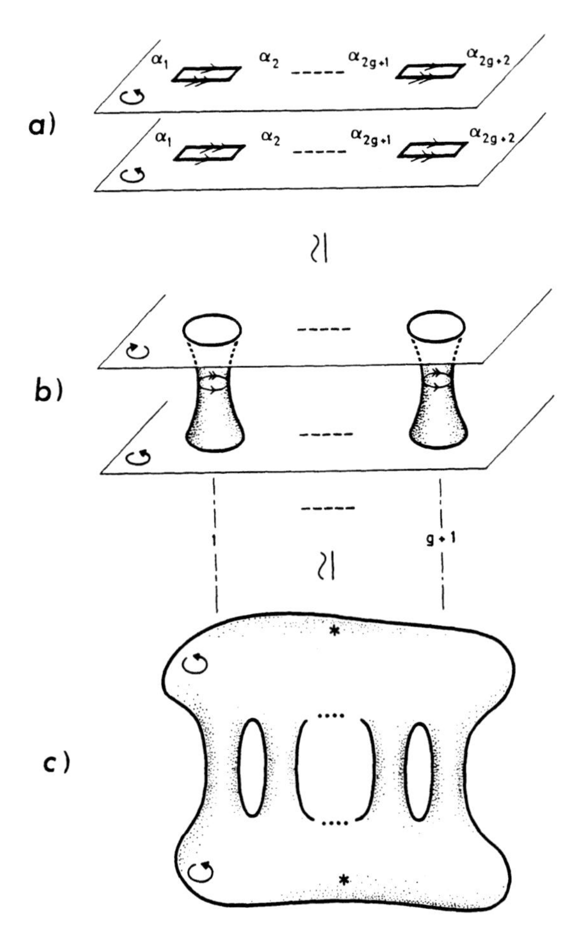

Fig. 3. a) Two separate copies of $\mathbb{C}$, each with $g+1$ cuts. b) The upper copy has been turned upside down and the sides of the cuts have been glued according to the arrows. c) The surface made compact by adding one point at infinity on each sheet.

is a Riemann surface (observe that $P(x,y) = 0 \Rightarrow \left(\frac{\partial P}{\partial X}(x,y), \frac{\partial P}{\partial Y}(x,y)\right) = (-Q'(x), 2y) \neq 0$ since Q has only simple roots).

Suppose $Q(X)$ is of even degree, say

$$Q(X) = \prod_{i=1}^{2g+2} (X - \alpha_i).$$

Figure 2 depicts the set $C_P \cap \mathbb{R}^2$ of real points of C_P when the roots α_i are real. The whole curve C_P appears as a two-sheeted covering of $\mathbb{C}$, ramified at $\alpha_1, \ldots, \alpha_{2g+2}$, *via* the map $(x,y) \mapsto x$. Topologically, C_P can be described as two copies of the complex plane suitably sewn along cuts between α_1 and α_2, α_3 and $\alpha_4, \ldots, \alpha_{2g+1}$ and α_{2g+2}; this proves that C_P has the topology of a compact surface of genus g with two points deleted (see figure 3; *cf.* Appendix A for the definition of the genus).

iii) Let $P(x,y) = X^n + Y^n - 1$ ($n \in \mathbb{N}^*$). Then C_P is the 'affine Fermat curve':

$$C_P = \left\{(x,y) \in \mathbb{C}^2 \mid x^n + y^n = 1\right\}.$$

The reader may enjoy proving that, topologically, C_P is a compact surface of genus $\frac{1}{2}(n-1)(n-2)$ with n points deleted.

I.2.3. Quotients of Riemann surfaces.

A third way to get Riemann surfaces is to construct them as *quotients* under group actions of some other Riemann surfaces.

For instance, an elliptic curve $E_\tau = \mathbb{C}/(\mathbb{Z} + \tau\mathbb{Z})$ is the quotient of the complex plane by the action by translation of the lattice $\mathbb{Z} + \tau\mathbb{Z}$.

In general, one can prove the following theorem:

Theorem I.2.1. *Let X be a Riemann surface and let Γ be a discrete group acting on X such that:*

i) for any $\gamma \in \Gamma$, the map $x \mapsto \gamma \cdot x$ from X to itself is holomorphic (hence an isomorphism of Riemann surfaces);

ii) the action of Γ on X is proper*: if $(z_n)_{n\geq 1}$ is a sequence in X and $(\gamma_n)_{n\geq 1}$ is a sequence in Γ such that the z_n's belong to a compact subset of X and the γ_n's are pairwise distinct, then $\gamma_n \cdot z_n$ goes to infinity in X* [5] *when n goes to infinity.*

Then the quotient space X/Γ possesses a natural structure of Riemann surface, characterized by the following property:

Let $\pi : X \to X/\Gamma$ denote the canonical map. For any open subset U of X/Γ, $\pi^{-1}(U)$ is an open subset of X [6] *and a function $f : U \to \mathbb{C}$ is holomorphic iff $f \circ \pi : \pi^{-1}(U) \to \mathbb{C}$ is holomorphic.*

Observe that according to ii), X/Γ is a nice Hausdorff space. Moreover, when Γ *acts freely on* X, *i.e.*, when for any $\gamma \in \Gamma - \{e\}$ the automorphism $z \mapsto \gamma \cdot z$ has no fixed point, this theorem is a simple formal consequence of the definition of a Riemann surface, and still holds if 'Riemann surface' is replaced by 'n-dimensional C^∞ (or complex) manifold'. The point is that, contrary to what occurs with quotients of manifolds of dimension > 1 [7], even if some

[5] *i.e.*, for any compact subset K of X, $\gamma_n x_n \notin K$ for n large enough.

[6] This is the definition of open subsets in X/Γ.

[7] For instance the quotient of $\mathbb{C}^2$ by the action of the involution $(x_1, x_2) \mapsto (-x_1, -x_2)$ is isomorphic with the cone $xy = z^2$ in $\mathbb{C}^3$ (put $x = x_1^2, y = x_2^2, z = x_1 x_2$), which is not smooth at $(0,0,0)$.

$\gamma \in \Gamma - \{e\}$ acts with fixed points, X/Γ is a *smooth* Riemann surface: there is no 'quotient singularity' in the world of Riemann surfaces.

Examples. i) Let X be the unit disc $D = \{z \in \mathbb{C} \mid |z| < 1\}$ and let Γ be the group $\mathbb{Z}_n$ of n-th roots of unity, acting by multiplication on $\mathbb{C}$. Then the Γ-invariant function on D $z \mapsto z^n$ induces an isomorphism

$$D/\mathbb{Z}_n \simeq D.$$

This example shows how a quotient X/Γ may be smooth even when some elements in $\Gamma\backslash\{e\}$ have fixed points. In fact it is the 'generic' example of this phenomenon: under the hypothesis of Theorem I.2.1, for any point $P \in X$, its stabilizer in Γ, namely

$$\Gamma_P = \{\gamma \in \Gamma \mid \gamma \cdot P = P\},$$

is a finite cyclic group $\mathbb{Z}_n$, and there exists a chart (U, z) of X such that

$$P \in U, z(P) = 0, z(U) = D,$$

$$\gamma \in \Gamma_P \Rightarrow \gamma \cdot U = U,$$
$$\gamma \in \Gamma\backslash\Gamma_P \Rightarrow \gamma \cdot U \cap U = \emptyset$$

and such that, read in this chart, the action of Γ_P on U is the action of n-th roots of unity on D. Then in X/Γ, the open set $\pi(U)$ may be identified with U/Γ_P, *i.e.*, with $D/\mathbb{Z}_n \simeq D$ (see, for instance, [GN], §I.5.(3)).

ii) The group $\mathbb{Z}$ acts on $\mathbb{C}$ by translation and this action clearly satisfies the conditions i) and ii) above. The $\mathbb{Z}$-periodic function $\exp(2\pi i z)$ induces an isomorphism

$$\mathbb{C}/\mathbb{Z} \simeq \mathbb{C}^*.$$

iii) An action of the group

$$\mathrm{PSL}(2,\mathbb{R}) = \left\{ \begin{pmatrix} a & b \\ c & d \end{pmatrix} \in M_2(\mathbb{R}) \mid ad - bc = 1 \right\} / \{\pm I_2\}$$

on the upper half plane $\mathfrak{H}$ is defined by

$$\begin{pmatrix} a & b \\ c & d \end{pmatrix} \cdot z = \frac{az + b}{cz + d} \, .$$

Thus any subgroup Γ of $\mathrm{PSL}(2,\mathbb{R})$ acts on $\mathfrak{H}$, and clearly this action satisfies condition i) in Theorem I.2.1. Moreover, if we let $\mathrm{PSO}(2) = \mathrm{SO}(2)/\{\pm I_2\}$, we have a homeomorphism

$$\begin{aligned} \mathrm{SL}(2,\mathbb{R})/\mathrm{SO}(2) = \mathrm{PSL}(2,\mathbb{R})/\mathrm{PSO}(2) &\xrightarrow{\sim} \mathfrak{H} \\ [g] &\mapsto g \cdot i, \end{aligned}$$

and, since $\mathrm{PSO}(2)$ is compact, this easily implies that any discrete subgroup Γ of $\mathrm{PSL}(2,\mathbb{R})$ satisfies condition ii) in Theorem I.2.1. Finally, to any discrete subgroup Γ of PSL(2, $\mathbb{R}$) is attached a Riemann surface $\mathfrak{H}/\Gamma$. Any such Γ is called a *Fuchsian group*.

A simple way to get discrete subgroups of PSL(2, $\mathbb{R}$) is to take subgroups of $\mathrm{PSL}(2,\mathbb{Z}) := \left[\mathrm{SL}(2,\mathbb{R}) \cap M_2(\mathbb{Z})\right]/\{\pm I_2\}$. The quotients of $\mathfrak{H}$ by some of these subgroups are closely related to elliptic modular functions. For instance, the modular function j, defined in terms of g_2 and g_3 (*cf.* (0.3) and (0.4)) by the formula

$$j(\tau) = 1728 \frac{g_2(\tau)^3}{g_2(\tau)^3 - 27 g_3(\tau)^2},$$

induces an isomorphism

$$\begin{aligned} \underline{j} : \mathfrak{H}/\mathrm{PSL}(2,\mathbb{Z}) &\xrightarrow{\sim} \mathbb{C} \\ [\tau] &\mapsto j(\tau). \end{aligned}$$

Consider now the congruence subgroup

$$\Gamma(2) = \left\{ \begin{pmatrix} a & b \\ c & d \end{pmatrix} \in \mathrm{SL}(2,\mathbb{Z}) \mid a \equiv d \equiv 1 \bmod 2, b \equiv c \equiv 0 \bmod 2 \right\} / \{\pm I_2\}$$

It acts freely on $\mathfrak{H}$ and there is an isomorphism

$$\mathfrak{H}/\Gamma(2) \simeq \mathbb{C} - \{0,1\}$$

induced by the function $\lambda : \mathfrak{H} \to \mathbb{C} - \{0,1\}$ defined by

$$\lambda(\tau) = \frac{\theta(\tau/2,\tau)^4}{\theta(0,\tau)^4} = \frac{\left(\sum_{n=-\infty}^{+\infty} (-1)^n e^{\pi i n^2 \tau}\right)^4}{\left(\sum_{n=-\infty}^{+\infty} e^{\pi i n^2 \tau}\right)^4},$$

which is modular with respect to $\Gamma(2)$. (The function θ was defined in (0.5). Note that the j function also may be expressed in terms of θ; in fact we have

$$j = 256 \frac{(\lambda^2 - \lambda + 1)^3}{\lambda^2 (\lambda - 1)^2} \quad .)$$

I.2.4. Analytic continuation.

Finally we should mention, at least because of its historical importance, a last way to construct Riemann surfaces: analytic continuation.

Consider a point x_0 in $\mathbb{C}$ and a germ of holomorphic function at x_0, *i.e.*, a series

$$f_0(z) = \sum_{n=0}^{\infty} a_n (z - x_0)^n$$

with a positive radius of convergence. The Riemann surface of f_0 is 'the largest connected Riemann surface unramified over $\mathbb{C}$ on which the germ f may be extended as a holomorphic function'. We now describe a formal construction of this Riemann surface.

Consider the set $\mathcal{O}$ of pairs (x, f) where x is a point of $\mathbb{C}$ and f is a germ of holomorphic function at x, and consider the map

$$\begin{aligned} \pi : \quad \mathcal{O} & \to \mathbb{C} \\ (x, f) & \mapsto x. \end{aligned}$$

There exists a unique structure of (highly non-connected) Riemann surface on $\mathcal{O}$ which satisfies the following conditions: for any $(x, f) \in \mathcal{O}$, if the radius of convergence of f is R and if, for any $x' \in \overset{\circ}{D}_x (R)$ (the open disc of center x and radius R), $f_{x'}$ denotes the Taylor series of f at x', the map

$$\begin{aligned} \overset{\circ}{D}_x (R) & \to \mathcal{O} \\ x' & \mapsto (x', f_{x'}) \end{aligned}$$

is a biholomorphic map from $\overset{\circ}{D}_x (R)$ onto its image. This Riemann surface structure on $\mathcal{O}$ makes the map π a holomorphic map, which indeed is locally biholomorphic (*i.e.*, unramified). Moreover, the map

$$\begin{aligned} F : \quad \mathcal{O} & \to \mathbb{C} \\ (x, f) & \mapsto f(x) \end{aligned}$$

is easily seen to be holomorphic.

The Riemann surface X of the germ of holomorphic function f_0 is now defined as the connected component of (x_0, f_0) in $\mathcal{O}$. By construction, X is a Riemann surface which 'lies over $\mathbb{C}$'; indeed the map

$$\pi : X \to \mathbb{C}$$

is unramified. Moreover the holomorphic function F extends the germ f, which can be seen as a germ of holomorphic function on a neighbourhood of (x, f) in X (identified with a neighbourhood of x in $\mathbb{C}$ by π).

The link between this construction of the analytic continuation and its more classical description is made by the following observation: a germ (x', f') belongs to X iff there exists a finite sequence (x_i, f_i), $0 \leq i \leq N$, of germs such that $(x_0, f_0) = (x, f)$, $(x_N, f_N) = (x', f')$ and such that the open discs of convergence of f_{i-1} and f_i intersect and f_i and f_{i-1} coincide on this intersection.

A well known example of Riemann surface obtained by this construction is the 'Riemann surface of the logarithm', obtained by applying the preceding construction to

$$(x, f) = \left(1, \sum_{n=1}^{\infty} (-1)^{n+1} \frac{(z-1)^n}{n}\right).$$

The associated Riemann surface X is an unramified covering of $\mathbb{C}^*$, with an infinite number of sheets. It is biholomorphic with $\mathbb{C}$ *via* the map

$$\begin{aligned} \mathbb{C} &\to X \\ x &\mapsto \left(e^x, x + \sum_{n=1}^{\infty}(-1)^{n+1}\,\frac{e^{-nx}}{n}\,(z-e^x)^n\right). \end{aligned}$$

Another example of analytic continuation is provided by algebraic functions. Assume f is a germ of algebraic function, *i.e.*, that there exists a non-zero polynomial $P \in \mathbb{C}[X,Y]$ such that

$$P(x', f(x')) = 0$$

for any x' in a neighbourhood of x. The polynomial P may be supposed irreducible and then the map

$$\begin{aligned} X &\to \mathbb{C}^2 \\ p = (x,f) &\mapsto (\pi(p), F(p)) = (x, f(x)) \end{aligned}$$

establishes an isomorphism between X and the open subset of C_P (*cf.* §I.2.2) obtained by deleting from C_P the finite set of points at which the first projection $(z_1,z_2) \mapsto z_1$ is ramified (this is equivalent to $\frac{\partial P}{\partial Y}(z_1,z_2) = 0$).

I.3. Classification of Riemann surfaces

A deep theorem of the theory of Riemann surfaces asserts that any connected Riemann surface may be obtained as a quotient from $\mathbb{P}^1\mathbb{C}$, $\mathbb{C}$, or $\mathfrak{H}$. More precisely, we have the following classification:

Theorem I.3.1. *The class of connected Riemann surfaces may be partitioned into three subclasses:*

• Elliptic Riemann surfaces: *these are the Riemann surfaces isomorphic to the Riemann sphere $\mathbb{P}^1\mathbb{C}$. They are characterized as the connected Riemann surfaces X such that there is a non-constant holomorphic map $\mathbb{P}^1\mathbb{C} \to X$.*

In particular, if $X = \mathbb{P}^1\mathbb{C}$ in Theorem I.2.1, then Γ is finite and $X/\Gamma \simeq \mathbb{P}^1\mathbb{C}$.

• Parabolic Riemann surfaces: *these are the Riemann surfaces isomorphic either to $\mathbb{C}$, or to $\mathbb{C}^*$ or to an elliptic*[8] *curve $E_\tau = \mathbb{C}/(\mathbb{Z}+\tau\mathbb{Z}), \tau \in \mathfrak{H}$. They are characterized as the connected Riemann surfaces X which are not elliptic and such that there is a non-constant holomorphic map from $\mathbb{C}$ to X, or as the surfaces isomorphic to a quotient of the complex plane $\mathbb{C}$ by a group action satisfying the hypotheses of Theorem I.2.1, or as the surfaces isomorphic to a quotient of the complex plane by the action by translation of a discrete subgroup of $\mathbb{C}$.*

[8] At this point, the terminology appears quite awful: an elliptic curve is a parabolic Riemann surface. It is unfortunately well established.

Among the parabolic Riemann surfaces, $\mathbb{C}$ *is characterized, up to isomorphism, as simply connected and non-compact,* $\mathbb{C}^*$ *as non-simply connected and non-compact, and the elliptic curves* E_τ*, as compact. Furthermore, two elliptic curves* E_τ *and* $E_{\tau'}$ *are isomorphic as Riemann surfaces iff* τ *and* τ' *have the same class in* $\mathfrak{H}/\mathrm{PSL}(2,\mathbb{Z})$, i.e., *iff* $j(\tau) = j(\tau')$.

• Hyperbolic Riemann surfaces: *these are the Riemann surfaces isomorphic to a quotient* $\mathfrak{H}/\Gamma$*, where* Γ *is a discrete subgroup of* $\mathrm{PSL}(2,\mathbb{R})$ *acting freely on* $\mathfrak{H}$*. They are characterized as the connected Riemann surfaces* X *such that there is no non-constant holomorphic map from* $\mathbb{C}$ *to* X*. Furthermore, two such surfaces* $\mathfrak{H}/\Gamma$ *and* $\mathfrak{H}/\Gamma'$ *are isomorphic iff there exists* $g \in \mathrm{PSL}(2,\mathbb{R})$ *such that* $\Gamma' = g\Gamma g^{-1}$.

This 'trichotomy' may be rephrased in terms of universal coverings:
If X *is a connected Riemann surface and if* $\tilde{X}$ *denotes its universal covering* (*cf.* [Rey]), *then:*

· X *is elliptic iff* $\tilde{X} \simeq \mathbb{P}^1\mathbb{C}$ (*and then in fact* $X \simeq \mathbb{P}^1\mathbb{C}$)
· X *is parabolic iff* $\tilde{X} \simeq \mathbb{C}$
· X *is hyperbolic iff* $\tilde{X} \simeq \mathfrak{H}$.

In particular, the classification above asserts that *any simply connected Riemann surface is isomorphic either to* $\mathbb{P}^1\mathbb{C}$*, or to* $\mathbb{C}$*, or to* $\mathfrak{H}$. This statement is known as the *uniformization theorem* and is the main point in the proof of Theorem A.3.1.

The division in three classes of connected Riemann surfaces may be understood in terms of conformal structures:

Theorem I.3.2. *A connected Riemann surface is elliptic (resp. parabolic, resp. hyperbolic) iff it may be equipped with a complete Riemannian metric defining its conformal structure whose Gaussian curvature is* $+1$ *(resp.* 0*, resp.* -1*).*

The 'only if' part in Theorem I.3.2 is easily checked as follows:

• An elliptic Riemann surface is isomorphic to the Riemann sphere $\mathbb{P}^1\mathbb{C} = \mathbb{C} \cup \{\infty\}$, which clearly possesses a conformal metric of curvature 1. Explicitly, we can take

$$ds^2 = \frac{4|dz|^2}{(1+|z|^2)^2}.$$

• The flat metric on $\mathbb{C}$

$$ds^2 = |dz|^2$$

is translation invariant. Hence it defines a complete metric of zero curvature on any parabolic Riemann surface $X = \mathbb{C}/\Gamma$ (Γ discrete subgroup of $\mathbb{C}$).

• The *Poincaré metric*

$$ds^2 = \frac{|dz|^2}{(\mathrm{Im}\, z)^2}$$

on $\mathfrak{H}$ is a complete Riemannian metric of curvature -1, which is preserved by the action of $\mathrm{PSL}(2,\mathbb{R})$ on $\mathfrak{H}$. Hence it defines a complete metric of curvature

-1 on any Riemann surface of the form $\mathfrak{H}/\Gamma$, where Γ is a discrete subgroup of $\mathrm{PSL}(2,\mathbb{R})$ acting freely on $\mathfrak{H}$, *i.e.*, on any hyperbolic Riemann surface.

Recall that, equipped with this metric, $\mathfrak{H}$ is a model of hyperbolic two-dimensional non-Euclidean geometry (*cf.* [Cr]). In this model, the geodesics are represented by the semi-circles centered on the real axis and by the lines orthogonal to it. The group of direct isometries of the hyperbolic plane coincides with $\mathrm{PSL}(2,\mathbb{R})$ acting on $\mathfrak{H}$ by homographic transformations.

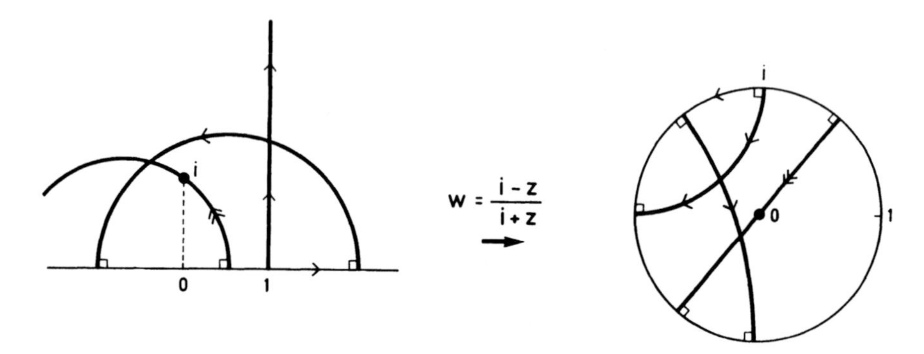

Fig. 4. Half plane and disc models of hyperbolic geometry.

It is often useful to use a variant of this model (*cf.* figure 4): we have a holomorphic diffeomorphism between $\mathfrak{H}$ and the unit disc

$$D = \{w \in \mathbb{C} \mid |w| < 1\}$$

which associate to $z \in \mathfrak{H}$ the point $w = \frac{i-z}{i+z} \in D$. Transported on D by this diffeomorphism, the Poincaré metric becomes

$$ds^2 = \frac{4|dw|^2}{(1-|w|^2)^2}.$$

The geodesics in this 'disc model' are the diameters and the arcs of circles orthogonal to the boundary circle $\partial D = \{w \in \mathbb{C} \mid |w| = 1\}$, and the group of direct isometries of the hyperbolic plane coincide with

$$\mathrm{PSU}(1,1) := \left\{ \begin{pmatrix} a & \bar{b} \\ b & \bar{a} \end{pmatrix} \; ; \; (a,b) \in \mathbb{C}^2 \; , \; |a|^2 - |b|^2 = 1 \right\} / \{\pm I_2\}$$

acting on D by homographic transformations.

Exercise: An element $g = (g_{ij})_{1 \le i,j \le 2}$ of $\mathrm{SL}(2,\mathbb{R}) - \{\pm I_2\}$ (resp. of $\mathrm{SU}(1,1) - \{\pm I_2\}$) acts without fixed point on $\mathfrak{H}$ (resp. on D) iff its trace $\mathrm{tr}\, g := g_{11} + g_{22}$ has modulus ≥ 2.

The uniformization theorem is one of the achievements of the mathematics in the last century. Its proof leaded to the development of rigorous methods in the study of elliptic differential equations and in potential theory, as well as in algebraic and differential topology. Uniformization establishes a fascinating interaction between these topics, the study of Riemann surfaces and hyperbolic geometry, which, still now, is far from being completely explored. Concerning this circle of ideas, we cannot refrain to quote the enthusiastic evocation of uniformization in the foreword of the first 'modern' book on Riemann surfaces, published by Hermann Weyl in 1913 ([Wey 1]):

> *... Die letzten Abschnitte endlich (§19-21) sind der von Klein und Poincaré in kühnem Riß entworfenen, von Koebe in jüngster Zeit auf ein breites Fundament gestellten* Theorie der Uniformisierung *gewidmet. Wir betreten damit den Tempel, in welchem die Gottheit (wenn ich dieses Bildes mich bedienen darf) aus der irdischen Haft ihrer Einzelverwirklichungen sich selber zurückgegeben wird: in dem Symbol des* zweidimensionalen Nicht-Euklidischen Kristalls *wird das Urbild der Riemannschen Flächen selbst, (soweit dies möglich ist) rein und befreit von allen Verdunklungen und Zufälligkeiten, erschaubar*[9]...

I.4. Compact Riemann surfaces

I.4.1. Compact Riemann surfaces with punctures.

We have just discussed the link between the construction of Riemann surfaces as quotients, and the construction from conformal structures. The interplay between these constructions and the construction from algebraic curves appears more clearly when one deals with *compact Riemann surfaces.*

In order to explain it, let us introduce a definition: we will call *compact Riemann surface with punctures* any Riemann surface X such that there exists an open subset $U \subset X$ such that:

(P1) there exists a biholomorphic map from U onto a disjoint finite union of punctured discs $\{0 < |z| < 1\}$;

(P2) $X - U$ is compact.

One easily checks that for any such X, one gets a new Riemann surface $\widehat{X}$ by gluing X and a disjoint finite union of discs $\{|z| < 1\}$ along the open set U (see figure 5). Clearly, $\widehat{X}$ is compact and contains X, and $\widehat{X} - X$ is finite. Moreover, $\widehat{X}$ is characterized by these properties and is therefore well defined.

It is useful to observe that if Σ is a finite subset of a Riemann surface X, X is a compact Riemann surface with punctures iff $X - \Sigma$ is such, and that, when this is true, $\widehat{X}$ may be identified with $\widehat{X - \Sigma}$.

---

[9] ... Finally the last Sections (§19-21) are devoted to the *uniformization theory,* which was sketched by Klein and Poincaré in an audacious breakthrough and was recently put on a firmer basis by Kœbe. Thus we get into the temple where the divinity (if I am allowed to use this image) is restored to itself, from the earthy jail of its particular realization: through the two dimensional *non-Euclidean crystal,* the archetype of the Riemann surface may be contemplated, pure and liberated from any obscurity or contingency (as far as it is possible)...

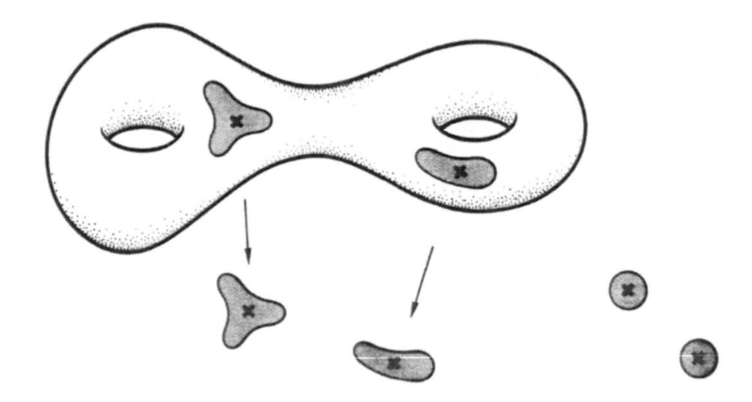

Fig. 5. A compact Riemann surface with punctures.

Example. Let F be a finite subset of $\mathbb{P}^1\mathbb{C}$, and let $\pi : X \to \mathbb{P}^1\mathbb{C} - F$ be an unramified covering of finite degree. Then X is a compact Riemann surface with punctures. Indeed, if we choose, for each $x \in F$, a neighbourhood D_x of x in $\mathbb{P}^1\mathbb{C}$ disjoint from $F - \{x\}$ and such that the D_x's are pairwise disjoint and biholomorphic to open discs, then the open subset of X

$$U = \bigcup_{x \in F} \pi^{-1}(D_x - \{x\})$$

of X clearly satisfies condition (P2), and satisfies condition (P1) as well, because any unramified covering of finite degree of a punctured disc is (biholomorphic to) a disjoint union of punctured discs (see [Rey]). Thus one gets a compact Riemann surface $\widehat{X}$. Moreover, the covering $\pi : X \to \mathbb{P}^1\mathbb{C} - F$ extends to a holomorphic map

$$\hat{\pi} : \widehat{X} \to \mathbb{P}^1\mathbb{C},$$

which may be ramified over $\pi^{-1}(F)$.

Exercise: Prove the following extension of the preceding example: let $f : X \to Y$ be a holomorphic map between Riemann surfaces which is proper and non constant on any connected component of X and whose ramification set is finite. If Y is a compact Riemann surface with punctures, then X also is a compact Riemann surface with punctures.

I.4.2. Algebraic curves and their normalizations.

Let $P \in \mathbb{C}[X, Y]$ be an irreducible polynomial, and let C_P be, as in §I.2.2, the set of smooth points of the curve of equation $P(x, y) = 0$. Let us assume that $P \notin \mathbb{C}[X]$ (*i.e.*, that C_P is not a line parallel to the Y-axis) and let n be the degree of Y in $P(X,Y)$. Then

$$F_0 = \{x \in \mathbb{C} \mid P(x,Y) \text{ has degree } < n \text{ or has a multiple root }\}$$

is a finite subset of $\mathbb{C}$. Indeed, it may be defined by the vanishing of the discriminant of $P(x,Y)$ as a polynomial in Y, and this discriminant is a polynomial function of x which does not vanishes identically, according to the hypothesis on P.

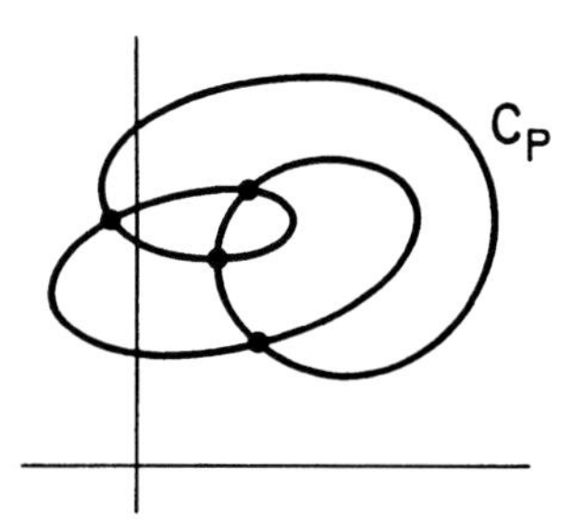

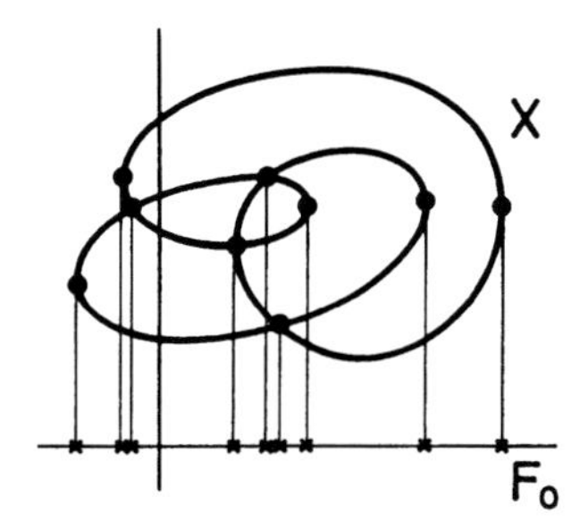

Fig. 6. The compact Riemann surface with punctures associated with an algebraic plane curve.

Let $X := \{(x,y) \in C_P \mid x \notin F_0\}$, $F := F_0 \cup \{\infty\}$, and let

$$\begin{aligned} \pi : \quad X &\to \mathbb{C} - F_0 = \mathbb{P}^1\mathbb{C} - F \\ (x,y) &\mapsto x. \end{aligned}$$

Then π is an unramified covering of degree n, and we are in the situation of the example in the preceding paragraph (see figure 6). Therefore X is a compact Riemann surface with punctures and we get a compact Riemann surface $\widehat{X}$ and a holomorphic map $\hat{\pi} : \widehat{X} \to \mathbb{P}^1\mathbb{C}$ which extends π.

As $X \subset C_P$ and $X - C_P$ is finite (it is included in $(F_0 \times \mathbb{C}) \cap \{P(x,y) = 0\}$ which is finite since P is irreducible and belongs to $\mathbb{C}[X,Y] - \mathbb{C}[X]$) we see that *$C_P$ itself is a compact Riemann surface with punctures, and that the map*

$$\begin{aligned} \pi : C_P &\to \mathbb{C} \\ (x,y) &\mapsto x \end{aligned}$$

extends to a holomorphic map

$$\hat{\pi} : \widehat{C}_P = \widehat{X} \to \mathbb{P}^1\mathbb{C}.$$

The compact Riemann surface $\widehat{C}_P$ is a compact Riemann surface canonically associated to the algebraic curve $\{P(x,y) = 0\}$. Moreover, one may prove that it is connected, as a consequence of the irreducibility of P.

Examples. Consider again the examples of §1.2.2. The reader can check that:

· if $P(X,Y) = X^2 - Y^3$, then $\widehat{C}_P$ is isomorphic to $\mathbb{P}^1\mathbb{C}$;

· if $P(X,Y) = Y^2 - \prod_{i=1}^{2g+2}(X - \alpha_i)$, where the α_i's are pairwise distinct, then $\widehat{C}_P$ is a compact surface of genus g; compact Riemann surfaces constructed in this way are called *hyperelliptic Riemann surfaces;*

· if $P(X,Y) = Y^2 - \prod_{i=1}^{2g+1}(X - \alpha_i)$, where the α_i's are pairwise distinct, then $\widehat{C}_P$ is also a hyperelliptic Riemann surface of genus g;

· if $P(X,Y) = X^n + Y^n - 1$, then $\widehat{C}_P$ may be identified with the 'Fermat curve', defined as the curve in $\mathbb{P}^2\mathbb{C}$ of homogeneous equation $X_1^n + X_2^n = X_0^n$. It is a compact surface of genus $\frac{1}{2}(n-1)(n-2)$.

In fact, by this construction, we can recover all compact connected Riemann surfaces:

Theorem I.4.2. *Let M be a compact connected Riemann surface.*
1) There exists a non-constant holomorphic map

$$f : M \to \mathbb{P}^1\mathbb{C}.$$

(i.e., a non-constant meromorphic function on M).
2) For any such f, there exists an irreducible polynomial $P \in \mathbb{C}[X,Y]$ and an isomorphism $M \simeq \widehat{C}_P$ such that, $\widehat{C}_P$ and M being identified through this isomorphism, f coincides with $\hat{\pi}$.

The assertion that on any compact Riemann surface, there is a non-constant meromorphic function is the difficult and important point in Theorem I.4.2. Let us emphasize that this property is special to *one-dimensional* compact complex manifolds. (On a compact complex surface, *i.e.*, on a compact two dimensional complex manifold, there may be no non-constant meromorphic function; *cf.* §III.2). This type of existence theorem goes back to Riemann, who gave a non-rigorous construction, based on the Dirichlet principle, of harmonic functions with prescribed singularities on ramified coverings of $\mathbb{P}^1\mathbb{C}$. Theorem I.4.2 was proved rigorously by methods closely related with the ones developed to prove the uniformization theorem, and, at the same time as this last theorem, has been at the origin of many developments in analysis around the term of last century.

Until now, we have considered only plane algebraic curves. To get a better understanding of the correspondence between compact Riemann surfaces and algebraic curves, it is important to deal with a more general notion of algebraic curves. This leads us to introduce a few definitions:

• an *algebraic subvariety* of complex dimension d of the affine space $\mathbb{C}^N$ (resp. of the projective space $\mathbb{P}^N\mathbb{C}$) is a subset V of $\mathbb{C}^N$ (resp. of $\mathbb{P}^N\mathbb{C}$) such that:

i) there exists a finite family of polynomials (resp. of homogeneous polynomials) $P_1, \ldots, P_k$ in $\mathbb{C}[X_1, \ldots, X_N]$ (resp. in $\mathbb{C}[X_0, X_1, \ldots, X_N]$) such that, for any $(x_1, \ldots, x_N) \in \mathbb{C}^N$ (resp. for any $(x_0 : \ldots : x_N) \in \mathbb{P}^N\mathbb{C}$), we have

$$(x_1, \ldots, x_N) \in V \Leftrightarrow P_1(x_1, \ldots, x_N) = \ldots = P_k(x_1, \ldots, x_N) = 0$$

$\Big($resp. :

$$\text{(I.4.1)} \quad (x_0 : \ldots : x_N) \in V \Leftrightarrow P_1(x_0, \ldots, x_N) = \ldots = P_k(x_0, \ldots, x_N) = 0\Big);$$

ii) the subset V_{reg} of V, formed by the points P of V which possess an open neighbourhood Ω in $\mathbb{C}^N$ (resp. in $\mathbb{P}^N\mathbb{C}$) such that $V \cap \Omega$ is a complex submanifold of Ω, is connected and of complex dimension d (by construction V_{reg} is an open subset[10] of V and a submanifold of $\mathbb{C}^N$ (resp. of $\mathbb{P}^N\mathbb{C}$)).

• an affine or projective algebraic variety V is said to be *smooth* if $V_{\text{reg}} = V$; then it is a complex submanifold of $\mathbb{C}^N$ or $\mathbb{P}^N$.

• an affine (resp. projective) *algebraic curve* is an affine (resp. projective) algebraic variety of dimension one.

Let V be any affine or projective algebraic curve. One may show that $V - V_{\text{reg}}$ *is finite* and that V_{reg} *is a compact Riemann surface with punctures* (the proof is a generalization of the proof for C_P: one shows that the linear projection from V_{reg} to a 'generic' line, restricted to the complement of a finite set of ramification points, is a proper finite unramified covering). Therefore we may consider the compact connected Riemann surface $\widehat{V}_{\text{reg}}$. In the projective case, the identity map from V_{reg} to itself extends to a holomorphic map from $\widehat{V}_{\text{reg}}$ to $V(\subset \mathbb{P}^N\mathbb{C})$. The Riemann surface $\widehat{V}_{\text{reg}}$ is called the *normalization* of V, and has been obtained from V by 'resolving its singularities' (the singular points of V are the points in $V - V_{\text{reg}}$). In the affine case—say $V \subset \mathbb{C}^N$—the closure $\overline{V}$ of V in $\mathbb{P}^N\mathbb{C}$ is an algebraic curve in $\mathbb{P}^N\mathbb{C}$, and $\widehat{V}_{\text{reg}}$ is nothing else than the normalization of $\overline{V}$. To summarize, *to any complex algebraic curve is canonically associated a compact connected Riemann surface.*

Examples: i) Let $P \in \mathbb{C}[X, Y]$ an irreducible polynomial of degree n (> 0). Then

$$V = \{(x, y) \in \mathbb{C}^2 \mid P(x, y) = 0\}$$

is an algebraic curve in $\mathbb{C}^2$, its closure $\overline{V}$ in $\mathbb{P}^2\mathbb{C}$ is the set of zeroes in $\mathbb{P}^2\mathbb{C}$ of the homogenized polynomial

$$\widetilde{P}(X_0, X_1, X_2) = X_0^n P\left(\frac{X_1}{X_0}, \frac{X_2}{X_0}\right),$$

and the normalization of $\overline{V}$ coincides with $\widehat{C}_P$.

ii) Any smooth projective algebraic curve—*e.g.*, the Fermat curve in $\mathbb{P}^2\mathbb{C}$, of equation $X_1^n + X_2^n = X_0^n$—is its own normalization.

iii) Let us give examples of smooth algebraic curves in $\mathbb{P}^3\mathbb{C}$ which are not contained in any projective plane of $\mathbb{P}^3\mathbb{C}$.

[10] In fact, one may prove that if V satisfies i), then V_{reg} is dense in V, and has only a finite number of connected components.

For instance we can take the image of the embedding

$$\begin{array}{ccc} \mathbb{P}^1\mathbb{C} & \to & \mathbb{P}^3\mathbb{C} \\ (x_0 : x_1) & \mapsto & (x_0^3 : x_0^2 x_1^1 : x_0^1 x_1^2 : x_1^3) \end{array}$$

the so-called *twisted cubic* in $\mathbb{P}^3\mathbb{C}$. It is an algebraic curve in $\mathbb{P}^3\mathbb{C}$ as it may be defined by the equations:

$$\begin{cases} X_0X_3 - X_1X_2 = 0 \\ X_2^2 - X_1X_3 = 0 \\ X_1^2 - X_0X_2 = 0. \end{cases}$$

A more sophisticated example is, for any $\tau \in \mathfrak{H}$, the image of the holomorphic map

$$\begin{aligned} \varphi_\tau : \mathbb{C}/(\mathbb{Z} + \tau\mathbb{Z}) &\to \mathbb{P}^3\mathbb{C} \\ [z] &\mapsto (\theta_{00}(2z,\tau) : \theta_{01}(2z,\tau) : \theta_{10}(2z,\tau) : \theta_{11}(2z,\tau)) \end{aligned}$$

defined by the theta functions 'with characteristics'

$$\theta_{ij}(z,\tau) = \sum_{n=-\infty}^{+\infty} \exp\left[\pi i(n+i/2)^2\tau + 2\pi i(n+i/2)(z+j/2)\right].$$

One may prove that φ_τ is an embedding and, using Riemann's quadratic relations between theta functions, that its image is defined by the quadratic equations (*cf.* [Mu4], I, p. 11–23):

$$\begin{cases} \theta_{00}(0)^2X_0^2 = \theta_{01}(0)^2X_1^2 + \theta_{10}(0)^2X_2^2 \\ \\ \theta_{00}(0)^2X_3^2 = \theta_{10}(0)^2X_1^2 - \theta_{01}(0)^2X_2^2. \end{cases}$$

A good reason to look at algebraic curves in higher dimensional projective spaces is that a general compact connected Riemann surface of genus $g \geq 2$ cannot be realized as a *smooth* algebraic curve in $\mathbb{P}^2\mathbb{C}$. Indeed, the genus of such a Riemann surface is $\frac{1}{2}(n-1)(n-2)$ where n is the degree of the irreducible homogeneous polynomial in $\mathbb{C}[X_0, X_1, X_2]$ which defines it, and clearly, not all positive integers are of this form. Moreover, no hyperelliptic Riemann surface of genus $g \geq 2$ can be embedded in $\mathbb{P}^2\mathbb{C}$. However, we have the following refined version of the existence theorem I.4.2., 1):

Theorem I.4.3. *For any compact connected Riemann surface X, there is a holomorphic embedding*

$$\varphi : X \to \mathbb{P}^3\mathbb{C}.$$

The crucial point now is that the image of such an embedding is an algebraic curve in $\mathbb{P}^3\mathbb{C}$ (observe the analogy with Theorem I.4.2. 2)). This is a (rather easy) special case of the following celebrated theorem of Chow:

Theorem I.4.4. *Any compact connected complex submanifold of $\mathbb{P}^N\mathbb{C}$ is an algebraic subvariety of $\mathbb{P}^N\mathbb{C}$.*

An important avatar of the 'algebraic character' of complex projective manifolds expressed by Chow's theorem is the fact that, on algebraic curves, 'meromorphic functions coincide with rational functions'. More precisely, we have:

Theorem I.4.5. *1) Let P be any irreducible polynomial in $\mathbb{C}[X,Y]$. For any meromorphic function f on $\widehat{C}_P$, there exist $R \in \mathbb{C}(X,Y)$ and a finite subset F of C_P such that R is defined and coincides with f on $C_P - F$.*

2) Let X be a compact connected Riemann surface embedded in $\mathbb{P}^N\mathbb{C}$. For any meromorphic function f on X, there exist homogeneous polynomials P, Q of the same degree in $\mathbb{C}[X_0, \ldots, X_N]$ and a finite subset F of X such that, for any $(x_0 : \ldots : x_N) \in X - F$,

$$Q(x_0, \ldots, x_N) \neq 0 \quad \text{and} \quad f(x_0 : \ldots : x_n) = \frac{P(x_0, \ldots, x_N)}{Q(x_0, \ldots, x_N)}.$$

Exercise: Prove assertion 2) of Theorem I.4.5 when $N = 1$ and $X = \mathbb{P}^1\mathbb{C}$; in other words, prove that any meromorphic function on $\mathbb{P}^1\mathbb{C}$ is a rational function. (Hint: use Appendix B.1.7).

Exercise: Prove that if C is the hyperelliptic Riemann surface $\widehat{C}_P$ defined by $P(X,Y) = Y^2 - \prod_{i=1}^{2g+2}(X - \alpha_i)$, and if x and y are the meromorphic functions on C defined by the coordinates, then any meromorphic function on C may be written in a unique way as $f = R(x) + S(x)y$, where R and S are rational functions (Hint: consider the involution on X which exchanges the two sheets of the covering $x : C \to \mathbb{P}^1\mathbb{C}$ and use the result of the preceding exercise).

I.4.3. Compact Riemann surfaces and uniformization.

Restricted to compact connected Riemann surfaces, the partition of Riemann surfaces discussed in §I.3 involves only their topology: a compact connected Riemann surface of genus g is elliptic (resp. parabolic, resp. hyperbolic) iff $g = 0$ (resp. $g = 1$, resp. $g \geq 2$). More precisely, uniformization theory tells us that the compact connected Riemann surfaces of genus 0 and 1 are, up to isomorphism, the Riemann sphere $\mathbb{P}^1\mathbb{C}$ and the elliptic curves $\mathbb{C}/(\mathbb{Z} + \tau\mathbb{Z})$, $\tau \in \mathfrak{H}$, and reduces the classification of compact connected Riemann surfaces of genus ≥ 2 to the classification of discrete subgroups Γ of $\mathrm{PSL}(2, \mathbb{R})$ acting freely on $\mathfrak{H}$ such that $\mathfrak{H}/\Gamma$ is compact (*cf.* Theorem I.3.1). One easily checks that, for any discrete subgroup Γ of $\mathrm{PSL}(2, \mathbb{R})$, this last condition is equivalent to the existence of a compact subset $K \subset \mathfrak{H}$ such that

$$\mathfrak{H} = \bigcup_{\gamma \in \Gamma} \gamma \cdot K,$$

or to the compactness of $\mathrm{PSL}(2,\mathbb{R})/\Gamma$. A discrete subgroup of $\mathrm{PSL}(2,\mathbb{R})$ which satisfies these conditions is said to be *cocompact.*

Moreover, any discrete subgroup Γ of $\mathrm{PSL}(2,\mathbb{R})$ may be shown to possess a subgroup Γ' of finite index which acts freely on $\mathfrak{H}$ (*cf.* [Sel]), and Γ' is easily shown to be cocompact iff Γ is.

To summarize, from the point of view of uniformization theory, *constructing compact Riemann surfaces of genus* ≥ 2 *is essentially equivalent to constructing discrete cocompact subgroups of* $\mathrm{PSL}(2,\mathbb{R})$.

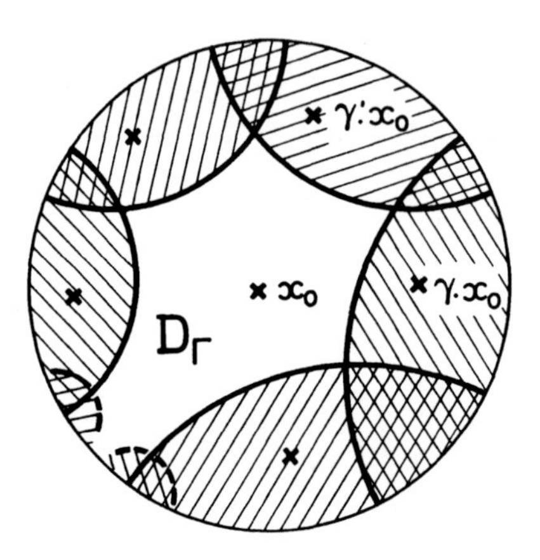

Fig. 7. The fundamental domain D_Γ.

It is possible to exhibit explicit examples of such subgroups, thanks to an arithmetic construction which goes back to Poincaré, based on the use of quaternion algebra (*cf.* [Ei], [Shi],[Vi]). Concretely, if p is any prime number and if n is any positive integer which is not a quadratic residue modulo p [11], then the set $\Gamma(n,p)$ of (classes modulo $\{\pm I_2\}$ of) matrices of the form

$$\begin{pmatrix} x_0 + x_1\sqrt{n} & x_2\sqrt{p} + x_3\sqrt{np} \\ x_2\sqrt{p} - x_3\sqrt{np} & x_0 - x_1\sqrt{n} \end{pmatrix} ,$$

where x_0, x_1, x_2, x_3 are integers such that

$$x_0^2 - nx_1^2 - px_2^2 + npx_3^2 = 1 ,$$

is a discrete cocompact subgroup of $\mathrm{PSL}(2,\mathbb{R})$. Furthermore, if $p \equiv 1 \pmod 4$, then $\Gamma(n,p)$ acts freely on $\mathfrak{H}$.

Another construction of discrete cocompact subgroups of $\mathrm{PSL}(2,\mathbb{R})$, which goes back to Poincaré and Klein, is based on hyperbolic geometry. It relies on the fact that $\mathrm{PSL}(2,\mathbb{R})$ is the group of direct isometries of $\mathfrak{H}$ equipped with

[11] *i.e.*, if there is no integer x such that $x^2 \equiv n \bmod p$.

the Poincaré metric and that, accordingly, such a subgroup may be seen as a 'discrete group of motion in the hyperbolic plane'. We now describe briefly this construction.

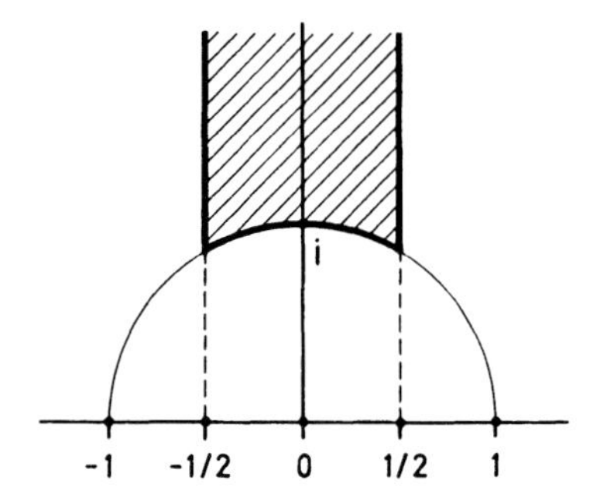

Fig. 8. A fundamental domain for PSL(2,$\mathbb{Z}$).

Remember that any lattice Λ in the Euclidean space $\mathbb{R}^N$ possesses a Dirichlet fundamental domain D_Λ, defined as

$$D_\Lambda = \bigcap_{\gamma \in \Lambda} \left\{ x \in \mathbb{R}^N \mid d(x,0) < d(x,\gamma) \right\},$$

where d denotes the Euclidean distance in $\mathbb{R}^N$ (*cf.* [Sen]). Similarly, for any discrete subgroup Γ of PSL$(2,\mathbb{R})$ acting freely on $\mathfrak{H}$ and to any base point $x_0 \in \mathfrak{H}$, we can attach the following subset of $\mathfrak{H}$

$$D_\Gamma = \bigcap_{\gamma \in \Gamma} \left\{ x \in \mathfrak{H} \mid d(x,x_0) < d(x,\gamma \cdot x_0) \right\},$$

where now d denotes the distance in the hyperbolic plane $\mathfrak{H}$ [12] (see figure 7). The subset D_Γ may be called a *fundamental domain for the action of Γ on $\mathfrak{H}$.* Indeed it enjoys the following properties:

(FD1) *D_Γ is a connected open subset of $\mathfrak{H}$;*

(FD2) *the subsets $\gamma \cdot D_\Gamma$, $\gamma \in \Gamma$, are pairwise disjoints;*

(FD3) $\bigcup_{\gamma \in \Gamma} \gamma \cdot \overline{D_\Gamma} = \mathfrak{H}$.

Moreover D_Γ is a convex polygon in the following sense:

· for any pair of points (A,B) in D_Γ, the geodesic segment between A and B lies entirely in D_Γ;

· the boundary ∂D_Γ of D_Γ is locally a piecewise geodesic curve.

Observe that a discrete subgroup Γ of PSL$(2,\mathbb{R})$ may have a fundamental domain D_Γ which satisfies all the preceding conditions but which is not

[12] *i.e.*, the geodesic distance associated to the Poincaré metric. The reader may prove that it is given by $d(z_1,z_2) = 2\,\sinh^{-1}\left(\frac{|z_1-z_2|}{2\sqrt{\operatorname{Im} z_1 \cdot \operatorname{Im} z_2}}\right)$.

obtained by the Dirichlet construction. It is the case of the well known fundamental domain of $\mathrm{PSL}(2,\mathbb{Z})$ depicted on figure 8.

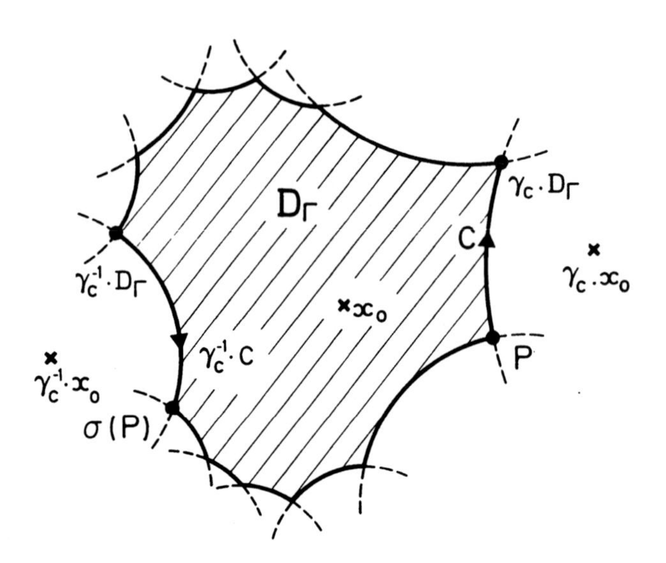

Fig. 9. The generators γ_C associated with the sides of D_Γ.

Let us return to the Dirichlet fundamental domain D_Γ. One easily checks that *Γ is cocompact in* $\mathrm{PSL}(2,\mathbb{R})$ *iff $\overline{D}_\Gamma$ is compact*, which we will assume from now on.

Then D_Γ is a convex hyperbolic polygon, with a finite number of vertices, which belong to $\mathfrak{H}$. Moreover, as explained on figure 9, to any side C of D_Γ is attached the unique element $\gamma_C \in \Gamma - \{I\}$ such that C is a common side of D_Γ and $\gamma_C \cdot D_\Gamma$. Then we clearly have:

(P1) *For any side C of D_Γ, $\gamma_C^{-1} \cdot C$ is a side of D_Γ.* Furthermore, ∂D_Γ may be oriented as the boundary of D_Γ, and *γ_C is an orientation reversing map from C to $\gamma_C \cdot C$.*

(P2) *If $C' = \gamma_C^{-1} \cdot C$, then $\gamma_{C'} = \gamma_C^{-1}$.*

Moreover, the γ_C's generate Γ. Indeed the set of polygons $\{\gamma \cdot D_\Gamma, \gamma \in \Gamma\}$ is a hyperbolic tiling of $\mathfrak{H}$ (see [Rey], figure 14, and the front cover of this book). Therefore, for any $\gamma \in \Gamma$, there exists a sequence $\gamma_0 \cdot D_\Gamma = D_\Gamma, \gamma_1 \cdot D_\Gamma, \ldots, \gamma_n \cdot D_\Gamma = \gamma \cdot D_\Gamma$ of 'tiles' such that $\gamma_{i+1} \cdot D_\Gamma$ and $\gamma_i \cdot D_\Gamma$ have a common side. Then $\gamma_{i+1}^{-1}\gamma_i \cdot D_\Gamma$ and D_Γ have a common side. Therefore $\gamma_{i+1}^{-1}\gamma_i$ is one of the γ_C's and γ is a product of γ_C's.

For any vertex P of D_Γ, we now define:

· its 'successor' $\sigma(P) = \gamma_C^{-1} \cdot P$, where C is the side of D_Γ which starts at P;

· the 'cycle' $\mathcal{C}(P)$ as the (finite) subset $\{\sigma^n(P)\}_{n \in \mathrm{N}}$ of the set of vertices

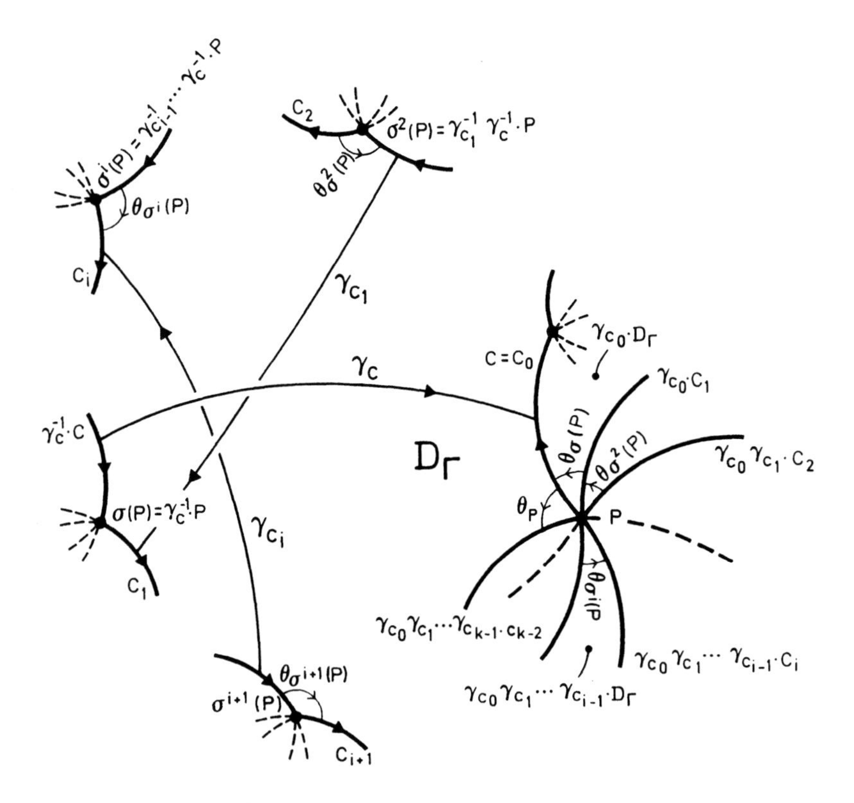

Fig. 10. The conditions (P3) and (P4).

of D_Γ[13];

· the angle θ_P of D_Γ at the vertex P.

Then, for any vertex P of D_Γ, we have, as shown on figure 10:

· *the 'cycle condition'*

(P3)
$$\sum_{Q \in \mathcal{C}(P)} \theta_Q = 2\pi$$

is satisfied;

· *if $\mathcal{C}(P) = \{P, \sigma(P), \ldots, \sigma^k(P)\}$ $(k \geq 1)$ and if C_i is the side of D_Γ starting at $\sigma^i(P)$, then*

(P4)
$$\gamma_{C_0}\gamma_{C_1} \cdots \gamma_{C_k} = I.$$

(Observe that

[13] In more concrete terms, if we glue every side C of D_Γ to $\gamma_C \cdot C$ using γ_C, then a vertex P of D_Γ is glued exactly with the elements of $\mathcal{C}(P)$.

$$\begin{aligned}
P &= \sigma^0(P) \\
&= \gamma_C \cdot \sigma(P) \\
&= \gamma_C \gamma_{C_1} \cdot \sigma^2(P) \\
&= \gamma_{C_0} \gamma_{C_1} \cdots \gamma_{C_{i-1}} \cdot \sigma^i(P) \\
&= \gamma_{C_0} \gamma_{C_1} \cdots \gamma_{C_k} \cdot \sigma^{k+1}(P) \\
&= \gamma_{C_0} \gamma_{C_1} \cdots \gamma_{C_k} \cdot (P)
\end{aligned}$$

and that Γ acts freely on $\mathfrak{H}$).

Poincaré's theorem is the following converse to this construction:

Theorem I.4.6. *Let D be any hyperbolic polygon in $\mathfrak{H}$,* i.e., *any connected open subset of $\mathfrak{H}$, whose closure $\overline{D}$ is compact in $\mathfrak{H}$ and whose boundary ∂D_Γ is a closed piecewise geodesic curve. Suppose that to any side C of D is attached an element $\gamma_C \in \mathrm{PSL}(2,\mathbb{R}) - \{I\}$ such that the conditions* (P1) *and* (P2) *are satisfied.* (This means essentially that the sides of D may be grouped in pairs of sides of the same length). *Then the definitions of $\sigma(P)$, $C(P)$ and θ_P still make sense, and if the condition* (P3) *is satisfied the γ_C's generate a cocompact discrete subgroup Γ of* $\mathrm{PSL}(2,\mathbb{R})$ *which acts freely on $\mathfrak{H}$ and possesses D as fundamental domain (i.e., D* satisfies the same conditions (FD1-3) as D_Γ).

Moreover, the relations (P4) *are satisfied and, together with* (P2), *provide a presentation of Γ.*

As a matter a fact, Theorem I.4.6 allows to construct easily many cocompact discrete subgroups of $\mathrm{PSL}(2,\mathbb{R})$.

Exercise. Prove that for any $g \geq 2$ there exists $r \in\,]0,1[$ such that the convex hyperbolic polygon D_g with vertices $r\, e^{\frac{\pi i k}{2g}}$, $1 \leq k \leq 4g$, (in the disc model) has all its angles equal to $\frac{\pi}{2g}$. Prove that one may choose elements γ_C's such that the conditions in Theorem I.4.6 are satisfied. (Hint: use §A.2.3 and figure 23.)

The construction of compact Riemann surfaces from cocompact discrete subgroups of $\mathrm{PSL}(2,\mathbb{R})$ possesses an important extension. A discrete subgroup Γ of $\mathrm{PSL}(2,\mathbb{R})$ is said to be of *finite covolume* when the area in the Poincaré metric of the fundamental domain D_Γ of Γ is finite:

$$\int_{D_\Gamma} \frac{dx\, dy}{y^2} < \infty.$$

This condition is equivalent to the existence of a Lebesgue measurable subset E of $\mathfrak{H}$ such that

$$\int_E \frac{dx\, dy}{y^2} < \infty$$

and

$$\Gamma \cdot E := \{\gamma \cdot x; \gamma \in \Gamma, x \in E\} = \mathfrak{H}.$$

For instance $\mathrm{PSL}(2,\mathbb{Z})$ satisfies this condition. Indeed if E is the fundamental domain of $\mathrm{PSL}(2,\mathbb{Z})$ depicted on figure 8 and if

$$E' = \{z \in E \mid \mathrm{Im} z \leq 1\}$$

and

$$E'' = \{z \in E \mid \mathrm{Im} z > 1\}$$

then clearly

$$\int_{E'} \frac{dx\, dy}{y^2} < \infty$$

and

$$\int_{E''} \frac{dx\, dy}{y^2} = \int_{\substack{-1/2 \leq x \leq 1/2 \\ y \geq 1}} \frac{dx\, dy}{y^2} = \int_1^\infty \frac{dy}{y^2} < \infty.$$

This easily implies that any subgroup of finite index in $\mathrm{PSL}(2,\mathbb{Z})$ also has finite covolume.

Theorem I.4.7. *For any discrete subgroup Γ of finite covolume in* $\mathrm{PSL}(2,\mathbb{R})$, *the Riemann surface $\mathfrak{H}/\Gamma$ is a compact Riemann surface with punctures.*

When $\Gamma = \mathrm{PSL}(2,\mathbb{Z})$, this is clear from the isomorphism

$$\underline{j} : \mathfrak{H}/\mathrm{PSL}(2,\mathbb{Z}) \simeq \mathbb{C},$$

since $\mathbb{C} = \mathbb{P}^1\mathbb{C} - \{\infty\}$ is a compact Riemann surfaces with punctures. When Γ is a subgroup of finite index in $\mathrm{PSL}(2,\mathbb{Z})$, this follows from the fact that the holomorphic map

$$\mathfrak{H}/\Gamma \to \mathfrak{H}/\mathrm{PSL}(2,\mathbb{Z})$$

is a ramified covering of finite degree, with a finite set of ramification points, since the map $j : \mathfrak{H} \to \mathbb{C} \simeq \mathfrak{H}/\mathrm{PSL}(2,\mathbb{Z})$ is ramified only over two points. For a general Γ, theorem I.4.7 is due to Siegel ([Siel]) and may be proved using hyperbolic geometry in $\mathfrak{H}$.

Finally, for any Γ with finite covolume, we get a compact Riemann surface $\widehat{\mathfrak{H}/\Gamma}$. The surfaces obtained by this construction, where Γ is a congruence subgroup of $\mathrm{PSL}(2,\mathbb{Z})$, *i.e.*, when there exists $N \in \mathbb{N}^*$ such that Γ contains

$$\Gamma(N) := \left\{ \begin{pmatrix} a & b \\ c & d \end{pmatrix} \in \mathrm{SL}(2,\mathbb{Z}) \mid a \equiv d \equiv 1 \bmod N \;;\; b \equiv c \equiv 0 \bmod N \right\},$$

are called *modular curves.* The points in the finite set $\widehat{\mathfrak{H}/\Gamma} - \mathfrak{H}/\Gamma$ are called the *cusps* of $\widehat{\mathfrak{H}/\Gamma}$.

I.4.4. Klein's quartic.

In the preceding Subsections, we discussed two very different descriptions of compact Riemann surfaces: any of them may be obtained either as an algebraic curve, or by a quotient construction involving a lattice in $\mathbb{C}$ or a cocompact or finite covolume subgroup in $\mathrm{PSL}(2,\mathbb{R})$. This double-faced aspect of one same class of mathematical objects is all the more remarkable that the explicit transition from one approach to the other is never trivial. For instance, for compact Riemann surfaces of genus one, this transition amounts to the theory of elliptic functions. In this Section, we briefly discuss a completely explicit example of compact Riemann surface, due to Klein ([Kl1]), which provides another illustration of this circle of ideas.

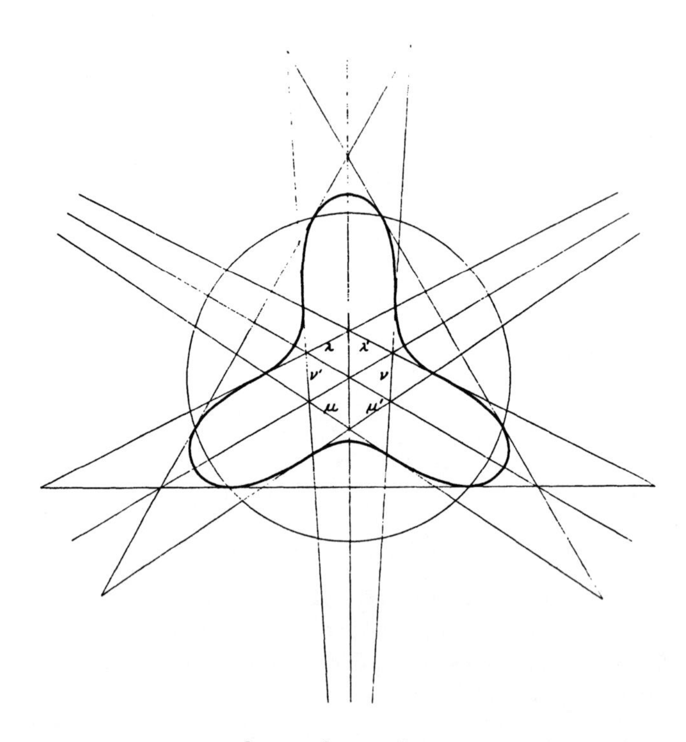

Fig. 11. The real points of $X^3Y + Y^3Z + Z^3X = 0$, according to Klein ([Kl1]).

Klein studies the modular curve $\widehat{\mathfrak{H}/\Gamma}(7)$, which nowadays is usually denoted $X(7)$, and obtains the following results:

• The congruence subgroup $\Gamma(7)$ is clearly a normal subgroup of $\mathrm{PSL}(2,\mathbb{Z})$, and the quotient group $G = \mathrm{PSL}(2,\mathbb{Z})/\Gamma(7)$ is isomorphic to $\mathrm{PSL}(2,\mathbb{Z}/7\mathbb{Z})$, a group with 168 elements. The group G acts on $\mathfrak{H}/\Gamma(7)$, hence on $X(7)$. In fact the holomorphic map

$$\tilde{j} : X(7) \to \mathbb{P}^1\mathbb{C}$$

which extends the map

$$\tilde{j} : \mathfrak{H}/\Gamma(7) \to \mathfrak{H}/\mathrm{PSL}(2;\mathbb{Z}) \overset{j}{\simeq} \mathbb{C}$$

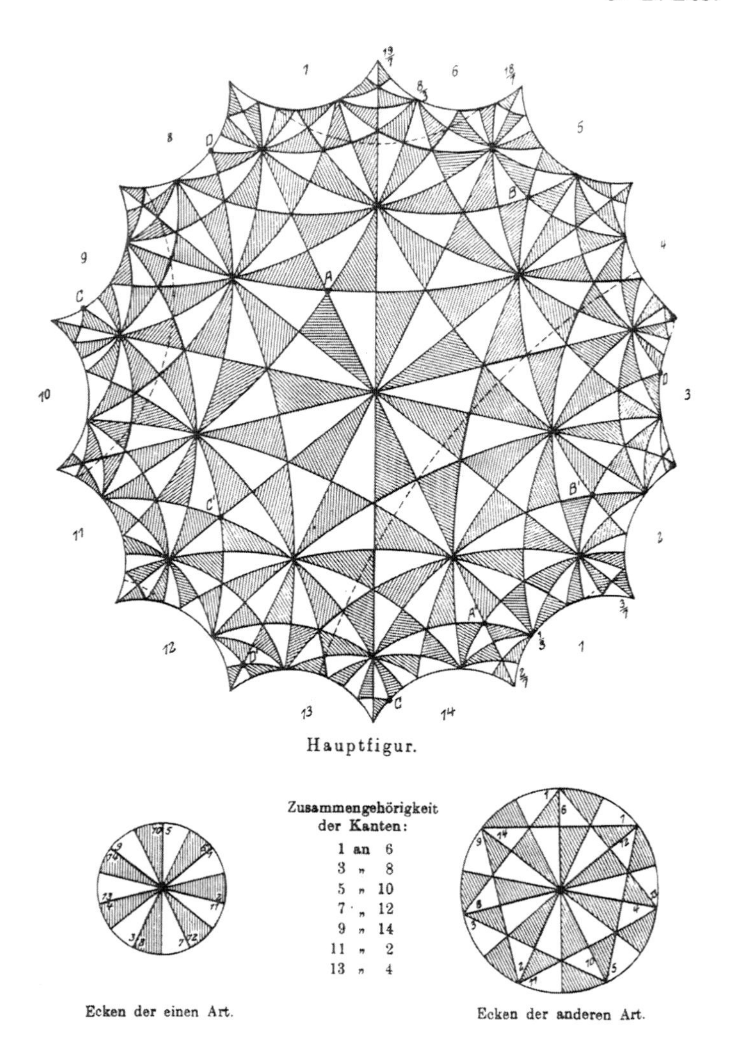

Fig. 12. Klein's 'Hauptfigur' ([Kl1]) : A fundamental domain for Γ.

by sending the cusps of $X(7)$ to ∞ is a Galois covering with Galois group G.

The map $\widetilde{j}$ is ramified only over ∞, 0 and 1728. The set $A = \widetilde{j}^{-1}(\infty)$ (resp. $B = \widetilde{j}^{-1}(0)$, resp. $C = \widetilde{j}^{-1}(1728)$) has cardinality 24 (resp. 56, resp. 84) and the ramification order of $\widetilde{j}$ at each point of this set is 7 (resp. 3, resp. 2). As $\widetilde{j}$ has degree 168, this implies by the Riemann-Hurwitz formula (*cf.* [Rey]) that the genus of $X(7)$ is 3.

- Klein proves that there exists a projective embedding

$$\varphi : X(7) \to \mathbb{P}^2\mathbb{C}$$

such that $\varphi(X(7))$ is the quartic curve of homogeneous equation

$$X^3Y + Y^3Z + Z^3X = 0 \tag{I.4.2}$$

(see figure 11 for Klein's picture of the real points of (I.4.2)). This embedding φ is defined in terms of holomorphic differential forms on $X(7)$ or, what amounts to the same, in terms of modular forms of weight 2 with respect to $\Gamma(7)$. The ramification points A (resp. B) are sent by φ on the inflection points of the quartic (I.4.2) (resp. on the contact points of its double tangents). As asserted in Theorem I.4.5., 2), the meromorphic function $\widetilde{j}$ may be written as a rational function of the homogeneous coordinates X, Y, Z of the projective embedding φ. Klein determines explicitly such a function. Namely, if

$$Q(X,Y,Z) = X^3Y + Y^3Z + Z^3X,$$

$$\nabla(X,Y,Z) = \frac{1}{54}\begin{vmatrix} \frac{\partial^2 Q}{\partial X^2} & \frac{\partial^2 Q}{\partial X \partial Y} & \frac{\partial^2 Q}{\partial X \partial Z} \\ \frac{\partial^2 Q}{\partial Y \partial X} & \frac{\partial^2 Q}{\partial Y^2} & \frac{\partial^2 Q}{\partial Y \partial Z} \\ \frac{\partial^2 Q}{\partial Z \partial X} & \frac{\partial^2 Q}{\partial Z \partial Y} & \frac{\partial^2 Q}{\partial Z^2} \end{vmatrix} = 5X^2Y^2Z^2 - (X^5Y + Y^5Z + Z^5X),$$

and

$$C(X,Y,Z) = \frac{1}{9}\begin{vmatrix} \frac{\partial^2 Q}{\partial X^2} & \frac{\partial^2 Q}{\partial X \partial Y} & \frac{\partial^2 Q}{\partial X \partial Z} & \frac{\partial \nabla}{\partial X} \\ \frac{\partial^2 Q}{\partial Y \partial X} & \frac{\partial^2 Q}{\partial Y^2} & \frac{\partial^2 Q}{\partial Y \partial Z} & \frac{\partial \nabla}{\partial Y} \\ \frac{\partial^2 Q}{\partial Z \partial X} & \frac{\partial^2 Q}{\partial Z \partial Y} & \frac{\partial^2 Q}{\partial Z^2} & \frac{\partial \nabla}{\partial Z} \\ \frac{\partial \nabla}{\partial X} & \frac{\partial \nabla}{\partial Y} & \frac{\partial \nabla}{\partial Z} & 0 \end{vmatrix} = X^{14} + Y^{14} + Z^{14} + \cdots,$$

then

$$\widetilde{j} = -\frac{C^3}{\nabla^7} \circ \varphi.$$

• Klein describes the uniformization of $X(7)$. The inverse image $\widetilde{j}^{-1}(\mathbb{P}^1\mathbb{R})$ of the real axis divides $X(7)$ in $2 \times 168 = 336$ pieces which, in the hyperbolic metric on $X(7)$, are geodesic triangles with angles $\frac{\pi}{7}$, $\frac{\pi}{3}$ and $\frac{\pi}{2}$ (these angles correspond to points in A, B and C, respectively). Therefore the cocompact Fuchsian group Γ such that $X(7) \simeq \mathfrak{H}/\Gamma$ possesses a fundamental domain built from 336 such hyperbolic triangles. Figure 12 reproduces Klein's 'Hauptfigur' which depicts such a fundamental domain in the disc model of the hyperbolic plane (the shaded triangles are the inverse images by $\widetilde{j}$ of the upper half sphere $\operatorname{Im} z \geq 0$ in $\mathbb{P}^1\mathbb{C}$).

This detailed investigation of $X(7)$ played an important role in the early history of uniformization. Klein's curve $X(7)$ was indeed the first compact Riemann surface of genus > 1 for which the uniformization theorem was known (see [Kl2], p.136).

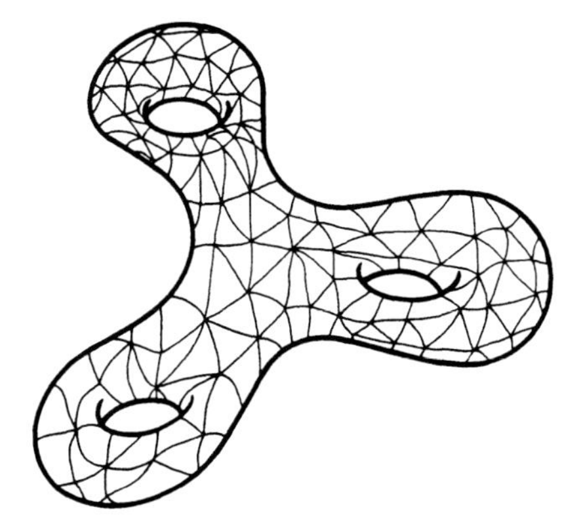

Fig. 13. A C^∞ triangulation of a Riemann surface.

A remarkable feature of Klein's curve is that it is isomorphic to an algebraic curve defined by a polynomial equation with *rational* coefficients. The link between this 'arithmetic' property and the 'geometric' construction of $X(7)$ is clarified by some recents work, initiated by a theorem of Belyi ([By]) which we discuss in the next Section.

I.5. Equilateral triangulations and algebraic curves defined over number fields

To conclude the first part of these notes, we describe a recent result which gives an 'arithmetic' counterpart to the various constructions of compact Riemann surfaces described in the preceding Section.

I.5.1. Euclidean triangulations of surfaces.

The construction of Riemann surfaces using conformal structures admits the following variant. Let X be an oriented compact connected topological surface and let $\mathcal{T}$ be a triangulation of X (see figure 13). A *Euclidean structure on* $\mathcal{T}$ is the data of a flat Riemannian metric on every triangle Δ of $\mathcal{T}$ such that:

i) equipped with this metric, Δ is isometric with a triangle (in the usual sense) in the Euclidean plane $\mathbb{R}^2$;

ii) on any edge E of $\mathcal{T}$, the metrics induced on E by the two triangles which contain E coincide.

A triangulation equipped with a Euclidean structure is called a *Euclidean triangulation.* An *equilateral triangulation* is a Euclidean triangulation whose all triangles are equilateral.

Let $V_\mathcal{T}$ be the set of vertices of a Euclidean triangulation $\mathcal{T}$ of X and let $E_\mathcal{T}$ be the union of the edges of $\mathcal{T}$. It is easily checked that their exists a unique structure of flat C^∞ Riemannian manifold on $X - E_{\mathcal{T}}$ compatible with the given metric on every triangle Δ of $\mathcal{T}$, and that near a vertex of $\mathcal{T}$, this Riemannian structure is 'conical' (see figure 14). This implies that $X - V_{\mathcal{T}}$ equipped with the holomorphic structure defined by the conformal class of this metric is a compact Riemann surface with punctures (the key point of the proof is pictorially explained on figure 15). In other words, the holomorphic structure on $X - V_{\mathcal{T}}$ extends to a holomorphic structure on X. To summarize, *a Euclidean triangulation of X defines a holomorphic structure on X*, hence a compact connected Riemann surface.

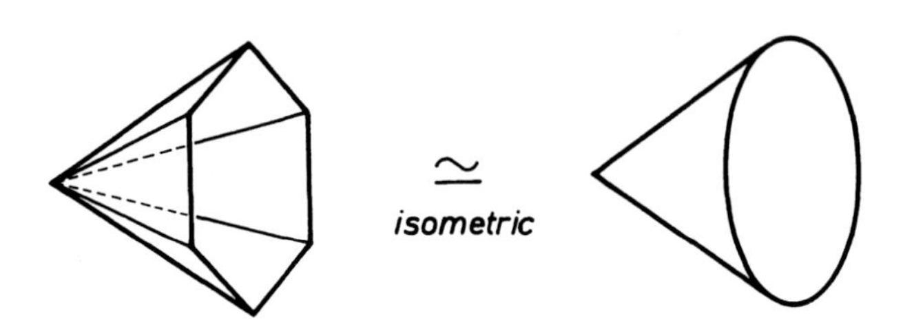

Fig. 14. The singularities of the metric defined through a Euclidean triangulation are conical.

Conversely, *the holomorphic structure of any compact connected Riemann surface M may be recovered* (up to isomorphism) *from some Euclidean triangulation $\mathcal{T}$ of M:*

• This is true if $M = \mathbb{P}^1\mathbb{C}$, using for instance the tetrahedral subdivision of $\mathbb{P}^1\mathbb{C}$ (see figure 16; the Riemann surface obtained from this Euclidean triangulation is isomorphic to $\mathbb{P}^1\mathbb{C}$ because it is compact and simply connected; *cf.* §I.3). Moreover, if $a_1, \ldots, a_r$ are given points in $\mathbb{P}^1\mathbb{C}$, we can assume, by refining the triangulation $\mathcal{T}$, that $a_1, \ldots, a_r$ are vertices of $\mathcal{T}$ (see figure 17).

• If M is any compact connected Riemann surface, there exists a ramified covering $\pi : M \to \mathbb{P}^1\mathbb{C}$ (*cf.* Theorem I.4.2). If $\mathcal{T}$ is a C^∞ triangulation of $\mathbb{P}^1\mathbb{C}$ such that $V_{\mathcal{T}}$ contains the ramification points of π, then there exists a unique C^∞ triangulation $\mathcal{T}'$ of M such that $V_{\mathcal{T}'} = \pi^{-1}(V_{\mathcal{T}})$ and $E_{\mathcal{T}'} = \pi^{-1}(E_{\mathcal{T}})$. If $\mathring{\Delta}$ is any open triangle of $\mathcal{T}$, $\pi^{-1}(\mathring{\Delta})$ is a disjoint union $\mathring{\Delta}_1 \cup \cdots \cup \mathring{\Delta}_r$ of open triangles of $\mathcal{T}'$. Moreover, there is a unique Euclidean structure on $\mathcal{T}'$ such that the maps $\pi : \mathring{\Delta}_i \xrightarrow{\sim} \mathring{\Delta}$ are isometries. Finally, the Euclidean triangulation $\mathcal{T}'$ defines the original holomorphic structure on M. Indeed, it defines this holomorphic structure on $M - V_{\mathcal{T}'}$, since $\pi : M - V_{\mathcal{T}'} \to \mathbb{P}^1\mathbb{C} - V_{\mathcal{T}}$ is locally isometric, hence conformal.

I.5.2. Characterization of Riemann surfaces defined over $\overline{\mathbb{Q}}$.

Let $\overline{\mathbb{Q}}$ be the field of algebraic numbers, *i.e.*, of complex numbers which

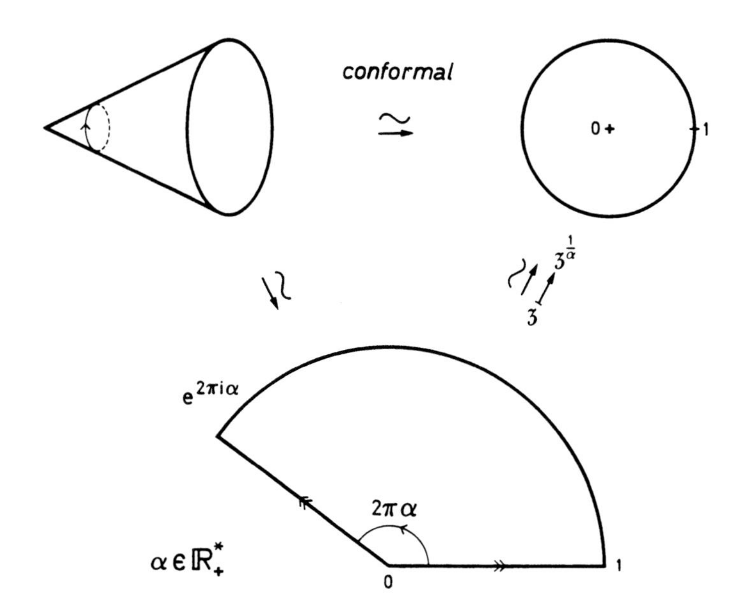

Fig. 15. A punctured cone is conformally equivalent to a punctured disc.

are roots of non-zero polynomials with rational coefficients. We will say that a compact connected Riemann surface M may be *defined over* $\overline{\mathbb{Q}}$ or equivalently, that M may be *defined over a number field*, if M is isomorphic to the Riemann surface $\widehat{C}_P$ associated with an irreducible polynomial P in $\overline{\mathbb{Q}}[X, Y]$. Moreover, for any compact connected Riemann surface M, a holomorphic map $\pi : M \to \mathbb{P}^1\mathbb{C}$ will be called a *Belyi map* if it is non-constant and unramified outside $\pi^{-1}(\{0, 1, \infty\})$.

The next theorem shows that the Riemann surfaces defined over $\overline{\mathbb{Q}}$ have remarkable characterizations from the various points of view on compact Riemann surfaces presented in preceding Sections.

Theorem I.5.2. *For any compact connected Riemann surface M, the following conditions are equivalent:*

i) M may be defined over $\overline{\mathbb{Q}}$;

ii) there exists a Belyi map $\pi : M \to \mathbb{P}^1\mathbb{C}$;

iii) M is isomorphic to $\widehat{X}$ where X is a finite unramified covering of $\mathbb{P}^1\mathbb{C} - \{0, 1, \infty\}$;

iv) M is isomorphic to $\widehat{\mathfrak{H}/\Gamma}$ where Γ is a subgroup of finite index in $\mathrm{PSL}(2, \mathbb{Z})$*;*

v) the holomorphic structure of M may be defined by an equilateral triangulation.

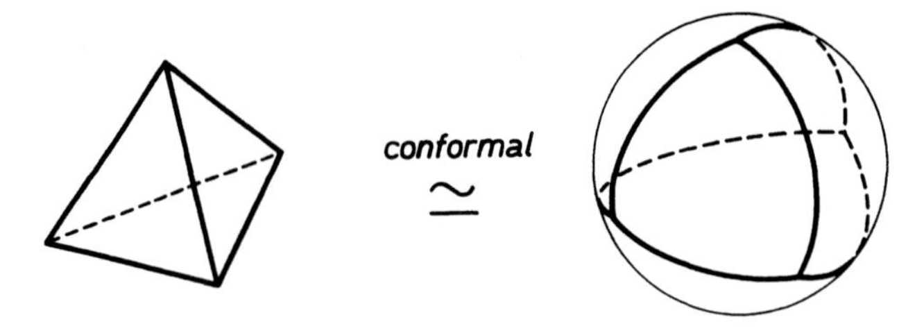

Fig. 16. The Riemann surface $\mathbb{P}^1\mathbb{C}$ can be obtained through a Euclidean triangulation.

The equivalence ii) ⇔ iii) is clear.

The implication iii) ⇒ iv) follows from the isomorphism (*cf.* §I.2.3)

$$\mathfrak{H}/\Gamma(2) \simeq \mathbb{P}^1\mathbb{C} - \{0, 1, \infty\}$$

and from the fact that, since $\Gamma(2)$ acts properly and freely on $\mathfrak{H}$ (*cf.* §I.2.3), which is simply connected, any finite unramified covering of $\mathfrak{H}/\Gamma(2)$ is isomorphic to a covering

$$\mathfrak{H}/\Gamma \to \mathfrak{H}/\Gamma(2)$$

where Γ is a subgroup of finite index in $\Gamma(2)$. The implication iv) ⇒ ii) follows from the fact that the map ρ defined by the following diagram, where the 'horizontal' map associates $\mathrm{PSL}(2,\mathbb{Z})\cdot z$ to $\Gamma \cdot z$ (*cf.* §I.2.3),

$$\begin{array}{ccc} \mathfrak{H}/\Gamma & \rightarrow & \mathfrak{H}/\mathrm{PSL}(2,\mathbb{Z}) \\ \rho \searrow & & \swarrow 1728^{-1}\underline{j} \\ & \mathbb{C} & \end{array}$$

is a proper map, which is unramified on $\rho^{-1}(\mathbb{C}\backslash\{0,1\})$ since $1728^{-1}j : \mathfrak{H} \to \mathbb{C}$ is ramified only over 0 and 1. Thus ρ extends to a Belyi map from $\widehat{\mathfrak{H}/\Gamma}$ to $\mathbb{P}^1\mathbb{C}$. (See [Ser], p.71).

The implication iii) ⇒ i) is a consequence of general results concerning the fields of definition of algebraic varieties and of their coverings (for instance, it is a consequence of the rationality criterion in [Wei 1]).

The most surprising fact in the theorem is the implication i) ⇒ ii). It is due to Belyi (*cf.* [Be]) and is proved as follows: by the very definition of a Riemann surface defined over $\overline{\mathbb{Q}}$, there exists a holomorphic map $\pi : X \to \mathbb{P}^1\mathbb{C}$ ramified only over points in $\overline{\mathbb{Q}} \cup \{\infty\}$. By composing π with polynomials with coefficients in $\mathbb{Q}$, it is possible to decrease the maximal degree[14] of these ramification values and ultimately to get a holomorphic map $\pi : X \to \mathbb{P}^1\mathbb{C}$ ramified only over $S \subset \mathbb{Q} \cup \{\infty\}$. By composing φ with some homographic transformation of

[14] By degree, we mean the degree over $\mathbb{Q}$ of algebraic numbers.

$\mathbb{P}^1\mathbb{C}$, we may assume that S contains $\{0,1,\infty\}$. At this point, Belyi observes that, for any $A, B \in \mathbb{Z}^*$ such that $A+B \neq 0$, the map

$$\varphi_{AB} : \mathbb{P}^1\mathbb{C} \to \mathbb{P}^1\mathbb{C}$$
$$z \mapsto \frac{(A+B)^{A+B}}{A^A B^B} z^A (1-z)^B,$$

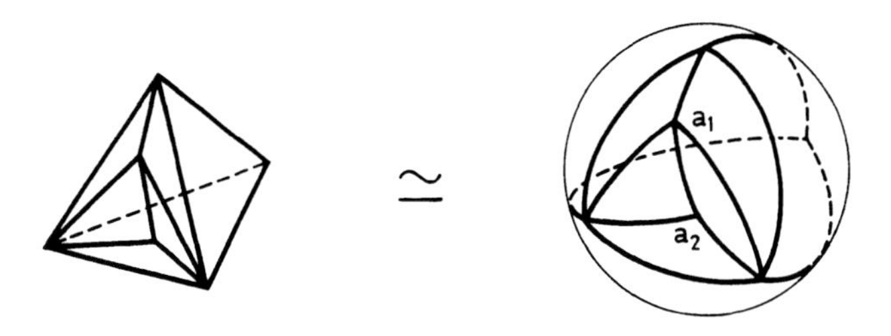

Fig. 17. A refinement of the subdivision shown in Figure 16.

is ramified at *four* points ($0, 1, \infty$ and $\frac{A}{A+B}$) but has only *three* ramification values (0,1 and ∞). If A, B are chosen such that $\frac{A}{A+B} \in S - \{0,1,\infty\}$, then $\varphi_{AB} \circ \pi$ is ramified only over $\{0,1,\infty\} \cup \pi(S - \{0,1,\infty, \frac{A}{A+B}\})$. As this set has a smaller cardinality than S, a simple decreasing induction on the cardinality of S proves the existence of a Belyi map.

The equivalence ii) $\Leftrightarrow$ v) is due to Shabat and Voevodsky ([SV1]) and is related to earlier work of Grothendieck (see [Gro] and [SV2]). The implication ii) $\Rightarrow$ v) follows from the construction of Euclidean triangulations from maps to $\mathbb{P}^1\mathbb{C}$ given in the preceding paragraph: if $\pi : M \to \mathbb{P}^1\mathbb{C}$ is a Belyi map and if $\mathcal{T}$ is an equilateral triangulation of $\mathbb{P}^1\mathbb{C}$ such that $V_\mathcal{T} \supset \{0,1,\infty\}$ [15], then the Euclidean triangulation $\mathcal{T}'$ obtained by 'pulling-back' $\mathcal{T}$ on M is an equilateral triangulation of M. To prove v) $\Rightarrow$ ii), one uses the fact that for any equilateral triangle Δ in $\mathbb{C}$, there exists a unique map $\beta_\Delta : \Delta \to \mathbb{P}^1\mathbb{C}$, holomorphic on a neighbourhood of Δ in $\mathbb{C}$ such that, restricted to any of the six triangles of the barycentric subdivision of Δ, β_Δ is a conformal diffeomorphism onto one of the two hemispheres of $\mathbb{P}^1\mathbb{C}$ bounded by $\mathbb{R} \cup \{\infty\}$ (see figure 18) and such that β_Δ send the vertices (resp. the midpoints of the edges, resp. the center) of Δ to 0 (resp. 1, resp. ∞). One easily checks that, if $\mathcal{T}$ is any equilateral triangulation of M, the map $\pi : M \to \mathbb{P}^1\mathbb{C}$ which coincide with β_Δ over any triangle Δ of $\mathcal{T}$ (*via* the identification of the oriented equilateral triangle Δ with an equilateral triangle in $\mathbb{C}$) is well defined and holomorphic and is a Belyi map.

[15] Such a $\mathcal{T}$ exists since by a suitable homographic transformation, any triple of distinct points of $\mathbb{P}^1\mathbb{C}$ - in particular three vertices of a given equilateral triangulation - may be sent to $(0,1,\infty)$.

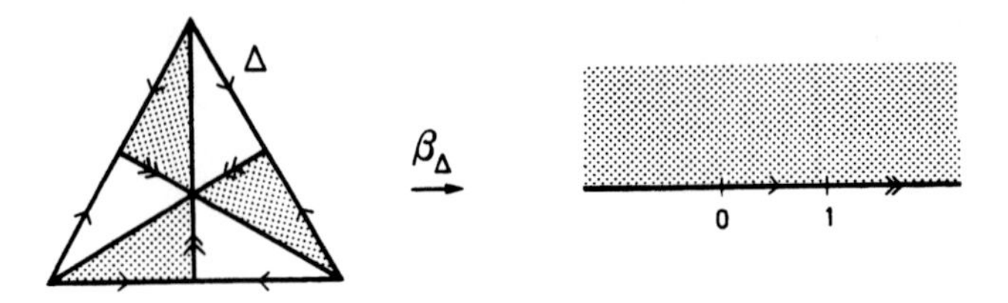

Fig. 18. The map β_Δ.

Exercise: Prove the existence of β_Δ:

i) by using the $\wp$ function associated to $\Gamma = \mathbb{Z} + e^{\pi i/3}\mathbb{Z}$;

ii) by considering the elliptic curve $E = \mathbb{C}/\Gamma$ and its quotient by $[z] \mapsto [-z]$;

iii) by considering the Euclidean 'triangulation' of the 2-sphere built from two congruent 30°-60°-90°-triangles glued along corresponding sides;

iv) by using the Schwarz-Christoffel formula for conformal mappings.

Exercise: Prove that an equilateral triangulation of Klein's quartic (*cf.* §I.4.3) may be defined by 'declaring equilateral' the 336 triangles of Klein's 'Hauptfigur' (figure 12).

Appendix A. The homology of oriented closed surfaces

A.1. The homology group $H_1(X;\mathbb{Z})$

The first homology group $H_1(X;\mathbb{Z})$ of a topological space X is the Abelian group defined by the following generators and relations:

- any loop c on X, *i.e.*, any continuous map $c : \mathbb{R}/\mathbb{Z} \to X$, defines a generator $[c]$ of $H_1(X;\mathbb{Z})$;
- if c and c' are two homotopic loops, *i.e.*, if there exists a continuous map $h : [0,1] \times \mathbb{R}/\mathbb{Z} \to X$ such that $h(0,t) = c(t)$ and $h(1,t) = c'(t)$ for any $t \in \mathbb{R}/\mathbb{Z}$, then we impose the relation $[c] = [c']$ (see figure 19);
- if c_1 and c_2 are two loops on X such that $c_1(0) = c_2(0)$ and if c is the loop defined by

$$c(t) = c_1(2t) \quad \text{if} \quad t \in [0,1/2]$$
$$c(t) = c_2(2t-1) \quad \text{if} \quad t \in [1/2,1],$$

then we impose the relation $[c] = [c_1] + [c_2]$ (see figure 20).

When X is arcwise connected, then for any $x_0 \in X$, every element of $H_1(X;\mathbb{Z})$ is the class $[c]$ of a loop $c : \mathbb{R}/\mathbb{Z} \to X$, such that $c(0) = x_0$, and the map which associates its class $[c]$ in $H_1(X,\mathbb{Z})$ to the homotopy class of such a

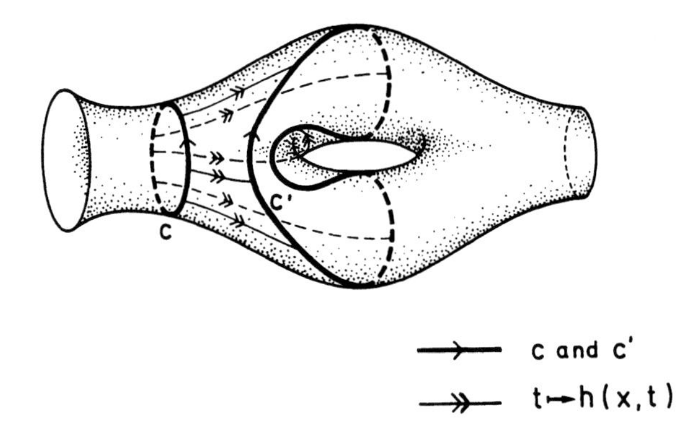

Fig. 19. Homotopy of loops.

loop c defines an isomorphism between the Abelianization[16] of the fundamental group $\pi_1(X; x_0)$ and $H_1(X; \mathbb{Z})$.

If X is a connected differentiable manifold, any element in $H_1(X; \mathbb{Z})$ may be represented by a C^∞ loop, *i.e.*, by a C^∞ map $c : \mathbb{R}/\mathbb{Z} \to X$. Furthermore, if η is any closed C^∞ 1-form over X, the integral $\int_0^1 c^*\eta$ of η along c depends only on the class of c in $H_1(X; \mathbb{Z})$. Moreover, it vanishes when η is an exact form and, consequently, depends only on the class of η in the first de Rham cohomology group of X with real coefficients, namely

$$H^1_{DR}(X; \mathbb{R}) \\ := \{\text{closed real } C^\infty \text{ 1-forms on } X\}/\{\text{exact real } C^\infty \text{ 1-forms on } X\}.$$

Hence, for any $(\gamma, \alpha) \in H_1(X; \mathbb{Z}) \times H^1_{DR}(X; \mathbb{R})$, we can define $\int_\gamma \alpha$ as the value of $\int_0^1 c^*\eta$ for any smooth loop c representing γ and for any closed C^∞ 1-form η representing α. In that way, one defines a bilinear map

$$\int : H_1(X; \mathbb{Z}) \times H^1_{DR}(X; \mathbb{R}) \to \mathbb{R}.$$

Similarly, using complex 1-forms instead of real 1-forms, one defines the first de Rham cohomology group of X with complex coefficients $H^1_{DR}(X; \mathbb{C})$, which may be identified with the complexification of $H^1_{DR}(X; \mathbb{R})$, and one may extend this pairing to a pairing

$$\int : H_1(X; \mathbb{Z}) \times H^1_{DR}(X; \mathbb{C}) \to \mathbb{C}.$$

[16] *i.e.*, the quotient by its commutator subgroup.

A.2. The homology of oriented closed surfaces

Suppose now that X is an *oriented closed surface*, *i.e.*, an oriented compact connected C^∞ manifold of dimension 2 (*e.g.* a compact connected Riemann surface!). In the next Sections we will use the following facts concerning the homology and the cohomology of X.

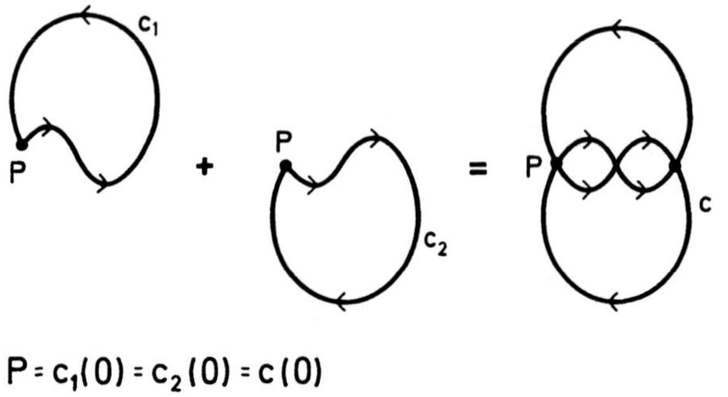

$P = c_1(0) = c_2(0) = c(0)$

Fig. 20. Addition of homology classes.

A.2.1. There exists an integer $g \geq 0$ such that

$$H_1(X;\mathbb{Z}) \simeq \mathbb{Z}^{2g}$$

and

$$H^1_{DR}(X;\mathbb{R}) \simeq \mathbb{R}^{2g}.$$

This integer is called the *genus* of X and completely classifies oriented closed surfaces, in the following sense: two such surfaces are diffeomorphic iff they have the same genus. Moreover, for any integer $g \geq 0$, there exists an oriented closed surface g of genus X: when $g = 0$, we can take $X = \mathbb{S}^2$ or $X = \mathbb{P}^1\mathbb{C}$; when $g = 1$, we can take as X the torus $\mathbb{R}^2/\mathbb{Z}^2$ or any complex elliptic curve; when $g \geq 2$, we can take as X a 'sphere with g handles' or the hyperelliptic Riemann surface defined as the compactification of the algebraic curve

$$y^2 = P(x)$$

where P is a polynomial of degree $2g+1$ or $2g+2$ with distinct roots (*cf.* §I.4.2).

A.2.2. The pairing

$$\int : H_1(X;\mathbb{Z}) \times H^1_{DR}(X;\mathbb{R}) \to \mathbb{R}$$

is a 'perfect duality' in the following sense: for any basis $(\gamma_1, \ldots, \gamma_{2g})$ of $H_1(X, \mathbb{Z})$, there exists a dual basis $(\omega_1, \ldots, \omega_{2g})$ of $H^1_{DR}(X)$ such that, for any $i, j = 1, \ldots, 2g$

$$\int_{\gamma_i} \omega_j = \delta_{ij}.$$

A.2.3. For any two classes γ_1, γ_2 in $H_1(X; \mathbb{Z})$, one can find representatives c_1, c_2 which are C^∞ loops on X which satisfy the following conditions:

- $c_1(\mathbb{R}/\mathbb{Z}) \cap c_2(\mathbb{R}/\mathbb{Z})$ is finite;
- for any $P \in c_1(\mathbb{R}/\mathbb{Z}) \cap c_2(\mathbb{R}/\mathbb{Z})$, there exist a unique $t_1 \in \mathbb{R}/\mathbb{Z}$ and a unique $t_2 \in \mathbb{R}/\mathbb{Z}$ such that

$$c_1(t_1) = c_2(t_2) = P,$$

and the couple of vectors $(\frac{\partial c_1}{\partial t}(t_1), \frac{\partial c_2}{\partial t}(t_2))$ is a basis of the tangent space of X at P.

If this basis is (resp. is not) compatible with the given orientation on X, we set

$$\varepsilon(P) = +1 \quad (\text{resp. } \varepsilon(P) = -1)$$

(see figure 21).

One may prove that the integer

$$\sum_{P \in c_1(\mathbb{R}/\mathbb{Z}) \cap c_2(\mathbb{R}/\mathbb{Z})} \varepsilon(P)$$

depends only on the classes γ_1 and γ_2. It is called the *intersection number* of γ_1, and γ_2, and is denoted $\gamma_1 \# \gamma_2$.

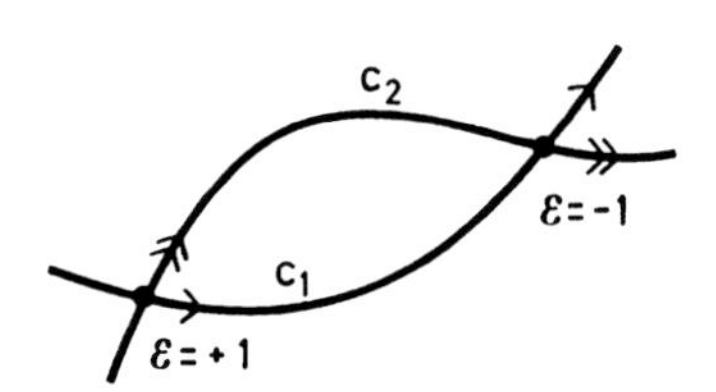

Fig. 21. Sign conventions for the intersections of loops.

The intersection product

$$\begin{aligned} H_1(X; \mathbb{Z}) \times H_1(X; \mathbb{Z}) &\to \mathbb{Z} \\ (\gamma_1, \gamma_2) &\mapsto \gamma_1 \# \gamma_2 \end{aligned}$$

so defined is bilinear and antisymmetric. Moreover, $H_1(X; \mathbb{Z})$ possesses bases which are *symplectic* with respect to this intersection product, *i.e.*, bases $(a_1, \ldots, a_g, b_1, \ldots, b_g)$ such that for any $i, j = 1, \ldots, g$:

$$a_i \# a_j = 0 \ ; \ b_i \# b_j = 0 \ ; \ a_i \# b_j = \delta_{ij}.$$

(see figure 22).

A.2.4. For any symplectic basis $(a_1, \ldots, a_g, b_1, \ldots, b_g)$ of $H_1(X;\mathbb{Z})$ and any $P \in X$, one can find loops $A_1, \ldots, A_g, B_1, \ldots, B_g$ over X which represent the homology classes $a_1, \ldots, a_g$, $b_1, \ldots, b_g$ and such that:

- the maps $A_i, B_i : [0,1] \to X$ are C^∞ and the vectors $\frac{\partial A_i}{\partial t}(\tau)$ and $\frac{\partial B_i}{\partial t}(\tau)$ do not vanish for $\tau \in [0,1]$.
- $A_i(0) = B_i(0) = P$ for any $i = 1, \ldots, g$; the maps $A_i, B_i : [0,1[\to X$ are injective and the images $A_1(]0,1[), \ldots, A_g(]0,1[), B_1(]0,1[), \ldots, B_g(]0,1[)$ are pairwise disjoint; the $4g$ vectors $\frac{\partial A_1}{\partial t}(0), \ldots, \frac{\partial A_g}{\partial t}(0)$, $\frac{\partial B_1}{\partial t}(0), \ldots, \frac{\partial B_g}{\partial t}(0)$, $\frac{\partial A_1}{\partial t}(1), \ldots, \frac{\partial A_g}{\partial t}(1)$, $\frac{\partial B_1}{\partial t}(1), \ldots, \frac{\partial B_g}{\partial t}(1)$ are pairwise non-collinear.

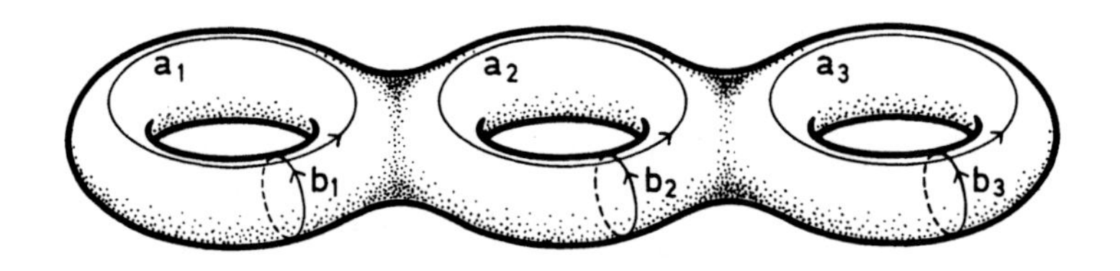

Fig. 22. A symplectic basis of $H_1(X;\mathbb{Z})$.

(See figure 23, a)-b)). Moreover, if the loops $A_1, \ldots, A_g, B_1, \ldots, B_g$ satisfy all these conditions, then the complement D in X of their images is diffeomorphic with an open disc. More precisely, if Δ is a $4g$-gon in the plane, there exists a regular C^∞ map[17] $\varphi : \Delta \to X$ such that:

- φ is surjective, $\varphi(\mathring{\Delta}) \subset D$ and $\varphi : \mathring{\Delta} \to D$ is a diffeomorphism (here $\mathring{\Delta}$ denotes the interior of Δ, which is diffeomorphic to an open disc);
- the restriction of φ to any edge of Δ coincides, modulo an affine identification between $[0,1]$ and this edge, with one of the maps A_i, B_i. (This implies that $\varphi^{-1}(P)$ is the set of vertices of Δ.) In that way, each of the maps A_i, B_i is recovered twice, and these maps occur in the way depicted on figure 23 c).

Roughly speaking, this means that one recovers X by gluing the sides of a $4g$-gon according to the prescription of figure 23 c). Figure 24 tries to show why the gluing of four consecutives sides of the $4g$-gon provides a handle.

Such a data (Δ, φ) is called a *canonical dissection of* X, to which the symplectic basis $(a_1, \ldots, a_g, b_1, \ldots, b_g)$ is said to be associated.

---

[17] This means that there exists an open neighbourhood U of Δ in the plane and an extension $\widetilde{\varphi} : U \to X$ of φ which is a local C^∞ diffeomorphism.

a)

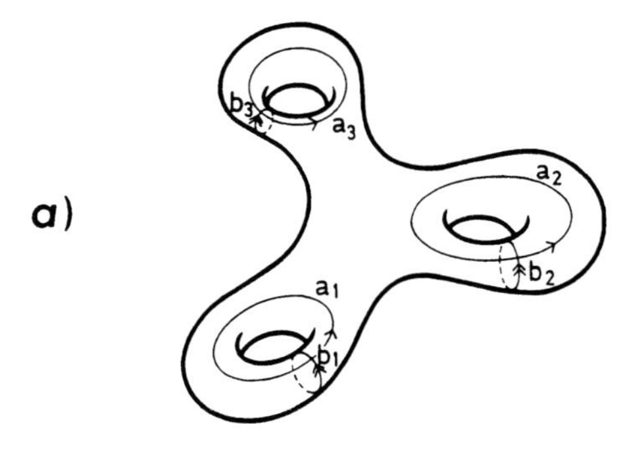

b)

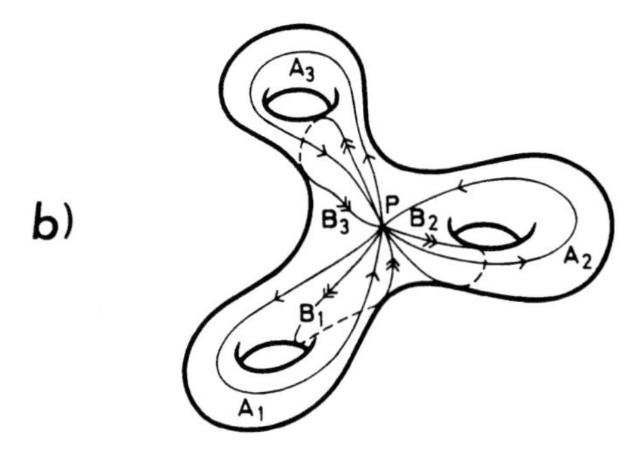

c)

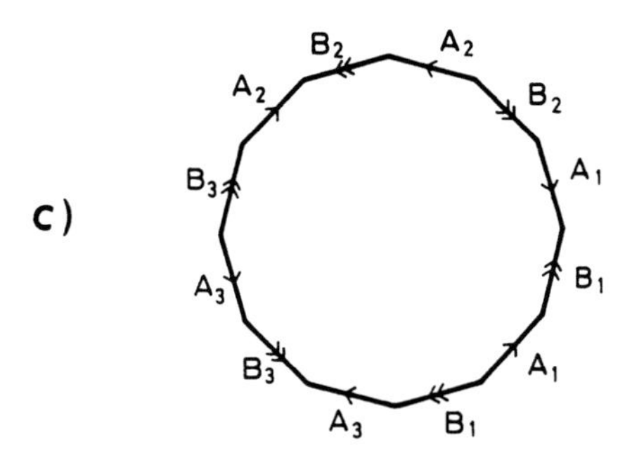

Fig. 23. A canonical dissection of X.

A.2.5. For any symplectic basis $(a_1, \ldots, a_g, b_1, \ldots, b_g)$ of $H_1(X; \mathbb{Z})$ and for any two closed 1-forms η and η' on X, we have:

$$\int_X \eta \wedge \eta' = \sum_{i=1}^{g} \left(\int_{a_i} \eta \cdot \int_{b_i} \eta' - \int_{a_i} \eta' \cdot \int_{b_i} \eta \right).$$

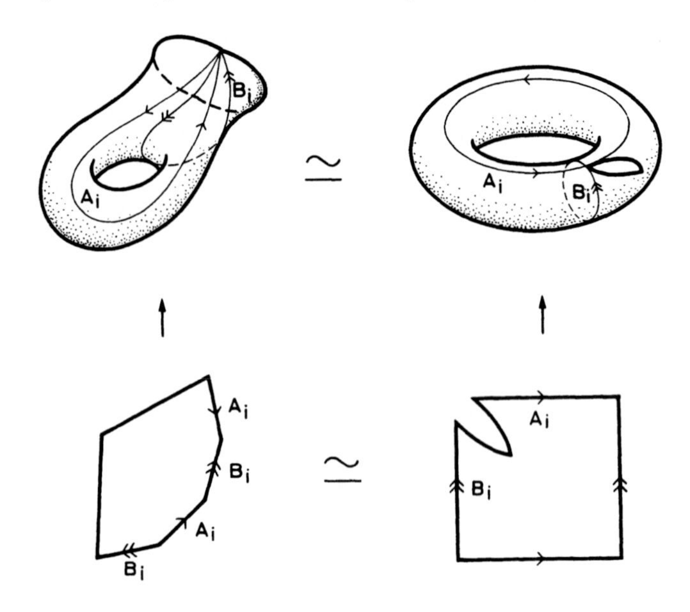

Fig. 24. Gluing four consecutives sides of the $4g$-gon provides a handle.

This lemma is a simple consequence of the compatibility between the de Rham isomorphism and the Poincaré duality for the (co-)homology of X. Let us sketch an elementary proof. The symplectic basis $(a_1, \ldots, a_g, b_1, \ldots, b_g)$ of $H_1(X; \mathbb{Z})$ is associated to some canonical dissection of X (*cf.* figure 23 and figure 25): X is obtained from a $4g$-gon Δ with sides $A_1, B_1, A_1', B_1', \ldots, A_g, B_g, A_g', B_g'$ by identifying A_i with A_i' *via* the map φ_i, and B_i with B_i' *via* the map ψ_i $(i = 1, \ldots, g)$; after this identification A_i and A_i' (resp. B_i and B_i') give a_i (resp. b_i). The forms η and η' may be identified with forms on Δ and the integrals over X (resp. a_i, b_i) may be computed over D (resp. A_i, B_i).

Over Δ, which is simply connected, there exists a function f such that $df = \eta$. Then we have for any $x \in A_i$ and for any $y \in B_i$ (see figure 25)

$$f \circ \varphi_i(x) - f(x) = \int_{b_i(x)} df = \int_{b_i} \eta \tag{A.2.1}$$

and

$$f(y) - f \circ \psi_i(y) = \int_{a_i(y)} df = \int_{a_i} \eta. \tag{A.2.2}$$

Now we get

$$\int_X \eta \wedge \eta' = \int_D \eta \wedge \eta' = \int_D df \cdot \eta' = \int_D d(f \cdot \eta')$$
$$= \int_{\partial_D} f \cdot \eta' \qquad \text{(Stokes formula)}$$
$$= \sum_{i=1}^{g} \int_{A_i + B_i - A'_i - B'_i} f\eta',$$

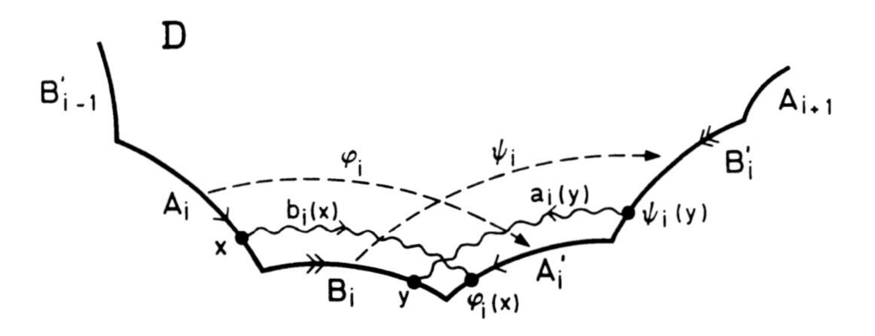

Fig. 25. Proof of equations (A.2.1) and (A.2.2).

and according to (A.2.1) and (A.2.2), we have

$$\int_{A_i - A'_i} f\eta' = \int_{A_i} (f - f \circ \varphi_i)\eta' = -\int_{b_i} \eta \cdot \int_{a_i} \eta'$$

and

$$\int_{B_i - B'_i} f\eta' = \int_{B_i} (f - f \circ \psi_i)\eta' = \int_{a_i} \eta \cdot \int_{b_i} \eta',$$

which proves the required identity.

Appendix B. Holomorphic line bundles on compact Riemann surfaces

B.1. C^∞ and holomorphic line bundles

This first Section presents a series of definitions, without motivations. These should be provided by the next two Sections.

B.1.1. Let X be a C^∞ manifold. Recall that a *C^∞ (complex) line bundle $\mathcal{L}$ on X* consists of a family $\{\mathcal{L}_x\}_{x \in X}$ of one-dimensional complex vector spaces parametrized by X, 'which depend on x in a C^∞ way'. A *section s* of $\mathcal{L}$ over a subset $E \subset X$ is a map which attaches a vector $s(x) \in \mathcal{L}_x$ to any point $x \in E$. When $s(x) \neq 0$ for any $x \in E$, the section s is said to be a *frame* over E.

A rigorous way to define the C^∞-structure of the line bundle $\mathcal{L}$ is to give a covering $\mathcal{U}$ of X by open sets and to give, for any $U \in \mathcal{U}$, a frame s_U over U in such a way that the *transition functions*

$$\varphi_{VU} : U \cap V \to \mathbb{C}^*,$$

defined on the non-empty pairwise intersections of open sets in $\mathcal{U}$ by the relation

$$s_V(x) = \varphi_{VU}(x)\ s_U(x), \tag{B.1.1}$$

are C^∞ functions. Then an arbitrary section s of $\mathcal{L}$ over an open subset Ω of X is C^∞ (resp. continuous) if, for any $U \in \mathcal{U}$, the function

$$f_U : U \cap \Omega \to \mathbb{C}^*$$

defined by

$$s(x) = f_U(x)\ s_U(x)$$

is C^∞ (resp. continuous).
The set of C^∞ (resp. continuous) sections of $\mathcal{L}$ over Ω will be denoted $C^\infty(\Omega; \mathcal{L})$ (resp. $C^0(\Omega; \mathcal{L})$). It is clearly a vector space. Moreover the pointwise product $f \cdot s$ of a function $f \in C^\infty(\Omega; \mathbb{C})$ and of a section $s \in C^\infty(U; \mathcal{L})$ clearly belongs to $C^\infty(U; \mathcal{L})$.

Two C^∞ structures on $\mathcal{L}$ coincide, by definition, if they have the same C^∞ sections. If $\mathcal{V}$ is another open covering of X, a family $\{t_V\}_{V \in \mathcal{V}}$ of frames of $\mathcal{L}$ on the open sets in $\mathcal{V}$ defines the same C^∞ structure on $\mathcal{L}$ as $(\mathcal{U}, \{s_U\}_{U \in \mathcal{U}})$ iff the t_V's are C^∞ sections of $\mathcal{L}$. A C^∞ frame of $\mathcal{L}$ on an open subset Ω is also called a *trivialization of* $\mathcal{L}$ *on* Ω. For instance, the s_U's are trivializations of $\mathcal{L}$.

B.1.2. Observe that the transition functions satisfy the following *cocycle condition*: if U, V and W are open sets in $\mathcal{V}$ with a non-empty intersection, then

$$\varphi_{WU} = \varphi_{WV} \cdot \varphi_{VU} \quad \text{on} \quad U \cap V \cap W. \tag{B.1.2}$$

Conversely, any family (φ_{UV}) of functions, $\varphi_{UV} \in C^\infty(U \cap V; \mathbb{C}^*)$, parametrized by the pairs (U, V) of open sets in $\mathcal{U}$ with a non-empty intersection, arises from a C^∞ line bundle $\mathcal{L}$ on X, provided these functions satisfy the cocycle condition. Indeed, starting from such a family (φ_{UV}) we may define a line bundle $\mathcal{L}$ on X and sections s_U on U, $U \in \mathcal{U}$, as follows: for any $x \in X$, let

$$\mathcal{U}_x = \{U \in \mathcal{U}\ ,\ x \in U\};$$

define $\mathcal{L}_x$ as the one-dimensional complex vector space obtained as the quotient of $\mathcal{U}_x \times \mathbb{C}$ by the equivalence relation $\sim$ such that

$$(U, \lambda) \sim (V, \mu) \Leftrightarrow \mu = \varphi_{UV}(x)\lambda,$$

and define s_U by $s_U(x) = [(U, 1)]$. Then the s_U's satisfy the relations (B.1.1) by construction. Therefore $\mathcal{L}$ is a C^∞ line bundle and (φ_{UV}) is the associated family of transition functions.

B.1.3. A C^∞ morphism ℓ from a C^∞ line bundle $\mathcal{L}$ on X to another line bundle $\mathcal{L}'$ on X is the data, for every $x \in X$, of a linear map

$$\ell(x) : \mathcal{L}_x \to \mathcal{L}'_x$$

such that for any open subset U of X and any $s \in \mathcal{C}^\infty(U; \mathcal{L})$, the section

$$\ell \cdot s : x \mapsto \ell(x) \cdot s(x)$$

of $\mathcal{L}'$ on U is C^∞. If $\mathcal{U}$ is an open covering of X and if $(s_U)_{U \in \mathcal{U}}$ and $(s'_U)_{U \in \mathcal{U}}$ are families of C^∞ frames of $\mathcal{L}$ and $\mathcal{L}'$ over the open sets in $\mathcal{U}$, then we may define functions $L_U \in \mathcal{C}^\infty(U; \mathbb{C})$, $U \in \mathcal{U}$, by the relations

$$\ell \cdot s_U = L_U \cdot s'_U. \tag{B.1.3}$$

Moreover, if (φ_{UV}) and (φ'_{UV}) are the transition functions attached to (s_U) and (s'_U), we have

$$\varphi_{VU}\, L_U = L_V\, \varphi'_{VU} \quad \text{on} \quad U \cap V. \tag{B.1.4}$$

Conversely, for any family $(L_U)_{U \in \mathcal{U}}$, of functions $L_U \in \mathcal{C}^\infty(U; \mathbb{C})$, satisfying (B.1.4), there exists a unique C^∞ morphism ℓ from $\mathcal{L}$ to $\mathcal{L}'$ such that (B.1.3) holds.

Morphisms of line bundles clearly may be composed. A C^∞ morphism $\ell : \mathcal{L} \to \mathcal{L}'$ is called an isomorphism if, for any $x \in X$, the map $\ell(x)$ is an isomorphism; this is equivalent to the non-vanishing of the L_U's defined in (B.1.3) or to the existence of a C^∞ morphism of line bundles from $\mathcal{L}'$ to $\mathcal{L}$ inverse of ℓ. Continuous morphisms and isomorphisms between line bundles are similarly defined.

B.1.4. The *trivial line bundle over* X is the line bundle such that $\mathcal{L}_x = \mathbb{C}$ for any $x \in X$ and whose C^∞ structure is defined by $\mathcal{U} = \{X\}$, $s_X = 1$. Its (C^∞) sections may be identified with (C^∞) $\mathbb{C}$-valued functions. There is an obvious notion of restriction of a C^∞ line bundle on X to an open subset Ω of X. A C^∞ line bundle $\mathcal{L}$ on X is said to be *trivial on* Ω if its restriction to Ω is isomorphic to the trivial line bundle over Ω. This is easily seen to be equivalent to the existence of a trivialization of $\mathcal{L}$ on Ω.

B.1.5. If L and L' are two one-dimensional complex vector spaces and if $v \in L$ and $v' \in L'$, we will write $v \cdot v'$ instead of $v \otimes v'$ for their tensor product, which is an element of the one-dimensional vector space $L \otimes L'$. If $v \in L - \{0\}$, we will denote v^{-1} the linear form on L taking the value 1 on v; it is an element of the one-dimensional vector space L^*.

The *tensor product* $\mathcal{L} \otimes \mathcal{L}'$ of two C^∞ line bundles $\mathcal{L}$ and $\mathcal{L}'$ over X is the C^∞ line bundle over X such that $(\mathcal{L} \otimes \mathcal{L}')_x = \mathcal{L}_x \otimes \mathcal{L}'_x$ and whose C^∞

structure is such that, for any open subset Ω of X, any $s \in \mathcal{C}^\infty(\Omega;\mathcal{L})$ and any $s' \in \mathcal{C}^\infty(\Omega;\mathcal{L}')$, the section

$$s \cdot s' : x \mapsto s(x) \cdot s'(x)$$

of $\mathcal{L} \otimes \mathcal{L}'$ over Ω is C^∞. The *external tensor product* $\mathcal{L} \boxtimes \mathcal{L}'$ is the C^∞ line bundle over $X \times X$ such that $(\mathcal{L} \boxtimes \mathcal{L}')_{(x,y)} = \mathcal{L}_x \otimes \mathcal{L}'_y$ and whose C^∞ structure is such that, for any two open subsets Ω and Ω' of X, any $s \in \mathcal{C}^\infty(\Omega;\mathcal{L})$ and $s' \in \mathcal{C}^\infty(\Omega';\mathcal{L}')$, the section

$$s \boxtimes s' : (x,y) \mapsto s(x) \cdot s'(y)$$

of $\mathcal{L} \boxtimes \mathcal{L}'$ over $\Omega \times \Omega'$ is C^∞. The *dual line bundle* $\mathcal{L}^*$ is the C^∞ line bundle over X such that $(\mathcal{L}^*)_x = (\mathcal{L}_x)^*$ and whose C^∞ structure is such that, for any open subset Ω of X and any C^∞ frame $s \in \mathcal{C}^\infty(\Omega;\mathcal{L})$, the section

$$s^{-1} : x \mapsto s(x)^{-1}$$

of $\mathcal{L}^*$ over Ω is C^∞.

B.1.6. When X is a Riemann surface (or more generally, a complex manifold of any dimension), we can define *holomorphic line bundles* over X and extend to them all the preceding statements by simply replacing 'C^∞' by 'holomorphic' whenever it occurs.

Clearly, if $\mathcal{L}$ is a holomorphic bundle over a Riemann surface X, it is automatically a C^∞ line bundle over X considered as a C^∞ surface. Conversely, if $\mathcal{M}$ is a C^∞ line bundle on X, the data of a structure of holomorphic line bundle on $\mathcal{M}$ compatible with its C^∞ structure is essentially equivalent to the data of an open covering $\mathcal{U}$ of X and of C^∞ frames s_U over U for every $U \in \mathcal{U}$ such that, for any two intersecting U and V in $\mathcal{U}$, φ_{UV} defined by (B.1.1) is holomorphic on $U \cap V$.

Observe also that the notion of meromorphic function extends easily to sections of holomorphic line bundles: a section of a holomorphic line bundle $\mathcal{L}$ is called *meromorphic* if, locally, it can be written as the product of a holomorphic frame of $\mathcal{L}$ by a meromorphic function.

B.1.7. A major difference between the C^∞ and the holomorphic situations regards the existence of global sections. For any C^∞ line bundle $\mathcal{L}$ on a C^∞ manifold X, the space $\mathcal{C}^\infty(X;\mathcal{L})$ of its global C^∞ sections is 'very large'. Indeed, if s is a C^∞ frame of $\mathcal{L}$ on an open set U in X and if ρ is any C^∞ function on X which vanishes outside a compact subset of U, then we define a global section $\rho \cdot s \in \mathcal{C}^\infty(X;\mathcal{L})$ by setting

$$\begin{aligned}(\rho \cdot s)(x) &= \rho(x) \cdot s(x) \quad \text{if} \quad x \in U \\ &= 0 \qquad\qquad\quad \text{if} \quad x \notin U.\end{aligned}$$

On the contrary, if $\mathcal{L}$ is a holomorphic line bundle over a Riemann surface X, the space of holomorphic sections of $\mathcal{L}$ over X, usually denoted $H^0(X;\mathcal{L})$, may be 'very small'. For instance, suppose that $\mathcal{L}$ is the trivial holomorphic bundle $\mathcal{O}$ over a compact connected Riemann surface X. Then *the space* $H^0(X;\mathcal{O})$ *is the space of constant* $\mathbb{C}$*-valued functions on* X, hence isomorphic to $\mathbb{C}$. Indeed, an element f of $H^0(X;\mathcal{O})$ is nothing else than a holomorphic function from X to $\mathbb{C}$. By compactness of X, $|f|$ assumes a maximum value at some point P of X. The maximum modulus principle shows that f is constant on a neighbourhood of P, hence, by analytic continuation, on X which is connected.

B.1.8. A C^∞ Hermitian metric $\| \ \|$ on a C^∞ line bundle $\mathcal{L}$ over a C^∞ manifold X is the data for any $x \in X$ of a Hermitian norm $\| \ \|$ on $\mathcal{L}_x$ such that, for any open subset U of X and any $s \in \mathcal{C}^\infty(U;\mathcal{L})$, the function

$$\| s \|: x \mapsto \| s(x) \|^2,$$

from U to $\mathbb{R}_+$, is C^∞.

A C^∞ Hermitian metric on a holomorphic line bundle is a C^∞ Hermitian metric on the underlying C^∞ line bundle.

B.1.9. Let X be a C^∞ manifold of dimension n.

We will denote $\wedge^n_X$ the C^∞ line bundle of complex differential forms of degree n over X. By definition, for any $x \in X$

$$\wedge^n_{X,x} = \wedge^n \, T^*_{X,x} \otimes_{\mathbb{R}} \mathbb{C}$$

and the C^∞ sections of $\wedge^n_X$ are the C^∞ complex differential forms of degree n. The line bundle $\wedge^n_X$ may also be defined as follows: let $\{(U,\psi_U)\}_{U\in\mathcal{U}}$ be a family of C^∞ coordinate charts on X such that $\mathcal{U}$ is a covering of X; for any intersecting U and U' in $\mathcal{U}$, denote

$$\psi_U = (x_i)_{1\leq i\leq n} \quad \text{and} \quad \psi_{U'} = (x'_i)_{1\leq i\leq n}$$

and

$$\varphi_{U'U} = \det\left(\frac{\partial x'_i}{\partial x_j}\right)_{1\leq i,j\leq n}; \tag{B.1.5}$$

the function $\varphi_{U'U}$ belongs to $C^\infty(U \cap U^*;\mathbb{R}^*)$ and the $\varphi_{U'U}$'s satisfy the cocycle condition (B.1.2) and are easily checked to define the line bundle $\wedge^n_X$.

Finally, we will denote $|\wedge|_X$ the C^∞ line bundle of *complex densities* over X. It is defined as the C^∞ line bundle associated to the transition functions $|\varphi_{UU'}|$, which are the absolute values of the transition functions (B.1.5) defining $\wedge^n_X$. If X is oriented, then $|\wedge|_X$ may be identified with $\wedge^n_X$ (indeed, we may assume that the (U,ψ_U)'s are oriented charts; then $|\varphi_{U'U}| = \varphi_{U'U}$). More generally, a local choice of orientation on X determines locally an identification between $\wedge^n_X$ and $|\wedge|_X$ and two different choices of orientation give rise to opposite identifications. Recall that, if X is oriented, for any compactly supported continuous section ω of $\wedge^n_X$, the integral $\int_X \omega$ makes sense. These observations show that for any compactly supported continuous section ω of $|\wedge|_X$, the integral $\int_X \omega$ makes sense, without any orientation assumption on X.

B.2. Holomorphic and meromorphic differential forms

B.2.1. Let X be a Riemann surface.

Let α be a complex C^∞ 1-form defined on an open subset U of X. For any holomorphic coordinate chart (Ω, z), $\Omega \subset U$ there exist C^∞ functions φ and ψ on Ω, such that

$$\alpha = \varphi\, dz + \psi\, d\bar{z}$$

on Ω. The functions φ and ψ are uniquely defined by this equation. Moreover, if we let

$$\alpha^{(1,0)} = \varphi\, dz \quad \text{and} \quad \alpha^{(0,1)} = \psi\, d\bar{z},$$

the forms $\alpha^{(1,0)}$ and $\alpha^{(0,1)}$ are easily shown not to depend on the choice of the local holomorphic coordinate z and therefore to be C^∞ forms defined globally on U. The form $\alpha^{(1,0)}$ (resp. $\alpha^{(0,1)}$) is called the $(1,0)$-*part* (resp. $(0,1)$-*part*) of α. The 1-form α is called a *form of type* $(1,0)$ (resp. a *form of type* $(0,1)$) iff its $(0,1)$-part (resp. its $(1,0)$-part) vanishes, *i.e.*, iff locally it may be written

$$\alpha = \varphi\, dz$$
$$(\text{resp. } \alpha = \psi\, d\bar{z}).$$

These definitions may be interpreted as follows in terms of line bundles.

Let T_X be the *holomorphic tangent bundle* of X. It is the holomorphic line bundle defined as follows: let $\{(U, z_U)\}_{U \in \mathcal{U}}$ be a family of holomorphic charts on X such that $\mathcal{U}$ is a covering of X; for any intersecting U and V in $\mathcal{U}$, define

$$\varphi_{VU} = \frac{dz_U}{dz_V}\ ;$$

it is a non-vanishing holomorphic function on $U \cap V$. The φ_{VU}'s satisfy the cocycle condition (B.1.2) and T_X is the associated holomorphic line bundle. Using a local holomorphic coordinate z, a section s of T_X may be written locally

$$s = f(z)\frac{\partial}{\partial z}$$

and is C^∞ or holomorphic iff f is.

Let ω_X be the dual line bundle T_X^* of T_X. It is a holomorphic line bundle on X which can be defined by the transition functions

$$\varphi_{VU} = \frac{dz_V}{dz_U}.$$

Using a local holomorphic coordinate z, a section s of ω_X may be written locally

$$s = f(z)\ dz$$

and is C^∞ or holomorphic iff f is. The C^∞ sections of ω_X may be identified with the C^∞ forms of type $(1,0)$. A holomorphic (resp. meromorphic) section of ω_X on an open subset Ω of X is called a *holomorphic* (resp. *meromorphic*) *differential* on X. The vector space $H^0(X;\omega_X)$ of holomorphic differentials on X is denoted $\Omega^1(X)$.

Let $\overline{\omega}_X$ be the 'complex conjugate' of the line bundle ω_X. It is the C^∞ line bundle on X defined by the transition functions

$$\varphi_{VU} = \overline{\left(\frac{dz_V}{dz_U}\right)}.$$

Using a local holomorphic coordinate z, a section s of $\overline{\omega}_X$ may be written locally

$$s = f(z)\, d\bar{z}$$

and is C^∞ iff f is. The C^∞ sections of $\overline{\omega}_X$ may be identified with the C^∞ forms of type $(0,1)$. Complex conjugation transforms C^∞ sections of ω_X into C^∞ sections of $\overline{\omega}_X$. The sections of $\overline{\omega}_X$ which are complex conjugate of holomorphic differentials on X are called *antiholomorphic differentials.* Their vector space will be denoted $\overline{\Omega^1}(X)$.

The following statement may be proved by a simple local computation using the Cauchy-Riemann equations and will be left as an exercise to the reader (see also *infra* (B.7.1)).

Proposition B.2.1. *For any Riemann surface X, a C^∞ section of ω_X* (resp. *of $\overline{\omega}_X$*) *over X belongs to $\Omega^1(X)$* (resp. *$\overline{\Omega^1}(X)$*) *iff, considered as a C^∞ complex 1-form over X, it is closed.*

Observe that there exists a canonical isomorphism of C^∞ line bundles

$$\omega_X \otimes \overline{\omega}_X \simeq \wedge^2_X \tag{B.2.1}$$

which, locally, sends $dz \otimes d\bar{z}$ to $dz \wedge d\bar{z}$.

Finally, the data of a Hermitian metric $\| \ \|$ on $\overline{\omega}_X$ is equivalent to the data of a Riemannian metric ds^2 on X compatible with its holomorphic structure (*cf.* §I.2.1). Namely, for any local holomorphic coordinate $z = x + iy$ on X, ds^2 and $\| \ \|$ are determined by each other *via* the relation

$$ds^2 = \| dz \|^{-2} (dx^2 + dy^2) .$$

The volume form μ associated to ds^2 is then given by

$$\mu = \frac{i}{2} \| dz \|^2 \cdot dz \wedge d\bar{z} . \tag{B.2.2}$$

B.2.2. From now on, we restrict our attention to holomorphic and meromorphic differentials over compact connected Riemann surfaces. We start with a few examples.

Examples. i) *There is no non-zero element in $\Omega^1(\mathbb{P}^1\mathbb{C})$.* Indeed let $\alpha \in \Omega^1(\mathbb{P}^1\mathbb{C})$. There is an entire function $f(z)$

$$f(z) = \sum_{n=0}^{\infty} a_n \, z^n$$

such that, over $\mathbb{C}$

$$\alpha = f(z) \, dz.$$

Since α is holomorphic on a neighbourhood of ∞ in $\mathbb{P}^1\mathbb{C}$, the differential

$$f\left(\frac{1}{z}\right) d\left(\frac{1}{z}\right) = -\sum_{n=0}^{\infty} a_n \, z^{-n-2} \, dz$$

is holomorphic near 0. This clearly implies that all the coefficients a_n vanish.

ii) Let $X = \mathbb{C}/(\mathbb{Z} + \tau\mathbb{Z})$ be an elliptic curve. Then dz is a non-vanishing holomorphic differential on X. Therefore ω_X is a trivial holomorphic line bundle on X, and $\Omega^1(X)$ is a one-dimensional vector space generated by dz (*cf.* §B.1.7).

iii) Let C be the hyperelliptic Riemann surface defined by the polynomial

$$P(X,Y) = Y^2 - \prod_{i=1}^{2g+2} (X - \alpha_i)$$

where the α_i's are pairwise distinct complex numbers (*cf.* §I.4.2). The coordinates X and Y defines meromorphic functions x and y on C. *The vector space $\Omega^1(C)$ has dimension g, and the following holomorphic differentials build a basis of $\Omega^1(C)$:*

$$\omega_i = \frac{x^{i-1}dx}{y} \qquad i = 1, \ldots, g.$$

Indeed, near the points at infinity of C, $\frac{1}{x}$ may be used as local coordinate and $\frac{1}{y} \sim \pm\left(\frac{1}{x}\right)^{g+1}$; near the points $(\alpha_i, 0)$, y may be used as local coordinate and $x - \alpha_i \sim \lambda y^2$; near any other point of C, x may be used as local coordinate and y does not vanish. This immediately implies that the ω_i, $1 \leq i \leq g$, are holomorphic differential forms on C and that $\omega_1 = \frac{dx}{y}$ has a zero of order $g-1$ at the two points at infinity and does not vanish elsewhere on C. Therefore any $\alpha \in \Omega^1(C)$ may be written

$$\alpha = f \, \omega_1$$

where f is a meromorphic function on C which is holomorphic everywhere except at the points at infinity where the orders of its poles are at most $g-1$. On the other hand, the function f may be written

$$f = R(x) + S(x)y$$

where R and S belong to $\mathbb{C}(X)$ (*cf.* §I.4.2, Exercise). Since f is holomorphic at finite distance, R and S must be polynomials. It is now easy to deduce that $S = 0$ and that R is a polynomial of degree at most $g-1$ by examining the local behaviour of α near the points at infinity.

iv) Let F be an irreducible homogeneous polynomial of degree $d \geq 3$ in $\mathbb{C}[X_0, X_1, X_2]$ such that the algebraic curve X in $\mathbb{P}^2\mathbb{C}$ of equation

$$F(X_0, X_1, X_2) = 0$$

is smooth (*e.g.* $F(X_0, X_1, X_2) = X_1^d + X_2^d - X_0^d$; *cf.* §I.4.2). The coordinates (X_0, X_1, X_2) define meromorphic functions $x = \frac{X_1}{X_0}$ and $y = \frac{X_2}{X_0}$ on X. Elementary computations show that, for any polynomial P in $\mathbb{C}[X_1, X_2]$ of degree $\leq d-3$, the meromorphic differential form on X

$$\alpha(P) := \left(\frac{\partial F}{\partial X_2}(1,x,y)\right)^{-1} P(x,y)dx = -\left(\frac{\partial F}{\partial X_1}(1,x,y)\right)^{-1} P(x,y)dy$$

is holomorphic. Moreover, it may be proved that the map $P \mapsto \alpha(P)$ is an isomorphism from the space of polynomials of degree $\leq d-3$ in $\mathbb{C}[X_1, X_2]$ onto $\Omega^1(X)$.

More generally, if X is the normalization of an algebraic curve, it is always possible to describe $\Omega^1(X)$ in terms of the algebraic data defining this curve.

v) Let Γ be a cocompact discrete group in PSL(2, $\mathbb{R}$), acting freely on $\mathfrak{H}$ and let $X = \mathfrak{H}/\Gamma$. Then $\Omega^1(X)$ may be identified with the modular forms of weight two with respect to Γ (*cf.* [Z]).

B.2.3. Let (U, z) be a holomorphic coordinate chart on a Riemann surface and let $P \in U$. If $\omega = f\ dz$ is a holomorphic differential on $U - \{P\}$, its *residue* at P is defined as the coefficient of $(z - z(P))^{-1}$ in the Laurent series expansion of ψ in term of $(z - z(P))$. It is noted $\mathrm{Res}_P\ \omega$ and is also the integral

$$\frac{1}{2\pi i}\int_{\partial D} \omega$$

where D is a small disc in U which contains P (this expression shows that $\mathrm{Res}_P\ \omega$ is independent on the choice of the local coordinate z).

Proposition B.2.2. *For any meromorphic differential ω on a compact Riemann surface X, we have*

$$\sum_{x \in X} \mathrm{Res}_x\ \omega = 0.$$

(This sum is finite since the set of poles of ω is finite.)

Indeed, let $\{P_j\}_{1 \leq j \leq N}$ be the poles of ω and let $\{D_j\}_{1 \leq j \leq N}$ be a family of disjoint discs on X such that $D_j \ni P_j$. Consider $U = X - \bigcup_{j=1}^{N} D_j$. It is a surface with boundary, bounded by $\bigcup_{j=1}^{N} \partial D_j$. Moreover, restricted to U, ω is a holomorphic differential, hence a closed C^∞ 1-form. Therefore we have:

$$\sum_{x \in X} \mathrm{Res}_x\ \omega = \sum_{j=1}^{N} \mathrm{Res}_{P_j}\ \omega$$

$$
\begin{aligned}
&= \sum_{j=1}^{N} \frac{1}{2\pi i} \int_{\partial D_j} \omega \\
&= -\frac{1}{2\pi i} \int_{\partial U} \omega \\
&= -\frac{1}{2\pi i} \int_{U} d\omega \qquad \text{(Stokes formula)} \\
&= 0.
\end{aligned}
$$

Consider now a meromorphic function f on a Riemann surface X and a point P of X such that f is not constant on a neighbourhood of X (if X is connected and non-constant, this holds for any $P \in X$). Let (U, z) be a holomorphic chart on X such that $P \in U$ and $z(P) = 0$. The *multiplicity* with which f takes the value $f(P)$ at P is the positive integer n defined by the following conditions:

· if $f(P) \neq \infty$, there exists $a \in \mathbb{C}^*$ such that, for $x \in U$

$$f(x) = f(P) + az(x)^n + O\left(z(x)^{n+1}\right)$$

· if $f(P) = \infty$, there exists $a \in \mathbb{C}^*$ such that, for $x \in U$

$$f(x) = az(x)^{-n} + O\left(z(x)^{-n+1}\right).$$

For any $\lambda \in \mathbb{C}$, the meromorphic differential form $(f-\lambda)^{-1}df$ is holomorphic at P iff $f(P) \notin \{\lambda, \infty\}$ and has a simple pole with residue n (resp. $-n$) if $f(P) = \lambda$ (resp. $f(P) = \infty$). Therefore Proposition B.2.2 applied to $(f-\lambda)^{-1}df$ implies:

Proposition B.2.3. *For any non-constant meromorphic function f on a compact connected Riemann surface X, the number (counted with multiplicities) of preimages under f of an element $a \in \mathbb{P}^1\mathbb{C}$ is independent on a.*

This number is easily seen to be the degree of f (*cf.* [Mil]).

In the sequel, we will use the following consequence of Proposition B.2.3:

Corollary B.2.4. *If on a connected compact Riemann surface X there exists a meromorphic function f which has exactly one pole and if this pole is simple, then f establishes an isomorphism from X onto $\mathbb{P}^1\mathbb{C}$.*

B.2.4. We end this Section with a remarkable result - a special case of *Hodge decomposition:*

Theorem B.2.5. *For any (connected) compact Riemann surface X, the map*

$$
\begin{aligned}
i: \quad \Omega^1(X) \oplus \overline{\Omega^1}(X) &\to H^1_{DR}(X; \mathbb{C}) \\
\alpha \oplus \beta &\mapsto [\alpha + \beta]
\end{aligned}
$$

is an isomorphism.

(Observe that $\alpha + \beta$ is a closed complex 1-form on X by Proposition B.2.1 and has a well defined class in the de Rham cohomology group $H^1_{DR}(X;\mathbb{C})$; *cf.* §A.1.)

We will prove Theorem B.2.5 in §C.5. It has the following consequence:

Corollary B.2.6. *If X is a connected compact Riemann surface of genus g, then the complex vector space $\Omega^1(X)$ has dimension g.*

When X is the Riemann sphere or a hyperelliptic Riemann surface, the corollary follows from the explicit description of $\Omega^1(X)$ given above combined with the determination of their genus (*cf.* Figure 3 and Examples in §I.2.2 and §I.4.2). In general, it shows that $\dim \Omega^1(X)$ which *a priori* depends on the holomorphic structure of X is in fact a topological invariant of X. Conversely, when X is realized as an algebraic curve, Corollary B.2.6 provides an algebraic interpretation of the genus of X.

B.3. Divisors and holomorphic line bundles on Riemann surfaces

One of the most important questions in the theory of Riemann surfaces is the existence of meromorphic functions on a given Riemann surface with prescribed zeros and poles.

For instance, consider a compact connected Riemann surface X. We have seen that there exist non-constant meromorphic functions on X (*cf.* Theorem I.4.2); more precisely, according to Theorem I.4.3. 2), there are enough meromorphic functions on X to separate the points of X and even to provide a projective embedding of X. On the other hand, any holomorphic function on X is constant (see Section B.1.7). A natural question is then the following:

Let $P_1, \ldots, P_k$ be distinct points on X and let $m_1, \ldots, m_k$ be positive integers. Is there any non-constant meromorphic function on X which is holomorphic on $X - \{P_1, \ldots, P_k\}$ and whose pole at P_i has order at most m_i[18] *for any $i = 1, \ldots, k$?*

A variant of this question is to consider some other points $Q_1, \ldots, Q_\ell$ on X and positive integers $n_1, \ldots, n_\ell$ and to ask for *meromorphic functions on X which satisfy the preceding conditions and moreover have a zero of order at least n_i at Q_i.*

Observe that if X is a plane algebraic curve and the P_i's are points at infinity on X, by Theorem I.4.5, this essentially amounts to finding a polynomial $P(X,Y)$ which satisfies some growth conditions at ∞ and with zeroes of order n_i at Q_i $(1 \leq i \leq \ell)$. Geometrically, it is equivalent to find another plane algebraic curve of bounded degree, which meets X at Q_i with a multiplicity at least n_i $(1 \leq i \leq \ell)$ and has a suitable behaviour at infinity.

[18] By definition, this condition is satisfied by any meromorphic function holomorphic at P_i.

In this Section, we explain how such questions may be translated into questions concerning the existence of holomorphic sections of some holomorphic line bundles on X.

Let X be a compact Riemann surface. A *divisor* on X is an element of the free Abelian group whose generators are the points of X. In other words, a divisor on X is a finite formal sum

$$D = \sum_{i=1}^{k} n_i P_i$$

of (distinct) points P_i of X, affected with multiplicities $n_i \in \mathbb{Z}$. Such a divisor is called *effective* if all the n_i's are non-negative. More generally, for any two divisors D_1 and D_2 on X, one writes $D_1 \geq D_2$ when $D_1 - D_2$ is effective. The *multiplicity* $n_P(D)$ of a point P of X in D is n_i if $P = P_i$ and is 0 if P is not one of the P_i's. The finite subset $|D|$ of X formed by the points $P \in X$ such that $n_P(D) \neq 0$ is called the *support* of D. The group of divisors on X will be denoted by $\mathrm{Div}(X)$.

Let us now explain how a divisor is attached to any non-zero meromorphic section of a holomorphic line bundle on X.

Let X be any Riemann surface. For any $P \in X$ and any non-zero meromorphic function defined on an open neighbourhood of P in X, the *valuation* of f at P is the integer $v_P(f)$ defined as follows:

· if f is holomorphic at P and has a zero of order n at P, then $v_P(f) = n$;

· if f has a pole of order $n > 0$ at P, then $v_P(f) = -n$.

Clearly, we have:

i) $v_P(f) = 0$ iff f is holomorphic and does not vanish at P

ii) if f and g are non-zero meromorphic functions on a neighbourhood of P, then

$$v_P(fg) = v_P(f) + v_P(g). \tag{B.3.1}$$

and

$$v_P(f^{-1}) = -v_P(f). \tag{B.3.2}$$

Let $\mathcal{L}$ be a holomorphic line bundle on X. Consider a point $P \in X$ and a holomorphic trivialization t of $\mathcal{L}$ over an open neighbourhood of P. The preceding properties show that, if s is any non-zero meromorphic section of $\mathcal{L}$ on an open neighbourhood V of P and if we define a meromorphic function f on $U \cap V$ by

$$s = ft,$$

then $v_P(f)$ does not depend on t. This integer is called the *valuation of s at P*, and is denoted by $v_P(s)$. When $\mathcal{L}$ is the trivial line bundle, this definition is clearly consistent with the preceding one. Moreover, the properties i) and ii) are still true if now f and g are non-zero meromorphic sections of two holomorphic line bundles $\mathcal{L}$ and $\mathcal{M}$; then fg is a section on $\mathcal{L} \otimes \mathcal{M}$ and f^{-1} is a section of $\mathcal{L}^*$.

From now on, suppose that X is compact and connected. Then, if s is any non-zero meromorphic section of $\mathcal{L}$ on X, the points of X such that $v_P(s) \neq 0$ are isolated, hence form a finite set. Thus, we can define the *divisor of* s as

$$\operatorname{div} s = \sum_{P \in X} v_P(s) P;$$

this sum is indeed a finite sum. Observe that $\operatorname{div} s$ is effective iff s is holomorphic.

Consider now the 'set' E of pairs $(\mathcal{L}, s)$ where $\mathcal{L}$ is a holomorphic line bundle on X and s is a non-zero meromorphic section of $\mathcal{L}$ on X, and let us say that two such pairs $(\mathcal{L}, s)$ and $(\mathcal{M}, t)$ are isomorphic iff there exists a holomorphic isomorphism $\varphi : \mathcal{L} \xrightarrow{\sim} \mathcal{M}$ such that

$$\varphi \cdot s = t.$$

This isomorphism relation, which we will denote by $\sim$, clearly is an equivalence relation on E.

The following Proposition makes precise the link between holomorphic line bundles and divisors on X provided by divisors of meromorphic sections.

Proposition B.3.1. *If two pairs* $(\mathcal{L}, s)$ *and* $(\mathcal{M}, t)$ *in* E *are isomorphic, then* $\operatorname{div} s = \operatorname{div} t$. *Moreover the map*

$$\begin{aligned} \operatorname{div} : E/\sim &\longrightarrow \operatorname{Div}(X) \\ [(\mathcal{L}, s)] &\longmapsto \operatorname{div} s \end{aligned}$$

is a bijection.

Clearly, $(\mathcal{L}, s) \sim (\mathcal{M}, t)$ iff $t \cdot s^{-1}$ is a non-vanishing holomorphic section of $\mathcal{M} \otimes \mathcal{L}^*$. This holds iff $\operatorname{div}(t \cdot s^{-1}) = 0$ and, according to (B.3.1) and (B.3.2), this is equivalent to $\operatorname{div} t = \operatorname{div} s$. This proves the first assertion of proposition (B.3.1) and the injectivity of div.

Let us now prove its surjectivity. Let

$$D = \sum_{i=1}^{k} n_i P_i$$

be a divisor on X, where the P_i's are distinct. For each i, let us choose a local variable z_i at P_i and a small enough $\varepsilon_i > 0$, such that the discs

$$D_i = \{M \in X \ ; \ z_i(M) \text{ is defined and } |z_i(M)| < \varepsilon_i\}$$

are pairwise disjoint subsets of X. Let $\mathcal{L}$ be the line bundle defined by the covering $\{U_i\}_{0 \leq i \leq k}$, where

$$U_0 = X - \{P_1, \ldots, P_k\}$$
$$U_i = D_i \text{ if } i = 1, \ldots, k,$$

and by the transition functions

$$\varphi_{U_0 U_i} = z_i^{n_i}$$

defined for $i = 1, \ldots, k$ on the intersections[19]

$$U_0 \cap U_i = \{M \in X; z_i(M) \text{ is defined and } 0 < |z_i(M)| < \varepsilon_i\}.$$

Then $\mathcal{L}$ is trivial on $X - \{P_1, \ldots, P_k\}$ by construction. Moreover, the constant function 1, seen as an element of $H^0(X - \{P_1, \ldots, P_k\}; \mathcal{L})$, defines a meromorphic section s of $\mathcal{L}$ over X, holomorphic and non-vanishing on $X - \{P_1, \ldots, P_k\}$, and, by definition of the valuation v_{P_i}, we have for any $i = 1, \ldots, k$:

$$v_{P_i}(s) = v_{P_i}(\varphi_{U_0 U_i}) = n_i.$$

This shows that

$$\operatorname{div} s = \sum_{i=1}^{k} n_i P_i = D$$

and finally we get

$$D = \operatorname{div}[(\mathcal{L}, s)].$$

By the injectivity of div, the pair $(\mathcal{L}, s)$ we have just defined is well defined, up to isomorphism, by this last relation. It is often denoted by $(\mathcal{O}(D), 1_{\mathcal{O}(D)})$.

Observe that, again by the injectivity of div, for any holomorphic line bundle $\mathcal{L}$ on X and *for any non-zero meromorphic section s of $\mathcal{L}$ over X, we have an isomorphism of holomorphic line bundles*

$$\mathcal{L} \simeq \mathcal{O}(\operatorname{div} s). \tag{B.3.3}$$

The reader will check easily that, for any $D \in \operatorname{Div}(X)$ and any open subset $U \subset X$, the trivialization of $\mathcal{O}(D)$ on $X - |D|$ defines an isomorphism

$$H^0(U; \mathcal{O}(D)) \simeq \{f \in \mathcal{M}(U) \mid \forall P \in U, v_P(f) \geq -m_P(D)\}. \tag{B.3.4}$$

In particular

$$H^0(X; \mathcal{O}(D)) \simeq \{f \in \mathcal{M}(X) \mid \operatorname{div}(f) \geq -D\}, \tag{B.3.5}$$

and the space of meromorphic functions considered at the beginning of this Section is nothing else than $H^0\left(X; \mathcal{O}\left(\sum_{i=1}^{k} m_i P_i - \sum_{j=1}^{\ell} n_j Q_j\right)\right)$.

Observe finally that the tensor product provides a commutative group law on $E/\sim$, defined by

[19] Observe that ($i < j$ and $U_i \cap U_j \neq \emptyset$) $\Rightarrow$ ($i = 0$ and $j \geq 1$). Thus we only need to consider the k transition functions attached to the intersections $U_0 \cap U_1, \ldots, U_0 \cap U_k$. Moreover, since $i < j < k \Rightarrow U_i \cap U_j \cap U_k = \emptyset$, no cocycle condition needs to be checked.

$$[(\mathcal{L}, s)] + [(\mathcal{M}, t)] = [(\mathcal{L} \otimes \mathcal{M}, st)],$$

and that div is then an isomorphism of groups. In particular, for any two divisors D_1 and D_2 on X, we have isomorphisms

$$\mathcal{O}(D_1) \otimes \mathcal{O}(D_2) \simeq \mathcal{O}(D_1 + D_2) \tag{B.3.6}$$

$$\mathcal{O}(D_1)^* \simeq \mathcal{O}(-D_1). \tag{B.3.7}$$

B.4. The degree of a line bundle

In this Section, we discuss how the classification of C^∞ line bundles over a closed oriented surface may be realized by means of a single numerical invariant, the *degree*.

Consider a C^∞ line bundle $\mathcal{L}$ over an oriented C^∞ surface X, P a point of X and s a continuous section of $\mathcal{L}$ over a neighbourhood of P which does not vanish on $U - \{P\}$. Let us choose a trivialization of $\mathcal{L}$ on a neigbourhood V of P and let Γ be a small closed contour in $U \cap V$ which goes once around P, anticlockwise under the orientation of X. Using this trivialization, the restriction of s to Γ may be identified with a continuous map $\Gamma \to \mathbb{C}^*$, and the winding number of this map is easily seen not to depend on the choices of the trivialization and of Γ. It is called the *index* of s at P and is denoted $\mathrm{Ind}_P s$. It is non-zero only if $s(P) = 0$. See figure 26 for examples where $X = \mathbb{C}$ and $\mathcal{L}$ is T_X, the holomorphic tangent bundle.

Proposition -Definition B.4.1. *Let X be a closed oriented surface (cf. Appendix A) and let $\mathcal{L}$ be a C^∞ line bundle over X. There exist (infinitely many) sections $s \in C^\infty(X; \mathcal{L})$ which vanish only on a finite subset of X and the integer*

$$\sum_{\substack{P \in X \\ s(P)=0}} \mathrm{Ind}_P s$$

associated to any such section is independent of s. It is called the degree *of $\mathcal{L}$ and is denoted* $\deg_X \mathcal{L}$.

By definition, the trivial line bundle has degree 0 and two isomorphic C^∞ line bundles have the same degree. If $\mathcal{L}_1$ and $\mathcal{L}_2$ are two C^∞ line bundles and if s_1 and s_2 are sections of $\mathcal{L}_1$ and $\mathcal{L}_2$ with finite sets F_1 and F_2 of zeros, then $s_1 \otimes s_2$ is a section of $\mathcal{L}_1 \otimes \mathcal{L}_2$ which vanishes only on $F_1 \cup F_2$ and for any P, we have

$$\mathrm{Ind}_P s_1 \otimes s_2 = \mathrm{Ind}_P s_1 + \mathrm{Ind}_P s_2.$$

This implies that

$$\deg_X \mathcal{L}_1 \otimes \mathcal{L}_2 = \deg_X \mathcal{L}_1 + \deg_X \mathcal{L}_2. \tag{B.4.1}$$

This identity applied to $\mathcal{L}_1 = \mathcal{L}$ and $\mathcal{L}_2 = \mathcal{L}^*$ shows that

$$\deg_X \mathcal{L}^* = -\deg_X \mathcal{L}. \tag{B.4.2}$$

The following proposition explains the significance of the degree of C^∞ line bundles over a closed surface.

Proposition B.4.2. *Let X be an oriented closed surface and $\mathcal{L}_1$ and $\mathcal{L}_2$ be two C^∞ line bundles over X. The following three conditions are equivalent:*

i) $\mathcal{L}_1$ and $\mathcal{L}_2$ are isomorphic as topological line bundles; (*i.e.*, there exists a continuous isomorphism from $\mathcal{L}_1$ to $\mathcal{L}_2$.)

ii) $\mathcal{L}_1$ and $\mathcal{L}_2$ are isomorphic as C^∞ line bundles; (*i.e.*, there exists a C^∞ isomorphism from $\mathcal{L}_1$ to $\mathcal{L}_2$.)

iii) $\deg_X \mathcal{L}_1 = \deg_X \mathcal{L}_2$.

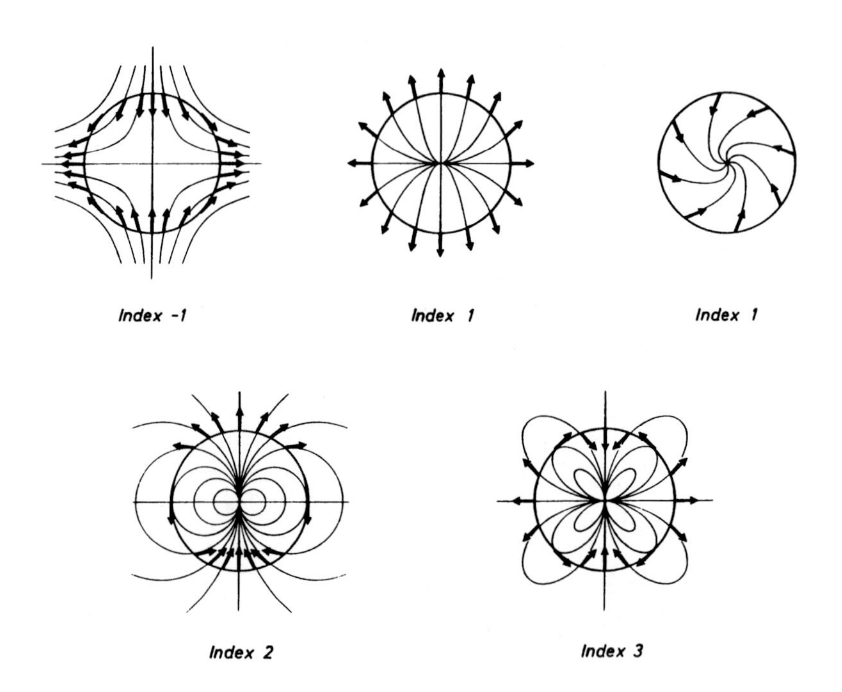

Fig. 26. The index of some vector fields.

According to (B.4.1), the $\mathbb{Z}$-valued map $\deg_X$, defined on the set of isomorphism classes of C^∞ line bundles over X, is in fact an isomorphism of Abelian groups, when the set of isomorphism classes of C^∞ line bundles over X is equipped with the structure of Abelian group defined by the tensor product[20].

In the sequel, we will consider the degrees only of holomorphic line bundles over a compact connected Riemann surface.

[20] The only non-trivial point in the verification of the group axioms is the existence of an inverse to the class of a line bundle $\mathcal{L}$: it is given by the class of $\mathcal{L}^*$, since $\mathcal{L}^* \otimes \mathcal{L}$ is isomorphic to the trivial bundle.

If X is such a Riemann surface, we may for instance consider the integers $\deg_X \omega_X$ and $\deg_X T_X$. According to (B.4.2), we have

$$\deg_X \omega_X = -\deg_X T_X, \tag{B.4.3}$$

and according to the *Poincaré-Hopf formula*, $\deg_X T_X$ may be expressed in terms of the genus g of X:

$$\deg_X T_X = 2 - 2g. \tag{B.4.4}$$

Finally, the degree of the line bundle associated to some divisor on X is easily computed:

Proposition B.4.3. *Let $D = \sum_{i=1}^{d} n_i P_i$ be a divisor on a compact connected Riemann surface. Then the degree of the line bundle $\mathcal{O}(D)$ is $\sum_{i=1}^{d} n_i$.*

Indeed consider the section $1_{\mathcal{O}(D)}$ of $\mathcal{O}(D)$ whose divisor is D and choose a C^∞ Hermitian metric $\| \ \|$ on $\mathcal{O}(D)$. An elementary computation shows that

$$s := \frac{1_{\mathcal{O}(D)}}{1 + \| 1_{\mathcal{O}(D)} \|^2}$$

is a C^∞ section of $\mathcal{O}(D)$ which vanishes exactly on $|D|$ and whose index at $P \in |D|$ is the multiplicity of P in D.

We will call *degree of the divisor* D the degree $\sum_{i=1}^{d} n_i$ of line bundle $\mathcal{O}(D)$.

Exercise. Use (B.4.3), (B.4.4) and Proposition B.4.3 to compute the genus of the Fermat curve (*cf.* §I.4.2).

Together with the isomorphism (B.3.3), Proposition B.4.3 has the following immediate consequence:

Corollary B.4.4. *A holomorphic line bundle on a compact connected Riemann surface X whose degree is negative has no non-zero holomorphic section over X.*

B.5. The operators $\overline{\partial}$ and $\overline{\partial}_{\mathcal{L}}$

Let U be an open subset of $\mathbb{C}$. For any function $f \in C^\infty(U;\mathbb{C})$, one defines

$$\frac{\partial f}{\partial \overline{z}}(x+iy) = \frac{1}{2}\left(\frac{\partial f(x+iy)}{\partial x} + i\frac{\partial f(x+iy)}{\partial y}\right)$$

and

$$\overline{\partial} f := \frac{\partial f}{\partial \overline{z}} \cdot d\overline{z}; \tag{B.5.1}$$

thus, $\overline{\partial} f$ is an element of $C^\infty(U;\overline{\omega}_U)$.

The operator $\overline{\partial}$ so defined clearly satisfies the Leibniz formula

$$\overline{\partial}(f_1 \cdot f_2) = f_1 \cdot \overline{\partial} f_2 + \overline{\partial} f_1 \cdot f_2. \tag{B.5.2}$$

Moreover $\overline{\partial} f$ vanishes iff f satisfies the Cauchy-Riemann equation

$$\frac{\partial f}{\partial x} + i\frac{\partial f}{\partial y} = 0,$$

i.e., iff f is holomorphic. These remarks show that

$$\overline{\partial}(h \cdot f) = h \cdot \overline{\partial} f \qquad \text{if } h \text{ is holomorphic.} \tag{B.5.3}$$

Let now $\varphi : U \to V$ be a holomorphic map between two open subsets of $\mathbb{C}$. A direct computation based on the Cauchy-Riemann equation satisfied by φ shows that, for any $f \in \mathcal{C}^\infty(V;\mathbb{C})$, we have

$$\overline{\partial}(f \circ \varphi) = \varphi^*(\overline{\partial} f). \tag{B.5.4}$$

This implies that the differential operator $\overline{\partial}$ may be defined on the C^∞ functions over any Riemann surface, in such a way that (B.5.1) holds locally for any holomorphic coordinate z. A direct intrinsic definition of $\overline{\partial} f$ is simply:

$$\overline{\partial} f = df^{(0,1)}.$$

Clearly the relations (B.5.2) and (B.5.3) are still true on any Riemann surface, as well as the characterization of holomorphic functions as the C^∞ functions f such that $\overline{\partial} f = 0$.

Consider now a holomorphic line bundle $\mathcal{L}$ on a Riemann surface X. Let U be an open set in X such that, over U, $\mathcal{L}$ possesses a holomorphic trivialization t (in other words, t is a non-vanishing holomorphic section of $\mathcal{L}$ over U). Then any $s \in \mathcal{C}^\infty(U,\mathcal{L})$ may be written $s = ft$ where $f \in \mathcal{C}^\infty(U;\mathbb{C})$ and the relation (B.5.3) easily implies that the element $\overline{\partial} f \cdot t$ of $\mathcal{C}^\infty(U;\mathcal{L}\otimes\overline{\omega}_X)$ does not depend on the choice of t. This shows that there exists a unique first order differential operator acting on the section of $\mathcal{L}$ with values in the sections of $\mathcal{L}\otimes\overline{\omega}_X$ which sends s to $\overline{\partial} f \cdot t$ for any such U, s, t and f. It is called the $\overline{\partial}$-operator with coefficients in $\mathcal{L}$ and denoted by $\overline{\partial}_{\mathcal{L}}$.

By construction of $\overline{\partial}_{\mathcal{L}}$, for any $s \in \mathcal{C}^\infty(U;\mathcal{L})$, we have

$$\overline{\partial}_{\mathcal{L}} s = 0 \quad \Leftrightarrow \quad s \text{ is holomorphic on } U.$$

In particular $H^0(X;\mathcal{L})$ is the kernel of

$$\overline{\partial}_{\mathcal{L}} : \mathcal{C}^\infty(X;\mathcal{L}) \to \mathcal{C}^\infty(X;\mathcal{L}\otimes\overline{\omega}_X).$$

The cokernel of this map, *i.e.*, the quotient vector space

$$\mathcal{C}^\infty(X;\mathcal{L}\otimes\overline{\omega}_X)/\overline{\partial}_{\mathcal{L}}(\mathcal{C}^\infty(X;\mathcal{L})),$$

is called the *first Dolbeault cohomology group of* $\mathcal{L}$ and is denoted by $H^1(X;\mathcal{L})$ ($H^0(X;\mathcal{L})$ is also called the zero-th cohomology group of $\mathcal{L}$).

Contrary to $H^0(X;\mathcal{L})$, the significance of the group $H^1(X;\mathcal{L})$ is not very intuitive *a priori.* The various theorems stated in the next Sections should make clear its great usefulness. Let us only say that it will appear as a vector space 'measuring the obstructions to build holomorphic sections of $\mathcal{L}$ over X'.

Suppose now that $\mathcal{L}$ and ω_X are equipped with C^∞ Hermitian metrics $\| \ \|_{\mathcal{L}}$ and $\| \ \|_{\omega_X}$. These metrics determine a C^∞ Hermitian metric on $\mathcal{L}\otimes\overline{\omega}_X$, such that

$$\| \ell\otimes\overline{\alpha} \|_{\mathcal{L}\otimes\overline{\omega}_X} = \| \ell \|_{\mathcal{L}} \cdot \| \alpha \|_{\omega_X}$$

for any $\ell \in \mathcal{L}_P$ and any $\alpha \in \omega_{X,P}$, and a positive 2-form μ on X given by (B.2.2). Using μ, $\| \cdot \|_{\mathcal{L}}$ and $\| \ \|_{\omega_X}$, we can define on $\mathcal{C}^\infty(X;\mathcal{L})$ and on $\mathcal{C}^\infty(X;\mathcal{L}\otimes\overline{\omega}_X)$ some L^2 Hermitian norms $\| \ \|_{L^2}$: for any $s\in\mathcal{C}^\infty(X;\mathcal{L})$ and any $t\in\mathcal{C}^\infty(X;\mathcal{L}\otimes\overline{\omega}_X)$,

$$\| s \|^2_{L^2} := \int_X \| s(x) \|^2 \, \mu(x)$$

$$\| t \|^2_{L^2} := \int_X \| t(x) \|^2 \, \mu(x).$$

These L^2 structures on $\mathcal{C}^\infty(X;\mathcal{L})$ and $\mathcal{C}^\infty(X;\mathcal{L}\otimes\overline{\omega}_X)$ allow to define the adjoint $\overline{\partial}^*_{\mathcal{L}}$ of $\overline{\partial}_{\mathcal{L}}$ by the identity

$$\langle \overline{\partial}_{\mathcal{L}} s, t\rangle = \langle s, \overline{\partial}^*_{\mathcal{L}} t\rangle$$

where $\langle \ , \ \rangle$ denote the scalar products associated with the Hermitian norms $\| \ \|_{L^2}$.

This being set, it is possible to give a heuristic interpretation of $H^1(X;\mathcal{L})$ which may be appealing to some mathematical physicists: *formally*, $H^1(X;\mathcal{L})$ may be identified with the kernel of $\overline{\partial}^*_{\mathcal{L}}$ - the 'zero-modes of the adjoint of $\overline{\partial}_{\mathcal{L}}$' in the language of physicists. Indeed, in the finite dimensional case, if $T: E\to F$ is a linear map between two finite dimensional vector spaces E and F endowed with Hermitian scalar products, then the kernel of the adjoint T^* of T, defined by

$$\langle Tx, y\rangle = \langle x, T^*y\rangle$$

for any $(x,y)\in E\times F$, is isomorphic with coker $T := F/T(E)$, *via* the map

$$\begin{aligned} \ker T^* &\to \operatorname{coker} T \\ y &\mapsto [y]. \end{aligned}$$

B.6. The finiteness theorem

The following theorem is a most basic fact concerning holomorphic sections of holomorphic line bundles on a compact Riemann surface.

Theorem B.6.1 (Finiteness Theorem). *Let X be a compact Riemann surface and let $\mathcal{L}$ be a holomorphic line bundle on X. The Dolbeault cohomology groups $H^0(X;\mathcal{L})$ and $H^1(X;\mathcal{L})$ are finite dimensional vector spaces.*

To show the significance of the finite dimensionality of $H^1(X;\mathcal{L})$, let us prove the following

Proposition B.6.2. *Let us keep the notations of Theorem B.6.1. For any $P \in X$, there exists a meromorphic section s of $\mathcal{L}$ which is holomorphic on $X - \{P\}$ and which has a pole at P of order at least 1 and at most $1 + \dim H^1(X;\mathcal{L})$.*

Consider a local holomorphic coordinate z at P, a non-vanishing holomorphic section t of $\mathcal{L}$ on an open neighbourhood U of P and a function $\rho \in C^\infty(X;\mathbb{C})$ such that $\rho \equiv 1$ near P and such that ρ vanishes outside a compact subset of U. For any positive integer i, the section $\overline{\partial}\rho \cdot z^{-i}t$ (resp. $\rho z^{-i}t$) of $\mathcal{L} \otimes \overline{\omega}_X$ (resp. of $\mathcal{L}$) over $U - \{P\}$ extended by zero defines an element of $\mathcal{C}^\infty(X;\mathcal{L}\otimes\overline{\omega}_X)$ (resp. of $\mathcal{C}^\infty(X-\{P\},\mathcal{L})$). Moreover, on $X - \{P\}$, we have

(B.6.1) $$\overline{\partial}_{\mathcal{L}}(\rho z^{-i}t) = \overline{\partial}\rho \cdot z^{-i}t.$$

Linear algebra shows that, if $d = \dim H^1(X;\mathcal{L})$, there exists $(\lambda_1, \ldots, \lambda_{d+1}) \in \mathbb{C}^{d+1} - \{0\}$ such that the class of

$$\alpha = \sum_{i=1}^{d+1} \lambda_i \overline{\partial}\rho \cdot z^{-i}t$$

in $H^1(X;\mathcal{L})$ vanishes. Then there exists $\beta \in \mathcal{C}^\infty(X;\mathcal{L})$ such that

(B.6.2) $$\alpha = \overline{\partial}_{\mathcal{L}}\beta,$$

and, according to (B.6.1) and (B.6.2),

$$s = \sum_{i=1}^{d+1} \lambda_i \rho z^{-i}t - \beta$$

is an element of $\mathcal{C}^\infty(X-\{P\},\mathcal{L})$ such that $\overline{\partial}_{\mathcal{L}}s = 0$. Thus s is holomorphic on $X - \{P\}$ and its 'polar part' at P is

$$\sum_{i=1}^{d+1} \lambda_i z^{-i}t$$

which is non-zero and of order at most $d+1$ by construction.

Observe that, according to (B.3.3), Proposition B.6.2 implies the following

Corollary B.6.3. *Any holomorphic line bundle $\mathcal{L}$ on a compact connected Riemann surface X is isomorphic, as a holomorphic line bundle, to the line bundle $\mathcal{O}(D)$ associated to some divisor D on X.*

Observe also that, applied to the trivial line bundle $\mathcal{O}$, Proposition B.6.2 shows that *for any* $P \in X$ *there exists a non-constant meromorphic function* f *on* X *which is holomorphic on* $X - \{P\}$ *and has a pole of order at most* $1 + \dim H^1(X;\mathcal{O})$ *at* P. In particular, this prove that *meromorphic functions separate the points of* X (compare with Theorem I.4.3).

B.7. $H^1(X;\mathcal{L})$ and polar parts of meromorphic sections

The construction used to prove Proposition B.6.2 may be extended to provide an alternative description of the cohomology group $H^1(X;\mathcal{L})$ in terms of meromorphic sections of $\mathcal{L}$ and of their 'polar parts' (hence a purely 'holomorphic' description). To do this we need to introduce a few notations:

• $\mathcal{M}(X;\mathcal{L})$ will denote the vector space of meromorphic sections of $\mathcal{L}$ over X;

• for any $x \in X$, $\mathcal{P}(x;\mathcal{L})$ is defined as the quotient of the set $\mathcal{M}(x;\mathcal{L})$ of pairs (U,s), where U is an open neighbourhood of x in X and s is meromorphic section of $\mathcal{L}$ on U, by the equivalence relation v defined by

$$(U,s) \sim (V,t) \Leftrightarrow t - s \text{ is holomorphic at } x.$$

The space $\mathcal{P}(x;\mathcal{L})$ is the space of 'polar parts at x' of meromorphic sections of $\mathcal{L}$. Indeed, if t is a non-vanishing holomorphic section of $\mathcal{L}$ on a neighbourhood of x and if z is a local coordinate at x such that $z(x) = 0$, then one defines an isomorphism between the vector space of polynomials in z^{-1} without constant term and $\mathcal{P}(x;\mathcal{L})$ by sending $\sum_{i=1}^{d} a_i z^{-i}$ to the class of $\sum_{i=1}^{d} a_i z^{-i} t$.

• $\mathcal{P}(X;\mathcal{L})$ will denote the subspace of $\prod_{x\in X}\mathcal{P}(x;\mathcal{L})$ defined by the condition

$$(p_x)_{x\in X} \in \mathcal{P}(X;\mathcal{L}) \Leftrightarrow p_x \neq 0 \text{ only for a finite set of } x'\text{s}.$$

• $P_{\mathcal{L}}$ will denote the linear map from $\mathcal{M}(X;\mathcal{L})$ to $\mathcal{P}(X;\mathcal{L})$ which associates to a meromorphic section s of $\mathcal{L}$ over X the element $(p_x)_{x\in X}$ of $\mathcal{P}(X;\mathcal{L})$, where p_x is the class of s in $\mathcal{P}(x;\mathcal{L})$. (In other words $P_{\mathcal{L}}(s)$ is the family of 'polar parts' of s.)

Consider now $(p_x)_{x\in X} \in \mathcal{P}(X;\mathcal{L})$ and let

$$\{P_1,\ldots,P_n\} = \{x \in X \mid p_x \neq 0\}.$$

Choose for each $i = 1,\ldots,n$ a meromorphic section f_i on some neighbourhood U_i of P_i whose class in $\mathcal{P}(P_i;\mathcal{L})$ is p_{P_i} and a function $\rho_i \in \mathcal{C}^\infty(X;\mathbb{C})$ such that $\rho_i \equiv 1$ near P_i and such that ρ_i vanishes outside some compact subset of U_i. Then, the section $f_i \cdot \bar\partial\rho_i$ of $\mathcal{L}\otimes\overline{\omega}_X$ over $U_i - \{P_i\}$ extended by zero defines an element of $\mathcal{C}^\infty(X;\mathcal{L}\otimes\overline{\omega}_X)$. One easily checks that the class of $\sum_{i=1}^{n} f_i \cdot \bar\partial\rho_i$ in $H^1(X;\mathcal{L})$ only depends on the class of $(p_x)_{x\in X}$ in

$$\operatorname{coker} P_{\mathcal{L}} := \mathcal{P}(X;\mathcal{L})/P_{\mathcal{L}}(\mathcal{M}(X;\mathcal{L})).$$

(This space is called the space of *repartition classes* of X). In that way, one defines a linear map

$$I_{\mathcal{L}} : \operatorname{coker} P_{\mathcal{L}} \to H^1(X;\mathcal{L})$$
$$[(p_x)_{x\in X}] \mapsto \left[\sum_{i=1}^{x} f_i \cdot \overline{\partial}\rho_i\right].$$

Theorem B.7.1. *The kernel of $P_{\mathcal{L}}$ is $H^0(X;\mathcal{L})$. The map $I_{\mathcal{L}}$ is an isomorphism from the cokernel of $P_{\mathcal{L}}$ onto $H^1(X;\mathcal{L})$.*

The first assertion is clear. The second one means that $H^1(X;\mathcal{L})$ 'measures the obstruction' to finding meromorphic sections of $\mathcal{L}$ with prescribed polar parts. See §C.4 for a proof.

Observe that, when $\mathcal{L} = \omega_X$, $C^\infty(X;\mathcal{L})$ is a subspace of the space of C^∞ complex 1-forms over X, $C^\infty(X;\mathcal{L}\otimes\overline{\omega}_X)$ may be identified with the space of C^∞ complex 2-forms over X (*cf.* (B.2.1)), and up to a sign, $\overline{\partial}_{\mathcal{L}}$ is nothing else than the exterior differential d acting from 1-forms to 2-forms (indeed, in local coordinates we have

$$\begin{aligned} d(f dz) &= df \wedge dz \\ &= \overline{\partial} f^{(0,1)} \wedge dz \end{aligned} \tag{B.7.1}$$

because $\partial f \wedge dz = 0$, since $dz \wedge dz = 0$). In particular, the image of $\overline{\partial}_{\mathcal{L}}$ is formed by exact two forms and their integral over X vanishes. This shows that the following map is well defined:

$$\operatorname{Res} : H^1(X;\omega_X) \to \mathbb{C}$$
$$[\alpha] \mapsto \frac{1}{2\pi i}\int_X \alpha.$$

The insertion of the factor $(2\pi i)^{-1}$ as well as the notation Res are explained by the following result, which we leave as an exercise.

Lemma B.7.2. *For any $(\alpha_x)_{x\in X} \in \mathcal{P}(X;\omega_X)$, we have*

$$\operatorname{Res}\circ P_{\omega_X}\left([(\alpha_x)_{x\in X}]\right) = \sum_{x\in X} \operatorname{Res}_x \alpha_x.$$

(In this equality $[(\alpha_x)_{x\in X}]$ denotes the class of $(\alpha_x)_{x\in X}$ in $\operatorname{coker} P_{\mathcal{L}}$ and $\operatorname{Res}_x\alpha_x$ the residue at x of any meromorphic differential representing α_x).

B.8. Serre duality

An important complement to the Finiteness Theorem B.6.1 is the Serre duality Theorem. To formulate it, we need a few preliminaries.

The Leibniz rule (B.5.2) generalizes immediately as follows: for any two holomorphic line bundles $\mathcal{L}_1$ and $\mathcal{L}_2$ on X, and for any $f_1 \in \mathcal{C}^\infty(X;\mathcal{L}_1)$ and any $f_2 \in \mathcal{C}^\infty(X;\mathcal{L}_2)$, then

$$\text{(B.8.1)} \quad \overline{\partial}_{\mathcal{L}_1\otimes\mathcal{L}_2}(f_1 \cdot f_2) = f_1 \cdot \overline{\partial}_{\mathcal{L}_2} f_2 + \overline{\partial}_{\mathcal{L}_1} f_1 \cdot f_2 \quad (\in \mathcal{C}^\infty(X;\mathcal{L}_1 \otimes \mathcal{L}_2 \otimes \overline{\omega}_X)).$$

In particular, if $f_1 \in H^0(X,\mathcal{L}_1)$, we have

$$\overline{\partial}_{\mathcal{L}_1\otimes\mathcal{L}_2}(f_1 \cdot f_2) = f_1 \cdot \overline{\partial}_{\mathcal{L}_2} f_2.$$

This shows that the bilinear map

$$\begin{aligned} H^0(X;\mathcal{L}_1) \times \mathcal{C}^\infty(X;\mathcal{L}_2 \otimes \overline{\omega}_X) &\to \mathcal{C}^\infty(X;\mathcal{L}_1 \otimes \mathcal{L}_2 \otimes \overline{\omega}_X) \\ (f_1, \alpha_2) &\mapsto f_1 \cdot \alpha_2 \end{aligned}$$

yields a quotient map

$$\text{(B.8.2)} \quad \begin{aligned} H^0(X;\mathcal{L}_1) \times H^1(X;\mathcal{L}_2) &\to H^1(X;\mathcal{L}_1 \otimes \mathcal{L}_2) \\ (f_1, [\alpha_2]) &\mapsto f_1 \cdot [\alpha_2] := [f_1 \cdot \alpha_2]. \end{aligned}$$

We can now state:

Theorem B.8.1 (Serre duality). *Let X be a compact Riemann surface and let $\mathcal{L}$ be a holomorphic line bundle on X. The bilinear map*

$$\langle \cdot, \cdot \rangle : H^0(X;\mathcal{L}^* \otimes \omega_X) \times H^1(X;\mathcal{L}) \to \mathbb{C}$$

obtained by composing the product (B.8.2)

$$H^0(X;\mathcal{L}^* \otimes \omega_X) \times H^1(X;\mathcal{L}) \to H^1(X;\mathcal{L}^* \otimes \mathcal{L} \otimes \omega_X) \simeq H^1(X;\omega_x)$$

with the map Res *is a perfect pairing*[21].

Observe that, by the definitions of Res and of the product (B.8.2), for any $s \in H^0(X;\mathcal{L})$ and $\alpha \in \mathcal{C}^\infty(X;\mathcal{L}^* \otimes \omega_X \otimes \overline{\omega}_X)$, we have

$$\langle s, [\alpha] \rangle = \frac{1}{2\pi i} \int_X s\alpha,$$

where $s\alpha$ is a section of $\mathcal{L} \otimes \mathcal{L}^* \otimes \omega_X \otimes \overline{\omega}_X$, identified with $(\Lambda^2 T^* X)_{\mathbb{C}}$. Theorem B.8.1 asserts that this pairing defines an isomorphism

$$H^1(X;\mathcal{L}) \simeq H^0(X;\mathcal{L}^* \otimes \omega_X)^*$$

and implies that

$$\text{(B.8.3)} \quad \dim H^1(X;\mathcal{L}) = \dim H^0(X;\mathcal{L}^* \otimes \omega_X).$$

[21] Recall that, if E and F are two finite dimensional vector spaces, a bilinear map $\langle \cdot, \cdot \rangle : E \times F \to \mathbb{C}$ is called a perfect pairing when the map $(E \to F^*, x \mapsto \langle x, \cdot \rangle)$, or equivalently the map $(F \to E^*,\ y \mapsto \langle \cdot, y \rangle)$, is bijective.

Corollary B.8.2. *Let X be a compact connected Riemann surface of genus g.*

1) There exists a canonical isomorphism

$$H^1(X;\mathcal{O}) \simeq H^0(X;\omega_X)^* = \Omega^1(X)^*.$$

In particular

$$\dim H^1(X;\mathcal{O}) = g \; . \tag{B.8.4}$$

Moreover the map

$$\overline{\Omega^1}(X) \to H^1(X;\mathcal{O})$$

deduced from the inclusion $\overline{\Omega^1}(X) \hookrightarrow \mathcal{C}^\infty(X;\overline{\omega}_X)$ is an isomorphism. In other words

$$\mathcal{C}^\infty(X;\overline{\omega}_X) = \overline{\Omega^1}(X) \oplus \overline{\partial}\, \mathcal{C}^\infty(X;\mathbb{C}).$$

2) The map

$$\mathrm{Res} : H^1(X;\omega_X) \to \mathbb{C}$$

is an isomorphism.

The first assertion in 1) (resp. the assertion 2)) follows from Theorem B.8.1 applied to $\mathcal{L} = \mathcal{O}$ (resp. $\mathcal{L} = \omega_X$). The equality (B.8.4) is then a consequence of Corollary (B.2.6). The last assertion in 1) is implied by the equality of dimensions

$$\dim \overline{\Omega^1}(X) = \dim \Omega^1(X)^* = \dim H^1(X;\mathcal{O})$$

and by the injectivity of the map $\overline{\Omega^1}(X) \to H^1(X;\mathcal{O})$, which follows from the identity

$$\langle \overline{\beta}, [\beta] \rangle = \frac{1}{2\pi i} \int_X \overline{\beta} \wedge \beta < 0$$

for any $\beta \in \overline{\Omega^1}(X)$.

Exercise: Deduce from Corollary B.2.4, Proposition B.6.2, and (B.8.4) that any compact connected Riemann surface of genus 0 is biholomorphic to $\mathbb{P}^1\mathbb{C}$.

Exercise: Use the Serre duality to prove that the heuristic interpretation of $H^1(X;\mathcal{L})$ given at the end of §B.5 is indeed correct.

B.9. Riemann-Roch theorem

Recall that, for any two vector spaces E and F, a linear map $T : E \to F$ is said to *have an index* if the kernel $\ker T$ and the cokernel $\operatorname{coker} T := F/T(E)$ of T are finite dimensional vector spaces. The *index* of T is then defined as the integer

$$\mathrm{ind}\; T = \dim\; \ker T - \dim\; \operatorname{coker}\; T$$

(compare [Bel] §1.4).

According to the very definition of the cohomology groups $H^0(X;\mathcal{L})$ and $H^1(X;\mathcal{L})$, the Finiteness Theorem B.6.1 precisely asserts that the operator

$$\overline{\partial}_{\mathcal{L}} : \mathcal{C}^\infty(X;\mathcal{L}) \to \mathcal{C}^\infty(X;\mathcal{L}\otimes\overline{\omega}_X)$$

has an index. Its index, namely

$$\dim \ker\overline{\partial}_{\mathcal{L}} - \dim \operatorname{coker}\overline{\partial}_{\mathcal{L}} = \dim H^0(X;\mathcal{L}) - \dim H^1(X;\mathcal{L}),$$

is given by the famous Riemann-Roch theorem:

Theorem B.9.1. *Let X be a compact connected Riemann surface of genus g and let $\mathcal{L}$ be a holomorphic line bundle on X. Then:*

$$\text{(B.9.1)} \qquad \dim H^0(X;\mathcal{L}) - \dim H^1(X;\mathcal{L}) = 1 - g + \deg_X \mathcal{L}.$$

Corollary B.9.2 (Riemann's inequality). *Let $D = \sum_{i=1}^k n_i P_i$ be a divisor on X. Then*

$$\text{(B.9.2)} \qquad \dim\{f \in \mathcal{M}(X) \mid \operatorname{div} f \geq -D\} \geq 1 - g + \sum_{i=1}^k n_i.$$

This follows immediately from (B.9.1) applied to $\mathcal{L} = \mathcal{O}(D)$ and from (B.3.5) and Proposition (B.4.3). Observe that Corollary B.9.2 is an existence statement concerning meromorphic functions with prescribed zeroes and poles, which answers the question discussed at the beginning of §B.3. Indeed (B.9.2) implies that if $\sum_{i=1}^k n_i > g$, there exists a non-constant meromorphic function f on X such that $\operatorname{div} f \geq -D$.

Classically, Riemann-Roch formula (B.9.1) is combined with the equality (B.8.3) and is written as

$$\text{(B.9.3)} \qquad \dim H^0(X;\mathcal{L}) - \dim H^0(X;\mathcal{L}^*\otimes\omega_X) = 1 - g + \deg_X \mathcal{L}.$$

Observe that, applied to $\mathcal{L} = \omega_X$, this equality becomes

$$g - 1 = 1 - g + \deg_X \mathcal{L}$$

since $\dim H^0(X;\omega_X) = g$ and $\dim H^0(X;\mathcal{O}) = 1$ (*cf.* Corollary B.2.6 and §B.1.7). This shows that *the degree of ω_X is $2g-2$* as mentioned above (see (B.4.3) and (B.4.4)).

Consequently, *if* $\deg_X \mathcal{L} > 2g - 2$, *then* $\deg_X \mathcal{L}^* \otimes \omega_X < 0$ *and* $H^0(X;\mathcal{L}^*\otimes\omega_X) = 0$ (by (B.4.1), (B.4.2) and Corollary B.4.4) *hence* $H^1(X;\mathcal{L}) = 0$ (by (B.8.3)) *and Riemann-Roch formula reads*

$$\text{(B.9.4)} \qquad \dim H^0(X;\mathcal{L}) = 1 - g + \deg_X \mathcal{L}.$$

Of course, this equality has an interpretation in terms of divisors, analogous to (B.9.2).

At this point of our discussion, the proof of Riemann-Roch Theorem is not difficult. Since any holomorphic line bundle over X is isomorphic to the line bundle $\mathcal{O}(D)$ associated to some divisor D, a simple induction shows that to prove (B.9.1), we only need to prove that:

i) (B.9.1) is true for $\mathcal{L} = \mathcal{O}$;

ii) for any holomorphic line bundle over X, (B.9.1) is true for $\mathcal{L} = \mathcal{N}$ iff (B.9.1) is true for $\mathcal{L} = \mathcal{N} \otimes \mathcal{O}(-P)$.

However the validity of (B.9.1) rewritten as (B.9.3) is clear when $\mathcal{L} = \mathcal{O}$, since dim $H^0(X;\mathcal{O}) = 1$ and $\dim H^0(X;\omega_X) = g$. As for ii), since

$$\deg_X \mathcal{N} \otimes \mathcal{O}(-P) = \deg_X \mathcal{N} + \deg_X \mathcal{O}(-P) = \deg_X \mathcal{N} - 1,$$

it amounts to proving that

$$\begin{aligned}\dim H^0(X;\mathcal{N} \otimes \mathcal{O}(-P)) - \dim H^1(X;\mathcal{N} \otimes \mathcal{O}(-P)) + 1 \\ = \dim H^0(X;\mathcal{N}) - \dim H^1(X;\mathcal{N})\end{aligned}$$

that is, according to Theorem B.7.1,

$$\text{ind} P_{\mathcal{L}\otimes\mathcal{O}(-P)} + 1 = \text{ind} P_{\mathcal{L}}. \tag{B.9.5}$$

The proof of (B.9.5) will be based on the following lemma of linear algebra, which we leave as an exercise[22].

Lemma B.9.3. *Let E, F and G be vector spaces and let $u : E \to F$ and $v : F \to G$ be linear maps with index. Then $v \circ u : E \to G$ is a linear map with index and*

$$\text{ind } v \circ u = \text{ind } u + \text{ind } v.$$

Observe that the meromorphic (resp. holomorphic) sections of $\mathcal{N} \otimes \mathcal{O}(-P)$ over an open subset $U \subset X$ may be identified with the meromorphic sections of $\mathcal{N}$ over U (resp. with the holomorphic sections of $\mathcal{N}$ over U which vanish at P if $P \in U$). These identifications provide linear bijections

$$\Phi^0 : \mathcal{M}(X;\mathcal{N} \otimes \mathcal{O}(-P)) \xrightarrow{\sim} \mathcal{M}(X;\mathcal{N})$$

and

$$\mathcal{M}(x;\mathcal{N} \otimes \mathcal{O}(-P)) \xrightarrow{\sim} \mathcal{M}(x;\mathcal{N})$$

and linear maps

$$\mathcal{P}(x;\mathcal{N} \otimes \mathcal{O}(-P)) \to \mathcal{P}(x;\mathcal{N})$$

and

$$\Phi^1 : \mathcal{P}(X;\mathcal{N} \otimes \mathcal{O}(-P)) \to \mathcal{P}(X;\mathcal{N}).$$

The following properties of Φ^0 and Φ^1 are easily checked:

- the following diagram

[22] We will use Lemma B.9.3 only when u is an isomorphism or when v is onto and dim ker $v = 1$. In these two cases, the proof of Lemma B.9.3 is very simple.

$$
\text{(B.9.6)} \qquad \begin{array}{ccc}
\mathcal{M}(X;\mathcal{N}\otimes\mathcal{O}(-P)) & \xrightarrow{P_{\mathcal{N}\otimes\mathcal{O}(-P)}} & \mathcal{P}(X;\mathcal{N}\otimes\mathcal{O}(-P)) \\
\downarrow \Phi^0 & & \downarrow \Phi^1 \\
\mathcal{M}(X;\mathcal{N}) & \xrightarrow{P_{\mathcal{N}}} & \mathcal{P}(X;\mathcal{N})
\end{array}
$$

is commutative (*i.e.*, $\Phi^1 \circ P_{\mathcal{N}\otimes\mathcal{O}(-P)} = P_{\mathcal{N}} \circ \Phi^0$).

• Φ^1 is onto and the kernel of Φ^1 is generated by the class of $(p_x)_{x\in X}$, where $p_x = 0$ if $x \neq P$ and p_P has a simple pole at P. In particular ind $\Phi^1 = 1$.

Thus the four maps which occur in (B.9.6) are maps with index and Lemma B.9.3 shows that

$$
\begin{aligned}
\mathrm{ind} P_{\mathcal{N}\otimes\mathcal{O}(-P)} + 1 &= \mathrm{ind}\ P_{\mathcal{N}\otimes\mathcal{O}(-P)} + \mathrm{ind}\ \Phi^1 \\
&= \mathrm{ind}\ \Phi^1 \circ P_{\mathcal{N}\otimes\mathcal{O}(-P)} \\
&= \mathrm{ind}\ P_{\mathcal{N}} \circ \Phi^0 \\
&= \mathrm{ind}\ P_{\mathcal{N}} + \mathrm{ind}\ \Phi^0 \\
&= \mathrm{ind}\ P_{\mathcal{N}},
\end{aligned}
$$

as was to be proved.

Finally, let us give a consequence of the results of this Appendix which will play a key role in the construction of the Jacobian embedding of compact Riemann surfaces.

Proposition B.9.4. *Let X be a compact connected Riemann surface of genus $g \geq 1$. Then for any $P \in X$, there exists $\omega \in \Omega^1(X)$ such that $\omega(P) \neq 0$.*

Assume the contrary. Then the injection

$$
H^0(X;\omega_X \otimes \mathcal{O}(-P)) \cong \{s \in H^0(X;\omega_X), \mid s(P) = 0\} \hookrightarrow H^0(X;\omega_X)
$$

is a bijection and since

$$
\deg_X \omega_X \otimes \mathcal{O}(-P) = \deg_X \omega_X - 1,
$$

Riemann-Roch formula (B.9.1) and Corollary B.8.2, 2) show that

$$
\dim\ H^1(X;\omega_X \otimes \mathcal{O}(-P)) = \ \dim\ H^1(X;\omega_X) + 1 = \ 2.
$$

Therefore, by Serre duality dim $H^0(X;\mathcal{O}(P)) = 2$ and there exists a non-constant meromorphic function on X, whose only pole is P, and is a simple pole. By Corollary B.2.4, this implies that X is isomorphic to $\mathbb{P}^1\mathbb{C}$, hence has genus 0.

Appendix C. Analysis on compact Riemann surfaces

This appendix is devoted to the proofs of the Finiteness Theorem B.6.1, of the Serre duality Theorem B.8.1, of Theorem B.7.1 (isomorphism between H^1 and repartition classes) and of Theorem B.2.5 (Hodge decomposition).

C.1. Regularizing operators

Let $\mathcal{L}_1$ and $\mathcal{L}_2$ be two C^∞ line bundles on a compact manifold X. Let $k \in \mathcal{C}^\infty(X \times X; \mathcal{L}_2 \boxtimes (\mathcal{L}_1^* \otimes |\wedge|_X))$ and $f \in \mathcal{C}^\infty(X; \mathcal{L}_1)$. For any $x \in X$, $y \mapsto k(x,y)f(y)$ is a C^∞ section of $\mathcal{L}_{1,x} \otimes |\wedge|_X$, and the integral

$$Kf(x) := \int_{y \in X} k(x,y)f(y) \tag{C.1.1}$$

is a well defined element of $\mathcal{L}_{1,x}$. Moreover, the section Kf of $\mathcal{L}_1$ over X so defined is C^∞. (This is clear by the elementary differentiability properties of integrals depending on a parameter when f is supported by a chart on which $\mathcal{L}_2$ is trivial. In general, partitions of unity reduce to this case). The section k is easily seen to be uniquely determined by the operator

$$K : \mathcal{C}^\infty(X; \mathcal{L}_1) \to \mathcal{C}^\infty(X; \mathcal{L}_2)$$

and is called its *kernel*. The linear maps associated in this way to kernels in $\mathcal{C}^\infty(X \times X; \mathcal{L}_2 \boxtimes (\mathcal{L}_1^* \otimes |\wedge|_X))$ are called the *regularizing operators* from $\mathcal{C}^\infty(X; \mathcal{L}_1)$ to $\mathcal{C}^\infty(X; \mathcal{L}_2)$.

A special class of regularizing operators are the *regularizing operators of finite rank*, which are defined by kernels k given by finite sums

$$k(x,y) = \sum_{i=1}^{N} \varphi_i(x)\psi_i(y), \tag{C.1.2}$$

where $\varphi_i \in \mathcal{C}^\infty(X; \mathcal{L}_1)$ and $\psi_i \in \mathcal{C}^\infty(X; \mathcal{L}_2^* \otimes |\wedge|_X)$, and have the form

$$f \mapsto \sum_{i=1}^{N} \left(\int_X \psi_i \cdot f \right) \varphi_i.$$

The Finiteness Theorem B.6.1 and the Serre duality Theorem B.8.1 are simple consequences of the following two propositions:

Proposition C.1.1. *Let $\mathcal{L}$ be a C^∞ line bundle on a compact manifold X and let*

$$K : \mathcal{C}^\infty(X; \mathcal{L}) \to \mathcal{C}^\infty(X; \mathcal{L})$$

be a regularizing operator. Then:

i) $\mathrm{Id} + K$ *is an operator with index from* $\mathcal{C}^\infty(X; \mathcal{L})$ *to itself.*

ii) For any linear map $\lambda : \mathcal{C}^\infty(X;\mathcal{L}) \to \mathbb{C}$ such that $\lambda \circ (\mathrm{Id} + K) = 0$, there exists $\ell \in \mathcal{C}^\infty(X;\mathcal{L}^* \otimes |\wedge|_X)$ such that, for any $f \in \mathcal{C}^\infty(X;\mathcal{L})$,

$$\lambda(f) = \int_X \ell(x) f(x) dx.$$

Proposition C.1.2. *For any holomorphic line bundle $\mathcal{L}$ over a compact Riemann surface X, there exists a linear map*

$$P : \mathcal{C}^\infty(X;\mathcal{L} \otimes \overline{\omega}_X) \to \mathcal{C}^\infty(X;\mathcal{L})$$

and regularizing operators

$$K_1 : \mathcal{C}^\infty(X;\mathcal{L}) \to \mathcal{C}^\infty(X;\mathcal{L})$$

and

$$K_2 : \mathcal{C}^\infty(X;\mathcal{L} \otimes \overline{\omega}_X) \to \mathcal{C}^\infty(X;\mathcal{L} \otimes \overline{\omega}_X)$$

such that

$$P \circ \overline{\partial}_{\mathcal{L}} = \mathrm{Id} + K_1$$

and

$$\overline{\partial}_{\mathcal{L}} \circ P = \mathrm{Id} + K_2.$$

Proposition C.1.1 is a variant of Fredholm's theory. The operator P, whose existence is asserted in Proposition C.1.2, is traditionally called, after Hilbert, a *parametrix* of $\overline{\partial}_{\mathcal{L}}$.

Proof of Theorems B.6.1 and B.8.1 (taking Propositions C.1.1 and C.1.2 for granted):

• $\ker \overline{\partial}_{\mathcal{L}}$ is clearly contained in

$$\ker P \circ \overline{\partial}_{\mathcal{L}} = \ker(\mathrm{Id} + K_1).$$

This space is finite dimensional according to Proposition C.1.1 i), applied to K_1. Hence $H^0(X;\mathcal{L})$ is finite dimensional.

• $\overline{\partial}_{\mathcal{L}}(\mathcal{C}^\infty(X;\mathcal{L}))$ clearly contains

$$\overline{\partial}_{\mathcal{L}} \circ P(\mathcal{C}^\infty(X;\mathcal{L} \otimes \overline{\omega}_X)) = (\mathrm{Id} + K_2)(\mathcal{C}^\infty(X;\mathcal{L} \otimes \overline{\omega}_X)).$$

This space has finite codimension in $\mathcal{C}^\infty(X;\mathcal{L} \otimes \overline{\omega}_X)$ according to Proposition C.1.1 i) applied to $K = K_2$. Hence the same is true for $\overline{\partial}_{\mathcal{L}}(\mathcal{C}^\infty(X;\mathcal{L}))$, *i.e.*, $H^1(X;\mathcal{L})$ is finite dimensional.

This proves Theorem B.6.1.

• Consider an element λ in $H^1(X;\mathcal{L})^*$ *i.e.*, a linear map

$$\lambda : \mathcal{C}^\infty(X;\mathcal{L} \otimes \overline{\omega}_X) \to \mathbb{C}$$

which vanishes on the image of $\overline{\partial}_{\mathcal{L}}$. Then λ vanishes on the image of $\overline{\partial}_{\mathcal{L}} \circ P = \mathrm{Id} + K_2$. Therefore, according to Proposition C.1.1 ii) applied to $K = K_2$, there exists $\ell \in \mathcal{C}^\infty(X; (\mathcal{L} \otimes \overline{\omega}_X)^* \otimes |\wedge|_X)$ such that, for any $f \in \mathcal{C}^\infty(X; \mathcal{L} \otimes \overline{\omega}_X)$

$$\lambda(f) = \int_{x \in X} \ell(x) f(x). \tag{C.1.3}$$

The section ℓ may be seen as a C^∞ section of $\mathcal{L}^* \otimes \omega_X$. Indeed,

$$(\mathcal{L} \otimes \overline{\omega}_X)^* \otimes |\wedge|_X \simeq \mathcal{L}^* \otimes \overline{\omega}_X^* \otimes |\wedge|_X$$

and

$$\overline{\omega}_X^* \otimes |\wedge|_X \simeq \omega_X,$$

since (*cf.* §B.1.9 and (B.2.1))

$$|\wedge|_X \simeq \wedge_X^2 \simeq \omega_X \otimes \overline{\omega}_X.$$

The relation (C.1.3) uniquely determines ℓ, and the map $\lambda \mapsto \ell$ so defined is an isomorphism from $H^1(X; \mathcal{L})^*$ onto the subspace of $\mathcal{C}^\infty(X; \mathcal{L} \otimes \overline{\omega}_X)$ formed by the sections ℓ such that

$$\forall\, s \in \mathcal{C}^\infty(X; \mathcal{L}) \, , \int_X \ell \cdot \overline{\partial} s = 0.$$

This subspace is nothing else than $H^0(X; \mathcal{L}^* \otimes \omega_X)$, as follows from the following

Lemma C.1.3. *For any $s \in \mathcal{C}^\infty(X; \mathcal{L})$ and any $\ell \in \mathcal{C}^\infty(X; \mathcal{L}^* \otimes \omega_X)$, we have*

$$\int_X \ell \cdot \overline{\partial}_{\mathcal{L}}\, s = - \int_X \overline{\partial}_{\mathcal{L}^* \otimes \omega_X}\, \ell \cdot s.$$

Indeed

$$\begin{aligned}
\int_X \ell \cdot \overline{\partial}_{\mathcal{L}}\, s + \int_X \overline{\partial}_{\mathcal{L}^* \otimes \omega_X}\, \ell \cdot s &= \int_X \overline{\partial}(\ell \cdot s) && (\textit{cf.}\ (\text{B.8.1})) \\
&= \int_X d(\ell \cdot s) && (\textit{cf.}\ (\text{B.7.1})) \\
&= 0 && (\text{Stokes formula}).
\end{aligned}$$

This proves Theorem B.8.1, since the bijection

$$\begin{aligned} H^0(X; \mathcal{L}^* \otimes \omega_X) &\to H^1(X; \mathcal{L})^* \\ \ell &\mapsto \lambda \end{aligned}$$

coincides (up to a factor $2\pi i$) with the linear map $H^0(X; \mathcal{L}^* \otimes \omega_X) \to H^1(X; \mathcal{L})^*$ defined by the bilinear pairing $\langle \cdot, \cdot \rangle$, which therefore is non-degenerate.

C.2. Fredholm's theory

This Section is devoted to the proof of Proposition C.1.1. We will indicate the main lines of the proof and leave (easy) details to the reader.

We begin with a few observations:

• If K_1 and K_2 are two regularizing operators from $\mathcal{C}^\infty(X;\mathcal{L})$ to itself, with kernel k_1 and k_2, then $K_1 \circ K_2$ is a regularizing operator, whose kernel is

$$k_1 * k_2 : (x,y) \mapsto \int_{z\in X} k_1(x,z)k_2(z,y). \tag{C.2.1}$$

In particular, if K_1 or K_2 is a regularizing operator of finite rank, then $K_1 \circ K_2$ is also such an operator.

• Choose C^∞ Hermitian metrics $\| \ \|_{\mathcal{L}}$ and $\| \ \|_{|\wedge|_X}$ on the line bundles $\mathcal{L}$ and $|\wedge|_X$. By duality and tensor product, these metrics define a Hermitian metric $\| \cdot \|$ on $\mathcal{L} \boxtimes (\mathcal{L}^* \otimes |\wedge|_X)$. Using $\| \ \|$, we can define a norm $|\!|\!| \ |\!|\!|$ on the space $\mathcal{C}^0(X\times X;\mathcal{L}\boxtimes(\mathcal{L}^*\otimes|\wedge|_X))$ of continuous kernels as

$$|\!|\!|k|\!|\!| = \sup_{(x,y)\in X\times X} \| k(x,y) \| .$$

Equipped with this norm, $\mathcal{C}^0(X\times X;\mathcal{L}\boxtimes(\mathcal{L}^*\otimes|\wedge|_X))$ is a Banach space. Moreover, one easily checks that the composition of kernels (C.2.1) is still well defined for continuous kernels and that the following estimate holds, for any two continuous kernels k_1 and k_2:

$$|\!|\!|k_1 * k_2|\!|\!| \leq M|\!|\!|k_1|\!|\!| \ |\!|\!|k_2|\!|\!|, \tag{C.2.2}$$

where M is the integral over X of the density μ defined locally on X by

$$\mu = \frac{|dx_1 \wedge \cdots \wedge dx_n|}{\| dx_1 \wedge \cdots \wedge dx_n \|_{|\wedge|_X}} .$$

The proof of Proposition C.1.1 is based on three preliminary lemmas.

Lemma C.2.1. *The kernels defining the regularizing operators of finite rank* (i.e., *the kernels of the form* (C.1.2)) *are dense in* $\mathcal{C}^\infty(X\times X,\mathcal{L}\boxtimes(\mathcal{L}^*\otimes|\wedge|_X))$ *equipped with the norm* $|\!|\!| \ |\!|\!|$.

Using partitions of unity, charts and local trivializations of $\mathcal{L}$, this follows from the fact that, if D_1 and D_2 are two (closed) balls in $\mathbb{R}^n$, any C^∞ function on $D_1 \times D_2$ may be uniformly approximated by functions of the form

$$(x,y) \mapsto \sum_{i=1}^{N} f_i(x)g_i(y)$$

where $f_i \in \mathcal{C}^\infty(D_1)$ and $g_i \in \mathcal{C}^\infty(D_2)$. This fact may be deduced, for instance, from the existence of uniform polynomial approximations.

Lemma C.2.2. *Let $K : \mathcal{C}^\infty(X;\mathcal{L}) \to \mathcal{C}^\infty(X;\mathcal{L})$ be a regularizing operator whose kernel k satisfies*

(C.2.3)
$$|||k||| < \frac{1}{M}\,.$$

Then $\mathrm{Id} + K$ *is invertible and there exists a regularizing operator*

$$L : \mathcal{C}^\infty(X;\mathcal{L}) \to \mathcal{C}^\infty(X;\mathcal{L})$$

such that $(\mathrm{Id} + K)^{-1} = \mathrm{Id} + L$.

Let us define

$$k_n = k * \cdots * k \qquad (n \text{ factors}).$$

Then it follows from (C.2.2) that

$$M|||k_n||| \le (M|||k|||)^n$$

and from (C.2.3) that

$$\sum_{n=1}^{\infty} |||k_n||| < +\infty.$$

Therefore

$$\ell = \sum_{n=1}^{\infty} (-1)^n\, k_n$$

is a well defined element of $\mathcal{C}^0(X \times X; \mathcal{L} \boxtimes (\mathcal{L}^* \otimes |\wedge|_X))$. Moreover, the continuity of the composition product $*$ with respect to the norm $|||\ |||$ (*cf.* (C.2.2)) implies that

(C.2.4)
$$\begin{aligned} k + \ell * k &= k + \sum_{n=1}^{\infty} (-1)^n\, k_n * k \\ &= -\sum_{n=1}^{\infty} (-1)^n\, k_n \\ &= -\ell \end{aligned}$$

and similarly that

(C.2.5)
$$k + k * \ell = -\ell.$$

Inserting (C.2.4) in (C.2.5), we get:

$$\ell = -k + k * k + k * \ell * k\,.$$

On the other hand, the formula

$$k * \ell * k(x,y) = \int_{(z_1,z_2)\in X\times X} k(x,z_1)\ell(z_1,z_2)k(z_2,z_1)$$

shows that $k * \ell * k$ is C^∞. Therefore, ℓ is C^∞.

If L denotes the operator of kernel ℓ, the relations (C.2.4) and (C.2.5) are equivalent to

$$K + L + LK = 0 \quad \text{and} \quad K + L + KL = 0$$

that is to

$$(\mathrm{Id} + L)(\mathrm{Id} + K) = (\mathrm{Id} + K)(\mathrm{Id} + L) = \mathrm{Id}.$$

This proves Lemma C.2.2.

Lemma C.2.3. *Proposition* C.1.1 *is true when K is a regularizing operator of finite rank.*

Indeed, if $\varphi_i \in \mathcal{C}^\infty(X;\mathcal{L})$ and $\psi_i \in \mathcal{C}^\infty(X;\mathcal{L}^* \otimes |\wedge|_X)$, $i = 1, \ldots, N$, are such that the kernel of K is $(x,y) \mapsto \sum_{i=1}^{N} \varphi_i(x)\psi_i(y)$, then

$$\ker(\mathrm{Id} + K) = \{f \in \mathcal{C}^\infty(X;\mathcal{L}) \mid f = -Kf\}$$

is contained in the vector space spanned by the φ_i's, and $(\mathrm{Id} + K)(\mathcal{C}^\infty(X;\mathcal{L}))$ contains $\ker K$ and *a fortiori* the subspace of $\mathcal{C}^\infty(X;\mathcal{L})$ defined by the vanishing of the linear forms

$$f \mapsto \int_X \psi_i \cdot f \quad , \quad i = 1, \ldots, N.$$

This clearly shows that $\mathrm{Id} + K$ has an index. Moreover, if a linear form $\lambda : \mathcal{C}^\infty(X;\mathcal{L}) \to \mathbb{C}$ vanishes on the image of $\mathrm{Id} + K$, then

$$\begin{aligned} \lambda(f) &= -\lambda \circ K(f) \\ &= -\sum_{i=1}^{N} \lambda(\varphi_i) \int_X \psi_i \cdot f \\ &= \int_X \ell \cdot f \end{aligned}$$

where

$$\ell = -\sum_{i=1}^{N} \lambda(\varphi_i)\psi_i.$$

Finally, we can prove Proposition C.1.1.

Let $K : \mathcal{C}^\infty(X;\mathcal{L}) \to \mathcal{C}^\infty(X;\mathcal{L})$ be a regularizing operator. According to Lemma C.2.1, we can write $K = K_1 + K_2$ where K_2 is regularizing of finite rank and where $|||K_1||| < \frac{1}{M}$. Then Lemma C.2.2 shows the existence of a regularizing operator L such that

$$(\mathrm{Id} + K_1)^{-1} = \mathrm{Id} + L.$$

Now we get:

$$\begin{aligned}\mathrm{Id} + K &= (\mathrm{Id} + K_1 + K_2)(\mathrm{Id} + L)(\mathrm{Id} + K_1) \\ &= (\mathrm{Id} + K_2 + K_2 L)(\mathrm{Id} + K_1).\end{aligned} \tag{C.2.6}$$

According to Lemma C.2.3 applied to $K_2 + K_2 L$, which is regularizing of finite rank,

$$\ker(\mathrm{Id} + K) = (\mathrm{Id} + K_1)^{-1} (\ker(\mathrm{Id} + K_2 + K_2 L))$$

is therefore finite dimensional and

$$\mathrm{im}(\mathrm{Id} + K) = \mathrm{im}(\mathrm{Id} + K_2 + K_2 L)$$

has finite codimension in $\mathcal{C}^\infty(X; \mathcal{L})$. Finally, if λ is a linear form such that $\lambda \circ (\mathrm{Id} + K) = 0$, then (C.2.6) shows that

$$\lambda \circ (\mathrm{Id} + K_2 + K_2 L) = 0$$

and, again by Lemma C.2.3, λ is of the required form.

C.3. A parametrix for $\overline{\partial}_{\mathcal{L}}$

This Section is devoted to the proof of Proposition C.1.2. Our proof is based on the fact that an 'inverse' of $\overline{\partial}$ acting on C^∞ functions with compact supports on $\mathbb{C}$ may be given by an explicit formula:

Lemma C.3.1. i) *For any $\varphi \in \mathcal{C}_c^\infty(\mathbb{C})$ and any $z \in \mathbb{C}$, we have*

$$\varphi(z) = \int_{z' \in \mathbb{C}} \frac{dz'}{2\pi i (z' - z)} \wedge \overline{\partial}\varphi(z').$$

ii) *For any $\sigma \in \mathcal{C}_c^\infty(\mathbb{C}; \overline{\omega}_{\mathbb{C}})$ and any $z \in \mathbb{C}$, we have*

$$\sigma(z) = \overline{\partial}\left(\int_{z' \in \mathbb{C}} \frac{dz'}{2\pi i (z' - z)} \wedge \sigma(z')\right).$$

These integrals make sense since $\frac{1}{z}$ is integrable near 0. Moreover, if we make the change of variables $z'' = z' - z$ and if we write $\sigma(z) = \varphi(z) d\overline{z}$, we obtain that these integrals are C^∞ as functions of z and that i) and ii) are equivalent.

Let us prove i). We can assume that $z = 0$, by replacing φ by $w \mapsto \varphi(z + w)$. Thus we only have to prove that

$$\varphi(0) = \int_{\mathbb{C}} \frac{dz}{2\pi i z} \wedge \overline{\partial}\varphi. \tag{C.3.1}$$

Let D_ε be the disc in $\mathbb{C}$ of center 0 and radius ε. On $\mathbb{C} - D_\varepsilon$, we have

$$\begin{aligned} d\left(\varphi \frac{dz}{2\pi i z}\right) &= d\varphi \wedge \frac{dz}{2\pi i z} && \left(\tfrac{dz}{z} \text{ is closed}\right) \\ &= \overline{\partial}\varphi \wedge \frac{dz}{2\pi i z} && \left(\partial\varphi \wedge dz = \frac{\partial\varphi}{\partial z}\, dz \wedge dz = 0\right). \end{aligned}$$

Therefore, Stokes formula gives:

$$\int_{\mathbb{C}-D_\varepsilon} \overline{\partial}\varphi \wedge \frac{dz}{2\pi i z} = -\int_{\partial D_\varepsilon} \varphi \frac{dz}{2\pi i z} \tag{C.3.2}$$

(the minus sign comes from the fact that the boundary of $\mathbb{C}-D_\varepsilon$ is the boundary of D_ε with reversed orientation). The relation (C.3.1) follows from (C.3.2) by taking the limit $\varepsilon \to 0$.

Lemma C.3.1 ii), has the following straightforward consequence, which will be of use later:

Lemma C.3.2. *Let $\mathcal{L}$ be a holomorphic line bundle over a Riemann surface X. For any $P \in X$ and any C^∞ section t of $\mathcal{L} \otimes \overline{\omega}_X$ over an open neighbourhood U of P in X, there exists $s \in \mathcal{C}^\infty(X;\mathcal{L})$ such that*

$$\overline{\partial}_{\mathcal{L}}\, s = t$$

on a neighbourhood of P in U.

The next lemma shows that a kernel on $X \times X$ which has the same singularity near the diagonal $\Delta_X := \{(x,x), x \in X\}$ as the kernel $dz'/2\pi i(z-z')$ of Lemma C.3.1 provides the required parametrix.

Lemma C.3.3. *Let p be an element of $C^\infty(X \times X - \Delta_X; \mathcal{L} \boxtimes (\mathcal{L}^* \otimes \omega_X))$ which satisfies the following condition: for any $x \in X$, there exists a local holomorphic coordinate z and a holomorphic trivialization t of $\mathcal{L}$ defined on a neighbourhood U of x such that the section of $\mathcal{L} \boxtimes (\mathcal{L}^* \otimes \omega_X)$ over $U \times U - \Delta_U$ defined by*

$$(x_1, x_2) \mapsto p(x_1,x_2) - \frac{1}{2\pi i}\frac{t \boxtimes t^{-1} dz}{z(x_2)-z(x_1)}$$

extends to a C^∞ section over $U \times U$.

Then, for any $f \in \mathcal{C}^\infty(X;\mathcal{L})$ and any $x \in X$, the integral

$$Pf(x) := \int_{y\in X} p(x,y) f(y)$$

is convergent and defines a C^∞ section Pf of $\mathcal{L}$ over X.

Moreover, the section k_1 (resp. k_2) of

$$\mathcal{L} \boxtimes (\mathcal{L}^* \otimes \omega_X \otimes \overline{\omega}_X) \simeq \mathcal{L} \boxtimes (\mathcal{L}^* \otimes |\wedge|_X)$$

(resp. of

$$(\mathcal{L} \otimes \overline{\omega}_X) \boxtimes (\mathcal{L}^* \otimes \omega_X) \simeq (\mathcal{L} \otimes \overline{\omega}_X) \boxtimes ((\mathcal{L} \otimes \overline{\omega}_X)^* \otimes |\wedge|_X))$$

over $X \times X - \Delta_X$ defined as the image of ℓ by $-\overline{\partial}_{\mathcal{L}^* \otimes \omega_X}$ (resp. by $\overline{\partial}_{\mathcal{L}}$) acting on the second (resp. on the first) variable extends to a C^∞ section over $X \times X$. If

$$K_1 : \mathcal{C}^\infty(X;\mathcal{L}) \to \mathcal{C}^\infty(X;\mathcal{L})$$

(resp.

$$K_2 : \mathcal{C}^\infty(X;\mathcal{L}\otimes\overline{\omega}_X) \to \mathcal{C}^\infty(X;\mathcal{L}\otimes\overline{\omega}_X))$$

denotes the regularizing operator with kernel k_1 (resp. k_2), then we have

$$P \circ \overline{\partial}_{\mathcal{L}} = \mathrm{Id} + K_1$$

and

$$\overline{\partial}_{\mathcal{L}} \circ P = \mathrm{Id} + K_2.$$

In particular, P is a parametrix of $\overline{\partial}_{\mathcal{L}}$.

The assertions concerning the smoothness of ℓ_1 and ℓ_2 near the diagonal immediately follow from the hypothesis on p, since

$$\frac{\partial}{\partial\,\overline{z}_1}\,\frac{1}{z_1 - z_2} = \frac{\partial}{\partial\,\overline{z}_2}\,\frac{1}{z_1 - z_2} = 0.$$

Consider an element σ in $\mathcal{C}^\infty(X;\mathcal{L}\otimes\overline{\omega}_X)$. The relation

$$\overline{\partial}_{\mathcal{L}}(P\sigma) = \sigma + K_2\ (\sigma)$$

follows from the hypothesis on p and from Lemma C.3.1, ii) if f has sufficiently small support. Partitions of unity reduce the general case to this one. The relation

$$P\left(\overline{\partial}_{\mathcal{L}}\ f\right) = f + K_1 f$$

for $f \in \mathcal{C}^\infty(X;\mathcal{L})$ is proved in the same way by using Lemma C.3.1, i) and Lemma C.1.3.

Finally, to prove Proposition C.1.2, it is enough to prove the existence of

$$p \in \mathcal{C}^\infty\left(X \times X - \Delta_X; \mathcal{L} \boxtimes (\mathcal{L}^* \otimes \omega_X)\right)$$

satisfying the hypotheses of Lemma C.3.3.

Let $((U_i, z_i))_{1\leq i\leq N}$ be a family of holomorphic charts on X which cover X and such that, over each U_i, $\mathcal{L}$ possesses a non vanishing holomorphic section t_i. Choose functions $\varphi_i \in \mathcal{C}^\infty(X;\mathbb{C})$ such that φ_i vanishes outside a compact subset of $U_i \times U_i$ and such that

$$\sum_{i=1}^{N} \varphi_i = 1$$

on a neighbourhood of Δ_X in $X \times X$ [23], and define sections $\sigma_i \in \mathcal{C}^\infty(X \times X - \Delta_X; \mathcal{L}\otimes\ (\mathcal{L}^* \otimes \omega_X))$ by

[23] In other words, $\left(\varphi_1, \ldots, \varphi_i, \ldots, \varphi_N, 1 - \sum_{i=1}^N \varphi_i\right)$ is a partition of unity associated to the covering $(U_1 \times U_1, \ldots, U_i \times U_i, \ldots, U_N \times U_N, X \times X - \Delta_X)$ of $X \times X$.

$$\sigma_i(x_1,x_2) = \frac{1}{2\pi i}\,\varphi_i(x_1,x_2)\,\frac{t_i \boxtimes t_i^{-1}dz_i}{z_i(x_1)-z_i(x_2)} \qquad \text{if } (x_1,x_2)\in U_i\times U_i$$

$$= 0 \qquad \text{if } (x_1,x_2)\notin U_i\times U_i.$$

Then $p = \sum_{i=1}^{N} \sigma_i$ satisfies the hypotheses of Lemma C.3.3. This follows from the fact that the section of $\mathcal{L} \boxtimes (\mathcal{L}^* \otimes \omega_X)$ over $(U_i\times U_i)\cap(U_j\times U_j)-\Delta_X$ defined by

$$(x_1,x_2)\mapsto \frac{t_i \boxtimes t_i^{-1}dz_i}{z_i(x_1)-z_i(x_2)} - \frac{t_j \boxtimes t_j^{-1}dz_j}{z_j(x_1)-z_j(x_2)}$$

extends to a C^∞ (in fact holomorphic) section over $(U_i\times U_i)\cap(U_j\times U_j)$, as a consequence of the following elementary lemma, the proof of which we leave to the reader:

Lemma C.3.4. *Let U and V be two open subsets of $\mathbb{C}$, let $\varphi : U \to V$ be a biholomorphic map and let ℓ be a non-vanishing holomorphic function on U. Then the holomorphic function on $U\times U-\Delta_U$*

$$(x_1,x_2)\mapsto \frac{1}{x_1-x_2} - \frac{\ell(x_1)\ell(x_2)^{-1}\varphi'(x_2)}{\varphi(x_1)-\varphi(x_2)}$$

extends holomorphically over $U\times U$.

C.4. Proof of Theorem B.7.1

The fact that $I_\mathcal{L}$ is well defined and injective follows from easy computations. For instance, let us sketch a proof of the injectivity. With the notations of the definition of $I_\mathcal{L}$, if

$$\sum_{i=1}^{n} f_i\cdot\overline{\partial}\rho_i = \overline{\partial}t$$

for some $t\in C^\infty(X;\mathcal{L})$, then the expression $\sum_{i=1}^{n}\rho_i f_i - t$ defines an element f of $C^\infty(X-\{P_1,\ldots,P_n\};\mathcal{L})$ on which $\overline{\partial}_\mathcal{L}$ vanishes - hence f is holomorphic on $X-\{P_1,\ldots,P_n\}$ - and which has the same singularity at P_i as f_i. Therefore, f is an element of $\mathcal{M}(X,\mathcal{L})$ whose image by $P_\mathcal{L}$ is $(p_x)_{x\in X}$.

The proof of the surjectivity of $I_\mathcal{L}$ is based on the following fact: let P be any point of X; then, *if n is a large enough positive integer*

(C.4.1) $$H^1(X;\mathcal{L}\otimes\mathcal{O}(nP))=0.$$

Indeed (C.4.1) holds as soon as

$$n > 2g-2-\deg_X\mathcal{L},$$

as follows from Serre duality Theorem B.8.1, from the relation

$$\deg_X\left((\mathcal{L}\otimes\mathcal{O}(nP))^*\otimes\omega_X\right)=2g-2-n-\deg_X\mathcal{L}\,,$$

and from Corollary B.4.4.

Consider now an element s of $\mathcal{C}^\infty(X;\mathcal{L}\otimes\overline{\omega}_X)$. It may be seen as an element of $\mathcal{C}^\infty(X;\mathcal{L}\otimes\mathcal{O}(nP)\otimes\overline{\omega}_X)$. Therefore, according to (C.4.1), there exists $t\in\mathcal{C}^\infty(X;\mathcal{L}\otimes\mathcal{O}(nP))$ such that

$$s=\overline{\partial}_{\mathcal{L}\otimes\mathcal{O}(nP)}\,t.$$

On $X-\{P\}$, t may be seen as a C^∞ section of $\mathcal{L}$ such that

$$s=\overline{\partial}_{\mathcal{L}}\,t. \tag{C.4.2}$$

On the other hand, Lemma C.3.2 shows that there exists an open neighbourhood U of P in X and $u\in\mathcal{C}^\infty(X;\mathcal{L})$ such that, on U

$$s=\overline{\partial}_{\mathcal{L}}\,u. \tag{C.4.3}$$

Let ρ be an element of $\mathcal{C}^\infty(X;\mathbb{C})$ such that $\rho\equiv 1$ on a neighbourhood of P and ρ vanishes outside a compact subset of U. From (C.4.2) and (C.4.3) it follows that, on $X-\{P\}$,

$$s=\overline{\partial}\rho\cdot(t-u)+\overline{\partial}_{\mathcal{L}}\left[\rho u+(1-\rho)t\right]$$

and that $t-u$ defines a meromorphic section of $\mathcal{L}$ over U. Therefore the class of s in $H^1(X;\mathcal{L})$ coincides with the image by $I_{\mathcal{L}}$ of the class of the element $(p_x)_{x\in X}\in\prod_{x\in X}\mathcal{P}(x;\mathcal{L})$ defined by:

$$\begin{aligned} p_x&=0 \quad \text{if} \quad x\neq P;\\ p_P&=t-u. \end{aligned}$$

This proves the surjectivity of $I_{\mathcal{L}}$.

C.5. The operator $\partial\overline{\partial}$

In this Section, we show how the methods used in §C.1-3 to prove Theorems B.6.1 (Finiteness) and B.8.1 (Serre duality) also allow to prove very easily the basic properties of the Laplace operator on a compact connected Riemann surface X, in particular that its image is the space of functions whose integral on X vanishes. The reason for including these results is their importance in the classical approach to the analysis on compact Riemann surfaces. We will use them only to give another proof of Lemma C.5.4, which is implied by Corollary B.8.2, 1).

Let X be any Riemann surface. We define, for any $f\in\mathcal{C}^\infty(X;\mathbb{C})$ and any $\alpha\in\mathcal{C}^\infty(X;\overline{\omega}_X)$:

$$\partial f:=(df)^{(1,0)}$$

and

$$\partial\alpha := d\alpha.$$

Therefore, for any $f \in C^\infty(X;\mathbb{C})$

$$df = \partial f + \overline{\partial} f$$

and

$$\begin{aligned} \partial\overline{\partial} f &= d\,(\overline{\partial} f) \\ &= -d(\partial f) \quad \text{(since } d(df) = 0\text{)} \\ &= -\overline{\partial}_{\omega_X} \partial f \quad \text{(since } \overline{\partial}_{\omega_X} = d \text{ on } C^\infty(X;\omega_X); cf. \text{ §B.7).} \end{aligned}$$

The following identities, true for any $f \in C^\infty(X;\mathbb{C})$ and any $\alpha \in C^\infty(X;\overline{\omega}_X)$, are also easily checked:

$$\overline{\partial f} = \overline{\partial}\,\overline{f} \quad \text{and} \quad \overline{\partial\alpha} = \overline{\partial}\,\overline{\alpha}.$$

The operator

$$\partial\overline{\partial} : C^\infty(X;\mathbb{C}) \to C^\infty(X;\wedge^2)$$

is compatible with holomorphic changes of variables (compare (B.5.4)) and has the following expression in terms of a local holomorphic coordinate $z = x + iy$:

$$\partial\overline{\partial}\varphi = -\frac{i}{2}\left(\frac{\partial^2\varphi}{\partial x^2} + \frac{\partial^2\varphi}{\partial y^2}\right) dx \wedge dy.$$

More generally, if X is endowed with a C^∞ Riemannian metric compatible with its complex structure (*cf.* §I.2.1) and if Δ and dA respectively denote the (positive) Laplace operator and the area 2-form associated with this metric, we have

$$\partial\overline{\partial}\varphi = \frac{i}{2}\Delta\varphi\, dA. \tag{C.5.1}$$

This identity shows that a function φ is harmonic with respect to this Riemannian metric (*i.e.*, $\Delta\varphi = 0$) iff $\partial\overline{\partial}\varphi = 0$. In particular, the harmonicity of a function depends only on the complex structure of X. The identity (C.5.1) proves the conformal invariance of the *Dirichlet functional*, namely

$$\frac{1}{2}\int_X \overline{\varphi}\cdot\Delta\varphi\, dA = -i\int_X \overline{\varphi}\cdot\partial\overline{\partial}\varphi.$$

Our proof of Hodge decomposition (Theorem B.2.5) will be based on the following result:

Proposition C.5.1. *Let X be a compact connected Riemann surface. The kernel of the operator*

$$\partial\overline{\partial} : C^\infty(X;\mathbb{C}) \to C^\infty(X;\wedge^2_X)$$

is the subspace of dimension 1 of $C^\infty(X;\mathbb{C})$ formed by the constant functions. Its image is the subspace of codimension 1 in $C^\infty(X;\wedge^2_X)$ formed by the 2-forms σ such that $\int_X \sigma = 0$.

Thanks to the identity (C.5.1), this proposition may be seen as a statement concerning the Laplace operator acting on $\mathcal{C}^\infty(X;\mathbb{C})$: it asserts that the kernel (resp. the image) of Δ is the space of constant functions (resp. the subspace of $\mathcal{C}^\infty(X;\mathbb{C})$ formed by the functions φ such that $\int_X \varphi \, dA = 0$).

The fact that a function $\varphi \in \mathcal{C}^\infty(X;\mathbb{C})$ such that $\partial\overline{\partial}\varphi = 0$ is constant follows for instance from the following expressions for the Dirichlet functional:

$$\begin{aligned}-i\int \overline{\varphi}\, \partial\overline{\partial}\varphi &= i\int \partial\varphi \wedge \overline{\partial\varphi} \\ &= i\int \overline{\overline{\partial}\varphi} \wedge \overline{\partial}\varphi,\end{aligned}$$

which are easily deduced by integration by part and which show that the Dirichlet functional is positive unless $\partial\varphi = \overline{\partial}\varphi = 0$, *i.e.*, unless φ is constant.

To determine the image of $\partial\overline{\partial}$, we begin by proving the following preliminary result:

Lemma C.5.2. *The image of $\partial\overline{\partial} : \mathcal{C}^\infty(X;\mathbb{C}) \to \mathcal{C}^\infty(X;\wedge^2_X)$ is a subspace of finite codimension in $\mathcal{C}^\infty(X;\wedge^2_X)$. Moreover, any linear form $\lambda : \mathcal{C}^\infty(X;\wedge^2_X) \to \mathbb{C}$ which vanishes on this subspace may be written*

$$\lambda : \sigma \mapsto \int_X \ell \cdot \sigma$$

for some $\ell \in \mathcal{C}^\infty(X;\mathbb{C})$.

This lemma is a consequence of the existence of a parametrix P for $\partial\overline{\partial}$, *i.e.*, of a linear map

$$P : \mathcal{C}^\infty(X;\wedge^2_X) \to \mathcal{C}^\infty(X;\mathbb{C})$$

such that

$$P \circ \partial\overline{\partial} - \mathrm{Id} : \mathcal{C}^\infty(X;\mathbb{C}) \to \mathcal{C}^\infty(X;\mathbb{C})$$

and

$$\partial\overline{\partial} \circ P - \mathrm{Id} : \mathcal{C}^\infty(X;\wedge^2_X) \to \mathcal{C}^\infty(X;\wedge^2_X)$$

are regularizing operators (use Proposition C.1.1 as in the proof of Theorems B.6.1 and B.8.1 in §C.1). In §C.3, a parametrix for $\overline{\partial}_{\mathcal{L}}$ was constructed starting from the 'local parametrix' $dz_2/2\pi i(z_2 - z_1)$. Similarly, a parametrix for $\partial\overline{\partial}$ is easily constructed from the 'local parametrix' $-\frac{1}{2\pi i}\log|z_1 - z_2|^2$. Indeed, we have:

Lemma C.5.3. i) *For any $\varphi \in \mathcal{C}^\infty(\mathbb{C};\mathbb{C})$ with compact support and any $z \in \mathbb{C}$, we have*

$$\varphi(z) = -\frac{1}{2\pi i}\int_{z'\in\mathbb{C}} \log|z - z'|^2 \, \partial\overline{\partial}\varphi(z').$$

ii) *For any $\sigma \in \mathcal{C}^\infty(\mathbb{C};\wedge^2_{\mathbb{C}})$ with compact support and any $z \in \mathbb{C}$, we have*

$$\sigma(z) = \partial\overline{\partial}\left[-\frac{1}{2\pi i}\int_{z'\in\mathbb{C}} \log|z-z'|^2\ \sigma(z')\right].$$

This lemma, which may be easily deduced from Lemma C.3.1 by integration by part, is a reformulation of the well known fact that the fundamental solution of the operator $\frac{\partial^2}{\partial x^2} + \frac{\partial^2}{\partial y^2}$ is $\frac{1}{2\pi}\log\sqrt{x^2+y^2}$.

We leave the details of the construction of the parametrix P to the reader, who will have no difficulty to establish statements concerning $\partial\overline{\partial}$ analogous (but simpler) to Lemma C.3.3 and Lemma C.3.4.

Let us complete the proof of Proposition C.5.1. According to Lemma C.5.2, the image of $\partial\overline{\partial}$ is the intersection of the kernels of the linear forms λ on $\mathcal{C}^\infty(X;\wedge^2_X)$ which may be written as

$$\lambda : \alpha \mapsto \int_X \ell\cdot\alpha$$

for some $\ell \in \mathcal{C}^\infty(X;\mathbb{C})$ and which satisfy the condition

$$\lambda\circ\partial\overline{\partial} = 0.$$

The identity

$$\int_X \ell\cdot\partial\overline{\partial}\alpha = \int_X \partial\overline{\partial}\ell\cdot\alpha,$$

which is easily proved by two integrations by parts, and the fact that the kernel of $\partial\overline{\partial}$ is the space of constant functions show that these linear forms are proportional to $\alpha\mapsto\int_X\alpha$. Therefore, the image of $\partial\overline{\partial}$ is the kernel of the linear form $\alpha\to\int_X\alpha$, as was to be proved.

The following consequence of Proposition C.5.1 is known as *Weyl's lemma:*

Lemma C.5.4. *For any $\beta\in\mathcal{C}^\infty(X;\overline{\omega}_X)$, there exists $f\in\mathcal{C}^\infty(X;\mathbb{C})$ such that $\beta-\overline{\partial}f$ is a closed form.*

Indeed, by Stokes formula

$$\int_X d\beta = 0.$$

Therefore, according to Proposition C.5.1, there exists $f\in\mathcal{C}^\infty(X;\mathbb{C})$ such that

$$\partial\overline{\partial}f = d\beta.$$

Then we get

$$\begin{aligned} d\left(\beta-\overline{\partial}f\right) &= d\beta - d\overline{\partial}f \\ &= d\beta - \partial\overline{\partial}f \\ &= 0. \end{aligned}$$

Finally we can prove Theorem B.2.5:

• The map i is injective: Consider $\omega \in \Omega^1(X)$ and $\omega' \in \overline{\Omega^1(X)}$ such that $\omega \oplus \omega'$ belong to $\ker i$. By definition of $H^1_{DR}(X;\mathbb{C})$, there exists $f \in C^\infty(X;\mathbb{C})$ such that

$$\omega + \omega' = df.$$

This implies

$$\omega = \partial f \tag{C.5.2}$$

$$\omega' = \overline{\partial} f \tag{C.5.3}$$

and

$$\overline{\partial}\partial f = \overline{\partial}\omega = 0.$$

According to Proposition C.5.1, it follows that f is constant. Then (C.5.2) and (C.5.3) show that $\omega = 0$ and $\omega' = 0$.

• The map i is surjective: Let α be a closed C^∞ complex 1-form on X. According to Lemma C.5.4 there exists $f \in C^\infty(X;\mathbb{C})$ such that $\alpha^{(0,1)} - \overline{\partial} f$ is closed. Then $\alpha' = \alpha - df$ is a closed 1-form on X which has the same class in $H^1_{DR}(X;\mathbb{C})$ as α. Moreover

$$\alpha'^{(0,1)} = \alpha^{(0,1)} - \overline{\partial} f$$

and

$$\alpha'^{(1,0)} = \alpha' - \alpha^{(1,0)}$$

are closed forms of type $(0,1)$ and $(1,0)$, hence antiholomorphic and holomorphic (*cf.* Proposition B.2.1). Therefore $\alpha'^{(1,0)} \oplus \alpha'^{(0,1)}$ is an element of $\Omega^1(X) \oplus \overline{\Omega^1(X)}$ which i sends to the cohomology class of α.

II. The Jacobian variety of a compact Riemann surface

II.1. The inversion problem—or the Jacobian according to Jacobi

II.1.1. The inversion problem for hyperelliptic integrals.

In this Section, we try to explain how mathematicians of the beginning of the XIX-th century were led to consider what we call now Abelian varieties through the study of *Abelian integrals*, *i.e.*, of integrals of algebraic functions.

First recall how elliptic functions arise naturally from the study of elliptic integrals. The simplest of these elliptic integrals are the integrals of the form

$$\int \frac{dx}{\sqrt{P(x)}} \tag{II.1.1}$$

where P is a polynomial of degree 3 or 4 with simple roots (a so-called elliptic integral of the first kind). This integral can be written as $\int \frac{dx}{y}$, the integral of the algebraic differential form dx/y on the algebraic curve E of equation

$$y^2 = P(x).$$

After a projective transformation of the coordinates x and y, we can suppose that P takes the form

$$P(x) = 4x^3 - g_2 x - g_3.$$

As P has simple roots, there exists τ in the upper half plane $\mathfrak{H}$ such that $g_2 = g_2(\tau)$ and $g_3 = g_3(\tau)$. Hence, if $\wp$ denotes the Weierstrass function associated with τ (*cf.* [Coh]), the curve E possesses the parametrization

$$\begin{cases} x = \wp(z) \\ y = \wp'(z) \qquad z \in \mathbb{C}/(\mathbb{Z} + \tau\mathbb{Z}). \end{cases}$$

Using this parametrization, we get immediately

$$\int \frac{dx}{y} = z + c.$$

In other words, the elliptic function $\wp$ is an inverse of the (multivalued) function $t \mapsto \int_{t_0}^{t} \frac{dx}{\sqrt{P(x)}}$.

In fact, one of the major contributions of Abel to the theory of 'Abelian integrals' is the following principle:

To study elliptic integrals

$$\int_{t_0}^{t} \frac{dx}{\sqrt{P(x)}}$$

as functions of t, look at the inverse function.

Observe that, from this point of view, the double periodicity of the Weierstrass $\wp$ function reflects the fact that the integral $\int_{t_0}^{t} \frac{dx}{\sqrt{P(x)}}$ is multivalued. This comes from the double valued character of the square root $\sqrt{P(x)}$ and from the non vanishing of the *periods* of the differential form $\frac{dx}{\sqrt{P(x)}}$, *i.e.*, of the integrals

$$\int_{\Gamma} \frac{dx}{\sqrt{P(x)}},$$

where Γ is a closed curve in $\mathbb{C}$ (or rather in E).

If instead of a polynomial P of degree 3 or 4, we start with a polynomial P of degree 2, say $P(x) = 1 - x^2$ the integral (II.1.1) is $\int \frac{dx}{\sqrt{1-x^2}}$, and its inverse is the sine function. Thus the addition formula for sine

$$\begin{aligned} \sin(a+b) &= \sin a \cos b + \cos a \sin b \\ &= \sin a \sqrt{1 - \sin^2 b} + \sqrt{1 - \sin^2 a}\, \sin b, \end{aligned}$$

which may also be written

$$\arcsin A + \arcsin B = \arcsin\left(A\sqrt{1-B^2} + \sqrt{1-A^2}\,B\right),$$

appears as an equality of integrals of algebraic differential forms

$$\text{(II.1.2)} \qquad \int_0^A \frac{dx}{\sqrt{1-x^2}} + \int_0^B \frac{dx}{\sqrt{1-x^2}} = \int_0^{F(A,B)} \frac{dx}{\sqrt{1-x^2}},$$

where F is the algebraic function

$$F(A,B) = A\sqrt{1-B^2} + \sqrt{1-A^2}B.$$

The identity (II.1.2) may be proved directly, by elementary change of variables in the integrals. The elliptic integral (II.1.1) satisfies an analogous addition formula, discovered by Fagnano and Euler in the middle of the eighteenth century and proved by an elementary (but clever) change of variables:

$$\text{(II.1.3)} \qquad \int_0^A \frac{dx}{\sqrt{P(x)}} + \int_0^B \frac{dx}{\sqrt{P(x)}} = \int_0^{G(A,B)} \frac{dx}{\sqrt{P(x)}}$$

where

$$G(A,B) = \frac{1}{4}\left(\frac{\sqrt{P(A)} - \sqrt{P(B)}}{A-B}\right)^2 - A - B.$$

This identity may be interpreted as an addition formula for the function $\wp$:

$$\text{(II.1.4)} \qquad \wp(z_1+z_2) = \frac{1}{4}\left(\frac{\wp'(z_1) - \wp'(z_2)}{\wp(z_1) - \wp(z_2)}\right)^2 - \wp(z_1) - \wp(z_2).$$

Incidentally, using

$$(\wp'(z))^2 = 4\,(\wp(z))^3 - g_2\,\wp(z) - g_3,$$

this relation is easily seen to be equivalent to the more symmetric one:

$$\text{(II.1.5)} \qquad \begin{vmatrix} 1 & \wp(z_1) & \wp'(z_1) \\ 1 & \wp(z_2) & \wp'(z_2) \\ 1 & \wp(z_3) & \wp'(z_3) \end{vmatrix} = 0 \quad \text{if} \quad z_1 + z_2 + z_3 \in \mathbb{Z} + \tau\mathbb{Z}.$$

Observe also that, starting from some local determination of the function $\wp$ as the inverse of an Abelian integral of the first kind, the addition formula may be used to extend $\wp$ to the whole complex plane.

Abel and Jacobi not only considered elliptic integrals, but more generally, *hyperelliptic integrals*, *i.e.*, integrals of the form $\int \frac{R(x)}{\sqrt{P(x)}}dx$, where R is a polynomial or a rational function, and P is polynomial with simple roots. The expression $\omega = \frac{R(x)}{\sqrt{P(x)}}dx$ defines a meromorphic differential form on the Riemann surface C_P defined by the equation

$$y^2 = P(x)$$

which extends to a meromorphic differential form on the associated compact Riemann surface $\hat{C}_P$, whose genus is $g = \left[\frac{d+1}{2}\right] - 1$ (*cf.* §I.4.2). The differential

ω is holomorphic (*i.e.*, has no pole) on $\hat{C}_P$ iff R is polynomial of degree at most $g-1$ (*cf.* §B.2.2). Hence the differential forms

$$\omega_i = \frac{x^{i-1}dx}{y} = \frac{x^{i-1}dx}{\sqrt{P(x)}} \qquad i = 1, \ldots, g$$

and their integrals $\int \frac{x^i}{\sqrt{P(x)}}dx$ are the natural generalizations of the differential form $\frac{dx}{y}$ and of the elliptic integral of the first kind considered above.

To extend the construction of elliptic functions by inversion of elliptic integrals, one would attempt to invert the multivalued functions

$$t \mapsto \int_{t_0}^{t} \frac{R(x)dx}{\sqrt{P(x)}}$$

where R is a polynomial of degree $\leq g-1$. However, as soon as the degree of P is ≥ 5, the group of periods

$$\left\{ \int_{\Gamma} \frac{R(x)dx}{\sqrt{P(x)}} \, , \; \Gamma \text{ closed curve on } C_P \right\}$$

is dense in $\mathbb{C}$ (at least for a generic polynomial $R(x)$ of degree $g-1$). So you cannot get simple functions by inversion of a hyperelliptic integral.

This was discovered by Jacobi. However, he went further than this negative statement, and discovered that the 'good' inversion problem is the following:

Invert the function $(t_1, \ldots t_g) \mapsto (I_1, \ldots, I_g)$ *of* g *points of* C_P *with values in* $\mathbb{C}^g$ *given by*

$$I_1 = \int_{t_0}^{t_1} \frac{dx}{\sqrt{P(x)}} + \cdots + \int_{t_0}^{t_g} \frac{dx}{\sqrt{P(x)}}$$
$$I_i = \int_{t_0}^{t_1} \frac{x^{i-1}dx}{\sqrt{P(x)}} + \cdots + \int_{t_0}^{t_g} \frac{x^{i-1}dx}{\sqrt{P(x)}}$$
$$I_g = \int_{t_0}^{t_1} \frac{x^{g-1}dx}{\sqrt{P(x)}} + \cdots + \int_{t_0}^{t_g} \frac{x^{g-1}dx}{\sqrt{P(x)}} \, .$$

For simplicity, assume, as Jacobi did (*cf. infra* Figure 29), that $d = 6$. Then the inversion problem is to solve the system

$$\int_0^{t_1} \frac{dx}{\sqrt{P(x)}} + \int_0^{t_2} \frac{dx}{\sqrt{P(x)}} = \alpha$$

$$\int_0^{t_1} \frac{xdx}{\sqrt{P(x)}} + \int_0^{t_2} \frac{xdx}{\sqrt{P(x)}} = \beta$$

There are ambiguities in the definitions of these integrals:

- the square root $\sqrt{P(x)}$ has two determinations; this difficulty is solved by looking at t_1 and t_2 as points on the Riemann surface C_P;

• α (resp. β) is well defined only up to the addition of the periods $\int_\Gamma \frac{dx}{\sqrt{P(x)}}$ (resp. $\int_\Gamma \frac{xdx}{\sqrt{P(x)}}$), where γ denote a closed loop on C_P. This leads to introduce

$$\Lambda = \left\{ \left(\int_\Gamma \frac{dx}{\sqrt{P(x)}}, \int_\Gamma \frac{xdx}{\sqrt{P(x)}} \right), \Gamma \text{ closed loop on } C_P \right\} \subset \mathbb{C}^2 .$$

Then Λ *is a lattice in* $\mathbb{C}^2$ (see Proposition II.1.1 below) and the multivalued map $(t_1, t_2) \mapsto (\alpha, \beta)$ defines an analytic map

$$C_P \times C_P \to \mathbb{C}^2/\Lambda$$

which to a pair of points (t_1, t_2) on the Riemann surface associates the class of (α, β) in the complex torus $\mathbb{C}^2/\Lambda$. Clearly (t_1, t_2) and (t_2, t_1) have the same image under this map. Hence this map may be seen as a map defined on the symmetric product

$$\mathrm{Sym}_2 C_P := C_P \times C_P/\sim ,$$

where $\sim$ is the equivalence relation which identifies (t_1, t_2) and (t_2, t_1). Jacobi's great result on the inversion problem is that *the map*

$$\mathcal{J} : \mathrm{Sym}_2 C_P \to \mathbb{C}^2/\Lambda$$

we have just defined, *is onto and generically one to one.*

The complex torus $\mathbb{C}^2/\Lambda$ of dimension 2 is called the *Jacobian* of the hyperelliptic curve $\widehat{C}_P$ of genus 2. In the following Sections, we extend the construction of the Jacobian to compact connected Riemann surfaces of arbitrary genus, and we prove the preceding statement concerning Λ and $\mathcal{J}$ in this generalized context.

II.1.2. The Jacobian and the inversion problem for general compact Riemann surfaces.

In this Section, X denotes a compact connected Riemann surface of genus g.

The natural generalization of the family $\left(\frac{x^{i-1}dx}{y}\right)_{1\leq i\leq g}$ of differentials on the hyperelliptic curve C_P considered above is a basis $(\omega_i)_{1\leq i\leq g}$ of the space $\Omega^1(X)$ of holomorphic 1-forms on X. The natural generalization of the group Λ of periods of C_P is the following subgroup of $\mathbb{C}^g$:

$$\Lambda = \left\{ \left(\int_\gamma \omega_i \right)_{1\leq i\leq g}, \gamma \in H_1(X; \mathbb{Z}) \right\} .$$

Observe that the g-uple $\left(\int_\gamma \omega_i\right)_{1\leq i\leq g}$ is the set of coordinates in the basis dual to $(\omega_1, \ldots, \omega_g)$ of the complex linear form

$$\int_\gamma : \Omega^1(X) \to \mathbb{C}$$

$$\omega \mapsto \int_\gamma \omega.$$

This leads to a more intrinsic (*i.e.*, independent on the choice of a basis in $\Omega^1(X)$) definition of Λ, which appears as a subgroup of $\Omega^1(X)^*$:

Proposition -Definition II.1.1. *The period map*

$$p : H_1(X;\mathbb{Z}) \to \Omega^1(X)^*$$

$$\gamma \mapsto \int_\gamma$$

is injective and its image is a lattice Λ in $\Omega^1(X)^*$ (considered as a real vector space).

This lattice is called the period lattice *of X and the g-dimensional complex torus $\Omega^1(X)^*/\Lambda$ is called the* Jacobian variety *of X and is denoted* Jac(X).

(Recall that a lattice in a real vector space V of dimension d is the Abelian group, isomorphic to $\mathbb{Z}^d$, generated by a basis of V; see [Sen]).

Examples: i) If $X = \mathbb{P}^1\mathbb{C}$, $\Omega^1(X) = \{0\}$ and Jac(X) is reduced to one point.

ii) Let X be the elliptic curve $\mathbb{C}/(\mathbb{Z}+\tau\mathbb{Z})$. Then $\Omega^1(X)$ is a one-dimensional vector space, generated by the non-vanishing holomorphic differential dz, and the induced isomorphism

$$\begin{array}{ccc} \Omega^1(X)^* & \to & \mathbb{C} \\ \lambda & \mapsto & \lambda(dz) \end{array}$$

is easily seen to map the lattice Λ onto $\mathbb{Z} + \tau\mathbb{Z}$. Therefore X and Jac(X) are isomorphic Riemann surfaces.

Let $H^1_{DR}(X;\mathbb{R})$ be the first de Rham cohomology group of X with real coefficients (*cf.* §A.1) and let $H^1_{DR}(X;\mathbb{R})^\wedge$ denote the space of real linear forms on $H^1_{DR}(X;\mathbb{R})$. Proposition II.1.1 is a consequence of the fact that, according to A.2.2, the map

$$p' : H_1(X;\mathbb{Z}) \to H^1_{DR}(X;\mathbb{R})^\wedge$$

$$\gamma \mapsto \int_\gamma$$

injectively embeds $H_1(X;\mathbb{Z})$ as a lattice in $H^1_{DR}(X;\mathbb{R})$, combined with the Hodge decomposition (*cf.* Theorem B.2.5):

$$\Omega^1(X) \oplus \overline{\Omega^1}(X) \xrightarrow{\sim} H^1_{DR}(X;\mathbb{C}) \qquad \text{(II.1.6)}$$

$$\alpha \oplus \beta \mapsto [\alpha + \beta].$$

Indeed, the isomorphism (II.1.6) induces an isomorphism of real vector spaces

$$j : \Omega^1(X) \xrightarrow{\sim} H^1_{DR}(X;\mathbb{R})$$
$$\alpha \mapsto [\alpha + \overline{\alpha}],$$

and one easily checks that the map

$$j' : \Omega^1(X)^* \to H^1_{DR}(X;\mathbb{R})^\vee$$
$$\lambda \mapsto \lambda \circ j^{-1} + \overline{\lambda} \circ j^{-1}$$

also is an isomorphism of real vector spaces and that $j' \circ p = p'$.

In order to state the generalization to general compact Riemann surfaces of the results of Jacobi alluded to above, we need to introduce a few preliminary definitions.

For any positive integer n, $\mathrm{Sym}_n X$ will denote the symmetric product of n copies of X, *i.e.*, the quotient of X^n by the action of the symmetric group S_n defined by

$$\sigma \cdot (x_1, \ldots, x_i, \ldots, x_n) = (x_{\sigma^{-1}(1)}, \ldots, x_{\sigma^{-1}(i)}, \ldots, x_{\sigma^{-1}(n)}).$$

This quotient has a natural structure of n dimensional complex *manifold* (without singularity!) characterized by a property similar to the property introduced in Theorem I.2.1 (replace X by X^n and Γ by S_n). For instance, if $g = 2$, $\mathrm{Sym}_2 X$ is clearly a complex manifold outside the diagonal $\Delta_X = \{[x,x], x \in X\}$ and, near a point $[(P,P)]$ of Δ_X, local holomorphic coordinates are given by

$$\sigma_1([P_1,P_2]) = z(P_1) + z(P_2)$$
$$\sigma_2([P_1,P_2]) = z(P_1)z(P_2),$$

where z denotes a local holomorphic coordinate on a neighbourhood of P in X. This proof may be extended to arbitrary values of n by using elementary symmetric functions of the coordinates.

Let $P_0 \in X$. For any $P \in X$ and any path L from P_0 to P on X, the linear form on $\Omega^1(X)$

$$\int_L : \omega \mapsto \int_L \omega$$

has a class in $\mathrm{Jac}(X) = \Omega^1(X)^*/\Lambda$ which depends only on P and P_0. Indeed, if L' is another such path, we have

$$\int_{L'} \omega - \int_L \omega = \int_\gamma \omega$$

where γ denotes the closed path $L' - L$. The element of $\mathrm{Jac}(X)$ so defined will be denoted

$$\omega \mapsto \int_{P_0}^P \omega$$

or $J_{P_0}(P)$. In other words, if $\Omega^1(X)^*$ is identified with $\mathbb{C}^g$ by the choice of a basis $(\omega_1, \ldots, \omega_g)$ of $\Omega^1(X)$, we have

$$J_{P_0}(P) = \left(\int_{P_0}^{P} \omega_i\right)_{1\leq i\leq g} \mod \Lambda. \tag{II.1.6}$$

Finally, we define

$$\mathcal{J}_{P_0} : \mathrm{Sym}_g X \to \mathrm{Jac}\,(X)$$

$$[(P_1, \ldots, P_g)] \mapsto \sum_{i=1}^{g} J_{P_0}(P_i).$$

Jacobi's results on the inversion problem extend to arbitrary compact Riemann surfaces, as shown by the following theorem, which is the main result of the second part of these notes.

Theorem II.1.2. *Assume that the genus g of X is at least 1.*

1) The map $J_{P_0} : X \to \mathrm{Jac}(X)$ is a holomorphic embedding, called the Jacobian embedding.

2) The map $\mathcal{J}_{P_0} : \mathrm{Sym}_g X \to \mathrm{Jac}(X)$ is a birational holomorphic map[24].

The problem of understanding the map $\mathcal{J}_{P_0}$ is what is now called the *inversion problem.* It has been the subject of many works during the nineteenth century and was, in a sense, solved in general by Riemann in his paper [Ri1] of 1857 where he showed how theta functions on $\mathrm{Jac}(X)$ allow to construct the inverse of the map $\mathcal{J}_{P_0}$ (*cf. infra* §III.5). However, many questions related to the map $\mathcal{J}_{P_0}$ and its geometry are still unsolved and the inversion problem is still at the origin of difficult and interesting problems (see *e.g.* [Ke2]).

Observe that when X is an elliptic curve $\mathbb{C}/(\mathbb{Z} + \tau\mathbb{Z})$ and $P_0 = 0$, the maps J_{P_0} and $\mathcal{J}_{P_0}$ are nothing else than the identity map from X to $\mathrm{Jac}(X)$, canonically identified with X. Therefore, in that case, Theorem II.1.2 is clear. As a matter of fact, the case $g = 1$ of Theorem II.1.2 shows that any compact connected Riemann surface X of genus 1 is isomorphic with $\Omega^1(X)^*/\Lambda$, hence to an elliptic curve $\mathbb{C}/(\mathbb{Z} + \tau\mathbb{Z})$ (choose an oriented basis (ω_1, ω_2) of Λ and let $\tau = \frac{\omega_2}{\omega_1}$).

Now let us come to the proof of Theorem II.1.2.

The holomorphy of J_{P_0} is clear on formula (II.1.6). The holomorphy of $\mathcal{J}_{P_0}$ is a consequence of the holomorphy of J_{P_0} and of the definition of the holomorphic structure on $\mathrm{Sym}_g X$ as a quotient structure.

Let us prove the following lemma:

Lemma II.1.3. *1) J_{P_0} is an immersion (*i.e.*, for any $P \in X$, the differential*

$$DJ_{P_0}(P) : T_P X \to T_{J_{P_0}(P)}\,\mathrm{Jac}(X)$$

[24] Recall (*cf.* Appendix D) that it means that $\mathcal{J}_{P_0}$ is holomorphic, surjective and generically one-to-one; more precisely, there exists closed analytic subsets F (resp. F') of codimension ≥ 1 (resp. ≥ 2) in $\mathrm{Sym}_g X$ (resp. in $\mathrm{Jac}(X)$) such that $\mathcal{J}_{P_0}(F) = F'$ and $\mathcal{J}_{P_0}$ maps $\mathrm{Sym}_g X - F$ biholomorphically onto $\mathrm{Jac}(X) - F'$.

is injective).

2) There exists $Q = [(P_1, \dots, P_g)]$ *in* $\mathrm{Sym}_g X$ *such that the differential*

$$D\mathcal{J}_{P_0}(Q) : T_Q \,\mathrm{Sym}_g X \to T_Q \,\mathrm{Jac}(X)$$

is bijective.

Indeed, the tangent space to $\mathrm{Jac}(X) = \Omega^1(X)^*/\Lambda$ at any point may be identified with $\Omega^1(X)^*$ and the injectivity of $DJ_{P_0}(P)$ is equivalent to the surjectivity of its transpose

$${}^tDJ_{P_0}(P) : \Omega^1(X) \to (T_P X)^* \simeq \omega_{X,P}.$$

But the definition of J_{P_0} (see (II.1.6)) shows that this map is the evaluation at P of holomorphic differentials on X, which is known to be surjective when $g \geq 1$ (*cf.* Proposition B.9.4).

If the points $P_1, \dots, P_g$ are pairwise distinct, then Q possesses a neighbourhood in $\mathrm{Sym}_g X$ which may be identified with a neighbourhood of $(P_1, \dots, P_g)$ in X^g. Therefore $T_Q \,\mathrm{Sym}_g X$ may be identified with $T_{P_1} X \oplus \cdots \oplus T_{P_g} X$. The bijectivity of $D\mathcal{J}_{P_0}(Q)$ is equivalent to the bijectivity of its transpose

$${}^tD\mathcal{J}_{P_0}(Q) : \Omega^1(X) \to \omega_{X,P_1} \oplus \cdots \oplus \omega_{X,P_g}.$$

But this map is easily seen to be the evaluation map

$$\omega \mapsto (\omega(P_1), \dots, \omega(P_g)).$$

As $\Omega^1(X)$ has dimension g, the existence of pairwise distinct points $P_1, \dots, P_g$ such that this map is bijective follows from elementary linear algebra.

Thanks to Lemma II.1.3, a), it will clearly be enough to show that J_{P_0} is injective to complete the proof of Theorem II.1.2, 1). On the other hand, since $\mathrm{Sym}_g X$ and $\mathrm{Jac}(X)$ are compact connected complex manifolds of the same dimension g, Lemma II.1.3, b) implies that $\mathcal{J}_{P_0}$ is onto and that, to prove that $\mathcal{J}_{P_0}$ is birational, we only have to show that the preimage $\mathcal{J}_{P_0}^{-1}(x)$ of any $x \in \mathrm{Jac}(X)$ is connected (*cf.* Appendix D). The next two Sections are devoted to a further study of the Jacobian variety $\mathrm{Jac}(X)$ which will allow us to prove the injectivity of J_{P_0} and the connectedness of the fibers $\mathcal{J}_{P_0}^{-1}(x)$, and therefore to complete the proof of Theorem II.1.2.

II.2. The classification of holomorphic line bundles—or the Jacobian for field theory physicists

In this Section, we study the set of isomorphism classes of holomorphic line bundles over a compact Riemann surface X - the so-called Picard group of X. We follow an approach in the spirit of gauge theory, which makes the Jacobian $\mathrm{Jac}(X)$ appear naturally as parametrizing the set of isomorphism classes of topologically trivial holomorphic line bundles on X. In the next Section, we will use this interpretation of $\mathrm{Jac}(X)$ to complete the proof of Theorem II.1.2.

II.2.1. $\overline{\partial}$-connections and holomorphic structures on complex line bundles.

Let $\mathcal{L}$ be a C^∞ complex line bundle on X. We define a $\overline{\partial}$-*connection* on $\mathcal{L}$ as a linear map

$$\overline{\nabla} : \mathcal{C}^\infty(X;\mathcal{L}) \to \mathcal{C}^\infty(X;\overline{\omega}_X \otimes \mathcal{L})$$

such that, for any $f \in \mathcal{C}^\infty(X,\mathbb{C})$ and any $s \in \mathcal{C}^\infty(X;\mathcal{L})$, the following equality holds:

$$\overline{\nabla}(fs) = f\overline{\nabla}s + \overline{\partial}f \cdot s. \tag{II.2.1}$$

The set of $\overline{\partial}$-connections on $\mathcal{L}$ will be denoted $\mathcal{C}_\mathcal{L}$.

A $\overline{\partial}$-connection $\overline{\nabla}$ is easily seen to be a differential operator of first order which locally, after the choice of a local holomorphic trivialization of $\mathcal{L}$, takes the form

$$\overline{\nabla}s = \overline{\partial}s + As$$

for some local section A of ω_X. (This could have been taken as definition).

If $\overline{\nabla}$ belongs to $\mathcal{C}_\mathcal{L}$ and $\varphi \in \mathcal{C}^\infty(X;\mathbb{C}^*)$, we can define another element $\overline{\nabla}^\varphi$ of $\mathcal{C}_\mathcal{L}$ by setting

$$\overline{\nabla}^\varphi(s) = \varphi^{-1}\,\overline{\nabla}(\varphi s). \tag{II.2.2}$$

One immediately checks that, for any $\varphi_1, \varphi_2 \in \mathcal{C}^\infty(X;\mathbb{C}^*)$, we have

$$\left(\overline{\nabla}^{\varphi_1}\right)^{\varphi_2} = \overline{\nabla}^{\varphi_1\varphi_2}.$$

In other words, formula (II.2.2) defines an action on $\mathcal{C}_\mathcal{L}$ of the (infinite dimensional) group $\mathcal{G} := \mathcal{C}^\infty(X;\mathbb{C}^*)$.

Observe now that the $\overline{\partial}$-operator

$$\overline{\partial} : \mathcal{C}^\infty(X;\mathbb{C}) \to \mathcal{C}^\infty(X;\overline{\omega}_X)$$

is a $\overline{\partial}$-connection on the trivial line bundle $X \times \mathbb{C}$. In that case, condition (II.2.1) is nothing else than Leibniz rule for $\overline{\partial}$. More generally, if $\underline{\mathcal{L}}$ is any holomorphic line bundle on X whose C^∞ underlying line bundle is $\mathcal{L}$, the $\overline{\partial}$-operator $\overline{\partial}_{\underline{\mathcal{L}}}$ associated with $\underline{\mathcal{L}}$ (see §B.5) is a $\overline{\partial}$-connection on $\mathcal{L}$. Indeed, to check the validity of (II.2.1) for $\overline{\nabla} = \overline{\partial}_{\underline{\mathcal{L}}}$, we can work locally on X; then we are reduced to the case of the trivial holomorphic bundle and of $\overline{\nabla} = \overline{\partial}$. In fact (II.2.1) is then nothing else than Leibniz rule (B.8.1) for $\mathcal{L}_1 = \mathcal{O}$ and $\mathcal{L}_2 = \underline{\mathcal{L}}$. Moreover, the $\overline{\partial}$-connection $\overline{\partial}_{\underline{\mathcal{L}}}$ allows to recover the holomorphic structure on $\underline{\mathcal{L}}$: a section $s \in \mathcal{C}^\infty(X;\mathcal{L})$ is holomorphic on an open set $U \subset X$ iff $\overline{\partial}_{\underline{\mathcal{L}}}\, s = 0$ on U.

The following theorem asserts that this construction is essentially the only way to get $\overline{\partial}$-connections on $\mathcal{L}$.

Theorem II.2.1. *Let $\mathcal{L}$ be a complex C^∞ line bundle on X.*

1) The map which associates the $\overline{\partial}$-connection $\overline{\partial}_{\underline{\mathcal{L}}}$ to any holomorphic line bundle $\underline{\mathcal{L}}$ over X whose underlying C^∞ line bundle is $\mathcal{L}$ establishes a bijection *between the set of holomorphic structures on $\mathcal{L}$ and the set $C_{\mathcal{L}}$ of $\overline{\partial}$-connections on $\mathcal{L}$.*

2) The holomorphic line bundles $\underline{\mathcal{L}}_1$ and $\underline{\mathcal{L}}_2$ on X associated to two $\overline{\partial}$-connections $\overline{\nabla}_1$ and $\overline{\nabla}_2$ in $C_{\mathcal{L}}$ are isomorphic (as holomorphic line bundles) iff there exists $\varphi \in \mathcal{G}$ such that

$$\overline{\nabla}_2 = \overline{\nabla}_1^{\varphi}.$$

As an immediate consequence, we obtain:

Corollary II.2.2. *The quotient space $C_{\mathcal{L}}/\mathcal{G}$ gets identified to the set of (holomorphic) isomorphism classes of holomorphic line bundles whose underlying C^∞ line bundles are isomorphic to $\mathcal{L}$ by the map which sends the class in $C_{\mathcal{L}}/\mathcal{G}$ of a $\overline{\partial}$-connection $\overline{\nabla}$ to the class of the holomorphic structure on $\mathcal{L}$ defined by $\overline{\nabla}$.*

Let us prove Theorem II.2.1. As the $\overline{\partial}$-operator $\overline{\partial}_{\underline{\mathcal{L}}}$ allows to recover the local holomorphic sections of $\underline{\mathcal{L}}$, the map in 1) is clearly injective. Its surjectivity will follow from the following

Lemma II.2.3. *Let $\overline{\nabla}$ be a $\overline{\partial}$-connection on $\mathcal{L}$. For any $P \in X$, there exists a C^∞ non-vanishing section σ of $\mathcal{L}$ on an open neighbourhood U of P such that $\overline{\nabla} s = 0$.*

Indeed, this lemma provides an open covering $\{U_i\}_{i\in I}$ of X and non-vanishing sections $\sigma_i \in \mathcal{C}^\infty(U_i;\mathcal{L})$, $i \in I$, such that

$$\overline{\nabla}\sigma_i = 0 \quad \text{on} \quad U_i\ ,$$

which therefore satisfy

$$\overline{\partial}(\sigma_j^{-1}\cdot\sigma_i) = 0 \quad \text{on} \quad U_i \cap U_j.$$

According to §B.1.6, there exists a unique holomorphic structure $\underline{\mathcal{L}}$ on $\mathcal{L}$ such that σ_i is a holomorphic section of $\mathcal{L}$ over U_i. If s is a local C^∞ section of $\mathcal{L}$, say over an open subset of Ω of U_i, we have

$$s = f\sigma_i$$

for some $f \in C^\infty(\Omega;\mathbb{C})$. Thus, over Ω, we get

$$\begin{aligned}
\overline{\partial}_{\underline{\mathcal{L}}}\, s &= \overline{\partial}_{\underline{\mathcal{L}}}(f\sigma_i) \\
&= \overline{\partial} f\cdot\sigma_i \qquad \text{(since } \sigma_i \text{ is a holomorphic section of } \underline{\mathcal{L}}\text{)} \\
&= \overline{\nabla}(f\cdot\sigma_i) \qquad \text{(since } \overline{\nabla}\sigma_i = 0\text{)} \\
&= \overline{\nabla} s.
\end{aligned}$$

This shows that

$$\overline{\partial}_{\underline{\mathcal{L}}} = \overline{\nabla}.$$

To prove Lemma II.2.3, choose a non-vanishing section t of $\mathcal{L}$ over an open neighbourhood V of P and let $A = \overline{\nabla} f$. Then for any open neighbourhood U of P contained in V and any $f \in \mathcal{C}^\infty(U;\mathbb{C})$, we have on U:

$$\overline{\nabla}(f \cdot t) = \overline{\partial} f \cdot t + fA. \tag{II.2.3}$$

According to Lemma C.3.2 (applied to the trivial line bundle), we can find such a neighbourhood U and a function $\varphi \in \mathcal{C}^\infty(U;\mathbb{C})$ such that, on U:

$$\overline{\partial}\varphi = -t^{-1} \cdot A.$$

Then (II.2.3) shows that

$$\sigma = e^{\varphi} \cdot t$$

satisfies the conditions required in Lemma II.2.3.

The proof of assertion 2) in Theorem II.2.1 is very simple and will be left as an exercise.

As emphasized in Corollary II.2.2, Theorem II.2.1 shows that the set of isomorphism classes of holomorphic line bundles over X, which, as C^∞ line bundles, are isomorphic with $\mathcal{L}$, may be identified with $\mathcal{C}_{\mathcal{L}}/\mathcal{G}$. On the other hand, when X is compact and connected, C^∞ line bundles on X are completely classified by their degree (*cf.* Proposition B.4.2). Therefore, if $\mathcal{L}_d$ is a C^∞ line bundle of degree d on X, then the set $\mathrm{Pic}_d(X)$ of isomorphism classes of holomorphic line bundles of degree d over X may be identified with $\mathcal{C}_{\mathcal{L}_d}/\mathcal{G}$.

II.2.2. The classification of topologically trivial holomorphic line bundles on a compact Riemann surface.

We now suppose that $\mathcal{L}$ is the trivial line bundle $X \times \mathbb{C}$ and we write $\mathcal{C}$ instead of $\mathcal{C}_{\mathcal{L}}$.

One easily checks that $\mathcal{C}$ is the space of differential operators $\overline{\nabla}_A$ defined by

$$\overline{\nabla}_A f := \overline{\partial} f + Af$$

where A is an arbitrary element in $\mathcal{C}^\infty(X;\overline{\omega}_X)$. Thus $\mathcal{C}$ may be identified with $\mathcal{C}^\infty(X;\overline{\omega}_X)$. For any $A \in \mathcal{C}^\infty(X;\overline{\omega}_X)$ and any $\varphi \in \mathcal{C}^\infty(X;\mathbb{C}^*)$, we have

$$\left(\overline{\nabla}_A\right)^{\varphi} = \overline{\nabla}_{A^{\varphi}}$$

where

$$A^{\varphi} = A + \varphi^{-1} \cdot \overline{\partial}\varphi.$$

Therefore, the quotient $\mathcal{C}/\mathcal{G}$ of $\mathcal{C}$ under the action of the group $\mathcal{G} = \mathcal{C}^\infty(X;\mathbb{C}^*)$ may be identified with the quotient of $\mathcal{C}^\infty(X;\overline{\omega}_X)$ by the subgroup

$$\mathcal{R}(X) = \{\varphi^{-1} \cdot \overline{\partial}\varphi \ , \ \varphi \in \mathcal{C}^\infty(X;\mathbb{C}^*)\}.$$

All what precedes holds for any Riemann surface. When X is compact and connected, we have an explicit description of the quotient $\mathcal{C}/\mathcal{G}$:

Theorem II.2.4. *Assume that X is a compact connected Riemann surface. Let $\overline{\Omega^1}(X)$ be the space of antiholomorphic differential forms on X (cf. §B.2.1), let*

$$\Lambda = \left\{\left(\omega \mapsto \int_\gamma \omega\right) \in \Omega^1(X)^* \ ; \ \gamma \in H_1(X;\mathbb{Z})\right\}$$

be the 'lattice of periods' in $\Omega^1(X)^*$ and let $\operatorname{Jac}(X) = \Omega^1(X)^*/\Lambda$ be the Jacobian of X. Consider the maps

$$\alpha : \overline{\Omega^1}(X) \to \Omega^1(X)^*/\Lambda$$

and

$$\beta : \overline{\Omega^1}(X) \to \mathcal{C}/\mathcal{G}$$

such that $\alpha(A)$ is the class in $\Omega^1(X)^*/\Lambda$ of the linear form

$$\begin{aligned} j(A) : \Omega^1(X) &\to \mathbb{C} \\ \omega &\mapsto \frac{1}{2\pi i}\int_X \omega \wedge A \end{aligned}$$

and $\beta(A)$ is the class in $\mathcal{C}/\mathcal{G}$ of $\overline{\nabla}_A$.

Then there exists a unique map

$$I : \operatorname{Jac}(X) \to \mathcal{C}/\mathcal{G}$$

such that the following diagram commutes

$$\begin{array}{ccccc} & & \overline{\Omega^1}(X) & & \\ & \alpha \swarrow & & \searrow \beta & \\ \operatorname{Jac}(X) = \Omega^1(X)^*/\Lambda & & \xrightarrow{I} & & \mathcal{C}/\mathcal{G}. \end{array}$$

Moreover it is a bijection.

According to Corollary II.2.2, the quotient $\mathcal{C}/\mathcal{G}$ may be identified with the set of isomorphism classes of holomorphic line bundles over X, which, as C^∞ line bundles, are trivial. As explained in Proposition B.4.2, these are exactly the topologically trivial holomorphic line bundles, or the holomorphic line bundles of degree zero. Therefore Theorem II.2.4 shows that *the map I establishes a bijection between the Jacobian variety of X and the set of isomorphism classes of topologically trivial line bundles over X*. Thus we get an alternative description of the Jacobian of X, in the spirit of gauge field theory.

For later use, it is important to notice that when we compose the bijection

$$i_1 : \operatorname{Pic}_0(X) \simeq \mathcal{C}/\mathcal{G}$$

described at the end of §II.2.1 and the bijection

$$i_2 = I^{-1} : \mathcal{C}/\mathcal{G} \xrightarrow{\sim} \operatorname{Jac}(X)$$

we obtain a bijection

$$i_2 \circ i_1 : \mathrm{Pic}_0(X) \stackrel{\sim}{\rightarrow} \mathrm{Jac}(X)$$

which is an *isomorphism of Abelian groups*, when the complex torus $\mathrm{Jac}(X)$ is seen as an Abelian group and when $\mathrm{Pic}_0(X)$ is endowed with the group structure defined by the tensor product of line bundles (the sum of the isomorphism classes of two holomorphic line bundles over X is defined as the isomorphism class of their tensor product). This group $\mathrm{Pic}_0(X)$ is called the *Picard group* (of degree 0) of X. Indeed, we have:

Lemma II.2.5. *Let $\overline{\nabla}_{A'}$ and $\overline{\nabla}_{A''}$ be two $\overline{\partial}$-connections on the trivial bundle $\mathcal{L} = X \times \mathbb{C}$. Let $\mathcal{L}'$ and $\mathcal{L}''$ be the holomorphic line bundles defined by these $\overline{\partial}$-connections. Then the tensor product $\mathcal{L}' \otimes \mathcal{L}''$ is a holomorphic line bundle which still has the trivial line bundle $\mathcal{L}$ as underlying C^∞-line bundle, and the $\overline{\partial}$-operator on $\mathcal{L}' \otimes \mathcal{L}''$ is $\overline{\nabla}_{A'+A''}$.*

Lemma II.2.5 is a straightforward consequence of the Leibniz rule for $\overline{\partial}$-operators (see (B.8.1)).

The next two subsections are devoted to the proof of Theorem II.2.4. This proof is rather technical and should be skipped at first reading.

II.2.3. $\mathcal{C}/\mathcal{G}$ as a quotient of $\overline{\Omega^1}(X)$.

We have seen that $\mathcal{C}/\mathcal{G}$ may be identified with the quotient $\mathcal{C}^\infty(X;\overline{\omega}_X)/\mathcal{R}(X)$.

For any $\psi \in \mathcal{C}^\infty(X;\mathbb{C})$, we have:

$$\overline{\partial}\,\psi = e^{-\psi} \cdot \overline{\partial}\, e^{\psi}.$$

Therefore $\mathcal{R}(X)$ contains $\overline{\partial}\,\mathcal{C}^\infty(X;\mathbb{C})$. On the other hand, we know that

$$\mathcal{C}^\infty(X;\overline{\omega}_X) = \overline{\Omega^1}(X) \oplus \overline{\partial}\,\mathcal{C}^\infty(X;\mathbb{C})$$

(*cf.* Corollary B.8.2., 1)).

These two facts imply

Lemma II.2.6. *The map $\overline{\Omega^1}(X) \rightarrow \mathcal{C}^\infty(X;\overline{\omega}_X)/\mathcal{R}(X)$ induced by the inclusion $\overline{\Omega^1}(X) \hookrightarrow \mathcal{C}^\infty(X;\overline{\omega}_X)$ is onto.*

We are now going to describe its kernel. To do this, we will use the following auxiliary result, the proof of which we leave to the reader as an exercise:

Lemma II.2.7. *For any C^∞ complex 1-form ω on X, the following two conditions are equivalent:*

i) there exists $f \in \mathcal{C}^\infty(X;\mathbb{C}^*)$ such that $\omega = f^{-1} \cdot df$;

ii) $d\omega = 0$ and, for any $\gamma \in H_1(X;\mathbb{Z})$, $\int_\gamma \omega \in 2\pi i\mathbb{Z}$.

(**Hint:** to prove the implication ii) $\Rightarrow$ i), choose a base point P_0 in X and consider the multivalued function $P \mapsto \int_{P_0}^{P} \omega$ and its exponential).

Let us now consider the subgroup

$$U(1) = \{z \in \mathbb{C} \mid |z| = 1\}$$

of the multiplicative group $\mathbb{C}^*$. A map $f \in \mathcal{C}^\infty(X; U(1))$ is called harmonic if any local determination $\log f$ of the logarithm of f is harmonic (*i.e.*, satisfies the equation $\partial\overline{\partial}\log f = 0$; *cf.* §C.5).

Lemma II.2.8. *For any $f \in \mathcal{C}^\infty(X; U(1))$ the following two conditions are equivalent:*
 i) $f^{-1} \cdot \overline{\partial} f \in \overline{\Omega^1}(X)$;
 ii) f is harmonic.

Indeed i) is satisfied iff $\overline{\partial}\log f \in \overline{\Omega^1}(X)$, *i.e.*, iff locally $\partial\overline{\partial}\log f = 0$.

Lemma II.2.9. *Let $\alpha \in \overline{\Omega^1}(X)$. The following three conditions are equivalent:*
 i) there exists a (harmonic) function $f \in \mathcal{C}^\infty(X; U(1))$ such that $\alpha = f^{-1} \cdot \overline{\partial} f$;
 ii) there exists $g \in \mathcal{C}^\infty(X; \mathbb{C}^*)$ such that $\alpha = g^{-1} \cdot \overline{\partial} g$;
 iii) for any $\gamma \in H_1(X; \mathbb{Z})$, $\int_\gamma (\alpha - \overline{\alpha}) \in 2\pi i\mathbb{Z}$.
Moreover, if g satisfies ii), then $|g|$ is constant and $f = |g|^{-1} g$ satifies i).

If f satisfies i), then $\overline{f} = f^{-1}$ and

$$\overline{\alpha} = \overline{f}^{-1} \cdot \partial\overline{f} = f \cdot \partial f^{-1} = -f^{-1} \cdot \partial f.$$

Therefore

$$f^{-1} \cdot df = \alpha - \overline{\alpha}$$

and iii) follows from Lemma II.2.7. Conversely, if iii) is satisfied, Lemma II.2.7 shows the existence of $f_0 \in \mathcal{C}^\infty(X; \mathbb{C}^*)$ such that

$$\alpha - \overline{\alpha} = f_0^{-1} \cdot df_0.$$

We have

$$\begin{aligned} |f_0|^{-1} \cdot d|f_0| &= d\log|f_0| \\ &= \operatorname{Re} d\log f_0 \\ &= \operatorname{Re}(\alpha - \overline{\alpha}) \\ &= 0. \end{aligned}$$

Therefore $|f_0|$ is constant, and $f := |f_0|^{-1} \cdot f_0$ belongs to $\mathcal{C}^\infty(X; U(1))$ and satisfies

$$\alpha - \overline{\alpha} = f^{-1} \cdot df$$

and *a fortiori* i).

If g satisfies ii), then

$$\overline{\alpha} = \overline{g}^{-1} \cdot \partial\overline{g} = -g^{-1} \cdot \partial g + |g|^{-2} \cdot \partial|g|^2 \ ;$$

therefore,

$$|g|^{-2} \cdot \partial \log |g|^2 = \overline{\alpha} + g^{-1}\partial g = \overline{\alpha} - \alpha + g^{-1} \cdot dg$$

is a closed form and $\log |g|^2$ is harmonic (*cf.* §C.5), hence constant (*cf.* Proposition C.5.1). This clearly implies that $f = |g|^{-1} \cdot g$ satisfies i). Finally, i) obviously implies ii).

The subgroup of $\overline{\Omega^1}(X)$ satisfying the conditions in Lemma II.2.9 will be denoted $H_1(X;\mathbb{Z})^\perp$. It is clearly the kernel of the map in Lemma II.2.6. Therefore we have proved the following:

Proposition II.2.10. *There exists a unique map*

$$\overline{\Omega^1}(X)/H_1(X;\mathbb{Z})^\perp \to \mathcal{C}/\mathcal{G}$$

which sends the class in $\overline{\Omega^1}(X)/H_1(X;\mathbb{Z})^\perp$ of $A \in \overline{\Omega^1}(X)$ to the class in $\mathcal{C}/\mathcal{G}$ of $\overline{\nabla}_A$, and this map is a bijection.

In other words, any $\overline{\partial}$-connection in $\mathcal{C}$ is in the same $\mathcal{G}$-orbit as a $\overline{\partial}$-connection $\overline{\nabla}_A = \overline{\partial} + A$, where A is antiholomorphic, and any two $\overline{\partial}$-connections $\overline{\nabla}_A$ and $\overline{\nabla}_{A'}$ of this form are in the same $\mathcal{G}$-orbit iff there exists a harmonic $\varphi \in \mathcal{C}^\infty(X;U(1))$ such that $\overline{\nabla}_{A'} = \overline{\nabla}_A^\varphi$.

An important fact which may be deduced from Proposition II.2.10 is that $\mathcal{C}/\mathcal{G}$ is naturally a complex torus of dimension g. This follows from the next lemma, which is established in the same way as Proposition II.1.1 (using A.2.2 and Theorem B.2.5) and whose detailed proof will be left to the reader:

Lemma II.2.11. *$H_1(X;\mathbb{Z})^\perp$ is a lattice in $\overline{\Omega^1}(X)$. Moreover, the map*

$$\begin{aligned} \varphi : H_1(X;\mathbb{Z})^\perp &\to \operatorname{Hom}(H_1(X;\mathbb{Z}),\mathbb{Z}) \\ A &\mapsto \left(\gamma \mapsto \frac{1}{2\pi i}\int_\gamma (A - \overline{A})\right) \end{aligned}$$

is an isomorphism of Abelian groups.

II.2.4. Completion of the proof of Theorem II.2.4.

Consider the linear map

$$j : \overline{\Omega^1}(X) \to \Omega^1(X)^*$$

which sends $A \in \overline{\Omega^1}(X)$ to the $\mathbb{C}$-linear form on $\Omega^1(X)^*$

$$j(A) : \alpha \mapsto \frac{1}{2\pi i}\int_X \alpha \wedge A.$$

It is a $\mathbb{C}$-linear isomorphism (this follows, for instance, from the fact that $(\omega_1,\omega_2) \mapsto i\int_X \omega_1 \wedge \overline{\omega}_2$ is a Hermitian scalar product on $\overline{\Omega^1}(X)$).

Theorem II.2.4 follows immediately from Proposition II.2.10 together with the following

Proposition II.2.12. *The isomorphism j maps $H_1(X;\mathbb{Z})^\perp$ onto the period lattice Λ.*

Observe that Propositions II.1.1 and II.2.12 provide another proof of Lemma II.2.11.

Let ψ be the map

$$H_1(X;\mathbb{Z}) \to \operatorname{Hom}(H_1(X;\mathbb{Z}),\mathbb{Z})$$

which sends $\gamma \in H_1(X;\mathbb{Z})$ to the group morphism

$$\psi(\gamma) : \gamma' \mapsto \gamma\#\gamma',$$

defined using the intersection number $\#$ of homology classes (*cf.* A.2.3). The map φ is an isomorphism, as follows for instance from the existence of symplectic bases (*cf.* A.2.3).

Consider now the following diagram

$$\text{(II.2.4)}\qquad \begin{array}{ccccc} H_1(X;\mathbb{Z})^\perp & \xrightarrow[\varphi]{\sim} & \operatorname{Hom}(H_1(X;\mathbb{Z}),\mathbb{Z}) & \xleftarrow{\sim} & H_1(X;\mathbb{Z}) \\ \downarrow & & & & \downarrow p \\ \overline{\Omega^1}(X) & & \xrightarrow{j} & & \Omega^1(X)^* \end{array}$$

(the isomorphism φ is defined in Lemma II.2.11 and the period map p in Proposition II.1.1). As Λ is defined as the image of p, Proposition II.2.12 follows from the commutativity of (II.2.4), which amounts to the following: let $\gamma \in H_1(X;\mathbb{Z})$; if $A \in \overline{\Omega^1}(X)$ is such that for any $\gamma' \in H_1(X;\mathbb{Z})$

$$\gamma\#\gamma' = \frac{1}{2\pi i}\int_{\gamma'}(A - \overline{A}), \tag{II.2.5}$$

then, for any $\alpha \in \Omega^1(X)$

$$\int_\gamma \alpha = \frac{1}{2\pi i}\int_X \alpha \wedge A. \tag{II.2.6}$$

Clearly, it is enough to check it for γ belonging to some basis of $H_1(X;\mathbb{Z})$. We may even suppose that γ is the first element a_1 of a symplectic basis $(a_1,\ldots,a_g,b_1,\ldots,b_g)$. Then, according to (II.2.5), we have

$$\frac{1}{2\pi i}\int_{a_j}(A - \overline{A}) = 0 \qquad (j = 1,\ldots,g) \tag{II.2.7}$$

and

$$\text{(II.2.8)} \qquad \frac{1}{2\pi i}\int_{b_j}(A-\overline{A})=\delta_{1j}.$$

On the other hand, since $\alpha \wedge \overline{A} = 0$, we have

$$\frac{1}{2\pi i}\int_X \alpha\wedge A = \frac{1}{2\pi i}\int_X \alpha\wedge(A-\overline{A}),$$

and if we apply A.2.5, we get

$$\frac{1}{2\pi i}\int_X \alpha\wedge A = \sum_{j=1}^{g}\left\{\int_{a_j}\alpha\cdot\frac{1}{2\pi i}\int_{b_j}(A-\overline{A}) - \frac{1}{2\pi i}\int_{a_j}(A-\overline{A})\cdot\int_{b_j}\alpha\right\}.$$

If we insert (II.2.7) and (II.2.8) in this equality, we get (II.2.6).

II.3. Abel's theorem and the algebraic description of the Jacobian

In the previous Section, we interpreted the Jacobian $\mathrm{Jac}(X)$ of a compact connected Riemann surface X as the space of isomorphism classes of holomorphic line bundles of degree zero over X. This was derived using a description of these line bundles in terms of $\overline{\partial}$-connections. On the other hand, we saw in Appendix B that holomorphic line bundles over X could be described in terms of divisors on X (*cf.* §B.4 and Corollary B.6.3). Combining the two approaches provides another interpretation of the Jacobian, which we explain in this Section and which will allow us to complete the proof of Theorem II.1.2.

II.3.1. Isomorphism classes of holomorphic line bundles and divisors.

The subgroup of $\mathrm{Div}(X)$ formed by the divisors of non-zero meromorphic function on X will be denoted $R(X)$. Two divisors D_1 and D_2 in $\mathrm{Div}(X)$ such that $D_1 - D_2 \in R(X)$ are said to be *linearly equivalent.*

Proposition II.3.1.

1) Any two divisors D_1 and D_2 on X are linearly equivalent iff the holomorphic line bundles $\mathcal{O}(D_1)$ and $\mathcal{O}(D_2)$ are isomorphic.

2) The quotient group $\mathrm{Div}(X)/R(X)$ *is isomorphic with the group of isomorphic classes of line bundles over X, via the map which associates the isomorphism class of the line bundle $\mathcal{O}(D)$ to the linear equivalence class of a divisor D over X.*

To prove 1), observe that $\mathcal{O}(D_1)$ and $\mathcal{O}(D_2)$ are isomorphic iff $\mathcal{O}(D_1)\otimes\mathcal{O}(D_2)^*$ is isomorphic with the trivial line bundle $\mathcal{O}$. As $\mathcal{O}(D_1)\otimes\mathcal{O}(D_2)^* \simeq \mathcal{O}(D_1-D_2)$ (*cf.* (B.3.6)), we are reduced to prove that a divisor D lies in $R(X)$ iff $\mathcal{O}(D)$ is isomorphic with $\mathcal{O}$. This follows from the fact that a non-vanishing section of $\mathcal{O}(D)$ is exactly a non-zero meromorphic function f on X such that $\mathrm{div}(f^{-1}) = D$ (compare (B.3.5)).

The assertion 2) follows immediately from 1) and from the fact that any holomorphic line bundle over X is isomorphic to the line bundle associated to some divisor (*cf.* Corollary B.6.3).

For any $d \in \mathbb{Z}$, we will denote $\mathrm{Div}_d(X)$ the set of divisors of degree d on X. Proposition II.3.1, 1) and the definition of the degree imply that $R(X)$ is included in $\mathrm{Div}_0(X)$ (this is also a consequence of Proposition B.2.1, applied to the meromorphic differentials $f^{-1} \cdot df$, $f \in \mathcal{M}(X)^*$). The isomorphism of Proposition II.3.1, 2) identifies the image $\mathrm{Div}_d(X)/R(X)$ of $\mathrm{Div}_d(X)$ in $\mathrm{Div}(X)/R(X)$ with the set $\mathrm{Pic}_d(X)$ of isomorphism classes of holomorphic line bundles of degree d over X.

If we compose the isomorphism

$$i_3 : \mathrm{Div}_0(X)/R(X) \simeq \mathrm{Pic}_0(X),$$

provided by Proposition II.3.1, 2), and the isomorphisms

$$i_1 : \mathrm{Pic}_0(X) \simeq \mathcal{C}/\mathcal{G}$$

and

$$i_2 = I^{-1} : \mathcal{C}/\mathcal{G} \simeq \mathrm{Jac}(X)$$

described in §II.2.1 and II.2.2, we get an isomorphism

$$i_2 \circ i_1 \circ i_3 : \mathrm{Div}_0(X)/R(X) \simeq \mathrm{Jac}(X). \tag{II.3.1}$$

The existence of such an isomorphism provides an algebraic description of the Jacobian Jac(X). Indeed, if X is realized as an algebraic curve (*cf.* §I.4.2) then $\mathcal{M}(X)$ may be identified with the field of restrictions to X of rational functions defined on the ambient space (*cf.* Theorem I.4.5). Therefore $R(X)$ and consequently the group $\mathrm{Div}_0(X)/R(X)$ may be defined in terms of purely algebraic objects, whereas the definition of Jac(X), involving the periods of X, was transcendental.

II.3.2. Abel's Theorem.

We now give an explicit formula for the isomorphism (II.3.1). It is the content of the following theorem, which is known as Abel's Theorem:

Theorem II.3.2. *Let $D = \sum_{i=1}^{r} Q_i - \sum_{i=1}^{r} P_i$ be any divisor of degree zero on X and, for $i = 1, \ldots, r$, let L_i be an oriented path from P_i to Q_i.*

The image by $i_2 \circ i_1$ of the isomorphism class of the holomorphic line bundle $\mathcal{O}(D)$ is the class in $\Omega^1(X)^*/\Lambda$ of the linear form

$$\begin{aligned} \Omega^1(X) &\to \mathbb{C} \\ \omega &\mapsto \sum_{i=1}^{r} \int_{L_i} \omega. \end{aligned}$$

Observe that if we replace the paths L_i by other paths L'_i, this linear form is modified by the addition of the 'periods' of the cycle $\sum_{i=1}^{r} (L'_i - L_i)$, which belong to Λ. Therefore its class in Jac(X) does not depend on the choice of the L_i's. More precisely, it is easily seen to depend only on D. It will be denoted $J(D)$.

The proof of Theorem II.3.2 is based on the following two lemmas, which we leave to the reader (the first is a simple consequence of the definitions; the second is an exercise in integration by parts).

Lemma II.3.3. *Let $\mathcal{L}$ be any holomorphic line bundle of degree zero over X. If $s \in \mathcal{C}^\infty(X;\mathcal{L})$ has no zero on X, then $\mathcal{L}$ is isomorphic to the holomorphic line bundle defined by the $\overline{\partial}$-connection $\overline{\nabla}_A$ where*

$$A = s^{-1} \cdot \overline{\partial}_{\mathcal{L}}\, s.$$

Lemma II.3.4. *Let $\varepsilon > 0$ and denote*

$$D_\varepsilon = \{z \in \mathbb{C} \mid |z| < \varepsilon\}.$$

If ω is a holomorphic 1-form defined on a neighbourhood of $\overline{D}_\varepsilon$ in $\mathbb{C}$, if z_1, $z_2 \in D_\varepsilon$ and if L is a path from z_1 to z_2 in D_ε, then the following equality holds

$$\int_{\partial D_\varepsilon} \log \frac{z - z_1}{z - z_2} \cdot \omega = 2\pi i \int_L \omega \tag{II.3.2}$$

(in the left hand side of (II.3.2), ∂D_ε is oriented counterclockwise, as usual, and $\log \frac{z-z_1}{z-z_2}$ denotes any determination of the logarithm of $\frac{z-z_1}{z-z_2}$ continuous along ∂D_ε).

The correspondence between tensor products of holomorphic line bundles, addition of divisors and addition of the 'gauge field' (Lemma II.2.5) shows that, to prove Theorem II.3.2, we can consider only the case $r = 1$, *i.e.*, divisors D of the form $Q - P$. Moreover, an easy connexity argument shows that we can assume that there exists a holomorphic coordinate chart (U, z) on X and $\varepsilon > 0$ such that

$$z(U) = D_\varepsilon$$

and such that U contains P and Q (see figure 27).

Consider the following meromorphic function on U:

$$s_0 = \frac{z - z(P)}{z - z(Q)}.$$

It may be seen as a non-vanishing holomorphic section of $\mathcal{O}(Q - P)$ on U. Choose $\varepsilon' \in]0, \varepsilon[$ such that $U' := z^{-1}(D_{\varepsilon'})$ contains P and Q. There exists a

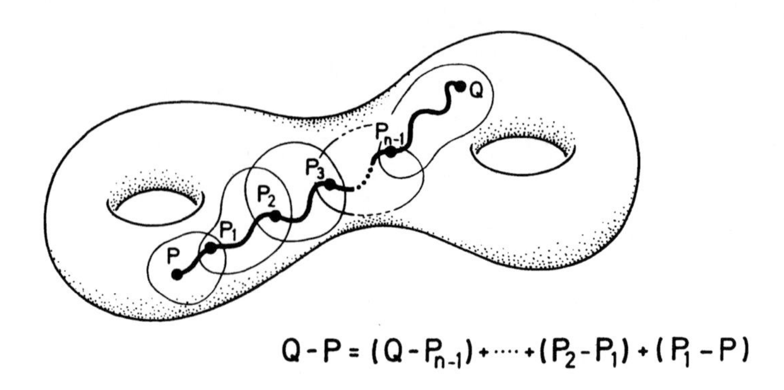

Fig. 27. Decomposition of divisors.

holomorphic determination $\log s_0$ of the logarithm of s_0 which is defined over $U - U'$. Choose also a function $\sigma \in \mathcal{C}^\infty(U;\mathbb{C})$ such that

$$\sigma = 1 \quad \text{on} \quad U' \tag{II.3.3}$$

and

$$\sigma = 0 \text{ on a neighbourhood of } \overline{X - U}. \tag{II.3.4}$$

Then define a function s on X as follows:

$$\begin{array}{lll} s & = s_0 & \text{on} \quad U' \\ s & = \exp\,(\sigma \cdot \log s_0) & \text{on} \quad U - U' \\ s & = 1 & \text{on} \quad X - U. \end{array}$$

This function is C^∞ and does not vanish on $X - \{P, Q\}$, and is meromorphic on U' and defines a C^∞ non-vanishing section of $\mathcal{O}(P - Q)$ over X. Therefore, according to Lemma II.3.3, the image by i_1 of the isomorphism class of $\mathcal{O}(Q-P)$ is the class in $\mathcal{C}/\mathcal{G}$ of

$$A = s^{-1} \cdot \overline{\partial}_{\mathcal{O}(Q-P)}\, s.$$

By construction of s, we have

$$A = 0 \quad \text{on} \quad U' \cup (X - U) \tag{II.3.5}$$

$$= \overline{\partial}(\sigma \cdot \log s_0) \quad \text{on} \quad \overline{U - U'}. \tag{II.3.6}$$

By definition, i_2 sends the class of A to the class in $\mathrm{Jac}(X)$ of the linear form on $\Omega^1(X)$ whose value on ω is $(2\pi i)^{-1} \int_X \omega \wedge A$. Now we get

$$\begin{aligned}
\frac{1}{2\pi i}\int_X \omega\cdot A &= \frac{1}{2\pi i}\int_{U-U'}\omega\wedge\overline{\partial}(\sigma\cdot\log s_0) && (cf. \text{(II.3.5) and (II.3.6)})\\
&= -\frac{1}{2\pi i}\int_{U-U'} d(\sigma\cdot\log s_0\cdot\omega) && (\text{since } \omega \text{ is a holomorphic differential})\\
&= -\frac{1}{2\pi i}\int_{\partial(U-U')}\sigma\cdot\log s_0\cdot\omega && (\text{by Stokes formula})\\
&= \frac{1}{2\pi i}\int_{\partial U'}\log s_0\cdot\omega && (cf. \text{(II.3.3) and (II.3.4)})
\end{aligned}$$

According to Lemma II.3.4, this expression equals $\int_L \omega$ for any path from P to Q in U'. This completes the proof of Theorem II.3.2.

In order to give a reformulation of Abel's Theorem, let us introduce a few notations. Let P_0 be any base point in X and let

$$J_{P_0} : X \to \mathrm{Jac}(X)$$

be the associated Jacobian embedding (*cf.* §II.1; in fact we do not have yet proved that J_{P_0} is an embedding). For any divisor $D = \sum_{i=1}^{N} n_i\ Q_i$ on X, let us define

$$J_{P_0}(D) = \sum_{i=1}^{N} n_i\ J_{P_0}(Q_i)$$

where, in the right hand side, the sum is computed in the Abelian group $\mathrm{Jac}(X)$. The maps

$$J_{P_0} : \mathrm{Div}_d(X) \to \mathrm{Jac}(X) \tag{II.3.7}$$

depend on the base point P_0 in a very simple way: for any other base point $P_0' \in X$ and any $D \in \mathrm{Div}_d(X)$, we have

$$J_{P_0'}(D) = J_{P_0}(D) - dJ_{P_0}(P_0').$$

In particular, the map $J_{P_0} : \mathrm{Div}_0(X) \to \mathrm{Jac}(X)$ does not depend on P_0 and is easily seen to coincide with the map J introduced just after the statement of Abel's Theorem.

The following assertions are direct consequences of Proposition II.3.1 and Theorem II.3.2, and constitute the classical version of Abel's theorem:

Corollary II.3.5. *Let $d \in \mathbb{Z}$. Any two divisors D_1 and D_2 in $\mathrm{Div}_d(X)$ are linearly equivalent iff*

$$J_{P_0}(D_1) = J_{P_0}(D_2).$$

Moreover the map

$$\begin{aligned}
\widetilde{J}_{P_0} : \mathrm{Div}_d(X)/R(X) &\to \mathrm{Jac}(X)\\
[D] &\mapsto J_{P_0}(D)
\end{aligned}$$

is a bijection, which coincides with $i_2 \circ i_1 \circ i_3$ when $d = 0$.

II.3.3. Completion of the proof of Theorem II.1.2.

Suppose now that X is a compact connected Riemann surface of genus $g \geq 1$. To complete the proof of Theorem II.1.2, we still need to prove the following facts (*cf.* §II.1.2):

i) the map $J_{P_0} : X \to \mathrm{Jac}(X)$ is injective;

ii) for any $z \in \mathrm{Jac}(X)$, the inverse image of z by

$$\mathcal{J}_{P_0} : \mathrm{Sym}_g X \to \mathrm{Jac}(X)$$

is connected.

To prove i), observe that if P and Q are points of X such that $J_{P_0}(P) = J_{P_0}(Q)$, then, according to Corollary II.3.5, there exists $f \in \mathcal{M}(X)^*$ such that

$$\mathrm{div}\, f = Q - P.$$

When X is not isomorphic with $\mathbb{P}^1\mathbb{C}$, hence when $g \geq 1$, this is impossible unless $Q = P$ (*cf.* Corollary B.2.4).

To prove ii), observe that $\mathrm{Sym}_g X$ may be seen as a subset of $\mathrm{Div}_g(X)$, namely the subset of effective divisors. Indeed the g-uple $(P_1, \ldots, P_g)$ up to permutation may be identified with the divisor $P_1 + \cdots + P_g$. Then the map $\mathcal{J}_{P_0}$ is nothing else than the restriction of $J_{P_0} : \mathrm{Div}_g(X) \to \mathrm{Jac}(X)$. Therefore Corollary II.3.5 shows that if $D_0 \in \mathrm{Sym}_g X$ is such that $\mathcal{J}_{P_0}(D_0) = z$, then $\mathcal{J}_{P_0}^{-1}(z)$ is the set of divisors $D' \in \mathrm{Sym}_g X$ such that the holomorphic line bundles $\mathcal{O}(D)$ and $\mathcal{O}(D_0)$ are isomorphic. The assertion ii) is now a particular case of the following general facts, which hold for any holomorphic line bundle $\mathcal{L}$ of degree $d > 0$ on a compact connected Riemann surface X:

• For any $D \in \mathrm{Div}_d(X)$, the following conditions are equivalent:

i) D is effective and the holomorphic line bundles $\mathcal{O}(D)$ and $\mathcal{L}$ are isomorphic;

ii) there exists $s \in H^0(X; \mathcal{L}) - \{0\}$ such that $\mathrm{div}(s) = D$.

This follows immediately from the basic facts concerning divisors of line bundles (*cf.* §B.3).

• The map $s \mapsto \mathrm{div}\, s$ from $H^0(X; \mathcal{L}) - \{0\}$ to the symmetric product $\mathrm{Sym}_d X$ (identified with effective divisors of degree d) is continuous. This follows easily from the compactness of $\mathrm{Sym}_d X$.

• Therefore, the set of divisors D which satisfy the condition i) above is connected, as the image of the connected set $H^0(X; \mathcal{L}) - \{0\}$ by a continuous map[25].

Exercise Use Riemann-Roch Theorem (*c.f.* §B.9) to show that any holomorphic line bundle $\mathcal{L}$ on X of degree g (resp. of degree 0) is isomorphic to a line bundle of the form $\mathcal{O}(\sum_{i=1}^g P_i)$ (resp. $\mathcal{O}(\sum_{i=1}^g P_i - gP_0)$). Combine this fact and Theorem II.3.2 to get another proof of the surjectivity of $\mathcal{J}_{P_0}$.

---

[25] More precisely, this set is easily seen to be isomorphic with the projective space $\mathbb{P}H^0(X; \mathcal{L}) = (H^0(X; \mathcal{L}) - \{0\})/\mathbb{C}^*$. In particular, the fibers of the map $\mathcal{J}_{P_0}$ are projective spaces (which, generically, are reduced to one point).

II.4. A historical digression.

The discussion of the preceding paragraph uses various modern concepts such as 'holomorphic line bundles', 'gauge groups', etc .... However, the main theorem asserting that the Jacobi map is essentially one to one may be formulated without these modern concepts and was indeed proved by Jacobi without them (at least for hyperelliptic curves). It may be of some interest to sketch the original approach of Abel to Abel's Theorem, which is quite simple and elegant, and which has been somewhat forgotten[26], contrary to the contribution of Riemann and his followers whose an account along the original lines may be found in modern textbooks.

To describe Abel's formulation, let us go back to the theory of elliptic integrals as it was developed at the end of the XVIII-th century. A general elliptic integral is an integral of the form

$$\int_a^x \frac{R(t)}{\sqrt{P(t)}}dt \tag{II.4.1}$$

where R is a rational function and P is a polynomial of degree 3 or 4. The basic example of such an integral—from which it derives its name—is the arc length on an ellipse: if we represent the ellipse

$$\frac{x^2}{a^2} + \frac{y^2}{b^2} = 1$$

parametrically by $x = a \sin\theta$ and $y = b\cos\theta$, then the arc length is given by

$$\int \sqrt{a^2\cos^2\theta + b^2\sin^2\theta}d\theta = \int a\sqrt{1-k^2\sin^2\theta}d\theta$$

where $k = \sqrt{1-b^2/a^2}$ is the eccentricity of the ellipse (we assume $a > b$). Setting $x = \sin\theta$, the integral becomes

$$a\int \frac{1-k^2x^2}{\sqrt{(1-x^2)(1-k^2x^2)}}dx$$

which indeed has the form (II.4.1).

As a function of x, an elliptic integral cannot be expressed in terms of elementary functions. However, it satisfies an addition formula, which generalizes (II.1.3):

$$\int_a^x \frac{R(t)}{\sqrt{P(t)}}dt + \int_a^y \frac{R(t)}{\sqrt{P(t)}}dt = \int_a^z \frac{R(t)}{\sqrt{P(t)}}dt + W(x,y) \tag{II.4.2}$$

where z is a rational function of $x, y, \sqrt{P(x)}$ and $\sqrt{P(y)}$, and where W is the sum of a rational function of $x, y, \sqrt{P(x)}$ and $\sqrt{P(y)}$ and of the logarithm

[26] See [Gr1] for a noteworthy counterexample to this last statement.

of such a function (according to (II.1.3), when $R(t)$ is constant, W is also constant).

In his work, Abel considers, instead of $\sqrt{P(x)}$, any 'algebraic function' $y(x)$ of x, defined implicitly by an irreducible polynomial equation[27]

$$\chi(x,y)=0, \tag{II.4.3}$$

and, instead of an elliptic integral, a general 'Abelian integral'

$$\psi(x)=\int_a^x f(t,y(t))dt\ ,$$

where f is any rational function of two variables. Moreover, instead of the sum of two elliptic integrals which occurs in the addition formula (II.4.2), he considers the sum $\psi(x_1)+\cdots+\psi(x_m)$ of the values of the Abelian integrals at a finite number of points $x_1,\ldots,x_m$.

In modern terms, the polynomial equation (II.4.3) defines an algebraic curve, hence by normalization a compact Riemann surface; the rational function f determines a meromorphic differential

$$\omega=f(x,y)dx$$

on X, and $\psi(x)$ is the multivalued function obtained by integration of ω along paths on $X-\Sigma$, where Σ denotes the finite set of points of X where ω is not holomorphic (observe that even considered as a function of $(x,y)\in X$, ψ is multivalued; indeed, for any two points A,B in $X-\Sigma$ to define the integral $\int_A^B\omega$, one needs to choose a path from A to B in $X-\Sigma$ and the value of the integral depends, in general, of the homology class of this path).

The key idea of Abel is to consider an auxiliary relation

$$\theta(x,y,a_1,\ldots,a_r)=0, \tag{II.4.4}$$

defined by a polynomial θ in (x,y) which depends rationally on parameters $(a_1,\ldots,a_r)$, and to study the sum

$$v=\psi(M_1)+\cdots+\psi(M_m)$$

when $M_1,\ldots,M_m$ are the intersections[28] of the curves defined by $\chi(x,y)=0$ and by (II.4.4).

Using the modern terminology, we can introduce the divisor on X

$$D(a_1,\ldots,a_r)=\sum_{i=1}^m M_i.$$

In general, it depends on $(a_1,\ldots,a_r)$. However, its class modulo linear equivalence is independent on $(a_1,\ldots,a_r)$. Indeed, at least for generic values of $(a'_1,\ldots,a'_r,a_1,\ldots,a_r)$, we have:

---

[27] We follow Abel's own notations.

[28] Possibly counted with some multiplicities.

$$D(a'_1, \ldots, a'_r) - D(a_1, \ldots, a_r) = \mathrm{div} g$$

where g is the meromorphic function on X defined by the rational function

$$\frac{\theta(x, y, a'_1, \ldots, a'_r)}{\theta(x, y, a_1, \ldots, a_r)}.$$

Abel's version of Abel's theorem is the statement that v *is a linear combination of a rational function of* $(a_1, \ldots, a_r)$ *and of the logarithms of such rational functions.*

Let us sketch Abel's proof[29]:

• Let n be the degree of $\chi(x, y)$ in the variable y, and let us denote by $y_1(x), \ldots, y_n(x)$ the n roots of $\chi(x, y) = 0$, considered as an equation in y. Without loss of generality, we may suppose that θ is a polynomial in $x, y, a_1, \ldots, a_r$. Then

$$\rho(x, a_1, \ldots, a_r) := \prod_{i=1}^{n} \theta(x, y_i(x), a_1, \ldots, a_r)$$

is easily seen to be a polynomial in $x, a_1, \ldots, a_r$. In fact ρ is nothing else than the resultant deduced from $\chi(x, y)$ and $\theta(x, y)$ by elimination of y. The relation

$$\rho(x, a_1, \ldots, a_r) = 0 \qquad \text{(II.4.5)}$$

characterizes the x's for which there exists y such that

$$\chi(x, y) = 0 \quad \text{and} \quad \theta(x, y, a_1, \ldots, a_r) = 0.$$

Moreover, if x satisfies (II.4.5), there exists a unique y such that this system is satisfied, which is given by a rational expression

$$y = \varphi(x, a_1, \ldots, a_r)$$

of x and $a_1, \ldots, a_r$.

• The resultant ρ may be factorized as $\rho = F_0 \cdot F$, where F_0 is the largest factor of ρ which does not involve the variable x. The relation $F(x, a_1, \ldots, a_r) = 0$ defines x as an algebraic function of $a_1, \ldots, a_r$, with several determinations $x_1, \ldots, x_m$ ($m =$ degree of F in x). Let

$$y_j = \varphi(x_j, a_1, \ldots, a_r)$$

be the corresponding values of y.

• To prove Abel's theorem, it is enough to prove that the differential

$$dv = \sum_{i=1}^{r} \frac{\partial v}{\partial a_i} \, da_i$$

[29] The following argument implicitly assumes some hypotheses of genericity on the choice of coordinates and on θ.

of v considered as a function of $(a_1, \ldots, a_r)$ has coefficients $\partial v/\partial a_i$ which are rational functions of $(a_1, \ldots, a_r)$ (this is clear when $r = 1$; see *e.g.* [Gr1], pp.328-329, for the general case).

By definition of v, we have the identity

$$dv = \sum_{j=1}^{m} f(x_j, y_j) dx_j, \tag{II.4.6}$$

where x_j and y_j are seen as function of $(a_1, \ldots, a_r)$. The differentials dx_j may be computed by differentiation of the relation

$$F(x_j, a_1, \ldots, a_r) = 0.$$

Indeed, if we denote

$$\delta F(x_j, a_1, \ldots, a_r) = \sum_{i=1}^{r} \frac{\partial F}{\partial a_i}(x_i, a_1, \ldots, a_r) da_i$$

and

$$F'(x_j, a_1, \ldots, a_r) = \frac{\partial F}{\partial x}(x_j, a_1, \ldots, a_r)$$

we get

$$dx_j = -\frac{\delta F(x_j, a_1, \ldots, a_r)}{F'(x_j, a_1, \ldots, a_r)}$$

and

$$f(x_j, y_j) dx_j = -\frac{f(x_j, \varphi(x_j, a_1, \ldots, a_r))}{F'(x_j, a_1, \ldots, a_r)} \delta F(x_j, a_1, \ldots, a_r).$$

Inserting these relations in (II.4.6), we get the identity

$$dv = \sum_{i=1}^{r} Q_i(x_1, \ldots, x_m, a_1, \ldots, a_r) da_i$$

for some rational functions Q_i of $(x_1, \ldots, x_m, a_1, \ldots, a_r)$ which clearly are symmetric in the x_j's. As these are the solutions of $F(x, a_1, \ldots, a_r) = 0$, any rational symmetric function of the x_j's is a rational function of the a_i's. Finally, we obtain that

$$dv = \sum_{i=1}^{r} R_i(a_1, \ldots, a_r) da_i$$

where the R_i's are rational functions, as was to be proved.

Abel also looks for conditions on ω which imply that v is constant. He proves that it is the case when

$$f(x, y) = \left(\frac{\partial \chi}{\partial y}\right)^{-1} g(x, y),$$

where $g(x,y)$ is a polynomial which satisfies a degree condition and some linear relations, which depend only on the polynomial χ (and not on θ). These conditions define a vector space of forms ω, of finite dimension γ. This vector space contains the space $\Omega^1(X)$ of holomorphic differential forms on X[30]. Consequently, if we apply Abel's results to the forms $\omega \in \Omega^1(X)$, we recover that, for any θ, the image of $D(a_1,\ldots,a_r)$ by the Jacobian embedding is independent of $(a_1,\ldots,a_r)$: this is essentially equivalent to the fact that the Jacobi map (II.3.7) is well defined on divisors modulo linear equivalence (*cf.* Corollary II.3.5).

In Abel's work, the genus of X appears in the following guise. Abel remarks that we can choose arbitrary r points $P_1,\ldots,P_r$ on X and that, in general, when θ depends linearly on $(a_1,\ldots,a_r)$, there exists a unique value of $(a_1,\ldots,a_r)$ such that $P_1,\ldots,P_r$ occur in the divisor $D(a_1,\ldots,a_r)$. If we denote by $Q_1,\ldots,Q_{m-r}$ the other points occurring in $D(a_1,\ldots,a_r)$, this implies that *one can associate to any* r *points* $P_1,\ldots,P_r$ *a finite set* $Q_1,\ldots,Q_{m-r}$ *of* $m-r$ *points which depend algebraically on* $P_1,\ldots,P_r$ *in such a way that a sum of* r *Abelian integrals* $\sum_{i=1}^r \int_{P_0}^{P_i}\omega$ *is equal to the sum* $-\sum_{i=1}^{m-r}\int_{P_0}^{Q_i}\omega$ *of* $m-r$ *Abelian integrals, up to some algebraic function.*

Abel proves that, for an arbitrary r, one may choose θ such that $m-r$ assumes a minimal value: this minimal value is independent of r, and defines an invariant p of the algebraic curve $\chi(x,y)=0$. This numerical invariant p is nothing else than the genus of X. The correspondence between $\mathrm{Sym}_r X$ and $\mathrm{Sym}_p X$ which associates $(Q_1,\ldots,Q_p)$ to $(P_1,\ldots,P_r)$ is indeed given by the inverse image of 0 by the following map

$$\mathrm{Sym}_r X \times \mathrm{Sym}_p X \to \mathrm{Jac}(X)$$

$$((P_1,\ldots,P_r),(Q_1,\ldots,Q_p)) \mapsto \sum_{i=1}^r J_{P_0}(P_i) + \sum_{i=1}^p J_{P_0}(Q_i).$$

These results of Abel are contained in his great work, *Mémoire sur une propriété générale d'une classe très étendue de fonctions transcendantes* ([A3]), which was presented at the Académie des Sciences de Paris in 1826 but was published only in 1841, after the death of Abel (1829) and even after the publication of the first edition of his collected works. Before 1841, only two short notes on this work had been published by Abel. They contain some of his results in the hyperelliptic case ([A1]), and the statement of Abel's theorem in his general form, with a sketch of proof ([A2]; see figure 28).

The work of Abel served as a foundation to the work of Jacobi on the 'inversion problem' for hyperelliptic integrals. We have seen that the Euler addition formula allows to extend the inverse function of an elliptic integral of the first kind, which is first only locally defined, to a holomorphic function

[30] In general we can have $\gamma > g(=\dim \Omega^1(X))$: some differential ω 'of the third kind' on X, *i.e.*, some meromorphic differential forms whose residue at any point of X vanishes, may give rise to constant functions v; see [Ho] pp.74 and 95 for a discussion and references on this point.

14.

Démonstration d'une propriété générale d'une certaine classe de fonctions transcendentes.

(Par Mr. *N. H. Abel.*)

Théorème. Soit y une fonction de x qui satisfait à une équation quelconque irréductible de la forme:

$$1.\quad 0 = p_0 + p_1.y + p_2.y^2 + \ldots. + p_{n-1}.y^{n-1} + y^n,$$

où $p_0, p_1, p_2, \ldots. p_{n-1}$ sont des fonctions entières de la variable x. Soit

$$2.\quad 0 = q_0 + q_1.y + q_2.y^2 + \ldots. + q_{n-1}.y^{n-1},$$

une équation semblable, $q_0, q_1, q_2, \ldots. q_{n-1}$ étant également des fonctions entières de x, et supposons variables les coëfficiens des diverses puissances de x dans ces fonctions. Nous désignerons ces coëfficiens par $a, a', a'', \ldots.$ En vertu des deux équations (1.) et (2.) x sera fonction de $a, a', a'', \ldots.$ et on en déterminera les valeurs en éliminant la quantité y. Désignons par:

$$3.\quad \varrho = 0$$

le résultat de l'élimination, en sorte que ϱ ne contiendra que les variables $x, a, a', a'', \ldots.$ Soit μ le degré de cette équation par rapport à x, et désignons par

$$4.\quad x_1, x_2, x_3, \ldots. x_\mu$$

ses μ racines, qui seront autant de fonctions de $a, a', a'', \ldots.$ Cela posé, si l'on fait

$$5.\quad \psi x = \int f(x, y).\partial x$$

où $f(x, y)$ désigne une fonction rationnelle quelconque de x et de y, je dis, que la fonction transcendente ψx jouira de la propriété générale exprimée par l'équation suivante:

$$6.\quad \psi x_1 + \psi x_2 + \ldots. + \psi x_\mu = u + k_1 \log v_1 + k_2 \log v_2 + \ldots. + k_n \log v_n,$$

$u, v_1, v_2, \ldots. v_n$ étant des fonctions rationnelles de $a, a', a'', \ldots.$, et $k_1, k_2, \ldots. k_n$ des constantes.

Démonstration. Pour prouver ce théorème il suffit d'exprimer la différentielle du premier membre de l'équation (6.) en fonction de $a, a', a'', \ldots.$; car il se réduira par la à une différentielle rationnelle, comme on le verra. D'abord les deux équations (1.) et (2.) donneront y en fonction rationnelle de $x, a, a', a'', \ldots.$ De même l'équation (3.): $\varrho = 0$ donnera pour ∂x une expression de la forme:

$$\partial x = \alpha.\partial a + \alpha'.\partial a' + \alpha''.\partial a'' + \ldots.,$$

où $\alpha, \alpha', \alpha'', \ldots.$ sont des fonctions rationnelles de $x, a, a', a'', \ldots.$ De là il suit que la différentielle $f(x, y).\partial x$ pourra être mise sous la forme:

$$f(x, y)\partial x = \varphi x.\partial a + \varphi_1 x.\partial a' + \varphi_2 x.\partial a'' + \ldots.,$$

où $\varphi x, \varphi_1 x, \ldots.$ sont des fonctions rationnelles de $x, a, a', a'', \ldots.$ En intégrant, il viendra:

$$\psi x = \int \{\varphi x.\partial a + \varphi_1 x.\partial a' + \ldots.\}$$

et de là on tire, en remarquant que cette équation aura lieu en mettant pour x les μ valeurs de cette quantité:

$$7.\quad \psi x_1 + \psi x_2 + \ldots. + \psi x_\mu$$

$$= \int [\{\varphi x_1 + \varphi x_2 + \ldots. + \varphi x_\mu\}\partial a + \{\varphi_1 x_1 + \varphi_1 x_2 + \ldots. + \varphi_1 x_\mu\}\partial a' + \ldots.]$$

Dans cette équation les coëfficiens des différentielles $\partial a, \partial a', \ldots.$ sont des fonctions rationnelles de $a, a', a'', \ldots.$ et de $x_1, x_2, \ldots. x_\mu$, mais d'ailleurs ils sont symétriques par rapport à $x_1, x_2, \ldots. x_\mu$; donc, en vertu d'un théorème connu, on pourra exprimer ces fonctions rationnellement par $a, a', a'', \ldots.$ et par les coëfficiens de l'équation $\varrho = 0$; mais ceux-ci sont eux-mêmes des fonctions rationnelles des variables $a, a', a'', \ldots.$, donc enfin les coëfficiens de $\partial a, \partial a', \partial a'', \ldots.$ de l'équation (7.) le seront également. Donc, en intégrant, on aura une équation de la forme (6.).

Je me propose de développer dans une autre occasion des nombreuses applications de ce théorème, qui jetteront un grand jour sur la nature des fonctions transcendentes dont il s'agit.

Christiania, le 6. Janvier 1829.

Fig. 28. Abel's original proof of 'Abel's Theorem' ([A2]).

defined on the whole complex plane. Around 1832, Jacobi discovered how Abel's theorem can be similarly used to 'invert' Abelian integrals.

First, in [J1], Jacobi derived from Abel's theorem an addition formula for hyperelliptic Abelian integrals. He proceeded as follows, in the case of Abelian integrals associated with hyperelliptic curves of genus 2, given by an equation

$$y^2 = P(x), \tag{II.4.7}$$

where P is a polynomial of degree 5 or 6. He introduced the two functions

$$\varphi(x) = \int_0^x \frac{dt}{\sqrt{P(t)}} \tag{II.4.8}$$

and

$$\varphi_1(x) = \int_0^x \frac{tdt}{\sqrt{P(t)}}. \tag{II.4.9}$$

Abel's theorem shows that, for any x, x', y and y', the system of two equations with unknown a and b

$$\varphi(a) + \varphi(b) = \varphi(x) + \varphi(y) + \varphi(x') + \varphi(y')$$

$$\varphi_1(a) + \varphi_1(b) = \varphi_1(x) + \varphi_1(y) + \varphi_1(x') + \varphi_1(y')$$

has solutions which may be expressed algebraically in terms of x, y, x' and y'. Now, if one sets

$$u = \varphi(x) + \varphi(y) \qquad v = \varphi_1(x) + \varphi_1(y)$$

and

$$u' = \varphi(x') + \varphi(y') \qquad v' = \varphi_1(x') + \varphi_1(y')$$

the preceding relations may be written

$$u + u' = \varphi(a) + \varphi(b)$$

$$v + v' = \varphi_1(a) + \varphi_1(b).$$

Therefore if we express x and y in terms of u and v, namely

$$x = \lambda(u, v), \qquad y = \lambda_1(u, v),$$

then we have:

$$x' = \lambda(u', v'), \qquad y' = \lambda_1(u', v'),$$

and

$$a = \lambda(u + u', v + v'), \qquad b = \lambda_1(u + u', v + v').$$

400 32. *C. G. J. Jacobi, considerationes generales de transcendentibus Abelianis.*

Theorema.

„*Designante X functionem integram rationalem ordinis quinti*
„*aut sexti, ponatur*

$$\int_0^x \frac{dx}{\sqrt{X}} = \Phi(x), \quad \int_0^x \frac{x\,dx}{\sqrt{X}} = \Phi_1(x);$$

„*sint porro*

$$x = \lambda(u, v), \quad y = \lambda_1(u, v)$$

„*functiones tales argumentorum u, v, ut simul sit:*

$$\Phi(x) + \Phi(y) = u, \quad \Phi_1(x) + \Phi_1(y) = v,$$

„*gaudebunt functiones illae*

$$\lambda(u, v), \quad \lambda_1(u, v)$$

„*proprietate ei simili, quae de functionibus trigonometricis et ellipticis*
„*in elementis proponitur, ut functiones illae argumentorum binominum*

$$u + u', \quad v + v'$$

„*algebraice exhibeantur per functiones, quae ad singula nomina*

$$u, v; \quad u', v'$$

„*pertinent; sive ut functiones*

$$\lambda(u + u', v + v'), \quad \lambda_1(u + u', v + v')$$

„*algebraice exhibeantur per functiones*

$$\lambda(u, v), \quad \lambda(u', v')$$
$$\lambda_1(u, v), \quad \lambda_1(u', v')."$$

Fig. 29. Jacobi's original work on the inversion of hyperelliptic integrals [J1].

In this way, Jacobi proved that *the functions* $\lambda(u+u', v+v')$ *and* $\lambda_1(u+u', v+v')$ *are algebraic functions of* $\lambda(u, v)$, $\lambda_1(u, v)$, $\lambda(u', v')$ *and* $\lambda_1(u', v')$ (see figure 29).

In modern terms, this algebraic addition formula is nothing else than the addition law on the Jacobian of the curve X defined by (II.4.7) expressed on S^2X (which is 'almost' isomorphic to Jac(X); *cf.* Theorem II.1.2, 2)).

Two years later, in 1834 (see [J2]), Jacobi came back to the study of the functions (II.4.8) and (II.4.9), met the problem of their multivaluedness, and discovered that λ and λ_1 admit four periods (in $\mathbb{C}^2$). As a function of one complex variable cannot have more than two independent periods[31], he was led to discover that the inversion problem, to be well posed, had to involve functions of several complex variables.

The inversion problem, for hyperelliptic Abelian integrals of genus 2, as posed by Jacobi, is to construct periodic entire functions on $\mathbb{C}^2$ with values in $\mathrm{Sym}_2 X$ (or rather in a variety birational to $\mathrm{Sym}_2 X$) which would define an inverse of the $\mathcal{J}$ map.

This problem was solved by Göpel and Rosenhain in 1847, who introduced for this purpose *theta functions of two variables*. The generalized problem concerning hyperelliptic curves of arbitrary genus was solved by Weierstrass, in

[31] It may be interesting to note that the basic fact that a discrete subgroup of $\mathbb{R}^2$ contains at most two rationally independent vectors was proven for the first time in this paper.

papers published from 1848 to 1856. These works motivated the study of theta functions of several variables, which were successfully used by Riemann in his great paper of 1857, *Theorie der Abel'schen Functionen* ([Ri1]), to give a solution of the inversion problem for an arbitrary algebraic curve. In the next Section, we try to explain Riemann's discoveries.

Appendix D. The surjectivity of some holomorphic maps

In §II.1.2, we used the following fact:

Let $f : X \to Y$ be a holomorphic map between compact connected complex manifolds of the same dimension. If there is a point $x_0 \in X$ such that the differential of f at x_0

$$Df(x_0) : T_{x_0}X \to T_{f(x_0)}Y$$

is an isomorphism, then f is onto.

Let us give a short proof of this statement, based on the elementary properties of the *degree* of C^∞ maps between compact manifolds of the same dimension (*cf.* [Mil]).

As complex manifolds, X and Y are naturally oriented, and, since f is holomorphic, for any $x \in X$ such that $Df(x)$ is bijective, $Df(x)$ is orientation preserving. This shows that, for any regular value $y \in Y$, the degree of f is the cardinality of $f^{-1}(y)$.

If $x_0 \in X$ is such that $Df(x_0)$ is bijective, then by the implicit function theorem, there exists an open neighbourhood U (resp. V) of x_0 (resp. of $f(x_0)$) in X (resp. Y) such that f maps bijectively and biholomorphically U onto V. The non-empty open subset V contains a regular value of f, whose inverse image is clearly non-empty and f has positive degree, hence is onto.

Suppose now that, moreover, *all the sets $f^{-1}(y)$, $y \in Y$, are connected.* Then f is of degree one; indeed, for any regular value y of f, $f^{-1}(y)$ is a connected finite set, *i.e.*, a one point set. Hence *there exists open dense subsets $U \subset X$ and $V \subset Y$ such that f sends biholomorphically U onto V*. In fact, it is possible to show that one can find U and V satisfying these conditions and such that $X - U$ is an analytic subset[32] of X of codimension ≥ 1 and $Y - V$ an analytic subset of Y of codimension ≥ 2 (see for instance [Mu3], §3B). Such a map f is called a *birational* holomorphic map.

[32] *i.e.*, a subset of X defined locally by the vanishing of a finite family of holomorphic functions.

III. Abelian varieties

III.1. Riemann bilinear relations

III.1.1. Integrals of products of closed 1-forms over a compact Riemann surface.

In the preceding Section, we associated to any compact connected Riemann surface X of genus g a complex torus $\mathrm{Jac}(X)$ of complex dimension g, the Jacobian of X. This complex torus can be described as the quotient $\Omega^1(X)^*/\Lambda$ where $\Omega^1(X)^*$ is the vector space dual to the space of holomorphic differentials on X and where Λ is the period lattice of X, defined as the image of the injection

$$\text{(III.1.1)} \qquad \begin{aligned} p : H_1(X, \mathbb{Z}) &\to \Omega^1(X)^* \\ \gamma &\mapsto (\omega \mapsto \int_\gamma \omega). \end{aligned}$$

In this Section and the following one, we will show that, as a lattice in the complex vector space $\Omega^1(X)^*$, Λ has very special properties. This will be a consequence of the following observations:

- *for any* $\omega, \omega' \in \Omega^1(X)$,

$$\text{(III.1.2)} \qquad \int_X \omega \wedge \omega' = 0\,, \qquad \text{(since } \omega \wedge \omega' = 0)$$

- *for any* $\omega \in \Omega^1(X) - \{0\}$,

$$\text{(III.1.3)} \qquad i \int_X \omega \wedge \overline{\omega} > 0\,,$$

together with the following lemma which was proved in A.2.5:

Lemma III.1.1. *Let X be any oriented compact connected differentiable surface of genus $g \geq 1$. For any symplectic basis $(a_1, \dots, a_g, b_1, \dots, b_g)$ of $H_1(X; \mathbb{Z})$ and for any two closed 1-forms η and η' on X, we have:*

$$\text{(III.1.4)} \qquad \int_X \eta \wedge \eta' = \sum_{i=1}^{g} \left(\int_{a_i} \eta \cdot \int_{b_i} \eta' - \int_{a_i} \eta' \cdot \int_{b_i} \eta \right).$$

III.1.2. Riemann bilinear relations.

Consider now a basis $(\omega_1, \dots, \omega_g)$ of $\Omega^1(X)$ and a symplectic basis $(a_1, \dots, a_g, b_1, \dots, b_g)$ of $H_1(X; \mathbb{Z})$.

Let P be the matrix of $(p(a_1), \dots, p(a_g), p(b_1), \dots, p(b_g))$ in the basis of $\Omega^1(X)^*$ dual to $(\omega_1, \dots, \omega_g)$. In other words

$$M = (A, B),$$

where A and B are the matrices in $M_g(\mathbb{C})$ defined by

$$A_{ij} = \int_{a_j} \omega_i \quad \text{and} \quad B_{ij} = \int_{b_j} \omega_i.$$

We can now state:

Theorem III.1.2. *1) The matrix A is invertible.*

2) The matrix $\Omega = A^{-1}B$ is symmetric and its imaginary part $\operatorname{Im}\Omega := (\operatorname{Im}\Omega_{ij})_{1\leq i,j\leq g}$ *is a positive*[33] *matrix.*

These conditions on the matrix (A, B) are known as *Riemann bilinear relations* (but were known long before Riemann in many special cases, *e.g.*, when X is hyperelliptic) and the subset $\mathfrak{H}_g$ of $M_g(\mathbb{C})$ defined by the conditions

$${}^t\Omega = \Omega \quad \text{and} \quad \operatorname{Im}\Omega > 0$$

is called nowadays the *Siegel upper half space.* When $g = 1$, it is nothing else than the Poincaré upper half plane:

$$\mathfrak{H}_1 = \mathfrak{H} = \{\tau \in \mathbb{C} \mid \operatorname{Im}\tau > 0\}.$$

The matrix Ω in Theorem III.1.2 is easily seen to be independent of the basis $(\omega_1, \ldots, \omega_g)$ of $\Omega^1(X)$. It is known as the *period matrix* of the Riemann surface X associated to the symplectic basis $(a_1, \ldots, a_g, b_1, \ldots, b_g)$ of $H_1(X;\mathbb{Z})$. This matrix Ω may be obtained from $(a_1, \ldots, a_g, b_1, \ldots, b_g)$ using the following recipe: according to Theorem III.1.2. 1), there exists a unique basis $(\omega_1, \ldots, \omega_g)$ of the space $\Omega^1(X)$ of holomorphic one forms on X such that

$$\int_{a_j} \omega_i = \delta_{ij}; \tag{III.1.7}$$

then we have

$$\Omega_{ij} = \int_{b_j} \omega_i. \tag{III.1.8}$$

If X is the elliptic curve $\mathbb{C}/(\mathbb{Z} + \tau\mathbb{Z})$ and if we choose as symplectic basis of $H_1(X;\mathbb{Z})$ the pair (a, b), where a (resp. b) is the class of the map from $\mathbb{R}/\mathbb{Z}$ to X which sends $[t] \in \mathbb{R}/\mathbb{Z}$ to $[t] \in X$ (resp. $[\tau t] \in X$), then the condition (III.1.7) reduces to $\int_a \omega_1 = 1$, implies that $\omega_1 = dz$, and (III.1.8) shows that Ω coincides with τ.

Observe that when $\Omega^1(X)^*$ is identified with $\mathbb{C}^g$ by using the basis dual to the basis $(\omega_1, \ldots, \omega_g)$ defined by (III.1.7), then Λ appears as the lattice $\mathbb{Z}^g + \Omega\mathbb{Z}^g$, and $\operatorname{Jac}(X)$ may be seen as the complex torus $\mathbb{C}^g/(\mathbb{Z}^g + \Omega\mathbb{Z}^g)$.

Let us prove Theorem III.1.2.

Let $\lambda = (\lambda_1, \ldots, \lambda_g) \in \mathbb{C}^g$ be such that

[33] *i.e.*, for any $v \in \mathbb{R}^g - \{0\}$, ${}^tv \cdot \operatorname{Im}\Omega \cdot v > 0$.

welche jedes Paar mit dem folgenden verbinden. Es möge c_ν von einem Punkte von b_ν nach einem Punkte von $a_{\nu+1}$ gehen. Das Schnittnetz wird als so entstanden betrachtet, dafs der $\overline{2\nu-1}$te Querschnitt aus $c_{\nu-1}$ und der von dem Endpunkte von $c_{\nu-1}$ zu diesem zurückgezogenen Linie a_ν besteht, und der 2νte durch die von der *positiven* auf die *negative* Seite von a_ν gezogene Linie b_ν gebildet wird. Die Begrenzung der Fläche besteht bei dieser Zerschneidung nach einer geraden Anzahl von Schnitten aus *einem*, nach einer ungeraden aus zwei Stücken.

Ein allenthalben endliches Integral w einer rationalen Function von s und z nimmt dann zu beiden Seiten einer Linie c denselben Werth an. Denn die ganze früher entstandene Begrenzung besteht, wie bemerkt, aus *einem* Stücke und bei der Integration längs derselben von der einen Seite der Linie c bis auf die andere wird $\int \partial w$ durch jedes früher entstandene Schnittelement zweimal, in entgegengesetzter Richtung, erstreckt. Eine solche Function ist daher in T' allenthalben aufser den Linien a und b stetig. Die durch diese Linien zerschnittene Fläche T möge durch T'' bezeichnet werden.

20.

Es seien nun $w_1, w_2, \ldots, w_p$ von einander unabhängige solche Functionen, und der Periodicitätsmodul von w_μ an dem Querschnitte a_ν gleich $A_\mu^{(\nu)}$ und an dem Querschnitte b_ν gleich $B_\mu^{(\nu)}$. Es ist dann das Integral $\int w_\mu\, dw_{\mu'}$, um die Fläche T'' positiv herum ausgedehnt, $=0$, da die Function unter dem Integralzeichen allenthalben endlich ist. Bei dieser Integration wird jede der Linien a und b zweimal, einmal in positiver und einmal in negativer Richtung durchlaufen, und es mufs während jener Integration, wo sie als Begrenzung des positiverseits gelegenen Gebiets dient, für w_μ der Werth auf der positiven Seite oder w_μ^+, während dieser der Werth auf der negativen oder w_μ^- genommen werden. Es ist also dies Integral gleich der Summe aller Integrale $\int (w_\mu^+ - w_\mu^-)\, dw_{\mu'}$ durch die Linien a und b. Die Linien b führen von der positiven zur negativen Seite der Linien a, und folglich die Linien a von der negativen zur positiven Seite der Linien b. Das Integral durch die Linie a_ν ist daher $= \int A_\mu^{(\nu)} dw_{\mu'} = A_\mu^{(\nu)} \int dw_{\mu'} = A_\mu^{(\nu)} B_{\mu'}^{(\nu)}$, und das Integral durch die Linie $b_\nu = \int B_\mu^{(\nu)} dw_{\mu'} = -B_\mu^{(\nu)} A_{\mu'}^{(\nu)}$. Das Integral $\int w\, dw_{\mu'}$, um die Fläche T'' positiv herum erstreckt, ist also $= \sum_\nu (A_\mu^{(\nu)} B_{\mu'}^{(\nu)} - B_\mu^{(\nu)} A_{\mu'}^{(\nu)})$, und diese Summe folglich $=0$. Diese Gleichung gilt für je zwei von den Functionen $w_1, w_2, \ldots, w_p$ und liefert also $\frac{p(p-1)}{1.2}$ Relationen zwischen deren Periodicitätsmoduln.

Nimmt man für die Functionen w die Functionen u oder wählt man sie so, dafs $A_\mu^{(\nu)}$ für ein von μ verschiedenes ν gleich 0 und $A_\nu^{(\nu)} = \pi i$ ist, so gehen diese Relationen über in $B_{\mu'}^{\mu}\pi i - B_\mu^{\mu'}\pi i = 0$ oder in $a_{\mu,\mu'} = a_{\mu',\mu}$.

21.

Es bleibt noch zu zeigen, dafs die Gröfsen a die zweite oben nöthig gefundene Eigenschaft besitzen.

Man setze $w = \mu + \nu i$ und den Periodicitätsmodul dieser Function an dem Schnitte a_ν gleich $A^{(\nu)} = \alpha_\nu + \gamma_\nu i$ und an dem Schnitte b_ν gleich $B^{(\nu)} = \beta_\nu + \delta_\nu i$. Es ist dann das Integral $\int \left(\left(\frac{\partial \mu}{\partial x}\right)^2 + \left(\frac{\partial \mu}{\partial y}\right)^2\right) dT$ oder $\int \left(\frac{\partial \mu}{\partial x}\frac{\partial \nu}{\partial y} - \frac{\partial \mu}{\partial y}\frac{\partial \nu}{\partial x}\right) dT$*) durch die Fläche T'' gleich dem Begrenzungsintegral $\int \mu\, d\nu$ um T'' positiv herum erstreckt, also gleich der Summe der Integrale $\int (\mu^+ - \mu^-)\, d\nu$ durch die Linien a und b. Das Integral durch die Linie a_ν ist $= \alpha_\nu \int d\nu = \alpha_\nu \delta_\nu$, das Integral durch die Linie b_ν gleich $\beta_\nu \int d\nu = -\beta_\nu \gamma_\nu$ und folglich

$$\int \left(\left(\frac{\partial \mu}{\partial x}\right)^2 + \left(\frac{\partial \mu}{\partial y}\right)^2\right) dT = \sum_{\nu=1}^{\nu=p} (\alpha_\nu \delta_\nu - \beta_\nu \gamma_\nu).$$

Diese Summe ist daher stets positiv.

Hieraus ergiebt sich die zu beweisende Eigenschaft der Gröfsen a, wenn man für w setzt $u_1 m_1 + u_2 m_2 + \cdots + u_p m_p$. Denn es ist dann $A_\nu = m_\nu \pi i$, $B_\nu = \sum_\mu a_{\mu,\nu} m_\mu$, folglich α_ν stets $=0$ und $\int \left(\left(\frac{\partial \mu}{\partial x}\right)^2 + \left(\frac{\partial \mu}{\partial y}\right)^2\right) dT = -\sum \beta_\nu \gamma_\nu$ $= -\pi \sum m_\nu \beta_\nu$ oder gleich dem reellen Theile von $-\pi \sum_{\mu,\nu} a_{\mu,\nu} m_\mu m_\nu$, welcher also für alle reellen Werthe der Gröfsen m positiv ist.

22.

Setzt man nun in der ϑ-Reihe (1.) §. 16 für $a_{\mu,\mu'}$ den Periodicitätsmodul der Function u_μ an dem Schnitt $b_{\mu'}$ und, durch $e_1, e_2, \ldots, e_p$ beliebige Constanten bezeichnend, $u_\mu - e_\mu$ für v_μ, so erhält man eine in jedem Punkte

*) Dies Integral drückt den Inhalt der Fläche aus, welche die Gesammtheit der Werthe, die w innerhalb T'' annimmt, auf der w-Ebene repräsentirt.

Fig. 30. Riemann's original proof of 'Riemann's bilinear relations' ([Ri1, Sections 20 and 21]).

$$\sum_{i=1}^{g} \lambda_i \, A_{ij} = 0. \qquad (1 \leq j \leq g)$$

Consider the holomorphic differential form

$$\omega = \sum_{i=1}^{g} \lambda_i \, \omega_i.$$

According to the definition of A, we have

$$\int_{a_j} \omega = 0. \qquad (1 \leq j \leq g)$$

This implies that

$$\int_{a_j} \overline{\omega} = 0 \qquad (1 \leq j \leq g)$$

and according to Lemma III.1.1, we get

$$\int_X \omega \wedge \overline{\omega} = 0.$$

Therefore $\omega = 0$ (*cf.* (III.1.3)) and $\lambda_1 = \cdots = \lambda_g = 0$. This proves the assertion 1) in Theorem III.1.2.

To prove 2), we use the basis $(\omega_1, \ldots, \omega_g)$ of $\Omega^1(X)$ defined by (III.1.7). Then, if we apply (III.1.4) to $\eta = \omega_i$ and $\eta' = \omega_j$ we get $\Omega_{ij} = \Omega_{ji}$. If we apply (III.1.4) to $\eta = \sum_{i=1}^{g} v_i \omega_i$ and $\eta' = \overline{\eta}$, with $v \in \mathbb{R}^g - \{0\}$, we get

$$^t v \cdot \operatorname{Im} \Omega \cdot v = \frac{i}{2} \int_X \eta \wedge \overline{\eta} > 0.$$

This proof of Theorem III.1.2, based on Lemma III.1.1, is essentially the original proof given by Riemann in 1857 in the Sections 20 and 21 of his paper on Abelian functions ([Ri1]; see figure 30).

Exercise: Let X be the hyperelliptic Riemann surface defined by the equation

$$y^2 = x^6 - x.$$

Figure 31 'shows' X as a two-sheeted covering of the 'x-plane', ramified over 0 and ζ_5^i, $0 \leq i \leq 4$, where $\zeta_5 = e^{2\pi i/5}$: the surface X may be realized by gluing two copies of $\mathbb{P}^1\mathbb{C}$ along the three segments $[0,1]$, $[\zeta_5, \zeta_5^2]$ and $[\zeta_5^3, \zeta_5^4]$. A symplectic basis (a_1, a_2, b_1, b_2) of $H_1(X; \mathbb{Z})$ is also shown; the parts of the loops in full line are on one sheet; the dotted parts are on the other one.

Prove the preceding assertions and compute the period matrix Ω of X associated to (a_1, a_2, b_1, b_2).

Hint: Consider the automorphism T of order 5 of X defined by

$$T(x, y) = (\zeta_5 x, \zeta_5^3 y)$$

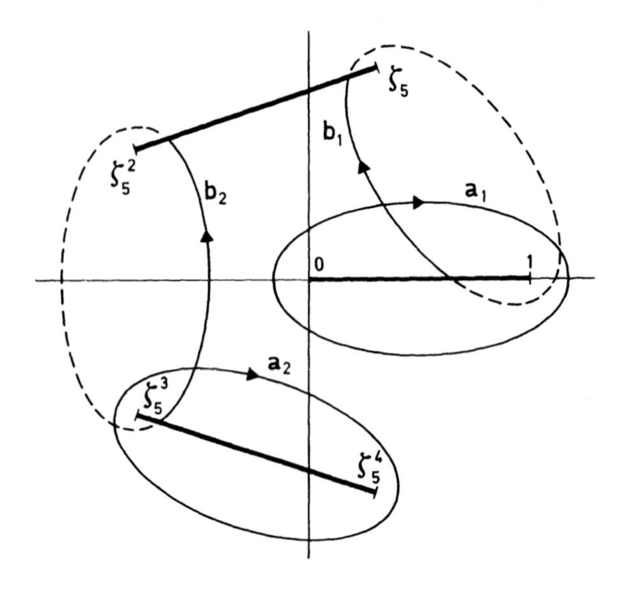

Fig. 31. The Riemann surface of $\sqrt{x^6 - x}$.

and the basis $(\omega_1, \omega_2) = \left(\frac{dx}{y}, \frac{x\,dx}{y}\right)$ of $\Omega^1(X)$. Express (a_1, a_2, b_1, b_2) in terms of $(a_1, T(a_1), T^2(a_1), T^3(a_1))$, and use the relation

$$\int_{T^i a_1} \omega = \int_{a_1} (T^i)^* \omega,$$

which is true for any $\omega \in \Omega^1(X)$, to compute the periods of ω_1 and ω_2 along a_1, a_2, b_1 and b_2 in terms of $\int_{a_1} \omega_1$ and $\int_{a_1} \omega_2$.

Answer:

$$\Omega = \begin{pmatrix} -\zeta_5^4 & \zeta_5^2 + 1 \\ \zeta_5^2 + 1 & \zeta_5^2 - \zeta_5^3 \end{pmatrix}.$$

III.1.3. The action of the symplectic group.

Before we proceed further, let us describe how Ω is transformed when changing the symplectic basis $(a_1, \ldots, a_g, b_1, \ldots, b_g)$. (Any general result on period matrices must be invariant under these transformations!)

Recall that $\mathrm{Sp}(2g, \mathbb{Z})$ denotes the group of symplectic matrices of size $2g$ with integer coefficients, *i.e.*, of matrices $M \in M_{2g}(\mathbb{Z})$ such that

$$ {}^tM \cdot \begin{pmatrix} 0 & I_g \\ -I_g & 0 \end{pmatrix} \cdot M = \begin{pmatrix} 0 & I_g \\ -I_g & 0 \end{pmatrix}. \tag{III.1.9}$$

If one writes

(III.1.10)
$$M = \begin{pmatrix} A & B \\ C & D \end{pmatrix},$$

with A, B, C, D in $M_g(\mathbb{Z})$ this condition is easily seen to be equivalent to

$$A \cdot {}^tB = B \cdot {}^tA \quad , \quad C \cdot {}^tD = D \cdot {}^tC \quad , \quad A \cdot {}^tD - B \cdot {}^tC = I_g.$$

The condition (III.1.9) means exactly that M preserves the standard symplectic form on $\mathbb{Z}^{2g}$. This implies that, given any 'reference' symplectic basis $\gamma = (\gamma_1, \ldots, \gamma_{2g})$ of $H_1(X; \mathbb{Z})$, one gets a bijection

$$\mathrm{Sp}(2g, \mathbb{Z}) \xrightarrow{\sim} \{\text{symplectic bases of } H_1(X; \mathbb{Z})\}$$

by associating the basis $\gamma \cdot M := \left(\sum_{i=1}^{2g} M_{ij}\gamma_i\right)_{1 \leq j \leq 2g}$ to $M = (M_{ij})_{1 \leq i,j \leq 2g} \in \mathrm{Sp}(2g, \mathbb{Z})$. Using this parametrization of symplectic bases, we can easily compute the period matrix Ω' associated to a symplectic basis $\gamma' = \gamma \cdot M$ in terms of the period matrix Ω associated to γ: using the block decomposition (III.1.10), one gets

(III.1.11)
$$\Omega' = ({}^tD\Omega + {}^tB)({}^tC\Omega + {}^tA)^{-1}.$$

The independence of the Riemann bilinear relations on the choice of a symplectic basis in $H_1(X; \mathbb{Z})$ comes from the existence of *an action of* $\mathrm{Sp}(2g, \mathbb{Z})$[34] *on* $\mathfrak{H}_g$ given by the following formula: *for any* $M = \begin{pmatrix} A & B \\ C & D \end{pmatrix} \in \mathrm{Sp}(2g, \mathbb{Z})$ *and any* $\Omega \in \mathfrak{H}_g$,

(III.1.12)
$$M \cdot \Omega = (A\Omega + B)(C\Omega + D)^{-1}.$$

This action generalizes the action of $\mathrm{SL}(2, \mathbb{Z})$ on $\mathfrak{H}$ considered in Section I.

Exercise: Check that $\det(C\Omega + D) \neq 0$, that $\gamma \cdot \Omega \in \mathfrak{H}_g$ and that if M' is another element of $\mathrm{Sp}(2g, \mathbb{Z})$, $M' \cdot (M \cdot \Omega) = (M'M) \cdot \Omega$.
Indeed, the relation (III.1.4) may be written

$$\Omega' = \widetilde{M} \cdot \Omega \ ,$$

where

$$\widetilde{M} = \begin{pmatrix} {}^tD & {}^tB \\ {}^tC & {}^tA \end{pmatrix}$$

is easily checked to belong to $\mathrm{Sp}(2g, \mathbb{Z})$.

[34] In fact of $\mathrm{Sp}(2g, \mathbb{R})$.

III.2. Complex Abelian varieties

Let us emphasize that the lattices $\mathbb{Z}^g + \Omega\mathbb{Z}^g$ attached to matrices $\Omega \in \mathfrak{H}_g$ are of a special kind, as soon as $g > 1$. This can be seen by the following dimension counting argument[35]. Consider the space $\mathcal{L}_g$ parametrizing the lattices in a g-dimensional complex vector space, *i.e.*, the space of pairs (V, Λ) where V is a complex vector space of dimension g and Λ is a lattice in V, modulo the equivalence relation $\simeq$ defined by:

$$(V, \Lambda) \simeq (V', \Lambda') \Leftrightarrow \text{there exists an isomorphism of complex vector spaces} \quad u : V \xrightarrow{\sim} V' \text{ such that } u(\Lambda) = \Lambda'.$$

This space $\mathcal{L}_g$ can also be described as the space of lattices Λ in $\mathbb{C}^g$ modulo the natural action of $\mathrm{GL}_g(\mathbb{C})$. Now $\mathrm{GL}_g(\mathbb{C})$ has complex dimension g^2, the space of lattices in $\mathbb{C}^g$ has complex dimension $2g^2$ (since they are defined, up to some discrete ambiguity, by $2g$ vectors in $\mathbb{C}^g$), and the isotropy in $\mathrm{GL}_g(\mathbb{C})$ of any such lattice is discrete. Hence the complex dimension of $\mathcal{L}_g$ is $2g^2 - g^2 = g^2$. On the other hand the pairs $(\mathbb{C}^g, \mathbb{Z}^g + \Omega\mathbb{Z}^g)$ with $\Omega \in \mathfrak{H}_g$ form in $\mathcal{L}_g$ a subspace of complex dimension $\frac{1}{2}g(g+1)$, since $\mathfrak{H}_g$ is an open subset of symmetric matrices in $M_g(\mathbb{C})$. If $g > 1$, we have $g^2 > \frac{1}{2}g(g+1)$, and a 'generic' (E, Λ) is not equivalent to a pair $(\mathbb{C}^g, \mathbb{Z}^g + \Omega\mathbb{Z}^g)$ with $\Omega \in \mathfrak{H}_g$.

The complex tori $\mathbb{C}^g/(\mathbb{Z}^g + \Omega\mathbb{Z}^g)$ defined by these special lattices have striking properties from the point of view of complex geometry. For instance, we have:

Theorem III.2.1. *Let V be a complex vector space of dimension g and let Λ be a lattice in V. The following two conditions are equivalent:*

i) the complex torus V/Λ can be embedded, as a complex manifold, in a complex projective space $\mathbb{P}^N(\mathbb{C})$;

ii) there exists a basis $(e_1, \ldots, e_g)$ of V (as a complex vector space) and a basis $(\lambda_1, \ldots, \lambda_{2g})$ of Λ such that the matrix of $(\lambda_1, \ldots, \lambda_{2g})$ with respect to $(e_1, \ldots, e_g)$ takes the form (Δ_δ, Ω) where $\Omega \in \mathfrak{H}_g$ and

$$\Delta_\delta = \begin{pmatrix} \delta_1 & & 0 & \cdots & 0 \\ & \ddots & & \ddots & \vdots \\ 0 & & \delta_i & & 0 \\ \vdots & \ddots & & \ddots & \\ 0 & \cdots & 0 & & \delta_g \end{pmatrix}, \quad \textit{with } \delta_i \in \mathbb{Z}\,, \ \delta_i > 0\,.$$

Many famous mathematicians of the last century contributed to this theorem (at least Hermite, Weierstrass, Riemann, Frobenius, Poincaré) and, in its final form, it is due to Lefschetz. Condition ii) is known as *Riemann condition.*

[35] We present this argument in an informal way; however it can easily be made rigorous.

Using elementary linear algebra, one may show that it can be reformulated in the following more intrinsic way:

ii)' there exists a Hermitian form $H : V \times V \to \mathbb{C}$ *which is positive definite* (i.e., $H(u,u) > 0$ *for all* $u \in V - \{0\}$*) and such that* $E = \operatorname{Im} H$ *is integer valued on* Λ (i.e., $E(\lambda_1, \lambda_2) \in \mathbb{Z}$, *for all* $\lambda_1, \lambda_2 \in \Lambda$*).*

Such a form H is called a *non-degenerate Riemann form on* V/Λ, or a *polarization* of V/Λ.

A complex torus V/Λ satisfying the equivalent conditions i), ii), or ii)' is called an *Abelian variety.* The Riemann bilinear relations show that the Jacobian Jac(X) of a compact connected Riemann surface satisfies the condition ii), hence is an Abelian variety.

One can also check that a Jacobian satisfies condition ii)': by definition, Jac(X) is the quotient V/Λ where $V = \Omega^1(X)^*$ and Λ is $H_1(X;\mathbb{Z})$, embedded in $\Omega^1(X)^*$ by the period map p (*cf.* (III.1.1)). The dual space $\Omega^1(X)^*$ may be identified with the space $\overline{\Omega^1}(X)$ of antiholomorphic differentials on X by the map

$$\overline{\Omega^1}(X) \to \Omega^1(X)^*$$
$$\alpha \mapsto (\omega \mapsto \int_X \alpha \wedge \omega).$$

Then a non-degenerate Riemann form on Jac(X) is given by

$$H : \overline{\Omega^1}(X) \times \overline{\Omega^1}(X) \to \mathbb{C}$$
$$(\alpha, \beta) \mapsto 2i \int_X \overline{\alpha} \wedge \beta.$$

Indeed, H is clearly positive definite and the restriction of its imaginary part $\operatorname{Im} H$ to $H_1(X;\mathbb{Z})$ is the intersection product (*cf.* Appendix A)

$$\# : H_1(X;\mathbb{Z}) \times H_1(X;\mathbb{Z}) \to \mathbb{Z}$$

(this is again a consequence of A.2.5, like Proposition II.2.12 or Theorem III.1.2).

An Abelian variety V/Λ equipped with a Riemann form H is called a *polarized* Abelian variety. When H is such that the alternating form

$$E = \operatorname{Im} H : \Lambda \times \Lambda \to \mathbb{Z}$$

is unimodular[36], then H is called a *principal polarization* and $(E/\Lambda, H)$ is said to be a *principally polarized* Abelian variety.

Elementary linear algebra shows that an Abelian variety V/Λ possesses a principal polarization iff condition ii) of Theorem III.2.1 is satisfied by the

---

[36] This means that there exists a basis $(\gamma_1, \dots, \gamma_{2g})$ of the lattice Λ such that $\det(E(\gamma_i, \gamma_j))_{1 \le i,j \le 2g} = 1$, or, equivalently, that there exists a basis $(a_1, \dots, a_g, b_1, \dots, b_g)$ of Λ such that for any $i, j = 1, \dots, g$, $E(a_i, a_j) = E(b_i, b_j) = 0$ and $E(a_i, b_j) = \delta_{ij}$.

quantities $\delta_1 = \cdots = \delta_g = 1$ and that any principally polarized Abelian variety $(V/\Lambda, H)$ of dimension g is isomorphic to a complex torus $\mathbb{C}^g/(\mathbb{Z}^g + \Omega\mathbb{Z}^g)$ equipped with the Riemann form

$$\text{(III.2.1)} \qquad \begin{aligned} H : \mathbb{C}^g &\to \mathbb{R}_+ \\ v &\mapsto {}^t v \cdot (\operatorname{Im} \Omega)^{-1} \cdot \overline{v}, \end{aligned}$$

for some $\Omega \in \mathfrak{H}_g$.

Of course, examples of principally polarized Abelian varieties are provided by Jacobians of compact Riemann surfaces.

Observe that, when $g = 1$, the Riemann condition is fulfilled by any lattice Λ in $\mathbb{C}$. In that case, the existence of a projective embedding of the elliptic curve $\mathbb{C}/\Lambda$ is a particular case of the existence of projective embeddings for compact Riemann surfaces. In fact, explicit embeddings of $\mathbb{C}/\Lambda$ may be given, by using the $\wp$ function or the theta functions associated to Λ (*cf.* §0.1 and §I.4.2).

In the next Section, we sketch a proof of the implication i) $\Rightarrow$ ii)' in Theorem III.2.1. Then we introduce theta functions (in several variables) and we explain how they provide projective embeddings of Abelian varieties of any dimension, which generalize the embeddings of elliptic curves we have just mentioned, and therefore allow to prove the implication ii) $\Rightarrow$ i).

III.3. The necessity of Riemann condition

In order to prove the implication i) $\Rightarrow$ ii)' in Theorem III.2.1, we need some basic facts concerning the Kähler geometry of the complex projective space $\mathbb{P}^N\mathbb{C}$.

First, recall that a Riemannian metric on a complex manifold M is said to be *Hermitian* if, in local holomorphic coordinates $(z^1, \ldots, z^n)$ it takes the form

$$ds^2 = \sum_{k,\ell=1}^{n} g_{k\ell}(z) dz^k d\overline{z}^\ell,$$

with

$$g_{k\ell}(z) = \overline{g_{\ell k}}(z).$$

The matrix $(g_{k\ell}(z))_{1 \leq k,\ell \leq n}$ is then the matrix of a Hermitian form. To such a Hermitian metric is associated the differential form of degree two[37] ω on M defined locally by

$$\omega = \frac{i}{2\pi} \sum_{k,\ell=1}^{n} g_{k\ell}(z) dz^k \wedge d\overline{z}^\ell.$$

The fact that $(g_{k\ell}(z))_{1 \leq k,\ell \leq n}$ is the matrix of a Hermitian form implies that ω is *positive*, *i.e.*, its restriction to any Riemann surface C holomorphically

[37] More precisely, of type $(1,1)$: it is a sum of terms, each of them involves 'only one dz and one $d\overline{z}$'.

embedded in M is a positive volume form on C. Conversely, any positive 2-form of type $(1,1)$ on M is associated to some Hermitian Riemannian metric on M.

Consider now the complex projective space $\mathbb{P}^N\mathbb{C}$. The unitary group $U(N+1)$ acts on $\mathbb{C}^{N+1}$, hence on

$$\mathbb{P}^N\mathbb{C} = (\mathbb{C}^{N+1} - \{0\})/\mathbb{C}^*.$$

There exists a unique Riemannian metric on $\mathbb{P}^N\mathbb{C}$ which is Hermitian and $U(N+1)$-invariant and which coincides with the usual Hermitian metric on $\mathbb{C}^N \simeq T_{(1:0:\ldots:0)}\mathbb{P}^N\mathbb{C}$. This metric may also be defined by giving the associated 2-form ω; explicitly, in terms of homogeneous coordinates $(z_0, \ldots, z_N)$ on $\mathbb{P}^N\mathbb{C}$, it is:

$$\text{(III.3.1)} \qquad \omega = \frac{i}{2\pi}\left[\frac{\sum_{i=0}^N dz_i \wedge d\overline{z}_i}{\sum_{i=0}^N |z_i|^2} - \frac{\sum_{i=0}^N \overline{z_i} dz_i \wedge \sum_{i=0}^N z_i d\overline{z_i}}{\left(\sum_{i=0}^N |z_i|^2\right)^2}\right]$$

$$= \frac{i}{2\pi}\partial\overline{\partial}\log\left(\sum_{i=0}^N |z_i|^2\right).$$

The last equality shows that ω *is a closed form.*

The 2-form ω is often called the *Fubini-Study 2-form* on $\mathbb{P}^N\mathbb{C}$, and the associated Hermitian metric, the *Fubini-Study metric*. Hermitian metrics which, like the Fubini-Study metric, have closed associated 2-forms, are known as *Kähler metrics.*

One easily checks that, if $\mathbb{P}^1\mathbb{C}$ is any complex projective line in $\mathbb{P}^N\mathbb{C}$, then

$$\int_{\mathbb{P}^1\mathbb{C}} \omega = 1.$$

As ω is closed and any oriented closed (real) surface in $\mathbb{P}^N\mathbb{C}$ is homologous to an integral multiple of $\mathbb{P}^1\mathbb{C}$ (see for instance [Gre], Theorem (19.21)), this implies that for any closed (real) surface S in $\mathbb{P}^N\mathbb{C}$, one has

$$\text{(III.3.2)} \qquad \int_S \omega \in \mathbb{Z}.$$

Consider now a complex torus $T = V/\Lambda$ embedded in $\mathbb{P}^N\mathbb{C}$. Then ω restricted to T is clearly a closed positive $(1,1)$-form such that for any oriented closed surface $S \subset T$, (III.3.2) holds.

The existence of such a 2-form on T entails condition ii)'. Indeed, the form $\widehat{\omega}$ on T defined as ω averaged by translation on T (*i.e.*, as the zero-th Fourier coefficient of ω) is by construction a positive translation invariant $(1,1)$-form on T; moreover it satisfies the same condition (III.3.1) as ω, since, for any oriented closed surface S in T, all the translated surfaces $S+x$, $x \in T$, are homologous and, consequently, the integrals $\int_{S+x}\omega$ and their average $\int_S \widehat{\omega}$ are equal to

$\int_S \omega$. The Hermitian metric on T associated to $\widehat{\omega}$ is translation invariant and corresponds to a Hermitian form H on V. A simple computation shows that, when S is a two-dimensional torus $(\mathbb{R}\lambda + \mathbb{R}\mu)/(\mathbb{Z}\lambda + \mathbb{Z}\mu)$, $\lambda, \mu \in \Lambda$, condition (III.3.2) amounts to the fact that

$$\operatorname{Im} H(\lambda, \mu) \in \pi\mathbb{Z}.$$

This shows that $\pi^{-1}H$ is a Riemann form for V/Λ.

III.4. Theta functions

We now come to the existence of projective embeddings for complex tori which possess a Riemann form (*i.e.*, to the implication ii) $\Rightarrow$ i) in Theorem II.2.1). To make things simpler, we will discuss the case of principally polarized Abelian varieties $\mathbb{C}^g/(\mathbb{Z}^g + \Omega\mathbb{Z}^g)$, $\Omega \in \mathfrak{H}_g$ (the general case where, in condition ii), the δ_i's may be different of 1, is only notationally more complicated).

To get a projective embedding of such an Abelian variety $A = \mathbb{C}^g/(\mathbb{Z}^g + \Omega\mathbb{Z}^g)$, one needs to produce meromorphic functions on A. A way to achieve this is to construct entire functions on $\mathbb{C}^g$, which are periodic with respect to the lattice $\mathbb{Z}^g + \Omega\mathbb{Z}^g$, up to some common factor of automorphy. Then the quotients of two of these entire functions will define a meromorphic function on A.

This procedure is well known when $g = 1$: elliptic functions may be written as quotients of theta series, which are entire functions 'almost periodic' with respect to a lattice in $\mathbb{C}$ (*cf.* [Coh], [Ge], [Z]). As a matter of fact, the construction of theta series generalizes to higher g's. The simplest of them is the Riemann theta function, defined for any $(z, \Omega) \in \mathbb{C}^g \times \mathfrak{H}_g$ as

$$\theta(z, \Omega) = \sum_{n \in \mathbb{Z}^g} \exp(\pi i\, {}^t n \cdot \Omega \cdot n + 2\pi i\, {}^t n \cdot z) \tag{III.4.1}$$

(compare with (0.5)). Observe that the series in (III.4.1) converges because $\operatorname{Im} \Omega > 0$: Riemann bilinear relations are exactly the appropriate conditions which allow to form theta-series associated to period matrices. This is one of Riemann's great discoveries published in [Ri1].

A simple computation shows that θ is indeed periodic up to some factor of automorphy: for any $(m, z, \Omega) \in \mathbb{Z}^g \times \mathbb{C}^g \times \mathfrak{H}_g$, we get that

$$\theta(z + m, \Omega) = \theta(z, \Omega) \tag{III.4.2}$$

and

$$\theta(z + \Omega \cdot m, \Omega) = e^{\alpha(m, z, \Omega)} \theta(z, \Omega), \tag{III.4.3}$$

where

$$\alpha(m, z, \Omega) = -\pi i\, {}^t m \cdot \Omega \cdot m - 2\pi i\, {}^t m \cdot z.$$

The functional equations (III.4.2) and (III.4.3) satisfied by θ may be generalized as follows. For any integer $\ell \geq 1$, we can define the *theta functions of weight ℓ associated to $\Omega \in \mathfrak{H}_g$* as the entire functions $f : \mathbb{C}^g \to \mathbb{C}$ such that, for any $(m, z) \in \mathbb{Z}^g \times \mathbb{C}^g$, the following identities hold:

$$f(z+m) = f(z) \tag{III.4.4}$$

$$f(z + \Omega \cdot m) = e^{\ell\alpha(m,z,\Omega)} f(z). \tag{III.4.5}$$

The vector space formed by these functions will be denoted by R_ℓ^Ω.

A slight generalization of the Riemann theta function is given by the *theta functions with characteristics* $\theta\left[\begin{smallmatrix} a \\ b \end{smallmatrix}\right]$, defined for any $(a,b) \in \mathbb{Q}^g \times \mathbb{Q}^g$ by

$$\begin{aligned}\theta\left[\begin{smallmatrix} a \\ b \end{smallmatrix}\right](z,\Omega) &= \sum_{n\in\mathbb{Z}^g} \exp[\pi i\,{}^t(n+a)\cdot\Omega\cdot(n+a) + 2\pi i\,{}^t(n+a)\cdot(z+b)] \\ &= \exp[\pi i\,{}^t a\cdot\Omega\cdot a + 2\pi i\,{}^t a\cdot(z+b)]\theta(z+\Omega\cdot a+b,\Omega).\end{aligned}$$

We can now state:

Proposition III.4.1.. *For any $\Omega \in \mathfrak{H}_g$ and any $\ell \geq 1$, R_ℓ^Ω is a vector space of dimension ℓ^g, which admits as bases*

$$\left(z \mapsto \theta\left[\begin{smallmatrix} a/\ell \\ 0 \end{smallmatrix}\right](\ell z, \ell\Omega)\right)_{a\in\{0,\ldots,\ell-1\}^g}$$

or

$$\left(z \mapsto \theta\left[\begin{smallmatrix} 0 \\ b/\ell \end{smallmatrix}\right](z, \ell^{-1}\Omega)\right)_{b\in\{0,\ldots,\ell-1\}^g}.$$

Theorem III.4.2. *For any $\Omega \in \mathfrak{H}_g$ and any $\ell \geq 3$, if $(f_1, \ldots, f_{\ell^g})$ is a basis of R_ℓ^Ω, then the functions f_i, $i = 1, \ldots, \ell^g$, have no common zero in $\mathbb{C}^g$ and define an embedding*

$$\begin{aligned}\mathbb{C}^g/(\mathbb{Z}^g + \Omega\mathbb{Z}^g) &\to \mathbb{P}^{\ell^g-1}\mathbb{C} \\ [z] &\mapsto (f_1(z) : \cdots f_{\ell^g}(z)).\end{aligned}$$

Observe that the 'periodicity relations' (III.4.4) and (III.4.5) show that, for any $z \in \mathbb{C}^g$, the point $(f_1(z) : \cdots : f_{\ell^g}(z))$ in $\mathbb{P}^{\ell^g-1}\mathbb{C}$ depends only on the class $[z]$ of z in $\mathbb{C}^g/(\mathbb{Z}^g + \Omega\mathbb{Z}^g)$.

Proposition III.4.1 may be proved by expanding the elements of R_ℓ^Ω in Fourier series (this is possible because of (III.4.4)) and by expressing condition (III.4.5) as a condition on their Fourier coefficients. The proof of Theorem III.4.2, the so-called *Lefschetz Embedding Theorem* is more involved, but still elementary. (See for instance [Mu1], §I.3, [GrH], pp. 317-324 or [Rob2]). Let us only explain why, as soon as $\ell \geq 2$, the functions of R_ℓ^Ω have no common zero: one immediately checks that, if $a_1, \ldots, a_\ell \in \mathbb{C}^g$ are such that

$$\sum_{i=1}^{\ell} a_i = 0, \tag{III.4.6}$$

then

$$f_{a_1,\ldots,a_\ell} : z \mapsto \prod_{i=1}^{\ell} \theta(z - a_i)$$

is an element of R_ℓ^Ω; as θ is not identically zero, for any $z_0 \in \mathbb{C}^g$ and any $\ell \geq 2$, there exist $a_1, \ldots, a_\ell$ satisfying (III.4.6) and such that $z_0 - a_1, \ldots, z_0 - a_\ell$ are not zeros of θ; then $f_{a_1,\ldots,a_\ell}$ is an element of R_ℓ^Ω which does not vanish at z_0.

III.5. Riemann's theorem

In this Section, we go back to the inversion problem for Abelian integrals and we describe how, according to Riemann, it can be 'solved' by using theta functions.

Let X be a compact connected Riemann surface of genus $g \geq 1$, let $(a_1, \ldots, a_g, b_1, \ldots, b_g)$ be a symplectic basis of $H_1(X;\mathbb{Z})$, and let $(\omega_1, \ldots, \omega_g)$ and Ω be the basis of $\Omega^1(X)$ and the period matrix attached to this symplectic basis (*cf.* (III.1.7) and (III.1.8)). The Jacobian variety Jac(X) of X is then identified with the complex torus $\mathbb{C}^g/(\mathbb{Z}^g + \Omega\mathbb{Z}^g)$.

Furthermore, let us choose a base point M_0 in X. This choice allows us to define Jacobi maps

$$J : X \to \mathrm{Jac}(X),$$

and

$$J_k : \mathrm{Sym}_k X \to \mathrm{Jac}(X)$$

by setting

$$J(M) = \int_{M_0}^{M} \vec{\omega} := \left(\int_{M_0}^{M} \omega_j\right)_{1\leq j\leq g} \quad \mathrm{mod}\ (\mathbb{Z}^g + \Omega\mathbb{Z}^g),$$

and

$$J_k(M_1 + \cdots + M_k) = J(M_1) + \cdots + J(M_k)$$

(compare §II.1 and §II.3.2).

Observe that the 'periodicity' relations (III.4.2) and (III.4.3) satisfied by θ imply the following facts:

- the hypersurface

$$\{z \in \mathbb{C}^g \mid \theta(z, \Omega) = 0\}$$

is invariant under translation by the lattice $\mathbb{Z}^g + \Omega\mathbb{Z}^g$; hence it defines a hypersurface Θ_0 in Jac(X);

- for any $u \in \mathbb{C}^g$ and any $P \in X$, the vanishing of

$$\theta\left(u - \int_{M_0}^{P} \vec{\omega}, \Omega\right) \tag{III.5.1}$$

does not depend on the path form M_0 to P chosen to compute $\int_{M_0}^P \vec{\omega}$; moreover, the order of vanishing at P of this function is also well defined.

We can now state Riemann's theorem:

Theorem III.5.1. *1) There exists $\Delta \in \mathrm{Jac}(X)$ such that the hypersurface $\Theta_0 + \Delta$ coincides with $J_{g-1}(\mathrm{Sym}_{g-1}X)$. This hypersurface will be called the* theta divisor, *and denoted by Θ.*

2) For a generic $z \in \mathbb{C}^g$,

$$[z] - J(X) := \{[z] - J(P), P \in X\}$$

is not contained in Θ. If $[z]$ satisfies this condition, $J_g^{-1}([z])$ consists of a unique element $P_1 + \cdots + P_g$ in $\mathrm{Sym}_g X$. The points $P_1, \ldots, P_g$ are the points P of X such that

$$[z] - J(P) \subset \Theta, \tag{III.5.2}$$

each of them counted with a multiplicity equal to the order of vanishing of

$$P \mapsto \theta\left(z - \Delta - \int_{M_0}^{P} \vec{\omega}, \Omega\right).$$

Roughly speaking, the second half of this theorem asserts that $J_g^{-1}([z])$ *is obtained by intersecting X, embedded by J in* $\mathrm{Jac}(X)$, *with the hypersurface* $[z] - \Theta$.

If $\tau \in \mathfrak{H}$ and if X is the elliptic curve $\mathbb{C}/(\mathbb{Z} + \tau\mathbb{Z})$, equipped with the usual symplectic basis (a, b) (*cf.* §III.1.2) and with the base point $M_0 = 0$, then the Jacobian $\mathrm{Jac}(X)$ coincides with X, and J is nothing else than the identity map. Then $\Theta = \{0\}$ (since any empty sum is zero), and 2) is obvious, while 1) asserts that $\theta(z, \tau)$, as a function of τ, has only one zero, Δ, modulo $\mathbb{Z} + \tau\mathbb{Z}$. In fact, one has

$$\Delta = \left[\frac{1+\tau}{2}\right].$$

When $g = 2$, $J_{g-1}(\mathrm{Sym}_{g-1}X) = J(X)$, and 1) shows that X is isomorphic with Θ_0. This shows that *the knowledge of* $\mathrm{Jac}(X)$, *as a principally polarized Abelian variety, allows us to recover the isomorphism class of the Riemann surface X.* As a matter of fact, this statement is true for Riemann surfaces of any genus and is known as *Torelli's theorem.*

The general philosophy examplified by Riemann's theorem is that, by using theta functions on their Jacobians, one can give constructive proofs of various results on compact connected Riemann surfaces. For instance, Riemann's theorem gives a constructive solution of the inversion problem. Here is another illustration of this principle:

Proposition III.5.2. *Let $e \in \mathbb{C}^g$ be such that $\theta(e, \Omega) = 0$ and $\theta(e + \int_P^Q \vec{\omega}, \Omega) \not\equiv 0$. For any divisor $D = \sum_{i=1}^k n_i\, P_i$ on X linearly equivalent to zero, one defines a meromorphic function φ on X such that $D = \operatorname{div}(\varphi)$ by setting*

$$\varphi(P) = \prod_{i=1}^{k} \theta\left(e + \int_{P_i}^{P} \vec{\omega}, \Omega\right)^{n_i}.$$

(See [Mu4], I, §I.3 for a proof).

In the next paragraphs, we sketch a proof of Riemann's theorem. This will illustrate the relations between theta functions and complex hypersurfaces on Abelian varieties. (For a detailed proof we refer to [Mu4] I, §II.3, [GrH] pp. 338-340, or [ACGH], §I.5; see also [Ke1] and [ACGH], VI for a remarkable extension due to Kempf).

To the Riemann form (III.2.1) on $A = \mathbb{C}^g/(\mathbb{Z}^g + \Omega\mathbb{Z}^g)$ is attached the following 2-form:

$$\omega = \frac{i}{2} \sum_{1 \leq i,j \leq g} Y_{ij} dz_i \wedge d\overline{z}_j$$

where

$$Y = (\operatorname{Im} \Omega)^{-1}.$$

It is a translation invariant form, which is Poincaré dual to Θ. In other words, ω is cohomologous to the current[38] δ_Θ of integration along Θ defined by the equality:

$$\int_A \delta_\Theta \wedge \alpha = \int_\Theta \alpha$$

for any $(2g-2)$-form on A. This follows for instance from the identity of currents

$$\omega - \delta_\Theta = \frac{1}{2\pi i} \partial\overline{\partial} \log \| \theta \|^2, \tag{III.5.3}$$

where

$$\partial\overline{\partial} := \sum_{i,j=1}^{g} \frac{\partial^2}{\partial z_i\, \partial \overline{z}_j}\, dz_i \wedge d\overline{z}_j$$

and where

$$\| \theta \|^2 (x + iy) := e^{-2\pi {}^t y \cdot Y \cdot y} |\theta(x + iy, \Omega)|^2$$

depends only on the class of $x + iy$ in A. The identity (III.5.3) is a refinement of the observation that $\log \| \theta \|$ is a function on A, which is C^∞ on $A - \Theta$ and has a logarithmic singularity along Θ[39], and whose Laplacian

---

[38] Recall that a current is a differential form whose coefficients are distributions. Here δ_Θ is a 2-form whose coefficients are distributions.

[39] *i.e.*, locally of the form $\log |\lambda| + \varphi$, where φ is C^∞ and where $\lambda = 0$ is a local holomorphic equation for Θ.

$$\Delta \log \parallel \theta \parallel = -\sum_{i=1}^{g} \left(\frac{\partial^2}{\partial x_i^2} + \frac{\partial^2}{\partial y_i^2} \right) \log \parallel \theta \parallel$$

is constant on $A - \Theta$.

Conversely, consider any complex hypersurface H in A. Applying a Green's kernel to δ_H produces a Green function f for H, *i.e.*, a function on A, C^∞ on $A - H$, with logarithmic singularities along H and whose Laplacian is constant on $A - H$.

Then $\partial\overline{\partial} f$ is easily seen to be the sum of a multiple $\lambda\delta_H$ of the integration current along H and of a C^∞ 2-form σ on A. Moreover, on $A - H$, we have:

$$\Delta\sigma = \Delta\partial\overline{\partial} f = \partial\overline{\partial}\Delta f = 0,$$

since Δf is constant on $A - H$. This implies that $\Delta\sigma = 0$ on A and therefore that σ is a translation invariant 2-form. Let $g := -2\pi i\lambda^{-1} f$ and $\sigma' := -\lambda^{-1}\sigma$. Then we have:

$$\frac{1}{2\pi i}\, \partial\overline{\partial} g = \sigma' - \delta_H \; .$$

In particular σ' is the translation invariant 2-form Poincaré dual to H.

Suppose now that $\sigma' = \omega$, *i.e.*, that H and ω are Poincaré dual, or, equivalently, that H and Θ are homologous hypersurfaces in A. Then the function

$$\begin{aligned} \widetilde{g} : \mathbb{C}^g &\to \mathbb{C} \\ x + iy &\mapsto g(x + iy) + 2\pi\, {}^t y \cdot Y \cdot y \end{aligned}$$

is pluriharmonic[40] on the complement of the inverse image $\widetilde{H}$ of H in $\mathbb{C}^g$ and has a logarithmic singularity along $\widetilde{H}$. Indeed we have

$$\frac{1}{2\pi i}\, \partial\overline{\partial}\, \widetilde{g} = -\delta_{\widetilde{H}}.$$

Using these facts, one shows easily that there exists an entire function $\varphi : \mathbb{C}^g \to \mathbb{C}$, vanishing to first order on H, such that

$$\widetilde{g} = \log |\varphi|^2$$

(see [P]). The periodicity of g with respect to $\mathbb{Z}^g + \Omega\mathbb{Z}^g$ implies that φ also is periodic, up to some automorphy factor, and finally using Proposition III.4.1 with $\ell = 1$, that φ has the form

$$\varphi(z) = \lambda\theta(z - a, \Omega) \qquad (\lambda \in \mathbb{C}^*, a \in \mathbb{C}^g).$$

This shows that

$$H = \{[z], \varphi(z) = 0\}$$

coincides with $\Theta_0 + [a]$.

[40] *i.e.*, its image by $\partial\overline{\partial}$ vanishes or, equivalently, its restriction to any complex line is harmonic.

This reduces the proof of Theorem III.5.1, 1) to the purely topological statement that ω and $\Theta := J_{g-1}(\mathrm{Sym}_{g-1}X)$ are Poincaré dual, which may be proved by a direct computation in de Rham cohomology.

Let us add that constructions similar to that of φ allow to prove that *any complex hypersurface on A may be defined as the zero set of some theta function* (possibly more general than the ones introduced in the preceding Section).

Concerning the second half of Theorem III.5.1, observe that if $P_1 + \cdots + P_g \in \mathrm{Sym}_g X$ and $[z] = J_g(P_1 + \cdot + P_g)$, then , for any $i = 1, \ldots, g$,

$$[z] - J(P_i) = J_{g-1}\left(\sum_{j \neq i} P_j\right) \in \Theta.$$

Therefore, the assertion to be shown follows from the fact that there are exactly (taking care of multiplicities) g points $P \in X$ satisfying (III.5.2), *i.e.*, that $[z] - \Theta$ and $J(X)$ meet in g points. As long as $J(X) \not\subset [z] - \Theta$, this is a purely topological statement which does not depend on $[z]$ and may be proved by a direct computation in de Rham cohomology. Indeed, the number of intersections, counted with multiplicities, of two closed oriented subvarieties of complementary dimensions in a closed (oriented) variety is invariant by deformation and may be computed as the integral of the products of the Poincaré dual de Rham cohomology classes.

III.6. Abelian varieties and algebraic geometry

In the preceding pages, we have described some of the basic facts about Jacobians and Abelian varieties. All of these were known at the turn of the century. In this last Section, we would like to give some hints on more recent developments.

III.6.1. Picard and Albanese varieties.

In this Section, we explain how the construction which attaches its Jacobian variety to a compact Riemann surface may be extended to higher dimensional complex manifolds.

Let us begin by a few preliminaries on complex tori.

Consider Λ a lattice in a finite dimensional real vector space V. Denote by $\widehat{V}$ the dual of V seen as a real vector space, *i.e.*, the real vector space of real linear forms on V. It contains the *dual lattice* Λ^* of Λ, defined by

$$\lambda \in \Lambda^* \Leftrightarrow (\forall x \in \Lambda\ ,\ \lambda(x) \in \mathbb{Z}).$$

The real torus $\widehat{V}/\Lambda^*$ is called the *dual torus* of the real torus V/Λ.

Suppose now that V is in fact a complex vector space. Then V/Λ is a complex torus. Furthermore, $\widehat{V}$ may also be equipped with a complex structure, defined by the equality:

$$(i\lambda)(x) = -\lambda(ix)$$

for any $(x, \lambda) \in V \times \widehat{V}$. Thus $\widehat{V}/\Lambda^*$ also appears as a complex torus, still called the dual torus of the complex torus V/Λ.

Exercise: Prove that the dual of the dual of a real (resp. complex) torus is canonically isomorphic to itself.

Suppose now that V/Λ is an Abelian variety. Then the dual torus $\widehat{V}/\Lambda^*$ is also an Abelian variety, called the *dual Abelian variety of* V/Λ. Indeed, if we denote by V^* the dual of V seen as a complex vector space (*i.e.*, the complex vector space of $\mathbb{C}$-linear forms with complex values on V) and by $\overline{V}^*$ the conjugate space (*i.e.*, the complex vector space of $\mathbb{C}$-antilinear forms with complex values on V), then any real linear form $\lambda \in \widehat{V}$ may be written uniquely as

$$\lambda = \mu + \overline{\mu}$$

for some $\mu \in \overline{V}^*$, and the isomorphism

$$\begin{aligned} i : \overline{V}^* &\to \widehat{V} \\ \mu &\mapsto \lambda \end{aligned}$$

so defined is $\mathbb{C}$-linear. Accordingly, if H is a polarization on V/Λ and if $h : V \to \overline{V}^*$ is defined by

$$h(v) : w \mapsto -\frac{i}{2} H(v, w),$$

then $i \circ h : V \to \widehat{V}$ is a $\mathbb{C}$-linear isomorphism. Moreover, for any $x, y \in \Lambda$, we have

$$(i \circ h(x))(y) = 2\mathrm{Re}(h(x)(y)) = \mathrm{Im} H(x, y) \in \mathbb{Z} .$$

Therefore $i \circ h(\Lambda)$ is a subgroup of Λ^*, of finite index since $i \circ h$ is injective. This shows that a multiple of the Hermitian scalar product

$$H^* : \widehat{V} \times \widehat{V} \to \mathbb{C}$$

defined by

$$H^*(i \circ h(v), i \circ h(w)) = H(v, w)$$

is a polarization on $\widehat{V}/\Lambda^*$.

One easily checks that, in general, V/Λ and $\widehat{V}/\Lambda^*$ are not isomorphic as complex manifolds, but, if H is a principal polarization, then $i \circ h(\Lambda) = \Lambda^*$ and the morphism from V/Λ onto $\widehat{V}/\Lambda^*$ defined by $i \circ h$ is an isomorphism between the polarized varieties $(V/\Lambda, H)$ and $(\widehat{V}/\Lambda^*, H^*)$.

In Section II, we have seen an instance of this isomorphism between principally polarized dual Abelian varieties. Indeed, we considered a compact connected Riemann surface X and we attached to it the lattice $\Lambda = H_1(X; \mathbb{Z})$, embedded in the dual of holomorphic 1-forms on $X, V = \Omega^1(X)^*$. The Jacobian of X was defined as the complex torus V/Λ, and was shown to be isomorphic with the space $\mathrm{Pic}_0(X)$ of isomorphism classes of holomorphic line bundles on X. This last space was first identified with the complex torus $\overline{\Omega^1}(X)/H_1(X; \mathbb{Z})^\perp$, which is nothing else than $\widehat{V}/\Lambda^* \simeq \overline{V}^*/i^{-1}(\Lambda^*)$. Then it was shown to be isomorphic with $\mathrm{Jac}(X)$, by an argument which is indeed closely related to the argument of the last paragraph. (Compare the proof of Proposition II.2.12 and the proof of Theorem II.1.2, which provides a principal polarization on $\mathrm{Jac}(X)$).

Up to the existence of principal polarizations, these constructions may be extended to higher dimensional projective varieties.

Let M be a compact complex manifold of complex dimension d, embedded in the projective space $\mathbb{P}^N\mathbb{C}$ (by Chow's theorem, such an M is in fact an algebraic subvariety of $\mathbb{P}^N\mathbb{C}$). Define $\Omega^1(M)$ as the space of holomorphic 1-forms on M, *i.e.*, of complex 1-forms on M, which, in any local holomorphic coordinates $(z_1, \ldots, z_d)$, may be written

$$\sum_{i=1}^{d} f_i(z) dz_i,$$

where the f_i's are holomorphic functions. One can show that any such form is closed and that $\Omega^1(M)$ is a finite dimensional vector space. Moreover, the first de Rham cohomology group of M, namely

$$\begin{aligned} &H^1_{DR}(M;\mathbb{C}) \\ &=\{\text{closed complex } C^\infty \text{ 1-forms on } M\}/\{\text{exact complex } C^\infty \text{ 1-forms on } M\}, \end{aligned}$$

possesses a Hodge decomposition, which generalizes Theorem B.2.5: the map

$$\begin{aligned} \Omega^1(M) \oplus \overline{\Omega^1}(M) &\to H^1_{DR}(M;\mathbb{C}) \\ \alpha \oplus \beta &\mapsto [\alpha + \beta] \end{aligned}$$

is an isomorphism.

This implies that any $\gamma \in H_1(M;\mathbb{Z})$ defines a linear form

$$\int_\gamma : \alpha \mapsto \int_\gamma \alpha$$

on $\Omega^1(M)$, and that the set $\{\int_\gamma \,;\ \gamma \in H_1(M;\mathbb{Z})\}$ of these linear forms is a lattice in $\Omega^1(M)^*$, which, abusively, we will still denote $H_1(M;\mathbb{Z})$.

According to our preliminary observations on complex tori, to this lattice $\Lambda = H_1(M;\mathbb{Z})$ in $V = \Omega^1(M)^*$ are naturally attached two complex tori, dual to each other, the *Albanese variety*

$$\mathrm{Alb}(M) = \Omega^1(M)^*/H_1(M;\mathbb{Z}) = V/\Lambda$$

and the *Picard variety*

$$\mathrm{Pic}_0(M) = \widehat{V}/\Lambda^* \simeq \overline{V}^*/i^{-1}(\Lambda^*) = \overline{\Omega^1}(M)/H_1(M;\mathbb{Z})^\perp$$

where $H_1(M;\mathbb{Z})^\perp$ is defined by

$$\alpha \in H_1(M;\mathbb{Z})^\perp \Leftrightarrow \forall \gamma \in H_1(M;\mathbb{Z}) \, , \, \int_\gamma (\alpha + \overline{\alpha}) \in \mathbb{Z}.$$

These complex tori are Abelian varieties. To get a polarization on $\mathrm{Pic}_0(M)$, consider ω the restriction to M of the Fubini-Study 2-form on $\mathbb{P}^N\mathbb{C}$ (*cf.* §III.3) and define, for any $(\alpha,\beta) \in \overline{\Omega^1}(M)$

$$H(\alpha,\beta) = 2i \int_M \overline{\alpha} \wedge \beta \wedge \omega^{d-1}.$$

When $d > 1$, this polarization is not necessarily principal, and the Albanese and Picard varieties of M are Abelian varieties dual to each other, but possibly not isomorphic. They have the following algebro-geometric interpretations:

• One may prove that the Picard variety $\mathrm{Pic}_0(M)$ still parametrizes the isomorphism classes of holomorphic line bundles on M which are topologically trivial.

• For any two points P, Q in M and any path L from P to Q in M, the class in $\Omega^1(X)^*/H_1(X;\mathbb{Z})$ of

$$\omega \mapsto \int_P^Q \omega$$

depends only on (P,Q). Therefore, for any base point $P_0 \in M$, we define a holomorphic map

$$j : M \to \mathrm{Alb}(M),$$

by setting

$$j(P) = \left(\omega \mapsto \int_{P_0}^P \omega\right) \mod H_1(X;\mathbb{Z}).$$

This map generalizes the Jacobian embedding (but is not always an embedding, *e.g.* when $\Omega^1(X) = 0$!).

If M is an Abelian variety V_0/Λ_0, then $\Omega^1(M)$ (resp. $\Omega^1(M)^*$, resp. $H_1(M;\mathbb{Z})$) is canonically isomorphic with V_0^* (resp. V_0^*, resp. Λ_0); hence $\mathrm{Alb}(M) \cong M$, and, if $P_0 = 0$, the map $j : M \to \mathrm{Alb}(M)$ is the identity.

More generally, any holomorphic map $f : M \to A$ with values in a complex torus may be factorized through j: there exists a unique holomorphic map $\widetilde{f} : \mathrm{Alb}(M) \to A$ such that the following diagram commute

$$\begin{array}{ccc} M & \xrightarrow{j} & \mathrm{Alb}(M) \\ & f \searrow & \downarrow \widetilde{f} \\ & & A \end{array}$$

A general Abelian variety A is not the Jacobian of a curve (*cf.* §III.6.3). However it is the Albanese or Picard variety of some smooth projective variety: according to the preceding discussion, it is its own Albanese variety, and the Picard variety of the dual Abelian variety. This may seem a little tautological. It is reassuring to know that, according to a theorem of Lefschetz, any Abelian variety is (isomorphic to) the Albanese variety of some smooth two-dimensional projective variety.

III.6.2. Abelian varieties as projective algebraic varieties.

We have seen in Theorem III.2.1 that Abelian varieties are complex manifolds which can be embedded in a complex projective space $\mathbb{P}^N\mathbb{C}$. Then Chow's theorem (Theorem I.4.4) shows that they may be considered as algebraic subvarieties of $\mathbb{P}^N\mathbb{C}$. As a matter of fact, a further consequence of Chow's theorem is that a complex Abelian variety may be defined as a projective variety $A \subset \mathbb{P}^N\mathbb{C}$ equipped with a structure of Abelian group

$$\begin{aligned} A \times A &\to A \\ (x, y) &\mapsto x + y \end{aligned}$$

which makes A an algebraic group, that is, roughly speaking, such that the homogeneous coordinates of $x+y$ (resp. of $-x$) are given by rational expressions in the homogeneous coordinates of x and y (resp. of x).

This algebraic definition leads to some refinements of the notion of Abelian variety. Let indeed K be any subfield of $\mathbb{C}$ (*e.g.* $\mathbb{Q}$ or $\overline{\mathbb{Q}}$). An algebraic subvariety $V \subset \mathbb{P}^N\mathbb{C}$ is said to be *defined over* K if it may be defined by polynomial equations with coefficients in K (*cf.* § I.4.2; this definition is compatible with the definition of § I.5.2: if M is a compact connected Riemann surface, M may be defined over $\overline{\mathbb{Q}}$ iff it may be embedded in some projective space $\mathbb{P}^N\mathbb{C}$ as an algebraic curve defined over $\overline{\mathbb{Q}}$). An *Abelian variety* $A \subset \mathbb{P}^N\mathbb{C}$ is said to be *defined over* K if it is defined over K as an algebraic variety, if the zero element of A has homogeneous coordinates in K, and if the addition and subtraction laws on A are given by rational expressions with coefficients in K. Furthermore, this algebraic definition of Abelian varieties may be extended to any field K, not necessarily included in $\mathbb{C}$, for example to finite fields.

Example: As in § 0.1, consider $\tau \in \mathfrak{H}$, E_τ the elliptic curve (=one-dimensional Abelian variety) $\mathbb{C}/(\mathbb{Z} + \tau\mathbb{Z})$, $\wp$ the Weierstrass function associated with the lattice $\mathbb{Z}+\tau\mathbb{Z}$ and $i : E_\tau \to \mathbb{P}^2\mathbb{C}$, the embedding defined by $(\wp, \wp')$. Embedded in $\mathbb{P}^2\mathbb{C}$ by i, E_τ appears as the algebraic curve of affine equation

$$y^2 = 4x^3 - g_2(\tau)x - g_3(\tau),$$

where $g_2(\tau)$ and $g_3(\tau)$ are defined by some Eisenstein series (*cf.* (0.3) and (0.4)). Moreover its zero element is the 'point at infinity' $(0:0:1)$ and the group law on E_τ is given by the following rules:

- $-(x, y) = (x, -y)$;
- if $(x', y') \neq (x, -y)$, let

$$\begin{aligned} m &= \frac{y' - y}{x' - x} && \text{if} \quad (x', y') \neq (x, y) \\ &= \frac{12x^2 - g_2}{2y} && \text{if} \quad (x', y') = (x, y); \end{aligned}$$

then

$$(x, y) + (x', y') = (x'', y'')$$

where

$$x'' = \frac{m^2}{4} - x - x'$$

and

$$y'' = m(x - x'') - y = m(x' - x'') - y'.$$

This follows from the identity $\wp(-z) = \wp(z)$ and from the addition law for $\wp$ (*cf.* (II.1.4)). See figure 32 for a geometric interpretation of these operations (compare with (II.1.5)).

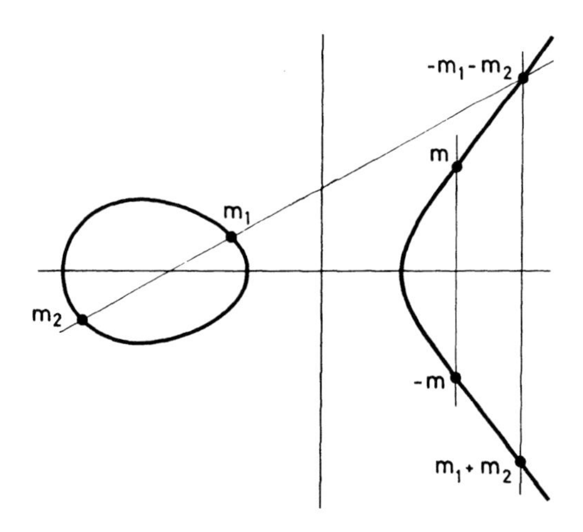

Fig. 32. The group operations on the elliptic curve $y^2 = (x - x_1)(x - x_2)(x - x_3)$.

These formulae prove that E_τ, embedded in $\mathbb{P}^2\mathbb{C}$ by i, is an Abelian variety defined over any subfield K of $\mathbb{C}$ which contains $g_2(\tau)$ and $g_3(\tau)$.

It is possible to develop the theory of Jacobians and Abelian varieties by purely algebraic means. (This is due initially to Weil, Chow, Matsusaka and Igusa). For instance, if X is a smooth projective algebraic curve, then Jac(X) may be constructed directly as a projective algebraic variety, without using any transcendental tool, such as complex analysis and theta functions. The starting point of these constructions is the algebraic description of the Jacobian we discussed in §II.3.1. Using these algebraic techniques, one may prove that, if X is defined over K, of genus g, then Jac(X) embedded in $\mathbb{P}^{\ell^g-1}\mathbb{C}$, $\ell \geq 3$ by means of theta functions (*cf.* Theorem III.4.2) is an Abelian variety defined over K. That type of result is crucial for the study of algebraic curves over number fields or over finite fields.

III.6.3. Modular forms and moduli spaces of Abelian varieties.

Two polarized Abelian varieties $(V/\Lambda, H)$ and $(V'/\Lambda', H')$ are said to be isomorphic iff there is a $\mathbb{C}$-linear isomorphism

$$\varphi : V \xrightarrow{\sim} V'$$

such that $\varphi(\Lambda) = \Lambda'$ and such that, for any $(x, y) \in V^2$,

$$H'(\varphi(x), \varphi(y)) = H(x, y).$$

We have seen that any principally polarized Abelian variety is isomorphic to some complex torus $\mathbb{C}^g/(\mathbb{Z}^g + \Omega\mathbb{Z}^g)$, $\Omega \in \mathfrak{H}_g$, equipped with the polarization

$$\begin{aligned} \mathbb{C}^g \times \mathbb{C}^g &\to \mathbb{C} \\ (x, y) &\mapsto {}^t\overline{x} \cdot (\operatorname{Im}\Omega)^{-1} \cdot y. \end{aligned}$$

Moreover, one can check that the principally polarized Abelian varieties associated to two matrices Ω and Ω' in $\mathfrak{H}_g$ are isomorphic iff Ω and Ω' belong to the same orbit under the action of $\mathrm{Sp}(2g, \mathbb{Z})$ on $\mathfrak{H}_g$ described in §III.1. In other words, the quotient space $\mathfrak{H}_g/\mathrm{Sp}(2g, \mathbb{Z})$ may be identified with the space $\mathcal{A}_g$ of isomorphic classes of principally polarized Abelian varieties. This space is often called the *moduli space of principally polarized Abelian varieties.*

The discrete group $\mathrm{Sp}(2g, \mathbb{Z})$ acts holomorphically and properly on $\mathfrak{H}_g$. This implies that the quotient $\mathfrak{H}_g/\mathrm{Sp}(2g, \mathbb{Z})$ is naturally endowed with a structure of complex analytic space (*i.e.*, roughly speaking, of complex manifold with singularities; the singularities come from the existence of points in $\mathfrak{H}_g$ fixed by the action of some elements of $\mathrm{Sp}(2g; \mathbb{Z}) - \{I_{2g}\}$ and are not really serious). It happens that this complex analytic space possesses a holomorphic embedding

$$i : \mathcal{A}_g \to \mathbb{P}^N(\mathbb{C})$$

in some complex projective space, and that its image $i(\mathcal{A}_g)$ is a quasi-projective algebraic variety, *i.e.*, the difference $X - Y$ of a projective algebraic variety X and of a projective algebraic variety Y included in X. The 'algebraicity' of $\mathcal{A}_g$ is closely related to the 'algebraicity' of Abelian varieties discussed in the preceding paragraph.

From a transcendental point of view, the most direct way to define a projective embedding of $\mathcal{A}_g$ is to use modular forms. Suppose $g \geq 2$. For any subgroup Γ of finite index in $\mathrm{Sp}(2g, \mathbb{Z})$ and any integer $k \geq 0$, a *modular form of weight k with respect to Γ on $\mathfrak{H}_g$* is by definition a holomorphic function $f : \mathfrak{H}_g \to \mathbb{C}$ such that[41] for any

$$\begin{pmatrix} A & B \\ C & D \end{pmatrix} \in \Gamma,$$

we have

[41] When $g = 1$, one must require an additional growth condition; *cf.* [Z].

$$f((A\Omega + B)(C\Omega + D)^{-1}) = \det(C\Omega + D)^k \cdot f(\Omega).$$

If Γ is the congruence subgroup Γ_n of $\mathrm{Sp}(2g, \mathbb{Z})$ defined by

$$\gamma \in \Gamma_n \Leftrightarrow \gamma \equiv I_{2g} \pmod{n}$$

then f is said to be a *modular form of level* n.

The quotient spaces $\mathfrak{H}_g/\Gamma$, Γ a finite index subgroup in $\mathrm{Sp}(2g, \mathbb{Z})$, are coverings of $\mathcal{A}_g$ which parametrize principally polarized Abelian varieties equipped with some additional structure. For instance $\mathfrak{H}_g/\Gamma_n$ parametrizes principally polarized Abelian varieties equipped with a basis of the subgroup of n-torsion points.

We can now state:

Theorem III.6.1. *Let Γ be a subgroup of finite index in $\mathrm{Sp}(2g, \mathbb{Z})$. For any integer $k \geq 0$, the space $[\Gamma, k]$ of modular forms of weight k with respect to Γ is finite dimensional. If k is large enough, the elements of $[\Gamma, k]$ have no common zero, and, for any basis $(f_1, \ldots, f_N)$ of $[\Gamma, k]$, the map*

$$\begin{aligned} \mathfrak{H}_g/\Gamma &\to \mathbb{P}^{N-1}\mathbb{C} \\ [x] &\mapsto (f_1(x) : \cdots : f_N(x)) \end{aligned}$$

is an embedding of $\mathfrak{H}_g/\Gamma$, whose image is a quasi-projective algebraic subvariety of $\mathbb{P}^{N-1}\mathbb{C}$.

As in the case $g = 1$, one may construct modular forms in g variables starting from theta series. For instance one may prove that, as a function of $\Omega \in \mathfrak{H}_g$, $\theta(0, \Omega)^2$ *is a modular form of weight 1 and level* 4. This is essentially equivalent to the following functional equation:

$$\theta(0, -\Omega^{-1})^2 = \det\left(\frac{\Omega}{i}\right) \cdot \theta(0, \Omega)^2$$

(*cf.* [C], §3.3 and [Z], §1.C when $g = 1$). More generally, we have:

Theorem III.6.2. *Any homogeneous polynomial of degree $2d$ in the 'Thetanullwerte' $\theta\left[\begin{smallmatrix} a \\ b \end{smallmatrix}\right](0, \Omega)$, $(a, b) \in \mathbb{Q}^g \times \mathbb{Q}^g$, is a modular form of weight d with respect to some congruence subgroup Γ.*

Conversely for any finite index subgroup Γ of $\mathrm{Sp}(2g, \mathbb{Z})$ any modular form with respect to Γ may be obtained as the quotient of two homogeneous polynomials in the 'Thetanullwerte'.

Taken together, Theorems III.6.1 and III.6.2 show that the 'Thetanullwerte' provide projective embeddings of the moduli spaces $\mathfrak{H}_g/\Gamma$. Thus they provide a far reaching generalization of the example, due to Klein, discussed in §I.4.4, where we described an embedding of $\mathfrak{H}_1/\Gamma(7)$ in $\mathbb{P}^2\mathbb{C}$.

With the notion of moduli space of Abelian varieties at hand, we can say a few words on the relation between Jacobians and general Abelian varieties.

If $g = 1$, any g-dimensional Abelian variety is isomorphic to a Jacobian (since it may be identified with its own Jacobian!). However, for $g \gg 0$, it is not the case. Indeed, consider the set $\mathcal{M}_g$ of isomorphism classes of compact connected Riemann surfaces of genus g. This set—the so-called *moduli space of smooth curves of genus g*—is somewhat similar to $\mathcal{A}_g$: it is endowed with a natural structure of complex space, with only mild singularities, and may be embedded in some projective space $\mathbb{P}^N\mathbb{C}$ as a quasi-projective subvariety. Moreover the (complex) dimension of $\mathcal{M}_g$ is $3g - 3$ (if $g \geq 2$) and one defines a holomorphic (in fact algebraic) map

$$j : \mathcal{M}_g \to \mathcal{A}_g$$

by sending the isomorphism class of a compact connected Riemann surface of genus g to the isomorphism class of its Jacobian (observe that by Torelli's Theorem - *cf.* §III.5 - the map j is injective). As the dimension of $\mathcal{A}_g$ is $\frac{1}{2}g(g+1)$, this shows that *the set $j(\mathcal{M}_g)$ of g-dimensional Jacobians is nowhere dense in* $\mathcal{A}_g$ when $\frac{1}{2}g(g+1) > 3g - 3$, *i.e.*, when $g \geq 4$.

The problem of describing explicitly $j(\mathcal{M}_g)$, *i.e.*, of characterizing the Abelian varieties which are Jacobians, is known as the *Schottky problem.* It has been much studied for one century and has known great progresses during the last ten years (*cf.* [Be], [D], [vG], [AD]).

On the other hand, we should mention that, for any Abelian variety A, there exists a Jacobian J such that A is a quotient of J (*i.e.*, such that there exists a surjective holomorphic map $\varphi : J \to A$). In fact, if A is a g-dimensional Abelian variety embedded in $\mathbb{P}^N\mathbb{C}$, we can take as J the Jacobian of the smooth curve obtained by intersecting A by a linear subspace of $\mathbb{P}^N\mathbb{C}$ of codimension $g-1$ in general position (see [Mi2], §10). This fact was classically used to prove general statements concerning Abelian varieties 'by reduction to Jacobians'.

Epilogue: Arithmetics on algebraic curves and Abelian varieties

The notions introduced in the last Sections allow us to state some results in 'arithmetic geometry', where the study of algebraic curves and Abelian varieties finally mixes with arithmetics.

If V is a projective algebraic variety defined over a subfield K of $\mathbb{C}$, it makes sense to consider the set $V(K)$ of points in V which are *rational over* K, *i.e.*, which have homogeneous coordinates in K. When V is an Abelian variety defined over K, the set of rational points $V(K)$ clearly is a subgroup of the group V.

The following theorems are two of the most remarkable results in arithmetic geometry.

Theorem . *Let A be any Abelian variety defined over a number field*[42] K. *The group of rational points $A(K)$ is finitely generated.*

Theorem . *Let X be any smooth algebraic projective curve, of genus ≥ 2, defined over a number field K. The set of rational points $X(K)$ is finite.*

The first of these theorems is due to Mordell ([Mo1], 1922) for elliptic curves and to Weil ([Weil], 1928) in general[43]. In the case of elliptic curves, with the notations of the example of §III.6.2, it asserts that if $g_2(\tau)$ and $g_3(\tau)$ belong to K, there exists a finite family of points of E_τ with coordinates in K such that any such point may be obtained from this finite family by iteration of the constructions depicted on figure 32.

The second theorem is due to Faltings ([Fa], 1983). Faltings' original proof, as well as subsequent proofs due to Masser, Wüstholz, Vojta, Faltings and Bombieri, use in a crucial way the interrelation between algebraic curves and Abelian varieties which was the subject of these lectures.

Bibliographical comments.

The literature on Riemann surfaces, algebraic curves, Jacobians, and Abelian varieties, is incredibly vast. We do not attempt at giving any sort of complete bibliography, but we simply list various sources where the reader may find more details on the topics touched in these notes.

A few general comments may be appropriate before we indicate the references concerning particular topics.

Reading original papers is still one of the best ways to a serious understanding of the topics discussed here. The collected papers of Abel ([A4]), Jacobi ([J4]), Riemann ([Ri2]) are still fascinating. Papers by Weierstrass, Dedekind, Weber, Frobenius, Hurwitz, Klein, Poincaré or Lefschetz, are written in a more contemporary style and may be more readable now. Houzel's paper ([Ho]) is very helpful to penetrate the developments of the studies on Abelian integrals till Riemann's time.

The first 'modern' book on Riemann surfaces is by Weyl ([Wey 1, 2]). It remains one of the best books on the subject. The book 'Curves and their Jacobians' by Mumford ([Mu2]) presents a highly readable overview on many developments of the topics discussed in these notes, and ought to be consulted by readers who want to go further. The book [ACGH], and the papers [Ros], [Mi1], [Mi2] should be very useful at a more advanced level.

---

[42] *i.e.*, an extension of $\mathbb{Q}$ generated by a finite family of algebraic numbers.

[43] In fact, in [Weil], Weil considers only Jacobians. But this easily implies the general case.

Reference list by sections.

Introduction
- Elliptic functions and elliptic curves: [Bm], [Coh], [Ge], [HC], [Jo], [L1], [Rob1]
- Modular forms of one variable: [Shi], [Z]
- See also references in [Coh], [Ge], [Z]

§ **I.1–2**
- General books on Riemann surfaces: [CGV], [Dy], [Fo], [FK], [GN], [Gu], [Sp], [Sie2]

§ **I.3**
- Uniformization: [Wey1–2], [FK], [Sie2], [Sp]

§ **I.4.2**
- Basic facts of algebraic geometry: [Abh], [Ko], [Mu3], [Sha]
- Algebraic curves: [Abh], [ACGH], [BK], [Cl], [Coo], [EC], [Fu], [Gr2], [GrH] Chap.2, [L2], [Wa]

§ **I.4.3**
- Fuchsian groups and quaternions algebras: [Ei], [Shi], [Vi]
- Poincaré's theorem for fundamental polygons: [Mas]
- Siegel's theorem on fundamental polygons: [Sie1]

§ **I.5**
- Equilateral triangulations and Belyi's theorem: [By], [Gro], [SV1–2]

Appendix A
- Homology: [Gre]
- Topology and classification of surfaces: [H], [SeiT]

§ **II.1**
- Abel's theorem and inversion of Abel's integrals: [Gr1], [Ke1]

§ **III.2–3, III.1**
- Jacobians of complex curves: [ACGH], [GrH] Chap.2, [Ke3], [Mi2], [Mu2], [Mu4] I,II, [Sie2]

§ **III.2–4**
- Abelian varieties and theta functions: [GrH] Chap.2, [I], [Ke4], [Mu1], [Rob2], [Ros], [Sie2] III, [Sw], [Wei3]

§**III.5**
- Riemann surfaces and theta functions: [ACGH], [Fay], [GrH] Chap.2, [Mu4] I,II

§**III.6.2**
- Abelian varieties as projective algebraic varieties: [Mu1], [Mil1]

§**III.6.3**
- Modular forms on $\mathfrak{H}_g$: [Fr], [I], [Mu4] I,II, [Sie2] III
- Schottky problem: [AD], [Be], [D], [vG]

Epilogue Arithmetics on algebraic curves and Abelian varieties:
- Original papers: [Mo], [Wei1], [Fa]
- Mordell-Weil theorem: [Mi1], [Mo2], [Mu1], [Ser]
- Mordell conjecture (= Faltings theorem): [Bo], [Maz]

References

[A1] Abel, N.H. Remarques sur quelques propriétés générales d'une certaine sorte de fonctions transcendantes, *Crelle J. reine u. angewandte Math.* **3** (1828), (=[A4],t. I, 444–456).

[A2] Abel, N.H. Démonstration d'une propriété générale d'une certaine classe de fonctions transcendantes, *Crelle J. reine u. angewandte Math.* **4** (1829), 200–201 (=[A4], t. I, 515–517).

[A3] Abel, N.H. Mémoire sur une propriété générale d'une classe très étendue de fonctions transcendantes, présenté à l'Académie des Sciences à Paris le 30 octobre 1826, *Mémoires présentés par divers savants*, t. VII, Paris, 1841 (=[A4], t. I, 145–211).

[A4] Abel, N.H. *Œuvres*, 2 vol., edited by Sylow and Lie, Christiania, Oslo, 1881.

[Abh] Abhyankar, S.S. *Algebraic Geometry for Scientists and Engineers*, American Mathematical Society, Providence, 1990.

[ACGH] Arbarello, E., Cornalba, M., Griffiths, P. and Harris, J. *Geometry of Algebraic Curves*, Springer-Verlag, New-York, 1985.

[AD] Arbarello, E. and De Concini, C. Geometrical aspects of the Kadomtsev-Petviashvili equation, in *CIME Lectures 1988*, Springer-Verlag Lecture Notes in Mathematics **1451** (1990), 95–137.

[Be] Beauville, A. Le problème de Schottky et la conjecture de Novikov, in Séminaire Bourbaki, Exposé n^{0}675, *Astérisque* **152–153** (1988), 101–112.

[Bel] Bellissard, J. Gap Labelling Theorems for Schrödinger Operators, *This volume*, Chapter 12.

[Bm] Bellman, R. *A Brief Introduction to Theta Functions*, Holt, Rinehart and Winston, 1961.

[By] Belyi, G.V. On Galois extensions of the maximal cyclotomic field. *Izvestiya AN SSSR, ser. mat.* **43**:2 (1979), 269–276. (=*Math. USSR Izvestiya*, Vol. **14** (1980), n°2, 247–256).

[Bo] Bombieri, E. The Mordell conjecture revisited, *Annali di Pisa* **17** (1990), 615–640.

[BK] Brieskorn, E. and Knörrer, H. *Ebene algebraische Kurven*, Birkhäuser, Basel, 1981.

[C] Cartier, P. An Introduction to Zeta Functions, *This volume*, Chapter 1.

[Ch] Chern, S.S. An elementary proof of the existence of isothermal parameters on a surface, *Proc. Amer. Math. Soc.* **6** (1955), 771–782.

[Cl] Clemens, C.H. *A Scrapbook of Complex Curve Theory*, Plenum Press, New York, 1980.

[Coh] Cohen, H. Elliptic curves, *This volume*, Chapter 3.

[Coo] Coolidge, J.L. *A Treatise on Algebraic Plane Curves*, Oxford University Press, Oxford, Dover reprint, New York, 1959.

[CGV] Cornalba, M., Gomez-Mont, X. and Verjovsky, A. (edited by) *Lectures on Riemann Surfaces*, World Scientific, Singapore, 1989.

[Cr] Coxeter, H.S.M. *Non-Euclidean Geometry*, The University of Toronto Press.

[D] Donagi, R. The Schottky problem, in *CIME Lectures 1985*, Montecatini, Springer-Verlag Lecture Notes in Mathematics **1337**, 1985.

[Dy] Douady, R. and A. Algèbre et Théories Galoisiennes, Cedic-Fernand Nathan, Paris, 1977.

[Ei] Eichler, M. Uber die Einheiten der Divisionsalgebren, *Math. Ann.* **114** (1937), 635–654.

[EC] Enriques, F. and Chisini, O. *Teoria geometrica delle equazioni e delle funzioni algebriche*, Zanichelli, Bologna, 1934.

[FK] Farkas, H. and Kra, I. *Riemann Surfaces*, Springer-Verlag, New York, 1980.

[Fa] Faltings, G. Endlichkeitsätze für abelsche Varietäten über Zahlkörpern, *Invent. Math.* **73** (1983), 349–366; corrigendum, *Invent. Math.* **75** (1984), 381.

[Fay] Fay, J. *Theta Functions on Riemann Surfaces*, Lecture Notes in Mathematics 332, Springer-Verlag, New York, 1973.

[Fo] Forster, O. *Lectures on Riemann Surfaces*, Springer-Verlag, New York, 1981.

[Fr] Freitag, E. *Siegelsche Modulfunktionen*, Springer-Verlag, New York, 1983.

[Fu] Fulton, W. *Algebraic Curves*, W.A. Benjamin, New York, 1969.

[Ge] Gergondey, R. Decorated Elliptic Curves: Modular Aspects, *This volume*, Chapter 5.

[Gre] Greenberg, M.J. *Lectures on Algebraic Topology*, Benjamin, Reading, 1967.

[Gr1] Griffiths, P. Variations on a theorem of Abel, *Inventiones Math.* **35** (1976), 321–390.

[Gr2] Griffiths, P. *Introduction to Algebraic Curves*, American Mathematical Society, Providence, 1989.

[GrH] Griffiths, P. and Harris, J. *Principles of Algebraic Geometry*, Wiley, New York, 1978.

[Gro] Grothendieck, A. *Esquisse d'un programme*, 1984.

[Gu] Gunning, R. C. *Lectures on Riemann Surfaces*, Princeton Math. Notes, Princeton U. Press, Princeton 1966.

[GN] Guenot, J. and Narasimhan, R. Introduction à la théorie des surfaces de Riemann, *Enseign. Math*, **21**, 1975, 123–328.

[H] Hirsch, M.W. *Differential Topology*, Springer-Verlag, New York, 1976.
[Ho] Houzel, C. Fonctions elliptiques et intégrales abéliennes, in *Abrégé d'histoire des mathématiques* 1700–1900, II, edited by J. Dieudonné, Hermann, Paris, 1978, 1–113.
[HC] Hurwitz, A. and Courant, R. *Vorlesungen über allgemeine Funktionentheorie und elliptische Funktionen*, Springer-Verlag, Berlin 1929.
[I] Igusa, J. *Theta Functions*, Springer-Verlag, New York, 1972.
[J1] Jacobi, C.G.J. Considerationes generales de transcendentibus abelianis, *Crelle J. reine u. angewandte Math.* **9** (1832), 394–403 (=[J4], 5–16).
[J2] Jacobi, C.G.J. De functionibus duarum variabilium quadrupliciter periodicis, quibus theoria transcendentium abelianarum innititur, *Crelle J. reine u. angewandte Math.* **13** (1834), 55–78 (=[J4], 23–50).
[J3] Jacobi, C.G.J. Zur Geschichte der elliptischen und abelschen Transcendenten, in [J4], 516–521.
[J4] Jacobi, C.G.J. *Gesammelte Werke*, vol. II, edited by K. Weierstrass, Berlin 1882; Chelsea, New York, 1969.
[Jo] Jordan, C. *Cours d'Analyse*, II, Gauthier-Villars, 1913.
[Ke1] Kempf, G. On the geometry of a theorem of Riemann, *Ann. of Math.* **98** (1973), 178–185.
[Ke2] Kempf, G. Inversion of Abelian integrals, Bull. of the A.M.S. **6** (1982), 25–32.
[Ke3] Kempf, G. *Abelian Integrals*, Monographias de Instituto de Matematicas, ♯13, UNAM, Mexico City, 1984.
[Ke4] Kempf, G.R. *Complex Abelian Varieties and Theta Functions*, Springer-Verlag, Berlin, 1991.
[Kl1] Klein, F. Uber die Transformation siebenter Ordnung der elliptischen Funktionen, *Math. Ann.* **14** B3 (1878), 428–471 (=[Kl2], p. 90–135).
[Kl2] Klein, F. *Gesammelte Mathematische Abhandlungen*, dritter Band, herausgegeben von R. Fricke, H. Vermeil und E. Bersel-Hagen, Verlag von Julius Springer, Berlin, 1923.
[Ko] Kollar, J. The structure of algebraic threefolds: an introduction to Mori's program, *Bull. of the A.M.S.* **17** (1987), 211–273.
[L1] Lang, S. *Elliptic Functions*, Addison Wesley, 1973.
[L2] Lang, S. *Introduction to Algebraic and Abelian Functions*, 2nd edn., Springer-Verlag, New York, 1982.
[Mas] Maskit, B. On Poincaré's theorem for fundamental polygons, *Adv. in Math.* **7** (1971), 219–230.
[Maz] Mazur, B. Arithmetic on curves, *Bull. of the A.M.S.* **14** (1986), 207–259.
[Mi1] Milne, J.S. Abelian varieties. *Arithmetic Geometry*, edited by G. Cornell and J.H. Silverman, Springer-Verlag, New York, 1986, 103–150.
[Mi2] Milne, J.S. Jacobian varieties. *Arithmetic Geometry*, edited by G. Cornell and J.H. Silverman, Springer-Verlag, New York, 1986, 167–212.
[Mil] Milnor, J. *Topology from a differential viewpoint*, University of Virginia Press, Charlottesville, 1965.
[Mo1] Mordell, L.J. On the rational solutions of the indeterminate equations of the third and fourth degrees, *Proc. Camb. Phil. Soc.* **21** (1922), 179–192.
[Mo2] Mordell, L.J. *Diophantine Equations*, Academic Press, London, 1969.
[Mu1] Mumford, D. *Abelian Varieties.* Oxford University Press, Oxford, 1970 (2nd edition, 1974).
[Mu2] Mumford, D. *Curves and their Jacobians.* University of Michigan Press, Ann Arbor, 1975.
[Mu3] Mumford, D. *Algebraic Geometry I: Complex Projective Varieties.* Springer-Verlag, New York, 1976.
[Mu4] Mumford, D. *Tata Lectures on Theta* I, II, III. Birkhäuser-Verlag, Basel, 1983, 1984, 1991.
[P] Poincaré, H. Sur les propriétés du potentiel et sur les fonctions abéliennes, *Acta Math.* **22** (1898), 89–178 [= Œuvres IV, 162–243].
[Rey] Reyssat, E. Galois Theory for Coverings and Riemann Surfaces, *This volume*, Chapter 7.

[Ri1] Riemann, B. Theorie der Abel'schen Functionen, Crelle J. reine u. angewandte Math. **54** (1857) (=[Ri2], p. 88–142).
[Ri2] Riemann, B. *Mathematische Werke*, B.G. Teubner, 1876, Dover Publications, 1953.
[Rob1] Robert, A. *Elliptic Curves*, Springer Lecture Notes in Mathematics **326**, 1973.
[Rob2] Robert, A. Introduction aux variétés abéliennes, *Enseign. Math.* **28** (1982), 91–137.
[Ros] Rosen, M. Abelian varieties over $\mathbb{C}$. *Arithmetic geometry*, edited by G. Cornell and J.H. Silverman, Springer-Verlag, New York, 1986, 79–101.
[SeiT] Seifert, H. and Threlfall, W. *Lehrbuch der Topologie*, Teubner, Leipzig, 1934, Dover, New York.
[Sel] Selberg, A. On discontinuous groups in higher-dimensional symmetric spaces, Internat. Colloq. Function Theory, Bombay 1960, 147–164, Tata Institute of Fundamental Research, Bombay, 1960.
[Ser] Serre, J.-P. *Lectures on the Mordell-Weil Theorem*, Vieweg, Wiesbaden, 1989.
[Sen] Senechal, M. Introduction to Lattice Geometry, *This volume*, Chapter 10.
[SV1] Shabat, G.B. and Voevodsky, V.A. Equilateral triangulations of Riemann surfaces and curves over algebraic number fields. *Doklady AN SSSR* **304** 2 (1989), 265–268 (=*Soviet Math. Dokl.*, Vol. **39** (1989), n°1, 38–41).
[SV2] Shabat, G.B. and Voevodsky, V.A. Drawing curves over number fields, *Grothendieck Festschrift*, Birkhäuser-Verlag.
[Sha] Shafarevich, I.R. *Basic Algebraic Geometry*, Springer-Verlag, New York, 1977.
[Shi] Shimura, G. *Introduction to the Arithmetic Theory of Automorphic Functions*, Math. Soc. of Japan n°11, Princeton University Press, Princeton, 1971.
[Sie1] Siegel, C.L. Some remarks on discontinuous groups, *Ann. of Math.* **46** (1945), 708–718.
[Sie2] Siegel, C.L. *Topics in Complex Function Theory*, I, II, III. Wiley, New York, 1972.
[Sp] Springer, G. *Introduction to Riemann surfaces*, Addison Wesley Publishing Co., Reading, 1957.
[Sw] Swinnerton-Dyer, H.P.F. *Analytic Theory of Abelian Varieties*, Cambridge University Press, Cambridge, 1974.
[vG] van Geemen, B. The Schottky problem, *Arbeitstagung Bonn* 1984, Springer-Verlag Lecture Notes in Mathematics, *1111* (1985), 385–406.
[Vi] Vignéras, M.-F. *Arithmétique des Algèbres de Quaternions*, Springer-Verlag Lecture Notes in Mathematics **800**, 1980.
[Wa] Walker, R.J. *Algebraic Curves*, Princeton University Press, Princeton, 1950, Dover reprint, New York, 1962.
[Weil] Weil, A. L'arithmétique sur les courbes algébriques, *Acta Math.* **52** (1928), 281–315.
[Wei2] Weil, A. The field of definition of a variety, *Amer. J. Math.* **78** (1956), 509–524.
[Wei3] Weil, A. *Introduction à l'Etude des Variétés Kähleriennes*, Hermann, Paris, 1971.
[Wey1] Weyl, H. *Die Idee der Riemannschen Fläche.* B.G. Teubner, Berlin, 1913.
[Wey2] Weyl, H. *The Concept of a Riemann Surface*, 3d. edition, Addison-Wesley, Reading, Mass., 1955.
[Z] Zagier, D. Introduction to Modular Forms, *This volume*, Chapter 4.

Chapter 3

Elliptic Curves

by Henri Cohen

1. Elliptic Curves

1.1 Elliptic Integrals and Elliptic Functions

The aim of this Chapter is to give a brief survey of results, essentially without proofs, about elliptic curves, class groups and complex multiplication. No previous knowledge of these subjects is required. Some excellent references are listed in the bibliography. The basic general reference books are (Borevitch and Shafarevitch 1966), (Shimura 1971A), (Silverman 1986 and 1994). The algorithms and tables in (Birch and Kuyk 1975) and (Cremona 1992) are invaluable.

Historically, the word elliptic (in the modern sense) came from the theory of elliptic integrals, which occur in many problems, for example in the computation of the length of an arc of an ellipse (whence the name), or in physical problems such as the movement of a pendulum. Such integrals are of the form

$$\int R(x,y)\,dx,$$

where $R(x,y)$ is a rational function in x and y, and y^2 is a polynomial in x of degree 3 or 4 having no multiple root. It is not our purpose here to explain the theory of these integrals (for this see e.g. Whittaker and Watson 1927, ch. XXII). However they have served as a motivation for the theory of elliptic *functions*, developed in particular by Abel, Jacobi and Weierstraß.

Elliptic functions can be defined as inverse functions of elliptic integrals, but the main property that interests us here is that these functions $f(x)$ are doubly periodic. More precisely we have:

(1.1.1) Definition. *An elliptic function is a meromorphic function $f(x)$ on the whole complex plane, which is doubly periodic, i.e. such that there exist complex numbers ω_1 and ω_2 such that $\omega_1/\omega_2 \notin \mathbb{R}$ and $f(x+\omega_1) = f(x+\omega_2) = f(x)$.*

If

$$L = \{m\omega_1 + n\omega_2 | m, n \in \mathbb{Z}\}$$

is the lattice generated by ω_1 and ω_2, it is clear that f is elliptic if and only if $f(x+\omega) = f(x)$ for all $x \in \mathbb{C}$ and all $\omega \in L$. The lattice L is called the period lattice of f. Furthermore it is clear that every element of $\mathbb{C}$ is equivalent modulo a translation by an element of L to a unique element of the set $F = \{x\omega_1 + y\omega_2,\ 0 \le x, y < 1\}$. Such a set will be called a *fundamental domain* for $\mathbb{C}/L$.

Standard residue calculations show immediately the following properties:

(1.1.2) Theorem. *Let $f(x)$ be an elliptic function with period lattice L, let $\{z_i\}$ be the set of zeros and poles of f in a fundamental domain for $\mathbb{C}/L$, and n_i be the order of f at z_i ($n_i > 0$ when z_i is a zero, $n_i < 0$ if z_i is a pole). Then*

1. *The sum of the residues of f in a fundamental domain is equal to 0.*
2. *$\sum_i n_i = 0$, in other words f has as many zeros as poles (counted with multiplicity).*
3. *If f is non-constant, counting multiplicity, f must have at least 2 poles (and hence 2 zeros) in a fundamental domain.*
4. *$\sum_i n_i z_i \in L$. This makes sense since z_i is defined modulo L.*

Note that the existence of non-constant elliptic functions is not *a priori* evident from definition 1.1.1. In fact, we have the following general theorem, due to Abel and Jacobi:

(1.1.3) Theorem. *Assume that z_i and n_i satisfy the above properties. Then there exists an elliptic function f with zeros and poles at z_i of order n_i.*

The simplest construction of non-constant elliptic functions is due to Weierstraß. One defines

$$\wp(z) = \frac{1}{z^2} + \sum_{\omega \in L \setminus \{0\}} \left(\frac{1}{(z+\omega)^2} - \frac{1}{\omega^2} \right),$$

and one easily checks that this is an absolutely convergent series which defines an elliptic function with a double pole at 0. Using the fact that non-constant elliptic functions must have poles, it is then a simple matter to check that if we define

$$g_2 = 60 \sum_{\omega \in L \setminus \{0\}} \frac{1}{\omega^4} \quad \text{and} \quad g_3 = 140 \sum_{\omega \in L \setminus \{0\}} \frac{1}{\omega^6},$$

then $\wp(z)$ satisfies the following differential equation:

$$\wp'^2 = 4\wp^3 - g_2\wp - g_3.$$

In more geometric terms, one can say that the map

$$z \mapsto \begin{cases} (\wp(z) : \wp'(z) : 1) & \text{for } z \notin L \\ (0:1:0) & \text{for } z \in L \end{cases}$$

from $\mathbb{C}$ to the projective complex plane gives an isomorphism between the torus $\mathbb{C}/L$ and the projective algebraic curve $y^2t = 4x^3 - g_2xt^2 - g_3t^3$. This is in fact a special case of a general theorem of Riemann which states that all compact Riemann surfaces are algebraic. Note that it is easy to prove that the field of elliptic functions is generated by $\wp$ and $\wp'$ subject to the above algebraic relation.

Since $\mathbb{C}/L$ is non-singular, the corresponding algebraic curve must also be non-singular, and this is equivalent to saying that the discriminant

$$\Delta = 16(g_2^3 - 27g_3^2)$$

of the cubic polynomial is non-zero. This leads directly to the definition of elliptic curves.

1.2 Elliptic Curves over a Field

From the preceding Section, we see that there are at least two ways to generalize the above concepts to an arbitrary field: we could define an elliptic curve as a curve of genus 1 or as a non-singular cubic. Luckily, the Riemann-Roch theorem shows that these two definitions are equivalent, hence we set:

(1.2.1) Definition. *Let K be a field. An elliptic curve over K is a non-singular projective cubic curve together with a point with coordinates in K.*

Up to suitable change of coordinates, it is a simple matter to check that such a curve can always be given by an equation of the following (affine) type:

$$y^2 + a_1xy + a_3y = x^3 + a_2x^2 + a_4x + a_6,$$

the point defined over K being the (unique) point at infinity.

This equation is not unique. However, over certain number fields K such as $\mathbb{Q}$, it can be shown that there exists an equation which is minimal, in a well defined sense. We will call it the minimal Weierstraß equation of the curve. Note that it does not necessarily exist for any number field K. For example, one can show (see Silverman 1986, page 226) that the elliptic curve $y^2 = x^3 + 125$ has no minimal Weierstraß equation over the field $\mathbb{Q}(\sqrt{-10})$.

(1.2.2) Theorem. *An elliptic curve over $\mathbb{C}$ has the form $\mathbb{C}/L$ where L is a lattice. In other words, if g_2 and g_3 are any complex numbers such that $g_2^3 - 27g_3^2 \neq 0$, then there exist ω_1 and ω_2 with* $\operatorname{Im}(\omega_1/\omega_2) > 0$ *and* $g_2 = 60\sum_{(m,n)\neq(0,0)}(m\omega_1 + n\omega_2)^{-4}$, $g_3 = 140\sum_{(m,n)\neq(0,0)}(m\omega_1 + n\omega_2)^{-6}$.

A fundamental property of elliptic curves is that they are commutative algebraic groups. This is true over any base field. Over $\mathbb{C}$ this follows immediately from theorem 1.2.2. The group law is then simply the quotient group law of $\mathbb{C}$ by L. On the other hand, it is not difficult to prove the addition theorem for the Weierstraß $\wp$ function, given by:

$$\wp(z_1+z_2)=\begin{cases} -\wp(z_1)-\wp(z_2)+\frac{1}{4}\left(\frac{\wp'(z_1)-\wp'(z_2)}{\wp(z_1)-\wp(z_2)}\right)^2, & \text{if } z_1\neq z_2; \\ -2\wp(z_1)+\frac{1}{4}\left(\frac{\wp''(z_1)}{\wp'(z_1)}\right)^2, & \text{if } z_1=z_2. \end{cases}$$

From this and the isomorphism given by the map $z\mapsto(\wp(z),\wp'(z))$, one obtains immediately:

(1.2.3) Proposition. *Let $y^2=4x^3-g_2x-g_3$ be the equation of an elliptic curve. The zero element for the group law is the point at infinity $(0:1:0)$. The inverse of a point (x_1,y_1) is the point $(x_1,-y_1)$ i.e. the symmetric point with respect to the real axis. Finally, if $P_1=(x_1,y_1)$ and $P_2=(x_2,y_2)$ are two non-opposite points on the curve, their sum $P_3=(x_3,y_3)$ is given by the following formulas. Set*

$$m=\begin{cases} \frac{y_1-y_2}{x_1-x_2}, & \textit{if } P_1\neq P_2; \\ \frac{12x_1^2-g_2}{2y_1}, & \textit{if } P_1=P_2. \end{cases}$$

Then

$$x_3=-x_1-x_2+m^2/4, \quad y_3=-y_1-m(x_3-x_1).$$

It is easy to see that this theorem enables us to define an addition law on an elliptic curve over any base field of characteristic zero, and in fact of characteristic different from 2 and 3. Furthermore, it can be checked that this law is indeed a group law.

One can more generally define such a law over any field, in the following way.

(1.2.4) Proposition. *Let*

$$y^2+a_1xy+a_3y=x^3+a_2x^2+a_4x+a_6$$

be the equation of an elliptic curve defined over an arbitrary base field. Define the zero element as the point at infinity $(0:1:0)$, the opposite of a point (x_1,y_1) as the point $(x_1,-y_1-a_1x_1-a_3)$. Finally, if $P_1=(x_1,y_1)$ and $P_2=(x_2,y_2)$ are two non-opposite points on the curve, define their sum $P_3=(x_3,y_3)$ by the following. Set

$$m=\begin{cases} \frac{y_1-y_2}{x_1-x_2}, & \textit{if } P_1\neq P_2; \\ \frac{3x_1^2+2a_2x_1+a_4-a_1y_1}{2y_1+a_1x_1+a_3}, & \textit{if } P_1=P_2, \end{cases}$$

and put

$$x_3 = -x_1 - x_2 - a_2 + m(m + a_1), \quad y_3 = -y_1 - a_3 - a_1 x_3 + m(x_1 - x_3).$$

Then these formulas define an (algebraic) Abelian group law on the curve.

The only non-trivial thing to check in this theorem is the associativity of the law. This can most easily be seen by interpreting the group law in terms of divisors, but we will not do this here.

The geometric interpretation of the formulas above is the following. Let P_1 and P_2 be points on the (projective) curve. The line D from P_1 to P_2 (the tangent to the curve if $P_1 = P_2$) intersects the curve at a third point R, say. Then if O is the point at infinity on the curve, the sum of P_1 and P_2 is the third point of intersection with the curve of the line from O to R. One checks easily that this leads to the above formulas.

1.3 Points on Elliptic Curves

Consider an abstract equation $y^2 + a_1xy + a_3y = x^3 + a_2x^2 + a_4x + a_6$, where the coefficients a_i are in $\mathbb{Z}$. One can consider this as a curve over any field K in the following manner. If K has characteristic zero, it contains an isomorphic copy of $\mathbb{Z}$ so the a_i can be considered as elements of K. If K has positive characteristic p, by reduction mod p one can consider the a_i as elements of the finite field $\mathbb{F}_p$, hence of K which contains an isomorphic copy. (Note that even if the initial curve was non-singular, the reduction mod p could be singular). We shall consider successively the case where $K = \mathbb{R}$, $K = \mathbb{Q}$, $K = \mathbb{F}_q$, where q is a power of p.

1.3.1 Elliptic curves over $\mathbb{R}$. In the case where the characteristic is different from 2 and 3, the general equation can be reduced to the following Weierstraß form:

$$y^2 = x^3 + a_4x + a_6.$$

(We could put a 4 in front of the x^3 as in the equation for the $\wp$ function, but this introduces unnecessary constant factors in the formulas). The discriminant of the cubic *polynomial* is $-(4a_4^3 + 27a_6^2)$, however the y^2 must be taken into account, and general considerations show that we must take

$$\Delta = -16(4a_4^3 + 27a_6^2)$$

as definition of the discriminant of the elliptic curve.

Several cases can occur. Let $Q(x) = x^3 + a_4x + a_6$ and Δ as given above.

1. $\Delta < 0$. Then the equation $Q(x) = 0$ has only one real root, and the graph of the curve has only one connected component.
2. $\Delta > 0$. Then the equation $Q(x) = 0$ has three distinct real roots, and the graph of the curve has two connected components: a non-compact one, which is the component of the zero element of the curve (i.e. the point at infinity), and a compact one, oval shaped.

From the geometric construction of the group law, one sees that the roots of $Q(x) = 0$ are exactly the points of order 2 on the curve (the points of order 3 correspond to the inflexion points).

3. $\Delta = 0$. The curve is not any more an elliptic curve, since it now has a singular point. This case subdivides into three subcases. Since the polynomial $Q(x)$ has at least a double root, write

$$Q(x) = (x-a)^2(x-b).$$

Note that $2a + b = 0$.

a) $a > b$. Then the curve has a unique connected component, which has a double point at $x = a$. The tangents at the double point have distinct real slopes.

b) $a < b$. Then the curve has two connected components: a non-compact one, and the single point of coordinates $(a, 0)$. In fact this point is again a double point, but with distinct *complex* tangents.

c) $a = b$. (In this case $a = b = 0$ since $2a + b = 0$). Then the curve has a cusp at $x = 0$, i.e. the tangents at the singular point are the same.

(The figures corresponding to these 5 cases and subcases can easily be hand drawn, or can be found in (Silverman 1986 page 47–48).)

In case 3, one says that the curve is a degenerate elliptic curve. One easily checks that the group law still exists, but on the curve minus the singular point. This leads to the following terminology: in cases 3a and 3b, the group law is isomorphic to a form of the group law of the multiplicative algebraic group $\mathbb{G}_m$. Hence they are called cases of multiplicative degeneracy. More precisely, 3a is called split multiplicative, 3b is non-split multiplicative. For analogous reasons, case 3c is called additive degeneracy. These notions can be used, not only for $\mathbb{R}$, but for any base field K. In that case, the condition $a > b$ is replaced by $a - b$ is a (non-zero) square in K.

1.3.2 Elliptic curves over $\mathbb{Q}$. From a number theorist's point of view, this is of course the most interesting base field. The situation in this case and in the case of more general number fields is much more difficult to study. The first basic theorem, due to Mordell (Mordell 1922) and later generalized by Weil (Weil 1930) to the case of number fields and of Abelian varieties, is as follows:

(1.3.1) Theorem. *Let E be an elliptic curve over $\mathbb{Q}$. The group of points of E with coordinates in $\mathbb{Q}$ (denoted naturally $E(\mathbb{Q})$) is a finitely generated Abelian group. In other words,*

$$E(\mathbb{Q}) \simeq E(\mathbb{Q})_{\text{tors}} \oplus \mathbb{Z}^r,$$

where r is a non-negative integer called the rank of the curve, and $E(\mathbb{Q})_{\text{tors}}$ is the torsion subgroup of $E(\mathbb{Q})$, which is a finite Abelian group.

The torsion subgroup of a given elliptic curve is easy to compute. On the other hand the study of possible torsion subgroups for elliptic curves over $\mathbb{Q}$

is a very difficult problem, solved only in 1977 by Mazur (1977 and 1978). His theorem is as follows:

(1.3.2) Theorem. *Let E be an elliptic curve over $\mathbb{Q}$. Then the torsion subgroup $E(\mathbb{Q})_{\text{tors}}$ of E can be isomorphic only to one of the 15 following groups:*

$$\mathbb{Z}/m\mathbb{Z} \quad \text{for } 1 \leq m \leq 10 \text{ or } m = 12,$$

$$\mathbb{Z}/2\mathbb{Z} \times \mathbb{Z}/2m\mathbb{Z} \quad \text{for } 1 \leq m \leq 4.$$

In particular, its cardinality is at most 16.

Note that all of the 15 groups above do occur for an infinite number of non-isomorphic elliptic curves and also that the corresponding theorem over a number field other than $\mathbb{Q}$ has recently been proved by Merel (Merel 1995) (with more possibilities for the groups of course).

The other quantity which occurs in Mordell's theorem is the rank r, and is a much more difficult number to compute, even for an individual curve. There is no known mathematically proven algorithm to compute r. Even the apparently simpler question of deciding whether r is zero or not (or equivalently whether the curve has a finite or an infinite number of rational points) is still not solved. This is the subject of active research, and we will come back in more detail to this question in Section 3.

Let us give an example of a down to earth application of the above definition. Consider the curve

$$y^2 = x^3 - 36x.$$

It is easy to show that the only torsion points are the points of order 1 or 2, i.e. the point at infinity and the three points $(0,0), (6,0), (-6,0)$. But the point $(-2,8)$ is also on the curve. Hence this point is not of finite order, so the curve has an infinite number of points, a fact which is not *a priori* evident. This curve is in fact closely related to the so-called congruent number problem, and the statement that we have just made means in that context that there exists an infinite number of non-equivalent right angled triangles with all three sides rational and area equal to 6, the simplest one (corresponding to the point $(-2,8)$) being the well known $(3,4,5)$ Pythagorean triangle.

As an exercise, the reader can check that the double of the point $(-2,8)$ is the point $\left(\frac{25}{4}, \frac{35}{8}\right)$, and that this corresponds to the right-angled triangle of area 6 with sides $\left(\frac{120}{7}, \frac{7}{10}, \frac{1201}{70}\right)$. See (Koblitz 1984) for the (almost) complete story on the congruent number problem.

1.3.3 Elliptic curves over a finite field. To study curves (or more general algebraic objects) over $\mathbb{Q}$, it is very useful to study first the reduction of the curve modulo primes. This leads naturally to elliptic curves over $\mathbb{F}_p$, and more generally over an arbitrary finite field $\mathbb{F}_q$, where q is a power of p. Note that when one reduces an elliptic curve mod p, the resulting curve over $\mathbb{F}_p$ may be singular, hence not any more an elliptic curve. Such p are called primes of bad reduction, and are finite in number since they must divide the discriminant of the curve. According to the terminology introduced in Section 1.3.1, we will say that the reduction mod p is (split or non-split) multiplicative or additive, according to the type of degeneracy of the curve over $\mathbb{F}_p$. The main theorem concerning elliptic curves over finite fields, due to Hasse, is as follows:

(1.3.3) Theorem. *Let p be a prime, and E an elliptic curve over $\mathbb{F}_p$. Then there exists an algebraic number α_p such that*

1. If $q = p^n$ then

$$|E(\mathbb{F}_q)| = q + 1 - \alpha_p{}^n - \overline{\alpha_p}^n$$

2.

$$\alpha_p \overline{\alpha_p} = p, \text{ or equivalently } |\alpha_p| = \sqrt{p}.$$

(1.3.4) Corollary. *Under the same hypotheses, we have*

$$|E(\mathbb{F}_p)| = p + 1 - a_p \quad \text{with } |a_p| < 2\sqrt{p}.$$

The numbers a_p are very important and are (conjecturally) coefficients of a modular form of weight 2. We will come back to this subject in Section 3.

2. Complex Multiplication and Class Numbers

In this Section, we will study maps between elliptic curves. We begin by the case of curves over $\mathbb{C}$.

2.1 Maps between Complex Elliptic Curves

Recall that a complex elliptic curve E has the form $\mathbb{C}/L$ where L is a lattice. Let $E = \mathbb{C}/L$ and $E' = \mathbb{C}/L'$ be two elliptic curves. A map ϕ from E to E' is by definition a holomorphic $\mathbb{Z}$-linear map from E to E'. Since $\mathbb{C}$ is the universal cover of E', ϕ lifts to a holomorphic $\mathbb{Z}$-linear map f from $\mathbb{C}$ to $\mathbb{C}$, and such a map has the form $f(z) = \alpha z$ for some non-zero complex number α, which induces a map from E to E' iff $\alpha L \subset L'$. Thus we have obtained:

(2.1.1) Proposition. *Let $E = \mathbb{C}/L$ and $E' = \mathbb{C}/L'$ be two elliptic curves over $\mathbb{C}$. Then:*

1. *E is isomorphic to E' if and only if $L' = \alpha L$ for a certain non-zero complex number α.*
2. *The set of maps from E to E' can be identified with the set of complex numbers α such that $\alpha L \subset L'$. In particular, the set* End(E) *of endomorphisms of E is a ring isomorphic to the set of α such that $\alpha L \subset L$.*

In terms of the Weierstraß equation of the curves, this theorem gives the following. Recall that the equation of E (resp E') is $y^2 = 4x^3 - g_2x - g_3$ (resp. $y^2 = 4x^3 - g_2'x - g_3'$) where

$$g_2 = 60 \sum_{\omega \in L \setminus \{0\}} \omega^{-4}, \quad g_3 = 140 \sum_{\omega \in L \setminus \{0\}} \omega^{-6},$$

and similarly for g_2' and g_3'. Hence the first part of the theorem says that if $E \simeq E'$ then there exists α such that

$$g_2' = \alpha^{-4} g_2, \quad g_3' = \alpha^{-6} g_3.$$

The converse is also clear from the Weierstraß equation. Now, since E is a non-singular curve, the discriminant $g_2^3 - 27g_3^2$ is non-zero, so we can define

$$j(E) = 1728 g_2^3/(g_2^3 - 27g_3^2),$$

and we obtain:

(2.1.2) Proposition. *The function $j(E)$ characterizes the isomorphism class of E over $\mathbb{C}$. More precisely, $E \simeq E'$ if and only if $j(E) = j(E')$.*

The quantity $j(E)$ is called the modular invariant of the elliptic curve E. The number $1728 = 12^3$ will be explained later. Although we have been working over $\mathbb{C}$, proposition 2.1.2 is still valid over any algebraically closed field of characteristic different from 2 and 3 (it is also valid in characteristic 2 or 3, for a slightly different definition of $j(E)$). On the other hand, it is false if the field is not algebraically closed (consider for example $y^2 = 4x^3 - 4x$ and $y^2 = 4x^3 + 4x$ over $\mathbb{R}$).

(2.1.3) *Remark.* It is easy to construct an elliptic curve with a given modular invariant j. Since we have not given the definition otherwise, we give the formulas when the characteristic is different from 2 and 3.

1. If $j = 0$, one can take $y^2 = x^3 - 1$.
2. If $j = 1728$, one can take $y^2 = x^3 - x$.
3. Otherwise, one sets $c = j/(j - 1728)$, and then one can take $y^2 = x^3 - 3cx + 2c$. (If one wants equations with a coefficient 4 in front of x^3, multiply by 4 and replace y by $y/2$.)

Now let $E = \mathbb{C}/L$ be an elliptic curve over $\mathbb{C}$. Then as a $\mathbb{Z}$-module, L can be generated by two $\mathbb{R}$-linearly independent complex numbers ω_1 and ω_2, and by suitably ordering them, we may assume that $\operatorname{Im} \tau > 0$, where $\tau = \omega_1/\omega_2$. Since multiplying a lattice by a non-zero complex number does not change the isomorphism class of E, we have $j(E) = j(E_\tau)$, where $E_\tau = \mathbb{C}/L_\tau$ and L_τ is the lattice generated by 1 and τ. By abuse of notation, we will write $j(\tau) = j(E_\tau)$. This defines a complex function j on the upper half-plane $\mathfrak{H}$:

$$\mathfrak{H} = \{\tau \in \mathbb{C}, \text{Im } \tau > 0\} .$$

If a, b, c and d are integers such that $ad - bc = 1$ (i.e. if $\begin{pmatrix} a & b \\ c & d \end{pmatrix} \in SL_2(\mathbb{Z})$), then the lattice generated by $a\tau + b$ and $c\tau + d$ is equal to L_τ. This implies the *modular invariance of* $j(\tau)$:

(2.1.4) Theorem. *For any* $\begin{pmatrix} a & b \\ c & d \end{pmatrix} \in SL_2(\mathbb{Z})$, *we have*

$$j\left(\frac{a\tau + b}{c\tau + d}\right) = j(\tau).$$

In particular, $j(\tau)$ is periodic of period 1. Hence it has a Fourier expansion, and one can prove the following theorem:

(2.1.5) Theorem. *There exist positive integers* c_n *such that, if we set* $q = e^{2i\pi\tau}$, *we have for all complex* τ *with* $\text{Im } \tau > 0$:

$$j(\tau) = \frac{1}{q} + 744 + \sum_{n \geq 1} c_n q^n.$$

Thus, the factor 1728 used in the definition of j has been put to avoid denominators in the Fourier expansion of $j(\tau)$, and more precisely to have a residue equal to 1 at infinity (the local variable at infinity being taken to be q). These theorems show that j is a meromorphic function on the compactification (obtained by adding a point at infinity) of the quotient $\mathfrak{H}/SL_2(\mathbb{Z})$, which is isomorphic as a Riemann surface to the Riemann sphere S^2. Under this isomorphism, we have:

(2.1.6) Proposition. *The function* j *is a one-to-one mapping from the sphere* S^2 *onto the projective complex plane* $\mathbb{P}_1(\mathbb{C})$. *In other words,* $j(\tau)$ *takes once and only once every possible value (including infinity) on* $\overline{\mathfrak{H}/SL_2(\mathbb{Z})}$.

This proposition is obtained essentially by combining remark 2.1.3 (surjectivity) with proposition 2.1.1 (injectivity).

Since the field of meromorphic functions on the sphere is the field of rational functions, we deduce that the field of *modular functions*, i.e. meromorphic functions invariant under $SL_2(\mathbb{Z})$, is the field of rational functions in j. In particular, modular functions which are holomorphic outside the point at infinity of the Riemann sphere are simply polynomials in j. Finally, if we want to have such a function which is one to one as in theorem 2.1.5, the only possibilities are linear polynomials $aj + b$. Now the constant 1728 has been chosen so that the residue at infinity is equal to one. If we want to keep this property, we must have $a = 1$. This leaves only the possibility $j + b$ for a function having essentially the same properties as j. In other words, the only freedom that we really have in the choice of the modular function j is the constant term 744 in its Fourier expansion.

Although it is a minor point, I would like to say that the normalization of j with constant term 744 is not the correct one for several reasons. The 'correct' constant should be 24, so the 'correct' j function should in fact be $j - 720$. Maybe the most natural reason is as follows: there exists a rapidly convergent series due to Rademacher for the Fourier coefficients c_n of j. For $n = 0$, this series gives 24, not 744. Other good reasons are due to Atkin and to Zagier (unpublished).

2.2 Isogenies

We now come back to the case of elliptic curves over an arbitrary field.

(2.2.1) Definition. *Let E and E' be two elliptic curves defined over a field K. An isogeny from E to E' is a map of algebraic curves from E to E' sending the zero element of E to the zero element of E'. The curves are said to be isogenous if there exists a non-constant isogeny from E to E'.*

The following theorem summarizes the main properties of non-constant isogenies:

(2.2.2) Theorem. *Let ϕ be a non-constant isogeny from E to E'. Then:*

1. *ϕ is a surjective map.*
2. *ϕ is a finite map, in other words the fiber over any point of E' is constant and finite.*
3. *ϕ preserves the group laws of the elliptic curves (note that this was not required in the definition), i.e. it is a map of algebraic groups.*

From these properties, one can see that ϕ induces an injective map from the corresponding function field of E' to that of E (over some algebraic closure of the base field). The degree of the corresponding field extensions is finite and called the degree of ϕ.

(2.2.3) *Remark.* If the above extension of fields is separable, for example if the base field has characteristic zero, then the degree of ϕ is also equal to the cardinality of a fiber, i.e. to the cardinality of its kernel $\phi^{-1}(O)$.

(2.2.4) Theorem. *Let E be an elliptic curve over a field K, and let m be a positive integer. Then the map $[m]$ (multiplication by m) is an endomorphism of E with the following properties:*

1. $\deg[m] = m^2$.

2. *If $E[m]$ denotes the kernel of $[m]$ in some algebraic closure of K, i.e. the group of points of order dividing m, and if the characteristic of K is prime to m (or if it is equal to 0), then we have*

$$E[m] \simeq (\mathbb{Z}/m\mathbb{Z}) \times (\mathbb{Z}/m\mathbb{Z}).$$

Another important point concerning isogenies is the following:

(2.2.5) Theorem. *There exists a unique isogeny $\hat{\phi}$ called the dual isogeny, such that*

$$\hat{\phi} \circ \phi = [m],$$

where m is the degree of ϕ. In addition, we also have

$$\phi \circ \hat{\phi} = [m]',$$

where $[m]'$ denotes multiplication by m on E'.

The following result also holds:

(2.2.6) Theorem. *Let E be an elliptic curve and Φ a finite subgroup of E. Then there exists an elliptic curve E' and an isogeny ϕ from E to E' whose kernel is equal to Φ. The elliptic curve E' is well defined up to isomorphism and is denoted E/Φ.*

We end this Section by giving a slightly less trivial example of an isogeny: Let E and E' be two elliptic curves over a field of characteristic different from 2, given by the equations

$$y^2 = x^3 + ax^2 + bx \quad \text{and } y^2 = x^3 - 2ax^2 + (a^2 - 4b)x,$$

where we assume that b and $a^2 - 4b$ are both non-zero. Then the map ϕ from E to E' given by

$$\phi(x, y) = \left(\frac{y^2}{x^2}, \frac{y(x^2 - b)}{x^2}\right)$$

is an isogeny of degree 2 with kernel $\{O, (0,0)\}$.

2.3 Complex Multiplication

Let E be an elliptic curve. To make life simpler, we will assume that the base field has characteristic zero. We have seen that the maps $[m]$ are elements of $\text{End}(E)$. Usually, they are the only ones, and since they are distinct, $\text{End}(E) \simeq \mathbb{Z}$. However it can happen that $\text{End}(E)$ is larger.

(2.3.1) Definition. *We say that E has complex multiplication if* $\text{End}(E)$ *contains elements other than $[m]$, i.e. if as a ring it is strictly larger than $\mathbb{Z}$.*

The theory of complex multiplication is vast, and we can just give a short glimpse at its contents. The first result is as follows:

(2.3.2) Proposition. *Let E be an elliptic curve defined over a field of characteristic zero, and assume that E has complex multiplication. Then the ring* $\operatorname{End}(E)$ *is an order in an imaginary quadratic field, i.e. has the form $\mathbb{Z} + \mathbb{Z}\tau$ where τ is a complex number with positive imaginary part and which is an algebraic integer of degree 2 (that is, satisfies an equation of the form*

$$\tau^2 - s\tau + n = 0,$$

with s and n in $\mathbb{Z}$ and $s^2 - 4n < 0$).

Proof. We shall give the proof in the case where the base field is $\mathbb{C}$. Then $E \simeq \mathbb{C}/L$ for a certain lattice L, and we know that $\operatorname{End}(E)$ is canonically isomorphic to the set of α such that $\alpha L \subset L$. After division by one of the generators of L, we can assume that L is generated by 1 and τ for a certain $\tau \in \mathfrak{H}$, where we recall that $\mathfrak{H}$ is the upper half-plane. Then if α stabilizes L, there must exist integers a, b, c and d such that $\alpha = a + b\tau$, $\alpha\tau = c + d\tau$. In other words, α is an eigenvalue of the matrix $\begin{pmatrix} a & b \\ c & d \end{pmatrix}$, hence is an algebraic integer of degree 2 (with $s = a + d$, $n = ad - bc$). Since $\alpha = a + b\tau$, this shows that $\mathbb{Q}(\tau) = \mathbb{Q}(\alpha)$ is a fixed imaginary quadratic extension k of $\mathbb{Q}$, and hence $\operatorname{End}(E)$ is (canonically isomorphic to) a subring of $\mathbb{Z}_k$, the ring of integers of k, and hence is an order in k if it is larger than $\mathbb{Z}$.

(2.3.3) Example. The curves $y^2 = x^3 - ax$ all have complex multiplication by $\mathbb{Z}[i]$ (map (x, y) to $(-x, iy)$). Similarly, the curves $y^2 = x^3 + b$ all have complex multiplication by $\mathbb{Z}[\rho]$, where ρ is a primitive cube root of unity (map (x, y) to $(\rho x, y)$). For a less trivial example, one can check that the curve

$$y^2 = x^3 - (3/4)x^2 - 2x - 1$$

has complex multiplication by $\mathbb{Z}[\omega]$, where $\omega = \frac{1+\sqrt{-7}}{2}$, multiplication by ω sending (x, y) to (u, v), where

$$u = \omega^{-2}\left(x + a + \frac{b}{x - a}\right)$$

$$v = \omega^{-3}y\left(1 - \frac{b}{(x - a)^2}\right),$$

where we have set $a = \frac{\omega-3}{4}$ and $b = -\frac{7}{16}(3\omega - 1)$ (I thank D. Bernardi for these calculations).

The next theorem concerning complex multiplication is as follows:

(2.3.4) Theorem. *Let τ be a quadratic algebraic number with positive imaginary part. Then the elliptic curve $E_\tau = \mathbb{C}/(\mathbb{Z}+\mathbb{Z}\tau)$ has complex multiplication by an order in the quadratic field $\mathbb{Q}(\tau)$, and the j-invariant $j(E_\tau) = j(\tau)$ is an algebraic integer.*

Note that although the context (and the proof) of this theorem involves elliptic curves, its statement is simply that a certain explicit function $j(\tau)$ on $\mathfrak{H}$ takes algebraic integer values at quadratic imaginary points.

(2.3.5) Example. Here are a few selected values of j.

$$\begin{aligned}
j((1+i\sqrt{3})/2) &= 0 \\
j(i) &= 1728 = 12^3 \\
j((1+i\sqrt{7})/2) &= -3375 = (-15)^3 \\
j(i\sqrt{2}) &= 8000 = 20^3 \\
j((1+i\sqrt{11})/2) &= -32768 = (-32)^3 \\
j((1+i\sqrt{19})/2) &= -884736 = (-96)^3 \\
j((1+i\sqrt{43})/2) &= -884736000 = (-960)^3 \\
j((1+i\sqrt{67})/2) &= -147197952000 = (-5280)^3 \\
j((1+i\sqrt{163})/2) &= -262537412640768000 = (-640320)^3 \\
j(i\sqrt{3}) &= 54000 = 2(30)^3 \\
j(2i) &= 287496 = (66)^3 \\
j((1+3i\sqrt{3})/2) &= -12288000 = -3(160)^3 \\
j(i\sqrt{7}) &= 16581375 = (255)^3 \\
j((1+i\sqrt{15})/2) &= (-191025-85995\sqrt{5})/2 = (1-\sqrt{5})/2((75+27\sqrt{5})/2)^3 \\
j((1+i\sqrt{23})/2) &= \theta = \alpha^3,
\end{aligned}$$

where θ is the largest negative root of the cubic equation $X^3 + 3491750X^2 - 5151296875X + 12771880859375 = 0$, and α is the largest negative root of the equation $X^3 + 155X^2 + 650X + 23375 = 0$.

The reason for the special values chosen will become clear later.

An amusing consequence of the above results is the following. We know that if $q = e^{2i\pi\tau}$ then $j(\tau) = 1/q + 744 + O(|q|)$. Hence when $|q|$ is very small (i.e. when the imaginary part of τ is large), it can be expected that $j(\tau)$ is well approximated by $1/q+744$. Taking the most striking example, this implies that $e^{\pi\sqrt{163}}$ should be close to an integer, and that $(e^{\pi\sqrt{163}} - 744)^{1/3}$ should be even closer. This is indeed what one finds:

$$e^{\pi\sqrt{163}} = 262537412640768743.99999999999925007259\ldots$$

$$(e^{\pi\sqrt{163}} - 744)^{1/3} = 640319.99999999999999999999999939031735\ldots$$

Note that by well known transcendence results, although these quantities are very close to integers, they cannot be integers and they are in fact transcendental numbers.

2.4 Class Numbers

To understand more precisely the statement of theorem 2.3.4, and the examples given after it, it is necessary to define the concept of class number. This can be done in full generality for any algebraic number field (see e.g. Borevitch-Shafarevitch 1966), but we need it here only for imaginary quadratic fields, and hence we can use the more explicit language of binary quadratic forms. Let $ax^2 + bxy + cy^2$ be a binary quadratic form. We will always consider forms which are positive definite (i.e. $D = b^2 - 4ac < 0$ and $a > 0$), with integral coefficients, and primitive, that is such that the gcd of a, b and c is equal to 1. On the set of such forms of given discriminant D, we can define an equivalence relation as follows: if $\begin{pmatrix} \alpha & \beta \\ \gamma & \delta \end{pmatrix} \in SL_2(\mathbb{Z})$ then we say that the form $a(\alpha x+\beta y)^2 + b(\alpha x+\beta y)(\gamma x+\delta y)+c(\gamma x+\delta y)^2$ is equivalent to $ax^2+bxy+cy^2$.

(2.4.1) Proposition. *The set of equivalence classes defined above is finite, and its cardinality is called the class number of D (or of the field $\mathbb{Q}(\sqrt{D})$ depending on the context), and denoted $h(D)$.*

Proof. The simplest way to prove this (and to compute $h(D)$ at the same time) is to introduce the notion of *reduced form*. A form is reduced if $|b| \le a \le c$ and furthermore if one of the inequalities is an equality then in addition we require $b \ge 0$. Then it is easy to show that in each equivalence class, there exists a unique reduced form, hence $h(D)$ is equal to the number of reduced forms, which is finite since it follows from the definition that $|b| \le a \le \sqrt{|D|/3}$.

From this proof, it is an easy exercise to compute $h(D)$ for any reasonable D. For example, if we denote by (a, b, c) the form $ax^2 + bxy + cy^2$, the reduced forms of discriminant $D = -23$ are the forms $(1, 1, 6)$, $(2, 1, 3)$ and $(2, -1, 3)$ hence $h(-23) = 3$. On the other hand, one checks that the only reduced form of discriminant $D = -163$ is the form $(1, 1, 41)$ hence $h(-163) = 1$.

Another, equivalent way, to get a hold on $h(D)$ is to consider the quadratic numbers associated to a binary quadratic form. The following lemma is easy to prove:

(2.4.2) Lemma. *Let τ be a quadratic algebraic number with positive imaginary part. Then there exists a unique (primitive, positive definite, binary quadratic form) $ax^2 + bxy + cy^2$ such that τ is a root of the equation $aX^2 + bX + c = 0$ (the converse is trivial). Furthermore, the quadratic form is reduced if and only if τ belongs to the standard fundamental domain $\mathcal{F}$ of $SL_2(\mathbb{Z})$, i.e. such that*

$$\overline{\mathcal{F}} = \{\tau \in \mathfrak{H}, |\tau| \geq 1, |\mathrm{Re}\ \tau| \leq \frac{1}{2}\},$$

$\mathcal{F}$ itself being the subset of $\overline{\mathcal{F}}$ where one excludes the points on the boundary with positive real part.

The discriminant of the form corresponding to τ will be called the discriminant of τ. Note that it is a square multiple of the discriminant of the quadratic field defined by τ, but does not have to be equal to it.

The link with complex multiplication is the following theorem, which is a more precise version of theorem 2.3.4:

(2.4.3) Theorem. *Let $\tau \in \mathfrak{H}$ be a quadratic imaginary number, and let D be its discriminant as just defined. Then $j(\tau)$ is an algebraic integer of degree exactly equal to $h(D)$. More precisely, the equation satisfied by $j(\tau)$ over $\mathbb{Z}$ is the equation $\prod(X - j(\alpha)) = 0$, where α runs over the quadratic numbers associated to the reduced forms of discriminant D.*

Note that $j(\tau)$ is indeed a root of this polynomial, since any quadratic form of discriminant D is equivalent to a reduced form, and since the j function is $SL_2(\mathbb{Z})$-invariant. The difficult part of this theorem is the fact that the polynomial has integral coefficients.

I can now explain the reason for the selection of j-values given in the preceding Section. From theorem 2.4.3, we see that $j(\tau)$ is rational (in fact integral) if and only if $h(D) = 1$ (we assume of course that τ is a quadratic number). Now it is a deep theorem, due to Baker and Stark (1967), that there exist only 13 values of D for which $h(D) = 1$. The first 9 correspond to field discriminants, and are -3, -4, -7, -8, -11, -19, -43, -67 and -163. There are 4 more corresponding to non-maximal orders: -12 and -27 (in the field $\mathbb{Q}(\sqrt{-3})$), -16 (in the field $\mathbb{Q}(\sqrt{-4})$), and -28 (in the field $\mathbb{Q}(\sqrt{-7})$).

Although it is known since Siegel that $h(D)$ tends to infinity with D, and even as fast as $|D|^{1/2-\epsilon}$ for any $\epsilon > 0$, this result is ineffective, and the explicit determination of all D with a given class number is very difficult. I have just stated that the class number 1 problem was only solved in 1967. The class number 2 problem was solved jointly by Baker and Stark in 1969: $D = -427$ is the largest discriminant (in absolute value) with class number 2. The general problem was solved in principle by Goldfeld, Gross and Zagier in 1983, who obtained an effective lower bound on $h(D)$ (Gross-Zagier 1983). However, the problem still needs some cleaning up, and only class numbers 3, 4, and all odd class numbers up to 23, have been explicitly finished, see (Arno et al. 1995). The last remaining j-values in our little table above are for $D = -15$ and $D = -23$, which are the first values for which the class number is 2 and 3.

Let D be the discriminant of a maximal order (i.e. either $D \equiv 1 \pmod 4$ and is squarefree, or $D \equiv 0 \pmod 4$, $D/4 \equiv 2$ or $3 \pmod 4$ and $D/4$ is squarefree), and let τ be a quadratic number of discriminant $D < 0$ (for example $\tau = (D + \sqrt{D})/2$). Set $K = \mathbb{Q}(\tau) = \mathbb{Q}(\sqrt{D})$. Then theorem 2.4.3 tells us that the field $H = K(j(\tau))$ obtained by adjoining $j(\tau)$ to K is an algebraic extension of degree $h(D)$ (this is not strictly true: it tells us this for $K = \mathbb{Q}$, but the statement holds nonetheless). Now in fact much more is true. One can define a group structure on binary quadratic forms called composition. This corresponds in the language of ideals to multiplication of ideals. Then for every D one has not only a class number, but a class group of cardinality $h(D)$. Then the field extension H/K possesses the following properties (see (Shimura 1971A) for the relevant definitions, which would carry us too far here): It is a Galois extension, with Abelian Galois group isomorphic to the class group. Furthermore, it is unramified, and it is the maximal Abelian unramified extension of K. By definition, such a field H is called the Hilbert class field of K. One sees that in the case of imaginary quadratic fields, the Hilbert class field can be obtained by adjoining a value of the j-function. This kind of construction is lacking for other types of fields (except of course for $\mathbb{Q}$).

A cursory glance at the table of j-values which we have given reveals many other interesting aspects. For example, in most cases, it seems that $j(\tau)$ is a cube. Furthermore, it can be checked that no big prime factors occur in the values of $j(\tau)$ (or of its norm when it is not in $\mathbb{Q}$). These properties are indeed quite general, with some restrictions. For example, if D is not divisible by 3, then up to multiplication by a unit, $j(\tau)$ is a cube in H. One can also check that $j(\tau) - 1728$ is a square if $D \equiv 1 \pmod 4$. Finally, not only the values of $j(\tau)$, but more generally the differences $j(\tau_1) - j(\tau_2)$ have only small prime factors (the case of $j(\tau_1)$ alone is recovered by taking $\tau_2 = \rho = (-1+\sqrt{-3})/2$). All these properties are proved in (Gross-Zagier 1985).

2.5 Modular Equations

Another remarkable property of the j-function, which is not directly linked to complex multiplication, but rather to the role that j plays as a modular invariant, is that the functions $j(N\tau)$ for N integral (or more generally rational) are algebraic functions of $j(\tau)$. The minimal equation of the form $\Phi_N(j(\tau), j(N\tau)) = 0$ satisfied by $j(N\tau)$ is called the modular equation of level N. This result is not difficult to prove. We will prove it explicitly in the special case $N = 2$. Set

$$P(X) = (X - j(2\tau))(X - j(\frac{\tau}{2}))(X - j(\frac{\tau+1}{2})) = X^3 - s(\tau)X^2 + t(\tau)X - n(\tau).$$

I claim that the functions s, t and n are polynomials in j. Since they are clearly meromorphic, and in fact holomorphic outside infinity, from Section 2.1 we know that it is enough to prove that they are modular functions (i.e. invariant under $SL_2(\mathbb{Z})$). Since the action of $SL_2(\mathbb{Z})$ on $\mathfrak{H}$ is generated by $\tau \mapsto \tau + 1$, and $\tau \mapsto -1/\tau$, it suffices to show the invariance of s, t and n under these transformations, and this is easily done using the modular invariance of j itself. This shows the existence of a cubic equation satisfied by $j(2\tau)$ over the field $\mathbb{C}(j(\tau))$. If one wants the equation explicitly, one must compute the first few coefficients of the Fourier expansion of $s(\tau)$, $t(\tau)$, and $n(\tau)$, using the Fourier expansion of $j(\tau)$:

$$j(\tau) = \frac{1}{q} + 744 + 198884q + 21493760q^2 + 864299970q^3 + \cdots$$

The result is as follows:

$$s = j^2 - 2^4 3 \cdot 31 j - 2^4 3^4 5^3,$$
$$t = 2^4 3 \cdot 31 j^2 + 3^4 5^3 4027 j + 2^8 3^7 5^6,$$
$$n = -j^3 + 2^4 3^4 5^3 j^2 - 2^8 3^7 5^6 j + 2^{12} 3^9 5^9.$$

This gives as modular polynomial of level 2 the polynomial

$$\begin{aligned}\Phi_2(X,Y) = X^3 + Y^3 - X^2Y^2 + 2^4 3 \cdot 31(X^2Y + XY^2) - 2^4 3^4 5^3 (X^2 + Y^2) \\ + 3^4 5^3 4027 XY + 2^8 3^7 5^6 (X + Y) - 2^{12} 3^9 5^9.\end{aligned}$$

As we can see from this example, the modular polynomials are symmetric in X and Y, and they have many other remarkable properties linking them closely to complex multiplication and class numbers, but we will not pursue this subject further here. See for example (Herrmann 1975) and (P. Cohen 1984) for results and more references on the polynomials Φ_N.

3. Rank and L-functions

We have seen in theorem 1.3.1 that if E is an elliptic curve defined over $\mathbb{Q}$, then

$$E(\mathbb{Q}) \simeq E(\mathbb{Q})_{\text{tors}} \oplus \mathbb{Z}^r,$$

where $E(\mathbb{Q})_{\text{tors}}$ is a finite group which is easy to compute for a given curve, and r is an integer called the rank. As has already been mentioned, r is very difficult to compute, even for a given curve. Most questions here have a conjectural answer, and very few are proved. In this short Chapter, we try to give some indications on the status of the subject at this time (May 1995).

3.1 L-functions

3.1.1 The zeta function of a variety. After clearing out the denominators of the coefficients, we can assume that our curve has coefficients in $\mathbb{Z}$. Now it is a classical technique to look at the equation modulo primes p, and to gather this information to obtain results on the equation over $\mathbb{Q}$ or over $\mathbb{Z}$. This can in fact be done in great generality for any smooth projective algebraic variety (and more general objects if needed), and not only for elliptic curves. Although it carries us away a little, I believe it worthwhile to do it in this context first. Let V be a (smooth projective) variety of dimension d, defined by equations with coefficients in $\mathbb{Z}$. For any prime p, we can consider the variety V_p obtained by reducing the coefficients modulo p (it may of course not be smooth any more). For any $n \geq 1$, let $N_n(p)$ be the number of points of V_p defined over the finite field $\mathbb{F}_{p^n}$ and consider the following formal power series in the variable T:

$$Z_p(T) = \exp(\sum_{n \geq 1} \frac{N_n(p)}{n} T^n).$$

Then we have the following very deep theorem, first conjectured by Weil (and proved by him for curves and Abelian varieties, see (Weil 1949)), and proved completely by Deligne in 1974 (Deligne 1974):

(3.1.1) Theorem. *Let V_p be a smooth projective variety of dimension d over $\mathbb{F}_p$. Then:*

1. *The series $Z_p(T)$ is a rational function of T, i.e. $Z_p(T) \in \mathbb{Q}(T)$.*
2. *There exists an integer e (called the Euler characteristic of V_p), such that*

$$Z_p(1/(p^d T)) = \pm p^{de/2} T^e Z_p(T).$$

3. *The rational function $Z_p(T)$ factors as follows:*

$$Z_p(T) = \frac{P_1(T) \cdots P_{2d-1}(T)}{P_0(T) P_2(T) \cdots P_{2d}(T)},$$

where for all i, $P_i(T) \in \mathbb{Z}[T]$, $P_0(T) = 1 - T$, $P_{2d}(T) = 1 - p^d T$, and for all other i,

$$P_i(T) = \prod_j (1 - \alpha_{ij} T) \quad \text{with } |\alpha_{ij}| = p^{i/2}.$$

The first assertion had been proved a few years before Deligne by B. Dwork using relatively elementary methods, but by far the hardest part in the proof of this theorem is the very last assertion, that $|\alpha_{ij}| = p^{i/2}$. This is called the Riemann hypothesis for varieties over finite fields.

Now given all the local $Z_p(T)$, we can form a global zeta function by setting for s complex with Re s sufficiently large:

$$\zeta(V, s) = \prod_p Z_p(p^{-s}).$$

This should be taken with a grain of salt, since there are some p (finite in number) such that V_p is not smooth.

Very little is known about this general zeta function. It is believed (can one say conjectured when so few cases have been closely examined?) that it can be analytically continued to the whole complex plane to a meromorphic function with a functional equation when the local factors at the bad primes p are correctly chosen, and that it satisfies the Riemann hypothesis, i.e. that apart from 'trivial' zeros and poles, all the other zeros and poles lie on certain vertical lines in the complex plane.

One recovers the ordinary Riemann zeta function by taking for V the single point 0. More generally, one can recover the Dedekind zeta function of a number field by taking for V the 0-dimensional variety defined in the projective line by $P(X) = 0$, where P is a monic polynomial with integer coefficients defining the field over $\mathbb{Q}$.

3.1.2 L-functions of elliptic curves. Let us now consider the special case where V is an elliptic curve E. In that case, Hasse's theorem 1.3.3 gives us all the information we need about the number of points of E over a finite field. This leads to the following corollary:

(3.1.2) Corollary. *Let E be an elliptic curve over $\mathbb{Q}$, and let p be a prime of good reduction (i.e. such that E_p is still smooth). Then*

$$Z_p(E) = \frac{1 - a_p T + pT^2}{(1-T)(1-pT)},$$

where a_p is as in theorem 1.3.3.

In fact, Hasse's theorem is simply the special case of the Weil conjectures for elliptic curves (and can be proved quite simply, see e.g. Silverman 1986 pp 134–136).

Ignoring for the moment the question of bad primes, the general definition of zeta functions gives us

$$\zeta(E,s) = \frac{\zeta(s)\zeta(s-1)}{L(E,s)},$$

where

$$L(E,s) = \prod_p (1 - a_p p^{-s} + p^{1-2s})^{-1}.$$

The function $L(E,s)$ will be called the Hasse-Weil L-function of the elliptic curve E. To give a precise definition, we also need to define the local factors at the bad primes p. This can be done, and finally leads to the following definition:

(3.1.3) Definition. *Let E be an elliptic curve over $\mathbb{Q}$, and let $y^2 + a_1xy + a_3y = x^3 + a_2x^2 + a_4x + a_6$ be a minimal Weierstraß equation for E (see 1.2.1). When E has good reduction at p, define $a_p = p + 1 - N_p$ where N_p is the number of (projective) points of E over $\mathbb{F}_p$. If E has bad reduction, define*

$$\epsilon(p) = \begin{cases} 1, & \textit{if } E \textit{ has split multiplicative reduction at } p; \\ -1, & \textit{if } E \textit{ has non-split multiplicative reduction at } p; \\ 0, & \textit{if } E \textit{ has additive reduction at } p. \end{cases}$$

Then we define the L-function of E as follows, for Re $s > 3/2$*:*

$$L(E,s) = \prod_{\text{bad } p} \frac{1}{1-\epsilon(p)p^{-s}} \prod_{\text{good } p} \frac{1}{1-a_p p^{-s}+p^{1-2s}}.$$

Note that in this definition it is crucial to take the minimal Weierstraß equation for E: taking another equation could increase the number of primes of bad reduction, and hence change a finite number of local factors. On the other hand, one can prove that $L(E,s)$ depends only on the isogeny class of E.

By expanding the product, it is clear that $L(E,s)$ is a Dirichlet series, i.e. of the form $\sum_{n\geq 1} a_n n^{-s}$ (this of course is the case for all zeta functions of varieties). We will set

$$f_E(\tau) = \sum_{n\geq 1} a_n q^n, \quad \text{where as usual } q = e^{2i\pi\tau}.$$

We can now state the first conjecture on L-functions of elliptic curves:

(3.1.4) Conjecture. *The function $L(E,s)$ can be analytically continued to the whole complex plane to an entire function. Furthermore, there exists a positive integer N, such that if we set*

$$\Lambda(E,s) = N^{s/2}(2\pi)^{-s}\Gamma(s)L(E,s),$$

then we have the following functional equation:

$$\Lambda(E,2-s) = \pm\Lambda(E,s).$$

In this case, the Riemann hypothesis states that apart from the trivial zeros at non-positive integers, the zeros of $L(E,s)$ all lie on the critical line $\operatorname{Re} s = 1$.

The number N occurring in conjecture 3.1.4 is a very important invariant of the curve. It is called the conductor of E, and can be defined without reference to any conjecture. It suffices to say that it has the form $\prod_p p^{e_p}$, where the product is over primes of bad reduction, and for $p > 3$, $e_p = 1$ if E has multiplicative reduction at p, $e_p = 2$ if E has additive reduction, while for $p \leq 3$, the recipe is more complicated (see Birch-Kuyk 1975).

3.1.3 The Taniyama-Weil Conjecture. Now a little acquaintance with modular forms reveals that the conjectured form of the functional equation of $L(E,s)$ is the same as the functional equation for the Mellin transform of a modular form of weight 2 over the group $\Gamma_0(N) = \left\{\begin{pmatrix} a & b \\ c & d \end{pmatrix} \in SL_2(\mathbb{Z}), c \equiv 0 \pmod{N}\right\}$ (see (Zagier 1990) in this volume for all relevant definitions about modular forms). Indeed, one can prove the following

(3.1.5) Theorem. *Let f be a modular cusp form of weight 2 on the group $\Gamma_0(N)$ (equivalently $f\frac{dq}{q}$ is a differential of the first kind on $X_0(N) = \overline{\mathfrak{H}/\Gamma_0(N)}$).*

Assume that f is a normalized eigenform of the Hecke operators with rational Fourier coefficients. Then there exists an elliptic curve E defined over $\mathbb{Q}$ such that $f = f_E$, i.e. such that the Mellin transform of $f(it/\sqrt{N})$ is equal to $\Lambda(E,s)$.

Such a curve E is called a modular elliptic curve, and is a natural quotient of the Jacobian of the curve $X_0(N)$. Since analytic continuation and functional equations are trivial consequences of the modular invariance of modular forms we obtain:

(3.1.6) Corollary. *Let E be a modular elliptic curve, and let $f = \sum_{n\geq 1} a_n q^n$ be the corresponding cusp form. Then conjecture 3.1.4 is true. In addition, it is known from Atkin-Lehner theory that one must have $f(-1/(N\tau)) = \epsilon N\tau^2 f(\tau)$ with $\epsilon = \pm 1$. Then the functional equation is*

$$\Lambda(E, 2-s) = -\epsilon\Lambda(E,s).$$

(Please note the minus sign, which is a cause for many confusions and mistakes in tables).

With theorem 3.1.5 in mind, it is natural to ask if the converse is true, i.e. whether any elliptic curve over $\mathbb{Q}$ is modular. This conjecture was first set forth by Taniyama. However, its full importance and plausibility was understood only after Weil proved the following theorem, which we state only in a vague form (the precise statement can be found e.g. in (Ogg 1969)):

(3.1.7) Theorem. *Let $f(\tau) = \sum_{n\geq 1} a_n q^n$, and for all primitive Dirichlet characters χ of conductor m set*

$$L(f,\chi,s) = \sum_{n\geq 1} \frac{a_n\chi(n)}{n^s},$$

$$\Lambda(f,\chi,s) = |Nm^2|^{s/2}(2\pi)^{-s}\Gamma(s)L(f,\chi,s).$$

Assume that these functions satisfy functional equations of the following form:

$$\Lambda(f,\chi,2-s) = w(\chi)\Lambda(f,\overline{\chi},s),$$

where $w(\chi)$ has modulus one, and assume that as χ varies, $w(\chi)$ satisfies certain compatibility conditions (this is where we are imprecise). Then f is a modular form of weight 2 over $\Gamma_0(N)$.

Because of this theorem, the above conjecture becomes much more plausible. The Taniyama-Weil conjecture is thus as follows:

(3.1.8) Conjecture. *Let E be an elliptic curve over $\mathbb{Q}$, let $L(E,s) = \sum_{n\geq 1} a_n n^{-s}$ be its L-series, and $f_E(\tau) = \sum_{n\geq 1} a_n q^n$, so that the Mellin transform of $f(it/\sqrt{N})$ is equal to $\Lambda(E,s)$. Then f is a cusp form of weight 2 on $\Gamma_0(N)$*

which is an eigenfunction of the Hecke operators. Furthermore there exists a morphism ϕ of curves from $X_0(N)$ to E, defined over $\mathbb{Q}$, such that the inverse image by ϕ of the differential $dx/(2y + a_1x + a_3)$ is the differential $c(2i\pi)f(\tau)d\tau = cf(\tau)dq/q$, where c is some constant.

The constant c, called Manin's constant, is conjectured to be always equal to ± 1 when ϕ is a "strong Weil parametrisation", see (Silverman 1986).

A curve satisfying the Taniyama-Weil conjecture has been called above a modular elliptic curve. However since this may lead to some confusion with modular curves (such as the curves $X_0(N)$) which are in general not elliptic, they are called Weil curves (which incidentally seems a little unfair to Taniyama).

The main theorem concerning this conjecture is Wiles's celebrated theorem, which states that when N is squarefree, the conjecture is true (Wiles and Taylor-Wiles 1995). This result has been generalized by Diamond (Diamond 1995) to the case where one only assumes that N is not divisible by 9 and 25. In addition, using Weil's Theorem 3.1.7, it was proved long ago by Shimura (Shimura 1971B and 1971C) that it is true for elliptic curves with complex multiplication.

There is also a recent conjecture of Serre (Serre 1987), which roughly states that any odd 2-dimensional representation of the Galois group $\mathrm{Gal}(\overline{\mathbb{Q}}/\mathbb{Q})$ over a finite field must come from a modular form. It can be shown that Serre's conjecture implies the Taniyama-Weil conjecture.

The Taniyama-Weil conjecture, and hence Wiles's proof, is mainly important for its own sake. However, it has attracted a lot of attention because of a deep result due to Ribet (Ribet 1990) saying that the Taniyama-Weil conjecture for squarefree N implies the full strength of Fermat's last "theorem" (FLT): if $x^n + y^n = z^n$ with x, y, z non-zero integers, then one must have $n \leq 2$. Thanks to Wiles, this is now really a theorem. Although it is not so interesting in itself, FLT has had amazing consequences on the development of number theory, since it is in large part responsible for the remarkable achievements of algebraic number theorists in the nineteenth century, and also as a further motivation for the study of elliptic curves, thanks to Ribet's result.

3.2 The Birch and Swinnerton-Dyer Conjecture

The other fundamental conjecture on elliptic curves was stated by Birch and Swinnerton-Dyer after doing quite a lot of computer calculations on elliptic curves (Birch-Swinnerton-Dyer 1963 and 1965). For the remaining of this paragraph, we assume that we are dealing with a curve E defined over $\mathbb{Q}$ and satisfying conjecture 3.1.4, for example a curve with complex multiplication, or more generally a Weil curve. (The initial computations of Birch and Swinnerton-Dyer were done on curves with complex multiplication).

Recall that we defined in a purely algebraic way the rank of an elliptic curve. A weak version of the Birch and Swinnerton-Dyer Conjecture (BSD) is that the rank is positive (i.e. $E(\mathbb{Q})$ is infinite) if and only if $L(E,1)=0$. This is quite remarkable, and illustrates the fact that the function $L(E,s)$ which is obtained by putting together local data for every prime p, conjecturally gives information on global data, i.e. on the rational points.

The precise statement of the Birch and Swinnerton-Dyer conjecture is as follows:

(3.2.1) Conjecture. *Let E be an elliptic curve over $\mathbb{Q}$, and assume that conjecture 3.1.4 (analytic continuation essentially) is true for E. Then if r is the rank of E, the function $L(E,s)$ has a zero of order exactly r at $s=1$, and in addition*

$$\lim_{s\to 1}(s-1)^{-r}L(E,s)=\Omega|\text{Ш}(E/\mathbb{Q})|R(E/\mathbb{Q})|E(\mathbb{Q})_{\text{tors}}|^{-2}\prod_p c_p,$$

where Ω is a real period of E (obtained by computing a complete elliptic integral), $R(E/\mathbb{Q})$ is the so-called regulator of E, which is an $r\times r$ determinant formed by pairing in a suitable way a basis of the non-torsion points, the product is over the primes of bad reduction, c_p are small easily computed integers, and Ш$(E/\mathbb{Q})$ is the most mysterious object, called the Tate-Shafarevitch group of E.

It would carry us much too far to explain in detail these quantities. However, it can be useful to give the corresponding result for the 0-dimensional case. The following theorem is due to Dirichlet.

(3.2.2) Theorem. *Let K be a number field, and let $\zeta_K(s)$ be the Dedekind zeta function of K. Recall that the group of units in K (algebraic integers whose inverses are also algebraic integers) form a finitely generated Abelian group. Call w the order of its torsion subgroup (this is simply the number of roots of unity in K), and let r be its rank (equal to r_1+r_2-1, where r_1 and $2r_2$ are the number of real and complex embeddings of K). Then at $s=0$ the function $\zeta_K(s)$ has a zero of order r, and one has:*

$$\lim_{s\to 0}s^{-r}\zeta_K(s)=-|Cl(K)|R(K)w^{-1},$$

where $Cl(K)$ is the class group of K and $R(K)$ its regulator.

This formula is very similar to the BSD formula, with the regulator and torsion points playing the same role, and with the class group replaced by the Tate-Shafarevitch group, the units of K being of course the analogues of the rational points.

(3.2.3) *Remark.* If one assumes Taniyama-Weil, BSD and a certain Riemann Hypothesis, one can give an algorithm to compute the rank of an elliptic curve. See e.g. (Mestre 1981).

Apart from numerous numerical verifications of BSD, few results have been obtained on BSD, and all are very deep. For example, it is only since 1987 that it has been proved by Rubin and Kolyvagin (Kolyvagin 1988, 1989) that Ш is finite for certain elliptic curves. The first result on BSD was obtained by Coates and Wiles (Coates-Wiles 1977) who showed that if E has complex multiplication and if $E(\mathbb{Q})$ is infinite, then $L(E,1)=0$. Further results have been obtained, in particular by Gross-Zagier, Rubin and Kolyvagin (see Gross-Zagier 1983 and 1986, and Kolyvagin 1988 and 1989). For example, the following is now known:

(3.2.4) Theorem. *Let E be a Weil curve. Then*

1. *If $L(E,1)\neq 0$ then $r=0$.*
2. *If $L(E,1)=0$ and $L'(E,1)\neq 0$ then $r=1$*

Furthermore, in both these cases $|Ш|$ is finite, and up to some trivial factors divides the conjectural $|Ш|$ involved in BSD.

The present status of BSD is essentially that very little is known when the rank is greater than or equal to 2.

Another conjecture about the rank is that it is unbounded. This seems quite plausible. The present record is due to Nagao, following ideas of Mestre (Mestre 1991, Nagao-Kouya 1994) who obtained an elliptic curve of rank 21.

References

Arno, S., Robinson, M. and Wheeler, F. Imaginary quadratic fields with small odd class number, to appear.

Birch, B. and Kuyk, W. eds. (1975) *Modular Forms of One Variable IV.* Lecture Notes in Math. **476**, Springer (1975).

Birch, B. and Swinnerton-Dyer, H. P. F. (1963) Notes on elliptic curves (I) and (II). *J. Reine Angew. Math.* **212** (1963), 7–25 and **218** (1965), 79–108.

Borevitch, Z. I. and Shafarevitch, I. R. (1966) *Number Theory.* Academic Press (1966).

Coates, J. and Wiles, A. (1977) On the conjecture of Birch and Swinnerton-Dyer. *Invent. Math.* **39** (1977), 223–251.

Cohen, P. (1984) On the coefficients of the transformation polynomials for the elliptic modular function. *Math. Proc. Camb. Phil. Soc.* **95** (1984), 389–402.

Cremona, J. (1992) *Algorithms for Modular Elliptic Curves.* Cambridge Univ. Press (1992).

Deligne, P. (1974) La conjecture de Weil. *Publ. Math. IHES* **43** (1974), 273–307.

Diamond, H. (1995) in preparation.

Faltings, G. (1983) Endlichkeitssätze für abelsche Varietäten über Zahlkörpern. *Invent. Math.* **73** (1983), 349–366.

Gross, B. and Zagier, D. (1983) Points de Heegner et dérivées de fonctions *L. C. R. Acad. Sc. Paris* **297** (1983), 85–87.

Gross, B. and Zagier, D. (1985) On singular moduli. *J. Reine Angew. Math.* **355** (1985), 191–219.

Gross, B. and Zagier, D. (1986) Heegner points and derivatives of L-functions. *Invent. Math.* **84** (1986), 225–320.

Hasse, H. (1933) Beweis des Analogons der Riemannschen Vermutung für die Artinschen u. F. K. Schmidtschen Kongruenzzetafunktionen in gewissen elliptischen Fällen. *Nachr. Gesell. Wissen. Göttingen I* **42** (1933), 253–262.

Herrmann, O. (1975) Über die Berechnung der Fouriercoefficienten der Funktion $j(\tau)$. *J. reine angew. Math.* **274/275** (1975), 187–195.
Koblitz, N. *Introduction to Elliptic Curves and Modular Forms.* Graduate Texts in Math. **97**, Springer (1984).
Kolyvagin, V. A. (1988) Finiteness of $E(\mathbb{Q})$ and $\text{Ш}(E/\mathbb{Q})$ for a class of Weil curves. *Izv. Akad. Nauk. SSSR* **52** (1988).
Kolyvagin, V. A. (1989) Euler systems, *in Progress in Math.* **87**, *Grothendieck Fesrschrift II*, Birkhäuser, Boston (1991), 435–483.
Mazur, B. (1977) Modular curves and the Eisenstein ideal. *Publ. Math. IHES* **47** (1977), 33–186.
Mazur, B. (1978) Rational isogenies of prime degree. *Invent. Math.* **44** (1978), 129–162.
Merel, L. (1995) in preparation.
Mestre, J.-F. (1981) Courbes elliptiques et formules explicites, *in* Sém. Th. Nombres Paris 1981–82. *Progress in Math.* **38**, Birkhäuser, 179–188.
Mestre, J.-F. (1982) Construction of an elliptic curve of rank ≥ 12, *C. R. Acad. Sc. Paris* **295** (1982), 643-644.
Mestre, J.-F. (1991) Courbes elliptiques de rang ≥ 12 sur $\mathbb{Q}(t)$. *C.R. Acad. Sci. Paris* (1991), 171–174.
Mordell, L. J. (1922) On the rational solutions of the indeterminate equations of the third and fourth degree. *Proc. Camb. Philos. Soc.* **21** (1922), 179–192.
Nagao, K. and Kouya, T. (1994) An example of elliptic curve over $\mathbb{Q}$ with rank ≥ 21. *Proc. Japan Acad.* **70** (1994), 104–105.
Ogg, A. (1969) *Modular Forms and Dirichlet Series.* Benjamin (1969).
Ribet, K. (1990) On modular representations of $\text{Gal}(\overline{\mathbb{Q}}/\mathbb{Q})$ arising from modular forms, *Invent. Math.* **100** (1990), 431–476.
Serre, J.-P. (1987) Sur les représentations modulaires de degré 2 de $\text{Gal}(\overline{\mathbb{Q}}/\mathbb{Q})$. *Duke Math. J.* **54** (1987), 179–230.
Shimura, G. (1971A) *Introduction to the arithmetic theory of automorphic functions.* Princeton Univ. Press (1971).
Shimura, G. (1971B) On the zeta-function of an Abelian variety with complex multiplication. *Ann. Math.* **94** (1971), 504–533.
Shimura, G. (1971C) On elliptic curves with complex multiplication as factors of the Jacobians of modular function fields. *Nagoya Math. J.* **43** (1971), 199–208.
Siegel, C.-L. (1936) Über die Classenzahl quadratischer Zahlkörper. *Acta Arith.* **1** (1936), 83–86.
Silverman, J. (1986) *The Arithmetic of Elliptic Curves.* Graduate Texts in Math. **106**, Springer (1986).
Silverman, J. (1994) *Advanced Topics in the Arithmetic of Elliptic Curves.* Graduate Texts in Math. **151**, Springer (1994).
Taniyama, Y. (1955), in the Problem session of the Tokyo-Nikko conference on number theory, problem 12 (1955).
Taylor, R. and Wiles, A. (1995) Ring theoretic properties of certain Hecke algebras, *Annals of Math.* (1995), to appear.
Weil, A. (1930) Sur un théorème de Mordell. *Bull. Sci. Math.* **54** (1930), 182–191.
Weil, A. (1949) Number of solutions of equations in finite fields. *Bull. AMS* **55** (1949), 497–508.
Wiles, A. (1995) Modular elliptic curves and Fermat's last theorem. *Annals of Math.* (1995), to appear.
Zagier, D. (1990) Introduction to modular forms, this volume.

Chapter 4

Introduction to Modular Forms

by Don Zagier

Table of Contents

1. A Supply of Modular Forms

The word 'modular' refers to the moduli space of complex curves (= Riemann surfaces) of genus 1. Such a curve can be represented as $\mathbb{C}/\Lambda$ where $\Lambda \subset \mathbb{C}$ is a lattice, two lattices Λ_1 and Λ_2 giving rise to the same curve if $\Lambda_2 = \lambda\Lambda_1$ for some non-zero complex number λ. (For properties of curves of genus 1, see the lectures of Cohen and Bost/Cartier in this volume.) A **modular function** assigns to each lattice Λ a complex number $F(\Lambda)$ with $F(\Lambda_1) = F(\Lambda_2)$ if $\Lambda_2 = \lambda\Lambda_1$. Since any lattice $\Lambda = \mathbb{Z}\omega_1 + \mathbb{Z}\omega_2$ is equivalent to a lattice of the form $\mathbb{Z}\tau + \mathbb{Z}$ with τ $(= \omega_1/\omega_2)$ a non-real complex number, the function F is completely specified by the values $f(\tau) = F(\mathbb{Z}\tau + \mathbb{Z})$ with τ in $\mathbb{C} \setminus \mathbb{R}$ or even, since $f(\tau) = f(-\tau)$, with τ in the complex upper half-plane $\mathfrak{H} = \{ \tau \in \mathbb{C} \mid \Im(\tau) > 0\}$. The fact that the lattice Λ is not changed by replacing the basis $\{\omega_1, \omega_2\}$ by the new basis $a\omega_1 + b\omega_2, c\omega_1 + d\omega_2$ $(a, b, c, d \in \mathbb{Z},\ ad - bc = \pm 1)$ translates into the **modular invariance property** $f(\frac{a\tau + b}{c\tau + d}) = f(\tau)$. Requiring that τ always belong to $\mathfrak{H}$ is equivalent to looking only at bases $\{\omega_1, \omega_2\}$ which are oriented (i.e. $\Im(\omega_1/\omega_2) > 0$) and forces us to look only at matrices $\left(\begin{smallmatrix} a & b \\ c & d \end{smallmatrix}\right)$ with $ad - bc = +1$; the group $PSL_2(\mathbb{Z})$ of such matrices will be denoted Γ_1 and called the (full) **modular group**. Thus a modular function can be thought of as a complex-valued function on $\mathfrak{H}$ which is invariant under the action $\tau \mapsto (a\tau + b)/(c\tau + d)$ of Γ_1 on $\mathfrak{H}$. Usually we are interested only in functions which are also holomorphic on $\mathfrak{H}$ (and satisfy a suitable growth condition at infinity) and will reserve the term 'modular function' for these. The prototypical example is the **modular invariant** $j(\tau) = e^{-2\pi i\tau} + 744 + 196884e^{2\pi i\tau} + \cdots$ which will be defined below (cf. Section **B**). However, it turns out that for many purposes the condition of modular invariance is too restrictive. Instead, one must consider functions on lattices which satisfy the identity $F(\Lambda_1) = \lambda^k F(\Lambda_2)$ when $\Lambda_2 = \lambda\Lambda_1$ for some integer k, called the **weight**. Again the function F is completely determined by its restriction $f(\tau)$ to lattices of the form $\mathbb{Z}\tau + \mathbb{Z}$ with τ in $\mathfrak{H}$, but now f must satisfy the **modular transformation property**

$$f\left(\frac{a\tau + b}{c\tau + d}\right) = (c\tau + d)^k f(\tau) \tag{1}$$

rather than the modular invariance property required before. The advantage of allowing this more general transformation property is that now there are functions satisfying it which are not only holomorphic in $\mathfrak{H}$, but also 'holomorphic at infinity' in the sense that their absolute value is majorized by a polynomial in $\max\{1, \Im(\tau)^{-1}\}$. This cannot happen for non-constant Γ_1-invariant functions by Liouville's theorem (the function $j(\tau)$ above, for instance, grows exponentially as $\Im(\tau)$ tends to infinity). Holomorphic functions $f : \mathfrak{H} \to \mathbb{C}$ satisfying (1) and the growth condition just given are called **modular forms** of weight k, and the set of all such functions—clearly a vector space over $\mathbb{C}$—is denoted by M_k or $M_k(\Gamma_1)$. The subspace of functions whose absolute value is majorized by a multiple of $\Im(\tau)^{-k/2}$ is denoted by S_k or $S_k(\Gamma_1)$, the space of **cusp forms** of weight k. It is a Hilbert space with respect to the **Petersson scalar product**

$$(f,g) = \iint_{\mathfrak{H}/\Gamma_1} v^k\, f(\tau)\,\overline{g(\tau)}\, d\mu \qquad (f,g \in S_k), \tag{2}$$

where we have written τ as $u + iv$ and $d\mu$ for the $SL(2,\mathbb{R})$-invariant measure $v^{-2}\,du\,dv$ on $\mathfrak{H}$.

The definition of modular forms which we have just given may not at first look very natural. The importance of modular forms stems from the conjunction of the following two facts:

(i) They arise naturally in a wide variety of contexts in mathematics and physics and often encode the arithmetically interesting information about a problem.

(ii) The space M_k is finite-dimensional for each k.

The point is that if $\dim M_k = d$ and we have more than d situations giving rise to modular forms in M_k, then we automatically have a linear relation among these functions and hence get 'for free' information—often highly non-trivial—relating these different situations. The way the information is 'encoded' in the modular forms is via the Fourier coefficients. From the property (1) applied to the matrix $\left(\begin{smallmatrix} a & b \\ c & d \end{smallmatrix}\right) = \left(\begin{smallmatrix} 1 & 1 \\ 0 & 1 \end{smallmatrix}\right)$ we find that any modular form $f(\tau)$ is invariant under $\tau \mapsto \tau + 1$ and hence, since it is also holomorphic, has a Fourier expansion as $\sum a_n e^{2\pi i n\tau}$. The growth conditions defining M_k and S_k as given above are equivalent to the requirement that a_n vanish for $n < 0$ or $n \le 0$, respectively (this is the form in which these growth conditions are usually stated). What we meant by (i) above is that nature—both physical and mathematical—often produces situations described by numbers which turn out to be the Fourier coefficients of a modular form. These can be as disparate as multiplicities of energy levels, numbers of vectors in a lattice of given length, sums over the divisors of integers, special values of zeta functions, or numbers of solutions of Diophantine equations. But the fact that all of these different objects land in the little spaces M_k forces the existence of relations among their coefficients. In these notes we will give many illustrations of this type of phenomenon and of the way in which modular forms are used. But to do this we first need to have a supply of modular forms on hand to work with. In this first part a number of constructions of modular forms will be given, the general theory being developed at the same time in the context of these examples.

A Eisenstein series

The first construction is a very simple one, but already here the Fourier coefficients will turn out to give interesting arithmetic functions. For k even and greater than 2, define the **Eisenstein series** of weight k by

$$G_k(\tau) = \frac{(k-1)!}{2(2\pi i)^k} \sum_{m,n}{}' \frac{1}{(m\tau + n)^k}, \tag{3}$$

where the sum is over all pairs of integers (m,n) except $(0,0)$. (The reason for the normalizing factor $(k-1)!/2(2\pi i)^k$, which is not always included in the definition, will become clear in a moment.) This transforms like a modular form of weight k because replacing $G_k(\tau)$ by $(c\tau+d)^{-k}G_k(\frac{a\tau+b}{c\tau+d})$ simply replaces (m,n) by $(am+cn, bm+dn)$ and hence permutes the terms of the sum. We need the condition $k>2$ to guarantee the absolute convergence of the sum (and hence the validity of the argument just given) and the condition k even because the series with k odd are identically zero (the terms (m,n) and $(-m,-n)$ cancel).

To see that G_k satisfies the growth condition defining M_k, and to have our first example of an arithmetically interesting modular form, we must compute the Fourier development. We begin with the **Lipschitz formula**

$$\sum_{n\in\mathbb{Z}}\frac{1}{(z+n)^k}=\frac{(-2\pi i)^k}{(k-1)!}\sum_{r=1}^{\infty}r^{k-1}e^{2\pi irz}\qquad(k\in\mathbb{Z}_{\geq 2},\ z\in\mathfrak{H}),$$

which is proved in Appendix **A**. Splitting the sum defining G_k into the terms with $m=0$ and the terms with $m\neq 0$, and using the evenness of k to restrict to the terms with n positive in the first and m positive in the second case, we find

$$\begin{aligned}G_k(\tau)&=\frac{(k-1)!}{(2\pi i)^k}\sum_{n=1}^{\infty}\frac{1}{n^k}+\sum_{m=1}^{\infty}\left(\frac{(k-1)!}{(2\pi i)^k}\sum_{n\in\mathbb{Z}}\frac{1}{(m\tau+n)^k}\right)\\&=\frac{(-1)^{k/2}(k-1)!}{(2\pi)^k}\zeta(k)+\sum_{m=1}^{\infty}\sum_{r=1}^{\infty}r^{k-1}e^{2\pi irm\tau},\end{aligned}$$

where $\zeta(s)=\sum\limits_{n=1}^{\infty}\frac{1}{n^s}$ is Riemann's zeta function. The number $\frac{(-1)^{k/2}(k-1)!}{(2\pi)^k}\zeta(k)$ is rational and in fact equals $-\frac{B_k}{2k}$, where B_k denotes the kth Bernoulli number ($=$ coefficient of $\frac{x^k}{k!}$ in $\frac{x}{e^x-1}$); it is also equal to $\frac{1}{2}\zeta(1-k)$, where the definition of $\zeta(s)$ is extended to negative s by analytic continuation (for all of this, cf. the lectures of Bost and Cartier). Putting this into the formula for G_k and collecting for each n the terms with $rm=n$, we find finally

$$G_k(\tau)=-\frac{B_k}{2k}+\sum_{n=1}^{\infty}\sigma_{k-1}(n)q^n=\frac{1}{2}\zeta(1-k)+\sum_{n=1}^{\infty}\sigma_{k-1}(n)q^n,\tag{4}$$

where $\sigma_{k-1}(n)$ denotes $\sum_{r|n}r^{k-1}$ (sum over all positive divisors r of n) and we have used the abbreviation $q=e^{2\pi i\tau}$, **a convention that will be used from now on.**

The right-hand side of (4) makes sense also for $k=2$ (B_2 is equal to $\frac{1}{6}$) and will be used to define a function $G_2(\tau)$. It is not a modular form (indeed, there can be no non-zero modular form f of weight 2 on the full modular group, since $f(\tau)\,d\tau$ would be a meromorphic differential form on the Riemann surface $\mathfrak{H}/\Gamma_1 \cup \{\infty\}$ of genus 0 with a single pole of order ≤ 1, contradicting the residue theorem). However, its transformation properties under the modular group can be easily determined using **Hecke's trick**: Define a function G_2^* by

$$G_2^*(\tau) = \frac{-1}{8\pi^2} \lim_{\epsilon \searrow 0} \left(\sum_{m,n}{}' \frac{1}{(m\tau+n)^2 |m\tau+n|^\epsilon} \right).$$

The absolute convergence of the expression in parentheses for $\epsilon > 0$ shows that G_2^* transforms according to (1) (with $k=2$), while applying the Poisson summation formula to this expression first and then taking the limit $\epsilon \searrow 0$ leads easily to the Fourier development $G_2^*(\tau) = G_2(\tau) + (8\pi v)^{-1}$ ($\tau = u + iv$ as before). The fact that the non-holomorphic function G_2^* transforms like a modular form of weight 2 then implies that the holomorphic function G_2 transforms according to

(5) $$G_2\left(\frac{a\tau+b}{c\tau+d}\right) = (c\tau+d)^2 G_2(\tau) - \frac{c(c\tau+d)}{4\pi i} \qquad \left(\begin{pmatrix} a & b \\ c & d \end{pmatrix} \in \Gamma_1 \right).$$

The beginnings of the Fourier developments of the first few G_k are given by

$$\begin{aligned}
G_2(\tau) &= -\frac{1}{24} + q + 3q^2 + 4q^3 + 7q^4 + 6q^5 + 12q^6 + 8q^7 + 15q^8 + \cdots \\
G_4(\tau) &= \frac{1}{240} + q + 9q^2 + 28q^3 + 73q^4 + 126q^5 + 252q^6 + \cdots \\
G_6(\tau) &= -\frac{1}{504} + q + 33q^2 + 244q^3 + 1057q^4 + \cdots \\
G_8(\tau) &= \frac{1}{480} + q + 129q^2 + 2188q^3 + \cdots \\
G_{10}(\tau) &= -\frac{1}{264} + q + 513q^2 + \cdots \\
G_{12}(\tau) &= \frac{691}{65520} + q + 2049q^2 + \cdots \\
G_{14}(\tau) &= -\frac{1}{24} + q + 8193q^2 + \cdots
\end{aligned}$$

Note that *the Fourier coefficients appearing are all rational numbers*, a special case of the phenomenon that M_k in general is spanned by forms with rational Fourier coefficients. It is this phenomenon which is responsible for the richness of the arithmetic applications of the theory of modular forms.

B The discriminant function

Define a function Δ in $\mathfrak{H}$ by

(6) $$\Delta(\tau) = q \prod_{r=1}^{\infty} (1-q^r)^{24} \qquad (\tau \in \mathfrak{H},\ q = e^{2\pi i \tau}).$$

Then

$$\begin{aligned}\frac{\Delta'(\tau)}{\Delta(\tau)} &= \frac{d}{d\tau}\left(2\pi i\tau + 24\sum_{r=1}^{\infty}\log(1-q^r)\right)\\ &= 2\pi i\left(1 - 24\sum_{r=1}^{\infty}\frac{rq^r}{1-q^r}\right)\\ &= -48\pi i\left(-\frac{1}{24} + \sum_{n=1}^{\infty}\Big(\sum_{r\mid n} r\Big)q^n\right) = -48\pi i G_2(\tau).\end{aligned}$$

The transformation formula (5) gives

$$\frac{1}{(c\tau+d)^2}\,\frac{\Delta'(\frac{a\tau+b}{c\tau+d})}{\Delta(\frac{a\tau+b}{c\tau+d})} = \frac{\Delta'(\tau)}{\Delta(\tau)} + 12\frac{c}{c\tau+d}$$

or

$$\frac{d}{d\tau}\Big(\log\Delta\Big(\frac{a\tau+b}{c\tau+d}\Big)\Big) = \frac{d}{d\tau}\log\big(\Delta(\tau)(c\tau+d)^{12}\big).$$

Integrating, we deduce that $\Delta(\frac{a\tau+b}{c\tau+d})$ equals a constant times $(c\tau+d)^{12}\Delta(\tau)$. Moreover, this constant must always be 1 since it is 1 for the special matrices $\left(\begin{smallmatrix} a & b\\ c & d\end{smallmatrix}\right) = \left(\begin{smallmatrix} 1 & 1\\ 0 & 1\end{smallmatrix}\right)$ (compare Fourier developments!) and $\left(\begin{smallmatrix} a & b\\ c & d\end{smallmatrix}\right) = \left(\begin{smallmatrix} 0 & -1\\ 1 & 0\end{smallmatrix}\right)$ (take $\tau = i$!) and these matrices generate Γ_1. Thus $\Delta(\tau)$ satisfies equation (1) with $k = 12$. Multiplying out the product in (6) gives the expansion

$$(7)\qquad \Delta(\tau) = q - 24q^2 + 252q^3 - 1472q^4 + 4830q^5 - 6048q^6 + 8405q^7 - \cdots$$

in which only positive exponents of q occur. Hence Δ is a cusp form of weight 12.

Using Δ, we can determine the space of modular forms of all weights. Indeed, there can be no non-constant modular form of weight 0 (it would be a non-constant holomorphic function on the compact Riemann surface $\mathfrak{H}/\Gamma_1 \cup \{\infty\}$), and it follows that there can be no non-zero modular form of negative weight (if f had weight $m < 0$, then $f^{12}\Delta^{|m|}$ would have weight 0 and a Fourier expansion with no constant term). Also, M_k is empty for k odd (take $a = d = -1$, $b = c = 0$ in (1)), as is M_2. For k even and greater than 2, we have the direct sum decomposition $M_k = \langle G_k\rangle \oplus S_k$, since the Eisenstein series G_k has non-vanishing constant term and therefore subtracting a suitable multiple of it from an arbitrary modular form of weight k produces a form with zero constant term. Finally, S_k is isomorphic to M_{k-12}: given any cusp form f of weight k, the quotient f/Δ transforms like a modular form of weight $k-12$, is holomorphic in $\mathfrak{H}$ (since the product expansion (6) shows that Δ does not vanish there), and has a Fourier expansion with only nonnegative powers of q (since f has an expansion starting with a strictly positive power of q and Δ an expansion starting with q^1). It follows that M_k has finite dimension given by

k	<0	0	2	4	6	8	10	12	14	16	18	...	k	...	$k+12$	...
$\dim M_k$	0	1	0	1	1	1	1	2	1	2	2	...	d	...	$d+1$	...

It also follows, since both G_k and Δ have rational coefficients, that M_k has a basis consisting of forms with rational coefficients, as claimed previously; such a basis is for instance the set of monomials $\Delta^l G_{k-12l}$ with $0 \le l \le (k-4)/12$, together with the function $\Delta^{k/12}$ if k is divisible by 12. We also get the first examples of the phenomenon, stressed in the introduction to this part, that non-trivial arithmetic identities can be obtained 'for free' from the finite-dimensionality of M_k. Thus both G_4^2 and G_8 belong to the one-dimensional space M_8, so they must be proportional; comparing the constant terms gives the proportionality constant as 120 and hence the far from obvious identity

$$\sigma_7(n) = \sigma_3(n) + 120 \sum_{m=1}^{n-1} \sigma_3(m)\sigma_3(n-m).$$

Similarly, $(240G_4)^3 - (504G_6)^2$ and Δ are both cusp forms of weight 12 and hence must be proportional. (Cf. Cohen's lectures for the interpretation of this identity in terms of elliptic curves.) In fact, one deduces easily from what has just been said that every modular form is (uniquely) expressible as a polynomial in G_4 and G_6.

Comparing the Fourier expansions of the first few G_k as given in the last section and the dimensions of the first few M_k as given above, we notice that S_k is empty exactly for those values of k for which the constant term $-B_k/2k$ of G_k is the reciprocal of an integer (namely, for $k = 2, 4, 6, 8, 10$ and 14). This is not a coincidence: one knows for reasons going well beyond the scope of these lectures that, if there are cusp forms of weight k, there must always be congruences between some cusp form and the Eisenstein series of this weight. If this congruence is modulo a prime p, then p must divide the numerator of the constant term of G_k (since the constant term of the cusp form congruent to G_k modulo p is zero). Conversely, for any prime p dividing the numerator of the constant term of G_k, there is a congruence between G_k and some cusp form. As an example, for $k = 12$ the numerator of the constant term of G_k is the prime number 691 and we have the congruence $G_{12} \equiv \Delta \pmod{691}$ (e.g. $2049 \equiv -24 \pmod{691}$) due to Ramanujan.

Finally, the existence of Δ allows us to define the function

$$\begin{aligned} j(\tau) = \frac{(240G_4)^3}{\Delta} &= \frac{(1+240q+2160q^2+\cdots)^3}{q-24q^2+252q^3+\cdots} \\ &= q^{-1} + 744 + 196884q + 21493760q^2 + \cdots \end{aligned}$$

and see (since G_4^3 and Δ are modular forms of the same weight on Γ_1) that it is invariant under the action of Γ_1 on $\mathfrak{H}$. Conversely, if $\phi(\tau)$ is any modular function on $\mathfrak{H}$ which grows at most exponentially as $\Im(\tau) \to \infty$, then the function $f(\tau) = \phi(\tau)\Delta(\tau)^m$ transforms like a modular form of weight $12m$ and (if m is large enough) is bounded at infinity, so that $f \in M_{12m}$; by what we saw above, f is then a homogeneous polynomial of degree m in G_4^3 and Δ, so $\phi = f/\Delta^m$ is a polynomial of degree $\le m$ in j. This justifies calling $j(\tau)$ 'the' modular invariant function.

C Theta Functions

We will be fairly brief on this topic, despite its great importance and interest for physicists, because it is treated in more detail in the lectures of Bost and Cartier. The basic statement is that, given an r-dimensional lattice in which the length squared of any vector is an integer, the multiplicities of these lengths are the Fourier coefficients of a modular form of weight $\frac{r}{2}$. By choosing a basis of the lattice, we can think of it as the standard lattice $\mathbb{Z}^r \subset \mathbb{R}^r$; the square-of-the-length function then becomes a quadratic form Q on $\mathbb{R}^r$ which assumes integral values on $\mathbb{Z}^r$, and the modular form in question is the **theta series**

$$\Theta_Q(\tau) = \sum_{x \in \mathbb{Z}^r} q^{Q(x)}.$$

In general this will not be a modular form on the full modular group $\Gamma_1 = PSL_2(\mathbb{Z})$, but on a subgroup of finite index. As a first example, let $r = 2$ and Q be the modular form $Q(x_1, x_2) = x_1^2 + x_2^2$, so that the associated theta-series, whose Fourier development begins

$$\Theta_Q(\tau) = 1 + 4q + 4q^2 + 0q^3 + 4q^4 + 8q^5 + 0q^6 + 0q^7 + 4q^8 + \cdots,$$

counts the number of representations of integers as sums of two squares. This is a modular form of weight 1, not on Γ_1 (for which, as we have seen, there are no modular forms of odd weight), but on the subgroup $\Gamma_0(4)$ consisting of matrices $\left(\begin{smallmatrix} a & b \\ c & d \end{smallmatrix}\right)$ with c divisible by 4; specifically, we have

$$\Theta_Q\left(\frac{a\tau + b}{c\tau + d}\right) = (-1)^{\frac{d-1}{2}}(c\tau + d)\Theta_Q(\tau)$$

for all $\left(\begin{smallmatrix} a & b \\ c & d \end{smallmatrix}\right) \in \Gamma_0(4)$. To prove this, one uses the Poisson summation formula to prove that $\Theta_Q(-1/4\tau) = -2i\tau\Theta_Q(\tau)$; together with the trivial invariance property $\Theta_Q(\tau+1) = \Theta_Q(\tau)$, this shows that Θ_Q is a modular form of weight 1 with respect to the group generated by $\left(\begin{smallmatrix} 0 & -\frac{1}{2} \\ 2 & 0 \end{smallmatrix}\right)$ and $\left(\begin{smallmatrix} 1 & 1 \\ 0 & 1 \end{smallmatrix}\right)$, which contains $\Gamma_0(4)$ as a subgroup of index 2.

More generally, if $Q : \mathbb{Z}^r \to \mathbb{Z}$ is any positive definite integer-valued quadratic form in r variables, r even, then Θ_Q is a modular form of weight $r/2$ on some group $\Gamma_0(N) = \{\left(\begin{smallmatrix} a & b \\ c & d \end{smallmatrix}\right) \in \Gamma_1 \mid c \equiv 0 \pmod{N}\}$ with some character χ (mod N), i.e.

$$\Theta_Q\left(\frac{a\tau + b}{c\tau + d}\right) = \chi(d)(c\tau + d)^{r/2}\Theta_Q(\tau) \qquad \text{for all } \begin{pmatrix} a & b \\ c & d \end{pmatrix} \in \Gamma_0(N).$$

The integer N, called the **level** of Q, is determined as follows: write $Q(x) = \frac{1}{2}x^t Ax$ where A is an even symmetric $r \times r$ matrix (i.e., $A = (a_{ij})$, $a_{ij} = a_{ji} \in \mathbb{Z}$, $a_{ii} \in 2\mathbb{Z}$); then N is the smallest positive integer such that NA^{-1} is again even. The character χ is given by $\chi(d) = \left(\frac{D}{d}\right)$ (Kronecker symbol) with $D = (-1)^{r/2} \det A$. For the form $Q(x_1, x_2) = x_1^2 + x_2^2$ above, we have $A = \left(\begin{smallmatrix} 2 & 0 \\ 0 & 2 \end{smallmatrix}\right)$,

$N = 4$, $\chi(d) = (-1)^{(d-1)/2}$. As a further example, the two quadratic forms $Q_1(x_1, x_2) = x_1^2 + x_1x_2 + 6x_2^2$ and $Q_2(x_1, x_2) = 2x_1^2 + x_1x_2 + 3x_2^2$ have level $N = 23$ and character $\chi(d) = \left(\frac{-23}{d}\right) = \left(\frac{d}{23}\right)$; the sum $\Theta_{Q_1}(\tau) + 2\Theta_{Q_2}(\tau)$ is an Eisenstein series $3+2\sum_{n=1}^{\infty}\left(\sum_{d|n}\chi(d)\right)q^n$ of weight 1 and level 23 (this is a special case of Gauss's theorem on the total number of representations of a natural number by all positive definite binary quadratic forms of a given discriminant), and the difference $\Theta_{Q_1} - \Theta_{Q_2}$ is two times the cusp form $q\prod_{n=1}^{\infty}(1-q^n)(1-q^{23n})$, the 24th root of $\Delta(\tau)\Delta(23\tau)$.

If we want modular forms on the full modular group $\Gamma_1 = PSL_2(\mathbb{Z})$, then we must have $N = 1$ as the level of Q; equivalently, the even symmetric matrix A must be unimodular. This can happen only if the dimension r is divisible by 8 (for a proof using modular forms, cf. Section **D** of Part 2). In dimension 8 there is only one such quadratic form Q up to isomorphism (i.e., up to change of base in $\mathbb{Z}^8$), and Θ_Q is a multiple of the Eisenstein series G_4. In dimension 16 there are two equivalence classes of forms Q, in dimension 24 there are 24, and in dimension 32 there are known to be more than 80 million classes. A theorem of Siegel tells us that the sum of the theta-series attached to all the Q of a given dimension r, each one weighted by a factor $1/|\mathrm{Aut}(Q)|$, is a certain multiple of the Eisenstein series $G_{r/2}$. Notice the applicability of the principle emphasized in the introduction that the finite-dimensionality of the spaces M_k, combined with the existence of modular forms arising from arithmetic situations, gives easy proofs of non-obvious arithmetic facts. For instance, the theta-series of the unique form Q of dimension 8 must be $240G_4$ (since it has weight 4 and starts with 1), so that there are exactly $240\sigma_3(n)$ vectors $x \in \mathbb{Z}^8$ with $Q(x) = n$ for each $n \in \mathbb{N}$; and the two forms of dimension 16 must have the same theta-series (since $\dim M_8 = 1$ and both series start with 1), so they have the same number $(= 480\sigma_7(n))$ of vectors of length n for every n. This latter fact, as noticed by J. Milnor, gives examples of non-isometric manifolds with the same spectrum for the Laplace operator: just take the tori $\mathbb{R}^{16}/\mathbb{Z}^{16}$ with the flat metrics induced by the two quadratic forms in question.

Finally, we can generalize theta series by including **spherical functions**. If $Q : \mathbb{Z}^r \to \mathbb{Z}$ is our quadratic form, then a homogeneous polynomial $P(x) = P(x_1, \ldots, x_r)$ is called spherical with respect to Q if $\Delta_Q P = 0$, where Δ_Q is the Laplace operator for Q (i.e. $\Delta_Q = \sum_j \frac{\partial^2}{\partial y_j^2}$ in a coordinate system (y) for which $Q = \sum y_j^2$, or $\Delta_Q = 2(\frac{\partial}{\partial x_1}, \ldots, \frac{\partial}{\partial x_r})A^{-1}(\frac{\partial}{\partial x_1}, \ldots, \frac{\partial}{\partial x_r})^t$ in the original coordinate system, where $Q(x) = \frac{1}{2}x^tAx$). If P is such a function, say of degree ν, then the generalized theta-series

$$\Theta_{Q,P}(\tau) = \sum_{x\in\mathbb{Z}^r} P(x)\, q^{Q(x)}$$

is a modular form of weight $\frac{r}{2} + \nu$ (and of the same level and character as for $P \equiv 1$), and is a cusp form if $\nu > 0$. As an example, let

$$Q(x_1, x_2) = x_1^2 + x_2^2,\ \Delta_Q = \frac{\partial^2}{\partial x_1^2} + \frac{\partial^2}{\partial x_2^2},\ P(x_1, x_2) = x_1^4 - 6x_1^2x_2^2 + x_2^4\ ;$$

then $\frac{1}{4}\Theta_{Q,P} = q - 4q^2 + 0q^3 + 16q^4 - 14q^5 + \cdots$ belongs to the space of cusp forms of weight 5 and character $\left(\frac{-4}{\cdot}\right)$ on $\Gamma_0(4)$, and since this space is 1-dimensional it must be of the form

$$\Delta(\tau)^{1/6}\Delta(2\tau)^{1/12}\Delta(4\tau)^{1/6} = q\prod_{n=1}^{\infty}(1-q^n)^{2+2\gcd(n,4)}.$$

That $P(x)$ here is the real part of $(x_1+ix_2)^4$ is no accident: in general, all spherical polynomials of degree ν can be obtained as linear combinations of the special spherical functions $(\zeta^t Ax)^\nu$, where $\zeta \in \mathbb{C}^r$ is isotropic (i.e., $Q(\zeta) = \frac{1}{2}\zeta^t A\zeta = 0$). Still more generally, one can generalize theta series by adding congruence conditions to the summation over $x \in \mathbb{Z}^r$ or, equivalently, by multiplying the spherical function $P(x)$ by some character or other periodic function of x. As an example of a spherical theta series of a more general kind we mention Freeman Dyson's identity

$$\Delta(\tau) = \sum_{\substack{(x_1,\ldots,x_5)\in\mathbb{Z}^5 \\ x_1+\cdots+x_5=0 \\ x_i\equiv i \pmod 5}} \left(\frac{1}{288}\prod_{1\le i<j\le 5}(x_i - x_j)\right) q^{(x_1^2+x_2^2+x_3^2+x_4^2+x_5^2)/10}$$

for the discriminant function Δ of Section **B**.

D Eisenstein series of half-integral weight

In the last section, there was no reason to look only at quadratic forms in an even number of variables. If we take the simplest possible quadratic form $Q(x_1) = x_1^2$, then the associated theta-series

$$\theta(\tau) = \sum_{n\in\mathbb{Z}} q^{n^2} = 1 + 2q + 2q^4 + 2q^9 + \cdots$$

is the square-root of the first example in that section and as such satisfies the transformation equation

$$\theta\left(\frac{a\tau+b}{c\tau+d}\right) = \epsilon(c\tau+d)^{\frac{1}{2}}\theta(\tau) \quad \forall\begin{pmatrix} a & b \\ c & d\end{pmatrix} \in \Gamma_0(4)$$

for a certain number $\epsilon = \epsilon_{c,d}$ satisfying $\epsilon^4 = 1$ (ϵ can be given explicitly in terms of the Kronecker symbol $\left(\frac{c}{d}\right)$). We say that θ is a **modular form of weight** $\frac{1}{2}$. More generally, we can define modular forms of any half-integral weight $r+\frac{1}{2}$ ($r \in \mathbb{N}$). A particularly convenient space of such forms, analogous to the space M_k of integral-weight modular forms on the full modular group, is the space $M_{r+\frac{1}{2}}$ introduced by W. Kohnen. It consists of all f satisfying the transformation law $f(\frac{a\tau+b}{c\tau+d}) = (\epsilon_{c,d}(c\tau+d)^{\frac{1}{2}})^{2r+1}f(\tau)$ for all $\left(\begin{smallmatrix} a & b \\ c & d\end{smallmatrix}\right) \in \Gamma_0(4)$ (equivalently, f/θ^{2r+1} should be $\Gamma_0(4)$-invariant) and having a Fourier expansion of the form $\sum_{n\ge 0} a(n)q^n$ with $a(n) = 0$ whenever $(-1)^r n$ is congruent to 2 or 3 modulo

4. For $r \geq 2$ this space contains an Eisenstein series $G_{r+\frac{1}{2}}$ calculated by H. Cohen. We do not give the definition and the calculation of the Fourier expansion of these Eisenstein series, which are similar in principle but considerably more complicated than in the integral weight case. Unlike the case of integral weight, where the Fourier coefficients were elementary arithmetic functions, the Fourier coefficients now turn out to be number-theoretical functions of considerable interest. Specifically, we have

$$G_{r+\frac{1}{2}}(\tau) = \sum_{\substack{n=0 \\ (-1)^r n \equiv 0 \text{ or } 1 \pmod 4}}^{\infty} H(r,n)\, q^n$$

where $H(r,n)$ is a special value of some L-series, e.g. $H(r,0) = \zeta(1-2r) = -\frac{B_{2r}}{2r}$ (where $\zeta(s)$ is the Riemann zeta-function and B_m the mth Bernoulli number), $H(r,1) = \zeta(1-r)$, and more generally $H(r,n) = L_\Delta(1-r)$ if the number $\Delta = (-1)^r n$ is equal to either 1 or the discriminant of a real or imaginary quadratic field, where the L-series $L_\Delta(s)$ is defined as the analytic continuation of the Dirichlet series $\sum_{n=1}^{\infty} \left(\frac{\Delta}{n}\right) n^{-s}$. These numbers are known to be rational, with a bounded denominator for a fixed value of r. The first few cases are

$$G_{2\frac{1}{2}}(\tau) = \tfrac{1}{120} - \tfrac{1}{12}q - \tfrac{7}{12}q^4 - \tfrac{3}{5}q^5 - q^8 - \tfrac{25}{12}q^9 - 2q^{12} - 2q^{13} - \tfrac{55}{12}q^{16} - 4q^{17}\cdots$$
$$G_{3\frac{1}{2}}(\tau) = -\tfrac{1}{252} - \tfrac{2}{9}q^3 - \tfrac{1}{2}q^4 - \tfrac{16}{7}q^7 - 3q^8 - 6q^{11} - \tfrac{74}{9}q^{12} - 16q^{15} - \tfrac{33}{2}q^{16}\cdots$$
$$G_{4\frac{1}{2}}(\tau) = \tfrac{1}{240} + \tfrac{1}{120}q + \tfrac{121}{120}q^4 + 2q^5 + 11q^8 + \tfrac{2161}{120}q^9 + 46q^{12} + 58q^{13}\cdots$$
$$G_{5\frac{1}{2}}(\tau) = -\tfrac{1}{132} + \tfrac{1}{3}q^3 + \tfrac{5}{2}q^4 + 32q^7 + 57q^8 + \tfrac{2550}{11}q^{11} + \tfrac{529}{3}q^{12} + 992q^{15}\cdots .$$

In each of these four cases, the space $M_{r+\frac{1}{2}}$ is one-dimensional, generated by $G_{r+\frac{1}{2}}$; in general, $M_{r+\frac{1}{2}}$ has the same dimension as M_{2r}.

Just as the case of G_2, the Fourier expansion of $G_{r+1/2}$ still makes sense for $r = 1$, but the analytic function it defines is no longer a modular form. Specifically, the function $H(r,n)$ when $r = 1$ is equal to the **Hurwitz-Kronecker class number** $H(n)$, defined for $n > 0$ as the number of $PSL_2(\mathbb{Z})$-equivalence classes of binary quadratic forms of discriminant $-n$, each form being counted with a multiplicity equal to 1 divided by the order of its stabilizer in $PSL_2(\mathbb{Z})$ (this order is 2 for a single equivalence class of forms if n is 4 times a square, 3 for a single class if n is 3 times a square, and 1 in all other cases). Thus the form $G_{3/2} = \sum_n H(n) q^n$ has a Fourier expansion beginning

$$G_{\frac{3}{2}}(\tau) = -\frac{1}{12}+\frac{1}{3}q^3+\frac{1}{2}q^4+q^7+q^8+q^{11}+\frac{4}{3}q^{12}+2q^{15}+\frac{3}{2}q^{16}+q^{19}+2q^{20}+3q^{23}\cdots .$$

As with G_2 we can use 'Hecke's trick' (cf. Section **A**) to define a function $G^*_{3/2}$ which is not holomorphic but transforms like a holomorphic modular form of weight 3/2. The Fourier expansion of this non-holomorphic modular form differs from that of $G_{3/2}$ only at negative square exponents:

$$G^*_{\frac{3}{2}}(\tau) = \sum_{n=0}^{\infty} H(n)\,q^n + \frac{1}{16\pi\sqrt{v}} \sum_{f\in\mathbb{Z}} \beta(4\pi f^2 v)\,q^{-f^2}$$

where v denotes the imaginary part of τ and $\beta(t)$ the function $\int_1^\infty x^{-3/2}\,e^{-xt}\,dx$, which can be expressed in terms of the error function.

E New forms from old

The words 'new' and 'old' here are not being used in their technical sense—introduced in Part 2—but simply to refer to the various methods available for manufacturing modular forms out of previously constructed ones.

The first and obvious method is **multiplication**: the product of a modular form of weight k and one of weight l is a modular form of weight $k+l$. Of course we have already used this many times, as when we compared G_4^2 and G_8. We also found the structure of the graded ring $M_* = \bigoplus M_k$ of all modular forms on the full modular group Γ_1: it is the free $\mathbb{C}$-algebra on two generators G_4 and G_6 of weights 4 and 6. The modular forms on a subgroup $\Gamma \subset \Gamma_1$ of finite index also form a ring. For instance, for $\Gamma = \Gamma_0(2)$ this ring is the free $\mathbb{C}$-algebra on two generators $G_2^{(2)}$ and G_4 of weights 2 and 4, where

$$G_2^{(2)}(\tau) = G_2(\tau) - 2G_2(2\tau) = \frac{1}{24} + \sum_{n=1}^{\infty}\Bigg(\sum_{\substack{d|n\\ d \text{ odd}}} d\Bigg)q^n = \frac{1}{24} + q + q^2 + 4q^3 + \cdots$$

(this is a modular form because $G_2(\tau)-2G_2(2\tau)$ can also be written as $G_2^*(\tau)-2G_2^*(2\tau)$, and G_2^* transforms like a modular form of weight 2 on Γ_1). In general, the graded ring of modular forms on Γ will not be a free algebra, but must be given by more than 2 generators and a certain number of relations; it will be free exactly when the Riemann surface $\mathfrak{H}/\Gamma \cup \{\text{cusps}\}$ has genus 0. We also note that the ring of modular forms on Γ contains $M_*(\Gamma_1) = \mathbb{C}[G_4, G_6]$ as a subring and hence can be considered as a module over this ring. As such, it is always *free on n generators*, where n is the index of Γ in Γ_1. For instance, every modular form of weight k on $\Gamma_0(2)$ can be uniquely written as $A(\tau)G_2^{(2)}(\tau)+B(\tau)G_4(\tau)+C(\tau)G_4(2\tau)$ where $A \in M_{k-2},\ B,\, C \in M_{k-4}$ (example: $G_2^{(2)}(\tau)^2 = \frac{1}{12}G_4(\tau) + \frac{1}{3}G_4(2\tau)$).

The next method is to apply to two known modular forms f and g of weights k and l **H. Cohen's differential operator**

$$(8) \qquad F_\nu(f,g) = (2\pi i)^{-\nu} \sum_{\mu=0}^{\nu} (-1)^\mu \binom{k+\nu-1}{\mu}\binom{l+\nu-1}{\nu-\mu} f^{(\nu-\mu)} g^{(\mu)},$$

where ν is a nonnegative integer and $f^{(\mu)}$, $g^{(\mu)}$ denote the μth derivatives of f and g. As we will see in a moment, this is a modular form of weight $k+l+2\nu$ on the same group as f and g. For $\nu = 0$ we have $F_0(f,g) = fg$, so the new method is a generalization of the previous one. For $\nu = 1$ we have $F_1(f,g) = \frac{1}{2\pi i}[lf'g - kfg']$; this operation is antisymmetric in f and g and satisfies the Jacobi identity, so that it makes $M_{*-2} = \bigoplus_n M_{n-2}$ into a graded Lie algebra. For ν positive, $F_\nu(f,g)$ has no constant term, so that F_ν maps $M_k \bigotimes M_l$ to $S_{k+l+2\nu}$. The first non-trivial example is $F_1(G_4, G_6) = -\frac{1}{35}\Delta$, which gives the formula

$$\tau(n) = \frac{5\sigma_3(n) + 7\sigma_5(n)}{12} n - 35 \sum_{\substack{a,b>0 \\ a+b=n}} (6a - 4b)\sigma_3(a)\sigma_5(b)$$

for the coefficient $\tau(n)$ of q^n in Δ. (Notice that this identity involves only integers; in general, it is clear that F_ν maps functions with integral or rational Fourier coefficients to another such function.) As another example, observe that applying F_ν to two theta series Θ_{Q_j} associated to quadratic forms $Q_j : \mathbb{Z}^{r_j} \to \mathbb{Z}$ $(j = 1, 2)$ gives rise to a theta-series attached to the form $Q_1 \bigoplus Q_2 : \mathbb{Z}^{r_1+r_2} \to \mathbb{Z}$ and a spherical polynomial of degree ν. For instance, if $\theta(\tau) = \sum q^{n^2}$ is the basic theta-series of weight $\frac{1}{2}$ on $\Gamma_0(4)$, then one checks easily that $\frac{8}{3}F_2(\theta,\theta)$ is the function $\Theta_{Q,P} = \sum_{x_1,x_2\in\mathbb{Z}}(x_1^4 - 6x_1^2x_2^2 + x_2^4)q^{x_1^2+x_2^2}$ discussed at the end of Section **C**. Thus the construction of modular forms via theta-series with spherical functions is a special case of the use of the differential operator F_ν.

We now sketch the proof that F_ν maps modular forms to modular forms. If f is a modular form of weight k on some group Γ, then for $\left(\begin{smallmatrix} a & b \\ c & d \end{smallmatrix}\right) \in \Gamma$ and $\mu \in \mathbb{Z}_{\geq 0}$ the formula

$$(9) \qquad f^{(\mu)}\Big(\frac{a\tau+b}{c\tau+d}\Big) = \sum_{\lambda=0}^{\mu} \frac{\mu!(k+\mu-1)!}{\lambda!(\mu-\lambda)!(k+\lambda-1)!} c^{\mu-\lambda}(c\tau+d)^{k+\mu+\lambda} f^{(\lambda)}(\tau)$$

is easily proved by induction on μ (to get from μ to $\mu+1$, just differentiate and multiply by $(c\tau+d)^2$). These transformation formulas can be combined into the single statement that the generating function

$$(10) \qquad \tilde{f}(\tau, X) = \sum_{\mu=0}^{\infty} \frac{1}{\mu!(k+\mu-1)!} f^{(\mu)}(\tau) X^\mu \qquad (\tau \in \mathfrak{H}, X \in \mathbb{C})$$

satisfies

$$(11) \quad \tilde{f}\Big(\frac{a\tau+b}{c\tau+d}, \frac{X}{(c\tau+d)^2}\Big) = (c\tau+d)^k\, e^{cX/(c\tau+d)}\, \tilde{f}(\tau, X) \qquad \Big(\left(\begin{smallmatrix} a & b \\ c & d \end{smallmatrix}\right) \in \Gamma\Big).$$

Writing down the same formula for a second modular form g of weight l, we find that the product

$$\tilde{f}(\tau, X)\,\tilde{g}(\tau, -X) = \sum_{\nu=0}^{\infty} \frac{(2\pi i)^\nu}{(\nu+k-1)!(\nu+l-1)!} F_\nu(f,g)(\tau)\, X^\nu$$

is multiplied by $(c\tau+d)^{k+l}$ when τ and X are replaced by $\frac{a\tau+b}{c\tau+d}$ and $\frac{X}{(c\tau+d)^2}$, and this proves the modular transformation property of $F_\nu(f,g)$ for every ν.

Finally, we can get new modular forms from old ones by applying the **'slash operator'**

$$f(\tau) \mapsto (f|_k\gamma)(\tau) = (\det\gamma)^{k/2}(c\tau+d)^{-k}f\left(\frac{a\tau+b}{c\tau+d}\right)$$

to an f of weight k on Γ, where $\gamma = \left(\begin{smallmatrix} a & b \\ c & d\end{smallmatrix}\right)$ is a 2×2 integral matrix which does *not* belong to Γ (if $\gamma \in \Gamma$, of course, then $f|_k\gamma = f$ by definition). This will in general be a modular form on some subgroup of Γ of finite index, but often by combining suitable combinations of images $f|_k\gamma$ we can obtain functions that transform like modular forms on Γ or even on a larger group. Important special cases are the operators

$$V_m f(\tau) = m^{-\frac{k}{2}}\left(f|_k\begin{pmatrix} m & 0 \\ 0 & 1\end{pmatrix}\right) = f(m\tau), \qquad U_m f(\tau) = m^{\frac{k}{2}-1}\sum_{j=1}^{m}\left(f|_k\begin{pmatrix} 1 & j \\ 0 & m\end{pmatrix}\right)(\tau)$$

$(m \in \mathbb{N})$, which map $\sum a(n)q^n$ to $\sum a(n)q^{mn}$ and to $\sum a(mn)q^n$, respectively. Both map forms of weight k on $\Gamma_0(N)$ to forms of the same weight on $\Gamma_0(mN)$; if m divides N, then U_m even maps forms on $\Gamma_0(N)$ to forms on $\Gamma_0(N)$. Sometimes, applying U_m can even *reduce* the level, which is always a good thing. For instance, if $f = \sum a(n)q^n$ is a modular form of even weight k on $\Gamma_0(4)$, then $U_2 f = \sum a(2n)q^n$ is a modular form of weight k on $\Gamma_0(2)$, and if f has the additional property that $a(n) = 0$ whenever $n \equiv 2 \pmod 4$, then $U_4 f = \sum a(4n)q^n$ even belongs to $M_k = M_k(\Gamma_1)$. Such f occur, for instance, when one multiplies (or applies the operator F_ν to) two forms $g_1 \in M_{r_1+\frac{1}{2}}$, $g_2 \in M_{r_2+\frac{1}{2}}$ with $r_1 + r_2 = k-1$ (resp. $r_1 + r_2 = k - 2\nu - 1$), since then r_1 and r_2 have opposite parity and consequently one of the g's contains only powers q^n with $n \equiv 0$ or $1 \pmod 4$, the other only powers with $n \equiv 0$ or $3 \pmod 4$. This situation will arise in Part 3 in the derivation of the Eichler-Selberg trace formula.

Important operators which can be built up out of the V_m and U_m are the **Hecke operators**, which are the subject of the next part.

F Other sources of modular forms

We have described the main *analytic* ways to produce modular forms on Γ_1 and its subgroups. Another method comes from algebraic geometry: certain power series $\sum a(n)q^n$ whose coefficients $a(n)$ are defined by counting the number of points of algebraic varieties over finite fields are known or conjectured to be modular forms. For example, the famous 'Taniyama-Weil conjecture' says that to any elliptic curve defined over $\mathbb{Q}$ there is associated a modular form $\sum a(n)q^n$ of weight 2 on some group $\Gamma_0(N)$ with $p+1-a(p)$ equal to the number of points of the elliptic curve over $\mathbb{F}_p$ for every prime number p. However, this cannot really be considered a way of constructing modular forms, since one can usually only *prove* the modularity of the function in question if one has an independent, analytic construction.

In a similar vein, one can get modular forms from algebraic number theory by looking at Fourier expansions $\sum a(n)q^n$ whose associated Dirichlet series $\sum a(n)n^{-s}$ are zeta functions coming from number fields or their characters. For instance, a theorem of Deligne and Serre says that one can get all modular forms of weight 1 in this way from the Artin L-series of two-dimensional Galois representations with odd determinant satisfying Artin's conjecture (that the L-series is holomorphic). Again, however, the usual way of applying such a result is to construct the modular form independently and then deduce that the corresponding Artin L-series satisfies Artin's conjecture.

In one situation the analytic, algebraic geometric, and number theoretic approaches come together. This is for the special class of modular forms called 'CM' (complex multiplication) forms: analytically, these are the theta series $\Theta_{Q,P}$ associated to a *binary* quadratic form Q and an arbitrary spherical function P on $\mathbb{Z}^2$; geometrically, they arise from elliptic curves having complex multiplication (i.e., non-trivial endomorphisms); and number theoretically, they are given by Fourier developments whose associated Dirichlet series are the L-series of algebraic Hecke grossencharacters over an imaginary quadratic field. An example is the function $\sum_{x_1,x_2\in\mathbb{Z}}(x_1^4-6x_1^2x_2^2+x_2^4)q^{x_1^2+x_2^2} = q\prod_{n=1}^{\infty}(1-q^n)^{2+2(n,4)}$ which occurred in Section **C**. The characteristic property of these CM forms is that they have highly lacunary Fourier developments. This is because binary quadratic forms represent only a thin subset of all integers (at most $O(x/(\log x)^{1/2})$ integers $\leq x$).

Finally, modular forms in one variable can be obtained by restricting in various ways different kinds of modular forms in more than one variable (Jacobi, Hilbert, Siegel, ...), these in turn being constructed by one of the methods of this part. The Jacobi forms will be discussed in Part 4.

2. Hecke Theory

The key to the rich internal structure of the theory of modular forms is the existence of a commutative algebra of operators T_n ($n \in \mathbb{N}$) acting on the space M_k of modular forms of weight k. The space M_k has a canonical basis of simultaneous eigenvectors of all the T_n; these special modular forms have the property that their Fourier coefficients $a(n)$ are algebraic integers and satisfy the multiplicative property $a(nm) = a(n)a(m)$ whenever n and m are relatively prime. In particular, their associated Dirichlet series $\sum a(n)n^{-s}$ have Euler products; they also have analytic continuations to the whole complex plane and satisfy functional equations analogous to that of the Riemann zeta function. We will define the operators T_n in Section **A** and describe their eigenforms and the associated Dirichlet series in Sections **B** and **C**, respectively. The final section of this part describes the modifications of the theory for modular forms on subgroups of $PSL_2(\mathbb{Z})$.

A Hecke operators

At the beginning of Part 1 we introduced the notion of modular forms of higher weight by giving an isomorphism

$$\begin{aligned} &F(\Lambda) \mapsto f(\tau) = F(\mathbb{Z}\tau + \mathbb{Z}), \\ &f(\tau) \mapsto F(\Lambda) = \omega_2^{-k} f(\omega_1/\omega_2) \quad (\Lambda = \mathbb{Z}\omega_1 + \mathbb{Z}\omega_2, \quad \Im(\omega_1/\omega_2) > 0) \end{aligned} \tag{1}$$

between functions f in the upper half-plane transforming like modular forms of weight k and functions F of lattices $\Lambda \subset \mathbb{C}$ which are homogeneous of weight $-k$, $F(\lambda\Lambda) = \lambda^{-k}F(\Lambda)$. If we fix a positive integer n, then every lattice Λ has a finite number of sublattices Λ' of index n, and we have an operator T_n on functions of lattices which assigns to such a function F the new function

$$T_n F(\Lambda) = n^{k-1} \sum_{\substack{\Lambda' \subseteq \Lambda \\ [\Lambda:\Lambda'] = n}} F(\Lambda') \tag{2}$$

(the factor n^{k-1} is introduced for convenience only). Clearly $T_n F$ is homogeneous of degree $-k$ if F is, so we can transfer the operator to an operator T_n on functions in the upper half-plane which transform like modular forms of weight k. This operator is given explicitly by

$$T_n f(\tau) = n^{k-1} \sum_{\left(\begin{smallmatrix} a & b \\ c & d \end{smallmatrix}\right) \in \Gamma_1 \backslash \mathcal{M}_n} (c\tau + d)^{-k} f\left(\frac{a\tau + b}{c\tau + d}\right) \tag{3}$$

and is called the n**th Hecke operator** in weight k; here $\mathcal{M}_n$ denotes the set of 2×2 integral matrices of determinant n and $\Gamma_1 \backslash \mathcal{M}_n$ the finite set of orbits of $\mathcal{M}_n$ under left multiplication by elements of $\Gamma_1 = PSL_2(\mathbb{Z})$. Clearly this definition depends on k and we should more correctly write $T_k(n)f$ or (the standard notation) $f|_kT_n$, but we will consider the weight as fixed and write simply T_nf for convenience. In terms of the slash operator

$$(f|_k\gamma)(\tau) = \frac{(ad-bc)^{k/2}}{(c\tau+d)^k} f\left(\frac{a\tau+b}{c\tau+d}\right) \qquad \left(\gamma = \begin{pmatrix} a & b \\ c & d \end{pmatrix},\ a,b,c,d \in \mathbb{R},\ ad-bc>0\right)$$

introduced in Part 1E, formula (3) can be expressed in the form

$$T_nf(\tau) = n^{\frac{k}{2}-1} \sum_{\mu \in \Gamma_1 \backslash \mathcal{M}_n} f|_k\mu.$$

From the fact that $|_k$ is a group operation (i.e. $f|_k(\gamma_1\gamma_2) = (f|_k\gamma_1)|_k\gamma_2$ for γ_1, γ_2 in $GL_2^+(\mathbb{R})$), we see that T_nf is well-defined (changing the orbit representative μ to $\gamma\mu$ with $\gamma \in \Gamma_1$ doesn't affect $f|_k\mu$ because $f|_k\gamma = f$) and again transforms like a modular form of weight k on Γ_1 ($(T_nf)|_k\gamma = T_nf$ for $\gamma \in \Gamma_1$ because $\{\mu\gamma \mid \mu \in \Gamma_1\backslash\mathcal{M}_n\}$ is another set of representatives for $\Gamma_1\backslash\mathcal{M}_n$). Of course, both of these properties are also obvious from the invariant definition (2) and the isomorphism (1).

Formula (3) makes it clear that T_n preserves the property of being holomorphic. We now give a description of the action of T_n on Fourier expansions which shows that T_n also preserves the growth properties at infinity defining modular forms and cusp forms, respectively, and also that the various Hecke operators commute with one another.

Theorem 1. *(i) If $f(\tau)$ is a modular form with the Fourier expansion $\sum_{m=0}^{\infty} a_m q^m$ ($q = e^{2\pi i\tau}$), then the Fourier expansion of T_nf is given by*

$$T_nf(\tau) = \sum_{m=0}^{\infty} \Big(\sum_{d|n,m} d^{k-1}\, a\big(\frac{nm}{d^2}\big) \Big)\, q^m, \tag{4}$$

where $\sum_{d|n,m}$ denotes a sum over the positive common divisors of n and m. In particular, T_nf is again a modular form, and is a cusp form if f is one.
(ii) The Hecke operators in weight k satisfy the multiplication rule

$$T_nT_m = \sum_{d|n,m} d^{k-1}\, T_{nm/d^2}. \tag{5}$$

In particular, $T_nT_m = T_mT_n$ for all n and m and $T_nT_m = T_{nm}$ if n and m are coprime.

Proof. If $\mu = \left(\begin{smallmatrix} a & b \\ c & d \end{smallmatrix}\right)$ is a matrix of determinant n with $c \neq 0$, then we can choose a matrix $\gamma = \left(\begin{smallmatrix} a' & b' \\ c' & d' \end{smallmatrix}\right) \in PSL_2(\mathbb{Z})$ with $\dfrac{a'}{c'} = \dfrac{a}{c}$, and $\gamma^{-1}\mu$ then has the form $\left(\begin{smallmatrix} * & * \\ 0 & * \end{smallmatrix}\right)$. Hence we can assume that the coset representatives in (3) have the form $\mu = \left(\begin{smallmatrix} a & b \\ 0 & d \end{smallmatrix}\right)$ with $ad = n$, $b \in \mathbb{Z}$. A different choice $\gamma\left(\begin{smallmatrix} a & b \\ 0 & d \end{smallmatrix}\right)$ ($\gamma \in PSL_2(\mathbb{Z})$) of representative also has this form if and only if $\gamma = \pm\left(\begin{smallmatrix} 1 & r \\ 0 & 1 \end{smallmatrix}\right)$ with $r \in \mathbb{Z}$, in which case $\gamma\left(\begin{smallmatrix} a & b \\ 0 & d \end{smallmatrix}\right) = \pm\left(\begin{smallmatrix} a & b+dr \\ 0 & d \end{smallmatrix}\right)$, so the choice of μ is unique if we require a, $d > 0$ and $0 \le b < d$. Hence

$$T_n f(\tau) = n^{k-1} \sum_{\substack{a,d>0 \\ ad=n}} \sum_{b=0}^{d-1} d^{-k} f\left(\frac{a\tau + b}{d}\right).$$

Substituting into this the formula $f = \sum a(m)\, q^m$ gives (4) after a short calculation. The second assertion of (i) follows from (4) because all of the exponents of q on the right-hand side are ≥ 0 and the constant term equals $a(0)\sigma_{k-1}(n)$ ($\sigma_{k-1}(n)$ as in Part 1**A**), so vanishes if $a(0) = 0$. The multiplication properties (5) follow from (4) by another easy computation. □

In the special case when $n = p$ is prime, the formula for the action of T_n reduces to

$$T_p f(\tau) = \frac{1}{p} \sum_{j=0}^{p-1} f\left(\frac{\tau + j}{p}\right) + p^{k-1} f(p\tau) = \sum_{m=0}^{\infty} a(mp)\, q^m + p^{k-1} \sum_{m=0}^{\infty} a(m)\, q^{mp},$$

i.e., $T_p = U_p + p^{k-1} V_p$ where U_p and V_p are the operators defined in 1**E**. (More generally, (4) says that T_n for any n is a linear combination of products $U_d V_a$ with $ad = n$.) The multiplicative property (5) tells us that knowing the T_p is sufficient for knowing all T_n, since if $n > 1$ is divisible by a prime p then $T_n = T_{n/p} T_p$ if $p^2 \nmid n$, $T_n = T_{n/p} T_p - p^{k-1} T_{n/p^2}$ if $p^2 | n$.

To end this section, we remark that formula (4), except for the constant term, makes sense also for $n = 0$, the common divisors of 0 and m being simply the divisors of m. Thus the coefficient of q^m on the right is just $a(0)\sigma_{k-1}(m)$ for each $m > 0$. The constant term is formally $a(0) \sum_{d=1}^{\infty} d^{k-1} = a(0)\zeta(1-k)$, but in fact we take it to be $\frac{1}{2} a(0)\zeta(1-k) = -a(0)\dfrac{B_k}{2k}$. Thus we set

$$\text{(6)} \qquad T_0 f(\tau) = a(0)\, G_k(\tau) \qquad \left(f = \sum_{m=0}^{\infty} a(m)\, q^m \in M_k\right);$$

in particular, T_0 maps M_k to M_k and $T_0 f = 0$ if f is a cusp form.

B Eigenforms

We have seen that the Hecke operators T_n act as linear operators on the vector space M_k. Suppose that $f(\tau) = \sum\limits_{m=0}^{\infty} a(m)\, q^m$ is an eigenvector of all the T_n, i.e.,

$$\text{(7)} \qquad T_n f = \lambda_n f \qquad (\forall n)$$

for some complex numbers λ_n. This certainly sometimes happens. For instance, if $k = 4$, 6, 8, 10 or 14 then the space M_k is 1-dimensional, spanned by the

Eisenstein series G_k of Part 1**A**, so $T_n G_k$ is necessarily a multiple of G_k for every n. (Actually, we will see in a moment that this is true even if $\dim M_k > 1$.) Similarly, if $k = 12, 16, 18, 20, 22$ or 26 then the space S_k of cusp forms of weight k is 1-dimensional, and since T_n preserves S_k, any element of S_k satisfies (7). From (7) and (4) we obtain the identity

$$\lambda_n\, a(m) = \sum_{d|n,m} d^{k-1}\, a\Big(\frac{nm}{d^2}\Big) \tag{8}$$

by comparing the coefficients of q^m on both sides of (7). In particular, $\lambda_n a(1) = a(n)$ for all n. It follows that $a(1) \neq 0$ if f is not identically zero, so we can normalize f by requiring that $a(1) = 1$. We call a modular form satisfying (7) and the extra condition $a(1) = 1$ a **Hecke form** (the term 'normalized Hecke eigenform' is commonly used in the literature). From what we have just said, it follows that a Hecke form has the property

$$\lambda_n = a(n) \qquad (\forall n), \tag{9}$$

i.e., the Fourier coefficients of f are equal to its eigenvalues under the Hecke operators. Equation (5) or (8) now implies the property

$$a(n)\, a(m) = \sum_{d|n,m} d^{k-1}\, a\Big(\frac{nm}{d^2}\Big) \tag{10}$$

for the coefficients of a Hecke form. In particular, the sequence of Fourier coefficients $\{a(n)\}$ is **multiplicative**, i.e., $a(1) = 1$ and $a(nm) = a(n)a(m)$ whenever n and m are coprime. In particular, $a(p_1^{r_1} \dots p_l^{r_l}) = a(p_1^{r_1}) \dots a(p_l^{r_l})$ for distinct primes $p_1, \dots, p_l$, so the $a(n)$ are determined if we know the values of $a(p^r)$ for all primes p. Moreover, (10) with $n = p^r$, $m = p$ gives the recursion

$$a(p^{r+1}) = a(p)\, a(p^r) - p^{k-1} a(p^{r-1}) \qquad (r \geq 1) \tag{11}$$

for the coefficients $a(p^r)$ for a fixed prime p, so it in fact is enough to know the $a(p)$ (compare the remark following Theorem 1).

Examples. 1. The form $G_k = -\dfrac{B_k}{2k} + \sum\limits_{m=1}^{\infty} \sigma_{k-1}(m)\, q^m \in M_k$ is a Hecke form for all $k \geq 4$ with $\lambda_n = a(n) = \sigma_{k-1}(n)$ for $n > 0$ and $\lambda_0 = a(0) = -\dfrac{B_k}{2k}$ (cf. (6)). In view of (4), to check this we need only check that the coefficients $a(n)$ of G_k satisfy (10) if n or $m > 0$; this is immediate if n or m equals 0 and can be checked easily for n and m positive by reducing to the case of prime powers (for $n = p^\nu$, $\sigma_{k-1}(n)$ equals $1 + p^{k-1} + \dots + p^{\nu(k-1)}$, which can be summed as a geometric series) and using the obvious multiplicativity of the numbers $\sigma_{k-1}(n)$.

2. The discriminant function Δ of Part 1 belongs to the 1-dimensional space S_{12} and has 1 as coefficient of q^1, so it is a Hecke form. In particular,

(10) holds (with $k = 12$) for the coefficients $a(n)$ of Δ, as we can check for small n using the coefficients given in (7) of Part 1:

$$a(2)a(3) = -24\times 252 = -6048 = a(6)\,,\ a(2)^2 = 576 = -1472+2048 = a(4)+2^{11}.$$

This multiplicativity property of the coefficients of Δ was noticed by Ramanujan in 1916 and proved by Mordell a year later by the same argument as we have just given.

The proof that Δ is a simultaneous eigenform of the T_n used the property $\dim S_k = 1$, which is false for $k > 26$. Nevertheless, there exist eigenforms in higher dimensions also; this is Hecke's great discovery. Indeed, we have:

Theorem 2. *The Hecke forms in M_k form a basis of M_k for every k.*

Proof. We have seen that G_k is an eigenform of all T_n. Conversely, any modular form with non-zero constant term which is an eigenform of all T_n $(n \geq 0)$ is a multiple of G_k by virtue of equation (6) of Section **A**. In view of this and the decomposition $M_k = \langle G_k \rangle \oplus S_k$, it suffices to show that S_k is spanned by Hecke forms and that the Hecke forms in S_k are linearly independent. For this we use the Hilbert space structure on S_k introduced in the introduction of Part 1 (eq. (2)). One checks from the definition (3) that the T_n are self-adjoint with respect to this structure, i.e. $(T_n f, g) = (f, T_n g)$ for all $f,\ g \in S_k$ and $n > 0$. (For $n = 0$, of course, T_n is the zero operator on S_k by equation (6).) Also, the T_n commute with one another, as we have seen. A well-known theorem of linear algebra then asserts that S_k is spanned by simultaneous eigenvectors of all the transformations T_n, and we have already seen that each such eigenform is uniquely expressible as a multiple of a Hecke form satisfying (10). Moreover, for a Hecke form we have

$$\begin{aligned} a(n)\,(f,f) &= (a(n)f,f) = (\lambda_n f,f) = (T_n f,f) \\ &= (f,T_n f) = (f,\lambda_n f) = (f,a(n)f) = \overline{a(n)}\,(f,f) \end{aligned}$$

by the self-adjointness of T_n and the sesquilinearity of the scalar product. Therefore the Fourier coefficients of f are real. If $g = \sum b(n)\,q^n$ is a second Hecke form in S_k, then the same computation shows that

$$a(n)\,(f,g) = (T_n f,g) = (f,T_n g) = \overline{b(n)}\,(f,g) = b(n)\,(f,g)$$

and hence that $(f,g) = 0$ if $f \neq g$. Thus the various Hecke forms in S_k are mutually orthogonal and *a fortiori* linearly independent. □

We also have

Theorem 3. *The Fourier coefficients of a Hecke form $f \in S_k$ are real algebraic integers of degree $\leq \dim S_k$.*

Proof. The space S_k is spanned by forms all of whose Fourier coefficients are integral (this follows easily from the discussion in Part 1, Section **B**). By formula (4), the lattice L_k of all such forms is mapped to itself by all T_n. Let $f_1, \dots, f_d$ ($d = \dim_{\mathbb{C}} S_k = \mathrm{rk}_{\mathbb{Z}} L_k$) be a basis for L_k over $\mathbb{Z}$. Then the action of T_n with respect to this basis is given by a $d \times d$ matrix with coefficients in $\mathbb{Z}$, so the eigenvalues of T_n are algebraic integers of degree $\leq d$. By (9), these eigenvalues are precisely the Fourier coefficients of the d Hecke forms in S_k. That the coefficients of Hecke forms are real was already checked in proving Theorem 2. □

From the proof of the theorem, we see that the trace of T_n ($n > 0$) acting on M_k or S_k is the trace of a $(d+1) \times (d+1)$ or $d \times d$ matrix with integral coefficients and hence is an integer. This trace is given in closed form by the Eichler-Selberg trace formula, which will be discussed in Part **3D**.

Example. The space S_{24} is 2-dimensional, spanned by

$$\Delta(\tau)^2 = 0q + q^2 - 48\,q^3 + 1080\,q^4 + \cdots$$

and

$$(240G_4(\tau))^3\,\Delta(\tau) = q + 696\,q^2 + 162252\,q^3 + 12831808\,q^4 + \cdots$$

If $f \in S_{24}$ is a Hecke form, then f must have the form $(240G_4)^3\Delta + \lambda\Delta^2$ for some $\lambda \in \mathbb{C}$, since the coefficient of q^1 must be 1. Hence its second and fourth coefficients are given by

$$a(2) = 696 + \lambda, \qquad a(4) = 12831808 + 1080\,\lambda.$$

The property $a(2)^2 = a(4) + 2^{23}$ ($n = m = 2$ in (10)) now leads to the quadratic equation

$$\lambda^2 + 312\,\lambda - 20736000 = 0$$

for λ. Hence any Hecke form in S_{24} must be one of the two functions

$$f_1,\ f_2 \ = \ (240G_4)^3\,\Delta + \left(-156 \pm 12\sqrt{144169}\right)\Delta^2.$$

Since Theorem 2 says that S_{24} must contain exactly two Hecke forms, f_1 and f_2 are indeed eigenvectors with respect to all the T_n. This means, for example, that we would have obtained the same quadratic equation for λ if we had used the relation $a(2)a(3) = a(6)$ instead of $a(2)^2 = a(4) + 2^{23}$. The coefficients $a_1(n)$, $a_2(n)$ of f_1 and f_2 are conjugate algebraic integers in the real quadratic field $\mathbb{Q}(\sqrt{144169})$.

C L-series

The natural reflex of a number-theorist confronted with a multiplicative function $n \mapsto a(n)$ is to form the Dirichlet series $\sum_{n=1}^{\infty} a(n)\,n^{-s}$, the point being that the multiplicative property implies that $a(p_1^{r_1}\dots p_l^{r_l}) = a(p_1^{r_1})\dots a(p_l^{r_l})$ and hence that this Dirichlet series has an Euler product $\prod_{p \text{ prime}} \bigl(\sum_{r\geq 0} a(p^r)\,p^{-rs}\bigr)$. We therefore define the **Hecke L-series** of a modular form $f(\tau) = \sum_{m=0}^{\infty} a(m)\,q^m \in M_k$ by

$$L(f,s) = \sum_{m=1}^{\infty} \frac{a(m)}{m^s} \tag{12}$$

(notice that we have ignored $a(0)$ in this definition; what else could we do?). Thus if f is a Hecke form we have an Euler product

$$L(f,s) = \prod_{p \text{ prime}} \left(1 + \frac{a(p)}{p^s} + \frac{a(p^2)}{p^{2s}} + \cdots\right)$$

because the coefficients $a(m)$ are multiplicative. But in fact we can go further, because the recursion (11) implies that for each prime p the generating function $A_p(x) = \sum a(p^r)x^r$ satisfies

$$\begin{aligned} A_p(x) &= 1 + \sum_{r=0}^{\infty} a(p^{r+1})\,x^{r+1} \\ &= 1 + \sum_{r=0}^{\infty} a(p)\,a(p^r)\,x^{r+1} - \sum_{r=1}^{\infty} p^{k-1}\,a(p^{r-1})\,x^{r+1} \\ &= 1 + a(p)\,x\,A_p(x) - p^{k-1}\,x^2\,A_p(x) \end{aligned}$$

and hence that

$$A_p(x) = \frac{1}{1 - a(p)x + p^{k-1}x^2}.$$

Therefore, replacing x by p^{-s} and multiplying over all primes p, we find finally

$$L(f,s) = \prod_p \frac{1}{1 - a(p)p^{-s} + p^{k-1-2s}} \qquad (f \in M_k \text{ a Hecke form}).$$

Examples. 1. For $f = G_k$ we have

$$a(p^r) = 1 + p^{k-1} + \cdots + p^{r(k-1)} = \frac{p^{(r+1)(k-1)} - 1}{p^{k-1} - 1},$$

$$A_p(x) = \sum_{r=0}^{\infty} \frac{p^{(r+1)(k-1)} - 1}{p^{k-1} - 1} x^r = \frac{1}{(1 - p^{k-1}x)(1 - x)}$$

$$L(G_k, s) = \prod_p \frac{1}{1 - \sigma_{k-1}(p)p^{-s} + p^{k-1-2s}} = \prod_p \frac{1}{(1 - p^{k-1-s})(1 - p^{-s})}$$

$$= \zeta(s - k + 1)\zeta(s),$$

where $\zeta(s)$ is the Riemann zeta function. (Of course, we could see this directly: the coefficient of n^{-s} in $\zeta(s-k+1)\zeta(s) = \sum_{d,e\geq 1} \frac{d^{k-1}}{(de)^s}$ is clearly $\sigma_{k-1}(n)$ for each $n \geq 1$.)

2. For $f = \Delta$ we have

$$L(\Delta, s) = \prod_p \frac{1}{1 - \tau(p)p^{-s} + p^{11-2s}},$$

where $\tau(n)$, the Ramanujan tau-function, denotes the coefficient of q^n in Δ; this identity summarizes all the multiplicative properties of $\tau(n)$ discovered by Ramanujan.

Of course, the Hecke L-series would be of no interest if their definition were merely formal. However, these series converge in a half-plane and define functions with nice analytic properties, as we now show.

Theorem 4. *(i) The Fourier coefficients $a(m)$ of a modular form of weight k satisfy the growth estimates*

$$(13) \quad a(n) = O(n^{k-1}) \qquad (f \in M_k), \qquad\qquad a(n) = O(n^{\frac{k}{2}}) \qquad (f \in S_k).$$

Hence the L-series $L(f,s)$ converges absolutely and locally uniformly in the half-plane $\Re(s) > k$ in any case and in the larger half-plane $\Re(s) > \frac{k}{2} + 1$ if f is a cusp form.

(ii) $L(f,s)$ has a meromorphic continuation to the whole complex plane. It is holomorphic everywhere if f is a cusp form and has exactly one singularity, a simple pole of residue $\frac{(2\pi i)^k}{(k-1)!}a(0)$ at $s = k$, otherwise. The meromorphically extended function satisfies the functional equation

$$(2\pi)^{-s}\,\Gamma(s)\,L(f,s) \;=\; (-1)^{\frac{k}{2}}(2\pi)^{s-k}\,\Gamma(k-s)\,L(f,k-s).$$

Proof. (i) Since the estimate $a(n) = O(n^{k-1})$ is obvious for the Eisenstein series G_k (we have $\sigma_{k-1}(n) = n^{k-1}\sum_{d|n} d^{-k+1} < n^{k-1}\sum_{d=1}^{\infty} d^{-k+1} < 2n^{k-1}$ because $k > 2$), and since every modular form of weight k is a combination of G_k and a cusp form, we need only prove the second estimate in (13). If f is a cusp form then by definition we have $|f(\tau)| < Mv^{-k/2}$ for some constant $M > 0$ and all $\tau = u + iv \in \mathfrak{H}$. On the other hand, for any $n \geq 1$ and $v > 0$ we have

$$a(n) = \int_0^1 f(u+iv)\, e^{-2\pi i n(u+iv)}\, du.$$

Hence

$$|a(n)| \leq M\, v^{-k/2}\, e^{2\pi n v},$$

and choosing $v = 1/n$ gives the desired conclusion. (This argument, like most of the rest of this part, is due to Hecke.)

(ii) This follows immediately from the 'functional equation principle' in Appendix **B**, since the function

$$\phi(v) \;=\; f(iv) - a(0) = \sum_{n=1}^{\infty} a(n)\, e^{-2\pi n v} \qquad (v > 0)$$

is exponentially small at infinity and satisfies the functional equation

$$\phi\Big(\frac{1}{v}\Big) = f\Big(\frac{-1}{iv}\Big) - a(0) \;=\; (iv)^k\, f(iv) - a(0) \;=\; (-1)^{\frac{k}{2}} v^k \phi(v) + (-1)^{\frac{k}{2}} a(0)\, v^k - a(0)$$

and its Mellin transform $\int_0^\infty \phi(v)\, v^{s-1}\, dv$ equals $(2\pi)^{-s}\Gamma(s)L(f,s)$. □

The first estimate in (13) is clearly the best possible, but the second one can be improved. The estimate $a(n) = \mathrm{O}(n^{\frac{k}{2}-\frac{1}{5}+\epsilon})$ for the Fourier coefficients of cusp forms on Γ_1 was found by Rankin in 1939 as an application of the Rankin-Selberg method explained in the next part. This was later improved to $a(n) = \mathrm{O}(n^{\frac{k}{2}-\frac{1}{4}+\epsilon})$ by Selberg as an application of Weil's estimates of Kloosterman sums. The estimate

$$a(n) = \mathrm{O}(n^{\frac{k-1}{2}+\epsilon}) \qquad \Big(f = \sum a(n)\, q^n \;\in\; S_k\Big), \tag{14}$$

conjectured by Ramanujan for $f = \Delta$ in 1916 and by Petersson in the general case, remained an open problem for many years. It was shown by Deligne in 1969 to be a consequence of the Weil conjectures on the eigenvalues of the Frobenius operator in the l-adic cohomology of algebraic varieties in positive characteristic; 5 years later he proved the Weil conjectures, thus establishing (14). Using the form of the generating function $A_p(x)$ given above, one sees that (14) is equivalent to

$$|a(p)| \leq 2p^{(k-1)/2} \qquad (p \text{ prime}). \tag{15}$$

In particular, for the Ramanujan tau-function $\tau(n)$ (coefficient of q^n in Δ) one has

$$|\tau(p)| \leq 2p^{11/2} \qquad (p \text{ prime}). \tag{16}$$

The proof of (16) uses the full force of Grothendieck's work in algebraic geometry and its length, if written out from scratch, has been estimated at 2000 pages; in his book on mathematics and physics, Manin cites this as a probable record for the ratio 'length of proof : length of statement' in the whole of mathematics.

D Forms of higher level

In most of these notes, we restrict attention to the full modular group $\Gamma_1 = PSL_2(\mathbb{Z})$ rather than subgroups because most aspects of the theory can be seen there. However, in the case of the theory of Hecke operators there are some important differences, which we now describe. We will restrict attention to the subgroups $\Gamma_0(N) = \{\left(\begin{smallmatrix} a & b \\ c & d \end{smallmatrix}\right) \in \Gamma_1 \mid c \equiv 0 \pmod{N}\}$ introduced in Part 1.

First of all, the definition of T_n must be modified. In formula (3) we must replace Γ_1 by $\Gamma = \Gamma_0(N)$ and $\mathcal{M}_n$ by the set of integral matrices $\left(\begin{smallmatrix} a & b \\ c & d \end{smallmatrix}\right)$ of determinant n satisfying $c \equiv 0 \pmod{N}$ and $(a, N) = 1$. Again the coset representatives of $\Gamma\backslash\mathcal{M}_n$ can be chosen to be upper triangular, but the extra condition $(a, N) = 1$ means that we have fewer representatives than before if $(n, N) > 1$. In particular, for p a prime dividing N we have $T_p = U_p$ and $T_{p^r} = (T_p)^r$ rather than $T_p = U_p + p^{k-1}V_p$ and a 3-term recursion relation for $\{T_{p^r}\}$. For general n, the operation of T_n is given by the same formula (4) as before but with the extra condition $(d, N) = 1$ added to the inner sum, and similarly for the multiplicativity relation (5).

The other main difference with the case $N = 1$ comes from the existence of so-called 'old forms.' If N' is a proper divisor of N, then $\Gamma_0(N)$ is a subgroup of $\Gamma_0(N')$ and every modular form $f(\tau)$ of weight k on $\Gamma_0(N')$ is *a fortiori* a modular form on $\Gamma_0(N)$. More generally, $f(M\tau)$ is a modular form of weight k on $\Gamma_0(N)$ for each positive divisor M of N/N', since

$$\begin{aligned} \begin{pmatrix} a & b \\ c & d \end{pmatrix} \in \Gamma_0(N) &\Rightarrow \begin{pmatrix} a & bM \\ c/M & d \end{pmatrix} \in \Gamma_0(N') \\ &\Rightarrow f\Big(M\frac{a\tau+b}{c\tau+d}\Big) = f\Big(\frac{a(M\tau)+bM}{(c/M)(M\tau)+d}\Big) = (c\tau+d)^k f(M\tau). \end{aligned}$$

The subspace of $M_k(\Gamma_0(N))$ spanned by all forms $f(M\tau)$ with $f \in M_k(\Gamma_0(N'))$, $MN'|N$, $N' \neq N$, is called the space of **old forms**. (This definition must be modified slightly if $k = 2$ to include also the modular forms $\sum_{M|N} c_M G_2^*(M\tau)$ with $c_M \in \mathbb{C}$, $\sum_{M|N} M^{-1}c_M = 0$, where G_2^* is the non-holomorphic Eisenstein series of weight 2 on Γ_1 introduced in Part **1A**, as old forms, even though G_2^* itself is not in $M_2(\Gamma_1)$.) Since the old forms can be considered by induction on N as already known, one is interested only in the 'rest' of $M_k(\Gamma_0(N))$. The answer here is quite satisfactory: $M_k(\Gamma_0(N))$ has a canonical splitting as the direct sum of the subspace $M_k(\Gamma_0(N))^{\text{old}}$ of old forms and a certainly complementary space $M_k(\Gamma_0(N))^{\text{new}}$ (for cusp forms, $S_k(\Gamma_0(N))^{\text{new}}$ is just the orthogonal complement of $S_k(\Gamma_0(N))^{\mathrm{old}}$ with respect to the Petersson scalar product), and if we define a **Hecke form of level** N to be a form in $M_k(\Gamma_0(N))^{\mathrm{new}}$ which is an eigenvector of T_n for all n prime to N and with $a(1) = 1$, then the Hecke forms are in fact eigenvectors of all the T_n, they form a basis of $M_k(\Gamma_0(N))^{\mathrm{new}}$, and their Fourier coefficients are real algebraic integers as before. For the pth Fourier coefficient (p prime) of a Hecke form in $S_k(\Gamma_0(N))^{\mathrm{new}}$ we have the same estimate (15) as before if $p \nmid N$, while the eigenvalue with respect to T_p when $p|N$ equals 0 if $p^2|N$ and $\pm p^{(k-1)/2}$ otherwise. Finally, there is no overlapping between the new forms of different level or between the different lifts $f(M\tau)$ of forms of the same level, so that we have a canonical direct sum decomposition

$$M_k(\Gamma_0(N)) = \bigoplus_{MN'|N} \{f(M\tau) \mid f \in M_k(\Gamma_0(N'))^{\mathrm{new}}\}$$

and a canonical basis of $M_k(\Gamma_0(N))$ consisting of the functions $f(M\tau)$ where $M|N$ and f is a Hecke form of level dividing N/M.

As already stated, the Fourier coefficients of Hecke forms of higher level are real algebraic integers, just as before. However, there is a difference with the case $N = 1$: For forms of level 1, Theorem 3 apparently always is sharp: in all cases which have been calculated, the number field generated by the Fourier coefficients of a Hecke cusp form of weight k has degree equal to the full dimension d of the space S_k, which is then spanned by a single form and its algebraic conjugates (cf. the example $k = 24$ given above). For forms of higher level, there are in general further splittings. The general situation is that $S_k(\Gamma_0(N))^{\mathrm{new}}$ splits as the sum of subspaces of some dimensions $d_1, \ldots, d_r \geq 1$, each of which is spanned by some Hecke form, with Fourier coefficients in a totally real number field K_i of degree d_i over $\mathbb{Q}$, and the algebraic conjugates of this form (i.e. the forms obtained by considering the various embeddings $K_i \hookrightarrow \mathbb{R}$). In general the number r and the dimensions d_i are unknown; the known theory implies certain necessary splittings of $S_k(\Gamma_0(N))^{\mathrm{new}}$, but there are often further splittings which we do not know how to predict.

Examples. 1. $k = 2$, $N = 11$. Here $\dim M_k(\Gamma_0(N)) = 2$. As well as one old form, the Eisenstein series

$$G_2^*(\tau) - 11G_2^*(11\tau) = \frac{5}{12} + \sum_{n=1}^{\infty} \Bigl(\sum_{\substack{d|n \\ 11\nmid d}} d \Bigr) q^n$$

of weight 2, there is one new form

$$f(\tau) = \sqrt[12]{\Delta(\tau)\Delta(11\tau)} = q - 2q^2 - q^3 + 2q^4 + q^5 + 2q^6 - 2q^7 + \cdots,$$

with Fourier coefficients in $\mathbb{Z}$. This form corresponds (as in the Taniyama-Weil conjecture mentioned in Part 1**F**) to the elliptic curve $y^2 - y = x^3 - x^2$, i.e., $p - a(p)$ gives the number of solutions of $y^2 - y = x^3 - x^2$ in integers modulo any prime p.

2. $k = 2$, $N = 23$. Again $M_k(\Gamma_0(N))^{\text{old}}$ is 1-dimensional, spanned by $G_2^*(\tau) - NG_2^*(N\tau)$, but this time $M_k(\Gamma_0(N))^{\text{new}} = S_k(\Gamma_0(N))^{\text{new}}$ is 2-dimensional, spanned by the Hecke form

$$f_1 = q - \frac{1-\sqrt{5}}{2}q^2 + \sqrt{5}\,q^3 - \frac{1+\sqrt{5}}{2}q^4 - (1-\sqrt{5})\,q^5 - \frac{5-\sqrt{5}}{2}q^6 + \cdots$$

with coefficients in $\mathbb{Z} + \mathbb{Z}\dfrac{1+\sqrt{5}}{2}$ and the conjugate form

$$f_2 = q - \frac{1+\sqrt{5}}{2}q^2 - \sqrt{5}\,q^3 - \frac{1-\sqrt{5}}{2}q^4 - (1+\sqrt{5})\,q^5 - \frac{5+\sqrt{5}}{2}q^6 + \cdots$$

obtained by replacing $\sqrt{5}$ by $-\sqrt{5}$ everywhere in f_1.

3. $k = 2$, $N = 37$. Again $M_k(\Gamma_0(N))^{\text{old}}$ is spanned by $G_2^*(\tau) - NG_2^*(N\tau)$ and $M_k(\Gamma_0(N))^{\text{new}} = S_k(\Gamma_0(N))^{\text{new}}$ is 2-dimensional, but this time the two Hecke forms of level N

$$f_1 = q - 2q^2 - 3q^3 + 2q^4 - 2q^5 + 6q^6 - q^7 + \cdots$$

and

$$f_2 = q + 0q^2 + q^3 - 2q^4 + 0q^5 + 0q^6 - q^7 + \cdots$$

both have coefficients in $\mathbb{Z}$; they correspond to the elliptic curves $y^2 - y = x^3 - x$ and $y^2 - y = x^3 + x^2 - 3x + 1$, respectively.

4. $k = 4$, $N = 13$. Here $M_k(\Gamma_0(N))^{\text{old}}$ is spanned by the two Eisenstein series $G_4(\tau)$ and $G_4(N\tau)$ and $M_k(\Gamma_0(N))^{\text{new}} = S_k(\Gamma_0(N))^{\text{new}}$ is 3-dimensional, spanned by the forms

$$f_1,\, f_2 = q + \frac{1\pm\sqrt{17}}{2}q^2 + \frac{5\mp 3\sqrt{17}}{2}q^3 - \frac{7\mp\sqrt{17}}{2}q^4 + \cdots$$

with coefficients in the real quadratic field $\mathbb{Q}(\sqrt{17})$ and the form

$$f_3 = q - 5q^2 - 7q^3 + 17q^4 - 7q^5 + 35q^6 - 13q^7 - \cdots$$

with coefficients in $\mathbb{Q}$.

Finally, there are some differences between the L-series in level 1 and in higher level. First of all, the form of the Euler product for the L-series of a Hecke form must be modified slightly: it is now

$$L(f,s) = \prod_{p\nmid N} \frac{1}{1 - a(p)p^{-s} + p^{k-1-2s}} \prod_{p\mid N} \frac{1}{1 - a(p)p^{-s}}.$$

More important, $L(f,s)$, although it converges absolutely in the same half-plane as before and again has a meromorphic continuation with at most a simple pole at $s = k$, in general *does not have a functional equation for every* $f \in M_k(\Gamma_0(N))$, because we no longer have the element $\left(\begin{smallmatrix} 0 & -1 \\ 1 & 0 \end{smallmatrix}\right) \in \Gamma$ to force the symmetry of $f(iv)$ with respect to $v \mapsto \dfrac{1}{v}$. Instead, we have the **Fricke involution**

$$w_N : f(\tau) \mapsto w_N f(\tau) = N^{-\frac{k}{2}} \tau^{-k} f\Big(\frac{-1}{N\tau}\Big)$$

which acts on the space of modular forms of weight k on $\Gamma_0(N)$ because the element $\left(\begin{smallmatrix} 0 & -1 \\ N & 0 \end{smallmatrix}\right)$ of $GL_2^+(\mathbb{R})$ normalizes the group $\Gamma_0(N)$. This involution splits $M_k(\Gamma_0(N))$ into the direct sum of two eigenspaces $M_k^{\pm}(\Gamma_0(N))$, and if f belongs to $M_k^{\pm}(\Gamma_0(N))$ then

$$(2\pi)^{-s} N^{s/2} \Gamma(s) L(f,s) = \pm(-1)^{k/2} (2\pi)^{s-k} N^{(k-s)/2} \Gamma(k-s) L(f, k-s).$$

(For $N = 1$ we have $w_N \equiv \mathrm{Id}$ since $\left(\begin{smallmatrix} 0 & -1 \\ N & 0 \end{smallmatrix}\right) \in \Gamma_0(N)$ in this case, so $M_k^- = \{0\}$ for all k, but for all other values of N the dimension of $M_k^+(\Gamma_0(N))$ is asymptotically $\frac{1}{2}$ the dimension of $M_k(\Gamma_0(N))$ as $k \to \infty$.) The involution w_N preserves the space $M_k(\Gamma_0(N))^{\mathrm{new}}$ and commutes with all Hecke operators T_n there (whereas on the full space $M_k(\Gamma_0(N))$ it commutes with T_n only for $(n, N) = 1$). In particular, each Hecke form of level N is an eigenvector of w_N and therefore has an L-series satisfying a functional equation. In our example **3** above, for instance, the Eisenstein series $G_2^*(\tau) - 37G_2^*(37\tau)$ and the cusp form f_2 are anti-invariant under w_{37} and therefore have plus-signs in the functional equations of their L-series, while f_1 is invariant under w_{37} and has an L-series with a minus sign in its functional equation. In particular, the L-series of f_1 vanishes at $s = 1$, which is related by the famous Birch-Swinnerton-Dyer conjecture to the fact that the equation of the corresponding elliptic curve $y^2 - y = x^3 - x$ has an infinite number of rational solutions.

3. The Rankin-Selberg Method and its Applications

The Rankin-Selberg convolution method is one of the most powerful tools in the theory of automorphic forms. In this part we explain two principal variants of it—one involving non-holomorphic Eisenstein series and one involving only the holomorphic Eisenstein series constructed in Part 1. We will also give several applications, the most important one being a proof of the formula of Eichler and Selberg for the traces of Hecke operators acting on spaces of holomorphic modular forms. The essential ingredients of the Rankin-Selberg method are various types of Eisenstein series, and we begin by studying the main properties of some of these.

A Non-holomorphic Eisenstein series

For $\tau = u + iv \in \mathfrak{H}$ and $s \in \mathbb{C}$ define

$$G(\tau, s) = \frac{1}{2}\sum_{m,n}{}' \frac{\Im(\tau)^s}{|m\tau + n|^{2s}}, \tag{1}$$

(sum over $m, n \in \mathbb{Z}$ not both zero). The series converges absolutely and locally uniformly for $\Re(s) > 1$ and defines a function which is Γ_1-invariant in τ for the same reason that G_k in Part 1 was a modular form. As a sum of pure exponential functions, it is a holomorphic function of s in the same region, but, owing to the presence of $v = \Im(\tau)$ and the absolute value signs, it is not holomorphic in τ. The function $G(\tau, s)$ is known in the literature under both the names 'non-holomorphic Eisenstein series' and 'Epstein zeta function' (in general, the Epstein zeta function of a positive definite quadratic form Q in r variables is the Dirichlet series $\sum'_{x\in\mathbb{Z}^r} Q(x)^{-s}$; if $r = 2$, then this equals $2^{s+1}d^{-s/2}G(\tau, s)$ where $-d$ is the discriminant of Q and τ the root of $Q(z, 1) = 0$ in the upper half plane). Its main properties, besides the Γ_1-invariance, are summarized in

Proposition. *The function $G(\tau, s)$ can be meromorphically extended to a function of s which is entire except for a simple pole of residue $\frac{\pi}{2}$ (independent of τ!) at $s = 1$. The function $G^*(\tau, s) = \pi^{-s}\Gamma(s)G(\tau, s)$ is holomorphic except for simple poles of residue $\frac{1}{2}$ and $-\frac{1}{2}$ at $s = 1$ and $s = 0$, respectively, and satisfies the functional equation $G^*(\tau, s) = G^*(\tau, 1 - s)$.*

Proof. We sketch two proofs of this. The first is analogous to Riemann's proof of the functional equation of $\zeta(s)$. For $\tau = u + iv \in \mathfrak{H}$ let Q_τ be the positive definite binary quadratic form $Q_\tau(m, n) = v^{-1}|m\tau + n|^2$ of discriminant -4 and $\Theta_\tau(t) = \sum_{m,n\in\mathbb{Z}} e^{-\pi Q_\tau(m,n)t}$ the associated theta series. The Mellin transformation formula (cf. Appendix **B**) implies

$$G^*(\tau, s) = \frac{1}{2}\Gamma(s)\sum_{m,n}{}' \big[\pi Q_\tau(m, n)\big]^{-s} = \frac{1}{2}\int_0^\infty (\Theta_\tau(t) - 1)t^{s-1}\, dt.$$

On the other hand, the Poisson summation formula (cf. Appendix **A**) implies that $\Theta_\tau(\frac{1}{t}) = t\Theta_\tau(t)$, so the function $\phi(t) = \frac{1}{2}(\Theta_\tau(t) - 1)$ satisfies $\phi(t^{-1}) = -\frac{1}{2} + \frac{1}{2}t + t\phi(\frac{1}{t})$. The 'functional equation principle' formulated in Appendix **B** now gives the assertions of the theorem.

The second proof, which requires more calculation, but also gives more information, is to compute the Fourier development of $G(\tau, s)$. The computation is very similar to that for G_k in Part 1, so we can be brief. Splitting up the sum defining $G(\tau, s)$ into the terms with $m = 0$ and those with $m \neq 0$, and combining each summand with its negative, we find

$$G(\tau, s) = \zeta(2s)v^s + v^s \sum_{m=1}^{\infty} \left(\sum_{n=-\infty}^{\infty} |m\tau + n|^{-2s} \right) \qquad (\tau = u + iv).$$

Substituting into this formula (3) of Appendix **A**, we find

$$G(\tau, s) = \zeta(2s)v^s + \frac{\pi^{\frac{1}{2}}\Gamma(s - \frac{1}{2})}{\Gamma(s)} v^{1-s} \sum_{m=1}^{\infty} m^{1-2s}$$
$$+ \frac{2\pi^s}{\Gamma(s)} v^{\frac{1}{2}} \sum_{\substack{m \geq 1 \\ r \neq 0}} m^{\frac{1}{2}-s} |r|^{s-\frac{1}{2}} K_{s-\frac{1}{2}}(2\pi m|r|v)\, e^{2\pi i m r u},$$

where $K_\nu(t)$ is the K-Bessel function $\int_0^\infty e^{-t\cosh u} \cosh(\nu u)\, du$. Hence

$$G^*(\tau, s) = \zeta^*(2s)v^s + \zeta^*(2s-1)v^{1-s} + 2v^{\frac{1}{2}} \sum_{n \neq 0} \sigma^*_{s-\frac{1}{2}}(|n|) K_{s-\frac{1}{2}}(2\pi|n|v) e^{2\pi i n u},$$

where $\zeta^*(s)$ denotes the meromorphic function $\pi^{-s/2}\Gamma(s/2)\zeta(s)$ and $\sigma^*_\nu(n)$ the arithmetic function $|n|^\nu \sum_{d|n} d^{-2\nu}$. The analytic continuation properties of G^* now follow from the facts that $\zeta^*(s)$ is holomorphic except for simple poles of residue 1 and -1 at $s = 1$ and $s = 0$, respectively, that $\sigma^*_\nu(n)$ is an entire function of ν, and that $K_\nu(t)$ is entire in ν and exponentially small in t as $t \to \infty$, while the functional equation follows from the functional equations $\zeta^*(1 - s) = \zeta^*(s)$ (cf. Appendix **B**), $\sigma^*_{-\nu}(n) = \sigma^*_\nu(n)$, and $K_{-\nu}(t) = K_\nu(t)$. $\square$

As an immediate consequence of the Fourier development of G^* and the identity $K_{\frac{1}{2}}(t) = \sqrt{\pi/2t}\, e^{-t}$, we find

$$\lim_{s \to 1} \left(G^*(\tau, s) - \frac{1/2}{s-1} \right) = \frac{\pi}{6}v - \frac{1}{2}\log v + C + 2 \sum_{m,r=1}^{\infty} \frac{1}{m} \Re(e^{2\pi i m r \tau})$$
$$= \frac{\pi}{6}v - \frac{1}{2}\log v + C - \sum_{r=1}^{\infty} \log|1 - e^{2\pi i r \tau}|^2$$
$$= -\frac{1}{24}\log\left(v^{12}|\Delta(\tau)|^2\right) + C,$$

where $C = \lim_{s\to 1}(\zeta^*(s) - (s-1)^{-1})$ is a certain constant (in fact given by $\frac{1}{2}\gamma - \frac{1}{2}\log 4\pi$, where γ is Euler's constant) and $\Delta(\tau)$ the discriminant function of Part 1. This formula is called the **Kronecker limit formula** and has many applications in number theory. Together with the invariance of $G(\tau, s)$ under $PSL_2(\mathbb{Z})$, it leads to another proof of the modular transformation property of $\Delta(\tau)$.

B The Rankin-Selberg method (non-holomorphic case) and applications

In this section we describe the 'unfolding method' invented by Rankin and Selberg in their papers of 1939–40. Suppose that $F(\tau)$ is a smooth Γ_1-invariant function in the upper half-plane and tends to 0 rapidly (say, exponentially) as $v = \Im(\tau) \to \infty$. (In the original papers of Rankin and Selberg, $F(\tau)$ was the function $v^{12}|\Delta(\tau)|^2$.) The Γ_1-invariance of F implies in particular the periodicity property $F(\tau+1) = F(\tau)$ and hence the existence of a Fourier development $F(u+iv) = \sum_{n\in\mathbb{Z}} c_n(v)e^{2\pi i n u}$. We define the **Rankin-Selberg transform** of F as the Mellin transform (cf. Appendix **B**) of the constant term $c_0(v)$ of F:

$$R(F;s) = \int_0^\infty c_0(v)\, v^{s-2}\, dv \tag{1}$$

(notice that there is a shift of s by 1 with respect to the usual definition of the Mellin transform). Since $F(u+iv)$ is bounded for all v and very small as $v \to \infty$, its constant term

$$c_0(v) = \int_0^1 F(u+iv)\, du \tag{2}$$

also has these properties. Hence the integral in (1) converges absolutely for $\Re(s) > 1$ and defines a holomorphic function of s in that domain.

Theorem. *The function $R(F;s)$ can be meromorphically extended to a function of s and is holomorphic in the half-plane $\Re(s) > \frac{1}{2}$ except for a simple pole of residue $\kappa = \frac{3}{\pi}\iint_{\mathfrak{H}/\Gamma_1} F(\tau)\, d\mu$ at $s = 1$. The function $R^*(F;s) = \pi^{-s}\Gamma(s)\zeta(2s)R(F;s)$ is holomorphic everywhere except for simple poles of residue $\pm\frac{\pi}{6}\kappa$ at $s = 1$ and $s = 0$ and $R^*(F;s) = R^*(F;1-s)$.*

(Recall that $d\mu$ denotes the $SL(2,\mathbb{R})$-invariant volume measure $v^{-2}\, du\, dv$ on $\mathfrak{H}/\Gamma_1$ and that the area of $\mathfrak{H}/\Gamma_1$ with respect to this measure is $\pi/3$; thus κ is simply the average value of F in the upper half-plane.)

Proof. We will show that $\zeta(2s)R(F;s)$ is equal to the Petersson scalar product of $\overline{F}$ with the non-holomorphic Eisenstein series of Section **A**:

$$\zeta(2s)R(F;s) = \iint_{\mathfrak{H}/\Gamma_1} G(\tau,s)\, F(\tau)\, d\mu. \tag{3}$$

The assertions of the theorem then follow immediately from the proposition in that section.

To prove (3) we use the method called 'unfolding' (sometimes also referred to as the 'Rankin-Selberg trick'). Let Γ_∞ denote the subgroup $\{\pm\left(\begin{smallmatrix}1&n\\0&1\end{smallmatrix}\right) \mid n\in\mathbb{Z}\}$ of Γ_1 (the '∞' in the notation refers to the fact that Γ_∞ is the stabilizer in Γ_1 of infinity). The left cosets of Γ_∞ in Γ_1 are in 1:1 correspondence with pairs of coprime integers (c,d), considered up to sign: multiplying a matrix $\left(\begin{smallmatrix}a&b\\c&d\end{smallmatrix}\right)$ on the left by $\left(\begin{smallmatrix}1&n\\0&1\end{smallmatrix}\right)$ produces a new matrix with the same second row, and any two matrices with the same second row are related in this way. Also, $\Im(\gamma(\tau)) = v/|c\tau+d|^2$ for $\gamma=\left(\begin{smallmatrix}a&b\\c&d\end{smallmatrix}\right)\in\Gamma_1$. Finally, any non-zero pair of integers (m,n) can be written uniquely as (rc,rd) for some $r>0$ and coprime c and d. Hence for $\Re(s)>1$ we have

$$G(\tau,s)=\frac{1}{2}\sum_{r=1}^{\infty}\sum_{c,d\text{ coprime}}\frac{\Im(\tau)^s}{|r(c\tau+d)|^{2s}}=\zeta(2s)\sum_{\gamma\in\Gamma_\infty\backslash\Gamma_1}\Im\big(\gamma(\tau)\big)^s.$$

Therefore, denoting by $\mathcal{F}$ a fundamental domain for the action of Γ_1 on $\mathfrak{H}$, and observing that the sum and integral are absolutely convergent and that both F and $d\mu$ are Γ_1-invariant, we obtain

$$\begin{aligned}\zeta(2s)^{-1}\iint\limits_{\mathfrak{H}/\Gamma_1}G(\tau,s)\,F(\tau)\,d\mu &= \iint\limits_{\mathcal{F}}\sum_{\gamma\in\Gamma_\infty\backslash\Gamma_1}\Im(\gamma\tau)^s\,F(\gamma\tau)\,d\mu\\ &= \sum_{\gamma\in\Gamma_\infty\backslash\Gamma_1}\iint\limits_{\gamma\mathcal{F}}\Im(\tau)^s\,F(\tau)\,d\mu.\end{aligned}$$

Notice that we have spoiled the invariance of the original representation: both the fundamental domain and the set of coset representatives for $\Gamma_\infty\backslash\Gamma_1$ must be chosen explicitly for the individual terms in what we have just written to make sense. Now comes the unfolding argument: the different translates $\gamma\mathcal{F}$ of the original fundamental domain are disjoint, and they fit together exactly to form a fundamental domain for the action of Γ_∞ on $\mathfrak{H}$ (here we ignore questions about the boundaries of the fundamental domains, since these form a set of measure zero and can be ignored.) Hence finally

$$\zeta(2s)^{-1}\iint_{\mathfrak{H}/\Gamma_1}G(\tau,s)\,F(\tau)\,d\mu=\iint_{\mathfrak{H}/\Gamma_\infty}\Im(\tau)^s\,F(\tau)\,d\mu.$$

Since the action of Γ_∞ on $\mathfrak{H}$ is given by $u\mapsto u+1$, the right-hand side of this can be rewritten as $\int_0^\infty\big(\int_0^1 F(u+iv)\,du\big)\,v^{s-2}\,dv$, and in view of equation (2) this is equivalent to the assertion (3). A particularly pleasing aspect of the computation is that—unlike the usual situation in mathematics where a simplification at one level of a formula must be paid for by an increased complexity somewhere else—the unfolding simultaneously permitted us to replace the complicated infinite sum defining the Eisenstein series by a single term

$\Im(\tau)^s$ *and* to replace the complicated domain of integration $\mathfrak{H}/\Gamma_1$ by the much simpler $\mathfrak{H}/\Gamma_\infty$ and eventually just by $(0,\infty)$. □

We now give some applications of the theorem. The first application is to the Γ_1-invariant function $F(\tau) = v^k|f(\tau)|^2$, where $f = \sum a(n)q^n$ is any cusp form in S_k (in the original papers of Rankin and Selberg, as already mentioned, f was the discriminant function of Part 1**B**, $k = 12$). We have

$$F(u+iv) = v^k \sum_{n=1}^{\infty} \sum_{m=1}^{\infty} a(n)\,\overline{a(m)}\, e^{2\pi i(n-m)u}\, e^{-2\pi(n+m)v}$$

and hence $c_0(v) = v^k \sum_{n=1}^{\infty} |a(n)|^2\, e^{-4\pi n v}$. Therefore

$$R(F;s) = \sum_{n=1}^{\infty} |a(n)|^2 \int_0^{\infty} v^k\, e^{-4\pi n v}\, v^{s-2}\, dv = \frac{\Gamma(s+k-1)}{(4\pi)^{s+k-1}} \sum_{n=1}^{\infty} \frac{|a(n)|^2}{n^{s+k-1}}.$$

This proves the meromorphic continuability and functional equation of the 'Rankin zeta function' $\sum |a(n)|^2 n^{-s}$; moreover, applying the statement about residues in the theorem and observing that κ here is just $3/\pi$ times the Petersson scalar product of f with itself, we find

$$(4)\qquad (f,f) = \frac{\pi}{3}\frac{(k-1)!}{(4\pi)^k}\, \mathrm{Res}_{s=1}\left(\sum_{n=1}^{\infty} \frac{|a(n)|^2}{n^{s+k-1}}\right).$$

If f is a Hecke form, then the coefficients $a(n)$ real and $\sum a(n)^2 n^{-s-k+1} = \zeta(s)\sum a(n^2) n^{-s-k+1}$ by an easy computation using the shape of the Euler product of the L-series of f, so this can be rewritten in the equivalent form

$$(5)\qquad (f,f) = \frac{\pi}{3}\frac{(k-1)!}{(4\pi)^k} \sum_{n=1}^{\infty} \frac{a(n^2)}{n^s}\Bigg|_{s=k}.$$

As a second application, we get a proof different from the usual one of the fact that the Riemann zeta function has no zeros on the line $\Re(s) = 1$; this fact is one of the key steps in the classical proof of the prime number theorem. Indeed, suppose that $\zeta(1+i\alpha) = 0$ for some real number α (necessarily different from 0), and let $F(\tau)$ be the function $G(\tau, \frac{1}{2}(1+i\alpha))$. Since both $\zeta(2s)$ and $\zeta(2s-1)$ vanish at $s = \frac{1}{2}(1+i\alpha)$ (use the functional equation of ζ!), the formula for the Fourier expansion of $G(\tau,s)$ proved in the last section shows that $F(\tau)$ is exponentially small as $v \to \infty$ and has a constant term $c_0(v)$ which vanishes identically. Therefore the Rankin-Selberg transform $R(F;s)$ is zero for $\Re(s)$ large, and then by analytic continuation for all s. But we saw above that $R(F;s)$ is the integral of $F(\tau)$ against $G(\tau,s)$, so taking $s = \frac{1}{2}(1-i\alpha)$,

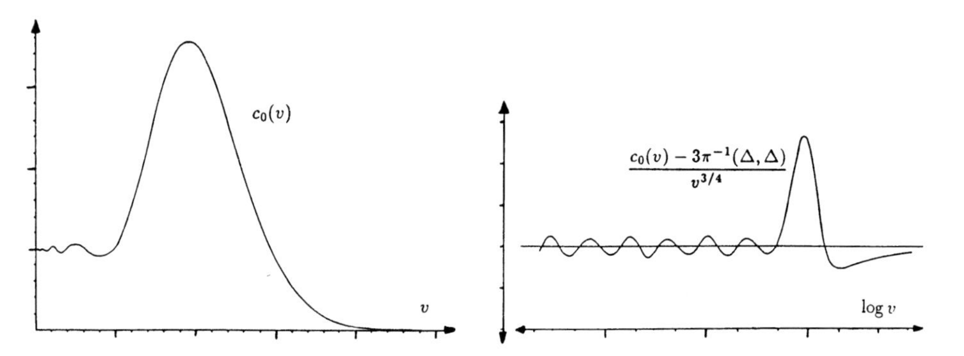

Fig. 1. The constant term $c_0(v) = v^{12} \sum_{n=1}^{\infty} \tau(n)^2 e^{-4\pi n v}$.

$G(\tau, s) = \overline{F(\tau)}$, we find that the integral of $|F(\tau)|^2$ over $\mathfrak{H}/\Gamma_1$ is zero. This is impossible since $F(\tau)$ is clearly not identically zero.

Finally, we can re-interpret the statement of the Rankin-Selberg identity in more picturesque ways. Suppose that we knew that the constant term $c_0(v)$ of F had an asymptotic expansion $c_0(v) = C_0 v^{\lambda_0} + C_1 v^{\lambda_1} + C_2 v^{\lambda_2} + \cdots$ as v tends to 0. Then breaking up the integral in the definition of $R(F; s)$ into the part from 0 to 1 and the part from 1 to infinity, and observing that the second integral is convergent for all s, we would discover that $R(F; s)$ has simple poles of residue C_j at $s = 1 - \lambda_j$ for each j and no other poles. Similarly, a term $C v^{\lambda} (\log v)^{m-1}$ would correspond to an mth order pole of $R(F; s)$ at $1 - \lambda$. But the theorem tells us that $R(F; s)$ has a simple pole of residue κ at $s = 1$ and otherwise poles only at the values $s = \frac{1}{2}\rho$, where ρ is a non-trivial zero of the Riemann zeta function. It is thus reasonable to think, and presumably under suitable hypotheses possible to prove, that $c_0(v)$ has an asymptotic expansion as $v \to 0$ consisting of one constant term κ and a sum of terms $C_\rho v^{1-\rho/2}$ for the various zeros of $\zeta(s)$. Assuming the Riemann hypothesis, these latter terms are of the form $v^{3/4}$ times an oscillatory function $A \cos(\frac{1}{2}\Im(\rho) \log v + \phi)$ for some amplitude A and phase ϕ. Figure 1 illustrates this behavior for the constant term $v^{12} \sum \tau(n)^2 e^{-4\pi n v}$ of $v^{12} |\Delta(\tau)|^2$; the predicted oscillatory behavior is clearly visible, and a rough measurement of the period of the primary oscillation leads to a rather accurate estimate of the imaginary part of the smallest non-trivial zero of $\zeta(s)$. In a related vein, we see that the difference between $c_0(v)$ and the average value κ of F for small v should be estimated by $\mathrm{O}(v^{\frac{1}{4}+\epsilon})$ if the Riemann hypothesis is true and by $\mathrm{O}(v^{\frac{1}{2}+\epsilon})$ unconditionally. Since $c_0(v)$ is simply the average value of $F(\tau)$ along the unique closed horocycle of length v^{-1} in the Riemannian manifold $\mathfrak{H}/\Gamma_1$, and since F is an essentially arbitrary function on this manifold, we can interpret this as a statement about the uniformity with which the closed horocycles on $\mathfrak{H}/\Gamma_1$ fill it up as their length tends to infinity.

C The Rankin-Selberg method (holomorphic case)

The calculations here are very similar to those of Section **B**, so we can be fairly brief. Let $f(\tau) = \sum_{n=1}^{\infty} a(n)q^n$ be a cusp form of weight k on Γ and $g(\tau) = \sum_{n=0}^{\infty} b(n)q^n$ a modular form of some smaller weight l. We assume for the moment that $k - l > 2$, so that there is a holomorphic Eisenstein series G_{k-l} of weight $k - l$. Our object is to calculate the scalar product of $f(\tau)$ with the product $G_{k-l}(\tau)g(\tau)$.

Ignoring convergence problems for the moment, we find (with $h = k - l$)

$$G_h(\tau) = \frac{(h-1)!}{(2\pi i)^h} \frac{1}{2} \sum_{m,n}{}' \frac{1}{(m\tau+n)^h} = \frac{(h-1)!}{(2\pi i)^h} \zeta(h) \sum_{\left(\begin{smallmatrix} \cdot & \cdot \\ c & d \end{smallmatrix}\right) \in \Gamma_\infty \backslash \Gamma_1} \frac{1}{(c\tau+d)^h},$$

whence

$$\begin{aligned} &\frac{(2\pi i)^h \, v^k}{(h-1)!\zeta(h)} f(\tau) \overline{G_h(\tau)\, g(\tau)} \\ &\qquad = \sum_{\left(\begin{smallmatrix} \cdot & \cdot \\ c & d \end{smallmatrix}\right) \in \Gamma_\infty \backslash \Gamma_1} \frac{v^k}{|c\tau+d|^{2k}} (c\tau+d)^k f(\tau) \overline{(c\tau+d)^l g(\tau)} \\ &\qquad = \sum_{\gamma \in \Gamma_\infty \backslash \Gamma_1} \Im(\gamma\tau)^k f(\gamma\tau) \overline{g(\gamma\tau)}, \end{aligned}$$

and consequently

$$\begin{aligned} \frac{(2\pi i)^h}{(h-1)!\zeta(h)} (f, G_h \cdot g) &= \iint_{\mathcal{F}} \sum_{\gamma \in \Gamma_\infty \backslash \Gamma_1} \Im(\gamma\tau)^k \, f(\gamma\tau)\, \overline{g(\gamma\tau)}\, d\mu \\ &= \sum_{\gamma \in \Gamma_\infty \backslash \Gamma_1} \iint_{\gamma\mathcal{F}} \Im(\tau)^k \, f(\tau)\, \overline{g(\tau)}\, d\mu \\ &= \int_0^\infty \left(\int_0^1 f(u+iv)\, \overline{g(u+iv)}\, du \right) v^{k-2}\, dv \\ &= \int_0^\infty \left(\sum_{n=1}^\infty a(n)\, \overline{b(n)}\, e^{-4\pi n v} \right) v^{k-2}\, dv \\ &= \frac{(k-2)!}{(4\pi)^{k-1}} \sum_{n=1}^\infty \frac{a(n)\,\overline{b(n)}}{n^{k-1}}. \end{aligned} \tag{1}$$

In other words, the scalar product of f and $G_h \cdot g$ is up to a simple factor equal to the value at $s = k - 1$ of the convolution of the L-series of f and $\overline{g}$.

The various steps in the calculation will be justified if $\iint_{\Gamma_\infty\backslash\mathfrak{H}} |f(\tau)g(\tau)|v^k\,d\mu$ converges. Since $f(\tau)=\mathrm{O}(v^{-k/2})$ and $g(\tau)=\mathrm{O}(v^{-l})$, this will certainly be the case if $k>2l+2$.

We can generalize the computation just done by replacing the product $G_h\cdot g$ by the function $F_\nu(G_h,g)$ defined in Section **E** of Part 1, where now $h+l+2\nu=k$. Here we find

$$\begin{aligned}\frac{(2\pi i)^h}{(h-1)!\zeta(h)}F_\nu(G_h,g) &= F_\nu\Big(\sum_{\gamma\in\Gamma_\infty\backslash\Gamma_1}\frac{1}{(c\tau+d)^h},\,g(\tau)\Big)\\ &= \Big(\frac{i}{2\pi}\Big)^\nu\sum_\gamma\sum_{\mu=0}^{\nu}\frac{(h+\nu-1)!(l+\nu-1)!}{\mu!(\nu-\mu)!(h-1)!(l+\mu-1)!}\,\frac{c^{\nu-\mu}g^{(\mu)}(\tau)}{(c\tau+d)^{h+\nu-\mu}}\\ &= \Big(\frac{i}{2\pi}\Big)^\nu\binom{h+\nu-1}{\nu}\sum_\gamma\frac{g^{(\nu)}(\gamma\tau)}{(c\tau+d)^k},\end{aligned}$$

where in the last line we have used formula (9) of Part 1, **E**. The same argument as before now leads to

$$\begin{aligned}\frac{(2\pi i)^h}{(h-1)!\zeta(h)}(f,F_\nu(G_h,g)) &= \frac{\binom{h+\nu-1}{\nu}}{(-2\pi i)^\nu}\int_0^\infty\int_0^1 f(\tau)\,\overline{g^{(\nu)}(\tau)}\,du\,v^{k-2}\,dv\\ &= \binom{h+\nu-1}{\nu}\frac{(k-2)!}{(4\pi)^{k-1}}\sum_{n=1}^\infty\frac{a(n)\,\overline{b(n)}}{n^{k-\nu-1}},\end{aligned}\tag{2}$$

the steps being justified this time if $k>2l+2\nu+2$ or $h>l+2$. Again the result is that the Petersson scalar product in question is proportional to a special value of the convolution of the L-series of f and g.

D Application: The Eichler-Selberg trace formula

Fix an even weight $k>0$ and let

$$t(n)=t_k(n)=\mathrm{Tr}(T(n),M_k),\qquad t^0(n)=t_k^0(n)=\mathrm{Tr}(T(n),S_k)$$

denote the traces of the nth Hecke operator $T(n)$ on the spaces of modular forms and cusp forms, respectively, of weight k. If we choose as a basis for M_k or S_k a $\mathbb{Z}$-basis of the lattice of forms having integral Fourier coefficients (which we know we can do by the results of Part 1), then the matrix representing the action of $T(n)$ with respect to this basis also has integral coefficients. Hence $t(n)$ and $t^0(n)$ are integers. The splitting $M_k=S_k\bigoplus\langle G_k\rangle$ and the formula $T_n(G_k)=\sigma_{k-1}(n)G_k$ for $k>2$ imply

$$t_k(n)=t_k^0(n)+\sigma_{k-1}(n)\qquad(n\ge 1,\ k>2).\tag{1}$$

Theorem. (Eichler, Selberg) *Let $H(N)$ $(N \geq 0)$ be the Kronecker-Hurwitz class numbers defined in* **D** *of Part 1 and denote by $p_k(t,n)$ the homogeneous polynomial*

$$p_k(t,n) = \sum_{0\leq r\leq \frac{k}{2}-1} \binom{k-2-r}{r}(-n)^r t^{k-2-2r} = \mathit{Coeff}_{X^{k-2}}\Big(\frac{1}{1-tX+nX^2}\Big)$$

of degree $\frac{k}{2}-1$ in t^2 and n (thus $p_2(t,n)=1$, $p_4(t,n)=t^2-n$, $p_6(t,n)=t^4-3t^2n+n^2$, etc.). Then

$$t_k(n) = -\frac{1}{2}\sum_{t\in\mathbb{Z},\, t^2\leq 4n} p_k(t,n)H(4n-t^2) + \frac{1}{2}\sum_{d|n}\max\{d,n/d\}^{k-1} \qquad (k\geq 2),$$

$$t_k^0(n) = -\frac{1}{2}\sum_{t\in\mathbb{Z},\, t^2\leq 4n} p_k(t,n)H(4n-t^2) - \frac{1}{2}\sum_{d|n}\min\{d,n/d\}^{k-1} \qquad (k\geq 4).$$

There is an analogous trace formula for forms of higher level (say, for the trace of $T(n)$ on $M_k(\Gamma_0(N))$ for n and N coprime), but the statement is more complicated and we omit it.

The equivalence of the two formulas (for $k>2$) follows from (1), since

$$\frac{1}{2}\sum_{d|n}\{\min\{d,n/d\}^{k-1} + \max\{d,n/d\}^{k-1}\} = \frac{1}{2}\sum_{d|n}\{d^{k-1} + (n/d)^{k-1}\} = \sigma_{k-1}(n).$$

Note also that $t_2(n)=0$ and $t_k^0(n)=0$ for $k\in\{2,4,6,8,10,14\}$ and all n, since the spaces M_2 and S_k are 0-dimensional in these cases. Equating to zero the expressions for $t_2(n)$ and $t_4^0(n)$ given in the theorem gives two formulas of the form

$$(2)\quad H(4n)+2H(4n-1)+\ldots=0, \qquad -nH(4n)-2(n-1)H(4n-1)+\ldots=0,$$

where the terms '$\ldots$' involve only $H(4m)$ and $H(4m-1)$ with $m<n$. Together, these formulas give a rapid inductive method of computing all the Kronecker-Hurwitz class numbers $H(N)$.

The importance of knowing $t^0(n)$ is as follows. Let $\mathbf{t}^0(\tau) = \mathbf{t}_k^0(\tau) = \sum_{n=1}^{\infty} t_k^0(n)q^n$. Then $\mathbf{t}^0$ is itself a cusp form of weight k on Γ_1 and its images under all Hecke operators (indeed, under $T(n_1),\ldots,T(n_d)$ for any $\{n_j\}_{j=1}^{d=\dim S_k}$ for which the n_1st, $\ldots$, n_dth Fourier coefficients of forms in S_k are linearly independent) generate the space S_k. To see this, let $f_i(\tau)=\sum_{n>0} a_i(n)q^n$ $(1\leq i\leq d)$ be the Hecke forms in S_k. We know that they form a basis and that the action of $T(n)$ on this basis is given by the diagonal matrix $\mathrm{diag}(a_1(n),\ldots,a_d(n))$. Hence the trace $t^0(n)$ equals $a_1(n)+\ldots+a_d(n)$ and $\mathbf{t}_k^0$ is just $f_1+\ldots+f_d$, which is indeed in S_k; the linear independence of the f_i

and the fact that the matrix $(a_i(n_j))_{1\le i,j\le d}$ is invertible then imply that the d forms $T(n_j)(\mathbf{t}^0) = \sum_{i=1}^d a_i(n_j) f_i$ are linearly independent and hence span S_k as claimed. The formula for $\mathrm{Tr}(T(n))$ thus gives an algorithm for obtaining all cusp forms of a given weight (and level).

We now sketch a proof of the Eichler-Selberg trace formula. The basic tool we will use is the 'holomorphic version' of the Rankin-Selberg method proved in the last section, but applied in the case when the Eisenstein series G_h and the modular form g have half-integral weight. The basic identities (1) and (2) of Section **C** remain true in this context with slight modifications due to the fact that the functions G_h and g are modular forms on $\Gamma_0(4)$ rather than $PSL_2(\mathbb{Z})$. They can be simplified by using the operator U_4 introduced in Section **E** of Part 1 and replacing $F_\nu(G_h, g)$ by $U_4(F_\nu(G_h, g))$, which belongs to $M_k(\Gamma_1)$ if $g \in M_l$, $h, l \in \mathbb{Z} + \frac{1}{2}$, $k = h + l + 2\nu \in 2\mathbb{Z}$ (cf. comments at the end of Part 1, **E**). In this situation, formula (2) of Section **C** still holds except for the values of the constant factors occurring. In particular, if $h = r + \frac{1}{2}$ with r odd and we take for g the basic theta-series $\theta(\tau) = 1 + 2\sum_{n=1}^\infty q^{n^2}$ of weight $\frac{1}{2}$ on $\Gamma_0(4)$, then we find

(3)

$$(f, U_4(F_\nu(G_{r+\frac{1}{2}}, \theta))) = c_{\nu,r} \sum_{n=1}^\infty \frac{a(n^2)}{(n^2)^{k-\nu-1}} \qquad (r > 1 \text{ odd}, \ \nu \ge 0, \ k = r + 2\nu + 1),$$

where $c_{\nu,r}$ is an explicitly known constant depending only on r and ν. We want to apply this formula in the case $r = 1$. Here the function $G_{3/2}$ is not a modular form and must be replaced by the function $G^*_{3/2}$ which was defined in Part 1**D**. The function $U_4(F_\nu(G^*_{3/2}, \theta))$ is no longer holomorphic, but we can apply the 'holomorphic projection operator' (cf. Appendix **C**) to replace it by a holomorphic modular form without changing its Petersson scalar product with the holomorphic cusp form f. Moreover, for $r = 1$ we have $\nu = \frac{1}{2}(k - r - 1) = \frac{k}{2} - 1$ and hence $2(k - \nu - 1) = k$, so the right-hand side of (3) is proportional to $\sum \frac{a(n^2)}{n^s}\big|_{s=k}$ and hence, by formula (5) of Section **B**, to (f, f) if f is a normalized Hecke eigenform. Thus finally

$$(f, \pi_{\mathrm{hol}}(U_4(F_\nu(G^*_{3/2}, \theta)))) = c_k\, (f, f)$$

for all Hecke forms $f \in S_k$, where c_k depends only on k (in fact, $c_k = -2^{k-1}\binom{\nu - \frac{1}{2}}{\nu}$). But since $\mathbf{t}^0(\tau)$ is the sum of all such eigenforms, and since distinct eigenforms are orthogonal, we also have $(f, \mathbf{t}^0) = (f, f)$ for all Hecke forms. It follows that

$$\pi_{\mathrm{hol}}(U_4(F_\nu(G^*_{3/2}, \theta))) = c_k\, \mathbf{t}(\tau) + c'_k G_k(\tau) \tag{4}$$

for some constant c'_k.

It remains only to compute the Fourier expansion of the function on the left of (4). We have

$$\theta(\tau) = \sum_{t\in\mathbb{Z}} q^{t^2}, \qquad G_{\frac{3}{2}}(\tau) = \sum_{m=0}^{\infty} H(m)q^m$$

and hence

$$\begin{aligned} F_\nu(\theta(\tau), G_{\frac{3}{2}}(\tau)) &= (2\pi i)^{-\nu} \sum_{\mu=0}^{\nu} (-1)^\mu \binom{\nu-\frac{1}{2}}{\mu}\binom{\nu+\frac{1}{2}}{\nu-\mu} \theta^{(\nu-\mu)}(\tau) G_{\frac{3}{2}}^{(\mu)}(\tau) \\ &= \sum_{\substack{m,t\in\mathbb{Z} \\ m\geq 0}} \sum_{\mu=0}^{\nu} (-1)^\mu \binom{\nu-\frac{1}{2}}{\mu}\binom{\nu+\frac{1}{2}}{\nu-\mu} t^{2(\nu-\mu)} m^\mu H(m) q^{m+t^2}, \end{aligned}$$

so

$$\begin{aligned} &U_4(F_\nu(\theta(\tau), G_{\frac{3}{2}}(\tau))) \\ &= \sum_{n=0}^{\infty} \sum_{t^2\leq 4n} \sum_{\mu=0}^{\nu} (-1)^\mu \binom{\nu-\frac{1}{2}}{\mu}\binom{\nu+\frac{1}{2}}{\nu-\mu} t^{2\nu-2\mu}(4n-t^2)^\mu H(4n-t^2) q^n \\ &= -\frac{1}{2} c_k \sum_{n=0}^{\infty} \sum_{t^2\leq 4n} p_k(t,n) H(4n-t^2) q^n \end{aligned}$$

(recall that $k = 2\nu + 2$). On the other hand, the difference of $G^*_{3/2}$ and $G_{3/2}$ is a linear combination of terms q^{-f^2} with coefficients which are analytic functions of $v = \Im(\tau)$. Hence the coefficient of q^n in $U_4(F_\nu(\theta, G^*_{3/2} - G_{3/2}))$ is a sum over all pairs $(t,f) \in \mathbb{Z}^2$ with $t^2 - f^2 = 4n$ of a certain analytic function of v. Applying π_{hol} means that this expression must be multiplied by $v^{k-2}e^{-4\pi n v}$ and integrated from $v = 0$ to $v = \infty$. The integral turns out to be elementary and one finds after a little calculation

$$\begin{aligned} &\text{coefficient if } q^n \text{ in } \pi_{\text{hol}}(U_4(F_\nu(\theta, G^*_{3/2} - G_{3/2}))) \\ &= \frac{1}{4} c_k \sum_{\substack{t,f\in\mathbb{Z} \\ t^2-f^2=4n}} \left(\frac{|t|+|f|}{2}\right)^{k-1} = \frac{1}{2} c_k \sum_{\substack{d|n \\ d>0}} \max(d, \frac{n}{d})^{k-1}. \end{aligned}$$

Adding this to the preceding formula, and comparing with (4), we find that the constant c'_k in (4) must be 0 and that we have obtained the result stated in the theorem.

4. Jacobi forms

When we introduced modular forms, we started with functions F of lattices $\Lambda \subset \mathbb{C}$ invariant under rescaling $\Lambda \mapsto \lambda\Lambda$ ($\lambda \in \mathbb{C}^*$); these corresponded via $f(\tau) = F(\mathbb{Z}\tau + \mathbb{Z})$ to modular functions. The quotient $\mathbb{C}/\Lambda$ is an elliptic curve, so we can think of F (or f) as functions of elliptic curves. It is natural to make them functions *on* elliptic curves as well, i.e., to consider functions Φ which depend both on Λ and on a variable $z \in \mathbb{C}/\Lambda$. The equations

$$\Phi(\lambda\Lambda, \lambda z) = \Phi(\Lambda, z), \quad \Phi(\Lambda, z+\omega) = \Phi(\Lambda, z) \qquad (\lambda \in \mathbb{C}^\times,\ \omega \in \Lambda)$$

correspond via $\phi(\tau, z) = \Phi(\mathbb{Z}\tau + \mathbb{Z}, z)$ to functions ϕ on $\mathfrak{H} \times \mathbb{C}$ satisfying

$$(1) \qquad \begin{aligned} &\phi\Big(\frac{a\tau+b}{c\tau+d}, \frac{z}{c\tau+d}\Big) = \phi(\tau, z), \\ &\phi(\tau, z+\ell\tau+m) = \phi(\tau, z) \qquad \Big(\begin{pmatrix} a & b \\ c & d \end{pmatrix} \in \Gamma_1,\ \ell, m \in \mathbb{Z}\Big). \end{aligned}$$

We call a meromorphic function ϕ on $\mathfrak{H} \times \mathbb{C}$ satisfying (1) a **Jacobi function.**

However, there can clearly never be a holomorphic Jacobi function, since by Liouville's theorem a holomorphic function on $\mathbb{C}$ invariant under all transformations $z \mapsto z + \omega$ ($\omega \in \Lambda$) must be constant. Thus, just as the concept of modular function was too restrictive and had to be extended to the concept of modular forms of weight k, corresponding to functions on lattices transforming under $\Lambda \mapsto \lambda\Lambda$ with a scaling factor λ^{-k}, the concept of Jacobi functions must be extended by incorporating appropriate scaling factors into the definition. The right requirements, motivated by examples which will be presented in Section **A**, turn out to be

$$(2) \qquad \phi\Big(\frac{a\tau+b}{c\tau+d}, \frac{z}{c\tau+d}\Big) = (c\tau+d)^k\, e^{\frac{2\pi i N c z^2}{c\tau+d}}\, \phi(\tau, z) \qquad \Big(\begin{pmatrix} a & b \\ c & d \end{pmatrix} \in \Gamma_1\Big)$$

and

$$(3) \qquad \phi(\tau, z+\ell\tau+m) = e^{-2\pi i N(\ell^2\tau + 2\ell z)}\phi(\tau, z) \qquad (\ell, m \in \mathbb{Z}),$$

where N is a certain integer. Finally, just as with modular forms, there must be a growth condition at infinity; it turns out that the right condition here is to require that ϕ have a Fourier expansion of the form

$$(4) \qquad \phi(\tau, z) = \sum_{n=0}^{\infty} \sum_{\substack{r\in\mathbb{Z} \\ r^2 \le 4Nn}} c(n, r)\, q^n\, \zeta^r \qquad \big(q = e^{2\pi i\tau},\ \zeta = e^{2\pi i z}\big)$$

(again, the rather odd-looking condition $r^2 \le 4Nn$ will be motivated by the examples). A function $\phi : \mathfrak{H} \to \mathbb{C}$ satisfying the conditions (2), (3) and (4) will be called a **Jacobi form** of **weight** k and **index** N.

Surprisingly, in most of the occurrences of modular forms and functions in physics—in particular, those connected with theta functions and with Kac-Moody algebras—it is actually Jacobi forms and functions which are involved. It is for this reason, and because the theory is not widely known, that we have devoted an entire part to these functions.

A Examples of Jacobi forms

The simplest theta series, namely the function

$$\theta(\tau)=\sum_{n\in\mathbb{Z}} q^{n^2}=1+2q+2q^4+2q^9+\cdots$$

introduced at the beginning of Section **1D**, is actually just the specialization to $z=0$ ('Thetanullwert') of the two-variable function

$$\theta(\tau,\,z)=\sum_{n\in\mathbb{Z}} q^{n^2}\zeta^{2n}=1+(\zeta^2+\zeta^{-2})\,q+(\zeta^4+\zeta^{-4})\,q^4+(\zeta^6+\zeta^{-6})\,q^9+\cdots,$$

and similarly the transformation equation

$$\theta\Big(\frac{a\tau+b}{c\tau+d}\Big)=\epsilon_{c,d}\,(c\tau+d)^{\frac12}\,\theta(\tau)\qquad\Big(\big(\begin{smallmatrix}a&b\\c&d\end{smallmatrix}\big)\in\Gamma_0(4),\ \epsilon_{c,d}^4=1\Big)$$

is just the specialization to $z=0$ of the more general transformation equation

$$\theta\Big(\frac{a\tau+b}{c\tau+d},\,\frac{z}{c\tau+d}\Big)=\epsilon_{c,d}\,(c\tau+d)^{\frac12}\,e^{\frac{2\pi icz^2}{c\tau+d}}\,\theta(\tau,\,z)\qquad\Big(\big(\begin{smallmatrix}a&b\\c&d\end{smallmatrix}\big)\in\Gamma_0(4)\Big).$$

It is also easily checked that θ satisfies

$$\theta(\tau,\,z+\ell\tau+m)=e^{-2\pi i(\ell^2\tau+2\ell z)}\,\theta(\tau,z)$$

(just replace n by $n+\ell$ in the summation defining θ), so that $\theta(\tau,z)$ is, with the obvious modifications in the definition given before, a Jacobi form of weight 1/2 and index 1 on the group $\Gamma_0(4)$. The function $\theta(\tau,z)$ is one of the classical Jacobi theta functions and this is the reason for the name 'Jacobi form.'

Just as for the one-variable theta functions discussed in Part 1, if we want to get forms of integral weight and on the full modular group, rather than of weight 1/2 and on $\Gamma_0(4)$, we must start with quadratic forms in an even number of variables and whose associated matrix has determinant 1. If $Q:\mathbb{Z}^{2k}\to\mathbb{Z}$ is a positive definite quadratic form in $2k$ variables given by an even symmetric unimodular matrix A (i.e. $Q(x)=\frac12 x^tAx$, $a_{ij}\in\mathbb{Z}$, $\frac12 a_{ii}\in\mathbb{Z}$, $\det A=1$), then for each vector $y\in\mathbb{Z}^{2k}$ the theta-function

$$\Theta_{Q,y}(\tau,z)=\sum_{x\in\mathbb{Z}^{2k}} q^{Q(x)}\,\zeta^{B(x,y)}\,,\tag{5}$$

where $\zeta=e^{2\pi iz}$ as before and $B(x,y)=x^tAy$ is the bilinear form associated to Q, is a Jacobi form of weight k and index $N=Q(y)$. The transformation law (2) is proved using the Poisson summation formula as for the special case $\Theta_Q(\tau) = \Theta_{Q,0}(\tau,0)$ studied in Part 1; the transformation law (3) is proved directly from the expansion (5) by making the substitution $x \mapsto x + \ell y$; and the form of the Fourier expansion required in (4) is clear from (5) and the Cauchy-Schwarz inequality $B(x,y)^2 \le 4Q(x)Q(y)$. (This motivates the inequality $r^2 \le 4Nn$ in (4), as promised.)

The next example is that of Eisenstein series. The Eisenstein series of Part 1 can be written as

$$G_k(\tau) = \frac{1}{2}\zeta(1-k) \sum_{\gamma\in\Gamma_\infty\backslash\Gamma_1} 1|_k\gamma(\tau),$$

where $|_k$ is the slash operator introduced in **1E** and the summation is over the cosets of $\Gamma_\infty = \{\pm\left(\begin{smallmatrix}1 & n\\ 0 & 1\end{smallmatrix}\right) \mid n \in \mathbb{Z}\}$ in $\Gamma_1 \subset PSL_2(\mathbb{Z})$ (cf. Part 3). In the Jacobi form context we must generalize the slash operator to a new operator $|_{k,N}$ defined by

$$(\phi|_{k,N}\gamma)(\tau,z) = (c\tau+d)^{-k}\, e^{-\frac{2\pi i N c z^2}{c\tau+d}}\, \phi\Big(\frac{a\tau+b}{c\tau+d}, \frac{z}{c\tau+d}\Big) \qquad \Big(\gamma = \left(\begin{smallmatrix}a & b\\ c & d\end{smallmatrix}\right) \in \Gamma_1\Big),$$

$$(\phi|_{k,N}[\ell,m])(\tau,z) = e^{2\pi i N(\ell^2\tau+2\ell z)}\, \phi(\tau, z+\ell\tau+m) \qquad (\ell,m \in \mathbb{Z})$$

(so that $\phi|_{k,N}\gamma = \phi|_{k,N}[\ell,m] = \phi$ if ϕ is a Jacobi form of weight k and index N). We then define an Eisenstein series

$$G_{k,N}(\tau,z) = \zeta(3-2k) \sum_{\gamma\in\Gamma_\infty\backslash\Gamma_1} \sum_{\ell\in\mathbb{Z}} \big((1|_{k,N}\gamma)|_{k,N}[\ell,0]\big)(\tau,z)$$

or more explicitly

$$G_{k,N}(\tau,z) = \frac{1}{2}\zeta(3-2k) \sum_{\substack{c,d,\ell\in\mathbb{Z}\\ (c,d)=1}} \frac{e^{2\pi i N(\ell^2\gamma_{c,d}(\tau) + \frac{2lz}{c\tau+d} - \frac{cz^2}{c\tau+d})}}{(c\tau+d)^k},$$

where $\gamma_{c,d}$ for each pair of coprime integers c, d denotes an element of $PSL_2(\mathbb{Z})$ with lower row $(c\ d)$. The series is convergent for $k > 2$ and defines a Jacobi form of weight k and index N. Moreover, its Fourier expansion can be computed by a calculation analogous to, though somewhat harder than, the one given in 1**A**. The result is that the Fourier coefficients are rational numbers of arithmetic interest, expressible in closed form in terms of the function $H(r,n)$ introduced in Section **D** of Part 1 in connection with Eisenstein series of half-integral weight. In particular, for $N = 1$ the result is simply

$$G_{k,1}(\tau,z) = \sum_{n=0}^{\infty} \sum_{|r|\le\sqrt{4n}} H(k-1, 4n-r^2)\, e^{2\pi i(n\tau+rz)}.$$

That the coefficient of $q^n\zeta^r$ depends only on $4n - r^2$ is not an accident: it is easily seen that the transformation equation (3) in the case $N = 1$ is equivalent to the condition that the Fourier coefficient $c(n,r)$ as defined in (4) depend only on $4n-r^2$, while for general N (3) is equivalent to the requirement that $c(n,r)$ depend only on $4n-r^2$ and on the residue of r modulo $2N$. The fact that the coefficients of the Jacobi Eisenstein series were essentially the same as the coefficients of Eisenstein series in one variable but of half-integral weight is also not accidental: There is in fact an intimate connection between Jacobi forms and modular forms of half-integral weight, obtained by associating to the Jacobi form ϕ the collection of functions $\sum_d C_\mu(d)\,q^d$, $\mu=1,2,\ldots,2N$, where $C_\mu(d)$ is the common value of the $c(n,r)$ with $4n-r^2=d$ and $r\equiv\mu \pmod{2N}$; each of these $2N$ functions is a modular form of weight $k-\frac{1}{2}$ with respect to some subgroup of Γ_1, and the entire $2N$-tuple satisfies a transformation law with respect to the whole group Γ_1. However, we do not elaborate on this here.

The beginnings of the Fourier expansions of the first few Jacobi Eisenstein series (of index 1) are

$$\begin{aligned}
G_{4,1}(\tau,z) = -\tfrac{1}{252} &+ \left(-\tfrac{1}{252}\zeta^2 - \tfrac{2}{9}\zeta - \tfrac{1}{2} - \tfrac{2}{9}\zeta^{-1} - \tfrac{1}{252}\zeta^{-2}\right) q \\
&+ \left(-\tfrac{1}{2}\zeta^2 - \tfrac{16}{7}\zeta - 3 - \tfrac{16}{7}\zeta^{-1} - \tfrac{1}{2}\zeta^{-2}\right) q^2 \\
&+ \left(-\tfrac{2}{9}\zeta^3 - 3\zeta^2 - 6\zeta - \tfrac{74}{9} - 6\zeta^{-1} - 3\zeta^{-2} - \tfrac{2}{9}\zeta^{-3}\right) q^3 + \cdots, \\
G_{6,1}(\tau,z) = -\tfrac{1}{132} &+ \left(-\tfrac{1}{132}\zeta^2 + \tfrac{2}{3}\zeta + \tfrac{5}{2} + \tfrac{2}{3}\zeta^{-1} - \tfrac{1}{132}\zeta^{-2}\right) q \\
&+ \left(\tfrac{5}{2}\zeta^2 + 32\zeta + 57 + 32\zeta^{-1} + \tfrac{5}{2}\zeta^{-2}\right) q^2 + \cdots, \\
G_{8,1}(\tau,z) = -\tfrac{1}{12} &+ \left(-\tfrac{1}{12}\zeta^2 - \tfrac{14}{3}\zeta - \tfrac{61}{2} - \tfrac{14}{3}\zeta^{-1} - \tfrac{1}{12}\zeta^{-2}\right) q + \cdots.
\end{aligned}$$

To get more examples, we can combine these in various ways. In particular, the two functions

$$\begin{aligned}
\phi_{10,1}(\tau,z) &= 882\,G_6(\tau)\,G_{4,1}(\tau,z) + 220\,G_4(\tau)\,G_{6,1}(\tau,z), \\
\phi_{12,1}(\tau,z) &= -840\,G_8(\tau)\,G_{4,1}(\tau,z) - 462\,G_6(\tau)\,G_{6,1}(\tau,z)
\end{aligned} \tag{6}$$

are Jacobi forms of index 1 and weights 10 and 12, respectively, and in fact are Jacobi *cusp* forms (i.e. $n>0$, $r^2<4Nn$ in (4)) with Fourier expansions starting

$$\begin{aligned}
\phi_{10,1}(\tau,z) &= \left(\zeta - 2 + \zeta^{-1}\right) q + \left(-2\zeta^2 - 16\zeta + 36 - 16\zeta^{-1} - 2\zeta^{-2}\right) q^2 \cdots, \\
\phi_{12,1}(\tau,z) &= \left(\zeta + 10 + \zeta^{-1}\right) q + \left(10\zeta^2 - 88\zeta - 132 - 88\zeta^{-1} + 10\zeta^{-2}\right) q^2 \cdots;
\end{aligned}$$

their ratio $\phi_{12,1}/\phi_{10,1}$ is $-3\pi^{-2}$ times the **Weierstrass $\wp$-function** $\wp(z;\mathbb{Z}\tau+\mathbb{Z})$ from the theory of elliptic functions (cf. the lectures of Cohen and Bost/Cartier in this volume).

Other important examples of Jacobi forms are obtained from the Fourier developments of Siegel modular forms on the symplectic group $Sp(2,\mathbb{Z})$, but we cannot go into this here since we have not developed the theory of Siegel modular forms.

B Known results

In this section we describe a few highlights from the theory of Jacobi forms.

(i) If $\phi \not\equiv 0$ is a Jacobi form of weight k and index N, then it is easily seen by integrating $\frac{d}{dz}\log\phi$ around a fundamental parallelogram for $\mathbb{C}/(\mathbb{Z}\tau+\mathbb{Z})$ that ϕ has exactly $2N$ zeros in this parallelogram (here we are considering τ as fixed and ϕ as a function of z alone). In particular, ϕ cannot have a zero of multiplicity greater than $2N$ at the origin, so in the Taylor expansion

$$\phi(\tau,z) = \chi_0(\tau) + \chi_1(\tau)\,z + \chi_2(\tau)\,z^2 + \cdots$$

the first $2N+1$ coefficients determine ϕ completely. On the other hand, one easily sees by differentiating (2) repeatedly with respect to z and then setting z equal to 0 that χ_0 is a modular form in τ of weight k (this, of course, is obvious), χ_1 a modular form of weight $k+1$, $\chi_2 - \frac{2\pi i N}{k}\chi_0'$ a modular form of weight $k+2$, and more generally

$$\xi_\nu(\tau) = \sum_{0\le\mu\le\nu/2} \frac{(-2\pi i N)^\mu\,(k+\nu-\mu-2)!}{(k+\nu-2)!\,\mu!}\,\chi_{\nu-2\mu}^{(\mu)}(\tau) \tag{7}$$

a modular form of weight $k+\nu$ for every integer $\nu \ge 0$. The fact that ϕ is determined by its first $2N+1$ Taylor coefficients means that we have an injective map from the space $J_{k,N}$ of Jacobi forms of weight k and index N into the direct sum $M_k \oplus M_{k+2} \oplus \cdots \oplus M_{k+2N}$ if k is even or $M_{k+1} \oplus M_{k+3} \oplus \cdots M_{k+2N-1}$ if k is odd. In particular, $J_{k,N}$ is finite dimensional, of dimension at most $\frac{1}{12}kN + \mathrm{O}(N^2)$.

The function ξ_ν defined by (7) has the Fourier expansion

$$\xi_\nu(\tau) = (2\pi i)^\nu\,\frac{(k-2)!}{(k+\nu-2)!}\sum_{n=1}^{\infty}\Bigl(\sum_{|r|\le\sqrt{4Nn}} p_{k-1,\nu}(r,Nn)\,c(n,r)\Bigr)\,q^n\,,$$

where the $c(n,r)$ are the coefficients defined by (4) and $p_{d,\nu}(a,b)$ denotes the coefficient of X^ν in $(1-aX+bX^2)^{-d}$. The fact that ξ_ν is a modular form is related to the heat equation operator $8\pi i N\frac{\partial}{\partial\tau} - \frac{\partial^2}{\partial z^2}$, and also to the formula (11) of 1**E**.

(ii) The bigraded ring of all Jacobi forms (of all weights and indexes) is not finitely generated, since the forms obtained as polynomials in any finite collection would have a bounded ratio of k to N and there is an Eisenstein series $G_{k,1}$ for all $k>2$. However, if we enlarge the space $J_{k,N}$ to the space $\tilde{J}_{k,N}$ of 'weak Jacobi forms,' defined as functions $\phi:\mathcal{H}\times\mathbb{C}\to\mathbb{C}$ satisfying the properties (2)–(4) but with the condition '$r^2\le 4Nn$' dropped in (4), then the bigraded ring $\bigoplus_{k,N}\tilde{J}_{k,N}$ is simply the ring of all polynomials in the four functions $G_4(\tau)$, $G_6(\tau)$, $\phi_{10,1}(\tau,z)/\Delta(\tau)$ and $\phi_{12,1}(\tau,z)/\Delta(\tau)$ (with $\phi_{10,1}$ and

$\phi_{12,1}$ as in (6) and Δ as in **1B**) of weight 4, 6, -2 and 0 and index 0, 0, 1 and 1, respectively. In particular, $\Delta(\tau)^N\phi(\tau,z)$ is a polynomial in $G_4(\tau)$, $G_6(\tau)$, $G_{4,1}(\tau,z)$ and $G_{6,1}(\tau,z)$ for any Jacobi form ϕ of index N.

(iii) There are no Jacobi forms of weight 1 on $PSL_2(\mathbb{Z})$, i.e., $J_{1,N}=\{0\}$ for all N.

(iv) One can define Hecke operators on the spaces $J_{k,N}$ and compute their traces. These turn out to be related to the traces of Hecke operators on the spaces of ordinary modular forms of weight $2k-2$ and level N. Using these, one can construct lifting maps from $J_{k,N}$ to a certain subspace $\mathfrak{M}_{2k-2}(N)\subset M_{2k-2}(\Gamma_0(N))$ which is canonically defined and invariant under all Hecke operators. Moreover, $J_{k,N}$ turns out to be isomorphic to the subspace of forms in $\mathfrak{M}_{2k-2}(N)$ whose Hecke L-series satisfy a functional equation with a minus sign, i.e., to the intersection of $\mathfrak{M}_{2k-2}(N)$ with $M_{2k-2}^{(-1)^k}(\Gamma_0(N))$ (cf. **2D**).

(v) There is another kind of Jacobi form, called **skew-holomorphic Jacobi forms**, for which statements analogous to those in (iv) hold but with the isomorphism now between the space of skew-holomorphic Jacobi forms and the subspace of forms in $\mathfrak{M}_{2k-2}(N)$ having a *plus* sign in the functional equation of their L-series. By definition, a skew-holomorphic Jacobi form of weight k and index N is a function ϕ on $\mathcal{H}\times\mathbb{C}$ which satisfies the transformation equations (2) and (3) but with $(c\tau+d)^k$ replaced by $(c\bar\tau+d)^{k-1}|c\tau+d|$ in (2) and which has a Fourier expansion like the one in (4) but with the condition $r^2\le 4Nn$ replaced by $r^2\ge 4Nn$ and with $q^n\zeta^r$ multiplied by $e^{-\pi(r^2-4Nn)v/N}$ $(v=\Im(\tau))$. Such a function is again holomorphic in z, but the Cauchy-Riemann condition $\frac{\partial\phi}{\partial\bar\tau}=0$ of holomorphy in τ is replaced by the heat equation $\frac{\partial\phi}{\partial\tau}=\frac{1}{8\pi i N}\frac{\partial^2\phi}{\partial z^2}$. The Fourier expansion together with the transformation property (3) can be written uniformly in the holomorphic and non-holomorphic case as

$$\phi(\tau,z)=\sum_{\substack{r,\Delta\in\mathbb{Z}\\ r^2\equiv\Delta\ (\mathrm{mod}\ 4N)}} C(\Delta,r)\,e^{2\pi i\left(\frac{r^2-\Delta}{4N}\Re(\tau)+i\frac{r^2+|\Delta|}{4N}\Im(\tau)+rz\right)} \tag{8}$$

where $C(\Delta,r)$ depends only on Δ and on $r\pmod{4N}$ and vanishes for $\Delta>0$ (holomorphic case) or $\Delta<0$ (non-holomorphic case).

(vi) There are explicit constructions of Jacobi and skew-Jacobi forms in terms of binary quadratic forms, due to Skoruppa. For instance, if we define $C(\Delta,r)=\sum\operatorname{sgn}(a)$, where the (finite) sum is over all binary quadratic forms $[a,b,c]=ax^2+bxy+cy^2$ of discriminant $b^2-4ac=\Delta$ with $a\equiv 0\pmod N$, $b\equiv r\pmod{4N}$ and $ac<0$, then (8) defines a skew-holomorphic Jacobi form of weight 2 and index N.

Appendices

The following appendices describe some analytic tools useful in the theory of modular forms.

A The Poisson summation formula

This is the identity

$$\sum_{n\in\mathbb{Z}} \varphi(x+n) = \sum_{r\in\mathbb{Z}} \left(\int_{\mathbb{R}} \varphi(t)e^{-2\pi irt}\,dt\right) e^{2\pi irx}, \tag{1}$$

where $\varphi(x)$ is any continuous function on $\mathbb{R}$ which decreases rapidly (say, at least like $|x|^{-c}$ with $c>1$) as $x\to\infty$. The proof is simple: the growth condition on φ ensures that the sum on the left-hand side converges absolutely and defines a continuous function $\Phi(x)$. Clearly $\Phi(x+1)=\Phi(x)$, so Φ has a Fourier expansion $\sum_{r\in\mathbb{Z}} c_r e^{2\pi irx}$ with Fourier coefficients c_r given by $\int_0^1 \Phi(x)e^{-2\pi irx}\,dx$. Substituting into this formula the definition of Φ, we find

$$\begin{aligned} c_r &= \int_0^1 \left(\sum_{n=-\infty}^{\infty} \varphi(x+n)e^{-2\pi ir(x+n)}\right) dx \\ &= \sum_{n=-\infty}^{\infty} \int_n^{n+1} \varphi(x)e^{-2\pi irx}\,dx = \int_{-\infty}^{\infty} \varphi(x)e^{-2\pi irx}\,dx, \end{aligned}$$

as claimed. If we write $\hat{\varphi}(t)$ for the **Fourier transform** $\int_{-\infty}^{\infty} \varphi(x)e^{-2\pi itx}\,dx$ of φ, then (1) can be written in the form $\sum_n \varphi(x+n) = \sum_r \hat{\varphi}(r)e^{2\pi irx}$, where both summations are over $\mathbb{Z}$. The special case $x=0$ has the more symmetric form $\sum_n \varphi(n) = \sum_r \hat{\varphi}(r)$, which is actually no less general since replacing $\varphi(x)$ by $\varphi(x+a)$ replaces $\hat{\varphi}(t)$ by $\hat{\varphi}(t)e^{2\pi ita}$; it is in this form that the Poisson summation formula is often stated.

As a first application, we take $\varphi(x) = (x+iy)^{-k}$, where y is a positive number and k an integer ≥ 2. This gives the **Lipschitz formula**

$$\sum_{n\in\mathbb{Z}} \frac{1}{(z+n)^k} = \frac{(-2\pi i)^k}{(k-1)!} \sum_{r=1}^{\infty} r^{k-1} e^{2\pi irz} \qquad (z\in\mathfrak{H},\ k\in\mathbb{Z}_{\geq 2}),$$

which can also be proved by expanding the right hand side of Euler's identity

$$\sum_{n\in\mathbb{Z}} \frac{1}{z+n} = \frac{\pi}{\tan \pi z} = -\pi i - 2\pi i \frac{e^{2\pi iz}}{1-e^{2\pi iz}}$$

as a geometric series in $e^{2\pi iz}$ and differentiating $k-1$ times with respect to z.

As a second application, take $\varphi(x) = e^{-\pi a x^2}$ with $a > 0$. Then $\hat{\varphi}(t) = a^{-\frac{1}{2}} e^{-\pi t^2/a}$, so we get

$$\text{(2)} \qquad \sum_{n=-\infty}^{\infty} e^{-\pi a (x+n)^2} = \sqrt{\frac{1}{a}} \sum_{r=-\infty}^{\infty} e^{-\pi a^{-1} r^2 + 2\pi i r x} \qquad (x \in \mathbb{R})$$

(the formula is actually valid for all $x \in \mathbb{C}$, as one sees by replacing $\varphi(x)$ by $\varphi(x+iy)$ with $y \in \mathbb{R}$). This identity, and its generalizations to higher-dimensional sums of Gaussian functions, is the basis of the theory of theta functions.

Finally, if s is a complex number of real part greater than 1, then taking $\varphi(x) = |x+iy|^{-s}$ with $y > 0$ leads to the following non-holomorphic generalization of the Lipschitz formula:

$$\sum_{n \in \mathbb{Z}} \frac{1}{|z+n|^s} = y^{1-s} \sum_{r=-\infty}^{\infty} k_{s/2}(2\pi r y) e^{2\pi i r x} \qquad (z = x+iy \in \mathfrak{H},\ \Re(s) > 1),$$

where $k_s(t) = \int_{-\infty}^{\infty} e^{-itx}(x^2+1)^{-s}\,dx$. The function $k_s(t)$ can be expressed in terms of the gamma function $\Gamma(s)$ and the **K-Bessel function** $K_\nu(t) = \int_0^\infty e^{-t\cosh u}\cosh(\nu u)\,du$ ($\nu \in \mathbb{C}$, $t > 0$) by

$$k_s(t) = \begin{cases} \frac{2\pi^{\frac{1}{2}}}{\Gamma(s)} \left(\frac{|t|}{2}\right)^{s-\frac{1}{2}} K_{s-\frac{1}{2}}(|t|) & \text{if } t \neq 0, \\ \frac{\pi^{\frac{1}{2}}\Gamma(s-\frac{1}{2})}{\Gamma(s)} & \text{if } t = 0 \end{cases}$$

(cf. Appendix **B**), so, replacing s by $2s$, we can rewrite the result as

(3)

$$\sum_{n \in \mathbb{Z}} \frac{1}{|z+n|^{2s}} = \frac{\pi^{\frac{1}{2}}\Gamma(s-\frac{1}{2})}{\Gamma(s)} y^{1-2s} + \frac{2\pi^s}{\Gamma(s)} y^{\frac{1}{2}-s} \sum_{r \neq 0} |r|^{s-\frac{1}{2}} K_{s-\frac{1}{2}}(2\pi|r|y)\, e^{2\pi i r x}$$

$$(z = x+iy \in \mathfrak{H},\ \Re(s) > \frac{1}{2}).$$

This formula is used for computing the Fourier development of the non-holomorphic Eisenstein series (Part **3A**).

B The gamma function and the Mellin transform

The integral representation $n! = \int_0^\infty t^n e^{-t}\,dt$ is generalized by the definition of the **gamma function**

$$\text{(1)} \qquad \Gamma(s) = \int_0^\infty t^{s-1} e^{-t}\,dt \qquad (s \in \mathbb{C},\ \Re(s) > 0).$$

Thus $n! = \Gamma(n+1)$ for n a nonnegative integer. Integration by parts gives the functional equation $\Gamma(s+1) = s\Gamma(s)$, generalizing the formula

$$(n+1)! = (n+1)n!$$

and also permitting one to define the Γ-function consistently for all $s \in \mathbb{C}$ as a meromorphic function with polar part $\frac{(-1)^n}{n!}\frac{1}{s+n}$ at $s=-n$, $n \in \mathbb{Z}_{\geq 0}$.

The integral (1) is a special case of the **Mellin transform**. Suppose that $\phi(t)$ $(t>0)$ is any function which decays rapidly at infinity (i.e., $\phi(t) = \mathrm{O}(t^{-A})$ as $t \to \infty$ for every $A \in \mathbb{R}$) and blows up at most polynomially at the origin (i.e., $\phi(t) = \mathrm{O}(t^{-C})$ as $t \to 0$ for some $C \in \mathbb{R}$). Then the integral

$$\mathbf{M}\phi(s) = \int_0^\infty \phi(t)\, t^{s-1}\, dt$$

converges absolutely and locally uniformly in the half-plane $\Re(s) > C$ and hence defines a holomorphic function of s in that region. The most frequent situation occurring in number theory is that $\phi(t) = \sum_{n=1}^\infty c_n e^{-nt}$ for some complex numbers $\{c_n\}_{n\geq 1}$ which grow at most polynomially in n. Such a function automatically satisfies the growth conditions just specified, and using formula (1) (with t replaced by nt in the integral), we easily find that the Mellin transform $\mathbf{M}\phi(s)$ equals $\Gamma(s)D(s)$, where $D(s) = \sum_{n=1}^\infty c_n n^{-s}$ is the Dirichlet series associated to ϕ. Thus the Mellin transformation allows one to pass between Dirichlet series, which are of number-theoretical interest, and exponential series, which are analytically much easier to handle.

Another useful principle is the following. Suppose that our function $\phi(t)$, still supposed to be small as $t \to \infty$, satisfies the functional equation

$$\phi(\frac{1}{t}) = \sum_{j=1}^J A_j t^{\lambda_j} + t^h \phi(t) \qquad (t>0), \tag{2}$$

where h, A_j and λ_j are complex numbers. Then, breaking up the integral defining $\mathbf{M}\phi(s)$ as $\int_0^1 + \int_1^\infty$ and replacing t by t^{-1} in the first term, we find for $\Re(s)$ sufficiently large

$$\begin{aligned}\mathbf{M}\phi(s) &= \int_1^\infty \left(\sum_{j=1}^J A_j t^{\lambda_j} + t^h\phi(t)\right) t^{-s-1}\, dt + \int_1^\infty \phi(t)\, t^{s-1}\, dt \\ &= \sum_{j=1}^J \frac{A_j}{s-\lambda_j} + \int_1^\infty \phi(t)\left(t^s + t^{h-s}\right)\frac{dt}{t}.\end{aligned}$$

The second term is convergent for all s and is invariant under $s \mapsto h-s$. The first term is also invariant, since applying the functional equation (2) twice shows that for each j there is a j' with $\lambda_{j'} = h - \lambda_j$, $A_{j'} = -A_j$. Hence we have the

Proposition. (Functional Equation Principle) *If $\phi(t)$ $(t>0)$ is small at infinity and satisfies the functional equation (2) for some complex numbers h, A_j and λ_j, then the Mellin transform $\mathbf{M}\phi(s)$ has a meromorphic extension to all s*

and is holomorphic everywhere except for simple poles of residue A_j at $s = \lambda_j$ $(j = 1, \ldots, J)$, and $\mathbf{M}\phi(h - s) = \mathbf{M}\phi(s)$.

This principle is used to establish most of the functional equations occurring in number theory, the first application being the proof of the functional equation of $\zeta(s)$ given by Riemann in 1859 (take $\phi(t) = \sum_{n=1}^{\infty} e^{-\pi n^2 t}$, so that $\mathbf{M}\phi(s) = \pi^{-s}\Gamma(s)\zeta(2s)$ by what was said above and (2) holds with $h = \frac{1}{2}$, $J = 2$, $\lambda_1 = 0$, $\lambda_2 = \frac{1}{2}$, $A_2 = -A_1 = \frac{1}{2}$ by formula (2) of Appendix **A**).

As a final application of the Mellin transform, we prove the formula for $k_s(t)$ stated in Appendix **A**. As we just saw, the function λ^{-s} $(\lambda > 0)$ can be written as $\Gamma(s)^{-1}$ times the Mellin transform of $e^{-\lambda t}$. Hence for $a \in \mathbb{R}$ we have $k_s(a) = \Gamma(s)^{-1}\,\mathbf{M}\phi_a(s)$ where

$$\phi_a(t) = \int_{-\infty}^{\infty} e^{-iax}\, e^{-(x^2+1)t}\, dx = \sqrt{\frac{\pi}{t}} e^{-t-a^2/4t}.$$

Hence $\pi^{-\frac{1}{2}}\Gamma(s)k_s(a) = \int_0^\infty e^{-t-a^2/4t}\, t^{s-\frac{3}{2}}\, dt$. For $a = 0$ this equals $\Gamma(s - \frac{1}{2})$, while for $a > 0$ it equals $2\left(\frac{a}{2}\right)^{s-\frac{1}{2}} \int_0^\infty e^{-a\cosh u} \cosh(s - \frac{1}{2})u\, du$, as one sees by substituting $t = \frac{1}{2}ae^u$.

C Holomorphic projection

We know that S_k has a scalar product $(\cdot,\cdot)$ which is non-degenerate (since $(f, f) > 0$ for every $f \neq 0$ in S_k). It follows that any linear functional $L : S_k \to \mathbb{C}$ can be represented as $f \mapsto (f, \phi_L)$ for a unique cusp form $\phi_L \in S_k$.

Now suppose that $\Phi : \mathfrak{H} \to \mathbb{C}$ is a function which is not necessarily holomorphic but transforms like a holomorphic modular form of weight k, and that $\Phi(\tau)$ has reasonable (say, at most polynomial) growth in $v = \Im(\tau)$ as $v \to \infty$. Then the scalar product $(f, \Phi) = \iint_{\mathfrak{H}/\Gamma_1} v^k f(\tau)\overline{\Phi(\tau)}\, d\mu$ converges for every f in S_k, and since $f \mapsto (f, \Phi)$ is linear, there exists a unique function $\phi \in S_k$ satisfying $(f, \phi) = (f, \Phi)$ for every $f \in S_k$. Clearly $\phi = \Phi$ if Φ is already in S_k, so that the operator π_{hol} which assigns ϕ to Φ is a projection from the infinite dimensional space of functions in $\mathfrak{H}$ transforming like modular forms of weight k to the finite dimensional subspace of holomorphic cusp forms. Our object is to derive a formula for the Fourier coefficients of $\pi_{\text{hol}}(\Phi)$.

To do this, we introduce the **Poincaré series**. For each integer $m \in \mathbb{N}$ set

$$P_m(\tau) = \sum_{\gamma = \left(\begin{smallmatrix} \cdot & \cdot \\ c & d \end{smallmatrix}\right) \in \Gamma_\infty \backslash \Gamma_1} \frac{e^{2\pi i m \gamma(\tau)}}{(c\tau + d)^k} \qquad \left(\gamma(\tau) = \frac{a\tau + b}{c\tau + d} \text{ for } \gamma = \left(\begin{smallmatrix} a & b \\ c & d \end{smallmatrix}\right)\right),$$

where the summation is over left cosets of $\Gamma_\infty = \left\{\pm\left(\begin{smallmatrix} 1 & b \\ 0 & 1 \end{smallmatrix}\right), b \in \mathbb{Z}\right\}$ in Γ_1. The series converges absolutely if $k > 2$ and defines a cusp form of weight k. The same unfolding argument as in the Rankin-Selberg method (Part 3, **B**) shows that for a form $f = \sum_1^\infty a(n)q^n \in S_k$ the Petersson scalar product (f, P_m) is given by

$$\begin{aligned}(f, P_m) &= \iint_{\mathfrak{H}/\Gamma_1} f(\tau)\,\overline{P_m(\tau)}\, v^k \,\frac{du\,dv}{v^2} \\ &= \iint_{\mathfrak{H}/\Gamma_\infty} f(\tau)\,\overline{e^{2\pi i m\tau}}\, v^k \,\frac{du\,dv}{v^2} \\ &= \int_0^\infty \left(\int_0^1 f(u+iv)\, e^{-2\pi i m u}\, du\right) e^{-2\pi m v} v^{k-2}\, dv \\ &= \int_0^\infty \left(a(m) e^{-2\pi m v}\right) e^{-2\pi m v} v^{k-2}\, dv \\ &= \frac{(k-2)!}{(4\pi m)^{k-1}} a(m).\end{aligned}$$

In other words, $(4\pi m)^{k-1} P_m(\tau)/(k-2)!$ is the cusp form dual to the operator of taking the mth Fourier coefficient of a holomorphic cusp form.

Now let $\sum_{n\in\mathbb{Z}} c_n(v) e^{2\pi i n u}$ denote the Fourier development of our function $\Phi(\tau)$ and $\sum_{n=1}^\infty c_n q^n$ that of its holomorphic projection to S_k. Then

$$\frac{(k-2)!}{(4\pi m)^{k-1}} c_m = (\pi_{\mathrm{hol}}(\Phi), P_m) = \overline{(P_m, \pi_{\mathrm{hol}}(\Phi))} = \overline{(P_m, \Phi)} = (\Phi, P_m)$$

by the property of P_m just proved and the defining property of $\pi_{\mathrm{hol}}(\Phi)$. Unfolding as before, we find

$$\begin{aligned}(\Phi, P_m) &= \int_0^\infty \left(\int_0^1 \Phi(u+iv) e^{-2\pi i m u}\, du\right) e^{-2\pi m v} v^{k-2}\, dv \\ &= \int_0^\infty c_m(v) e^{-2\pi m v} v^{k-2}\, dv\end{aligned}$$

provided that the interchange of summation and integration implicit in the first step is justified. This is certainly the case if the scalar product (Φ, P_m) remains convergent after replacing Φ by its absolute value and P_m by its majorant $\hat{P}_m(\tau) = \sum_{\Gamma_\infty\backslash\Gamma_1} |(c\tau+d)^{-k} e^{2\pi i m \gamma(\tau)}|$. We have

$$\begin{aligned}\hat{P}_m(\tau) &< |e^{2\pi i m\tau}| + \sum_{c\neq 0} \sum_{(d,c)=1} |c\tau+d|^{-k} \\ &= e^{-2\pi m v} + \frac{1}{\zeta(k)} v^{-k/2} \left[G(\tau, \frac{k}{2}) - \zeta(k) v^{k/2}\right],\end{aligned}$$

with $G(\tau, \frac{k}{2})$ the non-holomorphic Eisenstein series introduced in **A**, Part 3. The estimate there shows that $G(\tau, \frac{k}{2}) - \zeta(k) v^{k/2} = \mathrm{O}(v^{1-k/2})$ as $v \to \infty$, so $\hat{P}_m(\tau) = \mathrm{O}(v^{1-k})$. The convergence of $\iint_{\mathfrak{H}/\Gamma_1} |\Phi| \hat{P}_m v^{k-2}\, du\, dv$ is thus assured if $\Phi(\tau)$ decays like $\mathrm{O}(v^{-\epsilon})$ as $v \to \infty$ for some positive number ϵ. Finally, we can weaken the condition $\Phi(\tau) = \mathrm{O}(v^{-\epsilon})$ to $\Phi(\tau) = c_0 + \mathrm{O}(v^{-\epsilon})$ ($c_0 \in \mathbb{C}$) by the simple expedient of subtracting $c_0 \frac{-2k}{B_k} G_k(\tau)$ from $\Phi(\tau)$ and observing that G_k is orthogonal to cusp forms by the same calculation as above with $m = 0$ (G_k

is proportional to P_0). We have thus proved the following result, first stated by J. Sturm under slightly different hypotheses:

Lemma. (Holomorphic Projection Lemma) *Let $\Phi : \mathfrak{H} \to \mathbb{C}$ be a continuous function satisfying*

(1) *'(i)' $\Phi(\gamma(\tau)) = (c\tau + d)^k \Phi(\tau)$ for all $\gamma = \left(\begin{smallmatrix} a & b \\ c & d \end{smallmatrix}\right) \in \Gamma_1$ and $\tau \in \mathfrak{H}$; and*
(2) *'(ii)' $\Phi(\tau) = c_0 + O(v^{-\epsilon})$ as $v = \Im(\tau) \to \infty$,*

for some integer $k > 2$ and numbers $c_0 \in \mathbb{C}$ and $\epsilon > 0$. Then the function $\phi(\tau) = \sum_{n=0}^{\infty} c_n q^n$ with $c_n = \frac{(4\pi n)^{k-1}}{(k-2)!} \int_0^\infty c_n(v) e^{-2\pi n v} v^{k-2}\, dv$ for $n > 0$ belongs to M_k and satisfies $(f, \phi) = (f, \Phi)$ for all $f \in S_k$.

As an example, take $\Phi = (G_2^*)^2$, where G_2^* is the non-holomorphic Eisenstein series of weight 2 introduced in Part **1A**. Using the Fourier expansion $G_2^* = \frac{1}{8\pi v} + G_2 = \frac{1}{8\pi v} - \frac{1}{24} + \sum_1^\infty \sigma_1(n) q^n$ given there, we find

$$\begin{aligned} \Phi(\tau) =& \left(\frac{1}{576} - \frac{1}{96\pi v} + \frac{1}{64\pi^2 v^2}\right) \\ &+ \sum_{n=1}^{\infty} \left(-\frac{1}{12}\sigma_1(n) + \sum_{m=1}^{n-1} \sigma_1(m)\sigma_1(n-m) + \frac{1}{4\pi v}\sigma_1(n) \right) q^n, \end{aligned}$$

so that the hypotheses of the holomorphic projection lemma are satisfied with $k = 4$, $c_0 = \frac{1}{576}$, $\epsilon = 1$ and $c_n(v) = \left(-\frac{1}{12}\sigma_1(n) + \sum_{m=1}^{n-1} \sigma_1(m)\sigma_1(n-m) + \frac{1}{4\pi v}\sigma_1(n)\right) e^{-2\pi n v}$. The lemma then gives $\sum c_n q^n \in M_4$ with $c_n = -\frac{1}{12}\sigma_1(n) + \sum_{m=1}^{n-1} \sigma_1(m)\sigma_1(n-m) + \frac{1}{2} n \sigma_1(n)$ for $n \geq 1$. Since $\sum_0^\infty c_n q^n \in M_4 = \langle G_4 \rangle$, we must have $c_n = 240 c_0 \sigma_3(n) = \frac{5}{12}\sigma_3(n)$ for all $n > 0$, an identity that the reader can check for small values of n.

Similarly, if $f = \sum_0^\infty a_n q^n$ is a modular form of weight $l \geq 4$, then $\Phi = f G_2^*$ satisfies the hypotheses of the lemma with $k = l + 2$, $c_0 = -\frac{1}{24} a_0$ and $\epsilon = 1$, and we find that $\pi_{\text{hol}}(f G_2^*) = f G_2 + \frac{1}{4\pi i l} f' + \in M_{l+2}$.

References

We will not attempt to give a complete bibliography, but rather will indicate some places where the interested reader can learn in more detail about the theory of modular forms.

Three short introductions to modular forms can be especially recommended:

(i) the little book *Lectures on Modular Forms* by R.C. Gunning (Princeton, Ann. of Math. Studies **48**, 1962), which in 86 widely spaced pages describes the classical analytic theory and in particular the construction of Poincaré series and theta series,

(ii) Chapter 7 of J-P. Serre's 'Cours d'Arithmétique' (Presses Universitaires de France 1970; English translation: Springer, GTM **7**, 1973), which gives among other things a very clear introduction to the theory of Hecke operators and to the applications of theta series to the arithmetic of quadratic forms, and

(iii) the survey article by A. Ogg in *Modular Functions of One Variable. I* (Springer, Lecture Notes **320**, 1973), which treats some of the modern aspects of the theory and in particular the connection with elliptic curves. (The other volumes in this series, SLN **349**, **350**, **475**, **601**, and **627**, describe many of the developments of the years 1970–76, when the subject experienced a renascence after a long period of dormancy.)

Of the full-length books on the subject, the best introduction is probably Serge Lang's *Introduction to Modular Forms* (Springer, Grundlehren **222**, 1976), which treats both the analytic and the algebraic aspects of theory. It also includes a detailed derivation of the trace formula for Hecke operators on the full modular group (this is in an appendix by me and unfortunately contains an error, corrected in the volume SLN **627** referred to above). Other texts include Ogg's *Modular Forms and Dirichlet Series* (Benjamin 1969), which gives in great detail the correspondence between modular forms and Dirichlet series having appropriate functional equations, as well as an excellent presentation of the theory of theta series with spherical polynomial coefficients, G. Shimura's *Introduction to the Arithmetic Theory of Automorphic Functions* (Princeton 1971), which is more advanced and more heavily arithmetic than the other references discussed here, and the recent book *Modular Forms* by T. Miyake (Springer 1989), which contains a detailed derivation of the trace formula for the standard congruence subgroups of Γ_1. Another good book that treats the connection with elliptic curves and also the theory of modular forms of half-integral weight is N. Koblitz's *Introduction to Elliptic Curves and Modular Forms* (Springer, GTM **97**, 1984). Finally, anyone who really wants to learn the subject from the inside can do no better than to study Hecke's *Mathematische Werke* (Vandenhoeck 1959).

We also mention some books on subjects closely related to the theory of modular forms: for a classically oriented account of the theory of modular functions, Rankin's *Modular Forms and Functions* (Cambridge 1977) or Schoeneberg's *Elliptic Modular Functions: An Introduction* (Springer, Grundlehren **203**, 1974); for the theory of elliptic curves, Silverman's book *The Arithmetic of Elliptic curves* (Springer, GTM **106**, 1986); for the modern point of view on modular forms in terms of the representation theory of $GL(2)$ over the adeles of a number field, Gelbart's *Automorphic Forms on Adele Groups* (Princeton, Ann. of Math. Studies **83**, 1975) or, to go further, *Automorphic Forms, Representations, and L-Functions* (AMS 1979).

We now give in a little more detail sources for the specific subjects treated in these notes.

Part 1. The basic definitions of modular forms and the construction of the Eisenstein series G_k and the discriminant function Δ are given in essentially every introduction. Serre (op. cit.) gives a construction of Δ which is related to, but different from, the one given here: Instead of using the non-holomorphic modular form G_2^*, he uses G_2 itself but analyzes the effect on the value of the non-absolutely convergent series $\sum(m\tau+n)^{-2}$ of summing over m and n in different orders. This approach goes back to Eisenstein. The reader should beware of the fact that Serre normalizes the weight differently, so that, e.g., Δ has weight 6 instead of 12. The best treatment of theta series in the simplest case, namely when the underlying quadratic form is unimodular and there are no spherical coefficients, is also given in Serre's book, but for the general case one must go to Gunning's or (better) Ogg's book, as already mentioned. The Eisenstein series of half-integral weight are already a more specialized topic and are not to be found in any of the books mentioned so far. The construction of the Fourier coefficients of the particular Eisenstein series $G_{r+\frac{1}{2}}$ which we discuss (these are the simplest half-integral-weight series, but there are others) is due to H. Cohen (Math. Ann. **217**, 1975), for $r > 1$, while the construction of the series $G_{\frac{3}{2}}^*$ is contained in an article by Hirzebruch and myself (Inv. math. **36**, 1976, pp. 91–96). The development of the general theory of modular forms of half-integral weight, and in particular the construction of a 'lifting map' from these forms to forms of integral weight, is given in famous papers by G. Shimura (Ann. of Math. **97**, 1973 and in the above-mentioned Lecture Notes **320**); an elementary account of this theory is given in Koblitz's book cited above. Of the constructions described in Section **E**, the differential operator F_ν is constructed in the paper of H. Cohen just cited, but is in fact a special case of more general differential operators constructed by Rankin several years earlier, while the 'slash operators' and the operators V_m and U_m are treated in any discussion of Hecke operators for congruence subgroups of $SL(2,\mathbb{Z})$ and in particular in Chapter VII of Lang's book. Finally, the topics touched upon in Section **F** are discussed in a variety of places in the literature: the connection between modular forms of weight 2 and elliptic curves of weight 2 is discussed e.g. in Silverman's book or the Springer Lecture Notes **476** cited above; the theorem of Deligne and Serre appeared in Ann. Sc. Ec. Norm. Sup. 1974; and the theory of complex multiplication is discussed in Lang's book of the same name and in many other places.

Part 2. As already mentioned, the clearest introduction to Hecke operators for the full modular group is the one in Serre's book, the L-series and their functional equations are the main topic of Ogg's Benjamin book. The theory in the higher level case was first worked out by Atkin and Lehner (Math. Ann. **185**, 1970) and is presented in detail in Chapters VII-VIII of Lang's textbook. Some tables of eigenforms for weight 2 are given in the Lecture Notes volume **476** cited above.

Part 3. The classical reference for the function $G(\tau,s)$ and the Rankin-Selberg method is Rankin's original paper (Proc. Camb. Phil. Soc. **35**, 1939). However, the main emphasis there is on analytic number theory and the derivation of the estimate $a(n) = O\left(n^{\frac{k}{2}-\frac{1}{5}}\right)$ for the Fourier coefficients $a(n)$ of a cusp form f of weight k (specifically, Δ of weight 12). Expositions of the general method have been given by several authors, including the present one on several occasions (e.g. in two articles in *Automorphic Forms, Representation Theory and Arithmetic*, Springer 1981, and in a paper in J. Fac. Sci. Tokyo **28**, 1982; these also contain the applications mentioned in Section **B**). The proof of the Eichler-Selberg trace formula sketched in Section **D** has not been presented before. Standard proofs can be found in the books of Lang and Miyake, as already mentioned, as well, of course, as in the original papers of Eichler and Selberg.

Part 4. The theory of Jacobi forms was developed systematically in a book by M. Eichler and myself (Progress in Math.**55**, Birkhäuser 1985); special examples, of course, had been known for a long time. The results described in (iii), (v) and (vi) of Section **B** are due to N.-P. Skoruppa (in particular, the construction mentioned in (vi) is to appear in Inv. math. 1990), while the trace formula and lifting maps mentioned under (iv) are joint work of Skoruppa and myself (J. reine angew. Math. **393**, 1989, and Inv. math. **94**, 1988). A survey of these and some other recent developments is given in Skoruppa's paper in the proceedings of the Conference on Automorphic Functions and their Applications, Khabarovsk 1988.

Appendices. The material in Sections **A** and **B** is standard and can be found in many books on analysis or analytic number theory. The method of holomorphic projection was first given explicitly by J. Sturm (Bull. AMS **2**, 1980); his proof is somewhat different from the one we give.

Chapter 5

Decorated Elliptic Curves: Modular Aspects

by Robert Gergondey

Summary

Classical results concerning modular forms are interpreted by introducing moduli spaces for elliptic curves endowed with 'gadgets' that can be transported by isomorphisms. From both the arithmetical and the geometrical viewpoint, these moduli spaces and the corresponding universal decorated elliptic curves are 'nicer' objects, on which the various constructions turn out to be more natural than in the 'bare' situation.

Introduction

By 'modulating' a mathematical object of a given type, we mean, roughly speaking, parametrizing its essential variations, up to isomorphisms. Locally, this amounts to studying deformations. Globally, one is led to construct a family of such objects, that is both exhaustive and, as far as possible, without redundances. The space which parametrizes the family is endowed with the relevant structures (topological, differential, analytic, algebraic, arithmetical, ..., metric, Riemannian, Kählerian, ...). Such constructions are well-known, the classical paradigm being of course the case of genus-1 Riemann surfaces or, in other words, that of complex structures on the two-dimensional torus $\mathbb{T}^2$. As a matter of fact, the moduli of elliptic curves were introduced and studied in many ways. The fragrances of the 'garden of modular delights' have inebriated many a mathematician.

This Chapter is definitely not devoted to the theory of the moduli of elliptic curves, which has been presented in many books and articles (see the bibliography), and from many different viewpoints. Our oral exposition, which this text will follow rather closely, only aimed at illustrating – in forty minutes – the interest of several aspects, namely (a) considering (quasi-)elliptic modular functions, i. e. dealing with the 'universal' elliptic curve, (b) enriching elliptic curves by 'decorations', (c) considering as well the $\mathbf{C}^\infty$ and the $\mathbb{R}$-analytic cases, (d) studying the Eisenstein series with critical weight, and finally (e) considering modular, and even elliptic modular, derivations. Although each of these themes would have deserved a longer development than these few pages, we hope that this brief survey will nevertheless be useful. Let us finally emphasize that we do not claim any originality.

1. Hors-d'œuvre: Theta Functions

The function

$$\theta_3(\tau) = \sum_{n\in\mathbb{Z}} e^{i\pi n^2 \tau} \tag{1.1}$$

is the simplest of the theta 'constants'. It is holomorphic on the Poincaré upper half-plane $\mathfrak{H}_1 = \{\mathrm{Im}\tau > 0\}$, and obviously periodic with period 2. The following functional equation

$$\theta_3\left(-\frac{1}{\tau}\right) = \sqrt{\frac{\tau}{i}}\,\theta_3(\tau) \quad , \quad \mathrm{Re}\left(\sqrt{\frac{\tau}{i}}\right) > 0 \tag{1.2}$$

is one of the classical 'jewels' of Mathematics. This result can be proved in many ways, using either the Poisson formula, or residue calculus. The Mellin transform allows one to relate it to the functional equation obeyed by the Riemann function $\zeta(s) = \sum_{n\geq 1} n^{-s}$. The interest of the approach sketched here lies both in its simplicity and its very general structure.

The function

$$\theta_3(z|\tau) = \sum_{n\in\mathbb{Z}} e^{2i\pi(nz+\frac{n^2}{2}\tau)} \tag{1.3}$$

is analytic on $\mathbb{C}\times\mathfrak{H}_1$. For any fixed $\tau\in\mathfrak{H}_1$, the function $z\mapsto\theta_3(z|\tau)$ is an entire function which obeys

$$\begin{cases} \theta_3(z+1|\tau) &= \theta_3(z|\tau) \\ \theta_3(z+\tau|\tau) &= e^{-i\pi(2z+\tau)}\theta_3(z|\tau) \end{cases} \tag{1.4}$$

and which has simple zeros, located at

$$(m+1/2)+(n+1/2)\tau \quad , \quad (m,n)\in\mathbb{Z}^2 . \tag{1.5}$$

More generally, a function Θ analytic on $\mathbb{C}^n$ is called a theta function on $\mathbb{C}^n$, with respect to a lattice $L\subset\mathbb{C}^n$, if it obeys

$$\Theta(z+l) = e^{-i\pi\phi(z,l)}\Theta(z) \tag{1.6}$$

where $\phi(z,l)$ is a function on $\mathbb{C}^n\times L$, affine in z. The function θ_3 is thus a theta function with respect to the lattice $\mathbb{Z}\oplus\mathbb{Z}\tau\subset\mathbb{C}$.

Considering now τ as a variable, one can verify the *heat equation*

$$\frac{\partial^2 \theta_3}{\partial z^2} = 4i\pi \frac{\partial \theta_3}{\partial \tau} \tag{1.7}$$

Notice that

$$\Psi(z|\tau) = \theta_3\left(\frac{z}{\tau} \,\Big|\, -\frac{1}{\tau}\right) \tag{1.8}$$

is an entire function with the same simple zeros as $\theta_3(z|\tau)$. 'Periodicity' (up to a factor, see 1.4), allows one to show that

$$\Psi(z|\tau) = g(\tau)e^{i\pi z^2/\tau}\theta_3(z|\tau) \tag{1.9}$$

whereas the heat equation (1.7) implies that g assumes the form

$$g(\tau) = K\sqrt{\frac{\tau}{i}} \tag{1.10}$$

The special value $\tau = i$ fixes $K = 1$, yielding the transformation formula

$$\theta_3\left(\frac{z}{\tau} \,\Big|\, -\frac{1}{\tau}\right) = \sqrt{\frac{\tau}{i}}e^{i\pi z^2/\tau}\theta_3(z|\tau) \tag{1.11}$$

Specializing this result for $z = 0$, one recovers the functional equation for the 'constant' $\theta_3(\tau)$.

Some minor modifications of this approach allow one to obtain transformation formulas for the twisted theta functions

$$\theta_\chi(z|\tau) = \sum_{n\in\mathbb{Z}} \chi(n)e^{\frac{2i\pi}{N}(nz+\frac{n^2}{2}\tau)} \tag{1.12}$$

where $\chi : \mathbb{Z} \to \mathbb{C}$ is a primitive Dirichlet character mod N, that is

$$\begin{aligned} &\chi(n_1) = \chi(n_2) \quad \text{if} \quad n_1 \equiv n_2 \ (N) \\ &\chi(1) = 1 \quad ; \quad \chi(mn) = \chi(m)\chi(n) \\ &\chi(n) \neq 0 \quad \text{iff} \quad (n, N) = 1 \end{aligned} \tag{1.13}$$

One can thus show that the function

$$\eta(\tau) = q^{\frac{1}{24}} \sum_{n\in\mathbb{Z}} (-1)^n q^{\frac{3n^2+n}{2}} = \sum_{r\in\mathbb{Z}} (-1)^r q^{\frac{(6r+1)^2}{24}} = \theta_{\left(\frac{\cdot}{12}\right)}(\tau) \tag{1.14}$$

(where $q = e^{2i\pi\tau}$, and $\left(\frac{\cdot}{12}\right)$ is the Legendre-Jacobi symbol), obeys the very same functional equation as $\theta_3(\tau)$, namely

$$\eta\left(-\frac{1}{\tau}\right) = \sqrt{\frac{\tau}{i}}\,\eta(\tau) \tag{1.15}$$

Finally, the 'multi-dimensional' case of theta functions on $\mathbb{C}^n$, parametrized by the Siegel spaces $\mathfrak{H}_n$, can also be dealt with in an analogous fashion.

2. The Elliptic Modular Viewpoint

One of the underlying ideas of the previous Section, was the following: in a function such as $\theta_3(\tau)$, the variable τ parametrizes lattices of 'periods' for functions such as $\theta_3(z|\tau)$. The latter functions are thus related to the elliptic curves $E_{[\tau]} = {}^{\mathbb{C}}/_{L_\tau}$ (in the $\mathbb{C}$-analytic sense of toric Riemann surfaces with one distinguished point), where L_τ is the lattice of $\mathbb{C}$ generated over $\mathbb{Z}$ by 1 and τ. Any variation of τ corresponds to a variation of the curve $E_{[\tau]}$, so that θ_3 can be thought of as a rule which associates a number with an elliptic curve. The usual parametrization of the family of all elliptic curves (up to $\mathbb{C}$-analytic isomorphism) by the Poincaré upper half-plane is complete, in the sense that any elliptic curve E is $\mathbb{C}$-analytically isomorphic to at least one $E_{[\tau]}$. Moreover, we know that $E_{[\tau']}$ is $\mathbb{C}$-analytically isomorphic to $E_{[\tau]}$ if and only if

$$\tau' = \frac{a\tau + b}{c\tau + d} \quad \text{with} \quad \begin{pmatrix} a & b \\ c & d \end{pmatrix} \in \mathrm{SL}_2(\mathbb{Z}) \tag{2.1}$$

We are therefore led to consider the quotient

$$\mathfrak{M}_1 = {}_{\mathrm{PSL}_2(\mathbb{Z})}\backslash \mathfrak{H}_1 \tag{2.2}$$

as the moduli space, or 'universal parametrization', of elliptic curves.

This moduli space is nevertheless not as pleasant as one would wish, namely

- $\mathfrak{M}_1$ *is not* a *compact* topological space.
- The analytic structure obtained as a quotient of that of $\mathfrak{H}_1$ has *singularities*, since the action of $\mathrm{PSL}_2(\mathbb{Z})$ is not free: the curves $E_{[\sqrt{-1}]}$ and $E_{\left[\frac{-1+\sqrt{-3}}{2}\right]}$ possess automorphisms which differ from $\pm\mathrm{Id}$ (see Section 4).
- A function such as θ_3 *is not a function on* $\mathfrak{M}_1$, since it is not invariant under the action of $\mathrm{PSL}_2(\mathbb{Z})$.

A customary way of surrounding these difficulties is the following.

- One uses a *compactification* à la Alexandroff by adding one point at infinity ω, which can be interpreted as corresponding to the *singular degeneracy* of elliptic curves (namely a curve with a good singularity, and a regular part which is a multiplicative group). This degeneracy is obtained when the lattice L_τ degenerates into a rank-1 subgroup of $\mathbb{C}$ (see Section 5)

  $$\widehat{\mathfrak{M}}_1 = \mathfrak{M}_1 \cup \{\omega\} = {}_{\mathrm{PSL}_2(\mathbb{Z})}\backslash \widehat{\mathfrak{H}}_1 \tag{2.3}$$

  where $\widehat{\mathfrak{H}}_1 = \mathfrak{H}_1 \cup \mathbb{P}^1(\mathbb{Q})$ is endowed with the standard action of $\mathrm{PSL}_2(\mathbb{Z})$. For the analytic structure, ω is a logarithmic singular point, or a cusp. The *de-singularization* is relatively easy in one dimension: one chooses new uniformizing variables in the neighbourhood of the singular points. The Riemann surface obtained in this way is $\mathbb{C}$-analytically isomorphic to the Riemann sphere $\mathbb{P}^1(\mathbb{C}) = \widehat{\mathbb{C}}$, an explicit isomorphism being given by the modular function j. But we are still left with the *multi-valuedness* of the function θ_3.

As a matter of fact, such situations are generic when dealing with the moduli of mathematical structures, and compactification and de-singularization are rarely so easy as in the case of $\mathfrak{M}_1$. But a very fruitful way is open, namely to replace the modular problem under study by a rigidified problem, in which the objects, whose variations are to be parametrized, bear some additional structures chosen in an appropriate way. These superimposed structures, or decorations, which can be transported by isomorphism, will allow us to distinguish between isomorphic objects, and thus to break some symmetries in an artificial way. This is illustrated in figure 1, which shows how the distinction between formerly identified objects can de-singularize a modular problem.

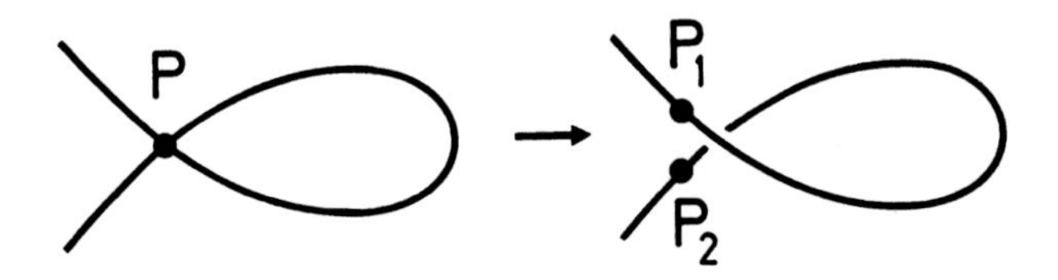

Fig. 1. The de-singularization

The 'compactification' problem, understood in terms of adding good degeneracies, will have to be interpreted in this rigidified context.

Finally, the multi-valued functions on the initial space will eventually be interpreted as univalued functions on the new moduli space. Moreover, these functions can be described in a natural fashion in terms of decorated objects. One can then go back to the original problem by controlling how the values of the functions depend on the decorations, in agreement with the spirit of both the Galois theory of ambiguity, and of gauge theories. This point will be illustrated on the example of elliptic curves. But there is clearly no reason to limit ourselves neither to elliptic curves, nor in this case to the types of decorations described below.

3. Decorations of Elliptic Curves

Let us start with the following

Definition. *An elliptic curve E defined on $\mathbb{C}$ is said to be (fully) decorated if it is endowed with*

– *a non-zero holomorphic 1-form $\alpha \in \Omega^{1,0}(E)$*

– *a positively oriented basis $\begin{bmatrix}\gamma_1\\ \gamma_2\end{bmatrix}$ of the Abelian group $H_1(E,\mathbb{Z})$ of oriented 1-cycles, up to homology. Here positively oriented means that the intersection index $i(\gamma_2,\gamma_1) = +1$.*

The above defined object will be denoted by $\left\{E;\alpha,\begin{bmatrix}\gamma_1\\ \gamma_2\end{bmatrix}\right\}$.

Definition. *A morphism $\phi : \left\{E;\alpha,\begin{bmatrix}\gamma_1\\ \gamma_2\end{bmatrix}\right\} \rightarrow \left\{E';\alpha',\begin{bmatrix}\gamma_1'\\ \gamma_2'\end{bmatrix}\right\}$ of decorated elliptic curves is a usual morphism $\phi :\ E \rightarrow E'$ such that*

$$\begin{aligned} &\phi^*\alpha' = \alpha \\ &\phi_*\gamma_\mu = \gamma_\mu' \qquad \mu = 1,2 \ , \end{aligned} \tag{3.1}$$

where ϕ^* is the inverse transport of differential forms, and ϕ_* is the direct transport of singular chains.

As a matter of fact, such a datum is entirely determined, up to isomorphism, by that of a positively oriented basis $\left\|\begin{bmatrix}\omega_1\\ \omega_2\end{bmatrix}\right\| = b$ of the $\mathbb{R}$-vector space $\mathbb{C}$, where positively oriented means $\omega_2 \wedge \omega_1 > 0$. With the notation $\left\lfloor \begin{bmatrix}\omega_1\\ \omega_2\end{bmatrix}\right. = \lfloor b$ for the lattice with basis $\left\|\begin{bmatrix}\omega_1\\ \omega_2\end{bmatrix}\right\|$, the correspondence is displayed in table 1, where dz stands, of course, for the (1,0) form defined by the (1,0) canonical form on $\mathbb{C}$, and $\widetilde{\omega}_\mu$ is the (geodesic) 1-cycle associated with ω_μ.

If we denote by $\mathfrak{B}^+$ the set of positively oriented $\mathbb{R}$-bases of $\mathbb{C}$, we realize that the correspondence described just above allows us to consider $\mathfrak{B}^+$ as the moduli space of decorated elliptic curves. This space is quite pleasant, in that it is a principal homogeneous space for the group $\mathrm{GL}_2^+(\mathbb{R})$ of real 2×2 matrices with (strictly) positive determinant, which can thus be endowed with the structure of a real analytic manifold of dimension four. $\mathfrak{B}^+$ can in fact also be thought of as the open set of $\mathbb{C}^2$ defined by the condition $\omega_2 \wedge \omega_1 > 0$. It is thus a complex analytic surface without singularities.

It is then possible to consider the trivial fiber bundle $\mathbb{C} \times \mathfrak{B}^+$ on $\mathfrak{B}^+$, and to perform in each fiber $\mathbb{C} \times \{b\}$ the quotient by the lattice $\lfloor b$, thus obtaining the elliptic curve E_b. The union $E_{\mathfrak{B}^+} = \cup_{b\in\mathfrak{B}^+} E_b$ is the quotient of $\mathbb{C} \times \mathfrak{B}^+$ by a gentle equivalence relation, which allows one to endow $E_{\mathfrak{B}^+}$ with the structure of a complex analytic manifold of dimension three, analytically fibered in (decorated) elliptic curves over $\mathfrak{B}^+$. We can thus refer to $E_{\mathfrak{B}^+}$ as the 'universal' decorated elliptic curve, making no mention of decorations, for the sake of modesty.

Table 1. Correspondence between elliptic curves and oriented bases

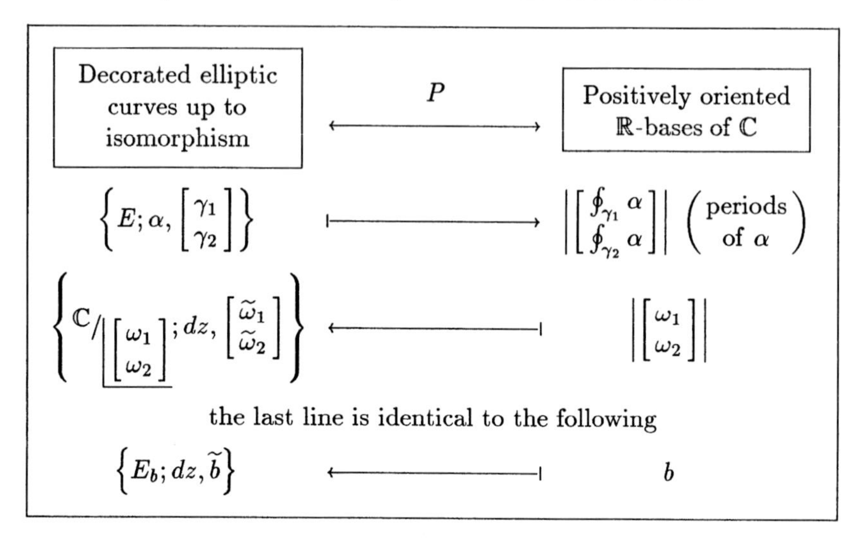

On an elliptic curve $\mathbb{C}/_L$ we define

- *additive functions*: meromorphic functions $\phi : \mathbb{C} \mapsto \mathbb{C} \cup \{\infty\}$ such that, for every $l \in L$, $\phi(z+l) - \phi(z)$ is independent of $z \in \mathbb{C}$. The case where $\phi(z+l) = \phi(z)$ is that of *elliptic* functions. Affine functions are said to be 'trivial' additive functions.
- *multiplicative functions*: entire functions $\theta : \mathbb{C} \mapsto \mathbb{C}$ such that, for every $l \in L$, $(\theta(z+l)/\theta(z))$ is an exponential (i.e. an invertible entire function). Exponential functions are said to be 'trivial' multiplicative functions.

The *theta functions* correspond to the case where the ratio $\theta(z+l)/\theta(z)$ is the exponential of an affine function; the exponentials of quadratic functions are said to be trivial among theta functions.

There exist, of course, twisted versions of these functions, associated with characters $\chi : L \mapsto \mathbb{C}^*$ of the lattice L. We may require e. g. $\phi(z+l) = \chi(l)\phi(z)$. But we want to emphasize the generalization to $\mathbb{R}$-analytic (or even only C^∞) functions over $\mathbb{C}$, that behaves similarly with respect to translations of L. In particular we may consider the effect of complex conjugation, and thus the $\overline{\mathbb{C}}$-analytic mirror image of what has just been exposed.

In the 'modular' case of the 'universal' elliptic curve $E_{\mathfrak{B}^+}$, it will be useful to deal with an $\mathbb{R}$-analytic structure, or with a mixed one: we will consider

$\mathbb{R}$-analytic functions $\Phi : \mathbb{C} \times \mathfrak{B}^+ \mapsto \mathbb{C} \cup \{\infty\}$, such that, for any $b \in \mathfrak{B}^+$, the partial function $z \mapsto \Phi(z|b)$ is of one of the types listed above, with respect to E_b.

Taking the example of the Weierstrass functions,

$$\xi(z|b) = -z^{-1} - \sum_{l \in \lfloor b^{\cdot}} \left[(z-l)^{-1} + l^{-1} + zl^{-2}\right] \qquad \lfloor b^{\cdot} = \lfloor b - \{0\} \tag{3.2}$$

is additive 'modular',

$$\wp(z|b) = z^{-2} + \sum_{l \in \lfloor b^{\cdot}} \left[(z-l)^{-2} - l^{-2}\right]$$

$$\wp'(z|b) = -2 \sum_{l \in \lfloor b} (z-l)^{-3} \tag{3.3}$$

are elliptic 'modular', and

$$\sigma(z|b) = z \prod_{l \in \lfloor b^{\cdot}} (1 - zl^{-1}) \exp\left(zl^{-1} + \frac{z^2 l^{-2}}{2}\right) \tag{3.4}$$

is theta 'modular'.

Consider now the more general functions

$$K^{\chi}_{s,t}(z|b) = \sum_{l \in \lfloor b} \chi(l)(z-l)^{-s}(\bar{z}-\bar{l})^{-t} \tag{3.5}$$

introduced by Kronecker, where χ is a character of $\lfloor b$, and $(s,t) \in \mathbb{C}^2$ obey $s \equiv t \ (\mathbb{Z})$ and $\mathrm{Re}(s+t) > 2$. When χ is trivial, these functions provide examples of 'modular' $\mathbb{R}$-analytic doubly periodic functions. When $t = 0$ (and thus s integer so that $s > 2$), elliptic 'modular' functions are recovered (essentially as derivatives of the Weierstrass $\wp$ function). Let us mention the fascinating study of the behavior of $K^{\chi}_{s,t}$ when (s,t) goes to the critical zone $\{\mathrm{Re}(s+t) = 2\}$.

4. Getting Rid of some Decorations

The decoration presented above can be either partially or totally dropped. This amounts to performing some identifications of indistinguishable objects, that we shall now to describe.

4.1. One may neglect the 1-form α in $\left\{E; \alpha, \begin{bmatrix}\gamma_1 \\ \gamma_2\end{bmatrix}\right\}$ thus considering elliptic curves endowed with a positively oriented basis of the free Abelian group $\mathrm{H}_1(E, \mathbb{Z})$ of 1-cycles up to homology, which will be called *marked* elliptic curves. Owing to the correspondence described in table 1, this amounts to identifying $b = \left\|\begin{bmatrix}\omega_1 \\ \omega_2\end{bmatrix}\right\|$ with $b\lambda = \left\|\begin{bmatrix}\omega_1\lambda \\ \omega_2\lambda\end{bmatrix}\right\|$ for $\lambda \in \mathbb{C}^*$, since any two holomorphic 1-forms on E are proportional. The relevant correspondence for marked elliptic curves therefore reads

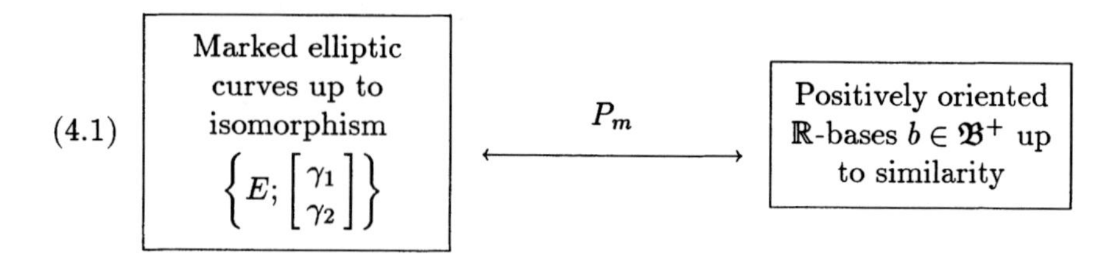

The set $\mathfrak{B}^+/_{\mathbb{C}^*}$ of the similarity classes of oriented bases is represented by the Poincaré half-plane $\mathfrak{H} = \{\tau \in \mathbb{C} | \mathrm{Im}\tau > 0\}$

$$
(4.2) \qquad \begin{array}{ccc} \mathfrak{B}^+/\mathbb{C}^* & \xrightarrow{\sim} & \mathfrak{H} \\ \left\| \begin{bmatrix}\omega_1\\ \omega_2\end{bmatrix} \right\| & \longmapsto & \tau = \dfrac{\omega_1}{\omega_2} \end{array}
$$

The upper half-plane is thus a moduli space for marked elliptic curves. The point made in the previous Section also holds in the present situation, so that a 'universal' marked elliptic curve $E_{\mathfrak{H}}$ can be defined.

4.2. One may neglect the marking $\begin{bmatrix}\gamma_1\\ \gamma_2\end{bmatrix}$, retaining just the 1-form α, thus considering *calibrated* elliptic curves $\{E; \alpha\}$. This amounts to identifying $b = \left\| \begin{bmatrix}\omega_1\\ \omega_2\end{bmatrix} \right\|$ and $gb = \left\| \begin{bmatrix}a\omega_1 + b\omega_2\\ c\omega_1 + d\omega_2\end{bmatrix} \right\|$, where $g = \begin{pmatrix} a & b\\ c & d\end{pmatrix} \in \mathrm{GL}_2^+(\mathbb{Z}) = \mathrm{SL}_2(\mathbb{Z})$, whence the following correspondence.

$$
(4.3) \qquad \boxed{\begin{array}{c}\text{Calibrated elliptic}\\ \text{curves } \{E;\alpha\} \text{ up}\\ \text{to isomorphism}\end{array}} \quad \xleftrightarrow{\;P_c\;} \quad \boxed{\begin{array}{c}\text{Lattices}\\ \text{in } \mathbb{C}\end{array}}
$$

Now the space $\mathcal{L} = \mathrm{GL}_2^+(\mathbb{Z}) \backslash \mathfrak{B}^+$ of the lattices in $\mathbb{C}$ is a moduli space for calibrated elliptic curves, and there exists a 'universal' calibrated elliptic curve $E_{\mathcal{L}}$.

4.3. One may finally neglect both the calibration α and the marking $\begin{bmatrix}\gamma_1\\ \gamma_2\end{bmatrix}$, thus recovering the original situation of bare elliptic curves. This amounts to identifying b with $gb\lambda$ for $g \in \mathrm{GL}_2^+(\mathbb{Z})$ and $\lambda \in \mathbb{C}^*$:

$$
(4.4) \qquad \boxed{\begin{array}{c}\text{Elliptic curves up}\\ \text{to isomorphism}\end{array}} \quad \xleftrightarrow{\;P_n\;} \quad \boxed{\begin{array}{c}\text{Lattices in } \mathbb{C} \text{ up to}\\ \text{similarity}\end{array}}
$$

The space $\mathfrak{M}_1 = \mathrm{GL}_2^+(\mathbb{Z})\backslash\mathfrak{B}^+/\mathbb{C}^*$ of lattices in $\mathbb{C}$, up to similarities, is the usual moduli space of elliptic curves. The universal elliptic curve $E_{\mathfrak{M}_1}$ lies above $\mathfrak{M}_1$.

The relationships between the various situations described so far are summarized by the diagram in table 2.

Table 2. Relations between various kinds of elliptic curves

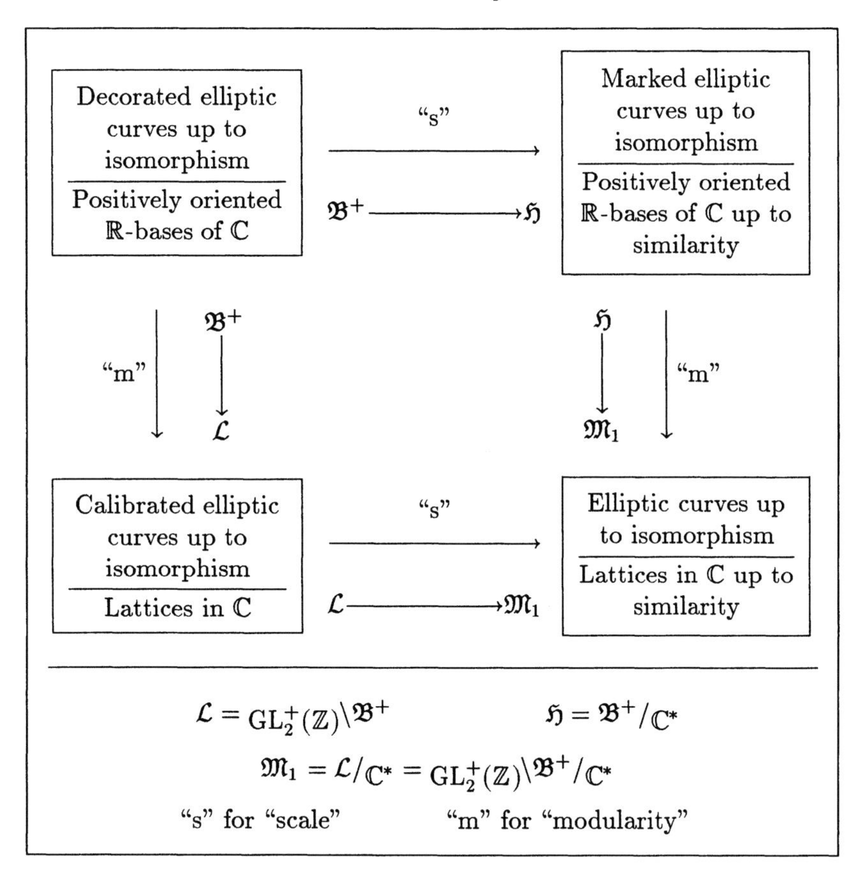

Remark. Both $\mathcal{L}$ and $\mathfrak{M}_1$ possess singularities for the quotient analytic structures, caused by the existence of lattices with non-trivial automorphisms, such as the square and the equilateral triangular lattices, shown on figure 2. These singularities are of a mild type, and one has learned how to deal with them.

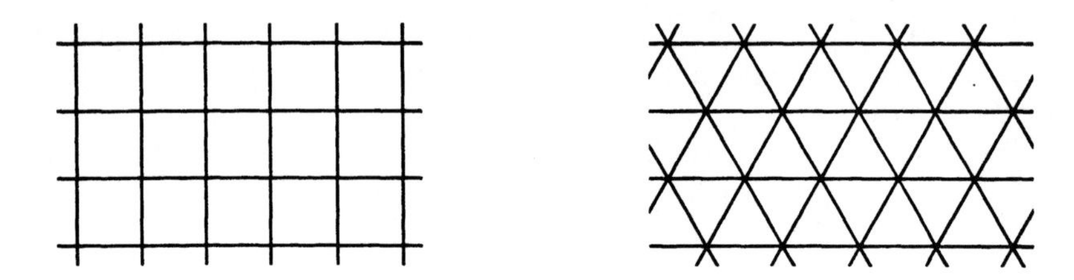

Fig. 2. Square and equilateral lattices

5. Degeneracies

We will content ourselves with a brief description of the following situation: for any analytic family $(E_t)_{t \in T}$ of elliptic curves parametrized by T, the allowed degeneracies when t goes to a limit value t_0 are the appearances of quadratic (nodal) singular points, as shown on figure 3.

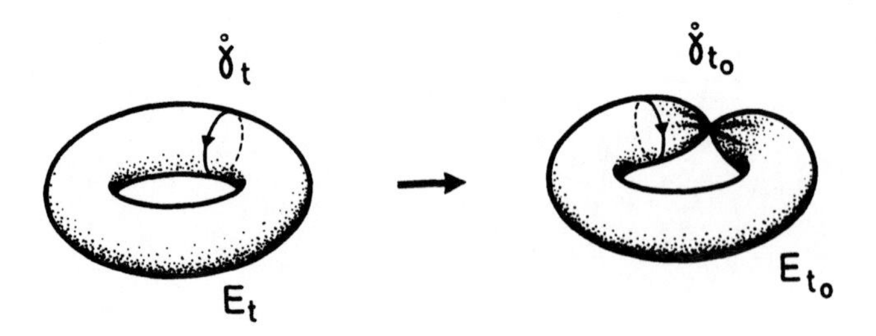

Fig. 3. Nodal singular point and vanishing cycle

The singular curve E_{t_0} can be viewed as a Riemann sphere on which two points have been identified in some well-defined fashion, and the non-singular part is this sphere minus these two points, or a cylinder, or the plane minus one point. What really matters is the existence of a non-trivial 1-cycle $\dot{\gamma}_t$ which becomes homologically trivial (vanishing) at the limit $\dot{\gamma}_{t_0}$. In order to implement this point on the periods $\oint_{\gamma_t} \alpha_t$, one must impose the behavior of the calibrations α_t in the neighbourhood of t_0 (the markings are locally rigid). In terms of the periods $\left|\left[\begin{matrix}\omega_{1,t}\\ \omega_{2,t}\end{matrix}\right]\right|$, one is facing either one of the following two cases

(a) The periods $\omega_{1,t}$ and $\omega_{2,t}$ have finite, $\mathbb{Q}$-dependent, not both vanishing limits ω_{1,t_0} and ω_{2,t_0}.

(b) One of the periods becomes infinite, whereas the other one has a finite, non-vanishing limit.

In both cases, the associated lattice in $\mathbb{C}$ (free Abelian rank-2 subgroup) degenerates into a free Abelian rank-1 subgroup.

The quotient of $\mathbb{C}$ by this subgroup is isomorphic to the multiplicative group $\mathbb{C}^*$. The singular point originates in the (metaphysically puzzling) identification of zero and infinity. Any loop around 0 (and ∞) is a vanishing cycle, whereas any path between 0 and ∞ is a non-vanishing cycle.

One is thus led to employ enlarged moduli spaces $\widehat{\mathfrak{B}}^+$, $\widehat{\mathcal{L}}$, $\widehat{\mathfrak{H}}$, $\widehat{\mathfrak{M}_1}$ of universal parametrizations for decorated, marked, calibrated, or bare elliptic curves, possibly degenerate.

6. Some Decorated and Calibrated Elliptic Curves

6.1 The standard curve

For $\tau \in \mathfrak{H}$, the Poincaré half-plane, consider the positively oriented $\mathbb{R}$-basis $b_\tau = \left|\left[\begin{matrix}2i\pi\tau\\ 2i\pi\end{matrix}\right]\right|$. It generates the lattice $L_\tau = \lfloor b_\tau$, and provides the decorated curve

$$\left\{\mathbb{C}/_{L_\tau}; dz; \left[\begin{matrix}\widetilde{2i\pi\tau}\\ \widetilde{2i\pi}\end{matrix}\right]\right\}, \tag{6.1}$$

We recall that $\widetilde{2i\pi\tau}$ (respectively $\widetilde{2i\pi}$) is the image in $\mathbb{C}/_{L_\tau}$ of the oriented segment joining 0 to $2i\pi\tau$ (respectively $2i\pi$). We will introduce the notation $E_{[\tau]} = \mathbb{C}/_{L_\tau}$. We have thus obtained a canonical section $\mathfrak{H} \overset{st}{\mapsto} \mathfrak{B}^+$ of the surjective map $\mathfrak{B}^+ \mapsto \mathfrak{B}^+/_{\mathbb{C}^*} = \mathfrak{H}$.

6.2 The 'Tate' curve

Let q be a varying element of $\mathbb{C}^*$, and $q_\tau = e^{2i\pi\tau}$ a fixed element of $\mathbb{C}$, with modulus less than 1, $\mathrm{Im}\tau > 0$, $\mathrm{Re}\tau \in [0,1[$ (notice that τ and q_τ are in one-to-one correspondence). The decorated curve $\left\{E_{[\tau]}; dz; \left[\begin{matrix}\widetilde{2i\pi\tau}\\ \widetilde{2i\pi}\end{matrix}\right]\right\}$ can be presented in the equivalent form

$$\left\{\mathbb{C}^*/_{q_\tau^{\mathbb{Z}}}; \frac{dq}{q}, \left[\begin{matrix}\gamma_1\\ \gamma_2\end{matrix}\right]\right\}, \tag{6.2}$$

where γ_1 is the image in $\mathbb{C}/_{q_\tau^{\mathbb{Z}}}$ of the arc $t \mapsto e^{2i\pi\tau t}$, $t \in [0,1]$ of a logarithmic spiral, and γ_2 is the image of the unit circle, described in the direct sense.

One goes from the standard curve to the Tate curve through the exponential map $z \mapsto e^z = q$, under which the periods $2i\pi n$ disappear. We will use the notation $E(q_\tau) = \mathbb{C}/_{q_\tau^{\mathbb{Z}}}$. Notice that the marking jumps when q_τ crosses the cut along the segment $[0,1]$: the marked curve does not depend on q_τ in an analytic way, whereas the calibrated curve does. The interest of this 'Tate curve' lies mainly in the study of moduli near degeneracies ($q_\tau \to 1$).

6.3 The 'Weierstrass' curve

This is a canonical representative for calibrated curves. If L denotes a lattice in $\mathbb{C}$, we define, with the usual normalization coefficients

$$g_2(L) = 60 \sum_{l \in L^{\cdot}} l^{-4} \quad L^{\cdot} = L - \{0\}$$

$$g_3(L) = 140 \sum_{l \in L^{\cdot}} l^{-6} \,. \tag{6.3}$$

One has $\Delta(L) = (g_2(L))^3 - 27(g_3(L))^2 \neq 0$, and the curve $E_{(L)} \subset \mathbb{P}^2(\mathbb{C})$ with equation

$$y^2 t = 4x^3 - g_2(L)xt^2 - g_3(L)t^3 \tag{6.4}$$

is an elliptic curve with neutral element $(0,1,0)$, and the 1-form $\alpha_L = \frac{dx}{y}$ is holomorphic.

Conversely, on any curve of equation

$$y^2 t = 4x^3 - g_2 xt^2 - g_3 t^3 \tag{6.5}$$

with $g_2^3 - 27g_3^2 \neq 0$, which is hence non-singular, hence elliptic, the holomorphic 1-form $\alpha = \frac{dx}{y}$ admits a lattice L of periods, such that we have $g_2(L) = g_2$ and $g_3(L) = g_3$. We have thus exhibited a one-to-one correspondence between Weierstrass calibrated curves represented by $\{(y^2t = 4x^3 - g_2xt^2 - g_3t^3); \frac{dx}{y}\}$ and the lattices of $\mathbb{C}$. One can therefore identify the space $\mathcal{L}$ of lattices in $\mathbb{C}^2$ and the complement in $\mathbb{C}$ of the curve $g_2^3 - 27g_3^2 = 0$.

7. Some Functions on the Set of Positively Oriented Bases

In Section 3 we have emphasized the role of $\mathbb{R}$-analytic functions defined on $\mathbb{C} \times \mathfrak{B}^+$. We will now focus our attention onto the associated 'constants', i. e. the ($\mathbb{R}$- or $\mathbb{C}$-) analytic functions defined on $\mathfrak{B}^+$.

$$\begin{array}{rcl} F : \mathfrak{B}^+ & \longrightarrow & \mathbb{C} \\ b = \left\| \begin{bmatrix} \omega_1 \\ \omega_2 \end{bmatrix} \right\| & \longmapsto & F(b) = F\left(\left\| \begin{bmatrix} \omega_1 \\ \omega_2 \end{bmatrix} \right\| \right) \end{array} \tag{7.1}$$

7.1 Effect of the scaling action: homogeneity

Definition. *The function $F : \mathfrak{B}^+ \mapsto \mathbb{C}$ is said to be homogeneous of weight $(s,t) \in \mathbb{C}^2$ if $F(b\lambda) = \lambda^{-s}\bar{\lambda}^{-t}F(b)$.*

Remark. This definition only makes sense for $s \equiv t\ (2\mathbb{Z})$. If F is $\mathbb{C}$-analytic, only the weights $(k,0)$ with $k \in 2\mathbb{Z}$ are possible.

Example. Eisenstein series

The Eisenstein series

$$G_{s,t}(b) = \sum_{l \in \lfloor b^{\cdot}} l^{-s}\bar{l}^{-t} \tag{7.2}$$

is a function of weight (s,t) whenever it is defined, i. e. for $s \equiv t \quad (2\mathbb{Z})$ and $\mathrm{Re}(s+t) > 2$. For $s = t$, Epstein's zeta functions are recovered. The function

$$G_{2,0}(b) = \sum_{m_1} \sideset{}{'}\sum_{m_2}^{s} (m_1\omega_1 + m_2\omega_2)^{-2} \tag{7.3}$$

has weight $(2,0)$ (notice that this double series is not summable: one has to perform first a symmetric summation over m_2, then a summation over m_1, avoiding the term with $(m_1, m_2) = (0,0)$). The function

$$G^*_{2,0}(b) = \lim_{\epsilon \to 0, \epsilon > 0} G_{2+\epsilon,\epsilon}(b) \tag{7.4}$$

also has weight $(2,0)$, and the function

$$G^*_{1,1}(b) = \lim_{\epsilon \to 0, \epsilon > 0} G_{1+\epsilon,1+\epsilon}(b) \tag{7.5}$$

has weight $(1,1)$. Finally

$$V(b) = \text{covolume of } \lfloor b = \omega_2 \wedge \omega_1 = \mathrm{Im}(\bar{\omega}_2\omega_1) \tag{7.6}$$

has weight $(-1,-1)$.

7.2 Effect of the action of the modular group: modularity

Definition. *A function $F : \mathfrak{B}^+ \to \mathbb{C}$ is said to be lattice-modular (abbreviated in l-modular) if it is $GL_2^+(\mathbb{Z})$-invariant, i. e. if, for any $g \in GL_2^+(\mathbb{Z})$, one has $F(gb) = F(b)$.*

The l-modular functions on $\mathfrak{B}^+$ thus define functions on the quotient space $\mathcal{L} = \mathrm{GL}_2^+(\mathbb{Z})\backslash\mathfrak{B}^+$ of lattices. To any function F on $\mathfrak{B}^+$, its composition with the canonical section st (see Section 6.1) $st : \mathfrak{H} \to \mathfrak{B}^+$ associates a function f on $\mathfrak{H}$

$$f(\tau) = F\left(\left\|\begin{bmatrix} 2i\pi\tau \\ 2i\pi \end{bmatrix}\right\|\right) = F(b_\tau) \tag{7.7}$$

When F is a homogeneous function of weight (s,t), it can be reconstructed from f by

$$F\left(\left\|\begin{bmatrix} \omega_1 \\ \omega_2 \end{bmatrix}\right\|\right) = \lambda^s\bar{\lambda}^t f\left(\frac{\omega_1}{\omega_2}\right) \quad \text{where} \quad \lambda = \frac{2i\pi}{\omega_2} \tag{7.8}$$

Under these conditions, saying that the homogeneous function F of weight (s,t) is l-modular amounts to saying that f is such that

$$f\left(\frac{a\tau+b}{c\tau+d}\right) = (c\tau+d)^s(c\bar{\tau}+d)^t f(\tau) \quad , \quad \begin{pmatrix} a & b \\ c & d \end{pmatrix} \in \mathrm{GL}_2^+(\mathbb{Z})\,. \tag{7.9}$$

Such functions have to obey in particular

$$\begin{aligned} f(\tau+1) &= f(\tau) \\ f\left(-\frac{1}{\tau}\right) &= \tau^s\bar{\tau}^t f(\tau) \end{aligned} \tag{7.10}$$

and these conditions turn out to be sufficient. This is a good place to notice that functions such as θ_3 and η, mentioned in Section 1, do not fit yet in this framework. In order for them to do so, one would have to decorate the elliptic curves in a still better fashion. The first relation in eq. (7.10) suggests to represent f by Fourier-like expansions. In the simplest cases, these expansions read

$$f(\tau) = \sum_{n\geq n_0} a_n q_\tau^n = \phi(q_\tau) \quad \text{where} \quad q_\tau = e^{2i\pi\tau}\,. \tag{7.11}$$

An especially interesting example of a q_τ expansion is that associated with $G_{k,0}$ for k an even integer > 2:
(7.12)

$$\gamma_k(q_\tau) = \frac{1}{(k-1)!}\left[\zeta(1-k) + 2\sum_{n\geq 1}\sigma_{k-1}(n)q_\tau^n\right] \quad \text{where} \quad \sigma_r(n) = \sum_{d|n} d^r\,.$$

8. Periods of Second Kind

We recall the definition of the $\wp$ and ξ Weierstrass functions of z, for fixed $b = \left\| \begin{bmatrix} \omega_1 \\ \omega_2 \end{bmatrix} \right\| \in \mathfrak{B}^+$:

$$\wp(z|b) = z^{-2} + \sum_{l \in \lfloor b^{\cdot}} \left[(z-l)^{-2} - l^{-2} \right]$$

$$\xi(z|b) = -z^{-1} - \sum_{l \in \lfloor b^{\cdot}} \left[(z-l)^{-1} + l^{-1} + zl^{-2} \right] \tag{8.1}$$

Notice that $d\xi = \wp(z)dz$.

The periods ω_μ can be evaluated by $\omega_\mu = \oint_{\gamma_\mu} \alpha$, where $\alpha = dz$ is, up to a factor, the unique holomorphic 1-form invariant under the translations of $\lfloor b$ (form of first kind).

The 1-form $\beta = d\xi = \wp dz$ is meromorphic, and invariant under the translations of $\lfloor b$. Its poles are double, situated on $\lfloor b$, and without residues (form of second kind). Its periods can be defined unambiguously by

$$\eta_\mu = \oint_{\gamma_\mu} \beta \tag{8.2}$$

or else $\xi(z + \omega_\mu) = \xi(z) + \eta_\mu$, $\mu = 1, 2$.

The expression for ξ yields the η_μ immediately

$$\eta_1 = -\omega_1 G_{2,0} \left(\left\| \begin{bmatrix} -\omega_2 \\ \omega_1 \end{bmatrix} \right\| \right)$$

$$\eta_2 = -\omega_2 G_{2,0} \left(\left\| \begin{bmatrix} \omega_1 \\ \omega_2 \end{bmatrix} \right\| \right) \tag{8.3}$$

We notice once more that $G_{2,0}$ is not l-modular, we have indeed

$$G_{2,0} \left(\left\| \begin{bmatrix} \omega_1 \\ \omega_2 \end{bmatrix} \right\| \right) - G_{2,0} \left(\left\| \begin{bmatrix} -\omega_2 \\ \omega_1 \end{bmatrix} \right\| \right) = \frac{\eta_1 \omega_2 - \eta_2 \omega_1}{\omega_1 \omega_2} = \frac{2i\pi}{\omega_1 \omega_2} \tag{8.4}$$

from the Legendre relation $\eta_1 \omega_2 - \eta_2 \omega_1 = 2i\pi$.

It can then be shown that

$$G^*_{2,0} \left(\left\| \begin{bmatrix} \omega_1 \\ \omega_2 \end{bmatrix} \right\| \right) = G_{2,0} \left(\left\| \begin{bmatrix} \omega_1 \\ \omega_2 \end{bmatrix} \right\| \right) - \pi \frac{\bar{\omega}_1}{\omega_1} \frac{1}{V\left(\left\| \begin{bmatrix} \omega_1 \\ \omega_2 \end{bmatrix} \right\| \right)} \tag{8.5}$$

is l-modular (but not $\mathbb{C}$-analytic!). However, $G_{2,0}$ possesses a q_τ-expansion

$$\gamma_2(q_\tau) = -\frac{1}{12} + 2 \sum_{n \geq 1} \sigma_1(n) q_\tau^n \tag{8.6}$$

in agreement with the γ_k for $k > 2$, since $\zeta(-1) = -\frac{1}{12}$.

9. Vector Fields on $\mathfrak{B}^+$ and $\mathcal{L}$

9.0. The vector fields under consideration will be $\mathbb{R}$-analytic, and will therefore operate as differentiations on the algebra of $\mathbb{R}$-analytic functions on $\mathfrak{B}^+$ or $\mathcal{L}$.

(a) A vector field D on $\mathfrak{B}^+$ is said to be *l-modular* if it commutes with the action of $\mathrm{GL}_2(\mathbb{Z})$

$$D \cdot F_g = (D \cdot F)_g \quad \text{where} \quad F_g(b) = F(gb) \tag{9.1}$$

An l-modular field D on $\mathfrak{B}^+$ is thus in one-to-one correspondence with a vector field on $\mathcal{L}$: the D-derivative $D \cdot F$ of an l-modular function is l-modular.

(b) A vector field D on $\mathfrak{B}^+$ (or on $\mathcal{L}$, if it is l-modular) is said to be *homogeneous of weight* (k,l) (with the usual restrictions) if it transforms a homogeneous function of weight (s,t) into a homogeneous function of weight $(s+k, t+l)$:

$$D \cdot F_{[\lambda]} = \lambda^k \bar{\lambda}^l (D \cdot F)_{[\lambda]} \quad \text{where} \quad F_{[\lambda]}(b) = F(b\lambda)\ ,\ \lambda \in \mathbb{C} \tag{9.2}$$

In order to describe these fields, we will work with co-ordinates $\omega_1, \bar{\omega}_1, \omega_2, \bar{\omega}_2$, on $\mathfrak{B}^+$. We recall that the latter is the open set of $\mathbb{C}^2$ defined by the inequality $(1/2i)(\bar{\omega}_2\omega_1 - \bar{\omega}_1\omega_2) > 0$.

9.1 Euler vector fields. The field

$$E = \omega_1 \frac{\partial}{\partial \omega_1} + \omega_2 \frac{\partial}{\partial \omega_2} \tag{9.3}$$

is $\mathbb{C}$-analytic, and its conjugate

$$\overline{E} = \bar{\omega}_1 \frac{\partial}{\partial \bar{\omega}_1} + \bar{\omega}_2 \frac{\partial}{\partial \bar{\omega}_2} \tag{9.4}$$

is $\overline{\mathbb{C}}$-analytic. The fields E and $\overline{E}$ are l-modular, and homogeneous of weight $(0,0)$. The condition that F is homogeneous of weight (s,t) reads

$$E \cdot F = -sF \qquad \overline{E} \cdot F = -tF\ . \tag{9.5}$$

9.2 Halphen-Fricke vector fields. The field

$$H = \eta_1 \frac{\partial}{\partial \omega_1} + \eta_2 \frac{\partial}{\partial \omega_2} \tag{9.6}$$

is $\mathbb{C}$-analytic, and its conjugate

$$\overline{H} = \bar{\eta}_1 \frac{\partial}{\partial \bar{\omega}_1} + \bar{\eta}_2 \frac{\partial}{\partial \bar{\omega}_2} \tag{9.7}$$

is $\overline{\mathbb{C}}$-analytic. The fields H and $\overline{H}$ are l-modular, and homogeneous of respective weights $(2,0)$ and $(0,2)$. We have for instance

$$H \cdot \omega_\mu = \eta_\mu \qquad H \cdot \bar{\omega}_\mu = 0 \qquad H \cdot \left(\frac{\omega_1}{\omega_2}\right) = \frac{2i\pi}{\omega_2^2} \tag{9.8}$$

It turns out to be advantageous to express H in terms of q_τ-expansions. If $\phi = \phi(q_\tau)$ is the expansion of a $\mathbb{C}$-analytic l-modular function F, homogeneous of weight $(k,0)$, a straightforward calculation shows that the expansion of $H \cdot F$ reads

$$H_k \cdot \phi = q_\tau \frac{d\phi}{dq_\tau} + k\gamma_2 \phi \tag{9.9}$$

Since the field H is l-modular, it allows one to differentiate the functions defined on $\mathcal{L}$. The 'modular forms' are certain $\mathbb{C}$-analytic functions on $\mathcal{L}$, essentially those which admit a q_τ-expansion of the form $\sum_{n\geq 0} a_n q_\tau^n$. They constitute a well-known graded algebra, with the weight $(k,0)$ as a grading. One starts with the graded algebra of polynomials in g_2 (weight $(4,0)$) and g_3 (weight $(6,0)$), and one makes the discriminant $\Delta = g_2^3 - 27g_3^2$ invertible. It is then easily realized, with the help of the q_τ-expansions, that

$$H \cdot g_2 = 6g_3 \qquad H \cdot g_3 = \frac{g_2^2}{3} \ . \tag{9.10}$$

In other words, H operates on the algebra of modular forms as

$$H = 6g_3 \frac{\partial}{\partial g_2} + \frac{g_2^2}{3} \frac{\partial}{\partial g_3} \ , \tag{9.11}$$

which can be applied to the discriminant itself, yielding

$$H \cdot \Delta = H \cdot (g_2^3 - 27g_3^2) = 6g_2^2 g_3 - 6g_2^2 g_3 = 0 \ . \tag{9.12}$$

This implies that the q_τ-expansion δ of Δ obeys

$$q_\tau \frac{d\delta}{dq_\tau} + \left(-1 + 24 \sum_{n\geq 1} \sigma_1(n) q_\tau^n \right) \delta = 0 \ , \tag{9.13}$$

or else

$$\frac{q_\tau}{\delta} \frac{d\delta}{dq_\tau} = 1 - 24 \sum_{d\geq 1} \frac{dq_\tau^d}{1 - q_\tau^d} \tag{9.14}$$

and

$$\delta = C q_\tau \prod_{n\geq 1} (1 - q_\tau^n)^{24} \ . \tag{9.15}$$

The numerical prefactor C can be determined by means of the q_τ-expansions, namely

(9.16)
$$\delta = q_\tau \prod_{n\geq 1}(1-q_\tau^n)^{24}$$

a classical formula, which admits many alternative proofs.

9.3 Weil vector fields. Besides the E, $(\overline{E})$, H, $(\overline{H})$ fields, which are $\mathbb{C}$- $(\overline{\mathbb{C}})$-analytic (notice that the $\mathbb{C}$-analyticity can be expressed as $\overline{E}\cdot F = \overline{H}\cdot F = 0$), the following Weil field

(9.17)
$$W = \frac{\pi}{V\left(\left|\left[\begin{matrix}\omega_1\\ \omega_2\end{matrix}\right]\right|\right)}\left(\bar{\omega}_1\frac{\partial}{\partial\omega_1}+\bar{\omega}_2\frac{\partial}{\partial\omega_2}\right)$$

is also very useful. It is 1-modular and homogeneous of weight $(2,0)$. Among other properties, one has $W\cdot\omega_\mu = (\pi/V)\bar{\omega}_\mu$, $W\cdot\bar{\omega}_\mu = 0$, and, more important, $W\cdot V = 0$ hence $W^2\cdot\omega_\mu = 0$.

9.4 Relations between these fields. Even on purely dimensional grounds, one may suspect the existence of relations between the various vector fields. As a matter of fact, the essential relation, concerning the weight $(2,0)$,

(9.18)
$$H + G^*_{2,0}E + W = 0\ ,$$

is nothing else but equation (8.5) in a disguised form.

In order for our knowledge of modular differential operators to be more complete, we now introduce the Ramanujan differential equations. We recall the action of H on $G_{4,0}$ and $G_{6,0}$

(9.19)
$$\begin{aligned} H\cdot G_{4,0} &= 14G_{6,0}\\ H\cdot G_{6,0} &= \frac{60}{7}G_{4,0}^2\ . \end{aligned}$$

As far as $G^*_{2,0}$ is concerned, an analogous formula holds, namely

(9.20)
$$H\cdot G^*_{2,0} = 5G_{4,0} - (G^*_{2,0})^2\ .$$

In order to prove the latter result, it is sufficient to notice that

(9.21)
$$\frac{H\cdot\eta_\mu}{\omega_\mu} = \frac{H^2\cdot\omega_\mu}{\omega_\mu}$$

is modular of weight $(4,0)$. The values are identical for $\mu = 1$ and $\mu = 2$, as a consequence of the Legendre relation. This quantity is thus equal to $G_{4,0}$, up to a factor easily determined:

(9.22)
$$\frac{H^2\cdot\omega_\mu}{\omega_\mu} = 5G_{4,0}\ .$$

In terms of q_τ-expansions, this result reads

(9.23)
$$q_\tau\frac{d\gamma_2}{dq_\tau} + 12\gamma_2^2 = \gamma_4\ .$$

as stated by Ramanujan.

10. Non-Conclusion

The theme of this lecture has only been broached here, and it is rather arbitrary to stop at this point. Even if we restrict ourselves to the moduli of elliptic curves over $\mathbb{C}$, we are far from having exhausted all the possible decorations, among which the level structures, defined in terms of the points of finite order (division of periods), of isogenies, etc. For instance, the Hecke operators have thus been left out (see the contribution of Zagier to this volume). Another important aspect has not been discussed (although we would have liked very much to do so), namely the arithmetical theory of moduli. The moduli spaces, which we have mostly studied from the viewpoint of the $\mathbb{R}$-analytic structure, can also be thought of as algebraic, defined on number rings which are 'close to' $\mathbb{Z}$. This arithmetical rigidity explains why some particular integers are often encountered in the above formulas. There is also a geometrical approach to these structures of arithmetic (and Arakelov) modular manifolds. Finally, we have just mentioned, en passant, the question of critical points (s,t) with the condition $\mathrm{Re}(s+t) = 2$. Very subtle ambiguities show up in the summation of the divergent Eisenstein series. One would like to see some decoration revealing a hidden 'quantum' symmetry which would govern those various summation procedures.

Scruple and Acknowledgements. The terminology used above (in Section 3) is somewhat non-canonical. The expression 'marked Riemann surface' is used by Gunning in the above sense and seems to become standard. However, expressions as 'decorated' and 'calibrated' are used here for convenience, and may be in conflict with other conventions (for instance 'calibrated geometries' of Thurston). In Section 7, we preferred 'lattice modular' in order to avoid possible confusions which could result from using 'latticial' or shortly 'modular'. We hope that the reader will not be puzzled by these appellations.

This Chapter was initially intended to be an appendix to the chapters relating the lectures by P. Cartier and J. B. Bost. However, Cartier's and Bost's contributions were available only very late, time was lacking to introduce the suitable adaptations required by the situation. Nevertheless, the arithmetic theory of elliptic functions and modular forms is also presented in the contributions by Cohen and Zagier, so that the present 'little appendix ' does not remain an orphan. It remains only for me to thank the organizers of such an 'unforgettable' meeting, especially M. Waldschmidt, J. -M. Luck and P. Moussa for their friendly persistence and efficient help.

References

We first list a few standard textbooks, which include additional references:

Husemöller, D. (1987) Elliptic curves, G T M **111** Springer

Koblitz, N. (1984) Introduction to elliptic curves and modular forms, G T M **97** Springer

Lang, S. (1973) Elliptic functions, (2d ed. 1987), G T M **112** Springer
Lang, S. (1976) Introduction to modular forms, Grundlehren **222** Springer
Gunning, R. C. (1962) Lectures on modular forms, Ann. of. Math. Studies **48** Princeton University Press
Serre, J. P. (1968) Cours d'arithmétique, Presses Universitaires de France
Shimura, G. (1971) Introduction to the arithmetic theory of automorphic functions, Iwanami Shoten, and Princeton University Press

For the arithmetic theory of moduli, the unavoidable references are:

Deligne, P. and Rapoport, M. (1973) Les schémas de modules de courbes elliptiques, in: Modular functions in one variable II, Lectures Notes in Mathematics, **349**
Katz, N. and Mazur, B. (1985) Arithmetic moduli of elliptic curves, Ann. of Math. Studies **108** Princeton University Press

It is always stimulating to read 'old' treatises. The following include (among many other subjects) the theory of modular derivations:

Jordan, C. (1912) Cours d'analyse, II, chap. VII, Gauthier Villars
Fricke, R. (1916 and 1923) Die elliptische Funktionen und ihre Anwendungen, I and II, Teubner
Halphen, G. H. (1886) Traité des fonctions elliptiques et de leurs applications I, Gauthier Villars

Now a masterpiece:

Weil, A. (1976) Elliptic functions according to Eisenstein and Kronecker E M G **88** Springer

The previous reference is of course an invitation to a new reading of:

Eisenstein, G. Werke I, II, Chelsea, reprint
Kronecker, L. Werke I, II, III, IV, V, Chelsea, reprint

Chapter 6

Galois Theory, Algebraic Number Theory, and Zeta Functions

by Harold M. Stark

1. Galois Theory

1.1 Introduction

The goal of these lectures is to explain the fundamentals of Galois theory and algebraic number theory and to get the reader to the point that he or she can make routine calculations. The following material has little in the way of prerequisites. Many examples will be given along the way and sometimes the examples will serve instead of formal proofs. In the last Section, we introduce the zeta functions of algebraic number fields. These functions can be factored into products of L-functions according to representations of Galois groups. Especially here, we will proceed by example. These lectures have been abstracted from my long promised forthcoming book [2]. A complete treatment of any of the topics here is somewhat beyond the length requirements of these lectures although we will come surprisingly close in some instances. I would also recommend Hecke's book [1] which has been a classic for almost 70 years.

1.2 Algebraic Extensions

A field is a system of 'numbers' in which all the usual rules for arithmetic using $+, -, \times, \div$ that we learned in grade school hold. Among these laws is the commutative law for multiplication, $ab = ba$; thus the quaternions don't give us a field. Examples include the complex numbers, $\mathbb{C}$; the real numbers, $\mathbb{R}$; and the rational numbers, $\mathbb{Q}$. Fields containing $\mathbb{Q}$ are said to be of *characteristic zero.* All subfields of $\mathbb{C}$ are of characteristic zero. In these lectures, we will deal almost exclusively with fields of characteristic zero, and the most interesting fields, the 'algebraic number fields', will be subfields of $\mathbb{C}$. *We suppose throughout that we are either in characteristic zero or are dealing with finite fields.* Everything we say should be taken in this context. The reader who has not seen any of this before can safely think of subfields of $\mathbb{C}$ except where noted. Outside of a few brief moments, all examples will be concerned with fields of characteristic zero.

Galois theory concerns the relations between fields and certain bigger fields. Given a field k, there is a standard construction giving bigger fields. Suppose that x is not in k but is 'consistent' with the arithmetic of k. This means that x may be arithmetically combined with elements of k or itself and that the usual laws of arithmetic hold for these combinations. Almost always in these lectures k will be a subfield of $\mathbb{C}$ and x will be either a variable or a number in $\mathbb{C}$. We define two extensions of k:

$$k[x] = \{\text{polynomials in } x \text{ with coefficients in } k\} ,$$

(the set of polynomials in x with coefficients in k) and

$$k(x) = \left\{ \begin{array}{l} \text{ratios of polynomials in } x \text{ whose coefficients} \\ \text{are in } k \text{ and whose denominators are non-zero} \end{array} \right\} .$$

In $k[x]$, we always have the operations of $+, -, \times$ and in $k(x)$ we have all four operations $+, -, \times, \div$. Thus $k(x)$ is always a field. In the same manner, we can use more than one x. For example

$$k[x, y] = k[x][y]$$

consists of polynomials in x and y with coefficients in k and

$$k(x, y) = k(x)(y)$$

consists of ratios of polynomials in x and y with coefficients in k whose denominators are non-zero. Again, $k(x, y)$ is a field. In general, if a field K contains another field k, then we say that K is an *extension field* of k or sometimes just an *extension* of k. Thus $k(x)$ and $k(x, y)$ are both extensions of k.

We now define algebraic extension fields. We say that α is *algebraic* over k if $f(\alpha) = 0$ for some non-zero $f(x)$ in $k[x]$ where x is a variable. The field $k(\alpha)$ is called an *algebraic extension* of k. For example, $\mathbb{C} = \mathbb{R}(i)$ where i is a root of the equation $x^2 + 1 = 0$. In this example $\mathbb{C} = \mathbb{R}[i]$ as well. Of all the polynomials $f(x)$ which can be used for α, it is customary to take f of minimal degree. This is equivalent to saying that $f(x)$ is irreducible in $k[x]$. If

$$f(x) = a_m x^m + a_{m-1} x^{m-1} + \cdots + a_0 \tag{1}$$

in $k[x]$ is irreducible with $a_m \neq 0$ and $f(\alpha) = 0$, we can divide through by a_m and thus assume that α satisfies an equation $f(\alpha) = 0$ where $f(x)$ is irreducible in $k[x]$ and where $a_m = 1$. A polynomial in (1) with $a_m = 1$ is called a *monic polynomial.* We sometimes refer to the monic irreducible polynomial satisfied by α over k as the *defining polynomial* for α over k. The defining polynomial for α over k is unique since the difference of two such polynomials would give a polynomial equation for α of lower degree. If the defining polynomial for α over k is of degree m, we say that the *degree* of α over k is m and that $K = k(\alpha)$ is also of *degree m over k*. It is also common to say that m is the *relative degree* of *K over k*. The notation for this is

$$[K:k]=m\ .$$

The degree of $K=k(\alpha)$ over k does not depend upon which α is used to give the field $k(\alpha)$ since for an algebraic α,

$$k(\alpha)=k[\alpha]=k+k\alpha+k\alpha^2+\cdots+k\alpha^{m-1}$$

is a vector space over k with basis $1,\alpha,\alpha^2,\cdots,\alpha^{m-1}$. (We will show that $k(\alpha)=k[\alpha]$ shortly.) Thus m serves as the dimension of this space and is independent of the choice of generator of the field. Since degrees are dimensions of vector spaces, we see that there is a multiplicative relation for degrees of successive extensions. If M in turn is an extension of K of degree n, then

$$[M:k]=[M:K][K:k]\ .$$

Indeed, if $M=K(\theta)$ where θ is of relative degree n over K, then

$$\begin{aligned}M&=K+K\theta+K\theta^2+\cdots+K\theta^{n-1}\\&=(k+k\alpha+k\alpha^2+\cdots+k\alpha^{m-1})+(k+k\alpha+k\alpha^2+\cdots+k\alpha^{m-1})\theta\\&\qquad+\cdots+(k\alpha+k\alpha^2+\cdots+k\alpha^{m-1})\theta^{n-1}\ ,\end{aligned}$$

so that M is a vector space of dimension mn over k with basis elements $\{\alpha^i\theta^j\}_{0\le i\le m-1,0\le j\le n-1}$. The fact that $k(\alpha)=k[\alpha]$ is due to the fact that the inverse of a non-zero algebraic number is an algebraic number and is indeed a polynomial in the original number. We show this for α. If α is defined as a zero of the polynomial $f(x)$ in (1) , and we have already taken out all the spare factors of x so that $a_0\neq 0$, then α^{-1} is given by

$$\alpha^{-1}=-a_0^{-1}(a_m\alpha^{m-1}+a_{m-1}\alpha^{m-2}+\cdots+a_1)\ .$$

More generally, all sums, differences, products and quotients of algebraic numbers are algebraic numbers. We illustrate the general proof by showing that $\sqrt{2}+\sqrt{3}$ is algebraic. In general, if α is algebraic over k of degree m and β is algebraic over k of degree n, we begin with the column vector $(\alpha^i\beta^j)_{0\le i\le m-1,0\le j\le n-1}$ of mn entries. Here we have the vector $\begin{pmatrix}1\\\sqrt{2}\\\sqrt{3}\\\sqrt{6}\end{pmatrix}$. We multiply this vector by $\sqrt{2}+\sqrt{3}$ and write the result as a rational matrix times the original vector:

$$\begin{pmatrix}1\\\sqrt{2}\\\sqrt{3}\\\sqrt{6}\end{pmatrix}(\sqrt{2}+\sqrt{3})=\begin{pmatrix}0&1&1&0\\2&0&0&1\\3&0&0&1\\0&3&2&0\end{pmatrix}\begin{pmatrix}1\\\sqrt{2}\\\sqrt{3}\\\sqrt{6}\end{pmatrix}$$

Thus $\sqrt{2}+\sqrt{3}$ is an eigenvalue of the matrix

$$\begin{pmatrix} 0 & 1 & 1 & 0 \\ 2 & 0 & 0 & 1 \\ 3 & 0 & 0 & 1 \\ 0 & 3 & 2 & 0 \end{pmatrix}$$

and so is algebraic since it is a root of the eigenvalue equation. This process does not always give the polynomial of minimal degree. For example, the same column vector can be used for $\sqrt{2}\sqrt{3} = \sqrt{6}$ which is of degree two even though the eigenvalue equation is of degree four.

Let us explore this process further. Suppose that α is of degree m and that β is in $k(\alpha)$. Since any element of $k(\alpha)$ is a unique linear combination of the numbers $1, \alpha, \cdots, \alpha^{m-1}$ with coefficients in k, we may write

$$\begin{pmatrix} 1 \\ \alpha \\ \vdots \\ \alpha^{m-1} \end{pmatrix} \beta = M(\beta) \begin{pmatrix} 1 \\ \alpha \\ \vdots \\ \alpha^{m-1} \end{pmatrix},$$

where $M(\beta)$ is an $m \times m$ matrix with entries in k which is uniquely determined by β since the i-th row of this equation just expresses the number $\alpha^{i-1}\beta$ of $k(\alpha)$ as a unique linear combination of the basis elements. Suppose that β_1 and β_2 are in $k(\alpha)$. Then we have

$$\begin{aligned} M(\beta_1 + \beta_2) \begin{pmatrix} 1 \\ \alpha \\ \vdots \\ \alpha^{m-1} \end{pmatrix} &= \begin{pmatrix} 1 \\ \alpha \\ \vdots \\ \alpha^{m-1} \end{pmatrix} (\beta_1 + \beta_2) = \begin{pmatrix} 1 \\ \alpha \\ \vdots \\ \alpha^{m-1} \end{pmatrix} \beta_1 + \begin{pmatrix} 1 \\ \alpha \\ \vdots \\ \alpha^{m-1} \end{pmatrix} \beta_2 \\ &= M(\beta_1) \begin{pmatrix} 1 \\ \alpha \\ \vdots \\ \alpha^{m-1} \end{pmatrix} + M(\beta_2) \begin{pmatrix} 1 \\ \alpha \\ \vdots \\ \alpha^{m-1} \end{pmatrix} \\ &= (M(\beta_1) + M(\beta_2)) \begin{pmatrix} 1 \\ \alpha \\ \vdots \\ \alpha^{m-1} \end{pmatrix}. \end{aligned}$$

Because these matrices are unique, we get

$$M(\beta_1 + \beta_2) = M(\beta_1) + M(\beta_2).$$

In like manner, we get

$$M(\beta_1 \beta_2) = M(\beta_1) M(\beta_2) = M(\beta_2) M(\beta_1).$$

In particular, these matrices commute under multiplication.

The first row of $M(\beta)$ reconstructs β when multiplied by the column vector ${}^t(1, \alpha, \cdots, \alpha^{m-1})$ (the 't' denotes the transpose). Therefore the correspondence

$\beta \to M(\beta)$ gives an $m \times m$ matrix representation of the field $k(\alpha)$ with matrix entries in k. For example, for the extension $\mathbb{C} = \mathbb{R}(i)$, we have

$$\begin{pmatrix} 1 \\ i \end{pmatrix}(x+iy) = \begin{pmatrix} x & y \\ -y & x \end{pmatrix}\begin{pmatrix} 1 \\ i \end{pmatrix} .$$

The matrix $M(x+iy) = \begin{pmatrix} x & y \\ -y & x \end{pmatrix}$ gives the standard 2×2 matrix representation of the complex numbers over the reals. With somewhat more work, we could show directly from scratch that the collection of matrices $\{M(\beta)\}$ forms a field thereby showing that there is an extension field of k in which the equation $f(x) = 0$ has a root. This would free us from thinking only of subfields of complex numbers.

Now we are ready to introduce the concept of conjugate algebraic numbers. Suppose that $f(x)$ is an irreducible monic polynomial of degree n in $k[x]$ and that $f(\alpha) = 0$. We factor f as

$$f(x) = \prod_{i=1}^{n}(x - \alpha^{(i)}) .$$

(Our major examples come with k a subfield of $\mathbb{C}$; we may thus think of the factorization as taking place over $\mathbb{C}$.) Then α is one of the $\alpha^{(i)}$. The $\alpha^{(i)}$ are called the *algebraic conjugates* of α over k. Just as α gives an extension $K = k(\alpha)$ of k, corresponding to the $\alpha^{(i)}$ are the *conjugate fields* $K^{(i)} = k(\alpha^{(i)})$ of K over k. If β in $k(\alpha)$ is given by

$$\beta = \sum_{j=1}^{n} b_j \alpha^{j-1} ,$$

with the b_j in k, then we define the *field conjugate* $\beta^{(i)}$ of β in $K^{(i)}$ by

$$\beta^{(i)} = \sum_{j=1}^{n} b_j (\alpha^{(i)})^{j-1} .$$

The map $\beta \mapsto \beta^{(i)}$ gives an isomorphism between the fields K and $K^{(i)}$. This is easy to see since the matrix representation $M(\beta)$ for β depends only on the b_j and the defining polynomial for α, but not on the zero of $f(x)$ chosen to be α. Therefore all the $K^{(i)}$ are isomorphic to the field of matrices $\{M(\beta)\}$ and hence to each other. As an example, complex conjugation gives field conjugates of $\mathbb{C}$ over $\mathbb{R}$ and we have

$$\begin{pmatrix} 1 \\ -i \end{pmatrix}(x-iy) = \begin{pmatrix} x & y \\ -y & x \end{pmatrix}\begin{pmatrix} 1 \\ -i \end{pmatrix}$$

with the same matrix $\begin{pmatrix} x & y \\ -y & x \end{pmatrix}$. Thus the familiar statement that the conjugate of the sum (difference, product, quotient) of two complex numbers is the sum (difference, product, quotient) of the conjugates of the two numbers is an illustration of the concept of conjugate fields. Note in the conjugation process that all elements of k are preserved.

There is a potential confusion in the word 'conjugate' since we have both algebraic conjugates and field conjugates of numbers β. The confusion however is restricted to multiplicities: every field conjugate of β is an algebraic conjugate of β. This is clear because if $g(x)$ is in $k[x]$ and $g(\beta) = 0$, then $g(\beta^{(i)}) = 0$ as well since

$$g(\beta^{(i)}) = (g(\beta))^{(i)} = 0^{(i)} = 0 .$$

1.3 Normal Extensions and Galois Groups

Sometimes several of the $\alpha^{(i)}$ are all in the same field. The case of most interest is the case where all the $\alpha^{(i)}$ are in $k(\alpha)$ already. For instance, both i and $-i$ are in $\mathbb{C} = \mathbb{R}(i)$. In the case that every $\alpha^{(i)}$ is in $k(\alpha)$, we say that $k(\alpha)$ is a *normal extension* of k or that $k(\alpha)$ is *normal* over k. This is not always the case. An example where we do not have a normal extension is the field $K = \mathbb{Q}(2^{1/3})$ over $\mathbb{Q}$ where we use the real cube root of 2. Thus every number in K is real and hence the other two conjugates of $2^{1/3}$ (the complex cube roots of 2) cannot be in K and so K is not normal over $\mathbb{Q}$.

When $K = k(\alpha)$ is normal over k, the n conjugation maps $\beta \mapsto \beta^{(i)}$ are maps from K to K and so are automorphisms of the field K over k (the 'over k' part of this means that every element of k is fixed by the automorphism). By the same argument as above, if $f(x)$ is in $k[x]$ and $f(\alpha) = 0$ then an automorphism of K over k takes α to a number which also satisfies the equation $f(x) = 0$. Thus any automorphism of K over k takes α to one of the $\alpha^{(i)}$. This determines the automorphism completely. Hence any automorphism of K which fixes k must be one of the n conjugation automorphisms. Thus the automorphisms of K over k form a group of order $[K : k]$ which is called the *Galois group* of K over k. For example, for $\mathbb{C} = \mathbb{R}(i)$, the identity and complex conjugation are the two elements in the Galois group of $\mathbb{C}/\mathbb{R}$. ('$\mathbb{C}/\mathbb{R}$' is pronounced, '$\mathbb{C}$ over $\mathbb{R}$'.) For notational purposes, when K is normal over k, we will often write the Galois group of K/k as $\mathrm{Gal}(K/k)$ or even as $G(K/k)$.

In our examples, we will frequently encounter permutations. For example the cyclic permutation σ of four letters a, b, c, d given by

$$\sigma : a \mapsto b \mapsto c \mapsto d \mapsto a ,$$

is written as

$$\sigma = (a, b, c, d) .$$

We may write σ with any of the four letters first,

$$\sigma = (a, b, c, d) = (b, c, d, a) = (c, d, a, b) = (d, a, b, c)$$

since all four of these describe the same permutation. It is often said that (a, b, c, d) is a four *cycle*. Any permutation on a finite number of letters can be resolved into cycles. For example, the permutation on five letters given by

$$\rho : a \mapsto b,\ b \mapsto d,\ c \mapsto e,\ d \mapsto a,\ e \mapsto c$$

is a combination of the three cycle (a, b, d) and (c, e). These cycles are said to be *disjoint* which means that they have no letters in common. We write the permutation ρ as the product of these disjoint cycles,

$$\rho = (a, b, d)(c, e) = (c, e)(a, b, d) \ .$$

Given a permutation, the way we do this is to pick one of the letters and then follow what the permutation does to it until we come back to the starting letter. For example, if we start with a, ρ takes a to b, b to d and then d to a giving the three cycle (a, b, d). If there are letters that have not yet been accounted for, then we pick one of them and repeat the process. In our example, we have not yet accounted for c. We see that ρ takes c to e and then e to c. This gives the two cycle (c, e). The order of writing these cycles doesn't matter since they are disjoint: when we perform the operations indicated by the various cycles, it does not matter which we perform first; the result will be ρ in either order. This independence of order is only true when the cycles are disjoint as we will see shortly.

Among the cycles are the one cycles. These are rather special and are usually not even written. For example, the permutation τ of the four letters a, b, c, d which just interchanges b and d,

$$\tau : b \mapsto d \mapsto b,\ a \mapsto a,\ c \mapsto c \ ,$$

can be written as

$$\tau = (a)(b, d)(c) \ .$$

But the two one cycles (a) and (c) fix everything and we usually just write τ as

$$\tau = (b, d) \ .$$

This means that b is taken to d and d is taken to b and everything else is left fixed.

The two permutations σ and τ act on the same four letters, but they don't commute. Thus the notation $\sigma\tau = (a, b, c, d)(b, d)$ depends upon which permutation is performed first. Since I read from left to right, I will take $\sigma\tau$ to mean that σ is first and τ operates on the result. Thus for example

$$\sigma\tau = (a, d)(b, c) \ .$$

We use the recipe above to find the cycle structure of $\sigma\tau$: σ takes a to b followed by τ which takes b to d; σ takes d to a and τ takes a to a; σ takes b to c and τ takes c to c; σ takes c to d and τ takes d to b. In like manner,

$$\tau\sigma = (a,b)(c,d) \ .$$

The permutations σ and τ generate an eight element permutation group called the dihedral group of order eight. Geometrically, these permutations consist of the eight ways that a square can be picked up and set exactly upon itself.

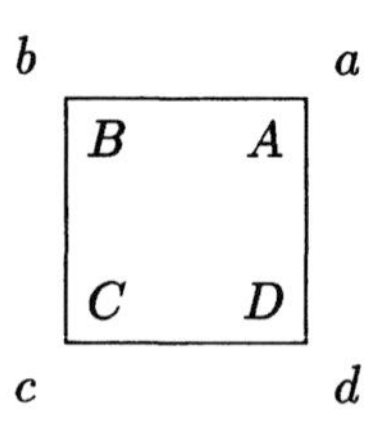

The letters a, b, c, d are fixed in the plane and serve to describe what is done to the square by a given operation. For example, σ represents a 90° counterclockwise rotation of the square about its center and τ turns the square over by rotating about the ac diagonal. In order to more easily keep track of successive operations, we will label the vertices of the square in its starting position by A, B, C, D. The labels will travel with the square while a, b, c, d remain fixed.

As an illustration, we calculate $\sigma\tau\sigma$. First we apply σ and then follow by τ (flipping about the ac diagonal and *not* about the AC diagonal) as shown.

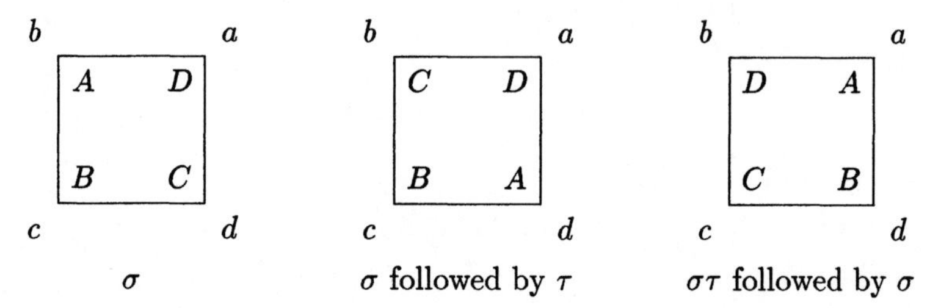

σ $\qquad$ σ followed by τ $\qquad$ $\sigma\tau$ followed by σ

Note that the effect of $\sigma\tau$ upon the original square is that of $(a,d)(b,c)$ which agrees with our permutation calculation above. A final 90° counterclockwise rotation gives $\sigma\tau\sigma$. The net result of all this is that $\sigma\tau\sigma$ has the same effect upon the original position of the square as just τ alone:

$$\sigma\tau\sigma = (b,d) = \tau \ .$$

We note that this is equivalent to

$$\tau\sigma = \sigma^{-1}\tau \ ,$$

which is very useful in making calculations. The dihedral group will appear in several of the examples of this Section.

Because Galois groups can be represented as permutation groups on the conjugates of field generators, if σ and τ are elements of $G(K/k)$, the notation

$\sigma\tau$ will mean σ is performed first and τ operates on the result. This in turn fixes our notation for a Galois group element acting on a number (the usual left-right difficulty). If σ is an element of $G(K/k)$ and β is in K, we will write β^σ (a minority will write this as $\beta \circ \sigma$) for the image of β under σ. Thus,

$$\beta^{\sigma\tau} = (\beta^\sigma)^\tau$$

Example. We take the field $K = \mathbb{Q}(\sqrt{2}, \sqrt{3})$ over $\mathbb{Q}$. Let $a = \sqrt{2}$, $b = -\sqrt{2}$, $c = \sqrt{3}$ $d = -\sqrt{3}$. The field K contains all the conjugates of its two generators, $\sqrt{2}$ and $\sqrt{3}$. K also contains two intermediate fields, $\mathbb{Q}(\sqrt{2})$ and $\mathbb{Q}(\sqrt{3})$:

$$\begin{array}{ccc} & K = \mathbb{Q}(\sqrt{2}, \sqrt{3}) & \\ & / \quad \backslash & \\ \mathbb{Q}(\sqrt{2}) & & \mathbb{Q}(\sqrt{3}) \\ & \backslash \quad / & \\ & \mathbb{Q} & \end{array}$$

Now K is normal over the subfield $\mathbb{Q}(\sqrt{2})$. The Galois group is of order 2 and the nontrivial element takes $\sqrt{3}$ to $-\sqrt{3}$. Thus

$$\mathrm{Gal}(K/\mathbb{Q}(\sqrt{2})) = \{1, (c, d)\} .$$

In like manner, K is normal over $\mathbb{Q}(\sqrt{3})$. The non-trivial element takes $\sqrt{2}$ to $-\sqrt{2}$ and so

$$\mathrm{Gal}(K/\mathbb{Q}(\sqrt{3})) = \{1, (a, b)\} .$$

We now have three elements of the four element group $(G(K/\mathbb{Q})$. We find the fourth element by multiplication to be $(a, b)(c, d)$ and so

$$\mathrm{Gal}(K/\mathbb{Q}) = \{1, (a, b), (c, d), (a, b)(c, d)\} .$$

Note also that $\sqrt{6} = \sqrt{2}\sqrt{3}$ is in K and so $\mathbb{Q}(\sqrt{6})$ is a third intermediate quadratic field in K. The corresponding Galois group is

$$\mathrm{Gal}(K/\mathbb{Q}(\sqrt{6})) = \{1, (a, b)(c, d)\} .$$

This example can be reached another way also. Let

$$e = \sqrt{2} + \sqrt{3},\ f = \sqrt{2} - \sqrt{3},\ g = -\sqrt{2} + \sqrt{3},\ h = -\sqrt{2} - \sqrt{3} .$$

These are the four conjugates of $\sqrt{2} + \sqrt{3}$. For example, the element (c, d) of $\mathrm{Gal}(K/\mathbb{Q})$ takes e to f. The numbers e, f, g, h are distinct and so e is a fourth degree algebraic number over $\mathbb{Q}$. Thus $K = \mathbb{Q}(\sqrt{2} + \sqrt{3})$. The three elements of $\mathrm{Gal}(K/\mathbb{Q})$ other than the identity can now be written as permutations on e, f, g, h:

$$\begin{array}{ll}
\sqrt{3} \mapsto -\sqrt{3},\ \sqrt{2} \to \ \ \sqrt{2} & \text{corresponds to } (e,f)(g,h) \\
\sqrt{3} \mapsto \ \ \sqrt{3},\ \sqrt{2} \to -\sqrt{2} & \text{corresponds to } (e,g)(f,h) \\
\sqrt{3} \mapsto -\sqrt{3},\ \sqrt{2} \mapsto -\sqrt{2} & \text{corresponds to } (e,h)(f,g)
\end{array}$$

and

$$\mathrm{Gal}(K/\mathbb{Q}) = \{1,\ (e,f)(g,h),\ (e,g)(f,h),\ (e,h)(f,g)\}\ .$$

Abstractly, this group is the same group as before, the elementary abelian $(2,2)$ group, also sometimes called the Klein four group. However, as a permutation group, the permutation structure has changed. It is only as an abstract group that the Galois group has a unique meaning.

1.4 The Fundamental Theorem of Galois Theory

We remind the reader that *we are assuming throughout that we are either in characteristic zero or are dealing with finite fields.* Outside of these restrictions, Galois theory has a different flavor, best left to graduate algebra courses. All algebraic number fields are of characteristic zero. We state without proof the fundamental theorem from which all other results in this Subsection will easily follow.

Theorem. *Let K be a normal algebraic extension of k with Galois group $G = \mathrm{Gal}(K/k)$. There is a 1-1 correspondence between all the subgroups H of G and all the fields L between K and k such that if H corresponds to L, then $H = \mathrm{Gal}(K/L)$ and L is the set of all elements of K fixed by every element of H.*

Remark. We call L the *fixed field* of H, but it is very important to remember that this means every element of L is fixed. For example, every automorphism of H takes K to K, but the non-identity elements of H do not preserve K elementwise. The fixed field of $\{1\}$ is K and the fixed field of G is k.

The picture we draw looks like,

$$\left.\begin{array}{c} \left.\begin{array}{c} K \\ | \\ L \end{array}\right\} H \\ | \\ k \end{array}\right\} G$$

Since $H = \mathrm{Gal}(K/L)$, we see that $[K:L] = |H|$, the order of H. Therefore, $[L:k] = [K:k]/[K:L] = |G|/|H|$. This says that if we split G up into cosets of H, then the number of cosets is $[L:k]$. As a simple illustration of the fundamental theorem, we may take $K = \mathbb{Q}(\sqrt{2}, \sqrt{3})$. We have seen all the subgroups of $G = \text{Gal}(K/\mathbb{Q})$ above and how they correspond to the intermediate fields. We take the fundamental theorem as fact and now derive some of its consequences.

Theorem. *Suppose that L_1 corresponds to H_1 and L_2 corresponds to H_2 in the correspondence above. Then $L_1 \supset L_2$ if and only if $H_1 \subset H_2$.*

Proof. The Proof is easy. Suppose that $L_1 \supset L_2$. If h is an element of H_1 then h fixes every element of L_1. Since L_2 is contained in L_1, h fixes every element of L_2 also. Thus h is in H_2. Therefore $H_1 \subset H_2$. Conversely, suppose that $H_1 \subset H_2$. If β is in L_2, then β is fixed by every element of H_2. Since H_1 is contained in H_2, β is fixed by every element of H_1 also. Thus β is in L_1. Therefore $L_1 \supset L_2$. □

The picture for this situation is

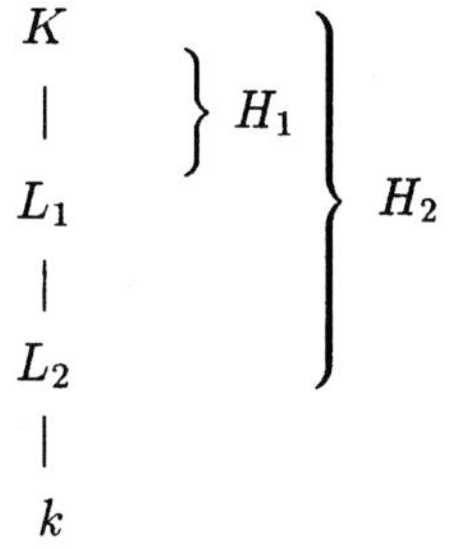

$$\begin{array}{c} K \\ | \\ L_1 \\ | \\ L_2 \\ | \\ k \end{array} \quad \begin{array}{l} \left.\begin{array}{c} \\ \\ \end{array}\right\} H_1 \\ \\ \end{array} \quad \left.\begin{array}{c} \\ \\ \\ \\ \end{array}\right\} H_2$$

The two extremes fit this pattern also. $G = \text{Gal}(K/k)$ contains all the subgroups of G while k is contained in all the intermediate fields between K and k. Also $\{1\} = \text{Gal}(K/K)$ is contained in all the subgroups of G and K contains all the intermediate fields. As an example of the applications, consider the situation,

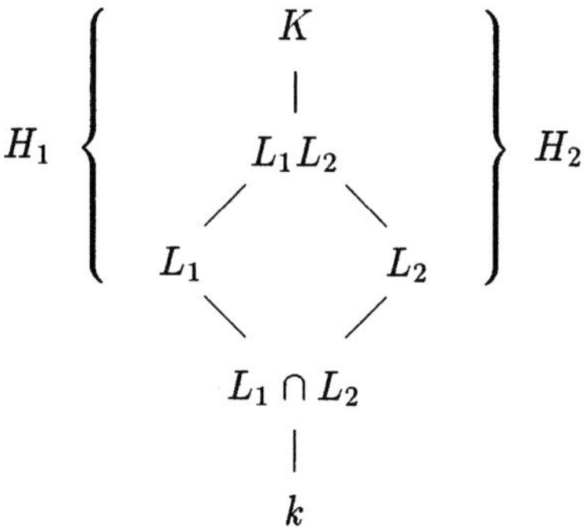

$$\begin{array}{ccccc} & & K & & \\ & & | & & \\ H_1 \Bigg\{ & & L_1 L_2 & & \Bigg\} H_2 \\ & / & & \backslash & \\ & L_1 & & L_2 & \\ & \backslash & & / & \\ & & L_1 \cap L_2 & & \\ & & | & & \\ & & k & & \end{array}$$

The field $L_1 L_2$ is by definition the field generated by the elements of L_1 and L_2 and is the smallest field containing L_1 and L_2. Thus $L_1 L_2$ corresponds to $H_1 \cap H_2$ which is the biggest subgroup of G contained in both H_1 and H_2. In like manner, $L_1 \cap L_2$ is the biggest field contained in both L_1 and L_2 and so corresponds to $\langle H_1, H_2\rangle$, the group generated by the elements of H_1 and H_2, which is the smallest subgroup of G containing both H_1 and H_2.

Now let us work out a more interesting example of the fundamental theorem. We take $K = \mathbb{Q}(2^{1/4}, i)$

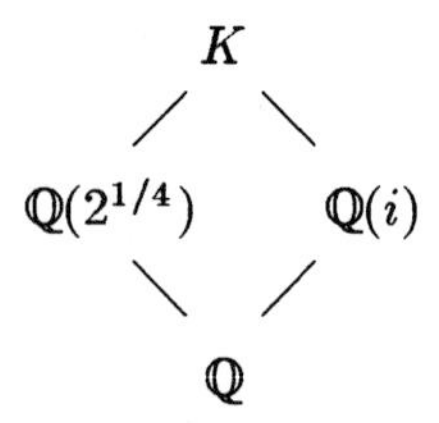

which is a normal eighth degree extension of $\mathbb{Q}$. Thus $G = \mathrm{Gal}(K/\mathbb{Q})$ is a group of order eight. We may present it as a permutation group on the conjugates of the generators. The four conjugates of $2^{1/4}$ are

$$a = 2^{1/4},\ b = i2^{1/4},\ c = -2^{1/4},\ d = -i2^{1/4}\ .$$

Once we know what an automorphism of K does to a and b, we know what it does to $i = b/a$ and so we may present G as a permutation group on the four letters a, b, c, d. We build up G as before by first finding subgroups of G. We begin with $K = \mathbb{Q}(i)(2^{1/4})$ which is an extension of $\mathbb{Q}(i)$ of degree four. The four conjugates of $2^{1/4}$ over $\mathbb{Q}(i)$ are a, b, c, d and any element of $\mathrm{Gal}(K/\mathbb{Q}(i))$ is determined completely by knowing to which of a, b, c, d the element sends $2^{1/4}$. In particular, there is an element σ of $\mathrm{Gal}(K/\mathbb{Q}(i))$ which sends $2^{1/4}$ to $i2^{1/4}$. In other words $a^\sigma = b$. It is easy to see what σ does to b, c, and d. We have $b^\sigma = (ia)^\sigma = i(a^\sigma)$ since i is fixed by σ. In other words, $b^\sigma = -a = c$. In like manner, $c^\sigma = d$ and $d^\sigma = a$. Therefore σ is represented by the four cycle, $\sigma = (a, b, c, d)$ and

$$\mathrm{Gal}(K/\mathbb{Q}(i)) = \{1, \sigma, \sigma^2, \sigma^3\}\ .$$

This gives us four elements of our group of order eight. As soon as we have a fifth element, we can generate the whole group. This is easy to get by considering K as a normal quadratic extension of $\mathbb{Q}(2^{1/4})$,

$$K = \mathbb{Q}(2^{1/4})(i)\ .$$

Thus there is an automorphism τ of $K/\mathbb{Q}(2^{1/4})$ which takes i to $-i$, but fixes all of $\mathbb{Q}(2^{1/4})$. This automorphism is just complex conjugation since $\mathbb{Q}(2^{1/4})$ is a real field. But now, since τ fixes $2^{1/4}$, it is clear that τ preserves a and c while interchanging b and d, $\tau = (b, d)$ and

$$\mathrm{Gal}(K/\mathbb{Q}(2^{1/4})) = \{1, \tau\}\ .$$

Since τ is not in the group generated by σ (this is either because the permutation (b, d) is clearly not a power of σ or because τ fixes $2^{1/4}$, while the only element of $\mathrm{Gal}(K/\mathbb{Q}(i))$ which fixes $2^{1/4}$ is the identity and τ is not the identity), we get another four elements from multiplying by τ. This gives us eight elements in all and thus we must now have all of G:

$$G = \{1, \sigma, \sigma^2, \sigma^3, \tau, \sigma\tau, \sigma^2\tau, \sigma^3\tau\}\,.$$

The group G is in fact the dihedral group of order 8 introduced above. The dihedral group is rich in subgroups. G has

One subgroup of order one:	$\{1\}$,
Five subgroups of order two:	$\{1, \sigma^2\}$ and four groups generated by the four flips about the axes and diagonals,
Three subgroups of order four:	$\{1, \sigma, \sigma^2, \sigma^3\}$ $\{1, \tau, \sigma^2, \sigma^2\tau\}$ $\{1, \sigma\tau, \sigma^2, \sigma^3\tau\}$,
One subgroup of order eight:	G.

Each of these subgroups corresponds to a field. Besides the groups of orders one and eight which correspond to K and $\mathbb{Q}$, the complete picture is

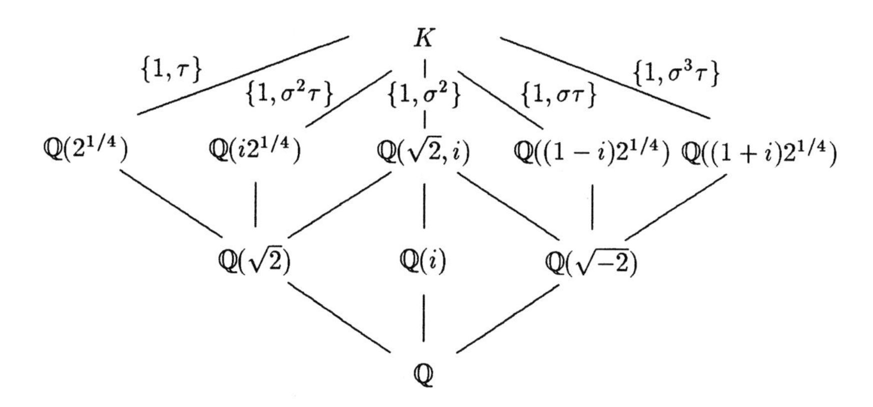

Each of these can be easily verified. For example, since $\sqrt{-2} = (i2^{1/4})(2^{1/4}) = ba$ and $\sigma\tau = (a, d)\,(b, c)$, we see that

$$(\sqrt{-2})^{\sigma\tau} = cd = (-2^{1/4})(-i2^{1/4}) = \sqrt{-2}$$

and so $\sigma\tau$ is in $\mathrm{Gal}(K/\mathbb{Q}(\sqrt{-2}))$. Likewise $\sigma\tau$ fixes $(1-i)2^{1/4} = 2^{1/4} - i2^{1/4}$ which is of degree two over $\mathbb{Q}(\sqrt{-2})$ (square it and see). The first three quartic fields in the list above were easy enough to come by. But when we don't know them beforehand, the last two are most easily found by Galois theory. Let α be any number in K. From α, we construct numbers in K fixed by $\sigma\tau$ by creating invariant combinations of conjugates of α such as $\alpha + \alpha^{\sigma\tau}$. Indeed, since $(\sigma\tau)^2 = 1$,

$$(\alpha + \alpha^{\sigma\tau})^{\sigma\tau} = \alpha^{\sigma\tau} + \alpha = \alpha + \alpha^{\sigma\tau} ,$$

and so $\alpha + \alpha^{\sigma\tau}$ is preserved. With $\alpha = 2^{1/4}$, we get the generator $(1-i)2^{1/4}$ of the fourth quartic field in the list. Not every α in K will give a generator of $\mathbb{Q}((1-i)2^{1/4})$ in this way, and in some situations for other fields K, no analogue of α will work. We will explore this process of using Galois theory to describe the intermediate fields in a later example.

1.5 Conjugate fields

In the last example, not every field between K and $\mathbb{Q}$ was normal over $\mathbb{Q}$. We now look at the conjugate fields more closely from the point of view of Galois theory. Suppose that K is a normal extension of k with Galois group $G(K/k)$. Let L be an intermediate field and suppose L corresponds to the subgroup H of G via the fundamental theorem so that $H = G(K/L)$. If g is in G, then L^g is one of the conjugate fields of L. Indeed, if $L = k(\alpha)$, then $L^g = k(\alpha^g)$. Further, the mapping from L to L^g depends only on the coset Hg of H since for all h in H, $\alpha^{hg} = \alpha^g$. As we run through the $[G : H]$ cosets of H in G, we run through all the conjugates of α. This is because $\prod(x - \alpha^{Hg})$ is invariant under G and so is divisible by the defining polynomial of α; since the degrees are the same, this is the defining polynomial of α. Therefore the L^{Hg} run through all the conjugate fields of L. We see easily that L^{Hg} is fixed by $g^{-1}Hg$:

$$(L^{Hg})^{g^{-1}Hg} = (L^g)^{g^{-1}Hg} = L^{gg^{-1}Hg} = L^{Hg} ,$$

and that L^g corresponds to the subgroup $g^{-1}Hg$ of G via the fundamental theorem. The subgroup $g^{-1}Hg$ of G is said to be *conjugate* to the subgroup H. Thus conjugate subgroups correspond to conjugate fields.

This is an excellent point to introduce relative norms and traces. If β is in L, we define the *relative norm* of β *from* L *to* k to be the product of all the field conjugates of β over k, and the *relative trace* of β from L to k to be the sum of all the field conjugates of β over k. Up to a plus or minus sign these numbers are the constant and x^{n-1} coefficients of the nth degree polynomial $\prod(x-\beta^{Hg})$. Since the whole polynomial is invariant under G, all of the coefficients are in k. We write $\mathbb{N}_{L/k}(\beta)$ and $\mathrm{tr}_{L/k}(\beta)$ for the relative norm and trace of β. When $k = \mathbb{Q}$, it is customary to just speak of the *norm* and *trace* of β and when the field L being dealt with is clear, just write $\mathbb{N}(\beta)$ and $\mathrm{tr}(\beta)$.

In the case that L/k is normal, all the conjugate fields of L are the same and thus all the $g^{-1}Hg$ must be the same as well:

$$g^{-1}Hg = H \qquad \text{for all } g \text{ in } G .$$

Alternatively, this condition may be stated as

$$Hg = gH \qquad \text{for all } g \text{ in } G .$$

In this form, the condition says in words that every right coset of H is a left coset also (and we don't even have to know which of Hg and gH is the right coset). Such a subgroup of G is called a *normal subgroup* of G. The converse is also true, if H is normal then all the conjugate fields $L^{Hg} = L$ and so L is normal over k. Thus in the correspondence of the fundamental theorem, intermediate normal fields correspond precisely to the normal subgroups of G. We see further in the case that L/k is normal that $G(L/k) = G/H$, the quotient group of cosets Hg with the group structure $(Hg_1)(Hg_2) = H(g_1g_2)$. The whole group G, and the identity subgroup $\{1\}$ are always normal subgroups of G; they correspond to k/k which is always normal over k and K/k which is normal over k by hypothesis.

In our last example, $\mathbb{Q}(2^{1/4}, i)/\mathbb{Q}$, some of the intermediate fields are normal over $\mathbb{Q}$ and some aren't. All three subgroups of order four are normal subgroups of G and correspond to the three quadratic fields $\mathbb{Q}(\sqrt{2})$, $\mathbb{Q}(i)$ and $\mathbb{Q}(\sqrt{-2})$ which are normal over $\mathbb{Q}$. (Indeed, this gives a Galois theory version of the easy group theory fact that a subgroup of index two of a finite group G is a normal subgroup: it corresponds to a quadratic extension of the ground field and quadratic extensions are visibly always normal extensions. This is even a rigorous proof: every finite group is a Galois group for a relative normal algebraic extension of some ground field. This is because every group is a subgroup of a symmetric group of permutations and the symmetric groups are all known to be Galois groups over $\mathbb{Q}$. By the fundamental theorem, all subgroups of symmetric groups are relative Galois groups as well.)

Of the five subgroups of G of order two, only one is normal. Namely the subgroup $\{1, \sigma^2\}$ which corresponds to the intermediate field, $\mathbb{Q}(\sqrt{2}, i)$, which is also visibly normal over $\mathbb{Q}$, since it contains all the conjugates of its generators, $\sqrt{2}$ and i.

1.6 Finite Fields

Our next example concerns finite field extensions. The ground field will be the finite field of p elements, $\mathbb{F}_p$. One way this field arises is via congruences of integers modulo p. Two integers a and b are said to be *congruent modulo p* if $a - b$ is divisible by p. The notation for this is

$$a \equiv b \pmod{p} .$$

The set of all integers congruent to a given integer a is called the *congruence class* of $a \pmod{p}$ or the *residue class* of $a \pmod{p}$. The reason for the terminology 'residue class' is that if a is divided by p giving a quotient q and remainder (residue) r, $a = pq + r$, then $a \equiv r \pmod{p}$. There are p possible remainders and so p residue classes $(\operatorname{mod} p)$. Suppose we denote the residue class of $a \pmod{p}$ by $\bar{a}$. As Gauss discovered, the arithmetic of integers gives rise to arithmetic of residue classes via the rules,

$$\bar{a} + \bar{b} = \overline{(a+b)} , \qquad \bar{a}\bar{b} = \overline{(ab)} ;$$

the calculations can be done using any representatives of the residue classes and the resulting residue class will always be the same. In the case when p is a prime, arithmetic of residue classes gives us the finite field of p elements, $\mathbb{F}_p$.

Let $k = \mathbb{F}_p$. Suppose that $g(x)$ is an irreducible polynomial in $k[x]$ of degree f and that a zero $\bar{\alpha}$ of g gives rise to an extension $K = k(\bar{\alpha})$ of k. (We write $\bar{\alpha}$ to match the residue class notation here and in Section 2.) The vector space version of K,

$$K = k + k\bar{\alpha} + \cdots + k(\bar{\alpha})^{f-1} ,$$

shows that $K = \mathbb{F}_q$ is a finite field of $q = p^f$ elements. As is well known, there is only one field of q elements. However, what this means is that two such fields are isomorphic and since conjugate fields are isomorphic, we do not instantly get the corollary that K/k is normal. However, this is the case and we will shortly prove it.

We begin with an example. The polynomial $x^2 + \bar{1}$ is an irreducible polynomial of degree two in $\mathbb{F}_3[x]$. Thus a zero $\bar{\imath}$ gives rise to the field $\mathbb{F}_9$ of nine elements

$$\mathbb{F}_9 = \{\bar{a} + \bar{b}\bar{\imath} \mid \bar{a}, \bar{b} \in \mathbb{F}_3\} .$$

This field is normal over $\mathbb{F}_3$ since the conjugate $-\bar{\imath}$ of $\bar{\imath}$ is clearly in $\mathbb{F}_9$. Thus there is a two element Galois group $G(\mathbb{F}_9/\mathbb{F}_3)$ whose non-identity element takes $\bar{\imath}$ to $-\bar{\imath}$. But there is a more interesting way to describe this automorphism. In $\mathbb{F}_3[x, y]$ we have the identities,

$$(x + y)^3 = x^3 + \bar{3}x^2y + \bar{3}xy^2 + y^3 = x^3 + y^3 \qquad (xy)^3 = x^3y^3 .$$

Thus raising every element of $\mathbb{F}_9$ to the third power provides an automorphism of $\mathbb{F}_9$. This automorphism preserves $\mathbb{F}_3$ since $\bar{0}, \bar{1}, \bar{2}$ cubed give $\bar{0}, \bar{1}, \bar{2}$ respectively. (In fact there are no non-trivial automorphisms of the finite field of p elements since $\bar{1}$ must be preserved and every other element of the field is a sum of $\bar{1}$'s and so must be preserved as well.) The polynomial equation $x^3 - x = \bar{0}$ has at most three roots in $\mathbb{F}_3$ (or any other field for that matter) and we have accounted for all three of them. Thus the map $\bar{\beta} \mapsto (\bar{\beta})^3$ must be the non-trivial automorphism of $G(\mathbb{F}_9/\mathbb{F}_3)$. For example,

$$(\bar{\imath})^3 = (\bar{\imath})^2\bar{\imath} = (-\bar{1})\bar{\imath} = -\bar{\imath} ,$$

as advertised. We will now show that this can be done in general for finite field extensions.

Again let $k = \mathbb{F}_p$ be the finite field of p elements where p is a prime and let K be an extension of k of degree f so that K has $q = p^f$ elements. It is known that the non-zero elements of K form a cyclic group of order $q-1$ under multiplication. Suppose that $\bar{\theta}$ is a generator of this group. Then $K = k(\bar{\theta})$ also since every non-zero element of K is already a power of $\bar{\theta}$ and so is certainly in $k(\bar{\theta})$. Furthermore, $(\bar{\theta})^{q-1} = \bar{1}$ and $q - 1$ is the minimal positive such power of $\bar{\theta}$ giving $\bar{1}$. Therefore $(\bar{\theta})^q = \bar{\theta}$ and for $1 < j < q$, $(\bar{\theta})^j \neq \bar{\theta}$. When $K = k$, this is a combination of the old Fermat theorem that $a^p \equiv a \pmod{p}$ for all integers a and the existence of a 'primitive root $\pmod{p}$'.

Now we are ready to find the automorphisms of K/k. We define a map σ of K to k by

$$\sigma : \bar{\beta} \mapsto (\bar{\beta})^p .$$

Then σ preserves every element of k and is an automorphism of K. The latter is because in a field of characteristic p,

$$(x+y)^p = x^p + y^p \qquad \text{and} \qquad (xy)^p = x^p y^p$$

are identities for variable x and y. The automorphism σ is called the *Frobenius automorphism* of K/k. This automorphism is the basis of a whole series of automorphisms and maps in number theory and geometry all of which are also called Frobenius automorphisms and Frobenius maps and are of fundamental importance. We will meet Frobenius automorphisms of number fields in the next two Sections.

The number $(\bar{\theta})^p$ is a conjugate of $\bar{\theta}$, but is not $\bar{\theta}$ itself. Indeed, by repeating the application of σ, we get several distinct conjugates of $\bar{\theta}$,

$$\bar{\theta}, (\bar{\theta})^\sigma = (\bar{\theta})^p, (\bar{\theta})^{\sigma^2} = (\bar{\theta})^{p^2}, \cdots, (\bar{\theta})^{\sigma^{f-1}} = (\bar{\theta})^{p^{f-1}} ,$$

until we finally we get to

$$(\bar{\theta})^{\sigma^f} = (\bar{\theta})^{p^f} = (\bar{\theta})^q = \bar{\theta}$$

again. Thus we have found f distinct conjugates of $\bar{\theta}$, all in K, and so K/k is normal. Further we have found all f automorphisms of K/k:

$$G(K/k) = \langle \sigma \rangle$$

is a cyclic group of order f generated by σ.

1.7 Jacobians in Genus Two

We now give one last example of the applications of Galois theory. It is an example that is not well known among mathematicians either. We will use Galois theory to construct the function field for the Jacobian of a Riemann surface of genus two. (If the words are meaningless, don't worry; the essence of the example has nothing to do with the terminology.) Let x and y be variables related by an equation of the form,

$$y^2 = f(x) , \tag{2}$$

where $f(x)$ is a polynomial in $\mathbb{C}[x]$ without repeated roots. We thus get a quadratic field extension $\mathbb{C}(x,y)$ of $\mathbb{C}(x)$. Let us suppose in this example that $f(x)$ is a quintic polynomial. In the language of complex variable theory, we are talking about a Riemann surface of genus two. In the language of algebraic geometry, we are talking about a hyperelliptic curve of genus two, this being the general case of curves of genus two. The word, 'hyperelliptic' is because when $f(x)$ is a polynomial of degree 3 or 4, we are dealing with a classical elliptic curve, and this terminology in turn is because the equation $y^2 = f(x)$ can be parameterized by elliptic functions in this case. Again, all the excess terminology will not actually appear in the example itself.

A point P on the curve is given by its coordinates, (x, y) where x and y satisfy the equation (2). For genus two, the Jacobian of the curve essentially consists of unordered pairs of points P_1, P_2 on the curve. (Technically, there is a collapse: all pairs of the shape (x, y), $(x, -y)$ correspond to just one point on the Jacobian. Because of this collapse, we are not dealing with a smooth version of the Jacobian. The collapse is able to take place because in two or more complex dimensions, a $0/0$ limit can have an infinite number of limiting values according to how the limit point is approached. In our case, this happens along the so-called theta divisor.) Just as a meromorphic function on the curve is just an element of $\mathbb{C}(x, y)$, so we have meromorphic functions on the Jacobian. They are the elements of $\mathbb{C}(x_1, y_1, x_2, x_2)$ which are symmetric in the points $P_1 = (x_1, y_1)$ and $P_2 = (x_2, y_2)$ on the curve. The set of such elements is a field. This field is called the *function field* of the Jacobian. For our purposes here, we may take this as the definition. It is the goal of this example to describe this field.

Let

$$u = x_1 + x_2, \ v = x_1 x_2, \ w = y_1 + y_2 \, .$$

Clearly u, v and w are in the function field for the Jacobian. We claim that the function field for the Jacobian is $\mathbb{C}(u, v)(w)$ which is a quartic extension of $\mathbb{C}(u, v)$. The quartic extension part of this claim is easy to see. First, we have

$$w^2 = y_1^2 + y_2^2 + 2y_1 y_2 = f(x_1) + f(x_2) + 2y_1 y_2 \, .$$

But $f(x_1) + f(x_2)$ is a symmetric function of the two independent variables x_1 and x_2 and so is a polynomial in the two base symmetric functions in x_1 and x_2: $x_1 + x_2$ and $x_1 x_2$. In other words $f(x_1) + f(x_2)$ is a polynomial in u and v. Thus $y_1 y_2$ is in $\mathbb{C}(u, v)(w)$ and this explains why it wasn't needed in the list of generators. We now repeat the squaring:

$$(w^2 - f(x_1) - f(x_2))^2 = (2y_1 y_2)^2 = 4f(x_1)f(x_2)$$

and again, $f(x_1)f(x_2)$ is also a polynomial in u and v. Thus we get a quartic equation for w over $\mathbb{C}(u, v)$. (The quartic equation is clearly irreducible. Given values for u and v we can determine x_1 and x_2 as an unordered pair, but x_1 and x_2 only determine $\pm y_1$ and $\pm y_2$ and thus there are four possible values for w which cannot be distinguished solely from the knowledge of x_1 and x_2, let alone from just u and v.)

However, while we were easily able to see that $y_1 y_2$ is in $\mathbb{C}(u, v)(w)$, it is not so obvious that every symmetric function of the coordinates of two points P_1 and P_2 must be in $\mathbb{C}(u, v)(w)$. For example, it is not immediately clear how to write $x_1 y_1 + x_2 y_2$ in terms of u, v and w. We will use Galois theory to show how to do it. The roots of

$$X^2 - (x_1 + x_2)X + x_1x_2 = 0$$

are just x_1 and x_2. Thus, $\mathbb{C}(x_1, x_2)$ is a quadratic extension of $\mathbb{C}(u, v)$. Again, this really is a quadratic extension of $\mathbb{C}(u, v)$ itself; the knowledge of just u and v can not possibly tell us which of the two roots is which. Given x_1, a further quadratic extension gives us y_1 and given x_2, another quadratic extension gives us y_2. Thus we have the chain of fields,

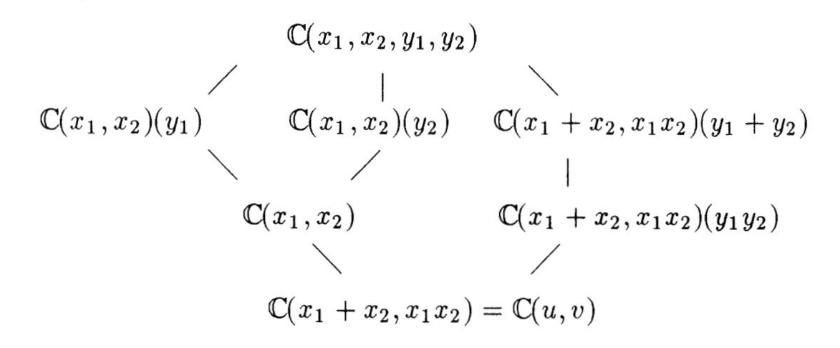

The top field, $\mathbb{C}(x_1, x_2, y_1, y_2)$ is clearly a normal extension of $\mathbb{C}(u, v)$ of degree 8 and once again, the Galois group

$$G = G(\mathbb{C}(x_1, x_2, y_1, y_2)/\mathbb{C}(u, v))$$

will turn out to be the dihedral group of order 8. We will express the group as permutations on the six letters $x_1, x_2, y_1, -y_1, y_2, -y_2$. Note that since $-y_1 = (-1)y_1$ and $-y_2 = (-1)y_2$, once we know the images of y_1 and y_2 under an element of G, we know what happens to $-y_1$ and $-y_2$. We begin by finding the four elements of the group which fix $\mathbb{C}(x_1, x_2)$. Besides the identity, there is clearly the element $(y_2, -y_2)$ which comes from looking at the quadratic extension $\mathbb{C}(x_1, x_2, y_1, y_2) = \mathbb{C}(x_1, x_2)(y_1)(y_2)$ of $\mathbb{C}(x_1, x_2)(y_1)$. Likewise, by looking at the quadratic extension $\mathbb{C}(x_1, x_2, y_1, y_2) = \mathbb{C}(x_1, x_2)(y_2)(y_1)$ of $\mathbb{C}(x_1, x_2)(y_2)$, we find the element $(y_1, -y_1)$ in our Galois group. The product of these two elements of order two gives the fourth desired element of G and we have

$$G(\mathbb{C}(x_1, x_2, y_1, y_2)/\mathbb{C}(x_1, x_2)) = \{1, (y_1, -y_1), (y_2, -y_2), (y_1, -y_1)(y_2, -y_2)\} .$$

We now have four elements of our total group, G. One more and we can find all eight by multiplication with these four. The remaining four elements don't fix the field $\mathbb{C}(x_1, x_2)$ elementwise. Thus the remaining four elements interchange x_1 and x_2; that is to say that they all have the two cycle (x_1, x_2) in their cycle decomposition. Pick one of these four elements. Either y_1 is sent to y_2, or y_1 is sent to $-y_2$. By multiplication with $(y_2, -y_2)$ if necessary, we can assume that our new element sends y_1 to y_2. It therefore also sends $-y_1$ to $-y_2$. Likewise, by multiplying through by $(y_1, -y_1)$ if necessary, we can assume that our new element sends y_2 to y_1 rather than $-y_1$. Again, this results in $-y_2$ being sent to $-y_1$. Thus we see that $(x_1, x_2)(y_1, y_2)(-y_1, -y_2)$ is in our whole Galois group. Therefore, the whole group contains the extra four elements,

$$(x_1, x_2)(y_1, y_2)(-y_1, -y_2) ,$$
$$(x_1, x_2)(y_1, y_2)(-y_1, -y_2)(y_1, -y_1) = (x_1, x_2)(y_1, y_2, -y_1, -y_2) ,$$
$$(x_1, x_2)(y_1, y_2)(-y_1, -y_2)(y_2, -y_2) = (x_1, x_2)(y_1, -y_2, -y_1, y_2) ,$$
$$(x_1, x_2)(y_1, y_2)(-y_1, -y_2)(y_1, -y_1)(y_2, -y_2) = (x_1, x_2)(y_1, -y_2)(y_2, -y_1) .$$

Therefore, the whole group is given by $G = G(\mathbb{C}(x_1, x_2, y_1, y_2)/\mathbb{C}(u, v))$, with the following eight elements,

$$G = \left\{ \begin{array}{l} 1,\ (y_1, -y_1),\ (y_2, -y_2),\ (y_1, -y_1)(y_2, -y_2) , \\ (x_1, x_2)(y_1, y_2)(-y_1, -y_2),\ (x_1, x_2)(y_1, y_2, -y_1, -y_2) , \\ (x_1, x_2)(y_1, -y_2, -y_1, y_2),\ (x_1, x_2)(y_1, -y_2)(y_2, -y_1) \end{array} \right\}$$

It is left to the reader to show that this is the dihedral group of order eight again. The dihedral group of order eight is the only one of the five groups of order eight which has exactly two elements of order four and no elements of order eight.

The critical question for us is, what is the subgroup H of G fixing $\mathbb{C}(x_1 + x_2, x_1 x_2)(y_1 + y_2) = \mathbb{C}(u, v)(w)$? The answer is that it is given by

$$H = \{1,\ (x_1, x_2)(y_1, y_2)(-y_1, -y_2)\} ,$$

since clearly the element $(x_1, x_2)(y_1, y_2)(-y_1, -y_2)$ of G fixes the generators $x_1 + x_2$, $x_1 x_2$, $y_1 + y_2$ of $\mathbb{C}(u, v)(w)$. But now we are done. Any rational function of the coordinates of two points $P_1 = (x_1, y_1)$ and $P_2 = (x_2, y_2)$ which is symmetric when P_1 and P_2 are interchanged is fixed by H and so belongs to $\mathbb{C}(u, v)(w)$. This was our original claim!

For example, $x_1 y_1 + x_2 y_2$ is fixed by H and so is in $\mathbb{C}(u, v)(w)$. We now show how Galois theory helps us write $x_1 y_1 + x_2 y_2$ in a form which is visibly in $\mathbb{C}(u, v)(w)$. It is now convenient to give $y_1 y_2$ a name. Let $t = y_1 y_2$. The field $\mathbb{C}(u, v)(w)$ is a quadratic extension of $\mathbb{C}(u, v)(t)$. Indeed, we saw above that w^2 is in $\mathbb{C}(u, v)(t)$. Quadratic extensions are always normal extensions. The non-trivial element of $G(\mathbb{C}(u, v)(w)/\ \mathbb{C}(u, v)(t))$ takes w to $-w$. Thus, if we write

$$\alpha = x_1 y_1 + x_2 y_2 = a + bw$$

where a and b are in $\mathbb{C}(u, v)(t)$, then the non-trivial element of $G(\mathbb{C}(u, v)(w)/\ \mathbb{C}(u, v)(t))$ takes α to $a - bw$. So, if we can find an expression for this in terms of x_1, x_2, y_1, y_2, we can find a and b. But, we can express $G(\mathbb{C}(u, v)(w)/\mathbb{C}(u, v)(t))$ in terms of elements of G. We already know the elements of

$$H = G(\mathbb{C}(x_1, x_2, y_1, y_2)/\mathbb{C}(u, v)(w)) .$$

It is easily checked that the group

$$J = G(\mathbb{C}(x_1, x_2, y_1, y_2)/\mathbb{C}(u, v)(t))$$

consists of the four elements

$$\{1, (x_1, x_2)(y_1, y_2)(-y_1, -y_2), (y_1, -y_1)(y_2, -y_2), (x_1, x_2)(y_1, -y_2)(y_2, -y_1)\}$$

all of which visibly fix t. Either of the two elements not in H (the last two elements) take w to its conjugate over $\mathbb{C}(u, v)(t)$, namely $-w$. We apply one of these to α. The result will be the same in either case, since it depends only on the coset of H used. We will calculate both of these and see. When we apply $(y_1, -y_1)(y_2, -y_2)$ to α, we get

$$x_1(-y_1) + x_2(-y_2) = -x_1y_1 - x_2y_2 = -\alpha ,$$

while when we apply $(x_1, x_2)(y_1, -y_2)(y_2, -y_1)$ to α we get

$$x_2(-y_2) + x_1(-y_1) = -x_2y_2 - x_1y_1 = -\alpha ,$$

the same result, as advertised. Thus we now have two equations for the two unknowns a and b:

$$a + bw = x_1y_1 + x_2y_2 ,$$
$$a - bw = -x_1y_1 - x_2y_2 .$$

Thus $a = 0$ and

$$b = \frac{x_1y_1 + x_2y_2}{w} = \frac{x_1y_1 + x_2y_2}{y_1 + y_2} . \tag{3}$$

We see that $\alpha = bw$ and that the four elements of $G(\mathbb{C}(x_1, x_2, y_1, y_2)/\mathbb{C}(u, v)(t))$ listed above really do preserve b so that b is in $\mathbb{C}(u, v)(t)$. However, b is not visibly in $\mathbb{C}(u, v)(t)$; that is, it is not obvious how b may be written solely in terms of u, v, and t.

To find the expression for b in terms of u, v and t, we now repeat what we just did. The field $\mathbb{C}(u, v)(t)$ is a quadratic extension of $\mathbb{C}(u, v)$ and again t^2 is in $\mathbb{C}(u, v)$ so that the non-trivial conjugate of t over $\mathbb{C}(u, v)$ is just $-t$. We write

$$b = c + dt$$

where c and d are in $\mathbb{C}(u, v)$. Then the non-trivial conjugate of b over $\mathbb{C}(u, v)$ is just $c - dt$. We want the expression for this in terms of x_1, x_2, y_1, y_2. But again, this is easy. It is found by applying any of the four elements of G which aren't in J.

For instance, we may apply the element $(y_2, -y_2)$ to b. When we do so, the result is,

$$\frac{x_1y_1 - x_2y_2}{y_1 - y_2} = c - dt . \tag{4}$$

The result will be the same if we use any of the other three possible group elements. We now have our two equations (3) and (4) for the two unknowns in c and d. We thus get

$$c = \frac{1}{2}\left[\frac{x_1y_1 + x_2y_2}{y_1 + y_2} + \frac{x_1y_1 - x_2y_2}{y_1 - y_2}\right]$$

and

$$d = \frac{1}{2y_1y_2}\left[\frac{x_1y_1 + x_2y_2}{y_1 + y_2} - \frac{x_1y_1 - x_2y_2}{y_1 - y_2}\right] .$$

Both c and d are supposed to be in $\mathbb{C}(u,v)$ and indeed, the reader may check that they are both fixed by all of G. However, we still don't *see* that this is so. But now at last, we are in position to see it. We deal with c first. Since c is supposed to be in $\mathbb{C}(u,v)$, and since u and v are written solely in terms of x_1 and x_2, it must be that both y_1 and y_2 are not really present in c. Thus if we combine the terms, both y_1 and y_2 deserve to vanish from the expression,

$$\begin{aligned} c &= \frac{1}{2}\left[\frac{x_1y_1 + x_2y_2}{y_1 + y_2} + \frac{x_1y_1 - x_2y_2}{y_1 - y_2}\right] \\ &= \frac{1}{2}\left[\frac{(x_1y_1 + x_2y_2)(y_1 - y_2) + (x_1y_1 - x_2y_2)(y_1 + y_2)}{y_1^2 - y_2^2}\right] \\ &= \frac{x_1y_1^2 - x_2y_2^2}{y_1^2 - y_2^2} \\ &= \frac{x_1f(x_1) - x_2f(x_2)}{f(x_1) - f(x_2)} . \end{aligned}$$

Indeed, the y_1 and y_2 contributions have canceled and we are left with a symmetric rational function of x_1 and x_2 which must be a rational function of u and v. (If we were to actually try and write this in terms of u and v, it would probably pay to continue to the point that both the numerator and denominator are symmetric in x_1 and x_2. This would be most easily accomplished by dividing the top and bottom by $x_1 - x_2$ which is a common factor and the resulting numerator and denominator are symmetric in x_1 and x_2 and so are polynomials in u and v.) The same thing may be done to d and we get,

$$\begin{aligned} d &= \frac{1}{2y_1y_2}\left[\frac{x_1y_1 + x_2y_2}{y_1 + y_2} - \frac{x_1y_1 - x_2y_2}{y_1 - y_2}\right] \\ &= \frac{1}{2y_1y_2}\left[\frac{(x_1y_1 + x_2y_2)(y_1 - y_2) - (x_1y_1 - x_2y_2)(y_1 + y_2)}{y_1^2 - y_2^2}\right] \\ &= \frac{x_2 - x_1}{y_1^2 - y_2^2} \\ &= \frac{x_2 - x_1}{f(x_1) - f(x_2)} . \end{aligned}$$

Thus d^{-1} turns out to be a polynomial in u and v.

2. Algebraic Number Theory

2.1 Unique Factorization Theory

An *algebraic number* is a root of an equation $f(x) = 0$ where $f(x)$ is a non-zero polynomial in $\mathbb{Q}[x]$. An algebraic extension of $\mathbb{Q}$ is called an *algebraic number field*. To do number theory, we now want to define algebraic integers. Naturally, we want the definition to accomplish certain things, if possible. First, and this is non-negotiable, we want the integers of a field K to form a ring: sums, differences and products of integers should be integers. Further, all ordinary integers should be algebraic integers so that we won't be confused. Also, the field should be the quotient field of this ring; in other words, every element of the field should be the quotient of two integers of the field just as every rational number is the ratio of two ordinary integers. Second, we would like there to be a unique factorization theory of integers in K.

This is a good place to say what a unique factorization theory would entail. Among the integers of a field, there would be two distinguished sets of integers. First and foremost would be the primes. Second would be the *units*, which are those integers whose inverses are also integers. In $\mathbb{Q}$, the units are ± 1. Because of units, we have to refine what we ordinarily think of as unique factorization. For example, in $\mathbb{Q}$, we may factor 20 completely as

$$20 = (2)(2)(5) = (-1)(2)(-2)(5) = (2)(-2)(-5) ,$$

among other ways. All of these factorizations are really the same and are complicated by the presence of units. We think of the primes 2 and -2 as essentially the same. The technical word for this is that the primes 2 and -2 are associates. In general, two algebraic numbers are *associates* if one is a unit times the other.

We will have a unique factorization theory for a field if every integer of the field factors uniquely into a product of units and primes up to associates and the order of the factors. If we have such a theory, we may pick from each class of associated primes a representative π. (This is the usual notation in the subject. We do our best to keep algebraic primes from appearing at the same time as $3.14159\cdots$, but failures have been known to occur.) If $\pi_1, \pi_2, \cdots$ run through a complete list of non-associated primes of K, then when we have unique factorization, every integer of K may be written uniquely in the form,

$$\alpha = \varepsilon \prod \pi_i^{a_i} , \tag{5}$$

where the a_i are ordinary non-negative integers all but finitely many of which are zero, and ε is a unit of K. For example, if our list of primes in $\mathbb{Q}$ were $2, -3, -5, 7, 11, -13, \cdots$, then the factorization of 20 would be $20 = -1 \cdot 2^2 \cdot (-5)$. In $\mathbb{Q}$, where we have positive and negative numbers and only two units, we usually take the positive member of each pair of associated primes as our representative. Since every number of K is supposed to be the quotient of two integers, this means in turn that every number of K has a unique factorization in the form (5) where now the a_i are allowed to be negative as well, but only finitely many are non-zero. The integers of K will be precisely those α with all the $a_i \geq 0$.

When there is unique factorization of integers, we also have a concept of greatest common divisors. To explain this, it will be convenient to define the concept of divisibility in a more general manner than usual, even for $\mathbb{Q}$. Suppose that α and δ are algebraic numbers with $\delta \neq 0$. Once algebraic integers have been defined, we will say that δ *divides* α (the notation is $\delta|\alpha$) if α/δ is an integer. For example, $\frac{1}{12}|\frac{1}{4}$, since the ratio is 3 which is an integer. Even in $\mathbb{Q}$, this takes some getting used to, but we will see that it is very handy. As a simple example, if α is an algebraic number then the condition that $1|\alpha$ is equivalent to the condition that α be an integer and the condition that $\alpha|1$ is equivalent to the condition that α^{-1} be an integer. Thus the units, which consist of those integers whose inverses are also integers, will be exactly those numbers which both are divisible by and divide 1.

When there is unique factorization, we can write the condition that $\delta|\alpha$ in terms of the factorizations of δ and α into primes: if

$$\alpha = \varepsilon \prod \pi_i^{a_i}, \qquad \delta = \eta \prod \pi_i^{d_i},$$

are the factorizations of α and δ into primes, with units ε and η, then $\delta|\alpha$ if and only if $d_i \leq a_i$ for all i. Now we are ready to introduce greatest common divisors. The notation will be that (α, β) denotes the greatest common divisor of the numbers α and β. In $\mathbb{Q}$, the words make us think of the largest number which divides the given numbers. This will not work in other fields where there often is no concept of largest, but greatest common divisors in $\mathbb{Q}$ have another property that we can use. The greatest common divisor is a common divisor divisible by all other common divisors.

As an example,

$$\left(\frac{1}{2}, \frac{1}{3}\right) = \frac{1}{6}.$$

This is to say that $\frac{1}{6}$ divides both $\frac{1}{2}$ and $\frac{1}{3}$ and that any rational number that divides both $\frac{1}{2}$ and $\frac{1}{3}$ divides $\frac{1}{6}$. This is not instantly obvious until we think of the prime factorization of the numbers in question. When there is unique factorization, as there is in $\mathbb{Q}$, we can easily see that there are greatest common divisors. In terms of the prime factorizations,

$$\left(\varepsilon \prod \pi_i^{a_i}, \eta \prod \pi_i^{b_i}\right) = \prod \pi_i^{\min(a_i, b_i)}.$$

For example,

$$\left(\frac{1}{2}, \frac{1}{3}\right) = (2^{-1} \cdot 3^0, 2^0 \cdot 3^{-1}) = 2^{-1} \cdot 3^{-1} = \frac{1}{6}.$$

Of course, any unit times the right hand side would work just as well. When we are dealing with numbers, greatest common divisors are unique only up to associates. In $\mathbb{Q}$, it has always been the convention to take the positive number of the pair of gcd's as the gcd. A special case arises often enough to be given a name. If $(\alpha, \beta) = 1$, we say that α and β are *relatively prime*. In $\mathbb{Q}$, this is exactly the usual concept; the prime factorizations show that the only way that two numbers can have greatest common divisor 1 is that both are integers without common prime factors.

2.2 Algebraic Integers

We denote the classical integers of $\mathbb{Q}$ by $\mathbb{Z}$. We will call these integers the *rational integers* because we are now ready to define algebraic integers. An *algebraic integer* is a zero of a monic polynomial in $\mathbb{Z}[x]$. Recall that a monic polynomial in $\mathbb{Z}[x]$ is a polynomial of the shape

$$f(x) = x^n + a_{n-1}x^{n-1} + \cdots + a_0 \tag{6}$$

where each a_j is in $\mathbb{Z}$. This is a rather strange definition at first glance. There are two major facts about it that make it reasonable.

The first fact about our definition is that the integers do form a ring; sums, differences and products of integers are integers. This is proved by exactly the argument that we used in the previous Section to show $\sqrt{2}+\sqrt{3}$ is algebraic. The new characteristic equation that we get by this technique is a monic polynomial in $\mathbb{Z}[x]$ when we start with algebraic integers. This was our non-negotiable first demand for a theory of integers. Note also that conjugates of integers are integers.

The second fact has to do with what we hope algebraic numbers will do for us. We have said that we hope there will be a nice multiplicative theory including unique factorization into primes. If there is such a theory, then a root of an equation $f(x) = 0$ where $f(x)$ in $\mathbb{Z}[x]$ is given by (6) must be an integer. The reason is as follows. Let α be a root of $f(x) = 0$. Every algebraic number will be the ratio of two 'integers' in a reasonable multiplicative theory: let $\alpha = \beta/\gamma$ where β and γ are 'integers'. If there is a unique factorization theory, then we may cancel out common 'prime' factors and so assume that β and γ are 'relatively prime'. From $\gamma^n f(\alpha) = 0$, we get

$$\beta^n + a_{n-1}\beta^{n-1}\gamma + \cdots + a_0\gamma^n = 0 ,$$

or

$$\beta^n = -a_{n-1}\beta^{n-1}\gamma - \cdots - a_0\gamma^n .$$

The right side is divisible by γ and so γ divides β^n. When there is unique factorization and β and γ have no common 'prime' factors, the only way γ can divide β^n is that γ have no prime factors at all. Such a number is a 'unit', an integer whose inverse is also an integer. Thus $\alpha = \beta\gamma^{-1}$ must be an integer.

Thus without our definition, there would be no hope for a unique factorization theory.

Now, let K be an algebraic number field. Having at last defined integers, we will let $\mathfrak{O}_K$ be the *ring of algebraic integers* in K. When a lower case letter such as k is used for the field, we frequently find a lower case $\mathfrak{o}$ used for the integers: $\mathfrak{o}_k$. For example, if $K = \mathbb{Q}(\sqrt{d})$ where d is a square-free rational integer (no repeated prime factors) other than 1, then the integers of K are given by

$$\mathfrak{O}_K = \mathbb{Z}[\sqrt{d}] = \{a + b\sqrt{d} \mid a \text{ and } b \text{ are in } \mathbb{Z}\} \qquad \text{when } d \not\equiv 1 \pmod 4$$

and

$$\mathfrak{O}_K = \mathbb{Z}\left[\frac{1+\sqrt{d}}{2}\right] = \{a + b\frac{1+\sqrt{d}}{2} \mid a \text{ and } b \text{ are in } \mathbb{Z}\} \quad \text{when } d \equiv 1 \pmod 4 .$$

We now look into the question of whether the integers do everything for us that we desire. First, every number in K is the quotient of two integers of K as we had hoped. This is easy to see. If we clear the denominators of coefficients in a non-zero polynomial in $\mathbb{Q}[x]$ with a zero α in K, we can put the equation for α in the form

$$a_n\alpha^n + a_{n-1}\alpha^{n-1} + \cdots + a_0 = 0 ,$$

where the a_j are in $\mathbb{Z}$. When we multiply through by a_n^{n-1}, we get

$$(a_n\alpha)^n + a_{n-1}(a_n\alpha)^{n-1} + \cdots + a_n^{n-1}a_0 = 0 ,$$

which shows that $a_n\alpha$ is an integer in K and α is the ratio of the two integers $a_n\alpha$ and a_n.

In order to discuss factorization of integers in a number field K, we need to know the units of K. The structure of the group of units of a field was found by Dirichlet and this is a starting point. Suppose that $K = \mathbb{Q}(\theta)$ is a degree n over $\mathbb{Q}$. Among the n conjugates of θ, suppose that r_1 conjugates are real and there are r_2 complex conjugate pairs so that $n = r_1 + 2r_2$. We will order the n conjugates of θ so that $\theta^{(j)}$ is real for $1 \le j \le r_1$ while $\theta^{(j+r_2)} = \overline{\theta^{(j)}}$ for $r_1 \le j \le r_1 + r_2$. We often say that the field K has r_1 *real conjugate fields* and $2r_2$ *complex conjugate fields.* It turns out that the subgroup of units of K of finite order is precisely the set of W-th roots of unity for some W. If $r_1 > 0$, then $W = 2$ and ± 1 are the only units of finite order. The units of infinite order are more interesting. The number $r = r_1 + r_2 - 1$ turns out to be the *rank* of the unit group.

Theorem. *There is a system of r units $\varepsilon_1, \cdots, \varepsilon_r$ of K such that every unit of K may be written uniquely in the form*

$$\varepsilon = \omega \prod_{i=1}^{r} \varepsilon_i^{a_i}$$

where ω is a W-th root of unity and the a_i are in $\mathbb{Z}$.

The set $\{\varepsilon_1, \cdots, \varepsilon_r\}$ is called a *system of fundamental units* of K. Actually finding such a system is not always easy. These numbers and their first r conjugates go into a very messy determinant called the regulator of K. Let

$$e_j = \begin{cases} 1 & \text{if } 1 \le j \le r_1 \\ 2 & \text{if } j \ge r_1 + 1 \,. \end{cases}$$

The *regulator* of K is the number

$$R = |\det(e_j \log|\varepsilon_i^{(j)}|)| \,.$$

As horrible as it is, the regulator turns out to be independent of the system of fundamental units used and is one of the principal invariants of the field K. It will appear again in Section 3. The reader should be warned that there are four competing versions of the regulator in print. One alternate eliminates the numbers e_j which just changes R by a power of 2. There are two other versions based on $(r+1) \times (r+1)$ determinants, but they are messier to write down. Again, they turn out to be simple rational multiples of R.

When $r = 0$, we say that $R = 1$ by definition. There are two ways this can happen. One way is when $r_1 = 1$ and $r_2 = 0$. Here $n = 1$ and so $K = \mathbb{Q}$. In this case $W = 2$ and ± 1 give the units. The second way $r = 1$ is possible when $r_1 = 0$ and $r_2 = 1$. In this case $n = 2$ and K is a *complex quadratic field.* Only two complex quadratic fields have roots of unity other than ± 1. The field $K = \mathbb{Q}(\sqrt{-1})$ contains the fourth roots of unity and $W = 4$. The field $K = \mathbb{Q}(\sqrt{-3})$ contains the sixth roots of unity and $W = 6$.

When we have a *real quadratic field* $(r_1 = 2, r_2 = 0)$, then $r = 1$ and there are infinitely many units. They are all of the form $\pm\varepsilon_1^a$ where a is in $\mathbb{Z}$ and ε_1 is a fundamental unit. There are four choices for ε_1, namely $\pm\varepsilon_1^{\pm 1}$. Any of these together with ± 1 also generates the unit group. Thanks to all the absolute value signs, the number R is the same for all four choices. It is customary in the literature to use the unique choice which is greater than one and call it *the fundamental unit.* For example when $K = \mathbb{Q}(\sqrt{2}), \varepsilon_1 = 1 + \sqrt{2}$ while for $K = \mathbb{Q}(\sqrt{5}), \varepsilon_1 = (1+\sqrt{5})/2$.

We return to the question of factorization of algebraic integers and present two examples. First we take the field $K = \mathbb{Q}(i)$. For this field, the ring of integers is

$$\mathfrak{O}_K = \mathbb{Z}[i] = \{a + bi \mid a \text{ and } b \text{ are in } \mathbb{Z}\} \,.$$

In this case, we have unique factorization into integers. Unfortunately, even though there are many fields with unique factorization (and quite likely infinitely many, although, amazingly enough, this has not yet been proved), not every field has unique factorization of integers. The standard example is in $\mathbb{Q}(\sqrt{-5})$. The integers here are

$$\mathbb{Z}[\sqrt{-5}] = \{a + b\sqrt{-5} \mid a \text{ and } b \text{ are in } \mathbb{Z}\} \,.$$

The units are just ± 1. The factorization,

$$6 = 2 \cdot 3 = (1 + \sqrt{-5}) \cdot (1 - \sqrt{-5}) \tag{7}$$

gives two fundamentally different factorizations of 6 into products of irreducible numbers. By a factorization into 'irreducible numbers', I mean that once we have factored 6 in either of the two ways, we can't factor any further except by introducing units and associates. These factorizations are really different; 2 clearly divides the product on the right, the quotient being 3, but 2 does not divide either factor since neither $(1 \pm \sqrt{-5})/2$ are integers (they are not in the form for integers of $\mathbb{Q}(\sqrt{-5})$ given above). This is incompatible with unique factorization: if 2 doesn't factor into pieces, 2 would have to be a prime and with unique factorization, a prime that divides a product of two integers divides at least one of them.

2.3 Divisor Theory

This has been a rapid introduction to integers. One of the great discoveries of the previous century in number theory was the restoration of the seemingly lost unique factorization. There are several equivalent ways to do this. We will take a route which does not require any additional knowledge of the reader to get started. This route is called *divisor theory.* We will begin with an example in $\mathbb{Q}$ that shows how we will repair the disaster in $\mathbb{Q}(\sqrt{-5})$ above. We will factor 60 in two ways as

$$60 = 4 \cdot 15 = 6 \cdot 10 \ .$$

This is analogous to the factorization of 6 in (7) above. Neither of the numbers 4 and 15 divides the numbers 6 and 10. In $\mathbb{Q}$ where we have unique factorization, we will refine both factorizations of 60 further into a common factorization. Since 4 does not divide 10, it follows that some piece of 4 divides 6. Likewise, since 4 does not divide 6, it follows that some piece of 4 also divides 10. Thus it is natural to look at the two greatest common divisors,

$$r = (4, 6) \text{ and } s = (4, 10) \ .$$

Likewise, we are motivated to look at

$$t = (15, 6) \text{ and } u = (15, 10) \ .$$

This leads us to the refined factorization,

$$60 = (4,6)(4,10)(15,6)(15,10) = rstu \ .$$

But $rs = 4$, $tu = 15$, $rt = 6$ and $su = 10$ and so this gives a further common refinement ot both factorizations of 60. Of course all this is easily checked since we can actually calculate all the greatest common divisors in the example but it is important to point out that we can arrive at the factorization, $60 = rstu$, which is a common refinement of the original two factorizations without numerically calculating any of the greatest common divisors. Indeed all we need to know is that one of the two factorizations of 60 is into relatively prime pieces.

As this also turns out to be the case in (7) where 2 and 3 are relatively prime, we now show that $60 = rstu$ gives a common refinement of the two factorizations of 60 without calculating any of r, s, t or u. We noted that some piece of 4 divides 6; the biggest such piece is by definition the greatest common divisor, $r = (4, 6)$. The rest of 4, namely $4/r$, must show up in 10. Thus $4/r$ is a factor of $s = (4, 10)$. Putting this together, we see that $4|rs$. In like manner, $15|tu$. Hence $60|rstu$. We now show the reverse: $rstu|60$; the result will be equality all along the line. Since 4 and 15 are relatively prime, the numbers $r = (4, 6)$ and $t = (15, 6)$ being pieces of 4 and 15 must be relatively prime. But when two relatively prime numbers such as r and t both divide a number such as 6, their product must divide the number as well. In other words, $rt|6$. Again in like manner $su|10$. Therefore $rtsu|60$ as advertised.

We will now create greatest common divisors for algebraic numbers. This will enable us to do the same thing for the two factorizations of 6 in $\mathbb{Q}(\sqrt{-5})$ above as we just did for the two factorizations of 60 in $\mathbb{Q}$. Suppose that $\alpha_1, \cdots, \alpha_m$ are numbers in K, not all zero. The 'greatest common divisor' of $\alpha_1, \cdots, \alpha_m$ is denoted by $(\alpha_1, \cdots, \alpha_m)$. We will call this symbol a *(fractional) divisor*. At the moment, this is just a symbol; it is up to us to make wise definitions so that this symbol will behave as greatest common divisors ought. We say that the divisor $(\alpha_1, \cdots, \alpha_m)$ is *generated* by $\alpha_1, \cdots, \alpha_m$. German letters are usually used to denote these divisors. An important class of divisors are the *principal (fractional)* divisors, (α) of just one non-zero number. For divisors to behave like gcd's, we have to have a notion of divisibility. To begin with, ordinary greatest common divisors have the property that the greatest common divisor of several numbers divides any integral linear combination of those numbers. We make this the definition in number fields: if β is an $\mathfrak{O}_K$-linear combination of the numbers $\alpha_1, \cdots, \alpha_m$ (not all zero) of K, say $\beta = \sum_{i=1}^{m} \gamma_i \alpha_i$ where the γ_i are in $\mathfrak{O}_K$ and $\mathfrak{a} = (\alpha_1, \cdots, \alpha_m)$, then we will say that $\mathfrak{a}$ *divides* β and write $\mathfrak{a}|\beta$.

Now we come to one of the key definitions. Since anything that divides each of several numbers should divide their greatest common divisor, if $\mathfrak{a} = (\alpha_1, \cdots, \alpha_m)$ and $\mathfrak{b} = (\beta, \cdots, \beta_n)$, we will say that $\mathfrak{a}$ *divides* $\mathfrak{b}$ if $\mathfrak{a}$ divides each β_j. In other words, we say that $(\alpha_1, \cdots, \alpha_m)$ divides $(\beta_1, \cdots, \beta_n)$ if each β_j is an integral linear combination of the α_i's. Thus, for example, $(\alpha) \,|\, (\beta_1, \cdots, \beta_n)$ means $\alpha|\beta_j$ for all j. Since greatest common divisors are supposed to be unique, we will go further and say that two divisors, $\mathfrak{a}$ and $\mathfrak{b}$ are *equal* if each divides the other, $\mathfrak{a}|\mathfrak{b}$ and $\mathfrak{b}|\mathfrak{a}$. As an illustration, in $\mathbb{Q}(\sqrt{-5})$, we have

$$(2, 1 + \sqrt{-5}) = (2, 1 - \sqrt{-5}) \tag{8}$$

since for example

$$1 - \sqrt{-5} = 1(2) + (-1)(1 + \sqrt{-5})$$

and so the left side divides the right side and the other direction goes the same way (in fact it is the conjugate relation).

We have defined what it means for one divisor to be divisible by another. it clearly becomes desirable to be able to perform the 'division'. However, as always, before we can learn to divide, we must learn to multiply.

Definition. $(\alpha_1, \cdots, \alpha_m) \cdot (\beta_1, \cdots, \beta_n) = (\{\alpha_i \beta_j\})$.

In other words, the product is the divisor generated by the mn $\alpha_i\beta_j$'s. Here is an example in $\mathbb{Q}$:

$$(2, 12) \cdot (24, 60) = (2 \cdot 24,\ 2 \cdot 60,\ 12 \cdot 24,\ 12 \cdot 60)\ .$$

It is clear that this must happen if we are to have a unique factorization theory. We see this by looking at the prime power factorization of both sides. For example in the relation above, the power of the prime 2 appearing in the first divisor $(2, 12)$ is 2^1 and in the second divisor $(24, 60)$ it is 2^2 while on the right side it is 2^{1+2} which occurs in the $2 \cdot 60$ term with at least this many 2's in the other three terms.

Theorem. *Multiplication of divisors satisfies the following usual properties,*
i) $\mathfrak{a}\mathfrak{b} = \mathfrak{b}\mathfrak{a}$,
ii) $\mathfrak{a}(\mathfrak{b}\mathfrak{c}) = (\mathfrak{a}\mathfrak{b})\mathfrak{c}$,
iii) $(1)\mathfrak{a} = \mathfrak{a}(1) = \mathfrak{a}$, *so that (1) serves as an identity.*

Proof. Let $\mathfrak{a} = (\alpha_1, \cdots, \alpha_m)$, $\mathfrak{b} = (\beta_1, \cdots, \beta_n)$, $\mathfrak{c} = (\gamma_1, \cdots, \gamma_r)$.
Property i) represents the fact that both sides are $(\{\alpha_i\beta_j\})$.
Property ii) is because both sides are $(\{\alpha_i\beta_j\gamma_k\})$, and property iii) is obvious. □

For example, if (β) is principal, then $(\beta\alpha_1, \cdots, \beta\alpha_m) = (\beta)(\alpha_1, \cdots, \alpha_m)$, a well known rule for greatest common divisors in the rational numbers. As an illustration,

$$\left(\frac{1}{2}, \frac{1}{3}\right) = \left(\frac{1}{6} \cdot 3, \frac{1}{6} \cdot 2\right) = \left(\frac{1}{6}\right) \cdot (3, 2) = \left(\frac{1}{6}\right)$$

It can be shown that there are multiplicative inverses. Given $\mathfrak{a}$, there is a unique $\mathfrak{b}$ such that $\mathfrak{a}\mathfrak{b} = (1)$. Of course, we write $\mathfrak{b} = \mathfrak{a}^{-1}$. This means that the divisors of K form a group under multiplication. This group I of all divisors has as a subgroup the group P of all principal divisors. The quotient group, I/P is called the *divisor class group* and can be shown to be finite. The order of this group is called the *class-number* of the field K. It plays a major role in the subject, but we will not treat it further here, except to note that the property of every divisor being principal in a field is equivalent to saying that the class-number of the field is one.

Now we have a multiplicative theory. What we still lack is the analog of integers for divisors so that we can begin to factorize them. In analogy to the fact that an algebraic number is an integer when it is divisible by 1, we make the equivalent definition here. We say that $\mathfrak{a} = (\alpha_1, \cdots, \alpha_m)$ is *integral* when $(1)|\mathfrak{a}$. According to the definition of divisibility, this means that each α_i is divisible by 1 which in this simple instance means that each α_i is integral. Thus an equivalent definition is that $\mathfrak{a}$ is integral when it has integral generators. We can now prove the natural alternate definition for divisibility.

Theorem. *For two fractional divisors $\mathfrak{a}$ and $\mathfrak{b}$, $\mathfrak{a}|\mathfrak{b}$ if and only if $\mathfrak{b}\mathfrak{a}^{-1}$ is integral.*

Proof. Let

$$\mathfrak{a} = (\alpha_1, \cdots, \alpha_m),\ \mathfrak{b} = (\beta_1, \cdots, \beta_n)$$

and set

$$\mathfrak{b}\mathfrak{a}^{-1} = \mathfrak{c} = (\gamma_1, \cdots, \gamma_r)\ .$$

First suppose that $\mathfrak{c}$ is integral so that all the γ_k are integral. Then from the relation $\mathfrak{b} = \mathfrak{a}\mathfrak{c}$,

$$(\beta_1, \cdots, \beta_n) = (\alpha_1, \cdots, \alpha_m) \cdot (\gamma_1, \cdots, \gamma_r)\ ,$$

we see that each β_j is an integral linear combination of the $\alpha_i\gamma_k$ and hence is an integral linear combination of the α_i. Therefore $\mathfrak{a}|\mathfrak{b}$.

Now for the converse. Suppose that $\mathfrak{a}|\mathfrak{b}$. We first show that if $\mathfrak{d} = (\delta_1, \cdots, \delta_s)$ is any other divisor, then $\mathfrak{a}\mathfrak{d}|\mathfrak{b}\mathfrak{d}$. Since $\mathfrak{b}\mathfrak{d}$ is generated by the ns numbers $\beta_j\delta_k$, we have to show that each $\beta_j\delta_k$ is an integral linear combination of the generators of $\mathfrak{a}\mathfrak{d}$. But β_j is an integral linear combination of the α_i and multiplying such a combination through by δ_k, we find that $\beta_j\delta_k$ is the same integral linear combination of the numbers $\alpha_i\delta_k$. Hence $\mathfrak{a}\mathfrak{d}|\mathfrak{b}\mathfrak{d}$, as claimed. Now we take $\mathfrak{d} = \mathfrak{a}^{-1}$. The result is that $(1)|\mathfrak{b}\mathfrak{a}^{-1}$. But this says precisely that $\mathfrak{c}$ is integral, as needed. Incidentally it is a consequence of the theorem just proved that $\mathfrak{a}|\mathfrak{b}$ if and only if $\mathfrak{a}\mathfrak{d}|\mathfrak{b}\mathfrak{d}$. □

Here is an example. Suppose that $\mathfrak{a}$ and $\mathfrak{b}$ are both integral and that

$$\mathfrak{a}\mathfrak{b} = (1)\ .$$

Then $(1)|\mathfrak{a}$ because $\mathfrak{a}$ is integral and $\mathfrak{a}|(1)$ because $(1)\mathfrak{a}^{-1} = \mathfrak{b}$ is integral. Therefore by definition, $\mathfrak{a} = (1)$ and of course, $\mathfrak{b} = (1)$ also. Thus there are no unit divisors other that (1) itself. This means that factorization into prime divisors will be free of the extra nuisance of units and associates.

It is time to get the unique factorization statement. Unique factorization in the rational numbers flows from the fact that there exist greatest common divisors. We now show that divisors have greatest common divisors.

Theorem. *Suppose that $\mathfrak{a} = (\alpha_1, \cdots, \alpha_m)$, $\mathfrak{b} = (\beta_1, \cdots, \beta_n)$ and let $\mathfrak{c} = (\alpha_1, \cdots, \alpha_m, \beta_1, \cdots, \beta_n)$. Then $\mathfrak{c}$ is the greatest common divisor of $\mathfrak{a}$ and $\mathfrak{b}$.*

Proof. This is easy to see and requires just talking through the words. First $\mathfrak{c}$ divides $\mathfrak{a}$ and $\mathfrak{b}$ since each α_i and each β_j is clearly an integral linear combination of all the α_i and β_j (all but one coefficient zero and the remaining coefficient one). Further, if $\mathfrak{d}$ is a common divisor of $\mathfrak{a}$ and $\mathfrak{b}$ then all the α_i and all the β_j are integral linear combinations of the generators of $\mathfrak{d}$ and hence the generators of $\mathfrak{c}$ are integral linear combinations of the generators of $\mathfrak{d}$. Thus $\mathfrak{d}$ divides $\mathfrak{c}$ also. This is practically magic! Needless to say, we write $\mathfrak{c} = (\mathfrak{a}, \mathfrak{b})$. □

Unique factorization now follows just as for $\mathbb{Q}$. The basic result leading to unique factorization is

Theorem. *Suppose that* $\mathfrak{a}$, $\mathfrak{b}$, $\mathfrak{c}$ *are integral divisors such that* $\mathfrak{a}|\mathfrak{bc}$ *and* $(\mathfrak{a}, \mathfrak{b}) = (1)$. *Then* $\mathfrak{a}|\mathfrak{c}$.

Proof. This is because

$$\mathfrak{c} = (1)\mathfrak{c} = (\mathfrak{a}, \mathfrak{b})\mathfrak{c} = (\mathfrak{ac}, \mathfrak{bc}) = \mathfrak{a}(\mathfrak{c}, \mathfrak{bca}^{-1})$$

and by hypothesis, $(\mathfrak{c}, \mathfrak{bca}^{-1})$ is integral. Thus $\mathfrak{ca}^{-1}$ is integral and $\mathfrak{a}|\mathfrak{c}$. □

In the special situation that $(\mathfrak{a}, \mathfrak{b}) = (1)$, we say that $\mathfrak{a}$ and $\mathfrak{b}$ are *relatively prime*. As with numbers, this can only happen when $\mathfrak{a}$ and $\mathfrak{b}$ are integral without any common non-trivial integral factors. We define a *prime divisor* to be an integral divisor other than (1) which has no integral divisors other than itself and (1).

Theorem. *Suppose that* $\mathfrak{P}$ *is a prime divisor which divides the product* $\mathfrak{ab}$ *of two integral divisors. Then either* $\mathfrak{P}$ *divides* $\mathfrak{a}$ *or* $\mathfrak{P}$ *divides* $\mathfrak{b}$.

Proof. Suppose that $\mathfrak{P}$ does not divide $\mathfrak{a}$. Since by definition a prime divisor either divides another integral divisor or is relatively prime to it, it follows that $(\mathfrak{a}, \mathfrak{P}) = (1)$. Therefore by the previous theorem, $\mathfrak{P}|\mathfrak{b}$. □

Without much further ado, we then get the unique factorization theorem for divisors which we just state for the record.

Theorem. *Every divisor* $\mathfrak{a}$ *of* K *has a unique representation in the form*

$$\mathfrak{a} = \prod \mathfrak{P}_j^{a_j} ,$$

where the $\mathfrak{P}_j$ *run through all the prime divisors of* K *and the* a_j *are in* $\mathbb{Z}$ *with all but finitely many being zero. Further* $\mathfrak{a}$ *is integral if and only if* $a_j \geq 0$ *for all* j.

Now we return to our example in $\mathbb{Q}(\sqrt{-5})$,

$$6 = 2 \cdot 3 = (1 + \sqrt{-5}) \cdot (1 - \sqrt{-5}) .$$

Both sides can be refined to the factorization in terms of divisors,

$$(6) = (2, 1 + \sqrt{-5}) \cdot (2, 1 - \sqrt{-5}) \cdot (3, 1 + \sqrt{-5}) \cdot (3, 1 - \sqrt{-5}) ,$$

where the product of the first and second divisors on the right gives (2), the third and fourth divisors on the right gives (3), the first and third divisors on the right multiply to $(1 + \sqrt{-5})$ and the second and fourth multiply to $(1 - \sqrt{-5})$. As we explained above when we factored 60 in $\mathbb{Q}$, the fact that (2) and (3) are relatively prime makes this clear without having to check the multiplications, but it is good practice to do so. For example,

$$
\begin{aligned}
(9) \qquad (2,\, 1+\sqrt{-5}) \cdot (2,\, 1-\sqrt{-5}) &= (4,\, 2-2\sqrt{-5}, 2+2\sqrt{-5}, 6) \\
&= (2)(2,\, 1-\sqrt{-5},\, 1+\sqrt{-5}, 3) \\
&= (2)(1) \\
&= 2\,.
\end{aligned}
$$

Since we have seen in (8) above that the two divisors on the left are in fact equal, this means that

$$
(10) \qquad (2) = (2,\, 1+\sqrt{-5})^2\,.
$$

On the other hand, in $\mathbb{Q}(\sqrt{2})$, we clearly have

$$
(11) \qquad (2) = (\sqrt{2})^2\,.
$$

Thus we are lead to the very strange conclusion,

$$
(12) \qquad (2,\, 1+\sqrt{-5}) = (\sqrt{2})\,!
$$

How shall we give meaning to this? The answer is that both (10) and (11) hold in $\mathbb{Q}(\sqrt{-5}, \sqrt{2})$ and hence by unique factorization (without units so that there are no $\pm$ problems), (12) must hold in this field as well. Thus the non-principal divisor $(2,\, 1+\sqrt{-5})$ in $\mathbb{Q}(\sqrt{-5})$ has become principal in $\mathbb{Q}(\sqrt{-5}, \sqrt{2})$. In fact, since the class-number of $\mathbb{Q}(\sqrt{-5})$ is 2, this carries with it the fact that every non-principal divisor of $\mathbb{Q}(\sqrt{-5})$ becomes principal in $\mathbb{Q}(\sqrt{-5}, \sqrt{2})$ by pure group theory arguments. This is a general principle. For every field K, there is a bigger field L for which every divisor of K becomes principal in L. Unfortunately, it can happen that L has further divisors which are still non-principal.

We turn next to the concept of the *norm* of a divisor. Our definition will be the exact analog of the norm of a number being the product of its field conjugates. Since we now know how to talk about a divisor as being in more than one field, if K is of degree n over $\mathbb{Q}$, we define

$$
(13) \qquad \mathbb{N}(\mathfrak{a}) = \prod_{i=1}^{n} \mathfrak{a}^{(i)}\,,
$$

where the product takes place in some larger field containing all the $K^{(i)}$. The result is a divisor of $\mathbb{Q}$, which is to say that the result has a system of generators all of which are in $\mathbb{Q}$. Therefore we may express the result in the form $\mathbb{N}(\mathfrak{a}) = N$ where $N > 0$ is unique. Because of this, it has become customary to simply write $\mathbb{N}(\mathfrak{a}) = N$. However, when we compute relative norms from one field to another, norms of divisors are divisors in the ground field and not numbers. In the case of a principal divisor (α), we see that

$$\mathbb{N}((\alpha)) = (\mathbb{N}(\alpha))$$

and according to our convention, as a number of $\mathbb{Q}$, this is the absolute value of $\mathbb{N}(\alpha)$. We note from the definition that

$$\mathbb{N}(\mathfrak{a}\mathfrak{b}) = \mathbb{N}(\mathfrak{a})\mathbb{N}(\mathfrak{b})$$

and that norms of integral divisors are integral divisors.

The fact that $\mathbb{N}(\mathfrak{a})$ is a divisor of $\mathbb{Q}$ is not obvious, but it is true. Since the product defining $\mathbb{N}(\mathfrak{a})$ in (13) is invariant under Galois actions, we might think that we can conclude from this that the product must be a divisor of the ground field. This is not sufficient. *Divisors can be preserved by every element of a Galois group without being definable in the ground field.* For example, in $K = \mathbb{Q}(\sqrt{-5})$, the divisor $(2, 1+\sqrt{-5})$ is preserved by $G(K/\mathbb{Q})$ as may be seen in (8) above. Even principal divisors may sometimes have this annoying property; for instance in K, the divisor $(\sqrt{-5})$ is preserved by $G(K/\mathbb{Q})$. It can be shown without too much difficulty that if a divisor of a normal extension K/k is preserved by all the elements of a Galois group $G(K/k)$, then the $[K:k]^{\text{th}}$ power of the divisor is a divisor of the ground field k. However we have no need for this rather esoteric fact in the rest of this paper.

The next theorem allows us to show that $\mathbb{N}(\mathfrak{a})$ is a divisor of $\mathbb{Q}$. To prepare for it, we need a definition. Suppose that $g(x_1, \cdots, x_m)$ is a non-zero polynomial in $K[x_1, \cdots, x_m]$. We define the *content* $\Im(g)$ of g to be the divisor of K generated by the coefficients of g.

Theorem (Kronecker's Content Theorem). *Suppose that $g_1, \cdots, g_r$ are non-zero polynomials in $K[x_1, \cdots, x_m]$. Then $\Im(g_1 \cdots g_r) = \Im(g_1) \cdots \Im(g_r)$.*

We will not give the proof although it is not too hard to show that the powers of every prime on both sides are the same. As a special case, let $\mathfrak{a} = (\alpha_1, \cdots, \alpha_m)$ be a divisor of K. Let $g(x_1, \cdots, x_m)$ be defined as the product of $n = [K:\mathbb{Q}]$ conjugate polynomials,

$$g(x_1, \cdots, x_m) = \prod_{i=1}^{n} [a_1^{(i)} x_1 + \cdots + \alpha_m^{(i)} x_m] \,. \tag{14}$$

Then according to Kronecker's content theorem and the definition of $\mathbb{N}(\mathfrak{a})$, the content of g is just $\mathbb{N}(\mathfrak{a})$. But g is a polynomial in $\mathbb{Q}[x_1, \cdots, x_m]$ by Galois theory since the factors giving g are permuted by any Galois action. Hence g is invariant under Galois actions and so the coefficients of g are all in $\mathbb{Q}$.

Not only does the content theorem give us a proof that relative norms of divisors are defined in the ground field, it allows us to do the computation in a reasonable manner. We illustrate by finding the norm of the divisor $\mathfrak{a} = (2,\, 1+\sqrt{-5})$ of $\mathbb{Q}(\sqrt{-5})$. In equation (9) above, we actually calculated the norm by the definition. By the content theorem, the norm is the content of the polynomial,

$$[2x + y(1+\sqrt{-5})][2x + y(1-\sqrt{-5})] = 4x^2 + 4xy + 6y^2 \ .$$

Since $(4,4,6) = (2)$, we see that $\mathbb{N}(\mathfrak{a}) = 2$. As another illustration, let $\mathfrak{a} = (3,\, 1+\sqrt{-5})$ in the same field. We have

$$[3x + y(1+\sqrt{-5})][3x + y(1-\sqrt{-5})] = 9x^2 + 6xy + 6y^2 \ ,$$

and so $\mathbb{N}(\mathfrak{a}) = 3$.

We have said that the inverse of a divisor exists. The content theorem gives us a way of finding it. Again, let $\mathfrak{a} = (\alpha_1, \cdots, \alpha_m)$ be a divisor of K. Then $\alpha_1 x_1 + \cdots + \alpha_m x_m$ is one of the n factors of the polynomial $g(x_1, \cdots, x_m)$ defined in (14). Hence

$$h(x_1, \cdots, x_m) = \frac{g(x_1, \cdots, x_m)}{\alpha_1 x_1 + \cdots + \alpha_m x_m}$$

is a polynomial in $K[x_1, \cdots, x_m]$. Let $\mathfrak{b}$ be the content of $h(x_1, \cdots, x_m)$. Thus $\mathfrak{b}$ is a divisor of K and by the content theorem,

$$\mathfrak{a}\mathfrak{b} = (N)$$

where $\mathbb{N}(\mathfrak{a}) = N$. Therefore, $\mathfrak{a}^{-1} = (N^{-1})\mathfrak{b}$. For example,

$$(3, 1+\sqrt{-5})^{-1} = (3^{-1})(3, 1-\sqrt{-5}) = \left(1, \frac{1+\sqrt{-5}}{3}\right) .$$

The content theorem also allows us to clear up several other loose ends.

Theorem. *Suppose that $f(x)$ is a monic polynomial with algebraic integer coefficients and that $f(x) = g(x)h(x)$ where $g(x)$ and $h(x)$ are monic polynomials. Then $g(x)$ and $h(x)$ have algebraic integer coefficients.*

Proof. Recall that any divisor divides its generators. Since f has integral coefficients, $\mathfrak{I}(f)$ has integral generators, one of which is 1 and so $\mathfrak{I}(f)$ is an integral divisor dividing 1. Hence $\mathfrak{I}(f) = (1)$. By the content theorem,

$$\mathfrak{I}(g)\mathfrak{I}(h) = \mathfrak{I}(f) = (1) \ .$$

But $\mathfrak{I}(g)$ and $\mathfrak{I}(h)$ both have 1 as one of their generators and so $\mathfrak{I}(g)|1$ and $\mathfrak{I}(h)|1$. Therefore $\mathfrak{I}(g)^{-1}$ and $\mathfrak{J}(h)^{-1}$ are integral. But their product is (1) and so each is (1). Thus

$$\mathfrak{I}(g) = \mathfrak{I}(h) = (1) \ .$$

But a divisor is integral only when all its generators are integral and hence g and h have integral coefficients. $\square$

We defined algebraic integers as zeros of monic polynomials, whether irreducible or not. One application of this last theorem is that the defining polynomial of an algebraic integer over any number field already has algebraic integer coefficients. For instance, we can now say that a rational number which is also an algebraic integer is already a rational integer, fortunately for the terminology. As another application, if $f(x)$ is a monic polynomial with algebraic integer coefficients and α is a zero of f, then α is an algebraic integer. This is because we can write $f(x)$ as a product of two monic polynomials, one of which is $(x - \alpha)$, and so α is an algebraic integer. This result is usually proved by looking at $\mathbb{N}(f(x))$ which by Galois theory is a monic polynomial in $\mathbb{Z}[x]$ and also has α as a zero. The Galois theory proof is more general, but I rather like the proof based on the content theorem.

We have introduced divisors as symbols with rather miraculous multiplicative properties. Especially in this paper it would be nice to have a physical model for divisors. Such a model is provided by ideals. If $\mathfrak{a} = (\alpha_1, \cdots, \alpha_m)$ is a fractional divisor of K, we let the *fractional ideal* generated by $\alpha_1, \cdots, \alpha_m$ be the set of $\mathfrak{O}_K$-linear combinations of the generators $\alpha_1, \cdots, \alpha_m$. In other words, the fractional ideal generated by $\alpha_1, \cdots, \alpha_m$ is the set of all numbers of K which are divisible by $\mathfrak{a}$. The ideal thus does not depend upon the generators of $\mathfrak{a}$ which are used. There is a $1-1$ correspondence between the set of divisors of K and the set of ideals of K. The notation $(\alpha_1, \cdots, \alpha_m)$ is also used for the fractional ideal and German letters are usually used to denote it; only the words are changed. When the divisor $\mathfrak{a}$ is integral, every element of the ideal is in $\mathfrak{O}_K$. In this case we have an *integral ideal* of K. This is precisely what algebra refers to as an ideal of the ring $\mathfrak{O}_K$, namely a subset of $\mathfrak{O}_K$ which is closed under addition and under multiplication by elements of $\mathfrak{O}_K$ (except that number theory doesn't allow the zero ideal). If $\mathfrak{P}$ is the ideal corresponding to a prime divisor which we also denote by $\mathfrak{P}$, and α and β are integers such that $\alpha\beta$ is in the ideal $\mathfrak{P}$, then the divisor $\mathfrak{P}$ divides the product $\alpha\beta$ and hence divides either α or β. Therefore either α or β is in the ideal $\mathfrak{P}$. In other words, in the language of algebra, $\mathfrak{P}$ is a *prime ideal.*

2.4 Prime Divisors

In preparation for dealing with zeta functions, we must discuss the prime divisors of number fields in greater detail. Let K/k be an extension of number fields of relative degree n. Suppose that $\mathfrak{P}$ is a prime divisor of K. We first show that there is a unique prime of k divisible by $\mathfrak{P}$. To see this, let $n = \mathbb{N}_{K/k}(\mathfrak{P})$. Considered as divisors of K, $\mathfrak{P}$ divides n since $\mathfrak{P}$ is one of the factors in the product defining the relative norm. On the other hand in k, n is an integral divisor which factors as a product of prime divisors. Unique factorization in K now says that $\mathfrak{P}$ must divide one of the primes of k in the factorization of n. Let $\mathfrak{p}$ be one such prime of k divisible by $\mathfrak{P}$. If $\mathfrak{p}'$ is another such prime of k divisible by $\mathfrak{P}$ then $\mathfrak{P}$ would divide $(\mathfrak{p}, p') = (1)$ which is impossible. Thus there can't be another such prime of k and $\mathfrak{p}$ is uniquely determined by $\mathfrak{P}$. As an ideal, $\mathfrak{p}$ consists exactly of those elements of k divisible by $\mathfrak{P}$ and so as an ideal, $\mathfrak{p} = \mathfrak{P} \bigcap k$.

Now we are ready to find the relative norm. Since $\mathfrak{P}|\mathfrak{p}$, it follows that $\mathbb{N}_{K/k}(\mathfrak{P})|\mathbb{N}_{K/k}(\mathfrak{p}) = \mathfrak{p}^n$ by definition of the relative norm and the fact that all conjugates of $\mathfrak{p}$ over k are just $\mathfrak{p}$. Hence

$$\mathbb{N}_{K/k}(\mathfrak{P}) = \mathfrak{p}^f$$

for some integer f in the range $1 \leq f \leq n$. We call f the *residue class degree of $\mathfrak{P}$ relative to K/k*, or sometimes just the *degree of $\mathfrak{P}$ relative to K/k*. When the choice of k is clear, we can also refer to f as the *relative residue class degree* of $\mathfrak{P}$, or even just the *relative degree* of $\mathfrak{P}$. When $k = \mathbb{Q}$, we simply call f the *degree* of $\mathfrak{P}$.

First degree primes are of particular importance in analytic treatments of the subject. Suppose that p is a prime of $\mathbb{Q}$ and that α is an integer in K such that $\mathbb{N}(\alpha)$ is *exactly divisible* by p (i.e. p divides $\mathbb{N}(\alpha)$ but p^2 does not). Unique factorization in K shows that there is a single first degree prime $\mathfrak{P}$ of K above p which exactly divides α. Hence

$$\mathfrak{P} = (p, \alpha) .$$

All first degree primes arise in this manner. For example, $\mathbb{N}(1 + \sqrt{-5}) = 6$ and so $(3, 1 + \sqrt{-5})$ is a first degree prime ideal of norm 3 in $\mathbb{Q}(\sqrt{-5})$. In case $\mathbb{N}(\alpha) = \pm p$, then $\mathfrak{P} = (\alpha)$ is a principal first degree prime. For instance in $\mathbb{Q}(2^{1/3}), \mathbb{N}(1+2^{1/3}) = 3$ and so $1+2^{1/3}$ generates a principal first degree prime in $\mathbb{Q}(2^{1/3})$ of norm 3.

The reason for the terminology, 'residue class degree' becomes apparent when we look at congruences. If α and β are in $\mathfrak{O}_K$, we say that α is congruent to β modulo $\mathfrak{P}$ and write

$$\alpha \equiv \beta (\operatorname{mod} \mathfrak{P})$$

when $\mathfrak{P}$ divides the difference $\alpha - \beta$. The set of all integers of K congruent to $\alpha (\operatorname{mod} \mathfrak{P})$ is called a *congruence class* $(\operatorname{mod} \mathfrak{P})$. The collection of all residue classes $(\operatorname{mod} \mathfrak{P})$ forms a finite field of p^f elements where it turns out that $\mathbb{N}_{K/\mathbb{Q}}(\mathfrak{P}) = p^f$. Thus the residue class degree is just the degree of the extension of finite fields $\mathfrak{O}_K/\mathfrak{P}$ over $\mathbb{F}_p$. Likewise, the relative residue class degree is just the degree of the finite field extension $\mathfrak{O}_K/\mathfrak{P}$ over $\mathfrak{o}_k/\mathfrak{p}$.

The picture we draw is

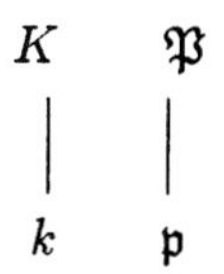

and we say that $\mathfrak{P}$ is *above* $\mathfrak{p}$. If we now factor $\mathfrak{p}$ in K, there will be a finite number of primes of K above $\mathfrak{p}$, say $\mathfrak{P}_1, \cdots, \mathfrak{P}_r$, where $\mathfrak{P}$ is one of the $\mathfrak{P}_j$ and we get a factorization,

(15)
$$\mathfrak{p} = \prod_{j=1}^{r} \mathfrak{P}_j^{e_j} .$$

We often say that $\mathfrak{p}$ *splits in* K into r prime power pieces. If $\mathrm{N}_{K/k}(\mathfrak{P}_j) = \mathfrak{p}^{f_j}$, then taking the norm of both sides gives,

$$\mathfrak{p}^n = \prod_{j=1}^{r} \mathfrak{p}^{f_j e_j} .$$

It follows therefore that

(16)
$$\sum_{j=1}^{r} f_j e_j = n .$$

An extreme case occurs when $r = n$. In this instance, we say that $\mathfrak{p}$ *splits completely* from k to K. Here we must have every $e_j = f_j = 1$.

All but finitely many $\mathfrak{p}$ have every $e_j = 1$. These are the primes of k which do not *ramify* in K. The finitely many $\mathfrak{p}$ of k which have a factorization (15) in K with at least one $e_j > 1$ ramify in K. Those $\mathfrak{P}$ of K which appear in such a factorization (15) with $e > 1$ are *ramified primes of* K *relative to* k and the exponent e is called the *ramification index of* $\mathfrak{P}$ *relative to* k. As always, when $k = \mathbb{Q}$, we drop the words, 'relative to k'. For example, by (10), 2 ramifies in $\mathbb{Q}(\sqrt{-5})$.

It is important to be able to determine the finitely many ramifying $\mathfrak{p}$ from k to K, if for no other reason than to avoid them since they are frequently the most difficult to deal with. It turns out that the primes of $\mathfrak{p}$ which ramify from k to K all divide the *relative discriminant* $D_{K/k}$ of K/k. This is an integral divisor (or if you wish, an ideal) of k which is a generalization of the concept of discriminants of polynomials. If $K = k(\alpha)$ where α is an integer and is a zero of a monic irreducible polynomial $g(x)$ in $\mathfrak{o}_k[x]$, then $D_{K/k}$ always divides the discriminant of the polynomial $g(x)$ and further, as divisors, the quotient is always a square. Often, but not always, there is an α such that $D_{K/k}$ is the divisor generated by the discriminant of the corresponding $g(x)$. This happens exactly when $\mathfrak{O}_K = \mathfrak{o}_k[\alpha]$. When $k = \mathbb{Q}$, we just refer to the *discriminant* D_K of K and tradition thinks of the discriminant as a number with the same sign as the polynomial discriminant. If $\mathfrak{p}$ in k has the factorization (15) in K, then each $\mathfrak{P}_j$ turns out to contribute at least $\mathfrak{p}^{e_j-1}$ to $D_{K/k}$ so that overall, there is always at least a factor of $\mathfrak{p}^{\sum(e_j-1)}$ in $D_{K/k}$. If $\mathfrak{p}$ does not divide the ramification index e_j, then the contribution $\mathfrak{p}^{e_j-1}$ to $D_{K/k}$ is exact. However, if $\mathfrak{p}|e_j$, then there is a contribution of at least $\mathfrak{p}^{e_j}$; however, the contribution can be greater.

For example, for $k = \mathbb{Q}(\sqrt{-5})$, the ring of integers is $\mathbb{Z}[\sqrt{-5}]$ and D_k is the discriminant of the polynomial x^2+5. For the reader's information, the discriminant of a trinomial $x^n + bx + c$ turns out to be

$$(-1)^{(n-1)(n-2)/2}(n-1)^{n-1}b^n + (-1)^{n(n-1)/2}n^n c^{n-1} .$$

Thus, $D_k = -20$, and the only ramifying primes are 2 and 5. Indeed, we have already seen that

$$(2) = (2,\, 1 + \sqrt{-5})^2 \qquad \text{and} \qquad (5) = (\sqrt{-5})^2 \,.$$

Note that (2) ramifies in k with ramification index 2 which is divisible by (2); hence there is a factor of at least 2^2 in D_k. Likewise, (5) ramifies in k with ramification index 2; hence there is a factor of exactly 5^1 in D_k.

For $K = \mathbb{Q}(2^{1/3})$, the ring of integers is $\mathbb{Z}[2^{1/3}]$ and D_k is the discriminant of the polynomial $x^3 - 2$. In other words, $D_k = -108$. Therefore the only ramifying primes are 2 and 3. In fact, it is clear that

$$(2) = (2^{1/3})^3 \,,$$

and it is true, although not quite so clear, that

$$(3) = (1 + 2^{1/3})^3 \,.$$

(The quotient of the right side by the left is integral, as is easily seen by simply cubing $1+2^{1/3}$ getting $3+3\cdot 2^{1/3}+3\cdot 2^{2/3}$; the integral ratio has norm 1 and so is (1). Hence as divisors both sides are equal.) Again, we note that (2) ramifies in k with ramification index 3 which is not divisible by (2) and so there is a factor of exactly 2^2 in D_k. Likewise (3) ramifies in k with ramification index 3 which is divisible by (3) and so there is a factor of at least 3^3 in D_k. Therefore D_k has a factor of at least 108 and since the polynomial discriminant is already this low, $d_k = -108$ as claimed. This verifies that the ring of integers of k really is $\mathbb{Z}[2^{1/3}]$.

If we do any sort of calculations with number fields, we have to be able to tell how primes split. To set up the general statement, let us take the example $k = \mathbb{Q}(\alpha)$ where $\alpha = 2^{1/3}$. As we have noted, the ring of integers of k is $\mathfrak{o}_k = \mathbb{Z}[\alpha]$. Let $g(x) = x^3 - 2$, one of whose zeros is α. We will find how the prime 5 splits in k. Suppose that $\mathfrak{p}$ is one of the prime factors of 5 in k and that $\mathfrak{p}$ is of degree f. Then the finite field $\mathfrak{o}_k/\mathfrak{p}$ is just $\mathbb{F}_5(\bar{\alpha})$ where $\bar{\alpha}$ is the congruence class of $\alpha (\operatorname{mod} \mathfrak{p})$. Therefore $\bar{\alpha}$ is a zero of an irreducible factor of $\bar{g}(x) = x^3 - \bar{2}$ in $\mathbb{F}_5[x]$. We check easily enough that $\bar{g}(x)$ has exactly one zero in $\mathbb{F}_5$ and hence factors completely into irreducible pieces over $\mathbb{F}_5[x]$ as

$$x^3 - \bar{2} = x^3 - \overline{27} = (x - \bar{3})(x^2 + \bar{3}x + \bar{9}). \tag{17}$$

Thus $\bar{\alpha}$ corresponds to one of the two irreducible polynomials on the right side and f is either 1 or 2. If $\bar{\alpha}$ corresponds to the first factor, then $\mathfrak{p}$ is of degree 1 while if $\bar{\alpha}$ corresponds to the second factor, then $\mathfrak{p}$ is of degree 2. We will now show that each factor on the right actually corresponds to a prime factor of 5 in k.

We may interpret the factorization (17) as saying that the two polynomials $x^3 - 2$ and $(x-3)(x^2+3x+9)$ differ by 5 times a polynomial in $\mathbb{Z}[x]$. Therefore when we set $x = \alpha$, we see that

$$5|(\alpha-3)(\alpha^2+3\alpha+9).$$

However, 5 does not divide either factor because we know the shape of all integers of k and neither $((\alpha-3)/5)$ nor $((\alpha^2+3\alpha+9)/5)$ are in this shape. (The numbers of k are uniquely expressible in the form $\mathbb{Q}+\mathbb{Q}\alpha+\mathbb{Q}\alpha^2$ and the integers are those numbers with coefficients in $\mathbb{Z}$). In other words, there is a piece of 5 in $\alpha-3$ and another piece of 5 in $\alpha^2+3\alpha+9$. This leads us to the two divisors

$$\mathfrak{p}_1=(5,\ \alpha-3) \qquad \text{and} \qquad \mathfrak{p}_2=(5,\ \alpha^2+3\alpha+9)$$

of 5. Slightly more work reveals that $\mathfrak{p}_1$ and $\mathfrak{p}_2$ are prime divisors whose product is 5. Thus we have found that 5 splits in k as $(5)=\mathfrak{p}_1\mathfrak{p}_2$ with the prime factors of 5 corresponding to the irreducible factors of $x^3-\bar{2}$ in $\mathbb{F}_5[x]$.

This is perfectly general. Suppose that $K=k(\alpha)$ where α is an integer which is a zero of a monic irreducible polynomial $g(x)$ of degree n. Suppose that $\mathfrak{p}$ is a prime of k divisible by the prime $\mathfrak{P}$ of K of relative degree f.

$$\begin{array}{ccccl} K & \mathfrak{P} & \mathfrak{O}_K/\mathfrak{P} & = & \text{finite field of } \mathbb{N}(\mathfrak{P})=\mathbb{N}(\mathfrak{p})^f \text{ elements} \\ | & | & | & & \\ k & \mathfrak{p} & \mathfrak{o}_k/\mathfrak{p} & = & \text{finite field of } \mathbb{N}(\mathfrak{p}) \text{ elements}. \end{array}$$

Again, if β is in $\mathfrak{O}_K$, we will denote the residue class of $\beta(\operatorname{mod}\mathfrak{P})$ as $\bar{\beta}$. Suppose for the sake of discussion that $\mathfrak{O}_K/\mathfrak{P}=\mathfrak{o}_k/\mathfrak{p}(\bar{\alpha})$. Then $\bar{\alpha}$ satisfies a reduced equation, $\bar{g}(\bar{\alpha})=\bar{0}$ and so $\bar{g}$ has an irreducible factor of degree f in $\mathfrak{o}_k/\mathfrak{p}$. As in our example, the factors correspond to the primes $\mathfrak{P}$ dividing $\mathfrak{p}$.

Theorem. *Suppose $g(x)$ factors into irreducible pieces $(\operatorname{mod}\mathfrak{p})$ as*

$$g(x)\equiv g_1(x)^{e_1}\cdots g_r(x)^{e_r}(\operatorname{mod}\mathfrak{p}).$$

If the power of $\mathfrak{p}$ in the polynomial discriminant of $g(x)$ is the same as the power of $\mathfrak{p}$ in $D_{K/k}$, then $\mathfrak{p}$ factors in K as

$$\mathfrak{p}=\mathfrak{P}_1^{e_1}\cdots\mathfrak{P}_r^{e_r}$$

where $\mathfrak{P}_j$ is of relative degree $\deg(g_j)$ and

$$\mathfrak{P}_j=(\mathfrak{p},g_j(\alpha)).$$

Three instances of this theorem should be pointed out. If $\mathfrak{p}$ doesn't divide the discriminant of $g(x)$ at all, then $\mathfrak{p}$ doesn't divide $D_{K/k}$ either; we are dealing with a prime which doesn't ramify from k to K and the theorem applies with all $e_j=1$. A variant of this instance is when $\mathfrak{p}$ exactly divides the discriminant of g. Since the polynomial discriminant differs from $D_{K/k}$ by a square factor, it must be that $\mathfrak{p}$ exactly divides $D_{K/k}$ as well. In this case, it must be that exactly one of the e_j is 2 and all the remaining $e_j = 1$. (Also, $\mathfrak{p}$ will never be a prime above 2 in this instance, because then there would be a factor of at least p^2 in $D_{K/k}$.) An example of this is given by the irreducible polynomial $g(x) = x^3 - x - 1$ in $\mathbb{Z}[x]$ whose discriminant is -23. If K is the cubic extension of $\mathbb{Q}$ generated by a zero of $g(x)$, then D_K must be -23 as well. We have the factorization

$$x^3 - x - 1 \equiv (x-3)(x-10)^2 (\bmod 23) ,$$

and so 23 splits in K as $\mathfrak{P}_1\mathfrak{P}_2^2$ as promised. Note that 23 ramifies in K, but that only one of the two primes over 23 is ramified.

A second instance of the theorem is when $g(x)$ is an *Eisenstein polynomial with respect to* $\mathfrak{p}$. This is a polynomial

$$g(x) = x^n + a_{n-1}x^{n-1} + \cdots + a_0$$

in $\mathfrak{o}_k[x]$ such that $\mathfrak{p}|a_j$ for $0 \le j \le n-1$ and $\mathfrak{p}$ exactly divides a_0. The *Eisenstein irreducibility criterion* says that such a polynomial is irreducible in $k[x]$ as an added bonus. It can also be shown that the power of $\mathfrak{p}$ in the discriminant of $g(x)$ matches that in $D_{K/k}$ and so the theorem gives $\mathfrak{p} = \mathfrak{P}^n$, where $\mathfrak{P} = (\mathfrak{p}, \alpha)$.

The third instance of the theorem is when we are lucky enough to know that $\mathfrak{O}_K = \mathfrak{o}_k[\alpha]$. In this case, $D_{K/k}$ is (the divisor generated by) the polynomial discriminant. Here, the theorem applies for all primes. For example,

$$x^2 + 5 \equiv (x+1)^2 (\bmod 2) \qquad \text{and} \qquad x^2 + 5 \equiv (x)^2 (\bmod 5)$$

gives the splitting of the ramifying primes 2 and 5 in $\mathbb{Q}(\sqrt{-5})$ above. Note that x^2+5 is an Eisenstein polynomial with respect to 5 and $(x+1)^2+5 = x^2+2x+6$ is an Eisenstein polynomial with respect to 2. Likewise,

$$x^3 - 2 \equiv (x)^3 (\bmod 2) \qquad \text{and} \qquad x^3 - 2 \equiv (x+1)^3 (\bmod 3)$$

gives the splitting of the ramifying primes 2 and 3 in $\mathbb{Q}(2^{1/3})$ above. Here again, $x^3 - 2$ is an Eisenstein polynomial with respect to 2 and $(x-1)^3 - 2 = x^3 - 3x^2 + 3x - 3$ is an Eisenstein polynomial with respect to 3. We repeat that in these two examples, the polynomial discriminants are the field discriminants.

As a simple illustration that care is needed when the hypotheses of the theorem aren't satisfied, look at the factorization

$$x^2 + 3 \equiv (x+1)^2 (\bmod 2)$$

which happens in spite of the fact that (2) is a second degree prime of $\mathbb{Q}(\sqrt{-3})$. The catch is that the discriminant of the polynomial $x^2 + 3$ is -12 while the discriminant of $\mathbb{Q}(\sqrt{-3})$ is just -3. So this gives an example where the powers of the prime in the two discriminants aren't the same and the splitting of the prime doesn't match the factorization of the polynomial. Note also that $(x+1)^2 + 3 = x^2 + 2x + 4$ is not an Eisenstein polynomial with respect to 2 since 2^2 divides the constant term.

Beginning here and continuing into the next Section on zeta functions we will follow two number fields as examples.

Example One: $k = \mathbb{Q}(\sqrt{-5})$. The discriminant here is -20; 2 and 5 are the ramifying primes. For the remaining primes of $\mathbb{Q}$, we have two possibilities.

(i) $(p) = \mathfrak{p}_1\mathfrak{p}_2$ with $f_1 = f_2 = 1$ from (16). For these primes p, the polynomial $x^2 + 5$ factors $\bmod p)$ into two linear factors. This is expressed in terms of the classical Legendre symbol by

$$\left(\frac{-5}{p}\right) = \left(\frac{-20}{p}\right) = 1$$

and is in essence the definition of the symbol. Since -5 and -20 are the same except for a factor of 4 which is a square, the two Legendre symbols are equal. We will see at the end of Section 3 that the -20 is often the better choice. Examples of such primes are $p = 3, 7, 29, 41$.

(ii) $(p) = \mathfrak{p}$ with $f = 2$. For these primes the polynomial $x^2 + 5$ does not factor $(\bmod p)$. In terms of the Legendre symbol, we write

$$\left(\frac{-5}{p}\right) = \left(\frac{-20}{p}\right) = -1 \,.$$

Examples of such primes are $p = 11, 13, 17, 19$.

Example Two: $k = \mathbb{Q}(\alpha)$ where $\alpha = 2^{1/3}$. The discriminant is -108. Only 2 and 3 ramify in k. From (16), the unramifying primes now come in three possible varieties.

(i) $(p) = \mathfrak{p}_1\mathfrak{p}_2\mathfrak{p}_3$, $f_1 = f_2 = f_3 = 1$. The prime $p = 31$ is the first of this variety. We have

$$x^3 - 2 \equiv (x-4)(x-7)(x-20)(\bmod 31)\,,$$

and so $(31) = \mathfrak{p}_1\mathfrak{p}_2\mathfrak{p}_3$ where

$$\mathfrak{p}_1 = (31,\ \alpha - 4)\,, \qquad \mathfrak{p}_2 = (31,\ \alpha - 7)\,, \qquad \mathfrak{p}_3 = (31,\ \alpha - 20)\,.$$

(ii) $(p) = \mathfrak{p}_1\mathfrak{p}_2$, $f_1 = 1$, $f_2 = 2$. The prime $p = 5$ is the smallest example. We factored (5) in k in our discussion of splitting above.

(iii) $(p) = \mathfrak{p}$, $f = 3$. The first instance here is $p = 7$. $x^3 - 2$ is irreducible $(\bmod 7)$ and we get

$$p = (7,\ \alpha^3 - 2) = (7, 0) = (7)\,.$$

2.5 Frobenius Automorphisms

Let $K/\mathbb{Q}$ be a normal algebraic extension with Galois group $G = G(K/\mathbb{Q})$ and suppose P is a prime of K lying over the prime p of $\mathbb{Q}$ with residue class degree f.

$$G\left\{\begin{array}{ccc} K & & \mathfrak{P} \\ | & & | \\ \mathbb{Q} & & p \end{array}\right.$$

Then all the conjugate divisors $\mathfrak{P}^g$ of $\mathfrak{P}$ are divisors of K and their norms all the same as the norm of P,

$$\mathbb{N}(\mathfrak{P}^g) = \prod_{g' \in G} (\mathfrak{P}^g)^{g'} = \prod_{g' \in G} (\mathfrak{P})^{g'} = \mathbb{N}(\mathfrak{P}) = p^f .$$

Thus all the conjugate prime divisors of $\mathfrak{P}$ have the same residue class degree. Also, since p^f is the product of all the $\mathfrak{P}^g$, the unique factorization theorem shows that the $\mathfrak{P}^g$ account for all the prime divisors of p. Let $\mathfrak{P} = \mathfrak{P}_1, \mathfrak{P}_2, \cdots, \mathfrak{P}_r$ denote the distinct divisors of p among the $\mathfrak{P}^g$. Then we have a factorization of p in K,

$$(p) = \prod_{j=1}^{r} \mathfrak{P}_j^{e_j} . \tag{18}$$

If we apply any g in G to this, we get

$$(p) = \prod_{j=1}^{r} (\mathfrak{P}_j^g)^{e_j} . \tag{19}$$

The unique factorization theorem says that the factorizations of p in (18) and (19) must be the same. By varying g, we see that all the e_j must be equal. Thus in a normal extension, we have a factorization

$$(p) = \prod_{j=1}^{r} \mathfrak{P}_j^e ,$$

where all the exponents are the same, all the $\mathfrak{P}_j$ have the same residue class degree f and (16) simplifies to

$$n = efr . \tag{20}$$

As an illustration of this, once Case ii) of Example Two occurs, as it does for $p = 5$, we see that the corresponding field is not normal over $\mathbb{Q}$ since $f_1 \neq f_2$. In fact, the condition that for any given p all the f_j related to primes above p are the same is a necessary and sufficient condition that a field be normal over $\mathbb{Q}$. We will sketch an analytic proof in Section 3.

There is now an interesting series of subgroups of G related to $\mathfrak{P}$ of which we will say a little. The first such subgroup we introduce is the *decomposition group* of $\mathfrak{P}$,

$$D_G = D_G(\mathfrak{P}) = \{g \in G \,|\, \mathfrak{P}^g = \mathfrak{P}\} .$$

The decomposition group tells us how p splits. If we write G in terms of cosets of D_G,

$$G = \bigcup D_G g_j ,$$

then the distinct conjugate divisors to $\mathfrak{P}$ are just the divisors $\mathfrak{P}_j = \mathfrak{P}^{g_j}$. Therefore the index of the decomposition group D_G in G is r, the number of primes of K into which p splits (decomposes). It now follows that the order of D_G is ef, For unramified primes, this is just f.

This reminds us of the finite field examples of Section 1 and we will now show that indeed D_G acts on the finite field $\mathfrak{O}_k/\mathfrak{P} = \mathbb{F}_q$ of $q = p^f$ elements. Indeed it is easy to see that the elements of D_G give automorphisms of $\mathbb{F}_q/\mathbb{F}_p$. This is because if g is in D_G and

$$\alpha \equiv \beta \pmod{\mathfrak{P}} ,$$

then

$$\alpha^g \equiv \beta^g \pmod{\mathfrak{P}^g} ,$$

and since $\mathfrak{P}^g = \mathfrak{P}$,

$$\alpha^g \equiv \beta^g \pmod{\mathfrak{P}} .$$

In other words, elements of D_G take congruence classes $(\operatorname{mod} \mathfrak{P})$ to congruence classes $(\operatorname{mod} \mathfrak{P})$. Therefore we get a homomorphism from D_G into the Galois group $G(\mathbb{F}_q/\mathbb{F}_p)$. It turns out that this map is onto. In particular there is an element of D_G which maps onto the Frobenius automorphism in $G(\mathbb{F}_q/\mathbb{F}_p)$. Naturally we call this element of D_G the Frobenius automorphism also. In other words, there is an automorphism

$$\sigma = \sigma(\mathfrak{P}) = \sigma\left(\frac{K/\mathbb{Q}}{\mathfrak{P}}\right)$$

of G called the *Frobenius automorphism* such that

$$\alpha^\sigma \equiv \alpha^p \pmod{\mathfrak{P}} \tag{21}$$

for all integers α in K. For unramified primes, D_G is the cyclic group of order f generated by the Frobenius automorphism and is isomorphic to $G(\mathbb{F}_q/\mathbb{F}_p)$. In this situation, the Frobenius automorphism

$$\sigma = \sigma\left(\frac{K/\mathbb{Q}}{\mathfrak{P}}\right)$$

is unique and is completely determined by the congruence (21).

In general, since the map from D_G to $G(\mathbb{F}_q/\mathbb{F}_p)$ is onto, we have an isomorphism of $G(\mathbb{F}_q/\mathbb{F}_p)$ with D_G/I_G where $I_G = I_G(\mathfrak{P})$ is the set of elements of D_G which act as the identity in $G(\mathbb{F}_q/\mathbb{F}_p)$. This is the set of g in D_G such that

$$\alpha^g \equiv \alpha \pmod{\mathfrak{P}} \qquad \text{for all } \alpha \text{ in } \mathfrak{O}_K .$$

The group I_G is called the *inertial group* of $\mathfrak{P}$. We have seen that the groups D_G/I_G is $G(\mathbb{F}_q/\mathbb{F}_p)$ and hence is the cyclic group of order f where f is the residue class degree of $\mathfrak{P}$. Since D_G has order ef, we see that I_G has order e where e is the ramification index of $\mathfrak{P}$. For the unramified primes $\mathfrak{P}$ of K, I_G consists just of the identity element and is harmless. However when dealing with ramified primes, there is a whole coset of Frobenius automorphisms in D_G. We will not explore this unpleasant situation further in this exposition nor will we go further into the reasons for the names of D_G and I_G. The reader is referred to [2] for more information.

We will need to know what the Frobenius automorphisms are for conjugate primes to $\mathfrak{P}$. They will turn out to be conjugate elements in G. If g is in G, and we apply g to (21), we get,

$$\alpha^{\sigma g} \equiv (\alpha^g)^p (\bmod\, \mathfrak{P}^g)$$

for all integers α in K. If we replace α by $\alpha^{g^{-1}}$, then for all α in $\mathfrak{O}_K$, we have

$$\alpha^{g^{-1}\sigma g} \equiv (\alpha)^p (\bmod\, \mathfrak{P}^g) .$$

Therefore,

$$\sigma\left(\frac{K/\mathbb{Q}}{\mathfrak{P}^g}\right) = g^{-1}\sigma\left(\frac{K/\mathbb{Q}}{\mathfrak{P}}\right) g ,$$

as claimed. The decomposition groups themselves now behave in the same way

$$D_G(\mathfrak{P}^g) = g^{-1} D_G(\mathfrak{P}) g . \tag{22}$$

Now suppose that k is a subfield of K with $H = G(K/k)$. Further suppose that $\mathfrak{P}$ is an unramified prime of K and let $\mathfrak{p}$ be the prime of k divisible by $\mathfrak{P}$. We let $f(\mathfrak{P})$ and $f(\mathfrak{p})$ be the residue class degrees of $\mathfrak{P}$ and $\mathfrak{p}$ respectively. We have the picture,

$$G\left\{\begin{array}{l} H\left\{\begin{array}{cc} K & \mathfrak{P} \\ | & | \\ k & \mathfrak{p} \end{array}\right. \\ \begin{array}{cc} | & | \\ \mathbb{Q} & p \end{array} \end{array}\right. \tag{23}$$

We may now introduce the *decomposition group* of $\mathfrak{P}$ *relative to* k,

$$D_H = D_H(\mathfrak{P}) = \{g \in H \mid (\mathfrak{P}^g) = \mathfrak{P}\} .$$

Clearly,

$$D_H = D_G \bigcap H\,.$$

But D_H is isomorphic to the group $G(\mathbb{F}_q/\mathbb{F}_{q'})$ where $q' = p^{f(\mathfrak{p})} = \mathbb{N}_{k/\mathbb{Q}}(\mathfrak{p})$, and this is the cyclic group of order $f(\mathfrak{P}/\mathfrak{p}) = f(\mathfrak{P})/f(\mathfrak{p})$ generated by the Frobenius automorphism

$$\sigma(\mathfrak{P}/\mathfrak{p}) = \sigma\left(\frac{K/k}{\mathfrak{P}}\right)$$

of H and is uniquely determined by the congruence

$$\alpha^{\sigma(\mathfrak{P}/\mathfrak{p})} \equiv \alpha^{\mathrm{N}(\mathfrak{p})} \pmod{\mathfrak{P}} \qquad \text{for all } \alpha \text{ in } \mathfrak{O}_K\,.$$

But we also have,

$$\alpha^{\sigma^{f(\mathfrak{p})}} \equiv \alpha^{p^{f(\mathfrak{p})}} \equiv \alpha^{\mathrm{N}(\mathfrak{p})} \pmod{\mathfrak{P}} \qquad \text{for all } \alpha \text{ in } \mathfrak{O}_K\,.$$

Hence we see that $\sigma(\mathfrak{P}/\mathfrak{p}) = \sigma^{f(\mathfrak{P})}$,

$$\sigma\left(\frac{K/k}{\mathfrak{P}}\right) = \sigma\left(\frac{K/\mathbb{Q}}{\mathfrak{P}}\right)^{f(\mathfrak{p})}\,.$$

Let us look at all this from the point of view of our two example fields. The field $\mathbb{Q}(\sqrt{-5})$ is normal over $\mathbb{Q}$. The Frobenius automorphism is the identity of the two element Galois group when we are dealing with a prime p which splits completely into two pieces ($f = 1$) and is complex conjugation when we are dealing with a prime p which remains prime in $\mathbb{Q}(\sqrt{-5})$ ($f = 2$). The field $k = \mathbb{Q}(2^{1/3})$ is much more interesting, since it is not normal over $\mathbb{Q}$. Let $K = \mathbb{Q}(2^{1/3}, e^{2\pi i/3})$ which is normal over $\mathbb{Q}$ of degree 6. We write,

$$\alpha = 2^{1/3}, \qquad \beta = e^{2\pi i/3}\cdot 2^{1/3}, \qquad \gamma = e^{4\pi i/3}\cdot 2^{1/3}\,.$$

An automorphism of K is determined by its action on α, β and γ and thus $G(K/\mathbb{Q})$ is the symmetric group on three letters:

$$G(K/\mathbb{Q}) = \{1, (\alpha\beta\gamma), (\alpha\gamma\beta), (\beta\gamma), (\alpha\beta), (\alpha\gamma)\}\,.$$

The splitting of primes from $\mathbb{Q}$ to K is governed by the behavior of Frobenius automorphisms. Likewise the splitting of primes from k to K is governed by Frobenius automorphisms. The combined information should allow us to tell how primes split from $\mathbb{Q}$ to k. Let $H = G(K/k) = \{1, (\beta\gamma)\}$ be the subgroup of G fixing k. We will write $\tau = (\beta\gamma)$; τ is just complex conjugation. Suppose that $\mathfrak{P}$ is an unramified prime of K which lies above the prime $\mathfrak{p}$ of k which in turn lies above the prime p of $\mathbb{Q}$ as indicated in the diagram (23). Let $\sigma(\mathfrak{P})$ be the Frobenius automorphism of $\mathfrak{P}$ relative to $\mathbb{Q}$ and of $\sigma(\mathfrak{P}/\mathfrak{p})$ be the Frobenius automorphism of $\mathfrak{P}$ relative to k and suppose that $\sigma(\mathfrak{P})$ is of order $f(\mathfrak{P})$ and $\sigma(\mathfrak{P}/\mathfrak{p})$ is of order $f(\mathfrak{P}/\mathfrak{p})$. Then $\sigma(\mathfrak{P}/\mathfrak{p}) = \sigma(\mathfrak{P})^{f(\mathfrak{p})}$ where $f(\mathfrak{P}) = f(\mathfrak{P}/\mathfrak{p})f(\mathfrak{p})$ as indicated above. We now look at the three conjugacy classes of G where the Frobenius automorphism $\sigma(\mathfrak{P})$ may lie.

i) $\sigma(\mathfrak{P}) = 1$. When the Frobenius automorphism is of order 1, there must be 6 distinct primes of K above p by (20) and so all 6 of the $\mathfrak{P}^g$ are distinct. In this case, $\sigma(\mathfrak{P})$ is in H and so $\sigma(\mathfrak{P}/\mathfrak{p}) = \sigma(\mathfrak{P})$ and there are two distinct primes of K above $\mathfrak{p}$, $\mathfrak{P}$ and $\mathfrak{P}^\tau$. Further $f(\mathfrak{P}/\mathfrak{p}) = f(\mathfrak{P}) = 1$ and $\mathfrak{p}$ is a first degree prime of k. When we put this all together, we see that the six first degree primes of K above p combine in pairs to give three first degree primes of k above p. The primes can thus be numbered so that the picture is,

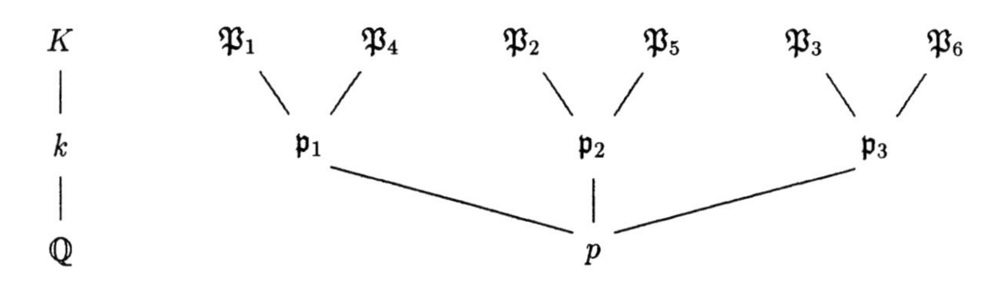

This is exactly the situation in Case i) of Example Two. Note that p not only splits completely in k, but also in K. We will see below that this is a general fact.

ii) $\sigma(\mathfrak{P})$ is in the conjugacy class $\{(\alpha\beta), (\alpha\gamma), (\beta\gamma)\}$ of elements of order two, one of which is τ. By choosing $\mathfrak{P}$ correctly among its conjugate primes, we may assume that $\sigma(\mathfrak{P}) = \tau$. Here $f = 2$ and there are three second degree primes $\mathfrak{P} = \mathfrak{P}_1, \mathfrak{P}_2, \mathfrak{P}_3$ above p. These primes behave differently with respect to k. We look at $\mathfrak{P}$ first. Since $\sigma(\mathfrak{P}) = \tau$ is in H, we see that $\sigma(\mathfrak{P}/\mathfrak{p}) = \sigma(\mathfrak{P}) = \tau$ also. Therefore $f(\mathfrak{P}/\mathfrak{p}) = 2$ and $f(\mathfrak{p}) = 1$. Thus the prime $\mathfrak{p} = \mathfrak{p}_1$ of k is a first degree prime of k which doesn't split in K; in K we have $\mathfrak{P} = \mathfrak{p}$. But the situation is different for the other two primes $\mathfrak{P}_2$ and $\mathfrak{P}_3$ where the Frobenius automorphisms work out to be the other two elements, $(\alpha\beta)$ and $(\alpha\gamma)$, of the conjugacy class. For these primes, $\sigma(\mathfrak{P})$ is not in H, but $\sigma(\mathfrak{P})^2 = 1$ is in H. Thus $f(\mathfrak{P}/\mathfrak{p}) = 1$ and $f(\mathfrak{p}) = 2$. This says that there are two primes of K (i.e. both $\mathfrak{P}_2$ and $\mathfrak{P}_3$ since there are no other choices) of relative degree one over a single prime $\mathfrak{p}_2$ of k which must be of degree two.

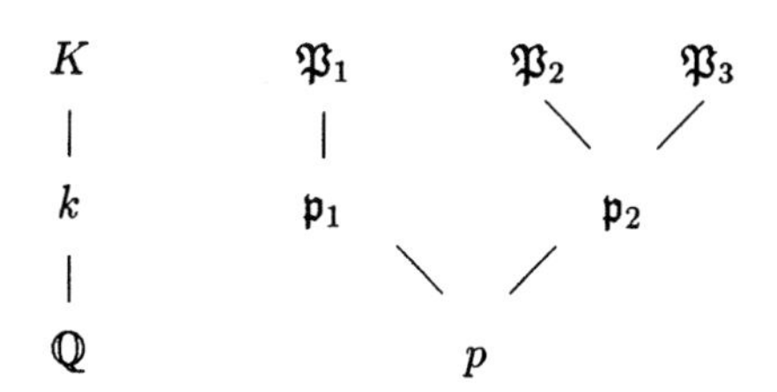

This corresponds precisely to the situation in Case ii) of Example Two. This is the most interesting case of all.

iii) $\sigma(\mathfrak{P})$ is in the conjugacy class $\{(\alpha\beta\gamma), (\alpha\gamma\beta)\}$ of elements of order three. Thus $f = 3$ and there are two third degree primes of K above p. We will denote them by $\mathfrak{P}_1$ and $\mathfrak{P}_2$. For one of these primes $\sigma(\mathfrak{P}) = (\alpha\beta\gamma)$ and for the other $\sigma(\mathfrak{P}) = (\alpha\gamma\beta)$. In each instance, $\sigma(\mathfrak{P})$ and $\sigma(\mathfrak{P})^2$ are not in H while $\sigma(\mathfrak{P})^3 = 1$ is in H. Thus in each instance $\sigma(\mathfrak{P}/p) = 1$. Thus there are two primes of K (which must be $\mathfrak{P}_1$ and $\mathfrak{P}_2$) of relative degrees one above a single prime $\mathfrak{p}$ of k which must be of degree three:

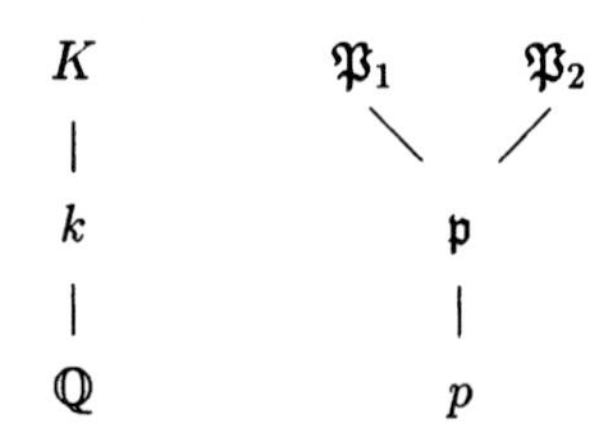

This gives us Case iii) of Example Two. We complete our crash course on Frobenius automorphisms with one last example. This example will be used in the next Section to show that the zeta function of a field determines the minimal normal extension of the field over $\mathbb{Q}$.

Theorem. *Suppose that p is a prime of $\mathbb{Q}$ which splits completely in each of two fields k_1 and k_2. Then p splits completely in the composite field k_1k_2. As a consequence, if p splits completely in a field k, then p also splits completely in the minimal normal extension of $\mathbb{Q}$ containing k.*

Proof. Let k be a field in which p is unramified. This is the case by definition for a prime which splits completely. Suppose that K is a normal extension of $\mathbb{Q}$ containing k and let $G = G(K/\mathbb{Q})$, $H = G(K/k)$. We will develop a group theoretical criterion for p to split completely in k. We will just do this for all p which don't ramify in K. (Since p is unramified in k by hypothesis, the criterion will actually hold for the finitely many remaining p once the ramification indices are put in.) Let $\mathfrak{P}$ be a prime of K above p and let $D_G = D_G(\mathfrak{P})$ be the corresponding decomposition group of order f where f is the degree of $\mathfrak{P}$. Likewise, let $\mathfrak{p}$ be a prime of k divisible by $\mathfrak{P}$. Then $\mathfrak{p}$ will be of first degree if and only if the decomposition group D_H is also of order f. Thus $\mathfrak{p}$ will be of first degree if and only if $D_H = D_G$. But this in turn happens if and only if D_G is contained in H. Therefore p splits completely in k if and only if every $D_G(\mathfrak{P}^g)$ is in H. Let J be the subgroup of G generated by all the $D_G(\mathfrak{P}^g)$ as g runs through G. Our criterion is then that *p splits completely in k if and only if J is contained in H*. For the record, since $D_G(\mathfrak{P}^g) = g^{-1}D_G g$, we see that J is a normal subgroup of G.

Now suppose that p splits completely in k_1 and k_2 and that $H_1 = G(K/k_1)$, $H_2 = G(K/k_2)$ are the corresponding Galois groups. Then J is contained in H_1 and in H_2 and so J is contained in $H_1 \bigcap H_2$. Therefore p splits completely in the corresponding fixed field, k_1k_2 . If p splits completely in k, then p splits completely in all the conjugate fields $k^{(i)}$ and hence p splits completely in the composite fields of all the $k^{(i)}$ which is just the minimal normal extension of $\mathbb{Q}$ containing k, as desired. □

3. Zeta Functions and L-Functions

3.1 Zeta Functions

For a number field k, the *zeta function* of k is

$$\zeta_k(s) = \sum_{\mathfrak{a}} \mathbb{N}(\mathfrak{a})^{-s} ,$$

where the sum is over all integral divisors of k. Often, $\zeta_k(s)$ is referred to as the *Dedekind zeta function.* This sum converges for complex s with $\mathrm{Re}(s) > 1$. (Analytic number theory has evolved the strange tradition of writing $s = \sigma + it$, but we will not do so in this paper because σ is rather heavily used already.) When $k = \mathbb{Q}$,

$$\zeta_{\mathbb{Q}}(s) = \zeta(s) = \sum_{n=1}^{\infty} n^{-s}$$

is just the *Riemann zeta function.* Euler had already discovered that $\zeta(s)$ has an *Euler product,*

$$\zeta(s) = \prod_p (1 + p^{-s} + p^{-2s} + p^{-3s} + \cdots) = \prod_p (1 - p^{-s})^{-1} .$$

For example, the product of the $p = 2$, 3 and 5 terms on the right gives the sum of all n^{-s} such that all of the prime factors of n lie among 2, 3 and 5. As an illustration, the 1500^{-s} terms arises from the products of the $2^{-2s}, 3^{-s}$ and 5^{-3s} terms. In the limit, the product over all p will pick up every n^{-s} exactly once thanks to the unique factorization theorem.

In like manner, the unique factorization theorem for divisors now gives us an *Euler product* for $\zeta_k(s)$:

$$\zeta_k(s) = \prod_{\mathfrak{p}} \left(\sum_{a=0}^{\infty} \mathbb{N}(\mathfrak{p})^{-as} \right) = \prod_{\mathfrak{p}} (1 - \mathbb{N}(\mathfrak{p})^{-s})^{-1} .$$

We will say that the *Euler p-factor* of $\zeta_k(s)$ is

$$\prod_{\mathfrak{p} | p} (1 - \mathbb{N}(\mathfrak{p})^{-s})^{-1} .$$

The series defining $\zeta_k(s)$ converges for $\mathrm{Re}(s) > 1$, but the function $\zeta_k(s)$ has an analytic continuation to the entire complex s-plane except for a first order pole at $s = 1$. The residue at $s = 1$ involves almost every field invariant we have introduced,

$$\mathrm{res}_{s=1} \zeta_k(s) = \frac{2^{r_1 + r_2} \pi^{r_2} hR}{W\sqrt{|D|}} ,$$

where r_1 is the number of real conjugate fields of k, $2r_2$ is the number of complex conjugate fields of k, n is the degree of k, h is the class-number of k,

R is the regulator of k, W is the number of roots of unity in k and D is the discriminant of k.

There is also a functional equation relating values at s to values at $1-s$. Let

$$\xi_k(s) = \left(\frac{|D|}{2^{2r_2}\pi^n}\right)^{s/2} \Gamma\left(\frac{s}{2}\right)^{r_1} \Gamma(s)^{r_2} \zeta_k(s) .$$

Then the *functional equation* for the Dedekind zeta function is,

$$\xi_k(1-s) = \xi_k(s) .$$

For instance, the functional equation relates the pole at $s=1$ to an r-th order zero at $s=0$, where $r = r_1 + r_2 - 1$ is the rank of the unit group of k and we find,

$$\left[s^{-r}\zeta_k(s)\right]\Big|_{s=0} = -\frac{hR}{W} . \tag{24}$$

For $k = \mathbb{Q}$, this gives the value of the Riemann zeta function at $s = 0$,

$$\zeta(0) = -\frac{1}{2} .$$

The reader is referred to either [1] or [2] for more information on the analytic continuation and functional equation.

When we take the logarithm of the Euler product for $\zeta_k(s)$, we get

$$\log \zeta_k(s) = -\sum_{\mathfrak{p}} \log(1 - \mathrm{N}(\mathfrak{p})^{-s}) = \sum_{\mathfrak{p}} \sum_{m=1}^{\infty} \frac{1}{m} \mathrm{N}(\mathfrak{p})^{-ms} . \tag{25}$$

This series also converges for $\mathrm{Re}(s) > 1$, but we now have a logarithmic singularity at $s = 1$. In fact, because $\zeta_k(s)$ has a first order pole at $s = 1$, we see that

$$\log \zeta_k(s) - \log\left(\frac{1}{s-1}\right)$$

is analytic at $s = 1$. We will say that $\log\left(\frac{1}{s-1}\right)$ is the singularity of $\log\zeta_k(s)$ at $s = 1$. We now want to look closely at which terms on the right of (25) actually contribute to the singularity at $s = 1$. Of course, any finite number of terms don't contribute at all, but there are infinite classes of terms that don't contribute either. Indeed, since there are at most $[k : \mathbb{Q}]$ primes $\mathfrak{p}$ in k above each rational prime p, and since $\sum p^{-2s}$ converges for $\mathrm{Re}(s) > 1/2$, we see that only the $m = 1$ terms in (25) matter. But further, even when $m = 1$, primes of second and higher degree don't contribute to the singularity for the same reason. Thus we have proved the

Theorem. *The function*

$$\sum_{\mathfrak{p}}{}' \mathbb{N}(\mathfrak{p})^{-s} - \log\left(\frac{1}{s-1}\right),$$

where the sum is over all first degree primes of k, is analytic at $s = 1$.

In other words, only the first degree primes of k contribute to the singularity at $s = 1$ in (25).

Definition. Let S be a set of primes of k such that

$$\lim_{s\to 1^+} \frac{\sum_{\mathfrak{p}\in S} \mathbb{N}(\mathfrak{p})^{-s}}{\log\left(\frac{1}{s-1}\right)} = \delta .$$

We will say that the primes of S have *analytic density* δ among the primes of k.

Except when $\delta = 0$, this may be phrased as the asymptotic statement,

$$\sum_{\mathfrak{p}\in S} \mathbb{N}(\mathfrak{p})^{-s} \sim \delta \log\left(\frac{1}{s-1}\right) \qquad \text{as } s \to 1^+ .$$

The notation $s \to 1^+$ means that s approaches 1 through real values from above. Clearly $0 \le \delta \le 1$. Again, only the first degree primes of S actually contribute to δ. The analytic density is the easiest of several types of densities that we could consider. The harder types actually count the elements of S versus the number of primes overall in some manner. When a harder way gives a density, it is the same as the analytic density.

Several sets of primes of interest to us can be easily shown to have analytic densities. The next theorem gives one instance.

Theorem. *Suppose that K is a normal extension of $\mathbb{Q}$ of degree n_K. The set of primes of $\mathbb{Q}$ which split completely in K has analytic density $1/n_K$.*

Proof. We have seen that with the exception of finitely many ramified primes, the first degree primes of K come in batches of n_K conjugate primes over rational primes of $\mathbb{Q}$ which split completely in K. Let S be the set of primes p in $\mathbb{Q}$ which split completely in K. Since the n_K primes $\mathfrak{P}$ of K above p all have p as their norm, we see from the last theorem that

$$\sum_{p\in S} p^{-s} \sim \frac{1}{n_K} \log\left(\frac{1}{s-1}\right) \qquad \text{as } s \to 1^+ .$$

In other words, the primes of $\mathbb{Q}$ which split completely in K have analytic density $1/n_K$. □

Let us look at this theorem from the point of view of our two example fields. In Example One, $k = \mathbb{Q}(\sqrt{-5})$ is normal over $\mathbb{Q}$ of degree two. Therefore the set of primes of $\mathbb{Q}$ which split completely in k has analytic density 1/2. Except for 2 and 5, the remaining primes of $\mathbb{Q}$ generate second degree primes in k. This set of primes of $\mathbb{Q}$ also has analytic density 1/2.

Example Two is much more interesting here because $k = \mathbb{Q}(2^{1/3})$ is not normal over $\mathbb{Q}$. We recall that there are three ways that a prime other than 2 or 3 can split in k. Let S_i denote the set of primes p of $\mathbb{Q}$ that split completely into three first degree primes of k, S_{ii} denote the set of those primes p of $\mathbb{Q}$ which split into two pieces in k, one first degree and one second degree and let S_{iii} denote the remaining primes of $\mathbb{Q}$ which are those that simply generate third degree primes of k. Let $K = k\left(e^{2\pi i/3}\right)$ which is the minimal normal extension of $\mathbb{Q}$ containing k. Since $[K : \mathbb{Q}] = 6$, 1/6 of the primes of $\mathbb{Q}$ split completely in K. But the primes of $\mathbb{Q}$ which split completely in K are exactly the primes of S_i. In other words, 1/6 of the primes of $\mathbb{Q}$ split completely in k.

This is not enough to account for all the first degree primes of k. Each prime of $\mathbb{Q}$ that splits completely in k splits into 3 first degree prime pieces in k. Therefore, we have only accounted for $3/6 = 1/2$ of the first degree primes of k. The primes of S_{ii} must account for the rest. Indeed, since each prime p of S_i gives rise to three first degree primes of k above p and each prime p of S_{ii} gives rise to one first degree prime of k above p, we have

$$\sum_{\mathfrak{p}}{}'' \mathrm{N}(\mathfrak{p})^{-s} = 3\sum_{p\in S_i} p^{-s} + \sum_{p\in S_{ii}} p^{-s}\,,$$

where $\sum''$ means that we sum over the first degree unramified primes of k. We now see that since the primes of S_i have density 1/6, the primes of S_{ii} must have density 1/2. In particular, there are primes of $\mathbb{Q}$ which don't split completely in k but which nevertheless have first degree primes of k above them.

We have accounted for $\frac{1}{6}+\frac{1}{2} = \frac{2}{3}$ of the primes of $\mathbb{Q}$ in S_i and S_{ii}. Therefore S_{iii} has the remaining $\frac{1}{3}$ of the primes of $\mathbb{Q}$. Thus all three of the sets S_i, S_{ii}, and S_{iii} have analytic densities. But more than that, the densities are related to the Galois group $G(K/\mathbb{Q})$. For a fixed prime p of $\mathbb{Q}$, the Frobenius automorphisms $\sigma(\mathfrak{P})$ of primes $\mathfrak{P}$ of K which divide p run through a conjugacy class of $G(K/\mathbb{Q})$. We saw in Subsection 2.5 that the primes of S_i all have $\sigma(\mathfrak{P}) = 1$, which is 1/6 of the elements of G. The primes of S_{ii} all have $\sigma(\mathfrak{P})$ in a conjugacy class of three elements; this conjugacy class consists of 1/2 of the elements of $G(K/\mathbb{Q})$. The primes of S_{iii} all have $\sigma(\mathfrak{P})$ in a conjugacy class of two elements; this conjugacy class consists of 1/3 of the elements of $G(K/\mathbb{Q})$. In other words, the primes of each of the three sets S_i, S_{ii} and S_{iii} have a density which is the density of the corresponding conjugacy classes of Frobenius automorphisms in $G(K/\mathbb{Q})$. This is a general theorem which goes by the name of the Chebotarev density theorem (a theorem with many spellings). We will show how to prove it in Subsection 3.3. For the moment, we just note that from the point of view of Frobenius automorphism, we found this same result in Example One. There are two conjugacy classes in $G(k/\mathbb{Q})$ of one element each and each class arises as a Frobenius automorphism half the time.

An analog of S_{ii} will arise in any non-normal extension k of $\mathbb{Q}$. There will always be infinitely many primes of $\mathbb{Q}$ which don't split completely in k, but yet have first degree primes in k above them. The reason is the same; the primes of $\mathbb{Q}$ which split completely in k have too low a density to account for all the first degree primes of k. When we factor any one of these primes in k and see different residue class degrees for the prime factors, we will be able to conclude that k is not normal over $\mathbb{Q}$.

The rest of this Section is devoted to proving that the zeta function of a field determines the minimal normal extension over $\mathbb{Q}$ containing the field. The next result provides the key fact.

Theorem. *Let K be a normal extension of $\mathbb{Q}$ and let M be another field such that with at most finitely many exceptions, any prime of $\mathbb{Q}$ which splits completely in K also splits completely in M. Then M is contained in K.*

Proof. The proof is analytic. Let S be the set of primes of $\mathbb{Q}$ which split completely in K, except that the potential finite number of such primes that don't split completely in M are to be excluded. Then S has analytic density $1/n_K$ where $n_K = [K : \mathbb{Q}]$. We saw in Subsection 2.5 that the primes of S split completely in KM. We write $n_{KM} = [KM : \mathbb{Q}]$ also. Since each prime of $\mathbb{Q}$ in S splits into n_{KM} first degree primes of KM, the analytic density of first degree primes of KM must be at least n_{KM}/n_K. But the maximum possible density is 1 and so,

$$n_{KM}/n_K \leq 1$$

Hence $n_{KM} \leq n_K$. Since $K \subset KM$, we now get $n_{KM} = n_K$. Therefore $KM = K$ and so M is a subfield of K. □

Theorem. *The zeta function of a field k determines the minimal normal extension of $\mathbb{Q}$ containing k.*

Proof. The Euler product shows us which primes of $\mathbb{Q}$ split completely in k and hence in the minimal normal extension K of $\mathbb{Q}$ containing k. The analytic density of this set gives us the degree of K. By the last theorem, if M is a normal extension of $\mathbb{Q}$ of the same degree as K and having the same primes of $\mathbb{Q}$ splitting completely in M, then $M = K$. Thus K is uniquely determined by $\zeta_k(s)$. □

Of course when we are just given $\zeta_k(s)$ as an analytic function, actually finding the field guaranteed by the last theorem is another matter entirely. In principle it can be done, but not yet in any reasonable manner. There are a whole series of conjectures regarding special values of zeta and L-functions that can be interpreted as being related to this problem. We might also ask whether $\zeta_k(s)$ determines k. The zeta function tells us the degree, the discriminant, the class-number times the regulator of the field; it determines the minimal normal extension of $\mathbb{Q}$ containing the field and of course, it tells us the splitting of every prime in the field. Naturally conjugate fields have the same zeta function, but it turns out that there are non-conjugate fields with the same zeta function! We will see how this can happen in Subsection 3.3.

3.2 Representations and Characters of Finite Groups

The zeta function of a field is like the atom of physics. In this and the next Section, we will show how to split it via group theory. We will do this by examining the individual Euler p-factors in the Euler product for a zeta function for primes which don't ramify. We begin by looking at the Euler p-factor in our two examples fields, $\mathbb{Q}(\sqrt{-5})$ and $\mathbb{Q}(2^{1/3})$.

Example One, $k = \mathbb{Q}(\sqrt{-5})$. There are two manners of factorization for the unramified primes p. Recall that one of the ways p can factor is that p splits into two pieces with

(i) $f_1 = f_2 = 1$. The Euler p-factor is then $(1-p^{-s})^{-1}(1-p^{-s})^{-1} = (1-p^{-s})^{-2}$. We can write this in matrix form as

$$(1-p^{-s})^{-2} = \det\left[I_2 - \begin{pmatrix} 1 & 0 \\ 0 & 1 \end{pmatrix} p^{-s}\right]^{-1} .$$

(Whether the determinant or the inverse is taken first, the result is the same, but the equality is easiest to see if the determinant is first.) The other way that p can factor is that p remains prime with

(ii) $f = 2$. The Euler p-factor is then $(1-p^{-2s})^{-1}$ and in matrix form, this is

$$(1-p^{-2s})^{-1} = \det\left[I_2 - \begin{pmatrix} 1 & 0 \\ 0 & -1 \end{pmatrix} p^{-s}\right]^{-1} .$$

Although this is the obvious choice of matrix, any similar matrix will do and we will find it more expedient to use the similar matrix $\begin{pmatrix} 0 & 1 \\ 1 & 0 \end{pmatrix}$:

$$(1-p^{-2s})^{-1} = \det\left[I_2 - \begin{pmatrix} 0 & 1 \\ 1 & 0 \end{pmatrix} p^{-s}\right]^{-1} .$$

The two matrices $\begin{pmatrix} 1 & 0 \\ 0 & 1 \end{pmatrix}$ and $\begin{pmatrix} 0 & 1 \\ 1 & 0 \end{pmatrix}$ form a group of order two, just as $G(k/\mathbb{Q})$ gives a group of order two. To explore this further, we turn to our second example.

Example Two, $k = \mathbb{Q}(2^{1/3})$. Here there are three ways an unramified prime can factor. The first way is that p splits completely into three prime divisors with

(i) $f_1 = f_2 = f_3 = 1$. The Euler p-factor is then

$$(1-p^{-s})^{-3} = \det\left[I_3 - \begin{pmatrix} 1 & 0 & 0 \\ 0 & 1 & 0 \\ 0 & 0 & 1 \end{pmatrix} p^{-s}\right]^{-1} .$$

Another way p can factor is into two pieces, one of first degree and one of second degree. Here we have

(ii) $f_1 = 1$, $f_2 = 2$. The Euler p-factor is $(1-p^{-s})^{-1}(1-p^{-2s})^{-1}$. We will put this in matrix form as

$$\begin{aligned}(1-p^{-s})^{-1}(1-p^{-2s})^{-1} &= \det\left[I_3 - \begin{pmatrix}0&1&0\\1&0&0\\0&0&1\end{pmatrix}p^{-s}\right]^{-1}\\ &= \det\left[I_3 - \begin{pmatrix}0&0&1\\0&1&0\\1&0&0\end{pmatrix}p^{-s}\right]^{-1}\\ &= \det\left[I_3 - \begin{pmatrix}1&0&0\\0&0&1\\0&1&0\end{pmatrix}p^{-s}\right]^{-1}.\end{aligned}$$

The eigenvalues of the three similar matrices here are $1, 1, -1$. Finally, p can simply remain prime in k with

(iii) $f = 3$. The Euler p-factor is $(1-p^{-3s})^{-1}$. We write this in matrix form as

$$\begin{aligned}(1-p^{-3s})^{-1} &= \det\left[I_3 - \begin{pmatrix}0&1&0\\0&0&1\\1&0&0\end{pmatrix}p^{-s}\right]^{-1}\\ &= \det\left[I_3 - \begin{pmatrix}0&0&1\\1&0&0\\0&1&0\end{pmatrix}p^{-s}\right]^{-1},\end{aligned}$$

the eigenvalues of the two similar matrices here are the three cube roots of unity.

Now, let us recall the Galois group related to Example 2. We write again,

$$\alpha = 2^{1/3}, \qquad \beta = e^{2\pi i/3}\cdot 2^{1/3}, \qquad \gamma = e^{4\pi i/3}\cdot 2^{1/3}.$$

The field $K = \mathbb{Q}(\alpha, e^{2\pi i/3}) = \mathbb{Q}(\alpha,\beta,\gamma)$ is normal over $\mathbb{Q}$ and we recall that $G(K/\mathbb{Q})$ is the symmetric group on three letters:

$$G(K/\mathbb{Q}) = \{1, (\alpha\beta\gamma), (\alpha\gamma\beta), (\alpha\beta), (\alpha\gamma), (\beta\gamma)\}.$$

We can *represent* this group by matrices in various ways. The appropriate way here is as permutation matrices on three letters: if g is in $G(K/\mathbb{Q})$, then we may write

$$\begin{pmatrix}\alpha\\\beta\\\gamma\end{pmatrix} g = M(g)\begin{pmatrix}\alpha\\\beta\\\gamma\end{pmatrix},$$

where $M(g)$ is a permutation matrix of zeros and ones. In this way, the six elements of our group correspond to six 3×3 matrices:

$$\begin{array}{ccc} 1 & (\alpha\beta\gamma) & (\alpha\gamma\beta) \\ | & | & | \\ \begin{pmatrix} 1 & 0 & 0 \\ 0 & 1 & 0 \\ 0 & 0 & 1 \end{pmatrix} & \begin{pmatrix} 0 & 1 & 0 \\ 0 & 0 & 1 \\ 1 & 0 & 0 \end{pmatrix} & \begin{pmatrix} 0 & 0 & 1 \\ 1 & 0 & 0 \\ 0 & 1 & 0 \end{pmatrix} \end{array}$$

$$\begin{array}{ccc} (\alpha\beta) & (\alpha\gamma) & \beta\gamma) \\ | & | & | \\ \begin{pmatrix} 0 & 1 & 0 \\ 1 & 0 & 0 \\ 0 & 0 & 1 \end{pmatrix} & \begin{pmatrix} 0 & 0 & 1 \\ 0 & 1 & 0 \\ 1 & 0 & 0 \end{pmatrix} & \begin{pmatrix} 1 & 0 & 0 \\ 0 & 0 & 1 \\ 0 & 1 & 0 \end{pmatrix} \end{array} \tag{26}$$

The map $g \to M(g)$ is a homomorphism of $G(K/\mathbb{Q})$ to 3×3 matrices: for g and h in $G(K/\mathbb{Q})$,

$$M(gh)\begin{pmatrix} \alpha \\ \beta \\ \gamma \end{pmatrix} = \begin{pmatrix} \alpha \\ \beta \\ \gamma \end{pmatrix} gh = M(g)\begin{pmatrix} \alpha \\ \beta \\ \gamma \end{pmatrix} h = M(g)M(h)\begin{pmatrix} \alpha \\ \beta \\ \gamma \end{pmatrix}$$

and so $M(gh) = M(g)M(h)$. The result is a *three dimensional representation* of the group $G(K/\mathbb{Q})$. Representations that arise in this manner are called *permutation representations.*

These six matrices are precisely the six matrices appearing in our expressions for the Euler p-factor in Example Two. In each case for the Euler p-factor, we can choose the matrix up to conjugacy. There are three conjugacy classes in this group. The conjugacy class of the identity alone (three one cycles) corresponds to an Euler p-factor with three first degree prime factors. The conjugacy class of elements of order two (one one cycle and one two cycle) corresponds to two factors, one first degree and one second degree. The conjugacy class of elements of elements of order three (one three cycle) corresponds to a single third degree prime. Further, we saw in Subsection 2.5 that these conjugacy classes are precisely the conjugacy classes of Frobenius automorphisms arising from primes which factor in the indicated manner.

It turns out that we can define an interesting function using any representation of $G(K/\mathbb{Q})$. In order to prepare for this, it will be useful to recall the basic facts about representations of finite groups. Let G be a finite group. Suppose that ρ is a finite dimensional representation of G. As we have just indicated, this means that ρ provides a homomorphism from G into the nonsingular $n \times n$ matrices, where n is the dimension of the representation. We usually call n the *degree* of the representation. In particular, $\rho(1) = I_n$, the $n \times n$ identity matrix. The *character* of a representation p is given by

$$\chi(g) = \operatorname{tr}(\rho(g)) \, .$$

We say that the *degree* of the character is n also and write $\deg(\chi) = n$. Note that $\chi(1) = \operatorname{tr}(I_n) = \deg(\chi)$. If h is another element of G, then

$$\rho(h^{-1}gh) = \rho(h)^{-1}\rho(g)\rho(h)$$

so that $\rho(h^{-1}gh)$ and $\rho(g)$ are similar matrices. Thus their traces are the same,

$$\chi(h^{-1}gh) = \chi(g) .$$

Hence a character is constant on conjugacy classes. A function on G which is constant on conjugacy classes is a *class function.* Characters are class functions; if C is a conjugacy class, we will sometimes write $\chi(C)$ for the common value of χ on the group elements in C.

Suppose that there is a fixed matrix A which relates two n-dimensional representations ρ_1 and ρ_2 of G in the following manner:

$$\rho_1(g) = A^{-1}\rho_2(g)A \quad \text{for all } \; g \text{ in } G .$$

We say that ρ_1 and ρ_2 are *equivalent representations* of G. Since similar matrices have the same trace, *equivalent representations have the same character.* Although far from obvious, the converse is also true. *If two representations have the same character, then they are equivalent.* If ρ_1 and ρ_2 are representations of G with characters χ_1 and χ_2, then

$$\rho(g) = \begin{pmatrix} \rho_1(g) & 0 \\ 0 & \rho_2(g) \end{pmatrix}$$

also gives a representation of G with character $\chi_1 + \chi_2$. We say that ρ is a *reducible representation.* We further say that any representation which is equivalent to a reducible representation is reducible. Any representation which is not reducible is *irreducible.* One irreducible representation is the *trivial representation* given by $\rho(g) = 1$ for all g in G. The corresponding character, to be denoted here from now on as χ_1, is identically one on G and is called the *trivial character.*

Suppose that G has r conjugacy classes. There turn out to be exactly r inequivalent irreducible representations. This allows us to form the *character table* of G. This is a table of the values of the r irreducible characters on the r conjugacy classes. It is customary to list the trivial character first in the list of characters and the identity conjugacy class first in the list of conjugacy classes. For example, the character table for S_3, the symmetric group on three letters from Example Two, is

	1	$(\alpha\beta\gamma), (\alpha\gamma\beta)$	$(\alpha\beta), (\alpha\gamma), (\beta\gamma)$
χ_1	1	1	1
χ_1'	1	1	-1
χ_2	2	-1	0

The entries in the first column give the degrees of the various characters. For S_3, the irreducible characters are of degrees $1, 1$ and 2. The charcter χ_1' is a character of the order two quotient group S_3/A_3 where $A_3 = \{1, (\alpha\beta\gamma), (\alpha\gamma\beta)\}$ is the alternating group on three letters. The character χ_2 is a second degree character corresponding to the two dimensional representation arising from the six linear transformations of a plane which carry an equilateral triangle centered at the origin onto itself.

The rows and columns of a character table satisfy *orthogonality relations.* For two irreducible characters χ and χ', the *row orthogonality relations* are,

$$\sum_{g\in G} \chi(g)\overline{\chi'(g)} = \begin{cases} |G| & \text{if } \chi = \chi' \\ 0 & \text{if } \chi \neq \chi' . \end{cases}$$

For two conjugacy classes C and C', the *column orthogonality relations* are,

$$\sum_{\chi} \chi(C)\overline{\chi(C')} = \begin{cases} |G|/|C| & \text{if } C = C' \\ 0 & \text{if } C \neq C' . \end{cases}$$

Because of the orthogonality relations, we have a recipe for writing any class function on G as a linear combination of the irreducible characters of G. This is most easily set up in terms of the following inner product on class functions, if $f(g)$ and $f'(g)$ are two class functions then we set

$$\langle f, f'\rangle = \langle f, f'\rangle_G = \frac{1}{|G|}\sum_{g\in G} f(g)\overline{f'(g)} .$$

The row orthogonality relations for two irreducible characters χ and χ' are thus,

$$\langle \chi, \chi'\rangle = \begin{cases} 1 & \text{if } \chi = \chi' \\ 0 & \text{if } \chi \neq \chi' . \end{cases}$$

Suppose that $f(g)$ is a class function. Then f can be written in the form,

$$f = \sum_{\chi} a_\chi \chi$$

where the sum is over the irreducible characters χ of G and our task is to find the coefficients a_χ. The row orthogonality relations show that

$$\langle f, \chi'\rangle = \sum_{\chi} a_\chi \langle \chi, \chi'\rangle = a_{\chi'} .$$

Therefore we may write an arbitrary class function f as

$$f = \sum_{\chi} \langle f, \chi\rangle \chi .$$

In particular, a class function f is a character if and only if every $\langle f, \chi\rangle$ is a non-negative rational integer.

We return to Example Two for an illustration. Again, set $G = G(K/\mathbb{Q})$. We look at the third degree character χ which corresponds to the permutation representation of G given in (26) and which by taking traces has character values,

	1	$(\alpha\beta\gamma), (\alpha\gamma\beta)$	$(\alpha\beta), (\alpha\gamma), (\beta\gamma)$
χ	3	0	1

The inner products are

$$\langle \chi, \chi_1 \rangle = 1 \,, \quad \langle \chi, \chi_1' \rangle = 0 \,, \quad \langle \chi, \chi_2 \rangle = 1 \,,$$

and thus we have,

$$\chi = \chi_1 + \chi_2 \,,$$

as may be easily checked. Thus χ splits into two irreducible pieces.

We have seen that $\zeta_k(s)$ is defined by a permutation representation with character χ given above. From the splitting of χ, we will find a splitting of $\zeta_k(s)$. To do this in general, we have to be able to deal with the analogs of χ and this will require us to find some sort of formula we can compute to get the values of χ on G. As a first step, we show that the permutation representation we constructed on the three letters α, β, γ corresponding to χ may also be thought of as a permutation representation of G on the three cosets of $H = G(K/k) = \{1, (\beta\gamma)\}$. Indeed since H is the subgroup of G which fixes α, the three cosets of H take α to its three conjugates, α, β and γ. Thus we can label the cosets Hg_α, Hg_β and Hg_γ so that the elements of the respective cosets applied to α give α, β and γ respectively. But now, we may write

$$\begin{pmatrix} \alpha \\ \beta \\ \gamma \end{pmatrix} = \begin{pmatrix} \alpha \circ Hg_\alpha \\ \alpha \circ Hg_\beta \\ \alpha \circ Hg_\gamma \end{pmatrix}$$

where $\alpha \circ g$ means α^g. So we see that the elements of the group G applied to the column ${}^t(\alpha, \beta, \gamma)$ act the same as the elements of G on the cosets of H.

We can create a permutation representation for any finite group on the cosets of any subgroup in the same manner. Let G be a finite group and H a subgroup. Suppose that we split G up into cosets of H as $G = \bigcup_{j=1}^{m} Hg_j$. Then we can create an m-th degree permutation representation of G on the cosets of H by defining a permutation matrix $M(g)$ of zeros and ones for each g in G such that

$$\begin{pmatrix} Hg_1 \\ \vdots \\ Hg_m \end{pmatrix} g = M(g) \begin{pmatrix} Hg_1 \\ \vdots \\ Hg_m \end{pmatrix} . \tag{27}$$

We want to find a formula now for the character χ which corresponds to this representation. Since $\chi(g)$ is the number of ones on the main diagonal of $M(g)$, we want to see how many ones there are on the diagonal. In the i-th row, there will be a one on the diagonal if and only if $Hg_ig = Hg_i$. This happens if and only if $Hg_igg_i^{-1} = H$ and in turn, this occurs if and only if $g_igg_i^{-1}$ is in H. Therefore,

$$\chi(g) = \sum_{\substack{i=1 \\ g_igg_i^{-1} \in H}}^{m} 1 . \tag{28}$$

Many readers will now recognize the construction for χ; χ is the character of G *induced* from the trivial character of H. In general, suppose that ψ is a character of H. We extend the definition of ψ to all of G by setting $\psi(g) = 0$ if g is not in H. The character ψ^* of G *induced* by ψ of H is then defined to be given by

$$\psi^*(g) = \sum_{i=1}^{m} \psi(g_igg_i^{-1}) . \tag{29}$$

We also say that ψ^* is *induced from H to G by* ψ. Another frequent notation is

$$\psi^* = \mathrm{Ind}_H^G \psi .$$

When several groups are in use at the same time, this is a convenient notation for keeping track of them.

There is another version of the formula (29) which does not depend on the choice of coset representatives. If h is in H, then $hg_igg_i^{-1}h^{-1}$ and $g_igg_i^{-1}$ are both in H or both not in H and they are conjugate by an element of H when they are both in H. Thus $\psi(hg_igg_i^{-1}h^{-1}) = \psi(g_igg_i^{-1})$. This shows that ψ^* does not depend upon the choice of coset representative. It also shows that we can sum over the whole coset and divide by the number of elements in the coset. This gives

$$\psi^*(g) = \frac{1}{|H|} \sum_{g' \in G} \psi\left(g'gg'^{-1}\right) . \tag{30}$$

Although the expression (29) is best for numerically computing the values of ψ^*, because it has fewer terms, (30) is easier to use theoretically because the sum ranges over the whole group. For instance, in (30), we can replace g' in G by g'^{-1} and get an equivalent expression,

$$\psi^*(g) = \frac{1}{|H|} \sum_{g' \in G} \psi\left(g'^{-1}gg'\right) \tag{31}$$

In the case that ψ is the trivial character of H, we see that (28) and (29) agree so that χ is induced from the trivial character of H as stated. While we constructed χ as the character of a representation of G, it is not obvious that ψ^* is a character for all ψ. There is an analogous construction which gives a representation of G whose character is ψ^*, but we won't go into it here. There is a theoretical way of seeing that ψ^* is a character which we will give because it also gives an efficient way of writing ψ^* in terms of the irreducible characters of G. This way is based on a fundamental calculational rule in character theory. To state it, we need one more definition. If χ is a character of a group G and H is a subgroup of G, we define the *restriction* $\chi|_H$ of χ to H by $\chi|_H(h) = \chi(h)$ for all h in H. This is a character of H (use the same representation restricted to H), but may not be irreducible on H even when χ is irreducible on G.

Theorem (The Frobenius Reciprocity Law). *If χ is a character of G and ψ is a character of a subgroup H of G, then*

$$\langle \psi^*, \chi \rangle_G = \langle \psi, \chi|_H \rangle_H \,.$$

Proof. We have

$$\begin{aligned} \langle \psi^*, \chi \rangle_G &= \frac{1}{|G|} \sum_{g \in G} \psi^*(g)\overline{\chi(g)} \\ &= \frac{1}{|G|} \sum_{g \in G} \frac{1}{|H|} \sum_{g' \in G} \psi(g'^{-1} g g')\overline{\chi(g)} \\ &= \frac{1}{|G||H|} \sum_{g' \in G} \sum_{g \in G} \psi(g)\overline{\chi(g' g g'^{-1})} \end{aligned}$$

But $\psi(g)$ is by definition 0 unless g is in H and χ is constant on conjugacy classes of G. Hence

$$\langle \psi^*, \chi \rangle_G = \frac{1}{|H|} \sum_{g \in H} \psi(g)\overline{\chi(g)} = \langle \psi, \chi|_H \rangle_H \,.$$

□

The Frobenius reciprocity law makes it clear that ψ^* is a character since for every irreducible character χ of G, $\langle \psi^*, \chi \rangle_G = \langle \psi, \chi|_H \rangle_H$ is a non-negative integer as is required.

We illustrate the reciprocity law with the groups corresponding to our two example fields. For the field $k = \mathbb{Q}(\sqrt{-5})$ of Example One, $G = G(k/\mathbb{Q})$ is a group with two elements, 1 and complex conjugation τ. The character table is

	1	τ
χ_1	1	1
χ_1'	1	-1

If we take $H = \{1\}$, then on H, χ_1 and χ_1' both restrict to the trivial character $\psi_1 = \psi_1(H)$. Therefore,

$$\psi_1^* = \chi_1 + \chi_1' \,.$$

As a result, $\psi_1^*(1) = 2$ and $\psi_1^*(\tau) = 0$. By just checking the traces, ψ_1^* is the character of the representation $M(g)$ of G given by

$$M(1) = \begin{pmatrix} 1 & 0 \\ 0 & 1 \end{pmatrix}, \qquad M(\tau) = \begin{pmatrix} 0 & 1 \\ 1 & 0 \end{pmatrix}.$$

which arose in looking at the Euler product for $\zeta_k(s)$. This should be no surprise, since we now see that $M(g)$ is exactly the permutation representation of G on its elements (i.e. on the cosets of H) and this is how we get induced trivial characters.

For Example Two, take the group $G = G(K/\mathbb{Q})$ of six elements again. Up to conjugation, there are three proper subgroups H of G. We will find the corresponding induced trivial characters $\psi_1(H)^*$ for each. We list the three subgroups and the character values necessary to apply the reciprocity law:

$H = H_1 = \{1\}$	
	1
ψ_1	1
$\chi_1\|_H$	1
$\chi_1'\|_H$	1
$\chi_2\|_H$	2

$H = H_2 = \{1, (\beta\gamma)\}$		
	1	$(\beta\gamma)$
ψ_1	1	1
$\chi_1\|_H$	1	1
$\chi_1'\|_H$	1	-1
$\chi_2\|_H$	2	0

$H = H_3 = \{1,\ (\alpha\beta\gamma),\ (\alpha\gamma\beta)\}$			
	1	$(\alpha\beta\gamma)$	$(\alpha\gamma\beta)$
ψ_1	1	1	1
$\chi_1\|_H$	1	1	1
$\chi_1'\|_H$	1	1	1
$\chi_2\|_H$	2	-1	-1

The reciprocity law yields the following induction formulas,

$$\psi_1(H_1)^* = \chi_1 + x_1' + 2\chi_2\,, \qquad \psi_1(H_2)^* = \chi_1 + \chi_2\,, \qquad \psi_1(H_3)^* = \chi_1 + \chi_1'\,.$$

For example, if $H = H_2 = \{1, \beta\gamma)\}$ is the subgroup of two elements which fixes k, then we find the following inner products,

$$\langle \psi_1, \chi_1|_H \rangle_H = 1\,, \qquad \langle \psi_1, \chi_1'|_H \rangle_H = 0\,, \qquad \langle \psi_1, \chi_2|_H \rangle_H = 1\,.$$

This gives $\psi_1(H_2)^* = \chi_1 + \chi_2$ which is what we found before. The result for H_1 illustrates a general fact for any finite group which we get from the Frobenius reciprocity law in the same manner:

$$\psi_1(\{1\})^* = \sum_\chi \deg(\chi)\chi\,.$$

Incidentally, the value at $g = 1$ is the order of the group, and so setting $g = 1$ gives the identity,

$$|G| = \sum_\chi \deg(\chi)^2\,.$$

The same result is also obtainable from the orthogonality relation of the first column with itself.

3.3 L-Functions

We now introduce the general Artin L-function. Let K be a normal algebraic extension of the number field k. Suppose that ρ is a finite dimensional representation of $G = G(K/k)$ with character χ. The *Artin L-function* corresponding to a character χ of $G(K/k)$ is given for $\mathrm{Re}(s) > 1$ by

$$L(s, \chi, K/k) \text{``}=\text{''} \prod_{\mathfrak{p}} \det\left\{I - \rho\left[\sigma\left(\frac{K/k}{\mathfrak{P}}\right)\right] \mathbb{N}(\mathfrak{p})^{-s}\right\}^{-1},$$

where "=" means that only the unramified $\mathfrak{p}$ are being given and where for each such $\mathfrak{p}$, we pick a prime $\mathfrak{P}$ of K above $\mathfrak{p}$. There are similar but messier expressions for the ramifying $\mathfrak{p}$. All identities that we give for L-functions will be exactly true and so we won't put quotation marks around the equal signs for these identities. However, the proofs will be incomplete because of the missing ramifying primes.

If $\mathfrak{p}$ is a prime of k which doesn't ramify in K, we will write $C(\mathfrak{p})$ for the conjugacy class in G of Frobenius automorphisms of primes in K above $\mathfrak{p}$. Since the various automorphisms in $C(\mathfrak{p})$ are conjugate, and determinants of similar matrices are the same, $L(s, \chi, K/k)$ does not depend upon the choice of automorphism in $C(\mathfrak{p})$. The explicit dependence on χ is shown by the exponential form,

$$L(s, \chi, K/k) \text{``}=\text{''} \prod_{\mathfrak{p}} \exp\left\{\sum_{n=1}^{\infty} \frac{1}{n} \chi\left[\sigma\left(\frac{K/k}{\mathfrak{P}}\right)^n\right] \mathbb{N}(\mathfrak{p})^{-ns}\right\}. \tag{32}$$

The exponential form is proved by looking at the eigenvalues of $\rho(\sigma)$ in the same way that the result $\det(e^A) = e^{\mathrm{tr}(A)}$ is proved for matrices. From the exponential form, we see that

$$L(s, \chi + \chi', K/k) = L(s, \chi, K/k) L(s, \chi', K/k).$$

We also see that when $\chi = \chi_1$ is the trivial character,

$$L(s, \chi_1, K/k) = \zeta_k(s).$$

Artin conjectured that if χ is an irreducible character other than χ_1, then $L(s, \chi, K/k)$ is an entire function of s, but this has not yet been proved in general. This and the Riemann hypothesis are the two most fundamental analytic questions in the subject. It has been proved that every such $L(s, \chi, K/k)$ has a meromorphic continuation to the entire complex s-plane with no zeros or poles for $\mathrm{Re}(s) \geq 1$.

All we need to prove the Chebotarev density theorem is the fact that for an irreducible character $\chi \neq \chi_1, (L(s, \chi, K/k)$ is continuable to $s = 1$ with neither a zero nor a pole at $s = 1$. We see from (32) that

$$\text{(33)} \qquad \log L(s,\chi,K/k) \text{``}=\text{''} \sum_{\mathfrak{p}} \sum_{n=1}^{\infty} \frac{1}{n} \chi\left[\sigma\left(\frac{K/k}{\mathfrak{P}}\right)^n\right] \mathbb{N}(\mathfrak{p})^{-ns}$$

As usual, the contribution for $n \geq 2$ is analytic at $s = 1$ and so for an irreducible character χ of $G(K/k)$, we get

$$\sum_{\mathfrak{p}} \chi(C(\mathfrak{p}))\mathbb{N}(\mathfrak{p})^{-s}$$

$$= \begin{cases} \log\left(\frac{1}{s-1}\right) + \text{ an analytic function of } s \text{ at } s=1 & \text{if } \chi = \chi_1 \\ \text{an analytic function of } s \text{ at } s = 1 & \text{if } \chi \neq \chi_1 \end{cases}$$

Theorem (Chebotarev Density Theorem). *Given a conjugacy class C of G, the set of primes of k such that $C(\mathfrak{p}) = C$ has analytic density $|C|/|G|$.*

Proof. According to the orthogonality relations,

$$\sum_{\chi} \left[\sum_{\mathfrak{p}} \chi(C(\mathfrak{p}))\mathbb{N}(\mathfrak{p})^{-s}\right] \overline{\chi(C)} = \sum_{\substack{\mathfrak{p} \\ C(\mathfrak{p})=C}} \frac{|G|}{|C|} \mathbb{N}(\mathfrak{p})^{-s} .$$

On the other hand, we have just seen that all $\chi \neq \chi_1$ contributions on the left side are analytic at $s = 1$ and that the $\chi = \chi_1$ contribution has a $\log\left(\frac{1}{s-1}\right)$ singularity. Hence,

$$\sum_{\substack{\mathfrak{p} \\ C(\mathfrak{p})=C}} \mathbb{N}(\mathfrak{p})^{-s} \sim \frac{|C|}{|G|} \log\left(\frac{1}{s-1}\right) \quad \text{as } s \to 1^+$$

and this proves the theorem. □

Corollary. *Every element of $G(K/k)$ is a Frobenius automorphism for infinitely many primes of K.*

We return to the field K of Example Two. We have already constructed a third degree character χ which corresponds to the permutation representation of $G = G(K/\mathbb{Q})$ on the cosets of $H = G(K/k)$. We have also seen that $\chi = \chi_1 + \chi_2$. Therefore,

$$\zeta_k(s) = L(s,\chi,K/\mathbb{Q}) = L(s,\chi_1,K/\mathbb{Q})L(s,\chi_2,K/\mathbb{Q}) = \zeta(s)L(s,\chi_2,K/\mathbb{Q}) .$$

Hence, we have factored $\zeta_k(s)$ into two pieces by representation theory. The clue to another key property of L-functions comes if we remember that $\chi = \psi_1(H)^*$. We have two different versions of $\zeta_k(s)$, one with respect to the trivial character ψ_1 on H and the other with respect to the character $\chi = \psi_1^*$ on G,

$$\zeta_k(s) = L(s,\psi_1^*,K/\mathbb{Q}) = L(s,\psi_1,K/k) .$$

This turns out to be true for all induced characters and is a fundamental result about Artin L-functions.

Theorem. *If K is a normal extension of $\mathbb{Q}$ and ψ is a character of a subgroup $H = G(K/k)$ of $G = G(K/\mathbb{Q})$, then*

$$L(s, \psi^*, K/\mathbb{Q}) = L(s, \psi, K/k) .$$

Remark. The theorem and its proof are true for any ground field and not just $\mathbb{Q}$. However, the notation would be messier, and we would have to invent a third German 'p' as well!

Proof. This theorem is the reason for much of the extensive development of Frobenius automorphisms in Section 2. The proof is still somewhat involved. We will take a single Euler p-factor in $L(s, \psi^*, K/\mathbb{Q})$ and show that it is the product over all $\mathfrak{p}$ of k above p of the Euler $\mathfrak{p}$-factors in $L(s, \psi, K/k)$. As always, we will not deal with the primes of $\mathbb{Q}$ which ramify in K. Let p be a prime of $\mathbb{Q}$ which doesn't ramify in K and let $\mathfrak{P}$ be one of the primes of K above p. We write $\sigma(\mathfrak{P}) = \sigma\left(\frac{K/\mathbb{Q}}{\mathfrak{P}}\right)$. The Euler p-factor of $L(s, \psi^*, K/\mathbb{Q})$ is

$$\exp\left\{\sum_{n=1}^{\infty} \frac{1}{n} \psi^*[\sigma(\mathfrak{P})^n] p^{-ns}\right\} .$$

By (31), we have

$$\begin{aligned}
\sum_{n=1}^{\infty} \frac{1}{n} \psi^*[\sigma(\mathfrak{P})^n] p^{-ns} &= \sum_{n=1}^{\infty} \frac{1}{n} \left[\frac{1}{|H|} \sum_{g\in G} \psi(g^{-1}\sigma(\mathfrak{P})^n g)\right] p^{-ns} \\
&= \sum_{n=1}^{\infty} \frac{1}{n} \left[\frac{1}{|H|} \sum_{g\in G} \psi\left((g^{-1}\sigma(\mathfrak{P})g)^n\right)\right] p^{-ns} \\
&= \sum_{n=1}^{\infty} \frac{1}{n} \left[\frac{1}{|H|} \sum_{g\in G} \psi\left(\sigma(\mathfrak{P}^g)^n\right)\right] p^{-ns} .
\end{aligned}$$

If $f(\mathfrak{P})$ is the common order of the $\sigma(\mathfrak{P}^g)$, then we know that as g runs through G, $\mathfrak{P}^g$ runs $f(\mathfrak{P})$ times through the set of primes of k above p. Hence

$$\begin{aligned}
\sum_{n=1}^{\infty} \frac{1}{n} \psi^*[\sigma(\mathfrak{P})^n] p^{-ns} &= \sum_{n=1}^{\infty} \frac{1}{n} \left[\frac{f(\mathfrak{P})}{|H|} \sum_{\mathfrak{P}|p} \psi\left(\sigma(\mathfrak{P})^n\right)\right] p^{-ns} \\
&= \frac{f(\mathfrak{P})}{|H|} \sum_{\mathfrak{P}|p} \sum_{n=1}^{\infty} \frac{1}{n} \psi\left(\sigma(\mathfrak{P})^n\right) p^{-ns} .
\end{aligned}$$

Each $\mathfrak{P}$ of K lies above a unique $\mathfrak{p}$ of k. We now take all the $\mathfrak{P}$ terms on the right for $\mathfrak{P}|\mathfrak{p}$ and show that they lead to exactly the Euler $\mathfrak{p}$-factor in $L(s, \psi, K/k)$. Thus we wish to look at

$$\frac{f(\mathfrak{P})}{|H|} \sum_{\mathfrak{P}|\mathfrak{p}} \sum_{n=1}^{\infty} \frac{1}{n} \psi\left(\sigma(\mathfrak{P})^n\right) p^{-ns} .$$

This is relatively easy since we have prepared the way in Subsection 2.5. The only n that contribute to the sum on the right are those n such that $\sigma(\mathfrak{P})^n$ is in H. These will all be multiples of the minimum such n and we saw in Subsection 2.5 that the minimum such n is exactly $f(\mathfrak{p})$, the residue class degree of $\mathfrak{p}$. Also $\sigma(\mathfrak{P})^{f(\mathfrak{p})} = \sigma(\mathfrak{P}/\mathfrak{p}) = \sigma\left(\frac{K/k}{\mathfrak{P}}\right)$ is the Frobenius automorphism with respect to K/k. Its order in H is $f(\mathfrak{P}/\mathfrak{p})$ where $f(\mathfrak{P}) = f(\mathfrak{p})f(\mathfrak{P}/\mathfrak{p})$. This number also is independent of which $\mathfrak{P}$ above $\mathfrak{p}$ is used. Hence,

$$\begin{aligned} \frac{f(\mathfrak{P})}{|H|} \sum_{\mathfrak{P}|\mathfrak{p}} \sum_{n=1}^{\infty} \frac{1}{n} \psi(\sigma(\mathfrak{P})^n) p^{-ns} &= \frac{f(\mathfrak{P})}{|H|} \sum_{\mathfrak{P}|\mathfrak{p}} \sum_{m=1}^{\infty} \frac{1}{f(\mathfrak{p})m} \psi\left(\sigma(\mathfrak{P})^{f(\mathfrak{p})m}\right) p^{-f(\mathfrak{p})ms} \\ &= \frac{f(\mathfrak{P}/\mathfrak{p})}{|H|} \sum_{\mathfrak{P}|\mathfrak{p}} \sum_{m=1}^{\infty} \frac{1}{m} \psi\left(\sigma(\mathfrak{P}/\mathfrak{p})^m\right) \mathrm{N}(\mathfrak{p})^{-ms} . \end{aligned}$$

The $\sigma(\mathfrak{P}/\mathfrak{p})$ for $\mathfrak{P}$ above $\mathfrak{p}$ are all conjugate in H and so for each m, the $\psi(\sigma(\mathfrak{P}/\mathfrak{p})^m)$ are all the same. Further, the number of $\mathfrak{P}$ above $\mathfrak{p}$ is exactly $|H|/f(\mathfrak{P}/\mathfrak{p})$. Therefore, if we just pick one particular $\mathfrak{P}$ of K above $\mathfrak{p}$, we have

$$\frac{f(\mathfrak{P})}{|H|} \sum_{\mathfrak{P}|\mathfrak{p}} \sum_{n=1}^{\infty} \frac{1}{n} \psi(\sigma(\mathfrak{P})^n) p^{-ns} = \sum_{m=1}^{\infty} \frac{1}{m} \psi(\sigma(\mathfrak{P}/\mathfrak{p})^m) \mathrm{N}(\mathfrak{p})^{-ms} .$$

The exponential of this is the Euler $\mathfrak{p}$-factor of $L(s, \psi, K/k)$ as claimed and this proves the Theorem. □

For the field $k = \mathbb{Q}(\sqrt{-5})$ of Example One, $G = G(k/\mathbb{Q})$ is a group with two elements, 1 and complex conjugation τ. We found characters and the character table in Subsection 3.2 as well as the induced trivial character from $\{1\}$ to G. Therefore

$$\begin{aligned} \zeta(s) = \zeta_{\mathbb{Q}}(s) &= L(s, \chi_1, k/\mathbb{Q}) , \\ \zeta_{\mathbb{Q}(\sqrt{-5})}(s) &= L(s, \chi_1, k/\mathbb{Q}) L(s, \chi_1', k/\mathbb{Q}) . \end{aligned}$$

We now extend Example One somewhat by looking at the field $K = \mathbb{Q}(\sqrt{-1}, \sqrt{-5})$ which contains k. As in Section 1 we write $a = \sqrt{-1}$, $b = -\sqrt{-1}$, $c = \sqrt{-5}$, $d = -\sqrt{-5}$. The field K contains all the conjugates of its two generators, $\sqrt{-1}$ and $\sqrt{-5}$ and so is a normal extension of $\mathbb{Q}$ of degree four. We find

$$G = \mathrm{Gal}(K/\mathbb{Q}) = \{1, (ab), (cd), (ab)(cd)\} .$$

Since G is abelian, there are four conjugacy classes and four irreducible characters, all of degree one. The character table is

	1	(ab)	(cd)	$(ab)(cd)$
χ_1	1	1	1	1
χ_1'	1	1	-1	-1
χ_1''	1	-1	1	-1
χ_1'''	1	-1	-1	1

There are three intermediate quadratic fields, $\mathbb{Q}(\sqrt{-5})$ fixed by $H' = \{1, (ab)\}$, $\mathbb{Q}(\sqrt{-1})$ fixed by $H'' = \{1, (cd)\}$, and $\mathbb{Q}(\sqrt{5})$ fixed by $H''' = \{1, (ab)(cd)\}$. Complex conjugation is $(ab)(cd)$. If H is any of the three subgroups of G of order two, then H is a normal subgroup of G and $\psi_1(H)^*$ is the sum of the two irreducible characters of G which are identically one on H, i.e. which are characters of G/H. Therefore we find

$$\begin{aligned}
\zeta(s) = \zeta_{\mathbb{Q}}(s) &= L(s, \chi_1, K/\mathbb{Q}) \,, \\
\zeta_{\mathbb{Q}(\sqrt{-5})}(s) &= L(s, \chi_1, K/\mathbb{Q}) L(s, \chi_1', K/\mathbb{Q}) \,, \\
\zeta_{\mathbb{Q}(\sqrt{-1})}(s) &= L(s, \chi_1, K/\mathbb{Q}) L(s, \chi_1'', K/\mathbb{Q}) \,, \\
\zeta_{\mathbb{Q}(\sqrt{5})}(s) &= L(s, \chi_1, K/\mathbb{Q}) L(s, \chi_1''', K/\mathbb{Q}) \,, \\
\zeta_K(s) &= L(s, \chi_1, K/\mathbb{Q}) L(s, \chi_1', K/\mathbb{Q}) L(s, \chi_1'', K/\mathbb{Q}) L(s, \chi_1''', K/\mathbb{Q}) \,.
\end{aligned}$$

The character χ_1' is a character of the quotient group $G/H' = G(k/\mathbb{Q})$ and the corresponding L-function $L(s, \chi_1', K/\mathbb{Q}) = L(s, \chi_1', k/\mathbb{Q})$ as seen from the two versions of $\zeta_k(s)$ which we have found. This too is a general fact about zeta functions corresponding to normal subgroups.

For the field $K = \mathbb{Q}(\alpha, \beta, \gamma)$ of Example 2, there are four subfields up to conjugacy. These are the fields $\mathbb{Q}$ fixed by all of $G, \mathbb{Q}(\sqrt{-3})$ fixed by $H_3 = \{1, (\alpha\beta\gamma), (\alpha\gamma\beta)\}, \mathbb{Q}(2^{1/3})$ fixed by $H_2 = \{1, (\beta\gamma)\}$, and K fixed by $H_1 = \{1\}$. We have worked out the induced trivial characters in the last Section. We therefore find

$$\begin{aligned}
\zeta(s) = \zeta_{\mathbb{Q}}(s) &= L(s, \chi_1, K/\mathbb{Q}) \,, \\
\zeta_{\mathbb{Q}(\sqrt{-3})}(s) &= L(s, \chi_1, K/\mathbb{Q}) L(s, \chi_1', K/\mathbb{Q}) \,, \\
\zeta_{\mathbb{Q}(2^{1/3})}(s) &= L(s, \chi_1, K/\mathbb{Q}) L(s, \chi_2, K/\mathbb{Q}) \,, \\
\zeta_K(s) &= L(s, \chi_1, K/\mathbb{Q}) L(s, \chi_1', K/\mathbb{Q}) L(s, \chi_2, K/\mathbb{Q})^2 \,.
\end{aligned} \tag{34}$$

We close this Section by returning to the question as to whether the zeta function of a field k determines the field. We have seen that $\zeta_k(s)$ determines the minimal normal extension K of $\mathbb{Q}$ containing k. Let $G = G(K/\mathbb{Q})$ and let $H = G(K/k)$. We now know that $\zeta_k(s) = L(s, \psi_1(H)^*, K/\mathbb{Q})$. Together with the fact that every element of G is a Frobenius automorphism, the exponential form of

$L(s, \psi_1(H)^*, K/\mathbb{Q})$ shows us that $\zeta_k(s)$ completely determines the character $\psi_1(H)^*$. Subgroups of G conjugate to H correspond to fields conjugate to k and give the same induced trivial character $\psi_1(H)^*$. The problem thus boils down to the question as to whether there can be non-conjugate subgroups of G giving the same induced trivial character. For some groups, the answer to this is yes. The best known example comes from the symmetric group on six letters, S_6 which is a Galois group for a field K over $\mathbb{Q}$. This group has an outer automorphism which changes the cycle structure of the elements of order three: elements of the type (abc) are interchanged with elements of the type $(abc)(def)$. Conjugation preserves cycle structures. Thus the subgroups of order three generated by (abc) and by $(abc)(def)$ will not be conjugate but will have the same induced trivial character. The subfields of K fixed by these two subgroups will be non-conjugate fields with the same zeta function.

3.4 The Reciprocity Law

Prior to the introduction of Artin L-functions in 1923, the algebraic and analytic study of *relative abelian extensions* (i.e. normal extensions with abelian Galois groups) was already very advanced. The subject was called *class field theory* because all such extensions turned out to correspond to congruence class groups in the ground field. There were also L-functions defined on these class groups and the zeta functions of the *class fields* were known to be the product of all the corresponding L-functions. In this Section, we will briefly introduce the congruence class groups and the corresponding L-functions.

We begin with congruence class groups. Let k be a number field and $\mathfrak{f}$ an integral divisor of k. We define congruences $(\operatorname{mod} \mathfrak{f})$ just as we did for prime divisors. If α and β are integers in k, we say that α is *congruent* to $\beta (\operatorname{mod} \mathfrak{f})$ and write $\alpha \equiv \beta (\operatorname{mod} \mathfrak{f})$ if $\mathfrak{f}$ divides $\alpha - \beta$. We will denote the congruence class of $\alpha (\operatorname{mod} \mathfrak{f})$ by $\bar{\alpha}$. Suppose that $\alpha \equiv \beta (\operatorname{mod} \mathfrak{f})$. Anything that divides β and $\mathfrak{f}$ divides $\alpha = \beta + (\alpha - \beta)$ and so divides $(\alpha, \mathfrak{f})$. Conversely anything that divides α and $\mathfrak{f}$ divides β and so divides $(\beta, \mathfrak{f})$. Therefore $(\alpha, \mathfrak{f}) = (\beta, \mathfrak{f})$. Thus we can talk about a congruence class $\bar{\alpha}$ being relatively prime to $\mathfrak{f}$ and write $(\bar{\alpha}, \mathfrak{f}) = (1)$ if $(\alpha, \mathfrak{f}) = (1)$. Over $\mathbb{Q}$, most elementary textbooks refer to these congruence classes as *reduced residue classes* $(\operatorname{mod} \mathfrak{f})$. The reduced residue classes form a group under multiplication $(\operatorname{mod} \mathfrak{f})$. This group is closely related to the congruence class groups of divisors that we wish to discuss. In all our examples, we will be able to describe the groups of divisors in terms of just these congruence classes, but in many fields, these congruence classes are not yet general enough due to sign problems with units.

We will deal with the set $I(\mathfrak{f})$ of all fractional divisors which are ratios of integral divisors relatively prime to $\mathfrak{f}$. In other words, a fractional divisor is in $I(\mathfrak{f})$ if all of the prime divisors occurring in its factorization don't divide $\mathfrak{f}$. The difficulty in defining divisor classes $(\operatorname{mod} \mathfrak{f})$ is that even a principal divisor has more than one generator and these generators may be in different congruence classes. It was also discovered that even over $\mathbb{Q}$, sign conditions on the generator of a divisor are also necessary to get all abelian extensions. The *principal ray class* $P(\mathfrak{f})(\bmod \mathfrak{f})$ will be the set of all principal divisors (α/β) such that α and β are relatively prime to $\mathfrak{f}$, $\alpha \equiv \beta(\bmod \mathfrak{f})$, and α/β is totally positive. If $k = \mathbb{Q}(\theta)$ and the conjugates of θ are ordered so that $\theta^{(1)}, \cdots, \theta^{(r_1)}$ are real while the remaining conjugates are complex, then we say that a number α in k is *totally positive* in k if $\alpha^{(i)} > 0$ for $1 \le i \le r_1$. Thus numbers of $\mathbb{Q}$ are totally positive when they are positive. On the other hand, in a complex quadratic field, any non-zero number is totally positive because $r_1 = 0$.

The quotient group $G(\mathfrak{f}) = I(\mathfrak{f})/P(\mathfrak{f})$ is called the *ray class group* $(\bmod \mathfrak{f})$. The elements of this group are the *ray classes* $(\bmod \mathfrak{f})$. Often, $P(\mathfrak{f})$ is referred to as the narrow principal ray class $(\bmod \mathfrak{f})$ and then the quotient group $G(\mathfrak{f})$ is called the narrow ray class group $(\bmod \mathfrak{f})$. When $\mathfrak{f} = (1)$, this group can still be bigger than the divisor class group of Section 2 because of the sign conditions on the principal class. $G(1)$ is often called the *narrow divisor class group* to distinguish it from the *wide class group* which is just I/P. However, in complex quadratic fields where many of our examples will come from, there are no sign conditions. Here the narrow and wide groups are always the same and we can also describe all the principal ray classes in terms of congruence classes. Likewise in $\mathbb{Q}$, every divisor is principal and has a unique positive generator, and so $G(f)$ is isomorphic to the group of reduced residue classes $(\bmod \mathfrak{f})$.

We now define the L-functions corresponding to these groups. Suppose that χ is a (first degree) character of the (abelian) group $G(\mathfrak{f})$. For an integral divisor $\mathfrak{a}$ relatively prime to $\mathfrak{f}$, we define $\chi(\mathfrak{a})$ to be χ of the ray class of $\mathfrak{a}$ in $G(\mathfrak{f})$. If $\mathfrak{a}$ and $\mathfrak{f}$ aren't relatively prime we set $\chi(\mathfrak{a}) = 0$. Our L-function is then

$$L(s,\chi) \text{``} = \text{''} \sum_{\mathfrak{a}} \chi(\mathfrak{a}) \mathbb{N}(\mathfrak{a})^{-s} ,$$

where the sum is over the integral divisors of k. As usual, there is an Euler product,

$$L(s,\chi) \text{``} = \text{''} \prod_{\mathfrak{p}} (1 + \chi(\mathfrak{p})\mathbb{N}(\mathfrak{p})^{-s} + \chi(\mathfrak{p})^2 \mathbb{N}(\mathfrak{p})^{-2s} + \cdots) = \prod_{\mathfrak{p}} (1 - \chi(\mathfrak{p})\mathbb{N}(\mathfrak{p})^{-s})^{-1} .$$

The reason for the " = " is that it is often convenient to alter the Euler $\mathfrak{p}$-factor for primes $\mathfrak{p}$ dividing $\mathfrak{f}$. For instance, when χ is the trivial character, the Euler $\mathfrak{p}$-factor is $(1 - \mathbb{N}(\mathfrak{p})^{-s})^{-1}$ for $\mathfrak{p}$ not dividing $\mathfrak{f}$ but is just 1 for $\mathfrak{p}|\mathfrak{f}$. It is best to change the $\mathfrak{p}$-factor to being $(1 - \mathbb{N}(\mathfrak{p})^{-s})^{-1}$ for all $\mathfrak{p}$ thereby getting $\zeta_k(s)$. This corresponds to $\mathfrak{f}$ not being the 'smallest' modulus of definition for the trivial character. The smallest modulus of definition for a character χ is called the *conductor* of the character and will be denoted by $\mathfrak{f}_\chi$. For example, the trivial character has conductor (1). We will not go further into this problem here; we merely think of the L-functions as defined via the Euler product for $\mathfrak{p}$ relatively prime to $\mathfrak{f}$ and leave the remaining primes to textbooks.

For non-trivial characters, these L-functions are known to be entire functions which are non-zero for $\mathrm{Re}(s) \ge 1$. The logarithm is given by

$$\log L(s,\chi)\text{``}=\text{''}\sum_{\mathfrak{p}}\sum_{n=1}^{\infty}\frac{1}{n}\chi(\mathfrak{p})^n\mathbb{N}(\mathfrak{p})^{-ns} \tag{35}$$

and as usual, the contribution for $n \geq 2$ is analytic at $s = 1$. Thus

$$\sum_{\mathfrak{p}}\chi(\mathfrak{p})\mathbb{N}(\mathfrak{p})^{-s}$$

$$= \begin{cases} \log\left(\frac{1}{s-1}\right) \;+\; \text{an analytic function of } s \text{ at } s=1 & \text{if } \chi=\chi_1 \\ \text{an analytic function of } s \text{ at } s=1 & \text{if } \chi\neq\chi_1 \end{cases}$$

Theorem. *The set of prime divisors in a given ray class of $G(\mathfrak{f})$ has analytic density* $1/|G(\mathfrak{f})|$.

Proof. In $\mathbb{Q}$, this is Dirichlet's theorem on primes in progressions. The proof is exactly the same as for the Chebotarev density theorem. Once we have stated the reciprocity law, we will see that in fact this theorem is the Chebotarev density theorem in the case of abelian groups. □

The general class groups which appear in this subject are quotient groups of $G(\mathfrak{f})$ by a subgroup H. Any (first degree) character χ of the (abelian) quotient group $G(\mathfrak{f})/H$ may be thought of as a character of $G(\mathfrak{f})$ which is identically one on H. By 1923, it was known that corresponding to a group $G = G(\mathfrak{f})/H$ is a unique *class field* K with many interesting properties. First, K is an abelian extension of k whose Galois group is isomorphic to G. From our point of view the fundamental fact is that

$$\zeta_K(s) = \prod_{\chi} L(s,\chi)$$

where the product is over all characters χ of $G = G(\mathfrak{f})/H$. The Euler product is therefore given by

$$\zeta_K(s)\text{``}=\text{''}\prod_{\mathfrak{p}\nmid f}(1-\mathbb{N}(\mathfrak{p})^{-f(\mathfrak{p})s})^{-|G|/f(\mathfrak{p})} \tag{36}$$

where $f(\mathfrak{p})$ is the order of the class of $\mathfrak{p}$ in G (i.e. the minimal power of the ray class of $\mathfrak{p}$ which is in H). This may be proved from the induced character point of view of Subsection 3.3, but can also be proved by summing (35) over χ and using the orthogonality relations to get the logarithm of (36) directly. The Euler product (36) allows us to guess that every prime divisor in a single class of $G = G(\mathfrak{f})/H$ splits in K in the same way into $|G|/f$ pieces, each of relative degree f where f is the order of the class. This too was a known fact. In particular, up to a finite number of exceptions due to using the wrong $\mathfrak{f}$, the prime divisors in H are exactly the primes of k which split completely in K. This uniquely determines the field K by the same sort of density argument as in Subsection 3.1. Finally, every abelian extension of k turns out to be a class field.

In an abelian group, conjugacy classes consist of single elements. Therefore each $\mathfrak{p}$ of k not dividing $\mathfrak{f}$ determines a unique Frobenius automorphism $\sigma(\mathfrak{p}) = \sigma(\mathfrak{P})$ of $G(K/k)$ which does not depend on the choice of $\mathfrak{P}$ in K above $\mathfrak{p}$. As we have just stated, it was already known that if $\mathfrak{p}$ is any prime divisor of k not dividing $\mathfrak{f}$, then the relative degree of primes of K above $\mathfrak{p}$ is the order of the class in $G = G(\mathfrak{f})/H$ containing $\mathfrak{p}$. This must be the order of the corresponding Frobenius automorphism $\sigma(\mathfrak{p})$ of $G(K/k)$. As we have also stated, it was even known that the Galois group $G(K/k)$ is isomorphic to $G(\mathfrak{f})/H$. However, it was not known or even conjectured that the groups are isomorphic because the Frobenius automorphism $\sigma(\mathfrak{p})$ which we get is the same element of $G(K/k)$ for all $\mathfrak{p}$ in the same congruence class. This conjecture was brought about by the introduction of the Artin L-functions which gave an apparently second way of factoring a zeta function of a class field. Since $G(K/k)$ and G are isomorphic, we can think of a first degree character χ as being a character of both groups. We then have two seemingly different L-functions corresponding to χ given by (33) and (35). Artin conjectured that these L-functions aren't different at all for the reason that the Frobenius automorphism provides an isomorphism between $G(K/k)$ and $G(\mathfrak{f})/H$. He proved this conjecture four years later. The result is,

Theorem (The Artin Reciprocity Law). *If $\mathfrak{p}$ is a prime of k relatively prime to $\mathfrak{f}$, then the Frobenius automorphism $\sigma(\mathfrak{p})$ of K corresponding to $\mathfrak{p}$ depends only on the class of $\mathfrak{p}$ in $G(\mathfrak{f})/H$ and this correspondence between elements of $G(K/k)$ and $G(\mathfrak{f})/H$ is an isomorphism.*

Remark. In any abelian group, the map $g \to g^{-1}$ is an automorphism of the group and hence there is a second isomorphism between the two groups of the theorem. Frequently the reciprocity map is cited in print in this latter form.

For the examples, one more fact about the class fields will be useful. It is the *conductor-discriminant theorem* due to Hasse which says among other things that the only ramifying primes from k to K are those dividing at least one of the $\mathfrak{f}_\chi$.

Theorem. *If K is the class field corresponding to the ray class group $G(\mathfrak{f})/H$ of k, then*

$$D_{K/k} = \prod_{\chi} \mathfrak{f}_\chi \,,$$

where the product is over all the characters of $G(\mathfrak{f})/H$. Further, the smallest $\mathfrak{f}$ is the least common multiple of the $\mathfrak{f}_\chi$.

We conclude this survey with several examples. First, we look at $k = \mathbb{Q}$. Corresponding to the multiplicative group $G(n)$ of reduced residue classes (mod n) is a class field K determined by the fact that the primes which split completely in K are exactly those congruent to 1 (mod n). Let

$$\omega = \omega_n = e^{2\pi i/n}$$

be a primitive n-th root of unity. Then $K = \mathbb{Q}(\omega)$ is the *cyclotomic field* of n-th roots of unity. The Frobenius automorphism associated to a prime p takes ω to ω^p by definition since ω^p really is conjugate to ω in this case. Thus the Frobenius automorphism is the identity if and only if $p \equiv 1 \pmod{n}$. In other words it is exactly the primes congruent to $1 \pmod{n}$ which split completely in $\mathbb{Q}(\omega)$ and this shows that $K = \mathbb{Q}(\omega)$. Since ω^p only depends upon $p \pmod{n}$, we also see the reciprocity law clearly holds here. Any abelian extension K' of $\mathbb{Q}$ is a class field and so corresponds to a class group $G(n)/H$ for some subgroup H of $G(n)$. Any prime $p \equiv 1 \pmod{n}$ is in H automatically and so the primes of $\mathbb{Q}$ which split completely in $\mathbb{Q}(\omega_n)$ also split completely in K'. Therefore K' is a subfield of $\mathbb{Q}(\omega_n)$. This is the Kronecker-Weber theorem that any abelian extension of $\mathbb{Q}$ is a subfield of a cyclotomic field. (Of course, from the point of view of the presentation here, the real result is that every abelian extension of $\mathbb{Q}$ is a class field, since the rest is easy.)

If d is the discriminant of the quadratic field $k = \mathbb{Q}(\sqrt{d})$, we have seen that the Legendre symbol

$$\chi_d(p) = \left(\frac{d}{p}\right)$$

tells us us how an odd prime p not dividing d splits in k. Quadratic extensions are always abelian extensions. Thus k is a class field over $\mathbb{Q}$ and according to the conductor-discriminant theorem k corresponds to a ray class group (mod $|d|$). The reciprocity law says that $\chi_d(p)$ is actually a congruence class character (mod $|d|$) of conductor $|d|$. When thought of this way, χ_d is often called a *Kronecker symbol.* A less precise way of interpreting this is to say that the splitting of primes in a quadratic field of discriminant d, is determined solely by the residue class of the prime (mod $|d|$).

Take for instance the field $\mathbb{Q}(\sqrt{-1}) = \mathbb{Q}(\sqrt{-4})$ of discriminant -4. The group $G(4)$ of congruence classes (mod 4) is a group of order two. The character table is,

m	1	3
$\chi_1(m)$	1	1
$\chi_{-4}(m)$	1	-1

where the second character must be χ_{-4} because that is all that is available. We thus recover the ancient fact that -1 is a square modulo primes which are 1 (mod 4) and -1 is a non-square modulo primes which are 3 (mod 4).

The next examples help to further explain the use of the words, 'reciprocity law'. Again, we take $\mathbb{Q}$ to be the ground field and look at the class field $k = \mathbb{Q}(\sqrt{5})$ of discriminant 5. The group $G(5)$ is of order 4 and the field k

corresponds to a quotient group $G(5)/H$ where H is a subgroup of $G(5)$ which also must be of order two. There is only one subgroup of order two, $H = \{\bar{1}, \bar{4}\}$. There can only be one non-trivial character on $G(5)/H$ and so we have found the character χ_5:

m	1	2	3	4
$\chi_5(m)$	1	-1	-1	1

Notice that H is the subgroup of squares (mod 5). Thus, $\chi_5(m)$ is also given by another Legendre symbol,

$$\chi_5(m) = \left(\frac{m}{5}\right) .$$

Together, the two versions of χ_5 for primes p are

$$\left(\frac{5}{p}\right) = \left(\frac{p}{5}\right) ,$$

which is nothing more than an instance of the quadratic reciprocity law.

Given an odd prime q, we can get the general quadratic reciprocity law in the same way. There are two cases. When $q \equiv 1 \pmod 4$, we look at the field $k = \mathbb{Q}(\sqrt{q})$ of discriminant q. Therefore, k is a class field corresponding to the group $G(q)/H$ where H is a subgroup of $G(q)$ of index 2. But $G(q)$ is the multiplicative subgroup of the finite field $\mathbb{F}_q$ and is cyclic. Therefore, $G(q)$ has a unique subgroup of index 2, namely the set of square classes $(\bmod q)$. This determines χ_q uniquely as the Legendre symbol $\chi_q(m) = \left(\frac{m}{q}\right)$. In other words we have

$$\left(\frac{q}{p}\right) = \left(\frac{p}{q}\right) \qquad (q \equiv 1 \pmod 4) . \tag{37}$$

When $q \equiv 3 \pmod 4$, we look at the field $k = \mathbb{Q}(\sqrt{-q})$ of discriminant $-q$. Again k is a class field corresponding to the group $G(q)/H$ where H is the subgroup of index 2 of square classes in $G(q)$. Again $G(q)$ is cyclic and χ_{-q} is uniquely determined as the Legendre symbol $\chi_{-q}(m) = \left(\frac{m}{q}\right)$. The result this time is

$$\left(\frac{-q}{p}\right) = \left(\frac{p}{q}\right) \qquad (q \equiv 3 \pmod 4) . \tag{38}$$

Since we have already found $\left(\frac{-1}{p}\right) = \chi_{-4}(p)$ above, we may rephrase (37) and (38) in the more classical shape

$$\left(\frac{p}{q}\right) = \begin{cases} -\left(\frac{q}{p}\right) & \text{if } p \equiv q \equiv 3 \pmod 4 \\ \left(\frac{q}{p}\right) & \text{otherwise} . \end{cases}$$

This completes the *quadratic reciprocity law.*

We return to the general quadratic field $k = \mathbb{Q}(\sqrt{d})$ of discriminant d. The zeta function of k factors as,

$$\zeta_k(s) = \zeta(s)L(s, \chi_d)\,.$$

Suppose that $d < 0$ so that k is complex quadratic. Since $\zeta(0) = -1/2$, according to (24),

$$L(0, \chi_d) = \frac{h}{W/2}\,,$$

where h is the class-number of k and W is the number of roots of unity in k. Let us try this out for $k = \mathbb{Q}(\sqrt{-1})$ where $d = -4$. We have

$$L(s, \chi_{-4}) = \chi_{-4}(1) \cdot 1^{-s} + \chi_{-4}(2) \cdot 2^{-s} + \chi_{-4}(3) \cdot 3^{-s} + \chi_{-4}(4) \cdot 4^{-s} + \cdots$$

so that

$$L(0, \chi_{-4}) = (1) + (0) + (-1) + (0) + (1) + (0) + (-1) + (0) + \cdots\,.$$

'Clearly' this gives

$$L(0, \chi_{-4}) = \frac{1}{2}\,.$$

Since $W = 4$, we find that $h = 1$.

In fact for any d,

$$\sum_{m=1}^{|d|} \chi_d(m) = 0\,, \tag{39}$$

by the orthogonality relations applied to the non-zero terms. This means that the sequence of partial sums for $L(0, \chi_d)$ is periodic and hence the series is Cesaro summable: $L(0, \chi_d)$ is the average of the partial sums,

$$\begin{aligned} L(0, \chi_d) &= \frac{1}{|d|} \sum_{M=1}^{|d|} \sum_{m=1}^{M} \chi_d(m) \\ &= \frac{1}{|d|} \sum_{m=1}^{|d|} (|d| + 1 - m)\chi_d(m)\,. \end{aligned}$$

By (39), this simplifies to

$$L(0, \chi_d) = -\frac{1}{|d|} \sum_{m=1}^{|d|} m\chi_d(m)\,.$$

This gives *Dirichlet's class-number formula for complex quadratic fields,*

$$h = -\frac{W/2}{|d|} \sum_{m=1}^{|d|} m\chi_d(m)\,. \tag{40}$$

The formula can be simplified further, but we will stop here.

When k is real quadratic, we find from (24) that

$$h\log(\varepsilon_1) = L'(0,\chi_d) = -\sum_{m=1}^{\infty}\chi_d(m)\log(m) = \log\left[\prod_{m=1}^{\infty} m^{-\chi_d(m)}\right],$$

where h is the class-number of k and ε_1 is the fundamental unit of k. It is not hard to make sense of this product. For example, when $d = 5$, we see that

$$h\log(\varepsilon_1) = \log\left[\prod_{n=0}^{\infty}\frac{(5n+2)(5n+3)}{(5n+1)(5n+4)}\right] = \log(1.618033989),$$

approximately. There is an even better arangement of the product which allows us to make use of the infinite product expansion for the sine function,

$$\begin{aligned} h\log(\varepsilon_1) &= \log\left[\frac{2}{1}\prod_{n=1}^{\infty}\frac{(5n+2)(5n-2)}{(5n+1)(5n-1)}\right] = \log\left[\frac{2\pi/5}{\pi/5}\prod_{n=1}^{\infty}\left(\frac{1-\frac{4/25}{n^2}}{1-\frac{1/25}{n^2}}\right)\right] \\ &= \log\left[\frac{\sin(2\pi/5)}{\sin(\pi/5)}\right]. \end{aligned}$$

The same thing works for any real quadratic field. When $d > 0$, the characters χ_d have the property that $\chi_d(d-m) = \chi_d(m)$ for all m. As a result, we find that

$$h\log(\varepsilon_1) = \log\left[\prod_{m\le d/2}\sin\left(\frac{m\pi}{d}\right)^{-\chi_d(m)}\right].$$

This is *Dirichlet's class-number formula for real quadratic fields.*

We finish by returning to our two examples. For Example One, the field $k = \mathbb{Q}(\sqrt{-5}) = \mathbb{Q}(\sqrt{-20})$ has discriminant -20 and so χ_{-20} is a congruence class character of conductor 20. Since χ_{-20} is defined by Legendre symbols, we see that

$$\chi_{-20}(p) = \left(\frac{-20}{p}\right) = \left(\frac{-4}{p}\right)\left(\frac{5}{p}\right) = \chi_{-4}(p)\chi_5(p). \tag{41}$$

The group $G(20) = \{\bar{1},\bar{3},\bar{7},\bar{9},\overline{11},\overline{13},\overline{17},\overline{19}\}$ is of order 8 and we have the four character values,

m	1	3	7	9	11	13	17	19
$\chi_1(m)$	1	1	1	1	1	1	1	1
$\chi_{-4}(m)$	1	−1	−1	1	−1	1	1	−1
$\chi_5(m)$	1	−1	−1	1	1	−1	−1	1
$\chi_{-20}(m)$	1	1	1	1	−1	−1	−1	−1

Thus k is the class field corresponding to the ray class group $G(20)/H$ where $H = \{\bar{1}, \bar{3}, \bar{7}, \bar{9}\}$. As an aside, now that we have the character values, we see from Dirichlet's class-number formula (40) that the class-number of k is

$$h = -\frac{1}{20}(1+3+7+9-11-13-17-19) = 2\,,$$

as we stated in Section 2.

The characters χ_{-4} and χ_5 also give rise to characters $\pmod{20}$; they just have smaller conductors. Along with the trivial character χ_1, we now have four characters of $G(20)$. These four characters are identically one on the group $H' = \{\bar{1}, \bar{9}\}$ and so we have the four characters of the ray class group $G(20)/H'$. Corresponding to this class group, there is a class field K which is a quartic extension of $\mathbb{Q}$. The field K is characterized by the fact that p splits completely in K if and only if p is congruent to 1 or 9 $\pmod{20}$. Any such prime splits in k and also in $\mathbb{Q}(\sqrt{-1})$ and so both of these fields are subfields of K. Hence we must have $K = \mathbb{Q}(\sqrt{-5}, \sqrt{-1})$. According to the conductor-discriminant theorem, up to a $\pm$ sign, D_K is $1 \cdot 4 \cdot 5 \cdot 20 = 20^2$. The zeta functions for k and K are given by

$$\begin{aligned} \zeta_k(s) &= \zeta(s)L(s,\chi_{-20})\,, \\ \zeta_K(s) &= \zeta(s)L(s,\chi_{-20})L(s,\chi_{-4})L(s,\chi_5)\,. \end{aligned} \tag{42}$$

We have seen the Artin L-function version of both of these when we dealt with Example One at the end of Subsection 3.3.

We continue with Example One. The field $K = k(\sqrt{-1})$ is a quadratic extension of k and so is a class field over k. Since $\sqrt{-1}$ is a zero of the polynomial x^2+1 of discriminant -4, $D_{K/k}|4$. But $K = k(\alpha)$ also where $\alpha = \frac{1 \pm \sqrt{5}}{2}$ and α is a zero of the polynomial $x^2 - x - 1$ of discriminant 5. Thus $D_{K/k}|5$ too. Hence $D_{K/k}$ divides $(4,5) = (1)$. Therefore $D_{K/k} = (1)$. In particular, there are no ramified primes from k to K. Hence the conductor-discriminant theorem says that K corresponds to a ray class group of the form $G((1))/H$ of order 2. But $G((1))$ is the divisor class group of k and is already of order two. Therefore $H = P(1)$, the principal divisor class. The class field corresponding to $\mathfrak{Q}/P$ is called the *Hilbert class field* of k; K is the Hilbert class field of k.

Besides the trivial character, there is another character ψ of $G((1))$. This character takes the values 1 on principal divisors and -1 on the non-principal divisors. We also have

$$\zeta_K(s) = \zeta_k(s)L(s,\psi)\,.$$

The Artin L-function form of $L(s,\psi)$ is a character of $G(K/k)$ which can be induced to a second degree character of $G(K/\mathbb{Q})$. From (42), we see that the result will be

$$L(s,\psi) = L(s,\chi_{-4})L(s,\chi_5)\,. \tag{43}$$

Let p be a prime of $\mathbb{Q}$ which splits in k. We know that the condition for this is that

$$\chi_{-20}(p) = \chi_{-4}(p)\chi_5(p) = 1\,. \tag{44}$$

We have seen that p is the product of two prime divisors in k and we even know how to find them: if we factor $x^2 + 5 \pmod p$ as

$$x^2 + 5 \equiv (x+a)(x-a) \pmod p$$

then $(p) = \mathfrak{p}_1\mathfrak{p}_2$ in k where

$$\mathfrak{p}_1 = (p, a + \sqrt{-5})\,, \qquad \mathfrak{p}_2 = (p, -a + \sqrt{-5}) = (p, a - \sqrt{-5}) = \overline{\mathfrak{p}_1}\,.$$

The Euler p-factor in (43) is

$$(1 - \psi(\mathfrak{p}_1)p^{-s})^{-1}(1 - \psi(\mathfrak{p}_2)p^{-s})^{-1} = (1 - \chi_{-4}(p)p^{-s})^{-1}(1 - \chi_5(p)p^{-s})^{-1}\,.$$

From (44), $\chi_{-4}(p)$ and $\chi_5(p)$ are both $+1$ or both -1. Hence the same is true of $\psi(\mathfrak{p}_1)$ and $\psi(\mathfrak{p}_2)$ and with the same values. But this is exactly what determines when $\mathfrak{p}_1$ and $\mathfrak{p}_2$ are principal. We get the long known

Theorem. *A prime p splits completely in k if and only if*

$$p \equiv 1, 3, 7, 9 \pmod{20}\,.$$

If $p \equiv 1, 9 \pmod{20}$ then p is the product of two principal divisors in k while if $p \equiv 3, 7 \pmod{20}$ then p is the product of two non-principal divisors in k.

For example, the already discussed prime divisor $(3, 1 + \sqrt{-5})$ above 3 is non-principal, while the prime divisor $\mathfrak{p} = (29, 16 + \sqrt{-5})$ above 29 is principal. Indeed $29 = (3 + 2\sqrt{-5})(3 - 2\sqrt{-5})$ does split as the product of two principal divisors. The only question is, which of the two factors generates $\mathfrak{p}$? One way to tell without dividing is to note that anything in $\mathfrak{p}$ is congruent to $0 \pmod{\mathfrak{p}}$. Hence both 29 and $16 + \sqrt{-5}$ are zero $(\bmod\, \mathfrak{p})$. Thus, $\sqrt{-5} \equiv -16 \equiv 13 (\bmod\, \mathfrak{p})$ and as a result, $3 + 2\sqrt{-5} \equiv 29 \equiv 0 \pmod{\mathfrak{p}}$. Therefore $\mathfrak{p} = (3 + 2\sqrt{-5})$.

We move on to Example Two. We set $k_2 = \mathbb{Q}(\sqrt{-3})$. The field K we have been considering is given by $K = k_2(\alpha)$ where $\alpha = 2^{1/3}$. K is normal over $\mathbb{Q}$ and so over k_2 as well. The Galois group $G(K/k_2)$ is cyclic of order 3 and so K is a class field with respect to k_2. Since α is a zero of the polynomial $x^3 - 2$ of discriminant -108, we see that D_{K/k_2} divides (108). Further, since (2) is prime in k_2, $x^3 - 2$ is an Eisenstein polynomial with respect to (2) in k_2 and so 2^2 is the correct power of (2) in D_{K/k_2}. However, the power of $(\sqrt{-3})$ in the polynomial discriminant is wrong and it turns out that $D_{K/k_2} = (36)$. [Translation of x by 1 leads to a polynomial which is not an Eisenstein polynomial because (3) is a square in k_2. Fortunately, the number $\theta = (\alpha + 1)^2/\sqrt{-3}$ is a zero of the polynomial $x^3 + \sqrt{-3}x^2 + 3x - \sqrt{-3}$ of discriminant -144 which is an Eisenstein polynomial with respect to $\sqrt{-3}$ in k_2. Therefore $(\sqrt{-3})^4$ is the correct power of $(\sqrt{-3})$ in D_{K/k_2}.]

According to class field theory, k is the class field for a ray class group $G(\mathfrak{f})/H$ which is cyclic of order three. The two non-trivial characters of this group are cubic characters and each is the square of the other. Hence each has the same conductor, as whenever one character is defined, its square is also. We take $\mathfrak{f}$ to be this common conductor. The trivial character of course has conductor (1). Therefore by the conductor-discriminant theorem, $(36) = 1 \cdot \mathfrak{f} \cdot \mathfrak{f}$. Unique factorization of divisors now gives us the conductor,

$$\mathfrak{f} = (6)$$

We look at congruences (mod 6). Since every divisor of k_2 is principal, this boils down to finding the reduced residue classes of numbers (mod 6) and then taking account of the units in k_2. The set of integers of k_2 is given by

$$\mathfrak{o} = \{a + b\lambda \mid a, b \text{ in } \mathbb{Z}\}$$

where

$$\lambda = \frac{-3 + \sqrt{-3}}{2}$$

is a zero of the polynomial $x^2 + 3x + 3$ of discriminant -3. Usually, the third root of unity,

$$\omega = \lambda + 1 = \frac{-1 + \sqrt{-3}}{2},$$

is used as a generator of the integers, but λ will be more useful since it is an associate of $\sqrt{-3}$ in k_2 and so in terms of divisors we have,

$$(3) = (\lambda)^2 .$$

Since

$$\frac{a + b\lambda}{6} = \frac{a}{6} + \frac{b}{6}\lambda ,$$

it follows that 6 divides an integer $a + b\lambda$ if and only if both a and b are divisible by 6. Therefore a system of representatives of the congruence classes of integers in $\mathfrak{o}$ (mod 6) is given by

$$\{a + b\lambda \mid 0 \leq a \leq 5,\ 0 \leq b \leq 5\} ,$$

as any integer of $\mathfrak{o}$ is clearly congruent (mod 6) to exactly one of these. Hence there are 36 congruence classes (mod 6). It is a general fact for integral divisors $\mathfrak{f}$ in number fields that there are $\mathbb{N}(\mathfrak{f})$ congruence classes of integers (mod $\mathfrak{f}$) in the field. This is often taken as the definition of the norm for an integral divisor, but when this is done, the concepts of norms of fractional divisors and relative norms make much less sense.

It turns out that 18 of 36 classes are relatively prime to 6. The general formula for the number $\phi(\mathfrak{f})$ of reduced residue classes (mod $\mathfrak{f}$) in a field is

$$\phi(\mathfrak{f}) = \mathbb{N}(\mathfrak{f}) \prod_{\mathfrak{p} \mid \mathfrak{f}} (1 - \mathbb{N}(\mathfrak{p}))^{-1} .$$

(Over $\mathbb{Q}$, this is the Euler ϕ-function.) In k_2 there are two prime divisors of (6) of norms 4 and 3 and this gives us 18 reduced classes. In this particular case, we can easily verify that there are 18 reduced residue classes by actually finding them. We are interested in those classes which are relatively prime to both λ and 2, since these are the only primes dividing 6. A number $a + b\lambda$ will be relatively prime to λ if and only if 3 doesn't divide a. In addition $a + b\lambda$ will not be divisible by 2 and so will be relatively prime to 2 when not both of a and b are even. Exactly 18 of the 36 class representatives are left: $a = 1$ or 5 with all 6 values of b, and $a = 2$ or 4 with $b = 1, 3, 5$.

There are six units in k_2, namely the six sixth roots of unity, $\pm 1, \pm\omega, \pm\omega^2$. It turns out that these units of k_2 are incongruent $(\text{mod}\, 6)$. In fact they are already incongruent $(\text{mod}\, 2\lambda)$ as is easily checked. Therefore the principal ray class $(\text{mod}\, 6)$ of k_2 uses up 6 of the 18 reduced residue classes $(\text{mod}\, 6)$. Thus the ray class group $G(6)$ of k_2 is of order 3. This means that K is the class field of k_2 corresponding to $G(6)$. The primes of k_2 which split completely in K are those which lie in the principal ray class $(\text{mod}\, 6)$. We will finish the algebraic side of this example by giving a nicer description of these primes.

We need to look at the group structure of the 18 reduced residue classes $(\text{mod}\, 6)$. It is expedient to first look at integers of $k_2\, (\text{mod}\, 2\lambda)$. There are 12 congruence classes $(\text{mod}\, 2\lambda)$ of which 6 are relatively prime to 2λ. Each of the 6 reduced residue classes $(\text{mod}\, 2\lambda)$ splits further into 3 reduced residue classes $(\text{mod}\, 6)$. Since all six units of k_2 are incongruent $(\text{mod}\, 2\lambda)$, the units give representatives of all the reduced residue classes $(\text{mod}\, 2\lambda)$. Therefore every divisor of k_2 which is relatively prime to 6 has a unique generator which is congruent to $1\, (\text{mod}\, 2\lambda)$. The reduced residue class congruent to $1\, (\text{mod}\, 2\lambda)$ splits into three reduced residue classes $(\text{mod}\, 6)$ with representatives $1, 1 + 2\lambda$, $1 + 4\lambda$ and these three classes actually form a subgroup of the group of 18 reduced residue classes which is isomorphic to $G(6)$.

The prime divisors (π) of k_2 which split completely from k_2 to K are precisely those with one of the six generators $\equiv 1\, (\text{mod}\, 6)$. Every second degree prime of k_2 other than (2) satisfies this condition. For $p \neq 2$, (p) is a second degree prime of k_2 when $p \equiv 2\, (\text{mod}\, 3)$ and thus when $p \equiv 5\, (\text{mod}\, 6)$. Therefore $-p$ is a generator of (p) which is $1\, (\text{mod}\, 6)$. This accounts for the half of the primes of $\mathbb{Q}$ which give three second degree primes of K. The interesting primes are the first degree primes.

Theorem. *Suppose that p is a prime of $\mathbb{Q}$. Then p splits completely in K if and only if p is represented by the quadratic form*

$$Q(x, y) = x^2 + 27y^2\ .$$

Proof. Note that 2 and 3 aren't represented by $Q(x, y)$. First suppose that p splits completely in K. Then p splits in k_2 as $p = (\pi)(\bar{\pi})$ and (π) must split completely from k_2 to K. Therefore, we may choose the generator π of (π) so that $\pi \equiv 1\, (\text{mod}\, 6)$. In other words, we may assume that π is of the form

$$\pi = x + 6y\lambda\ .$$

where in fact $x \equiv 1 \pmod 6$ as well. Hence

$$p = \mathbb{N}(\pi) = x^2 - 18xy + 108y^2 = (x - 9y)^2 + 27y^2 \tag{45}$$

is of the desired form.

Conversely, suppose p is of the desired form. By changing variables, we may write

$$p = \mathbb{N}(x + 6y\lambda) = (x + 6y\lambda)(x + 6y\bar{\lambda})\ .$$

Let $\pi = x + 6y\lambda$, $\bar{\pi} = x + 6y\bar{\lambda}$. Then p splits in k_2 as $(\pi)(\bar{\pi})$ and we wish to show that both (π) and $(\bar{\pi})$ split completely from k_2 to K. Since p is a prime other than 2 or 3, (45) makes it clear that neither 2 nor 3 divides x. Therefore x is relatively prime to 6 and so $x \equiv \pm 1 \pmod 6$. Thus one of $\pm(x + 6y\lambda)$ is congruent to $1 \pmod 6$ and generates (π) and hence (π) splits completely from k_2 to K. Likewise $(\bar{\pi})$ splits completely in K. □

To summarize a major part of what we have learned about Example Two, 1/6 of the primes of $\mathbb{Q}$ split completely from $\mathbb{Q}$ to $k = \mathbb{Q}(2^{1/3})$ and these are exactly the primes which can be written in the form $p = x^2 + 27y^2$. The first prime of this form is $p = 31$. Half the primes of $\mathbb{Q}$ split in k as the product of first and second degree prime factors; these are the primes $p \equiv 2 \pmod 3$. The remaining 1/3 of the primes of $\mathbb{Q}$ generate third degree primes of k.

We close by looking at the L-functions for the cubic characters of $G(6)$ in Example Two. For each integral divisor $\mathfrak{a}$ relatively prime to 6, pick the generator congruent to $1 \pmod{2\lambda}$. We can then describe the three characters of $G(6)$ from the table,

	1	$1+2\lambda$	$1+4\lambda$
ψ_1	1	1	1
ψ_3	1	ω	ω^2
$\overline{\psi}_3$	1	ω^2	ω

This gives rise to three L-functions, with the trivial character giving $\zeta_{k_2}(s)$. Hence

$$\zeta_K(s) = \zeta_{k_2}(s) L(s, \psi_3) L(s, \overline{\psi}_3)\ . \tag{46}$$

From the point of view of Artin L-functions, ψ_3 and $\overline{\psi}_3$ become characters of the order three subgroup $G(K/k_2)$ of $G(K/\mathbb{Q})$. The induced characters are

$$\psi_3^* = \overline{\psi}_3^{\,*} = \chi_2\ .$$

Therefore the two L-functions are actually the same function,

$$L(s, \psi_3) = L(s, \overline{\psi_3}) = L(s, \chi_2, K/\mathbb{Q})\ .$$

This lets us check that the expressions for $\zeta_K(s)$ in (34) and (46) agree. We also get a formula which was known to Dedekind and which was the first non-abelian precursor of Artin L-functions,

$$\zeta_k(s) = \zeta(s)L(s, \psi_3)\,.$$

We conclude by mentioning the tantalizing analytic determination of k from $G(6)$,

$$L'(0, \psi_3) = -\log(2^{1/3} - 1)\,,$$

which follows from (24) once we know that the class-number of k is one and that $(2^{1/3} - 1)^{-1}$ is the fundamental unit of k.

References

Hecke, E.: Vorlesungen über die Theorie der algebraischen Zahlen. Chelsea, New York 1948. Translated as Lectures on the Theory of Algebraic Numbers. Springer, New York Heidelberg Berlin 1981.
Stark, H.M.: The Analytic Theory of Algebraic Numbers. Springer, 1993

Chapter 7

Galois Theory for Coverings and Riemann Surfaces

by Eric Reyssat

1. Coverings

As part of algebraic number theory, Galois theory deals with greatest common divisors (gcd's, also called ideals) and their conjugates (Stark 1992). One studies how the automorphisms of a field K change a gcd, or a number, to a conjugate. The set of 'prime gcd's' of K may be viewed as a topological space, as is now classical in algebraic geometry. Without going into this, we mention it to motivate the fact that Galois theory may have a meaning in a topological context. And it does indeed, in the context of *coverings.*

Here is one example of a covering (fig.1):

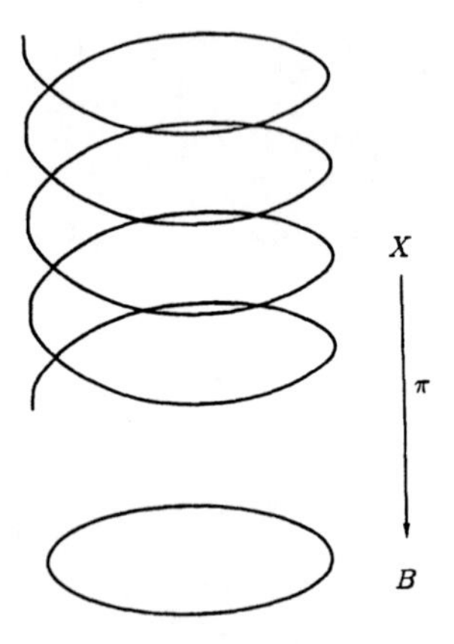

Fig. 1.

More precisely and generally, let X, B be two curves, or surfaces, or general topological spaces. We put one above the other, which means that we choose a map $X \xrightarrow{\pi} B$ (see the down arrow in fig.1). Now there is a local condition for our data (X, B, π) to form a *covering*. The condition is that above each small piece $\mathcal{U}$ of B lay just copies of $\mathcal{U}$; more precisely, the map $\pi : X \to B$ is a covering if B may be covered by open sets $B = \bigcup \mathcal{U}_i$ such that $\pi^{-1}(\mathcal{U}_i) \simeq \mathcal{U}_i \times F_i$ where each F_i is a non empty discrete topological space (think of it as being finite, or like $\mathbb{Z}$, but not continuous). The space B is called the *base space* of the covering. Note that F_i is independent of i provided X is connected (it is not the disjoint union of several domains). In the following, we will always assume that X is a connected *manifold* (small pieces look like pieces of $\mathbb{R}^n$, without any boundary point or singularities such as peaks or edges), and let $F = F_i$.

The simplest example of covering is the so called *trivial covering*

$$X = B \times F.$$

In this case, the local condition is even globally satisfied : X is the disjoint union of copies of B (see fig.2)

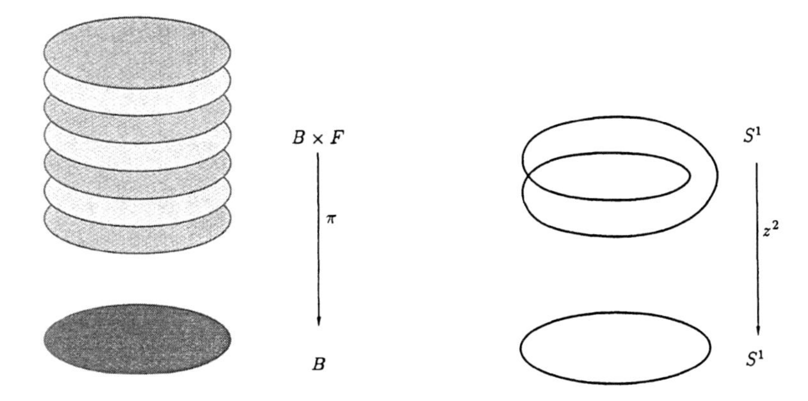

Fig. 2. Trivial covering **Fig. 3.** Squaring the circle !

A less trivial example is given by fig .1. : the real line rolled up above the circle $S^1 \subset \mathbb{C}$, the projection π being given by $\pi : t \longmapsto e^{2i\pi t}$ (the number π is not the projection π!). In this case, the set F is $\mathbb{Z}$. Another example is given by cutting the circle, turning one end once more and glueing it again (fig.3.). This is described by the map $\begin{matrix} z & \longmapsto & z^2 \\ S^1 & \longrightarrow & S^1 \end{matrix}$ from the unit circle onto itself, with the set $F = \{0, 1\}$.

The same construction works with the set of all non zero complex numbers instead of the unit circle, using the map $\begin{matrix} z & \longmapsto & z^2 \\ \mathbb{C}^* & \longrightarrow & \mathbb{C}^* \end{matrix}$. This is harder to draw : think of cutting the complex plane along the positive real axis, turning it once and glueing it again without self intersection (this is not possible when embedded in the 3-space). The figure may be like fig.4.

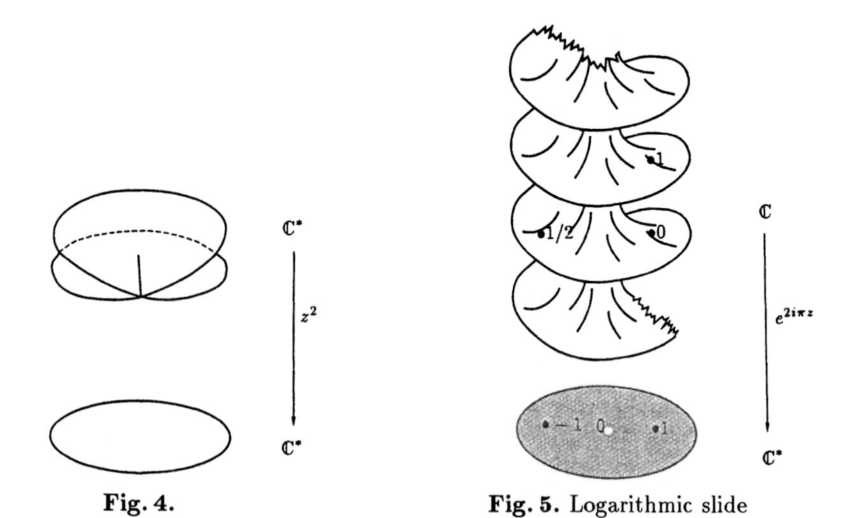

Fig. 4. **Fig. 5.** Logarithmic slide

Of course one can also complexify the exponential example by using the map $\overset{z}{\mathbb{C}} \overset{\longmapsto}{\longrightarrow} \overset{e^{2i\pi z}}{\mathbb{C}^*}$, and get an infinite slide (see fig.5). Above 1 lie all the integers, and the set F is still $\mathbb{Z}$.

Given a covering $X \xrightarrow{\pi} B$, there is above each point $b \in B$ a copy of F. This copy $\pi^{-1}(b)$ is called the *fiber* at b. One can think of its points as being *conjugate* to each other. For instance, the fiber at -1 for the above exponential example is the set $\{\frac{1}{2} + n \quad ; \quad n \in \mathbb{Z}\}$. The points $\frac{1}{2} + n$ are conjugates.

The *degree* of π is the cardinality of F (possibly infinite). One says that π is finite or infinite according to its degree. If π is of degree n, it is called an *n-sheeted covering*. Hence the exponential gives an infinite covering and $z \longmapsto z^n$ is an n-sheeted covering. Note that we usually speak of the *sheets* of X although one cannot enumerate them globally (except for a trivial covering). For instance, on the covering of fig.5, one cannot say on which sheet lie the points 0 and 1 of $\mathbb{C}$, but they definitely lie on different sheets. This is no more clear (nor interesting) for the points 1 and $\frac{1}{2}$ whose projections on $\mathbb{C}^*$ lie far apart.

In the case of the complex exponential, $\mathbb{C}^*$ is obtained as a quotient of $\mathbb{C}$ by a group $\mathbb{Z}$. In general if a group G acts on a manifold X in a good way, then the natural map $X \longrightarrow X/G$ is a covering (good means continuous — all $x \longmapsto gx$ are continuous — , properly discontinuous — each x has a neighbourhood U_x meeting only a finite number of conjugates gU_x — , and free — $gx \neq x$ for $g \neq 0$). The fundamental theorem of Galois theory (see below) implies that all coverings of a manifold are of this form.

For instance, a lattice Λ in $\mathbb{C}$ acts by translation and $\mathbb{C} \longrightarrow \mathbb{C}/\Lambda$ is an infinite covering. The result of taking the quotient gives a torus (see fig.6 : identify the slanted lines then the horizontal ones):

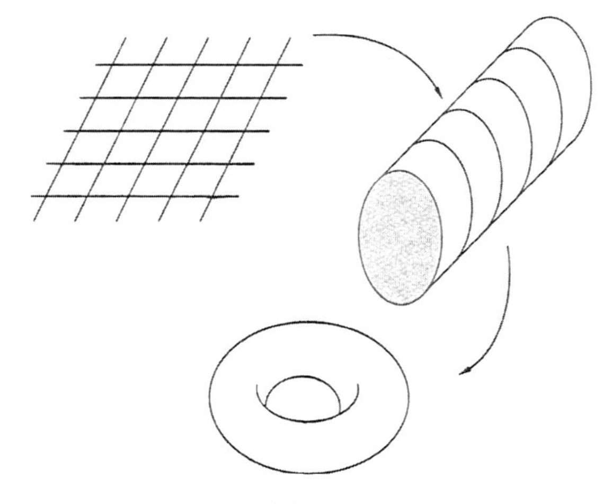

Fig. 6.

The multiplication by n on $\mathbb{C}$ induces an n^2-sheeted covering $\mathbb{C}/\Lambda \longrightarrow \mathbb{C}/\Lambda$; its fiber above x is

$$\left\{\frac{x + a\omega_1 + b\omega_2}{n} \quad ; \quad 0 \le a, b \le n-1\right\}$$

if

$$\Lambda = \{m_1\omega_1 + m_2\omega_2 \quad ; \quad m_1, m_2 \in \mathbb{Z}\}$$

is the lattice generated by ω_1 and ω_2.

2. Galois Theory

Roughly speaking, *Galois theory* deals with the problem of passing from one sheet to another. More precisely, one tries to permute the sheets, using automorphisms : an *automorphism* of the covering $\pi : X \to B$ is a bijective bicontinuous map $\phi : X \to X$ preserving fibers : $\pi \circ \phi = \pi$.

Example. The map $z \longmapsto iz$ is an automorphism of the covering $\begin{matrix} z & \longmapsto & z^4 \\ \mathbb{C}^* & \longrightarrow & \mathbb{C}^* \end{matrix}$: it doesn't change anything in the base.

The automorphisms form a group $\mathrm{Aut}_B\, X$ transforming a point $x \in X$ into some conjugates of it like a lift going up and down. But the best lifts must be able to lift us from any floor to any other one ! This is the essential property of Galois fields, in which the automorphisms may transform any number (or gcd) to any conjugate of it ; this is why we shall say that a covering $X \to B$ is *Galois* if any point in X may be transformed to any conjugate by some automorphism : $\mathrm{Aut}_B X$ acts transitively on each fiber. In this case, the groupe $G = \mathrm{Aut}_B X$ is also called the *Galois group* of the covering.

For instance :

1 The covering $\overset{z}{S^1} \overset{\longmapsto}{\longrightarrow} \overset{z^n}{S^1}$ is Galois since two points in a fiber differ only by an n-th root of unity ω, and the map $z \longmapsto \omega z$ is an automorphism.
2 Every 2-sheeted covering is Galois : The group of automorphisms is composed of the identity map and the exchange of sheets.

Now before stating the main theorem of Galois theory, we need the notion of a universal covering.

Given a covering $\pi : X \to B$, it is a trivial fact that any path drawn on the base B may be lifted to X : for any path γ on B from b_0 to b_1, and any point x_0 of X above b_0, there is a unique path Γ on X above γ starting from x_0 (see fig.7).

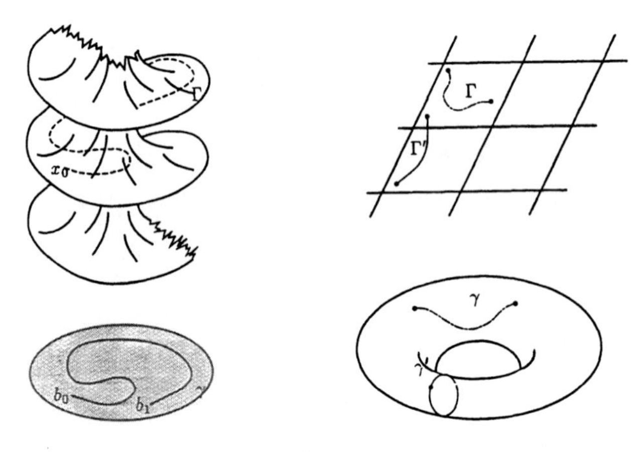

Fig. 7.

Looking at paths is a way to construct *the best* covering of a manifold. First recall that the fundamental group $\pi_1(B)$ of B is the set of closed paths (that is having same origin and end) on B, up to continuous deformation.

Example. In $\mathbb{C}^*$, one cannot transform continuously a path turning once around 0 into one which does not. One can only do it if the two paths turn the same (positive or negative) number of times around 0, so that $\pi_1(\mathbb{C}^*) = \mathbb{Z}$.

When one can contract *any* closed curve in X to a point (this means that $\pi_1(X) = 0$) then X is said to be *simply connected.* In the case of open sets of $\mathbb{C}$, this simply means that there is no 'hole' in X. For instance, $\mathbb{C}$ and all discs are simply connected, but $\mathbb{C}^*$ or a torus are not : the paths shown on fig.8 are not contractible to a point.

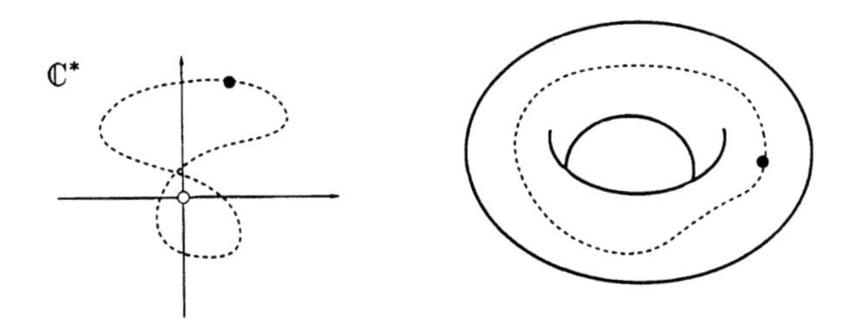

Fig. 8.

Theorem. *Every (connected) manifold B has a covering $\pi : \tilde{B} \to B$ which is connected and simply connected. It has the 'universal' property of being a covering of any other covering $p : X \to B$ of B in a natural way : there is a covering map $q : \tilde{B} \to X$ making the following diagram commutative :*

$$\begin{array}{ccc} & X & \\ {}^{q}\nearrow & & \searrow{}^{p} \\ \tilde{B} & \longrightarrow & B \end{array}$$

In particular, this covering is essentially unique (two of them are isomorphic since each one is a covering of the other).

The space $\tilde{B}$ can be viewed as the set of all curves on B starting from a fixed point $b_0 \in B$, up to continuous deformation.

Definition. *$\tilde{B}$ is called the universal covering of B.*

Example. Since $\mathbb{R}$ is simply connected, the covering $\mathbb{R} \to S^1$ is the universal covering of the circle S^1. In particular, it covers all the n-sheeted coverings $S^1 \to S^1$ given by the powers.

$$\begin{array}{ccccc} t \longmapsto z = e^{2i\pi t/n} & \longmapsto & z^n = e^{2i\pi t} \\ \mathbb{R} \longrightarrow \quad S^1 & \longrightarrow & S^1 \end{array} .$$

The heart of Galois theory for coverings lies in the following theorem, relating the automorphisms of $\tilde{B}$ and the fundamental group $\pi_1(B)$. Note that there is a natural action of $\pi_1(B)$ on each fiber of the covering $\pi : X \to B$ since the lifting of a closed path on B to a covering x of B starts and ends at points of the same fiber.

Theorem. (Fundamental theorem of Galois coverings) *Let B be a connected manifold, and $\tilde{B}$ its universal covering. Then :*

$\boxed{1}$ *$\tilde{B}$ is a Galois covering of B, and $G = \mathrm{Aut}_B(\tilde{B}) \simeq \pi_1(B)$.*

$\boxed{2}$ *The subgroups of G and the (connected) coverings of B are in natural 1-1 correspondence, given as follows : To any subgroup H of G corresponds the natural covering $\tilde{B}/H \to B$. Conversely, to a covering $X \to B$ corresponds the stabilizer H of any point x_0 (or any fiber) in $\pi_1(B)$. This group H is also isomorphic to the fundamental group $\pi_1(X)$.*

$\boxed{3}$ *The covering $\tilde{B}/H \to B$ is Galois if and only if the subgroup H is normal in G. In this case there is an isomorphism $\mathrm{Aut}_B(\tilde{B}/H) \simeq G/H$.*

Remark. At this stage, the comparison with the algebraic theory (see Stark's 1992; Section 1.4) shows clearly that the Galois theories for algebraic number fields and for topological coverings are but two different occurrences of one and the same concept. The analogue of the fundamental group $\pi_1(B)$ of some space B is the absolute Galois closure $\mathrm{Gal}(\overline{\mathbb{Q}}/\mathbb{Q})$. This group is far from being well understood and leads to interesting problems in number theory.

Example. We have seen that S^1 has $\mathbb{Z}$ as fundamental group and $\mathbb{R}$ as universal covering with the map $t \longmapsto e^{2i\pi t}$. Hence all coverings of S^1 are given by subgroups of $\mathbb{Z}$, which are the groups $n\mathbb{Z}$ ($n \geq 0$). For $n = 0$, the group $H = \{id\}$ corresponds to the covering $\mathbb{R} \to S^1$ itself, and for $n > 0$ we get the cyclic covering $\overset{z}{S^1} \overset{\longmapsto}{\longrightarrow} \overset{z^n}{S^1}$, with $\mathrm{Aut}_{S^1} S^1 = \mathbb{Z}/n\mathbb{Z}$. All coverings of S^1 are Galois.

For the same reason, the only coverings of the punctured unit disc

$$D^* = \{z \quad ; \quad 0 < |z| < 1\},$$

are the universal covering $\overset{z}{\mathfrak{H}} \overset{\longmapsto}{\longrightarrow} \overset{e^{2i\pi z}}{D^*}$ (where $\mathfrak{H}$ is the Poincaré upper half plane) and the cyclic coverings (which are Galois with cyclic Galois groups) $\overset{z}{D^*} \overset{\longmapsto}{\longrightarrow} \overset{z^n}{D^*}$.

We will see more examples of coverings in the next Section.

We end this Section by stating a basic property of universal coverings concerning the lifting of mappings : if Y is a simply connected manifold, and

$f : Y \to B$ is continuous, then f may be lifted to a map $F : Y \to \tilde{B}$ making the following diagram commutative :

$$\begin{array}{ccc} & & \tilde{B} \\ & \overset{F}{\nearrow} & \downarrow p \\ Y & \xrightarrow{f} & B \end{array}$$

Moreover, if we choose any point $y \in Y$, and any $\tilde{b} \in \tilde{B}$ above $f(y)$ (that is $p(\tilde{b}) = f(y)$) then there is a unique map $F : Y \to \tilde{B}$ with the above property and satisfying $F(y) = \tilde{b}$.

3. Riemann Surfaces

3.1 Generalities

A surface S is a space which can be covered by small discs looking like discs in $\mathbb{R}^2$ or $\mathbb{C}$ (see fig.9) : there are bicontinuous maps $z : \mathcal{D} \to D(0,1)$ from open sets $\mathcal{D}$ of S to the unit disc of $\mathbb{C}$.

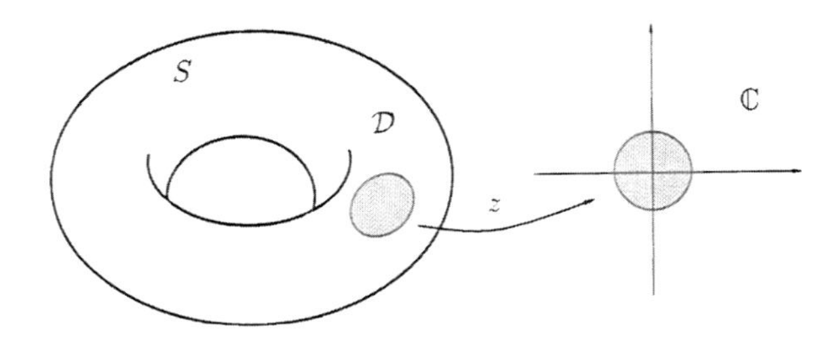

Fig. 9.

Such maps are called *charts* of S centered at $z^{-1}(0)$. Using a chart z, we can say that a function $f : S \to \mathbb{C}$ is analytic on $\mathcal{D}$ if $f \circ z^{-1}$ is. We can define in the same way the notion of a meromorphic function, a pole, the order of a zero or a pole, ... But to be consistent, these notions have to be independent of the choice of charts. For this, we cover S by charts with the property that on the overlapping of two discs, both meanings of analyticity are equal ; such a set of consistent charts is called an *analytic structure* or a structure of a *Riemann surface* on S. It is not always possible to choose a Riemann surface structure on any S. The surface S has to be orientable for this ; if S is compact, this is also sufficient, and if not it is still usually true (it is sufficient that S be orientable and triangulable). So each reasonable surface you can think of, except for the Klein bottle and the projective plane $\mathbb{P}_2(\mathbb{R})$, has a Riemann surface structure.

As an example, the quotient $\mathbb{C}/\Lambda$ of the complex plane by a lattice is a Riemann surface if one chooses the natural charts : we cover S by taking in $\mathbb{C}$ small discs with no two equivalent points (see fig.10).

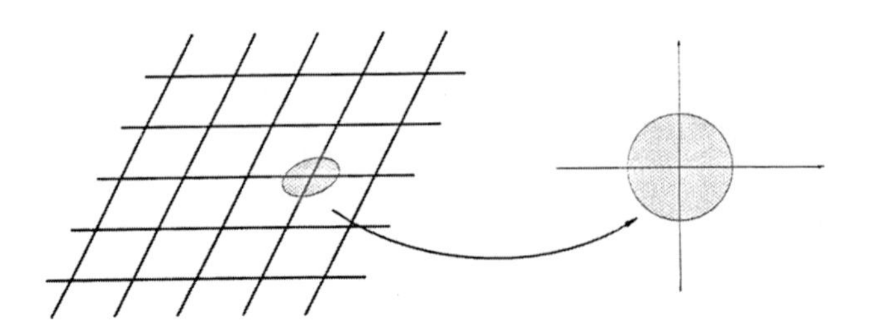

Fig. 10.

A meromorphic function on $\mathbb{C}/\Lambda$ is now a meromorphic function on $\mathbb{C}$, periodic with respect to Λ (each point in Λ is a period). A classical example of this is the Weierstrass $\wp$-function :

$$\wp(z) = \frac{1}{z^2} + \sum_{\Lambda \setminus 0} \left(\frac{1}{(z-\omega)^2} - \frac{1}{\omega^2} \right)$$

and its derivative $\wp'$ (see Cohen 1990 for details).

In the case of Riemann surfaces, the classification of the possible universal coverings is very simple (at least the result is) since there are *only three* possibilities :

Theorem. *The universal covering $\tilde{X}$ of a Riemann surface X is either $\mathbb{C}$, or the projective line $\mathbb{P}_1(\mathbb{C})$, or the unit disc D which is also isomorphic to the Poincaré upper half plane $\mathfrak{H} = \{z \in \mathbb{C} \,;\, \Im m\, z > 0\}$.*

In view of the fundamental theorem of Galois theory, every Riemann surface X is isomorphic to the quotient $\tilde{X}/G$ of its universal covering $\tilde{X}$ by some group G of automorphisms of $\tilde{X}$, isomorphic to the fundamental group of X. One shows that this group acts with discrete orbits and without fixed points. But the automorphisms of the three surfaces mentioned in the above theorem are well known :

$$\begin{aligned}\operatorname{Aut}\mathbb{C} &= \{z \longmapsto az+b \quad ; \quad a,b \in \mathbb{C}\},\\ \operatorname{Aut}\mathbb{P}_1(\mathbb{C}) &= PSL_2(\mathbb{C})\\ &= \{z \longmapsto \frac{az+b}{cz+d} \quad ; \quad a,b,c,d \in \mathbb{C},\ ad-bc=1\},\\ \operatorname{Aut}\mathfrak{H} &= PSL_2(\mathbb{R}).\end{aligned}$$

This allows to classify the Riemann surfaces X themselves, according to their universal covering :

- $\tilde{X} = \mathbb{P}_1(\mathbb{C})$. In this case, X itself is isomorphic to $\mathbb{P}_1(\mathbb{C})$.
- $\tilde{X} = \mathbb{C}$. Here G has to be a discrete group of translations, isomorphic to 0, $\mathbb{Z}$ or a lattice Λ, so that X is $\mathbb{C}$ or $\mathbb{C}/\mathbb{Z} \simeq \mathbb{C}^*$ or some torus $\mathbb{C}/\Lambda$.
- $\tilde{X} = D$. This is the case for all other Riemann surfaces. Each one corresponds to some subgroup of $PSL_2(\mathbb{R})$ acting discretely and without fixed points on D. They are called Fuchsian groups. Three of these surfaces are of particular interest : they are the unit disc D itself, the punctured unit disc D^* and the annuli $D_r = \{z \in \mathbb{C} \quad ; \quad 0 < |z| < r\}$.

The seven Riemann surfaces just mentioned (the last one is in fact a class of Riemann surfaces parametrized by r) are the only ones corresponding to an Abelian group G. This means that they have an Abelian fundamental group $\pi_1(X)$. Using the fundamental theorem, we see that all coverings of these seven (classes of) Riemann surfaces are Galois.

Before leaving general Riemann surfaces, let us prove one more application of this theory of coverings, namely the little theorem of Picard : by the examples given above, it is easy to see that if a, b are two complex numbers, the universal covering of $\mathbb{C}\backslash\{a,b\}$ has to be D, so that any analytic function $f : \mathbb{C} \to \mathbb{C}\backslash\{a,b\}$ lifts to an entire function from $\mathbb{C}$ to the unit disc D. Hence, this lifting is entire and bounded (since D is bounded) so is constant by Liouville theorem. As a consequence, f is also constant, which proves the classical *theorem of Picard* : a non-constant entire function must take every complex value with at most one exception.

3.2 Compact Riemann Surfaces

We go back to compact Riemann surfaces. Topologically, any compact orientable surface (recall that a Riemann surface is orientable) is known to be homeomorphic to a 'sphere with g handles', which is also a torus with g holes (see fig.11).

The fundamental group $\pi_1(X)$ of such a surface X is generated by $2g$ elements $a_1, \ldots, a_g, b_1, \ldots, b_g$ with the single relation

$$a_1 b_1 a_1^{-1} b_1^{-1} \ldots a_g b_g a_g^{-1} b_g^{-1} = 1.$$

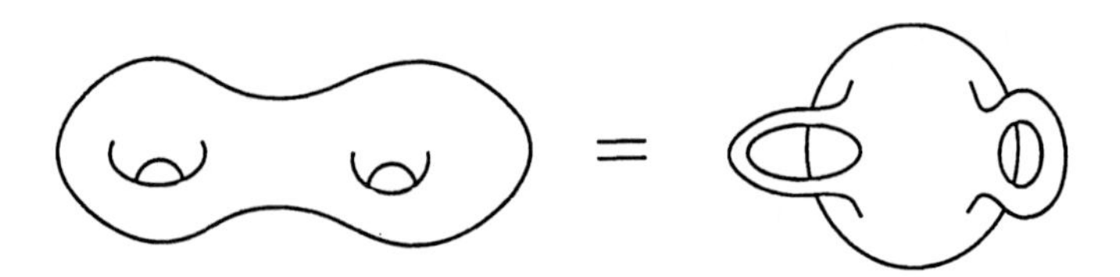

Fig. 11.

One may represent $a_1, \ldots, a_g$ as g cycles on X each turning once around one hole, and $b_1, \ldots, b_g$ as g cycles each passing through one hole. In particular $\pi_1(X)$ is an Abelian group if and only if $g \leq 1$; in this case X is either $\mathbb{P}^1$ or an elliptic curve, and all coverings of X are Galois. The number g is a very important topological invariant, called the *genus* of X. One way to compute it is to use any triangulation on X : if F, E, and V are the number of faces, edges and vertices of such a triangulation, the genus of X is given by the relation

$$2 - 2g = V - E + F. \tag{1}$$

Proposition. *Verify using this formula that the Riemann sphere has genus* 0 *and a torus has genus* 1.

Galois theory for Riemann surfaces requires the study of maps between Riemann surfaces. Let $f : X \to Y$ be such a map. If z, t are two charts at $x_0 \in X$ and $y_0 = f(x_0) \in Y$, we say that f is *analytic* around x_0 if the complex function $t \circ f$ is an analytic function of z in the usual sense : there is a converging power series satisfying

$$t \circ f(x) = \sum_{n_0}^{\infty} a_n z^n(x), \qquad a_{n_0} \neq 0$$

(note that $n_0 \geq 1$ since t is a chart at $f(x_0)$). We say that f is a morphism if it is everywhere analytic. The first exponent $n_0 = e_{x_0}(f)$ is called the *ramification index* of f at x_0, and f is said to be *ramified* at x_0, or x_0 is a *ramification point,* if $e_{x_0}(f) > 1$.

Theorem. *Let $f : X \to Y$ be a non constant morphism between two compact Riemann surfaces. Then a) there are only a finite number of ramification points $x_1, \ldots, x_k$. b) f defines (by removing the 'bad' fibers) an n-sheeted covering*

$$\pi : X \setminus \{f^{-1}(f(x_1)), \ldots, f^{-1}(f(x_k))\} \longrightarrow Y \setminus \{f(x_1), \ldots, f(x_k)\}.$$

c) The number of points on each fiber $\pi^{-1}(y)$ (bad or not) is independent of the fiber (hence equals the number n of sheets), provided that we count the multiplicity as the ramification index :

$$\sum_{x\in\pi^{-1}(y)} e_x(f) = n.$$

We say in this case that f is a *ramified covering*, or a *branched covering* of *degree n.*

The picture is given in fig.12.

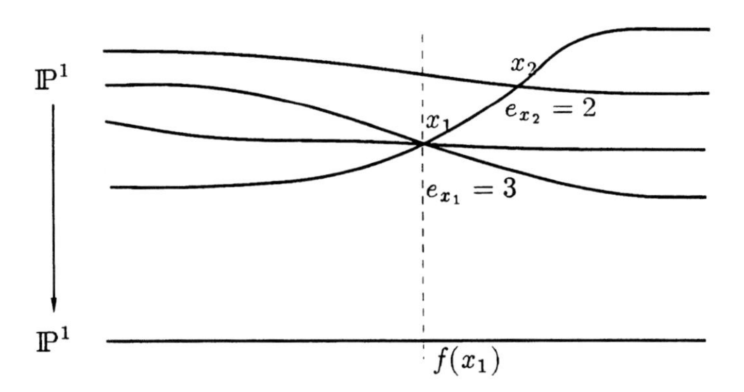

Fig. 12.

Remark. The terminology about coverings is somewhat dangerous. A branched covering is *not* a covering as defined earlier, unless there are no ramification points !

Example. Given a complex elliptic curve $E = \mathbb{C}/\Lambda$, the Weierstrass $\wp$ function defines a ramified covering $z \to \wp(z)$ of the projective line $\mathbb{P}^1$ by E of degree 2 : the elliptic curve E may be viewed as the set of points $(\wp(z), \wp'(z))$ in the projective plane $\mathbb{P}^2$, and to each value of $\wp(z)$ there usually correspond two points z and $-z$ on E since the $\wp$-function satisfies an equation of the form $\wp'^2(z) = 4\wp^3(z) - a\wp(z) - b$ and one point when $\wp' = 0$ or at 'infinity' (the point $0 \in \mathbb{C}/\Lambda$).

We fix as above a morphism $f : X \to Y$ of compact Riemann surfaces. If we define $X' = X \setminus \{\text{ramified fibers}\}$ and $Y' = Y \setminus f(\{\text{ramification points}\})$, then every automorphism of X' preserving the fibers of $\pi : X' \to Y'$ extends to an automorphism of X preserving fibers.

We say that $f : X \to Y$ is a *Galois* (ramified) covering if the restriction $f\,|_{X'}: X' \to Y'$ is a Galois covering. In this case, for every automorphism $\sigma \in \operatorname{Aut}_{Y'} X' = \operatorname{Aut}_Y X$ and every point x_0 on X, the ramification indices $e_{x_0}(f)$ and $e_{\sigma x_0}(f)$ are equal.

Proof. We choose charts z on X and t on Y such that $t \circ f(x) = z(x)^{e_{x_0}}$. If $u = \sigma(x)$, then

$$\begin{aligned} t \circ f(u) &= t \circ f\left(g^{-1}(u)\right) \qquad (\text{since } g^{-1} \in \operatorname{Aut}_Y X) \\ &= z\left(g^{-1}(u)\right)^{e_{x_0}} \end{aligned}$$

but zg^{-1} is a chart at u_0. □

Example. Let's look at the map $f : z \longmapsto z^3 - 3z$ from $X = \mathbb{P}^1$ to $Y = \mathbb{P}^1$. It induces a covering $X' = \mathbb{C} \setminus \{\pm 1, \pm 2\} \to Y' = \mathbb{C} \setminus \{\pm 2\}$ (see fig.13).

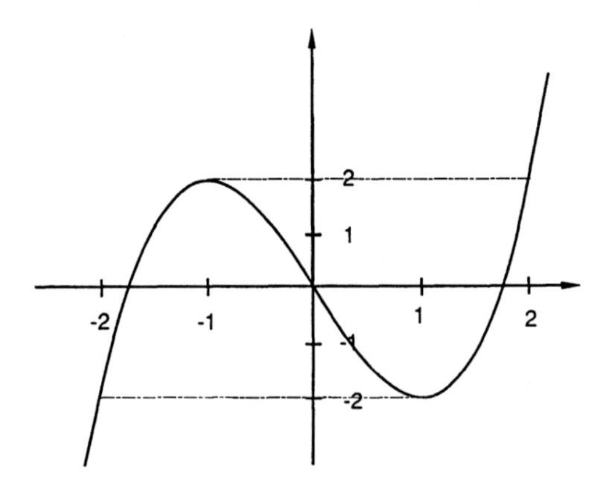

Fig. 13.

Any automorphism $\sigma \in \operatorname{Aut}_{X'} Y'$ is analytic. As an analytic function, it has no essential singularity since it is injective, hence extends to a map $\mathbb{P}^1 \to \mathbb{P}^1$ which is (almost everywhere) injective. Hence it is an automorphism of $\mathbb{P}^1$. But f is ramified at -1, and unramified at the point 2, which is in the same fiber. So there does not exist an automorphism g sending -1 on 2 (or a number near -1 or a number near 2.). We conclude that this covering is not a Galois covering (compare with example two of Stark's lecture (Stark 1992)).

3.3 The Riemann-Hurwitz Formula

Assume that $f : X \to Y$ is a non constant morphism of degree n. The Riemann-Hurwitz formula will allow us to compute the genus of X in terms of the genus of Y and the ramification of f. To do this, we choose any triangulation on Y. By adding new vertices and edges, we may assume that the image by f of its ramification points are vertices. Taking inverse images, we get a triangulation on X with $F_X = n F_Y$ faces and $E_X = n E_Y$ edges (since almost every point on Y has exactly n inverse images). The number of vertices on X is given by $V_X = n V_Y - \sum(e_x - 1)$ since a point above $y \in Y$ has $n - \sum_{f(x)=y}(e_x - 1)$ points above it. Computing the genus by the formula (1) given above finally shows

Theorem. (Riemann-Hurwitz) *Given a non constant morphism $f : X \to Y$ of degree n between two compact Riemann surfaces, the genus g_X of X and the genus g_Y of Y are related by the formula :*

$$2g_X - 2 = n(2g_Y - 2) + \sum_{x \in X} (e_x(f) - 1).$$

Example. The covering $z \to (\wp(z), \wp'(z))$ of $\mathbb{P}^1$ by an elliptic curve E discussed above is of degree 2, and has four ramification points, necessarily of index 2. These are the four 2-torsion points on E (see Cohen 1990). This is consistent with the Riemann-Hurwitz formula which reads

$$2.1 - 2 = 2(2.0 - 2) + 1 + 1 + 1 + 1$$

An application of this is the estimation of the number of automorphisms of a Riemann surface X of genus at least 2. The automorphism group $G = \operatorname{Aut} X$ may be shown to be finite (using its action on the so-called Weiertrass points, and the existence theorem of Riemann-Roch — see Reyssat 1989).

Let $G = \operatorname{Aut} X$. Then the quotient X/G is naturally endowed with a Riemann surface structure, and the map $f : X \to X/G$ is a branched Galois covering of degree $\operatorname{Card}(G)$. All points of a fiber $f^{-1}(y)$ have the same ramification index e_y, and there are $\frac{\operatorname{Card}(G)}{e_y}$ of these points. This implies that the total ramification of the fiber is $\frac{\operatorname{Card}(G)}{e_y}(e_y - 1)$ and the Riemann-Hurwitz formula gives

$$2(g_X - 1) = \operatorname{Card}(G) \left\{ 2g_{X/G} - 2 + \sum_{y \in Y} \left(1 - \frac{1}{e_y}\right) \right\}.$$

Now it is an exercise of elementary arithmetic to show that if $\gamma \geq 0$ and $e_i \geq 2$ are integers such that the number $t = 2\gamma - 2 + \sum \left(1 - \frac{1}{e_i}\right)$ is positive, then $t \geq \frac{1}{42}$.

We summarize all this in the

Theorem. (Hurwitz) *Let X be a compact Riemann surface of genus $g \geq 2$. Then the group of automorphisms of X is a finite group with at most $84(g - 1)$ elements.*

Remark. The equality $t = \frac{1}{42}$ in the preceding inequality is attained in only one case, when $\gamma = 0$ and the set of e_i's is $\{2,3,7\}$. This case occurs for some curves : it is attained for the modular curve denoted by $X(7)$ which is constructed by compactifying the quotient $\mathfrak{H}/\Gamma(7)$ of the Poincaré upper half plane $\mathfrak{H}$ by the modular group

$$\Gamma(7) = \left\{ \begin{pmatrix} ab \\ cd \end{pmatrix} \in SL_2(\mathbb{Z})\,;\ \begin{pmatrix} ab \\ cd \end{pmatrix} \equiv \begin{pmatrix} 10 \\ 01 \end{pmatrix} \pmod 7 \right\}.$$

This curve has genus 3 and 168 automorphisms. It is linked to the beautiful tessellation of the unit disc by hyperbolic triangles with angles $\pi/2$, $\pi/3$, and $\pi/7$. (see fig.14)

Fig. 14.

3.4 Curves and the Genus Formula

We will also apply the Riemann-Hurwitz formula to the genus formula which gives the genus of a plane algebraic curve. For this we have first to study compact Riemann surfaces as function fields (which means in terms of polynomials). For a compact Riemann surface X, we denote by $\mathcal{M}(X)$ the field of meromorphic functions on X which are also the morphisms from X to $\mathbb{P}^1$. It's a non trivial fact that every Riemann surface (in particular the compact ones) carries non constant meromorphic functions, and the celebrated theorem of Riemann-Roch 'counts' them in the compact case.

Theorem. (Siegel) *If f is a non constant meromorphic function on a compact Riemann surface X, let n be its degree as a ramified covering of $\mathbb{P}^1$. Then the field $\mathcal{M}(X)$ is an algebraic extension of $\mathbb{C}(f)$ of degree n.*

This means that $\mathcal{M}(X)$ is the field $\mathbb{C}(f,g)$ generated by f and a second function g algebraically related to f by an irreducible polynomial equation $P(f,g)=0$ of degree n.

Example. For the elliptic curve $E = \mathbb{C}/\Lambda$, we know the parametrisation by Weierstrass functions $\wp$ and $\wp'$ which are related by an equation

$$\wp' = 4\wp^3 - g_2\wp - g_3.$$

The theorem says that the field of all meromorphic functions on E is of degree 2 on $\mathbb{C}(\wp)$ (since $\wp$ has one double pole, the fiber above ∞ has two points counting multiplicity, hence the degree of $\wp$ is 2). But $\wp'$ is not in this field $\mathbb{C}(\wp)$ (since it is an odd function and $\wp$ is even), hence $\wp'$ generates the field $\mathcal{M}(E)$ over $\mathbb{C}(\wp)$. We obtain finally that $\mathcal{M}(E) = \mathbb{C}(\wp,\wp')$: all meromorphic functions on E (they are also called elliptic functions associated with the lattice Λ) are rational functions of $\wp$ and $\wp'$.

By the above theorem, we see that to each compact Riemann surface X is associated (non uniquely !) a polynomial P. Conversely, for any complex irreducible polynomial $P \in \mathbb{C}[z,w]$ the set of points where $P'_z(z,w) \neq 0$ or $P'_w(z,w) \neq 0$ is naturally a Riemann surface, and it gives a compact Riemann surface by adding a finite number of points. We see that to each compact Riemann surface X is associated its function field $\mathcal{M}(X)$, which in turn gives an algebraic plane curve of equation $P(f,g)=0$, and from such a polynomial we construct a new Riemann surface X_P. It is a fact that $X_P = X$ up to isomorphism. We thus close the loop . All this means that what can be stated in terms of compact Riemann surfaces (and morphisms between them) can also be stated in terms of function fields without loosing information, and conversely. The precise statement behind this idea is that there is an equivalence of categories between the compact Riemann surfaces and the function fields of one variable. As an example, let us state the following

Theorem. *Let $f : X \to Y$ be a non constant morphism of compact Riemann surfaces. It induces by composition an embedding $\mathcal{M}(X) \hookrightarrow \mathcal{M}(Y)$ so that we may speak of the extension $\mathcal{M}(Y)/\mathcal{M}(X)$. Then f is a Galois covering if and only if the extension $\mathcal{M}(Y)/\mathcal{M}(X)$ is Galois. This is also equivalent to the condition that the group of automorphisms* $\mathrm{Aut}_Y X$ *(which is isomorphic to the* *group* $\mathrm{Aut}_{\mathcal{M}(Y)}\,\mathcal{M}(X)$ *of field automorphisms) has cardinality* $n = degree(f)$, *which is also the degree of the extension* $\mathcal{M}(Y)/\mathcal{M}(X)$.

Now that we know that a compact Riemann surface 'is' also a curve X_P in the projective plane, we can state the genus formula. It can be shown that, by suitably choosing the functions f and g on X, the curve X_P can be constructed having only nodes as singularities : these are 'ordinary crossings' like on fig .15 and not 'cusps' or even worse singularities like on fig .16.

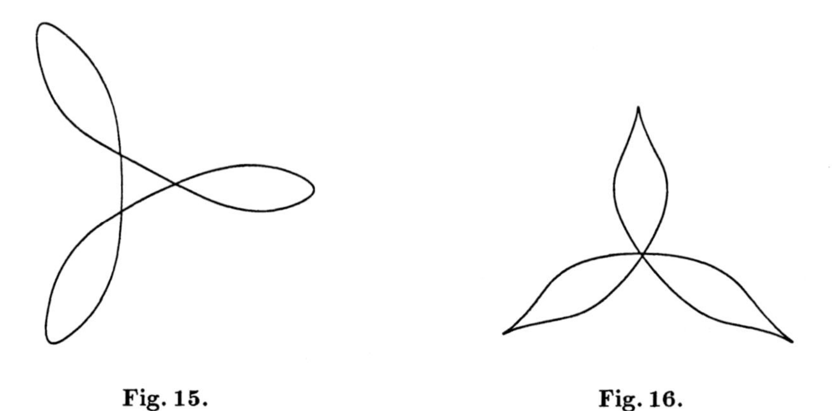

Fig. 15. **Fig. 16.**

Theorem. *If X is a plane curve of degree n in the projective plane $\mathbb{P}^2$ having only r nodes as singularities, then the associated Riemann surface is of genus*

$$g = \frac{(n-1)(n-2)}{2} - r.$$

To prove this, one applies the Riemann-Hurwitz formula to the morphism $f : \begin{array}{ccc} X & \longrightarrow & \mathbb{P}^1 \\ (f,g) & \longmapsto & f \end{array}$ of degree n (generically we see n points in each fiber). The ramification is given by the vanishing of the derivative f' giving the $\frac{(n-1)(n-2)}{2}$ term, and the nodes giving the $-r$ term (see Reyssat 1990). For instance, the well known Fermat curve of equation $x^n + y^n = 1$ is smooth (no singularity at all), hence is of genus $\frac{(n-1)(n-2)}{2}$.

Finally, coverings of Riemann surfaces also appear in the original definition of Riemann surfaces : we start with any power series $\sum a_n z^n$ at the origin. By analytic continuation, we can rewrite it at all the points of some domain of $\mathbb{C}$ that we take as large as possible. At some points the continuation is only possible if we use Puiseux series of the form $\sum b_n (z - z_0)^{n/k}$. Then this set of power and Puiseux series carries a Riemann surface structure, and (except for branch points where $k \geq 1$) it is a covering of a part of $\mathbb{C}$. Any Riemann surface is of this form (see Reyssat 1990). This theory appears in the study of multivalued analytic functions like solutions of differential equations, transformed by monodromy (Beukers 1990).

4. Constructing Galois Extensions with given Galois Group

We have seen how Galois theory may make sense for function fields of one variable as well as for number fields (see Stark 1992). Not only the theory has the same structure, but one can use one case to study the other. The geometric aspect of the function field case usually makes it easier, and gives information or intuition for the treatment of the algebraic case. One application of this connection is the problem of finding which finite groups occur as Galois group of number fields over $\mathbb{Q}$. We start with $\mathbb{C}(T) = \mathcal{M}(\mathbb{P}^1)$. If $t_1, \ldots, t_k \in \mathbb{P}^1$, then the fundamental group $\pi_1(\mathbb{P}^1 \setminus \cup\{t_i\})$ is the group generated by k elements $s_1, \ldots, s_k$ and the only relation $s_1 \ldots s_k = 1$. Since any finite group may be generated by $k-1$ elements for some k, we can view it as a quotient of this fundamental group. By Galois theory, this means that any finite group G is the group of automorphisms of some covering X of $\mathbb{P}^1 \setminus \Sigma$ for some finite set Σ, hence is the Galois group of some finite extension of $\mathbb{C}(T)$ (since the automorphisms extend to $\mathbb{P}^1$). Then the procedure is to descend to $\mathbb{Q}(T)$ and then to $\mathbb{Q}$ itself. This is not an easy task and nobody knows how to make it work in full generality, but it does work for many groups (see Serre 1988). One difficulty in obtaining Galois groups over $\mathbb{Q}(T)$ from extensions of $\mathbb{C}(T)$ lies in the fact that one must deal with extensions of $\mathbb{Q}(T)$ with ramification points which are not in $\mathbb{Q}$, as shown by the following theorem of Dèbes-Fried (Dèbes and Fried 1990) :

Theorem. *Let G be a finite group. For G to be the Galois group of a regular extension Y of $\mathbb{R}(T)$ (that is $Y \cap \mathbb{C} = \mathbb{R}$) with only real ramification points, it is necessary and sufficient that G be generated by elements of order 2.*

References

Beukers, F. (1990) Differential Galois theory ; this volume.
Cohen, H. (1990) Elliptic curves ; this volume.
Dèbes, P., Fried, M. (1990) Non rigid situations in constructive Galois theory ; preprint 1990.
Douady, R., Douady, A. (1979) Algèbre et théories galoisiennes ; Cedic Fernand Nathan (1979).
Reyssat, E. (1989) Quelques aspects des surfaces de Riemann ; Birkhäuser (1989).

Serre, J.-P. (1988) Groupes de Galois sur $\mathbb{Q}$; Séminaire Bourbaki n^{r}. 689, Astérisque **161-162** (1988) 73-85

Stark, H. (1992) Galois theory, algebraic number theory and zeta functions ; this volume.

Chapter 8

Differential Galois Theory

by Frits Beukers

1. Introduction

Perhaps the easiest description of differential Galois theory is that it is about algebraic dependence relations between solutions of linear differential equations. To clarify this statement, let us consider three examples. First consider the differential equation

$$z(1-z)y'' + (\frac{1}{2} - \frac{7}{6}z)y' + \frac{11}{3600}y = 0 \tag{1.1}$$

It is not at all obvious from its appearance that the solutions of Eq.(1.1) are algebraic functions over $\mathbb{C}(z)$. That is, any solution of (1.1) satisfies a polynomial equation with coefficients in $\mathbb{C}(z)$ (As always, $\mathbb{C}(z)$ stands for the set of rational functions with complex coefficients). As a second example consider

$$(z^4 - 34z^3 + z^2)y''' + (6z^3 - 153z^2 + 3z)y'' + (z^2 - 112z + 1)y' + (z-5)y = 0 \tag{1.2}$$

Again it may come as a surprise that any three independent solutions of (1.2) satisfy a homogeneous quadratic relation with coefficients in $\mathbb{C}$ (Beukers and Peters 1984). The third example is of a different kind. We know that the general solution of $\frac{d}{dz}y + f \cdot y = g,\ f, g \in \mathbb{C}(z)$ reads

$$y = \left(\int g \exp(\textstyle\int f dz) + C\right) \exp(-\textstyle\int f dz).$$

One might wonder whether or not the general solution of a second order linear differential equation can be written in a similar way, i.e. as a function involving only the coefficients of the differential equation, integrations and exponentiations. It will turn out that the answer is in general 'no' (see Theorem 2.4.3). The main point to our present story is that phenomena and questions such as the ones above were at the origin of differential Galois theory. By the end of the 19-th century such questions were studied by many people of whom we mention Halphén, (Fano 1900), (Picard 1898), (Vessiot 1892). One of the tools which slowly emerged was what we now call a differential Galois group. Here we give an informal description of what a differential Galois group is. For the precise definitions we refer to Section 2.1. Consider a linear differential equation

$$p_n(z)\partial^n y + \cdots + p_1(z)\partial y + p_0(z)y = 0, \; p_i(z) \in \mathbb{C}(z), \tag{1.3}$$

where $\partial = d/dz$. Let $y_1, \ldots, y_n$ be a basis of solutions and consider the field F obtained by adjoining to $\mathbb{C}(z)$ all functions $\partial^i y_j$ $(i = 0, \ldots, n-1; j = 1, \ldots, n)$. Notice that for any $g = (g_{ij}) \in \mathrm{GL}(n, \mathbb{C})$ (=invertible $n \times n$ matrices) the set of functions $\sum_{j=1}^n g_{ij} y_j (i = 1, \ldots, n)$ is again a basis of solutions. Let $\mathcal{J}$ be the set of all $Q \in \mathbb{C}(z)[X_1^{(0)}, \ldots, X_n^{(0)}, X_1^{(1)}, \ldots, X_n^{(n-1)}]$ such that $Q(\ldots, \partial^j y_i, \ldots) = 0$ (we substituted $\partial^j y_i$ for $X_i^{(j)}$). The group

$$\left\{ (g_{ik}) \in GL(n, \mathbb{C}) | \; Q(\ldots, \partial^j (\sum_{k=1}^n g_{ik} y_k), \ldots) = 0 \text{ for all } Q \in \mathcal{J} \right\}$$

will be called the differential Galois group of (1.1) and we shall denote it by $\mathrm{Gal}_\partial(F/\mathbb{C}(z))$. As we see, it respects all relations over $\mathbb{C}(z)$ which exist between the functions y_i and their derivatives and thus it plays a rôle as a kind of bookkeeping system of algebraic relations. In particular, if the $\partial^j y_i$ $(j = 0, \ldots, n-1; i = 1, \ldots, n)$ are algebraically independent over $\mathbb{C}(z)$, i.e. no Q's exist, we have $\mathrm{Gal}_\partial(F/\mathbb{C}(z)) = \mathrm{GL}(n, \mathbb{C})$ and the differential Galois group is maximal. On the other extreme, it will turn out that $\mathrm{Gal}_\partial(F/\mathbb{C}(z))$ is finite if and only if all solutions of (1.1) are algebraic over $\mathbb{C}(z)$. In that case $\mathrm{Gal}_\partial(F/\mathbb{C}(z))$ is actually isomorphic to the ordinary Galois group of the corresponding finite extension of $\mathbb{C}(z)$. Moreover, it turns out that a differential Galois group is a linear algebraic group, the standard example of a Lie group over $\mathbb{C}$ (see Section 3 for more precise definitions). Unfortunately, the study of linear algebraic groups was only at a very primitive stage in the 19-th century and could not be of any assistance. Nevertheless it did become clear that the differential Galois group is an important tool in algebraic dependence questions. Then, at the beginning of the 20-th century the study of these questions became more or less obsolete. It might be interesting to philosophize on the reasons for this silence. What matters for our story is that after preparatory work of (Ritt 1932), E.R.Kolchin published a paper in 1948 which marks the birth of modern differential Galois theory (Kolchin 1948). In this paper, and other papers as well, Kolchin took up the work of the 19-th century mathematicians and addressed questions such as existence and uniqueness of Picard-Vessiot extensions, and stressed the need for an approach which is entirely algebraic. Some years later I. Kaplansky (Kaplansky 1957) wrote a small booklet explaining the basics of the ideas of Ritt and Kolchin. For a quick and very pleasant introduction to differential Galois theory Kaplansky's book was practically the only reference up till now. In the meantime Kolchin had developed and generalized his ideas to a very large extent including systems of partial differential equations. This work culminated in two books (Kolchin 1973), (Kolchin 1985). Unfortunately, these books are very hard to read for a beginner. Remarkably enough, the development of differential Galois theory still remained in the hands of a small group of people until only a few years ago. It was then that Kolchin's ideas attracted attention from other fields of mathematics. One of the fields which could have benefited from differential Galois theory many years ago was transcendental number theory. Siegel (Siegel 1929) discovered that there exists a large class of functions, satisfying linear differential equations, for which one can establish algebraic independence results of their values at certain points provided that the functions themselves are algebraically independent over $\mathbb{C}(z)$. The latter problem could have been approached via the determination of differential Galois groups. However, apart from a few remarks, the first papers in which this connection is made explicitly are (Kolchin 1968) and later (Beukers, Brownawell and Heckman 1988). Very recently, N. M. Katz (Katz 1990) wrote a book on exponential sums and differential Galois theory, in which important parallels are drawn between l-adic representations of $\mathrm{Gal}(\bar{\mathbb{Q}}/\mathbb{Q})$ and differential Galois groups. Another such parallel can be drawn with the Mumford-Tate group of an Abelian variety. More generally, such parallels become clear if one views objects as differential equations, l-adic representations, etc. as examples of Tannakean categories (Deligne 1987). However, in this article we shall not go that far but restrict ourselves to giving a basic introduction to Galois theory of ordinary differential equations with some applications. We shall introduce differential Galois groups and their basic properties in Section 2. We formulate definitions and theorems in a fairly general form using differential fields of characteristic zero and algebraically closed constant field. However, the reader who is not interested in such generalities, is welcome to read $\mathbb{C}(z)$ or $\mathbb{C}((z))$ (that is the field of formal Laurent series in z) any time he or she meets the word differential field. The derivation ∂ then becomes ordinary differentiation and the field of constants is $\mathbb{C}$ in this case. In Section 3 we provide some background on linear algebraic groups and refer to the existing literature for proofs and more general definitions. Finally, in Section 4 we illustrate techniques for the computation of differential Galois groups by computing them for the generalized hypergeometric equation in one variable. The mathematically inclined reader may regret the absence of proofs for the main theorems of differential Galois theory and other theorems as well. However, in this article we have not attempted to be complete, but only to give proofs at those places where we thought they might be instructive. For some easily accessible proofs of the main theorems we refer to (Levelt 1990) in combination with Kaplansky's book. Finally we mention the survey (Singer 1989) and the book (Pommaret 1983). In the latter book the author takes up a theory for partial differential equations and gives some applications to physics, among which a claim for a new approach to gauge field theory.

2. Differential Fields and Their Galois Groups

2.1 Basic notions

The coefficients of our linear differential equation are chosen from a differential field. A *differential field* is a field F equipped with a map $\partial : F \to F$ satisfying the rules $\partial(f+g) = \partial f + \partial g$ and $\partial(fg) = f\partial g + g\partial f$. The map ∂ is called a *derivation* and we use the notation (F, ∂) for the differential field. The examples we shall mainly consider are $(\mathbb{C}(z), d/dz)$ and $(\mathbb{C}((z)), d/dz)$. Here, $\mathbb{C}((z))$ stands for the (not necessarily converging) Laurent series in z, and d/dz denotes ordinary differentiation with respect to z. A trivial and very uninteresting example is when F is any field and $\partial f = 0$ for all $f \in F$. The *constant field* of a differential field (F, ∂) is the subfield consisting of all elements of F whose derivation is zero. Notation: C_F or, if no confusion can arise, just C. In the remainder of this article **we shall assume that the characteristic of C is 0, and that C is algebraically closed.** For example, $C = \mathbb{C}$. Two differential fields (F_1, ∂_1) and (F_2, ∂_2) are said *to be differentially isomorphic* if there exists a field isomorphism $\phi : F_1 \to F_2$ such that $\phi \circ \partial_1 = \partial_2 \circ \phi$. The map ϕ is called a *differential isomorphism.* A differential isomorphism of a field to itself is called a *differential automorphism.* A differential field $(\mathcal{F}, \partial')$ is called a *differential extension* of (F, ∂) if $F \subset \mathcal{F}$ and ∂' restricted to F coincides with ∂. Usually, ∂' is again denoted by ∂, and we shall adopt this habit. For example, $(\mathbb{C}((z)), d/dz)$ is a differential extension of $(\mathbb{C}(z), d/dz)$. Conversely, (F, ∂) is called a *differential subfield* of $(\mathcal{F}, \partial)$. Let $u_1, \ldots, u_r \in \mathcal{F}$. The smallest differential subfield of $(\mathcal{F}, \partial)$ containing F and the elements $u_1, \ldots, u_r$ is denoted by $F < u_1, \ldots, u_r >$. It is actually obtained by adjoining to F the elements $u_1, \ldots, u_r$ together with all their derivatives. Let (F, ∂) be a differential field and consider the linear differential equation

$$\text{(2.1.1)} \qquad \mathcal{L}y = 0, \quad \mathcal{L} = \partial^n + f_1\partial^{n-1} + \cdots + f_n, \quad f_i \in F \quad (i = 1, \ldots, n)\,.$$

Usually, the solutions of (2.1.1) do not lie in F. So we look for differential extensions of (F, ∂) containing the solutions.

(2.1.1) Definition. *A differential extension $(\mathcal{F}, \partial)$ of (F, ∂) such that*

i. $C_{\mathcal{F}} = C_F$,

ii. *$\mathcal{F}$ contains n C_F-linear independent solutions $y_1, \ldots, y_n$ of* Eq.(2.1.1) *and* $\mathcal{F} = F < y_1, \ldots, y_n >$,

is called a Picard-Vessiot extension *of* Eq.(2.1.1).

(2.1.2) Theorem. (Kolchin). *Let (F, ∂) be a differential field. Assume, as we do throughout this article, that the characteristic of F is zero and that C_F is algebraically closed. Then to any linear differential equation there exists a Picard-Vessiot extension. Moreover, this extension is unique up to differential isomorphism.*

For a proof we refer to (Kolchin2 1948), (Kolchin 1973) or (Levelt 1990).

(2.1.3) *Remark.* Note the similarity between a Picard-Vessiot extension of a linear differential equation and the splitting field of a polynomial.

(2.1.4) Lemma. *Let (F, ∂) be a differential field and $u_1, \ldots, u_r \in F$. Then $u_1, \ldots, u_r$ are linearly dependent over C_F if and only if the determinant*

$$W(u_1, \ldots, u_r) = \begin{vmatrix} u_1 & u_2 & \cdots & u_r \\ \partial u_1 & \partial u_2 & \cdots & \partial u_r \\ \vdots & \vdots & \ddots & \vdots \\ \partial^{r-1} u_1 & \partial^{r-1} u_2 & \cdots & \partial^{r-1} u_r \end{vmatrix}$$

vanishes.

Proof. If $u_1, \ldots, u_r$ are linearly dependent over C_F then so are the columns of $W(u_1, \ldots, u_r)$, hence this determinant vanishes. Suppose conversely that $W(u_1, \ldots, u_r)$ vanishes. By induction on r we shall prove that $u_1, \ldots, u_r$ are linearly dependent over C_F. First notice the identity $W(vu_1, \ldots, vu_r) = v^r W(u_1, \ldots, u_r)$ for any $v \in F$. In particular, if we take $v = u_r^{-1}$ (and assuming $u_r \neq 0$), then

$$\begin{aligned} W(u_1, \ldots, u_r)/u_r^r &= W(u_1/u_r, \ldots, u_{r-1}/u_r, 1) \\ &= (-1)^{r-1} W(\partial(u_1/u_r), \ldots, \partial(u_{r-1}/u_r)). \end{aligned}$$

If $r = 1$, then our statement is obvious. Now suppose that $r > 1$. If $u_r = 0$ we have a linear dependence relation. So we assume $u_r \neq 0$. The vanishing of $W(u_1, \ldots, u_r)$ then implies the vanishing of $W(\partial(u_1/u_r), \ldots, \partial(u_{r-1}/u_r))$. By induction hypothesis there exist $a_1, a_2, \ldots, a_{r-1} \in C_F$, not all zero, such that $a_1 \partial(u_1/u_r) + \cdots + a_{r-1} \partial(u_{r-1}/u_r) = 0$, hence $a_1 u_1 + \cdots + a_{r-1} u_{r-1} = a_r u_r$ for some $a_r \in C_F$, as asserted. □

(2.1.5) *Remark.* The determinant $W(u_1, \ldots, u_r)$ is called the *Wronskian determinant* of $u_1, \ldots, u_r$. If $y_1, \ldots, y_n$ is an independent set of solutions of Eq.(2.1.1) one easily verifies that $\partial W = -f_1 W$, where $W = W(y_1, \ldots, y_n)$.

(2.1.6) Corollary. *Let $y_1, \ldots, y_n$ be a set of solutions of* Eq.(2.1.1), *linearly independent over C_F. Then any solution y of* Eq.(2.1.1) *is a C_F-linear combination of $y_1, \ldots, y_n$. In particular, the solutions of* Eq.(2.1.1) *form a linear vector space of dimension n over C_F.*

Proof. Since $y, y_1, \ldots, y_n$ all satisfy the same linear differential equation of order n, we have $W(y, y_1, \ldots, y_n) = 0$. Lemma 2.1.4 now implies that $y, y_1, \ldots, y_n$ are linearly dependent over C_F, and our corollary follows. □

2.2 Galois theory

(2.2.1) Definition. *Let* $(\mathcal{F}, \partial)$ *be a Picard-Vessiot extension of* Eq.(2.1.1). *The* differential Galois group *of* Eq.(2.1.1) *(or of* $\mathcal{F}/F$*) is defined as the group of all differential automorphisms* $\phi : \mathcal{F} \to \mathcal{F}$ *such that* $\phi f = f$ *for all* $f \in F$. *Notation:* $\mathrm{Gal}_\partial(\mathcal{F}/F)$.

Let $\phi : \mathcal{F} \to \mathcal{F}$ be an element of $\mathrm{Gal}_\partial(\mathcal{F}/F)$. Let y be any solution of Eq.(2.1.1). Since ϕ fixes F, we have $0 = \phi(\mathcal{L}y) = \mathcal{L}(\phi y)$. Hence ϕy is again a solution of Eq.(2.1.1). In other words, elements of $\mathrm{Gal}_\partial(\mathcal{F}/F)$ act as C_F-linear maps on the n-dimensional vector space V of solutions of Eq.(2.1.1).

Conversely, it follows from the definitions that any linear map in $GL(V)$ which respects all polynomial relations over F between the solutions of equation (2.1.1) and their derivatives, is an element of $\mathrm{Gal}_\partial(\mathcal{F}/F)$. So we find, just like in the introduction, the following statement

(2.2.2) Lemma. *Let notations be as above, and let* $\mathcal{J}$ *be the set of polynomials* $Q \in F[X_1^{(0)}, \ldots, X_n^{(0)}, X_1^{(1)}, \ldots, X_n^{(n-1)}]$ *such that* $Q(\ldots, \partial^j y_i, \ldots) = 0$ *(we have substituted* $\partial^j y_i$ *for* $X_i^{(j)}$*). Then*

$$\mathrm{Gal}_\partial(\mathcal{F}/F) = \left\{ g_{ik} \in GL(n, C_F) \mid Q(\ldots, \partial^j (\textstyle\sum_{k=1}^n g_{ik} y_k), \ldots) = 0 \quad \forall Q \in \mathcal{J} \right\}.$$

(2.2.3) *Remark.* The determinant of $\phi \in \mathrm{Gal}_\partial(\mathcal{F}/F)$ can be read off from its action on $W(y_1, \ldots, y_n)$, since $W(\phi y_1, \ldots, \phi y_n) = \det \phi W(y_1, \ldots, y_n)$.

We now state the principal theorems on differential Galois groups. Again, for their proofs we refer to (Kolchin1 1948) or (Kaplansky 1957).

(2.2.4) Theorem. (Kolchin). *The differential Galois group of a Picard-Vessiot extension is a linear algebraic group over the field of constants. Its dimension equals the transcendence degree of the Picard-Vessiot extension.*

The reader who is not familiar with algebraic groups might consult Section 3 for a bare minimum of definitions, examples and results. The more ambitious reader might consult (Humphreys 1972 and 1975) or (Freudenthal and de Vries 1969).

The following result is known as the Galois correspondence for differential fields.

(2.2.5) Theorem. (Kolchin). *Let* $(\mathcal{F}, \partial)$ *be a Picard-Vessiot extension of* (F, ∂) *with differential Galois group* $G = \mathrm{Gal}_\partial(\mathcal{F}/F)$. *Then,*

i. *If* $f \in \mathcal{F}$ *is such that* $\phi f = f$ *for all* $\phi \in G$, *then* $f \in F$.

ii. *Let H be an algebraic subgroup of G such that* $F = \{f \in \mathcal{F} | \phi f = f \text{ for all } \phi \in H\}$. *Then* $G = H$.

iii. *There is a one-to-one correspondence between algebraic subgroups H of G and intermediate differential extensions M of F (i.e. $F \subset M \subset \mathcal{F}$) given by*

$$H = \mathrm{Gal}_\partial(\mathcal{F}/M) \qquad M = \{f \in \mathcal{F} | \phi f = f \ \forall \phi \in H\} .$$

iv. *Under the correspondence given in iii) a normal algebraic subgroup H of G corresponds to a Picard-Vessiot extension M of F and conversely. In such a situation we have* $\mathrm{Gal}_\partial(M/F) = G/H$.

Of particular interest among the subgroups of G is the connected component of the identity, G^0 (see Section 3). Its fixed field is an algebraic extension of F, since G^0 is a subgroup of finite index. A particular case is when G is finite. Then $\dim G$ is zero and, by Theorem 2.2.4, the extension $\mathcal{F}/F$ is algebraic. Clearly, the converse also holds.

2.3 Examples

In two of the examples below, we have considered differential equations over any differential field F. As we said before, the reader who does not like such generalities is welcome to substitute $\mathbb{C}(z)$ for F and $\mathbb{C}$ for C_F, or any other familiar fields.

(2.3.1) Example.

$$\partial y = ay, \qquad a \in F$$

Let $\mathcal{F}$ be the Picard-Vessiot extension and u a non-trivial solution. Clearly, any element ϕ of $G = \mathrm{Gal}_\partial(\mathcal{F}/F)$ acts as $\phi : u \to \lambda u$ for some $\lambda \in C_F^\times$. One easily checks that any algebraic subgroup of $C_F^\times$ is either $C_F^\times$ itself, or a finite cyclic subgroup of order m, say. In the latter case we see that $\phi : u^m \to u^m$ for any $\phi \in G$. Hence $u^m \in F$ and u is algebraic over F.

(2.3.2) Example.

$$\partial y = a \ , \ a \in F \ , \ a \neq 0$$

This is obviously not a homogeneous equation, so instead we consider $a\partial^2 y = (\partial a)(\partial y)$. Letting u be a solution of $\partial y = a$, we easily see that $1, u$ form a basis of solutions of our homogeneous equation. Let $\mathcal{F} = F(u)$ be the Picard-Vessiot extension. Let $\phi \in G = \mathrm{Gal}_\partial(\mathcal{F}/F)$. Then $\phi u = \alpha u + \beta$ for some $\alpha, \beta \in C_F$. Since $\partial(\phi u) = \partial(\alpha u + \beta) = \alpha \partial u = \alpha a$ and $\partial(\phi u) = \phi(\partial u) = \phi a = a$ we see that $\alpha a = a$. Hence $\alpha = 1$. Thus G is a subgroup of the additive group $G_a = \{\left(\begin{smallmatrix} 1 & \lambda \\ 0 & 1 \end{smallmatrix}\right) \mid \lambda \in C_F\}$. One easily checks that an algebraic subgroup of G_a is either G_a itself or the trivial group. The latter case corresponds to $u \in F$. Notice in particular, that if $u \notin F$ then G is one-dimensional and so, by Theorem 2.2.4, u is transcendental over F.

(2.3.3) Example.

$$zy'' + \frac{1}{2}y' - \frac{1}{4}y = 0 \quad \text{with } (F, \partial) = (\mathbb{C}(z), \frac{d}{dz})$$

A basis of solutions reads $y_1 = e^{\sqrt{z}}$, $y_2 = e^{-\sqrt{z}}$. Hence the Picard-Vessiot extension is $\mathcal{F} = \mathbb{C}(\sqrt{z}, e^{\sqrt{z}})$. Let $\phi \in \text{Gal}_\partial(\mathcal{F}/F)$. Then $\phi y_1 = \alpha y_1 + \beta y_2$, $\phi y_2 = \gamma y_1 + \delta y_2$ for certain $\alpha, \beta, \gamma, \delta \in \mathbb{C}$ and $\alpha\delta - \beta\gamma \neq 0$. Moreover, $y_1 y_2 = 1$. Hence $\phi(y_1)\phi(y_2) = 1$, which immediately implies $1 = (\alpha y_1 + \beta y_2)(\gamma y_1 + \delta y_2) = \alpha\gamma y_1^2 + (\beta\gamma + \alpha\delta)y_1 y_2 + \beta\delta y_2^2$. Hence $\alpha\gamma = \beta\delta = 0$, $\beta\gamma + \alpha\delta = 1$. An easy computation now shows that

$$G \subset \left\{ \begin{pmatrix} \lambda & 0 \\ 0 & \lambda^{-1} \end{pmatrix}, \begin{pmatrix} 0 & \lambda \\ \lambda^{-1} & 0 \end{pmatrix} \middle| \; \lambda \in \mathbb{C}^\times \right\} \tag{2.3.1}$$

Since $e^{\sqrt{z}}$ is transcendental over $\mathbb{C}(z)$, we have dim G=1. Moreover, via Galois correspondence, the sequence of fields $\mathbb{C}(z) \subset \mathbb{C}(\sqrt{z}) \subset \mathcal{F}$ corresponds to the sequence of algebraic groups $G \supset G_1 \supset \{1\}$, where G_1 has index 2 in G. With all this information it is now a simple exercise to show that the inclusion sign in Eq.(2.3.1) is actually an equality sign. The connected component of the identity is precisely the group G_1, which equals $\{ \begin{pmatrix} \lambda & 0 \\ 0 & \lambda^{-1} \end{pmatrix} \mid \lambda \in \mathbb{C}^\times \}$.

(2.3.4) Example.

$$y'' + \frac{1}{z}y' + y = 0, \quad \text{with } (F, \partial) = (\mathbb{C}(z), \frac{d}{dz})$$

This is the Bessel-equation of order 0. A basis of solutions is formed by $J_0(z)$ and $Y_0(z) = J_0(z)\log z + f(z)$, where $f(z)$ is some power series in z. Both $J_0(z)$ and $f(z)$ have infinite radius of convergence. Let $\mathcal{F}$ be the Picard-Vessiot extension and G its differential Galois group . One easily verifies that $W(J_0, Y_0) = 1/z \in \mathbb{C}(z)$. Hence $G \subset SL(2, \mathbb{C})$. Secondly, G acts irreducibly on the space of solutions. This can be seen as follows. Suppose G acts reducibly, that is, there exists a solution y such that $\phi : y \to \lambda(\phi)y$ for any $\phi \in G$. This means that y'/y is fixed under G, hence $y'/y \in \mathbb{C}(z)$. This is certainly not possible if y contains $\log z$. Hence we can take $y = J_0(z)$, and $J_0'/J_0 \in \mathbb{C}(z)$. Again, this is impossible since J_0 is known to have infinitely many zeros and J_0'/J_0 would have infinitely many poles. So G acts irreducibly. Thirdly, $J_0(z)$ is transcendental over $\mathbb{C}(z)$ and $Y_0(z)$ is transcendental over $\mathbb{C}(z, J_0(z))$ for the very simple reason that $\log z$ is transcendental over the field of Laurent series in z. Hence the transcendence degree of $\mathcal{F}/\mathbb{C}(z)$ is at least two, implying that dim$G \geq 2$. It is a nice exercise to verify that an algebraic group $G \subset SL(2, \mathbb{C})$, acting irreducibly and of dimension ≥ 2 is actually equal to $SL(2, \mathbb{C})$. An alternative for the third argument is the following consideration. Let z describe a closed loop around the origin. After analytic continuation along this loop, we find that $J_0 \to J_0$ and $Y_0 \to Y_0 + 2\pi i J_0$. Hence G contains the element $\tau = \begin{pmatrix} 1 & 2\pi i \\ 0 & 1 \end{pmatrix}$. It is again a nice exercise to show that an algebraic group

$G \subset SL(2, \mathbb{C})$, acting irreducibly and which contains τ, is precisely $SL(2, \mathbb{C})$ itself.

2.4 Applications

(2.4.1) Theorem. *Let $\mathcal{F}/F$ be a Picard-Vessiot extension corresponding to* Eq.(2.1.1). *Let V be its C_F-vector space of solutions and $G = \mathrm{Gal}_\partial(\mathcal{F}/F)$ the differential Galois group . Then the following statements are equivalent:*

i. *There is a non-trivial linear subspace $W \subset V$ which is stable under G.*
ii. *The operator $\mathcal{L}$ factors as $\mathcal{L}_1\mathcal{L}_2$, where $\mathcal{L}_1$ and $\mathcal{L}_2$ are linear differential operators with coefficients in F and order strictly less than n.*

Moreover, $y \in W \Leftrightarrow \mathcal{L}_2 y = 0$.

Proof. ii)$\Rightarrow$i). Let $W \subset V$ be the space of solutions of $\mathcal{L}_2 y = 0$. Since $0 = \phi(\mathcal{L}_2 y) = \mathcal{L}_2(\phi y)$ for any $\phi \in G$, the space W is clearly stable under G. i)$\Rightarrow$ii). Let $v_1, \ldots, v_r$ be a basis of W. Consider the r-th order linear differential equation $\mathcal{L}_2 y = W(v_1, \ldots, v_r, y)/W(v_1, \ldots, v_r)$. Notice that the leading coefficient of $\mathcal{L}_2$ is 1 and the other coefficients are determinants in $v_1, \ldots, v_r$ and their derivatives divided by $W(v_1, \ldots, v_r)$. It is easy to see that the coefficients of $\mathcal{L}_2$ are fixed under G hence, by Theorem 2.2.5(i), they lie in F. Notice that $\mathcal{L}_2 v_i = 0$ for $i = 1, 2, \ldots, r$. We also have $\mathcal{L} v_i = 0$ for $i = 1, 2, \ldots, r$. By division with remainder of differential operators we find $\mathcal{L}_1$ and $\mathcal{L}_3$ such that $\mathcal{L} = \mathcal{L}_1\mathcal{L}_2 + \mathcal{L}_3$, where $\mathcal{L}_3$ has order less than r, However, we have automatically $\mathcal{L}_3 v_i = 0$ for $i = 1, 2, \ldots, r$. Since $v_1, \ldots, v_r$ are linearly independent, this implies $\mathcal{L}_3 = 0$. □

We shall call Eq.(2.1.1) *irreducible* over F if $\mathcal{L}$ does not factor over F. So Theorem 2.4.1 implies that Eq.(2.1.1) is irreducible if and only its differential Galois group acts irreducibly on the space of solutions.

(2.4.2) Corollary. *Let notations be as in the previous theorem. Then G^0, the component of the identity in G, acts irreducibly on V if and only if $\mathcal{L}$ does not factor over any finite extension of F.*

From ordinary Galois theory we know that the zeros of a polynomial P can be determined by repeatedly taking roots if and only if the Galois group of the splitting field of P is solvable. One of the nice applications of differential Galois theory is an analogue of this theorem for differential equations. A differential extension L of F is called a *Liouville extension* if there exists a chain of extensions $F = F_0 \subset F_1 \subset F_2 \subset \cdots \subset F_r = L$ such that $F_{i+1} = F_i < u_i >$ $(i = 0, \ldots, r-1)$, where u_i is a solution of an equation of the form $\partial y = a_i y$ or $\partial y = a_i$, $a_i \in F_i$. In other words, Liouville extensions arise by repeatedly solving first order differential equations (or, as some people say, by quadratures). From the next theorem it follows for example, that the Bessel equation of order zero (see Example 2.3.4) cannot be solved by quadratures.

(2.4.3) Theorem. *Let $\mathcal{L}y = 0$ be a linear differential equation over F and $\mathcal{F}/F$ its Picard-Vessiot extension. Then $\mathcal{F}$ is a Liouville extension if and only if* $\mathrm{Gal}_\partial(\mathcal{F}/F)$ *is solvable in the sense of algebraic groups.*

Proof. Suppose L/F is a Liouville extension. Then, via the differential Galois correspondence, the chain $F = F_0 \subset F_1 \subset \cdots \subset F_r = L$ corresponds to the chain of algebraic subgroups $\mathrm{Gal}_\partial(\mathcal{F}/F) = G_0 \supset G_1 \supset \cdots \supset G_n = \{\mathrm{id}\}$, and we have that G_{i+1} is normal in G_i for every $i, 0 \leq i < r$ because F_{i+1}/F_i is a Picard-Vessiot extension. According to Examples 2.3.1 and 2.3.2 the differential Galois group of F_{i+1}/F_i is Abelian, hence G_i/G_{i+1} is Abelian. Thus we see that $\mathrm{Gal}_\partial(\mathcal{F}/F)$ is solvable. Now suppose that $G = \mathrm{Gal}_\partial(\mathcal{F}/F)$ is solvable. Then, by the Lie-Kolchin theorem, we can find a basis of solutions $u_1, \ldots, u_n$ of $\mathcal{L}y = 0$ such that G acts on $u_1, \ldots, u_n$ by upper triangular matrices. In other words, to every $g \in G$ there exist $g_{ij} \in C_F$ such that $g : u_i \rightarrow g_{ii}u_i + g_{i,i+1}u_{i+1} + \cdots + g_{in}u_n$. In particular, G acts by multiplication with elements from C_F on u_n. Hence, $\partial u_n/u_n$ is fixed under G and thus, $\partial u_n = au_n$ for some $a \in F$. Put $v_i = \partial(u_i/u_n) \quad (i = 1, 2, \ldots, n-1)$. Note that we can recover the u_i from the v_i by integration and multiplication by u_n. Moreover, the group G acts on $v_1, \ldots, v_{n-1}$ again by upper triangular matrices. Hence we can repeat the argument and find that $\partial v_{n-1} = bv_{n-1}$ for some $b \in F$ and G acts by upper triangular matrices on $\partial(v_i/v_{n-1}) \quad (i = 1, \ldots, n-2)$. Repeating this argument n times we find the proof of our assertion. □

In what follows we shall often be interested in linear differential equations modulo some equivalence.

(2.4.4) Definition. *Two differential equations $\mathcal{L}_1y = 0$ and $\mathcal{L}_2y = 0$ of order n are called* equivalent *over F if there exists a linear differential operator $\mathcal{L}$ such that $\mathcal{L}u_1, \ldots, \mathcal{L}u_n$ is a basis of solutions of $\mathcal{L}_2y = 0$ whenever $u_1, \ldots, u_n$ is a basis of solutions of $\mathcal{L}_1y = 0$.*

(2.4.5) *Remark.* It is not hard to show that if such an $\mathcal{L}$ exists, then there exists an inverse differential operator $\mathcal{L}'$ which maps bases of solutions of $\mathcal{L}_2y = 0$ into bases of solutions of $\mathcal{L}_1y = 0$. Namely, if we carry out a 'left' version of the Euclidean algorithm to $\mathcal{L}$ and $\mathcal{L}_1$, we can find linear differential operators $\mathcal{L}'$, $\mathcal{L}_3$ over F such that $\mathcal{L}'\mathcal{L} + \mathcal{L}_3\mathcal{L}_1 = 1$. Notice, that for any solution y of $\mathcal{L}_1y = 0$ we have $\mathcal{L}'\mathcal{L}y = y$. Hence $\mathcal{L}'$ is the desired inverse.

2.5 Monodromy and local Galois groups

In this part we shall make some comments on the case of differential equations over $\mathbb{C}(z)$, which is the most common one in mathematics and physics. For a proper understanding of this Subsection some knowledge of the concepts of monodromy, (ir)regular singularity, local exponents is desirable. Since this is not the place to introduce them, we refer to the introductory books (Poole 1936) (Ince 1926) (Hille 1976). Let $\mathcal{L}y = 0$ be a linear differential equation over $\mathbb{C}(z)$. Let $y_1, \ldots, y_n$ be a basis of solutions around a regular point z_0. They are given by converging power series in $z - z_0$ and can be continued analytically along any path avoiding the singularities of the differential equation. After analytic continuation along a closed loop Γ beginning and ending in z_0, the functions $y_1, \ldots, y_n$ undergo a linear substitution, called the *monodromy substitution* corresponding to Γ. It is obvious that such a monodromy substitution is an element of the differential Galois group of the differential equation. Actually, for Fuchsian equations (having only regular singularities) more is true.

(2.5.1) Theorem. *Let $\mathcal{L}y = 0$ be a Fuchsian equation of order n over $\mathbb{C}(z)$ and $\mathcal{M}$ the group generated by the monodromy substitutions acting on the space of solutions. Let G be the differential Galois group of $\mathcal{L}y = 0$ and let $\overline{\mathcal{M}}$ be the Zariski-closure of $\mathcal{M}$ in $GL(n, \mathbb{C})$. Then $G = \overline{\mathcal{M}}$.*

Proof. First note that $\mathcal{M} \subset G$ and hence $\overline{\mathcal{M}} \subset G$. Thus it suffices to show that the field which is fixed by $\mathcal{M}$ is precisely $\mathbb{C}(z)$. However, this follows already from Riemann. Any function $f(z)$ of z, having trivial monodromy and such that at any $z_0 \in \mathbb{C} \cup \{\infty\}$ there exists $n \in \mathbb{N}$ such that $(z - z_0)^n f(z)$ is bounded near z_0 (denote $z - z_0 = 1/z$ if $z_0 = \infty$) is necessarily rational. The boundedness follows from the fact that we have only regular singularities. □

From the fact that $\mathcal{M} \subset G$ one often obtains elements of G which determine the possibilities for G to a large extent. Consider Example 2.3.4 of the Bessel equation. There we had found the monodromy element τ around $z = 0$. Together with the fact that G acts irreducibly in this case, this already yielded $SL(2, \mathbb{C}) \subset G$. An elegant way to study a differential equation over $\mathbb{C}(z)$ locally at a point z_0 is to consider it as a differential equation over $\mathbb{C}((z - z_0))$. Here, $\mathbb{C}((z - z_0))$ denotes the field of (not necessarily converging) Laurent expansions in $z - z_0$, and we replace $z - z_0$ by $1/z$ if $z_0 = \infty$. Let $\mathcal{F}$ be the Picard-Vessiot extension of $\mathcal{L}y = 0$ considered as a linear differential equation over $\mathbb{C}(z)$, and let $\mathcal{F}'$ be the smallest differential field containing both $\mathbb{C}((z - z_0))$ and $\mathcal{F}$. It is not hard to show that $\mathrm{Gal}_\partial(\mathcal{F}'/\mathbb{C}((z - z_0))) \subset \mathrm{Gal}_\partial(\mathcal{F}/\mathbb{C}(z))$. We note that $\mathcal{F}'$ can only be strictly larger than $\mathbb{C}((z - z_0))$ if z_0 is a singularity of $\mathcal{L}y = 0$. So it makes sense to look only locally at singular points. The nice thing is, that for linear differential equations over $\mathbb{C}((z - z_0))$ there is a complete classification theory, developed by many people of whom we mention Fuchs, Frobenius, (Turrittin 1955), (Levelt 1975) and very recently (Babbitt and Varadarajan 1989). In the following theorem we restrict ourselves to the case $z_0 = 0$, the general case being entirely similar.

(2.5.2) Theorem. *Consider an n-th order linear differential equation with coefficients in* $\mathbb{C}((z))$. *Then there exist* $d \in \mathbb{N}$, *a diagonal matrix* $P(X)$ *with entries in* $\mathbb{C}[X]$, *a constant matrix* A *in Jordan normal form and a vector* $(f_1, \ldots, f_n)$ *with entries in* $\mathbb{C}((z))$ *such that*

$$\begin{pmatrix} y_1 \\ \vdots \\ y_n \end{pmatrix} = e^{P(z^{1/d})} z^A \begin{pmatrix} f_1 \\ \vdots \\ f_n \end{pmatrix}$$

yields a basis of solutions $y_1, \ldots, y_n$ *of our differential equation.*

(2.5.3) *Remark.* The notation z^A stands for the matrix obtained by expanding $z^A = e^{A \log z}$ via the Taylor series for e^x. For example, if $A = \mathrm{diag}(a_1, \ldots, a_n)$ then $z^A = (z^{a_1}, \ldots, z^{a_n})$ and if $A = \left(\begin{smallmatrix} 0 & 1 \\ 0 & 0 \end{smallmatrix}\right)$ then $z^A = \left(\begin{smallmatrix} 1 & \log z \\ 0 & 1 \end{smallmatrix}\right)$.

(2.5.4) *Remark.* If $A = \mathrm{diag}(a_1, \ldots, a_n)$ then the theorem states that there exists a basis of solutions of the form $\exp(P_i(z^{1/d}))z^{a_i} f_i(z)$, $f_i(z) \in \mathbb{C}((z))$ $(i = 1, \ldots, n)$. Moreover, if $z = 0$ is a regular singularity, then all entries $P_i(X)$ of $P(X)$ are identically zero.

(2.5.5) *Remark.* Notice that the Picard-Vessiot extension $\mathcal{F}'$ of our differential equation can be obtained by adjoining to $\mathbb{C}((z))$ the elements $z^{1/d}, \exp(P_i(z^{1/d}))$ and the entries of z^A. A simple consideration shows that $\mathrm{diag}(e^{t_1}, \ldots, e^{t_n})$ is contained in $\mathrm{Gal}_\partial(\mathcal{F}'/\mathbb{C}((z)))$ for any n-tuple $t_1, \ldots, t_n \in \mathbb{C}$ satisfying $\sum_{i=1}^n a_i t_i = 0$ whenever $\sum_{n=1}^n a_i P_i(z^{1/d}) = 0$ with $a_1, \ldots, a_n \in \mathbb{Z}$. This element acts on the solutions y_i via $y_i \to e^{t_i} y_i$ $(i = 1, \ldots, n)$.

As an example take the Bessel equation again (Example 2.3.4). At $z = 0$, a regular singularity, we have two independent solutions, namely the Bessel function $J_0(z)$ and the function $Y_0(z) = \log z J_0(z) + f(z)$, where $f(z)$ is a power series in z with infinite radius of convergence (see (Erdélyi et al. 1953)). Clearly, $\mathrm{Gal}_\partial(\mathcal{F}'/\mathbb{C}((z))) = \{\left(\begin{smallmatrix} 1 & \lambda \\ 0 & 1 \end{smallmatrix}\right) \mid \lambda \in \mathbb{C}\}$. At $z = \infty$, an irregular singularity, we have the solutions $\exp(iz)f(1/z)$ and $\exp(-iz)f(-1/z)$, where $f(t)$ is an asymptotic expansion in t (see (Erdélyi et al. 1953). Clearly, $\mathrm{Gal}_\partial(\mathcal{F}'/\mathbb{C}((z))) = \{\left(\begin{smallmatrix} \lambda & 0 \\ 0 & \lambda^{-1} \end{smallmatrix}\right) \mid \lambda \in \mathbb{C}^\times\}$. Thus we find that $\mathrm{Gal}_\partial(\mathcal{F}/\mathbb{C}(z))$ contains a unipotent subgroup and a semisimple subgroup. It must be noted that the irreducibility of the action of $\mathrm{Gal}_\partial(\mathcal{F}/\mathbb{C}(z))$ cannot be proved by such local considerations. This is a global property which must be decided in other ways (see Example 2.3.4 and also Section 4). It turns out that in many cases the existence of a special element, found by local considerations, and the fact that $\mathrm{Gal}_\partial(\mathcal{F}/\mathbb{C}(z))^0$ acts irreducibly, found by other means, largely determine the group $\mathrm{Gal}_\partial(\mathcal{F}/\mathbb{C}(z))$. This is the basic principle used in Section 4, where we determine Galois groups of hypergeometric differential equations. In Section 3 the required tools from linear algebraic groups are provided. Finally we mention that instead of considering linear differential equations over $\mathbb{C}((z-z_0))$ we can consider them over the smaller field $\mathbb{C}_{an}((z-z_0))$, the field of locally converging Laurent series in $z-z_0$. In the neighbourhood of regular singularities this does not yield anything new. If z_0 is an irregular singularity however, it makes a big difference. Considerations over $\mathbb{C}_{an}((z-z_0))$ give a finer classification and by studying the so-called *Stokes phenomenon* which then arises and one can sometimes obtain more information on the differential Galois group (see (Martinet and Ramis 1989) or (Duval and Mitschi 1989)). Moreover, Stokes's matrices form an important ingredient in understanding the local to global behaviour of differential Galois groups of equations having irregular singularities.

3. Linear Algebraic Groups

3.1 Definitions and examples

In order to understand and be able to work with differential Galois groups, some knowledge of linear algebraic groups is indispensable. In fact, it is the strong classification theory of linear algebraic groups and the familiarity of their representations which lends its power to the study of linear differential equations. In this Section we collect some definitions and theorems on linear algebraic groups which will be useful in the explicit determination of differential Galois groups. Readers who are interested in a systematic account of linear algebraic groups and their representations should consult (Humphreys 1972 and 1975) or (Springer 1981). In this Section we let k be an algebraically closed field of characteristic zero, for example, $k=\mathbb{C}$.

(3.1.1) Definition. *A* linear algebraic group *over k is a subgroup $G \subset GL(n,k)$ with the property that there exist polynomials $p_r \in k[X_{11},\ldots,X_{nn}]$ $(r = 1,\ldots,m)$ in the n^2 variables X_{ij} $(i,j=1,\ldots,n)$ such that $G = \{(g_{ij}) \in GL(n,k) \mid p_r(g_{11},\ldots,g_{nn}) = 0$ for $r=1,\ldots,m\}$.*

Examples:

i. The special linear group $SL(n,k) = \{g \in GL(n,k) \mid \det g = 1\}$.
ii. The orthogonal group $O(n,k) = \{g \in GL(n,k) \mid {}^tgg = Id\}$ where ${}^tg$ denotes the transpose of g.
iii. The special orthogonal group $SO(n,k) = O(n,k) \cap SL(n,k)$.
iv. The symplectic group $Sp(2n,k) = \{g \in SL(2n,k) \mid {}^tgJg = J\}$ in even dimension, where J is any non degenerate anti-symmetric $2n \times 2n$ matrix.
v. The exceptional group $G_2 = \{g \in SL(7,k) \mid F(gx,gy,gz) = F(x,y,z)\}$ where F is a sufficiently general antisymmetric trilinear form.
vi. The group of upper triangular matrices $U = \{g \in GL(n,k) \mid g_{ij} = 0$ for all $i<j\}$.

vii. The unipotent group $U_1 = \{g \in U(n,k) \mid g_{ii} = 1 \text{ for all } i\}$. In fact, any algebraic subgroup of U_1 is called unipotent.

Notice that the groups mentioned above are actually Lie groups over k. This follows very quickly from the definition of linear algebraic groups. Of course, over $\mathbb{R}$ for example, there exist other Lie groups such as $SU(n,\mathbb{C})$. It should be emphasized here that this Lie group cannot be realized as the set of complex solutions of polynomial equations, since the definition of $SU(n,\mathbb{C})$ involves complex conjugation, which is not an algebraic operation over $\mathbb{C}$. It follows from the definition that a linear algebraic group is an algebraic variety. It may or may not be irreducible. To avoid confusion with 'irreducible' in the sense of representation theory we shall reserve, by slight abuse of terminology, the word '*connected*' for 'irreducible' in the sense of algebraic geometry. The connected component of our algebraic group which contains the identity element will be called the *component of the identity* and is denoted by G^0. The group G^0 is a normal algebraic subgroup of G of finite index and the connected components of G are precisely the cosets with respect to G^0. Moreover, any algebraic subgroup of G of finite index automatically contains G^0. An example is $SO(n,k)$ which is the component of the identity in $O(n,k)$. The coset decomposition consists of the determinant 1 and -1 matrices. Another example occurs in Example 2.3.3.

(3.1.2) Definition. *An m-dimensional* rational representation *of a linear algebraic group $G \subset GL(n,k)$ is a homomorphism $\rho : G \to GL(m,k)$ with the property that there exist polynomials $\rho_{ij} \in k[X_{11},\dots,X_{nn}]$ $(i,j = 1,\dots,m)$ and an $r \in \mathbb{Z}$ such that $(\rho(g))_{ij} = \det(g)^r \rho_{ij}(g_{11},\dots,g_{nn})$. In other words, up to a possible common factor of the form* $\det(g)^r$ $(r \in \mathbb{Z})$, *the homomorphism ρ is given by polynomials.*

Notice that if $G \subset SL(n,k)$, any rational representation of G is given by polynomials, since $\det(g) = 1$ for all $g \in G$. We shall assume that the reader is familiar with elementary concepts such as irreducible representations, equivalence of representations, invariant subspace, etc. Let $\rho : G \to GL(m,k)$ be a rational representation, which may even be the standard inclusion $G \subset GL(n,k)$. Then we have the dual representation $\rho^d : g \to {}^t\rho(g^{-1})$, again in dimension m. Let $\rho' : G \to GL(m',k)$ be another rational representation. Then the direct sum representation $\rho \oplus \rho'$ is the $m + m'$ dimensional representation obtained by simply writing

$$(\rho \oplus \rho')(g) = \begin{pmatrix} \rho(g) & \\ & \rho'(g) \end{pmatrix}.$$

The mm' dimensional tensor representation $\rho \otimes \rho'$ is obtained as follows. Replace each entry $\rho(g)_{ij}$ $(i,j = 1,\dots,m)$ in $\rho(g)$ by the $m' \times m'$ matrix $\rho(g)_{ij} \times \rho'(g)$. This yields an $mm' \times mm'$ matrix which we call $(\rho \otimes \rho')(g)$. One easily checks that this yields a rational representation. By repeating these constructions and taking subrepresentations of them, we obtain a wealth of new representations. For example, the r-th symmetric product and the r-th exterior power of a representation ρ are subrepresentations of $\rho \otimes \rho \otimes \cdots \otimes \rho$ (r times). The adjoint representation of G is a subrepresentation of $i \otimes i^d$, where $i : G \to GL(n,k)$ is a faithful representation. In this paper our main interest will be in reductive groups. Usually they are defined as algebraic groups whose unipotent radical is trivial. Here we prefer to use a definition which is more practical for our purposes.

(3.1.3) Definition. *Let $G \subset GL(n,k)$ be an algebraic group and $\rho : G \to GL(m,k)$ a faithful rational representation. We call G* reductive *if ρ is completely reducible, i.e. $k^m = V_1 \oplus \cdots \oplus V_r$, where the V_i are irreducible $\rho(G)$-invariant subspaces.*

That this definition does not depend on ρ is shown by the following theorem.

(3.1.4) Theorem. *Any rational representation of a reductive group is completely reducible.*

A proof that our definition is equivalent to the usual definition, can be found in (Beukers, Brownawell and Heckman 1988, Appendix). It relies heavily on the Lie-Kolchin theorem. It is clear that Theorem 3.1.4 greatly simplifies the study of representations of reductive groups. A reductive group is called *semi-simple* if its centre is finite. In particular, a reductive group $G \subset SL(n,k)$ which acts irreducibly on k^n is automatically semi-simple since, by Schur's lemma, elements of the centre of G are scalar, of which there exist only finitely many in $SL(n,k)$.

3.2 Theorems

The following statements give us very easy criteria to recognize algebraic groups from the occurrence of certain typical elements. They are taken from (Katz 1990), where the Lie algebra versions are given. In all theorems of this Subsection we let $G \subset GL(V)$ be a reductive, connected algebraic group acting irreducibly on the finite dimensional vector space V. The notation $C \cdot G$ stands for the group obtained by taking all products of elements of some suitable scalar group C with elements from G. The notation $\operatorname{diag}(d_1, \ldots, d_n)$ will stand for the $n \times n$ matrix having the elements $d_1, \ldots, d_n$ on the diagonal and zeros at all off-diagonal places.

(3.2.1) Theorem. (O. Gabber) *Let $D \subset GL(V)$ be a group of diagonal matrices such that $dGd^{-1} = G$ for all $d \in D$. Consider the diagonal group T consisting of all* $\operatorname{diag}(t_1, \ldots, t_n) \in GL(V)$ *such that $t_i t_j = t_k t_l$ whenever $d_i d_j = d_k d_l$ for all* $\operatorname{diag}(d_1, \ldots, d_n) \in D$. *Then $T^0 \subset C \cdot G$.*

The proof of this theorem is based on the fact that derivations of semi-simple Lie algebras are inner. The following three theorems rely heavily on the classification theory of semi-simple Lie algebras. Theorem 3.2.2 is a collection of results found by Gabber, Kostant, Zarhin, Kazhdan-Margulis, Beukers and Heckman and undoubtedly many other people, while Theorems 3.2.3 and 3.2.4 were found by O. Gabber with the purpose of determining differential Galois groups.

(3.2.2) Theorem. *Let $D \subset GL(V)$ be such that $dGd^{-1} = G$, $\forall d \in D$. Then*

i. *if D contains $\mathrm{diag}(\lambda, 1, \ldots, 1)$ with $\lambda \neq \pm 1$, then $G = C \cdot SL(V)$ (i.e. $G = SL(V)$ or $G = GL(V)$).*
ii. *if D contains $\mathrm{diag}(-1, 1, \ldots, 1)$ then $G = C \cdot SL(V)$ or $G = C \cdot SO(V)$.*
iii. *if D contains an element with 1 on all diagonal places and 0 at all other places with precisely one exception, then $G = C \cdot SL(V)$ or $G = C \cdot Sp(V)$.*
iv. *if $D = \{\mathrm{diag}(\lambda, \lambda^{-1}, 1, \ldots, 1) \mid \lambda \neq 0\}$ then $G = C \cdot SL(V)$ or $G = C \cdot Sp(V)$ or $G = C \cdot SO(V)$.*

(3.2.3) Theorem. (O. Gabber) *Let $D \subset GL(V)$ be such that $dGd^{-1} = G$, $\forall d \in D$. Then*

i. *if $D = \{\mathrm{diag}(\lambda, \lambda, \lambda^{-1}, \lambda^{-1}, 1, \ldots, 1) \mid \lambda \neq 0\}$, then we have the following possibilities:*

$G = C \cdot SL(V)$ or $C \cdot SO(V)$ or $C \cdot Sp(V)$.
$G = C \cdot (SL(2) \times (SL(k)$ or $SO(k)$ or $Sp(k)))$ with $k = \dim V - 2$ and in standard tensor representation.
$G = C \cdot G_2$ in 7-dimensional standard representation.
$G = C \cdot SO(7)$ in 8-dimensional spin representation.
$G = C \cdot SL(3)$ in 8-dimensional adjoint representation.
$G = C \cdot (SL(3) \times SL(3))$ in 9-dimensional standard tensor representation.

ii. *if $D = \{\mathrm{diag}(\lambda, \mu, \lambda\mu, \lambda^{-1}, \mu^{-1}, (\lambda\mu)^{-1}, 1, \ldots, 1)\}$, then we have the same possibilities as in the previous case, except that in the second possibility only $k = 4$ is allowed.*

(3.2.4) Theorem. (O. Gabber) *If $\dim V$ is a prime p, then G has the following possibilities, $C \cdot SL(V)$, $C \cdot SO(V)$, $C \cdot G_2$ (only when $p = 7$), $C \cdot \mathrm{symm}^{p-1}(SL(2, k))$, where symm^{p-1} stands for the $(p-1)$-the symmetric power of the standard representation of $SL(n, k)$.*

Notice that all the above criteria require that G be a connected group acting irreducibly on V. Without either of these requirements the theorems are false. When determining the differential Galois group of a given linear differential equation in practice, it is usually not too hard to decide that the differential Galois group acts irreducibly (see Theorem 2.4.1). However, connectedness is harder to check, and very often G is not even connected. So one usually studies G^0, the component of the identity. Now the problem is to check that G^0 acts irreducibly on V. In some cases, like the hypergeometric equation (see Section 4) this is doable, but in other cases it may be a very tedious job. The following theorem partly avoids these problems at the cost of determination of irreducibility of higher order differential equations.

(3.2.5) Theorem. *Let $G \subset GL(V)$ be a linear algebraic group such that G modulo its scalars is an infinite group. Then the following statements are equivalent,*

i. *G acts irreducibly on the symmetric square S^2V of V.*
ii. *$G = C \cdot SL(V)$ or $G = C \cdot Sp(V)$.*

A proof of this theorem can be found in (Beukers, Brownawell and Heckman 1988, Appendix).

4. Hypergeometric Differential Equations

4.1 Introductory remarks

Let $p, q \in \mathbb{N}$, $q \geq p$ and $\alpha_1, \ldots, \alpha_p; \beta_1, \ldots, \beta_q \in \mathbb{C}$. For the moment, take $\beta_q = 1$ and $\beta_i \notin \mathbb{Z}_{\leq 0}$ $(i = 1, \ldots, q)$. The generalized hypergeometric function in one variable z is defined by

$$(4.1.1) \qquad {}_pF_{q-1}\left(\begin{matrix}\alpha_1 & \cdots & \alpha_p \\ \beta_1 & \cdots & \beta_{q-1}\end{matrix}\,\middle|\, z\right) = \sum_{n=0}^{\infty} \frac{(\alpha_1)_n \cdots (\alpha_p)_n}{(\beta_1)_n \cdots (\beta_q)_n} z^n$$

where $(\alpha)_n = \alpha(\alpha+1)\cdots(\alpha+n-1)$ is the *Pochhammer symbol.* The adjective 'generalized' refers to the fact that it is a generalization of the classical case $q = 2$, which was already studied by Euler, Gauss and Riemann. See (Erdélyi et al. 1953), (Klein 1933) or (Gray 1986). Note, that if $\alpha_i \in \mathbb{Z}_{\leq 0}$ for some i, then ${}_pF_{q-1}$ is a polynomial. When $q > p$ the radius of convergence of (4.1.1) is infinite. When $p = q$ and ${}_pF_{q-1}$ is not a polynomial, the radius of convergence of (4.1.1) is one. When specializing α_i, β_j we obtain a large number of familiar functions which occur throughout mathematics and physics, particularly in the case when $q = 2$, $p \leq q$. We give some examples:

i.

$${}_0F_1\left(\begin{matrix}\cdot \\ 1\end{matrix}\,\middle|\, z^2\right) = J_0(2iz),$$

where J_0 is the Bessel function of order zero.

ii.

$${}_2F_1\left(\begin{matrix}1 & 1 \\ & 2\end{matrix}\,\middle|\, z\right) = -\frac{\log(1-z)}{z}.$$

iii.

$$ {}_2F_1\left(\begin{matrix}1 & \alpha \\ & 1\end{matrix}\middle| z\right) = (1-z)^{-\alpha} . $$

iv.

$$ {}_2F_1\left(\begin{matrix}-n & n+1 \\ & 1\end{matrix}\middle| \frac{1-x}{2}\right) = P_n(x), $$

where P_n is the Legendre polynomial of degree n.

v.

$$ {}_1F_1\left(\begin{matrix}-m \\ 1/2\end{matrix}\middle| x^2\right) = (-1)^m \frac{m!}{(2m)!} H_{2m}(x), $$

where $H_{2m}(x)$ is the Hermite polynomial of degree $2m$.

Let now $\alpha_i, \beta_j \quad (i = 1, \ldots, p\ ; j = 1, \ldots, q)$ be arbitrary complex numbers. Throughout this Section we write $\partial = z(d/dz)$. Consider the following differential equation of order q,

$$ \prod_{j=1}^{q} (\partial + \beta_j - 1)F = z \prod_{i=1}^{p} (\partial + \alpha_i)F \tag{4.1.2} $$

which we call the *hypergeometric differential equation* . Let $f(z)$ be any solution of the form

$$ z^\lambda \sum_{n=0}^{\infty} a_n z^n, \qquad a_0 = 1 \tag{4.1.3} $$

Substitution of (4.1.3) into (4.1.2) yields $\lambda = 1 - \beta_i$ for some i and the recursion relation

$$ a_{n+1} = \frac{\prod_{i=1}^{p}(n+\lambda+\alpha_i)}{\prod_{j=1}^{q}(n+\lambda+\beta_j)} a_n $$

We thus see that solution (4.1.3) is in fact

$$ z^{1-\beta_i}{}_pF_{q-1}\left(\begin{matrix}1+\alpha_1-\beta_i & \cdots & 1+\alpha_p-\beta_i \\ 1+\beta_1-\beta_i & .\vee. & 1+\beta_q-\beta_i\end{matrix}\middle| z\right) \tag{4.1.4} $$

where $^\vee$ denotes suppression of the term $1+\beta_i-\beta_i$. In particular, when $\beta_q = 1$ we see that

$$ {}_pF_{q-1}\left(\begin{matrix}\alpha_1 & \cdots & \alpha_p \\ \beta_1 & \cdots & \beta_{q-1}\end{matrix} \mid z\right) $$

is a solution. Whenever $\beta_j - \beta_i \in \mathbb{Z}$ for some i, j, the function (4.1.4) may not be well defined since the coefficients may become infinite. In that case the functions (4.1.4) do not give a basis of solutions of (4.1.2) and it turns out, that solutions containing $\log z$ also show up. If, on the other hand, $\beta_i - \beta_j \notin \mathbb{Z}$ for all i, j, then the functions (4.1.4) do form a basis of solutions of (4.1.2). There are some crucial differences between the cases $p = q$ and $p < q$. As remarked already, the radius of convergence of the hypergeometric function with $p = q$ is generally 1, and with $p < q$ it is infinite. A reason for this shows up if we write out (4.1.2) explicitly. In the case $p = q$ the equation has three singularities $0, 1, \infty$ and it is a Fuchsian equation. A solution around $z = 0$ usually does not converge beyond 1 and so its radius of convergence is 1. In case $p < q$ there are only the singularities $0, \infty$ and ∞ is an irregular singularity. The monodromy in this case is not very interesting since it is the image of the fundamental group $\pi_1(\mathbb{C} \setminus \{0\}) \cong \mathbb{Z}$, and thus cyclic. The differential Galois group will carry much more information in this case. When $p = q$ however, our equation is Fuchsian and according to Theorem 2.5.1 the monodromy group determines the differential Galois group . We also have the following theorem.

(4.1.1) Theorem. (Pochhammer) *Suppose $p = q$. Then there exist $p - 1$ linearly independent holomorphic solutions of* Eq.(4.1.2) *in a neighbourhood of $z = 1$.*

A proof can be found in (Beukers and Heckman 1989). In fact, Fuchsian hypergeometric equations are characterized by this property.

(4.1.2) Theorem. *Suppose we are given a Fuchsian differential equation over $\mathbb{C}(z)$ of order n which has the singularities $0, 1, \infty$ and no others. If this equation admits $n - 1$ linearly independent holomorphic solutions in a neighbourhood of $z = 1$ then it is equivalent over $\mathbb{C}(z)$ to an equation of the form Eq.(4.1.2) with $p = q = n$.*

A consequence of all the above is, that in the case $p = q$ the local monodromy matrix aroud $z = 1$ has $n - 1$ eigenvalues 1 with corresponding eigenvectors. Such a matrix is called a *pseudoreflection* (if the n-th eigenvalue would be -1 we would have had a true reflection). Its Jordan normal form is either $\mathrm{diag}(\lambda, 1, \ldots, 1)$ with $\lambda \neq 1$ or a matrix with 1 on the diagonal places, a non zero element at the place $1, 2$ and zeros everywhere else. The fact that the local monodromy around $z = 1$ is generated by a pseudoreflection is crucial in the determination of the differential Galois group in case $p = q$. The rest of this Section will be devoted to a sketch of a systematic treatment of the differential Galois group of a hypergeometric differential equation .

4.2 Reducibility and imprimitivity

Recall (Subsection 2.4) that a differential equation is called irreducible over $\mathbb{C}(z)$ if its corresponding differential operator does not factor over $\mathbb{C}(z)$. If $\mathcal{F}/\mathbb{C}(z)$ is the corresponding Picard-Vessiot extension, then it follows from Theorem 2.4.1 that a differential equation is irreducible over $\mathbb{C}(z)$ if and only if its differential Galois group $\mathrm{Gal}_\partial(\mathcal{F}/\mathbb{C}(z))$ acts irreducibly on the space of solutions.

(4.2.1) Lemma. Eq.(4.1.2) *is irreducible over $\mathbb{C}(z)$ if and only if $\alpha_i \neq \beta_j \pmod{\mathbb{Z}}$ for all i, j.*

Proof. Here we shall prove that $\alpha_i \not\equiv \beta_j \pmod{\mathbb{Z}}$ for all i,j implies irreducibility. An elementary proof of the converse can be found in (Katz 1990) or (Beukers and Heckman 1989). Suppose Eq.(4.1.2) has a non-trivial factorization $\mathcal{L}_1\mathcal{L}_2 y = 0$. Usual theory shows that $\mathcal{L}_2 y = 0$ has a solution of the form $z^\lambda \sum_{k=0}^{\infty} a_k z^k$. It is automatically a solution of Eq.(4.1.2) and the arguments from Subsection 4.1 show that it is of the form (4.1.4). This implies the existence of a hypergeometric function with parameters, say, $\mu_1, \ldots, \mu_p; \nu_1, \ldots, \nu_q$ $\nu_q = 1$ and $\mu_i - \nu_j \neq \mathbb{Z}$ $\forall i,j$ which satisfies a linear differential equation of order $< q$. Writing a_n for the coefficients of the power series of this function, we deduce such a differential equation from the existence of a non trivial recurrence for the a_n of the form
(4.2.1)

$$A_k(n)a_{n+k} + A_{k-1}(n)a_{n+k-1} + \cdots + A_1(n)a_{n+1} + A_0(n)a_n = 0, \quad \forall n \geq 0,$$

where the $A_i(n)$ are polynomials in n of degree $< q$ and $a_n \neq 0$ $\forall n \geq 0$. Since $a_n = (\mu_1)_n \cdots (\mu_p)_n / (\nu_1)_n \cdots (\nu_q)_n$ for all n, the quotients a_{n+i}/a_n are rational functions in n and Eq.(4.2.1) is equivalent to

$$A_k(x)\frac{\prod_{i=1}^{p}(\mu_i + x)\cdots(\mu_i + x + k - 1)}{\prod_{j=1}^{q}(\nu_j + x)\cdots(\nu_j + x + k - 1)} + \cdots + A_0(x) = 0$$

Here, x is some arbitrary variable replacing n. Since $\mu_i - \nu_j \notin \mathbb{Z}$ for all i,j and $\deg A_k < q$, the left most term has for some j a pole of the form $x = 1 - k - \nu_j$ which none of the other terms possesses. Hence relation (4.2.1) cannot exist, which implies our assertion. □

For the following lemma, recall Definition 2.4.4 on the equivalence of linear differential equations .

(4.2.2) Lemma. *Let $\mathcal{H}$ and $\mathcal{H}'$ be two irreducible (over $\mathbb{C}(z)$) hypergeometric equations with parameters $\alpha_1, \ldots, \alpha_p; \beta_1, \ldots, \beta_q$ and $\alpha'_1, \ldots, \alpha'_{p'}; \beta'_1, \ldots, \beta'_{q'}$ respectively. Then $\mathcal{H}$ and $\mathcal{H}'$ are equivalent over $\mathbb{C}(z)$ if and only if $p = p'$, $q = q'$ and, after renumbering if necessary, $\alpha_i \equiv \alpha'_i \pmod{\mathbb{Z}}$ $\beta_j \equiv \beta'_j \pmod{\mathbb{Z}}$ for $i = 1, \ldots, p; j = 1, \ldots, q$.*

Proof. We shall only show that $\alpha_i \equiv \alpha'_i \pmod{\mathbb{Z}}$, $\beta_j \equiv \beta'_j \pmod{\mathbb{Z}}$ for all i,j implies the equivalence. Let $V(\alpha_1, \ldots, \alpha_p; \beta_1, \ldots, \beta_q)$ be the solution space of $\mathcal{H}$. Notice that $(\partial + \alpha_k)f \in V(\alpha_1, \ldots, \alpha_k + 1, \ldots, \alpha_p; \beta_1, \ldots, \beta_q)$ for every $f \in V(\alpha_1, \ldots, \alpha_p; \beta_1, \ldots, \beta_q)$. Moreover, $\partial + \alpha_k$ is a $\mathbb{C}$-linear map. Its kernel, which is nothing but the one dimensional space spanned by $z^{-\alpha_k}$, is contained in $V(\alpha_1, \ldots, \alpha_p; \beta_1, \ldots, \beta_q)$ if and only if $\alpha_k = \beta_l - 1$ for some l. Hence, under our assumptions, the systems with parameters $\alpha_1, \ldots, \alpha_p; \beta_1, \ldots, \beta_q$ and $\alpha_1, \ldots, \alpha_k + 1, \ldots, \alpha_p; \beta_1, \ldots, \beta_q$ are equivalent. An isomorphism of the solution spaces is given by the differential map $\partial + \alpha_k$. Similarly, the map $\partial + \beta_k - 1$ maps $V(\alpha_1, \ldots, \alpha_p; \beta_1, \ldots, \beta_q)$ bijectively onto $V(\alpha_1, \ldots, \alpha_p; \beta_1, \ldots, \beta_k - 1, \ldots, \beta_q)$. So, by using the operators $\partial + \alpha_k$, $\partial + \beta_k - 1$ and their inverses one can shift the parameters α_i, β_j freely by integers. □

An important consequence of Lemma 4.2.2 is that the differential Galois group in the irreducible case depends only on the parameters mod $\mathbb{Z}$. From now on we shall assume that our hypergeometric differential equation is irreducible. We now go over to imprimitivity. Let V be a vector space and G a group acting on V in an irreducible way. We shall say that G is *imprimitive* on V if there exists a direct sum decomposition $V = V_1 \oplus \cdots \oplus V_m$ $(m > 1)$ such that G permutes the V_i. If such a decomposition does not exist we call G *primitive* on V. Let V be the vector space of solutions of Eq.(4.1.2) and G its differential Galois group . We have already assumed that G acts irreducibly on V. There are two more or less obvious ways in which G is imprimitive on V. They are known as Kummer induction and (inverse) Belyi induction, using Katz' terminology. Kummer induction arises as follows. Consider Eq.(4.1.2) and replace z by $\zeta z^{1/d}/d^{p-q}$, where d is a natural number and ζ any d-th root of unity. We obtain

$$\mathcal{A}F = \zeta z^{1/d}\mathcal{B}F \tag{4.2.2}$$

where

$$\mathcal{A} = \prod_{i=1}^{q}(\partial + \frac{\beta_i - 1}{d}), \quad \mathcal{B} = \prod_{j=1}^{p}(\partial + \frac{\alpha_j}{d}).$$

One easily verifies the following operator equation,

$$z^{\frac{d-1}{d}}\left\{\sum_{t=0}^{d-1}(\mathcal{A}z^{-\frac{1}{d}})^{d-t-1}z^{-\frac{t}{d}}(z^{\frac{1}{d}}\mathcal{B})^t\right\}(\mathcal{A} - z^{\frac{1}{d}}\mathcal{B}) = z(z^{-\frac{1}{d}}\mathcal{A})^d - (z^{\frac{1}{d}}\mathcal{B})^d$$

Notice that

$$z(z^{-\frac{1}{d}}\mathcal{A})^d = \prod_{k=0}^{d-1}\prod_{i=1}^{q}\left(\partial + \frac{\beta_i - 1}{d} - \frac{k}{d}\right)$$

$$(z^{\frac{1}{d}}\mathcal{B})^d = z\prod_{k=0}^{d-1}\prod_{j=1}^{p}\left(\partial + \frac{\alpha_j}{d} + \frac{k}{d}\right)$$

Hence, Eq.(4.2.2) is a factor of the hypergeometric differential equation $\mathcal{H}$ with parameters

$$\frac{\alpha_1}{d}, \frac{\alpha_1+1}{d}, \ldots, \frac{\alpha_1+d-1}{d}, \frac{\alpha_2}{d}, \ldots, \frac{\alpha_p+d-1}{d}; \frac{\beta_1}{d}, \frac{\beta_1+1}{d}, \ldots, \frac{\beta_q+d-1}{d}.$$

More precisely, the direct sum of the solution spaces of Eq.(4.2.1) taken over all d-th roots of unity ζ, is precisely the solution space of $\mathcal{H}$. In particular, if $f_1(z), \ldots, f_q(z)$ is a basis of solutions of Eq.(4.2.1), then $f_i(\zeta z^{1/d}/d^{p-q})$ $(i = 1, \ldots, q, \quad \zeta^d = 1)$ is a basis of solutions of $\mathcal{H}$. Moreover, G, the differential Galois group of $\mathcal{H}$ permutes the solution spaces of Eq.(4.2.1) for different ζ, since G acts on $z^{1/d}$ by multiplication with d-th roots of unity. So, G is imprimitive. Notice that the parameter set of $\mathcal{H}$ has the property that modulo $\mathbb{Z}$ this set does not change if we add $1/d$ to all parameters.

(4.2.3) Definition. *A hypergeometric differential equation with the parameters* $\alpha_1, \ldots, \alpha_p; \beta_1, \ldots, \beta_q$ *is called* Kummer induced *if it is irreducible and if there exists a number* $d \in \mathbb{N}$ *such that modulo* $\mathbb{Z}$ *we have equality of the following sets*

$$\left\{\alpha_1 + \frac{1}{d}, \ldots, \alpha_p + \frac{1}{d}\right\} \equiv \{\alpha_1, \ldots, \alpha_p\} \pmod{\mathbb{Z}}$$

$$\left\{\beta_1 + \frac{1}{d}, \ldots, \beta_q + \frac{1}{d}\right\} \equiv \{\beta_1, \ldots, \beta_q\} \pmod{\mathbb{Z}}$$

By the remarks made above, its differential Galois group is imprimitive. A study of the differential Galois group can now be carried out by studying the differential Galois group of the hypergeometric equation from which it is induced. Belyi induction arises as follows. Let $a, b \in \mathbb{N}$ and define the algebraic function $t(z)$ of z by $z = \gamma t^a(1-t)^b$, where $\gamma = (1+b/a)^a(1+a/b)^b$ is chosen such that branching takes place only above $z = 0, 1, \infty$. Now take $\lambda, \mu \in \mathbb{C}$. Then $t^\lambda(1-t)^\mu$ satisfies the hypergeometric differential equation of order $q = p = a+b$ in z with parameters $\lambda, \lambda + 1/a, \ldots, \lambda + (a-1)/a, \mu, \mu + 1/b, \ldots, \mu + (b-1)/b; \nu, \nu + 1/(a+b), \ldots, \nu + (a+b-1)/(a+b)$, where $\nu = (a\lambda + b\mu)/(a+b)$. Inverse Belyi induction is a small variation on the above theme. With the same notations, let $z^{-1} = \gamma t^a(1-t)^b$, Then the function $t^\lambda(1-t)^\mu$ satisfies almost the same hypergeometric differential equation , the difference being that the α- and β-parameters are now interchanged.

(4.2.4) Definition. *A hypergeometric differential equation is called* Belyi induced *if it is irreducible and if there exist* $\lambda, \mu, \nu \in \mathbb{C}$ *and* $a, b \in \mathbb{N}$ *such that* $a\lambda + b\mu = (a+b)\nu$ *and the parameter set modulo* $\mathbb{Z}$ *is given by*

$$\lambda, \lambda + \frac{1}{a}, \ldots, \lambda + \frac{a-1}{a}, \mu, \ldots, \mu + \frac{b-1}{b}; \nu, \nu + \frac{1}{a+b}, \ldots, \nu + \frac{a+b-1}{a+b}.$$

The equation is called inverse Belyi induced *if, with the same notations, its parameter set modulo* $\mathbb{Z}$ *reads*

$$\nu, \nu + \frac{1}{a+b}, \ldots, \nu + \frac{a+b-1}{a+b}; \lambda, \lambda + \frac{1}{a}, \ldots, \lambda + \frac{a-1}{a}, \mu, \ldots, \mu + \frac{b-1}{b}$$

(4.2.5) Theorem. *Let G be the differential Galois group of* Eq.(4.1.2), *which we assume to be irreducible. Then G is imprimitive if and only if* Eq.(4.1.2) *is either Kummer induced or Belyi induced or inverse Belyi induced.*

This theorem is not stated as such in either (Katz 1990) or (Beukers and Heckman 1989). The proof is fairly technical, but its ingredients are contained in the two references mentioned.

4.3 The primitive case

(4.3.1) Theorem. *Let G be the differential Galois group of* Eq.(4.1.2), *which we assume to be irreducible. Suppose G is primitive. Then either G^0 acts irreducibly or G^0 consists of scalars, in which case we have either $G^0 \simeq \mathbb{C}^\times$ or $G^0 = \{id\}$.*

(4.3.2) *Remark.* From the proof of this theorem one sees that the latter cases can only occur when $p = q$. In such a case we have that G modulo its centre is finite. Theorem 4.3.3 will describe precisely when this happens.

Proof. First assume $p < q$. Suppose G^0 acts reducibly. The fixed field $K/\mathbb{C}(z)$ corresponding to G^0 is algebraic, and since $0, \infty$ are the only singularities, it must be of the form $K = \mathbb{C}(z^{1/e})$ for some $e \in \mathbb{N}$. But then G/G^0 is cyclic. Letting W be an irreducible invariant subspace of G^0, one sees that a maximal subset of distinct subspaces in $\{gW|\ g \in G\}$ yields a system of imprimitivity for G. Hence G is imprimitive, contrary to our assumption. Thus we conclude that G^0 acts irreducibly. Now assume $p = q$. This case is more tedious due to the occurrence of equations which have only algebraic solutions. We sketch the proof here, and rely on results from (Beukers and Heckman 1989). Let H be the monodromy group of Eq.(4.1.2) and let G be its Zariski closure in $GL(n)$. In H we have the so-called reflection subgroup H_r which is generated by the elements $\{hh_1h^{-1}|h \in H\}$, where h_1 is the local monodromy element around the point $z = 1$. According to our remarks in Subsection 4.1, h_1 is a pseudo reflection. Let h_0 be the monodromy element around $z = 0$. It is well known that H is generated by h_1, h_0. So, H/H_r is generated by h_0, i.e. H/H_r is cyclic. In (Beukers and Heckman 1989, Theorem 5.14) it is shown that if H_r is imprimitive, then H_r is finite. Since H/H_r is cyclic this implies that $G^0 = \{id\}$ or $\mathbb{C}^\times$. Now suppose that H_r is primitive. Then (Beukers and Heckman Prop.6.3) states that $\overline{H}_r^0$, component of the identity of the Zariski closure of H_r, is either trivial or irreducible on V. This proves our assertion. □

The following theorem enables one to recognize the cases of Theorem 4.3.1. We say that two sets of points $A = \{a_1, \ldots, a_n\}$ and $B = \{b_1, \ldots, b_n\}$ with $|a_i| = |b_j| = 1$ for $i, j = 1, \ldots, n$ *interlace* on the unit circle $|z| = 1$ if, following this circle clockwise, one meets the points of A and B alternately.

(4.3.3) Theorem. *Suppose $p = q$. Then G modulo its centre is finite if and only if the following conditions hold,*

i. $\exists \delta \in \mathbb{C}$ *such that* $\mu_i = \alpha_i + \delta \in \mathbb{Q}$ *and* $\nu_j = \beta_j + \delta \in \mathbb{Q}$ *for all* i, j.

ii. *Letting N be the common denominator of μ_i, ν_j, the sets $\{\exp(2\pi i h\mu_k)|\ k = 1, \ldots, q\}$ and $\{\exp(2\pi i h\nu_l|\ l = 1, \ldots, q\}$ interlace on the unit circle $|z| = 1$ for all $h \in \mathbb{N}$ with $1 \le h \le N$,* $\gcd(h, N) = 1$.

In particular, if $\delta \in \mathbb{Q}$ then G is finite.

(4.3.4) *Remark.* In case $p = q = 2$ a complete list of all cases where G is finite was given by H.A.Schwarz (Schwarz 1873). For the case $p = q > 2$ a complete list was determined in (Beukers and Heckman 1989, Section 8). A proof of Theorem 4.3.3 can be found in the same reference in Section 4.

The following theorem gives a rough description of the differential Galois groups that occur for hypergeometric differential equations in case G^0 acts primitively (i.e. the general case as we know by now).The case $p \neq q$ was settled by O.Gabber and N.M.Katz in 1986 (Katz 1990) and the case $p = q$ by (Beukers and Heckman 1989).

(4.3.5) Theorem. *Let V be the solution space of* Eq.4.1.2 *and G its differential Galois group . Suppose G is primitive and that G^0 acts irreducibly on the space V. Then we have the following possibilities for G,*

If $p = q$, $\exp 2\pi i \sum_k(\alpha_k - \beta_k) \neq \pm 1$ *then* $G = C \cdot SL(V)$.
If $p = q$, $\exp 2\pi i \sum_k(\alpha_k - \beta_k) = 1$ *then* $G = C \cdot SL(V)$ *or* $C \cdot Sp(V)$.
If $p = q$, $\exp 2\pi i \sum_k(\alpha_k - \beta_k) = -1$ *then* $G = C \cdot SL(V)$ *or* $C \cdot SO(V)$.
If $q - p = 1$ *then* $G = GL(V)$.
If $q - p$ *is odd then* $G = C \cdot SL(V)$.
If $q-p$ *is positive and even, then* $G = C \cdot SL(V)$ *or* $C \cdot SO(V)$ *or* $C \cdot Sp(V)$ *or, in addition*

if $q = 7, p = 1$, $G = C \cdot G_2$
if $q = 8, p = 2$, $G = C \cdot SL(3)$, $SL(3)$ *in adjoint representation*
if $q = 8, p = 2$, $G = C \cdot (SL(2) \times SL(2) \times SL(2))$
if $q = 8, p = 2$, $G = C \cdot (SL(2) \times Sp(4))$
if $q = 8, p = 2$, $G = C \cdot (SL(2) \times SL(4))$
if $q = 9, p = 3$, $G = C \cdot (SL(3) \times SL(3))$.

Proof. Suppose $p = q$. We know from Section 4.1 that the local monodromy around the point $z = 1$ is generated by the pseudoreflection h (i.e. $h - Id$ has rank one) with special eigenvalue $\lambda = \exp 2\pi i \sum_k(\beta_k - \alpha_k)$. Clearly, the group hG^0h^{-1} is again connected and so $hG^0h^{-1} = G^0$, i.e. h normalizes G^0. We now obtain the first three assertions of our theorem by application of Theorem 3.2.3 to the normalizing element h and the group G^0. Suppose $q > p$. From classical references (Barnes 1906) or (Meijer 1946) we know that we have at ∞ a basis of formal solutions $y_1, \ldots, y_q$ given by

$$\begin{pmatrix} y_1 \\ \vdots \\ y_q \end{pmatrix} = \exp(P(z^{1/(q-p)}))z^A \begin{pmatrix} f_1 \\ \vdots \\ f_q \end{pmatrix},$$

where $P(X) = \mathrm{diag}(X, \zeta X, \ldots, \zeta^{q-p-1}X, 1, \ldots, 1)$, A is a constant $q \times q$ matrix in Jordan normal form and the f_i are formal Laurent expansions in $z^{-1/(q-p)}$. From the remarks made at the end of Section 2.5 we see that the local Galois group, and hence the group G, contains the torus D consisting of all

$\mathrm{diag}(d_1,\ldots,d_q)$ such that $d_i = 1$ for $i > q-p$ and $d_1^{a_1}\cdots d_{q-p}^{a_{q-p}} = 1$ whenever $\sum_{i=1}^{q-p} a_i\zeta^{i-1} = 0$, $a_i \in \mathbb{Z}$. Trivially, D normalizes G^0. We apply Gabber's Theorem 3.2.1. In order to do so, we must determine all $1 \le i,j,k,l \le q$ such that $d_id_j = d_kd_l$ for all elements of D. Since $d_i = 1$ for $i > q-p$ this means that we have to watch for non trivial relations of the form $d_id_j = d_kd_l$, $d_id_j = d_k$, $d_i = d_k$, $d_id_j = 1$, $d_i = 1$ with $1 \le i,j,k,l \le q-p$. From the definition of D this means that we must check for non trivial relations of the form $\zeta^i + \zeta^j = \zeta^k + \zeta^l$, $\zeta^i + \zeta^j = \zeta^k$ or $\zeta^i + \zeta^j = 0$. It is not hard to show that

i. $\zeta^i + \zeta^j = \zeta^k + \zeta^l \Rightarrow i = k, j = l$ or $q-p$ even, $i \equiv k + \frac{q-p}{2} \bmod q-p$, $j \equiv l + \frac{q-p}{2} \bmod q-p$ or the same possibilities with i and j interchanged.
ii. $\zeta^k = \zeta^i + \zeta^j \Rightarrow 6 \mid (p-q)$, $i \equiv k \pm \frac{q-p}{6} \pmod{q-p}$, $j \equiv k \mp \frac{q-p}{6} \pmod{q-p}$.
iii. $\zeta^i + \zeta^j = 0 \Rightarrow q-p$ even, $j \equiv i + \frac{q-p}{2} \pmod{q-p}$.

Thus, if $q-p$ is odd there are no restrictions and Theorem 3.2.1 implies that $\mathbb{C}\cdot G^0$ contains the torus

$$T = \{\mathrm{diag}(t_1,\ldots,t_{q-p},1,\ldots,1) \mid t_1\cdots t_{q-p} \neq 0\}.$$

In particular $\mathbb{C}\cdot G^0$ contains $T_1 = \{\mathrm{diag}(t,1,\ldots,1)|t \neq 0\}$ and Theorem 3.2.2 now implies $\mathbb{C}\cdot G^0 = GL(V)$. Hence $SL(V) \subset G$. Moreover, if $q-p=1$, we see that $D = T_1$ and so $G = GL(V)$ in this case. Now suppose $q-p$ is even, but $6 \not| (q-p)$. Then the relations are generated by $d_id_j = 1$, $i-j \equiv \frac{q-p}{2} \pmod{q-p}$ and Theorem 3.2.1 implies that $\mathbb{C}\cdot G^0$ contains

$$\{\mathrm{diag}(t_1,\ldots,t_{\frac{q-p}{2}},t_1^{-1},\ldots,t_{\frac{q-p}{2}}^{-1},1,\ldots,1) \mid t_1\cdots t_{\frac{q-p}{2}} \neq 0\}$$

In particular, $\mathbb{C}\cdot G^0$ contains the torus $\{\mathrm{diag}(t,1,\ldots,1,t^{-1},1,\ldots,1) \mid t \neq 0\}$ and Theorem 3.2.2 implies that $\mathbb{C}\cdot G^0$ is $GL(V)$, $\mathbb{C}\cdot Sp(V)$ or $\mathbb{C}\cdot SO(V)$. Finally, if $6 \mid (q-p)$ then, by the same arguments as above, we find that $\mathbb{C}\cdot G^0$ contains (after permutation of basis vectors, if necessary)

$$\{\mathrm{diag}(ts,t,s,(ts)^{-1},t^{-1},s^{-1},1,\ldots,1) \mid ts \neq 0\},$$

which is called the G_2-torus. Application of Theorem 3.2.3 yields the desired result. □

In the cases of Theorem 4.3.5 where we have a choice between $SL(V)$ and a smaller selfdual group, there is a very easy criterion to make this decision.

(4.3.6) Theorem. *Let V be the solution space of* Eq.4.1.2 *and G its differential Galois group . Then the following two statements are equivalent,*

i. *G is contained in $\mathbb{C}\cdot Sp(V)$ or $\mathbb{C}\cdot SO(V)$*
ii. *$\exists\delta$ such that $\{\alpha_1,\ldots,\alpha_p\} = \{\delta-\alpha_1,\ldots,\delta-\alpha_p\} (\mathrm{mod}\ \mathbb{Z})$ and $\{\beta_1,\ldots,\beta_q\} = \{\delta-\beta_1,\ldots,\delta-\beta_q\} (\mathrm{mod}\ \mathbb{Z})$.*

The proof, which is elementary, can be found in (Katz 1990). In this same book it is shown that there actually exist examples of hypergeometric differential equations having G_2 as differential Galois group .

(4.3.7) Theorem. (Gabber-Katz) *Let V be the solution space of* Eq.(4.1.2) *and G its differential Galois group . Then the following statements are equivalent,*

i. $G = C \cdot G_2$

ii. $q = 7, p = 1$ *and* $\exists \mu, \nu$ *with* $\mu \neq 0, \nu \neq 0, \mu + \nu \neq 0$ *such that after renumbering of indices, if necessary, the numbers* $\beta_1 - \alpha_1, \ldots, \beta_7 - \alpha_1$*, considered (mod $\mathbb{Z}$), equal* $\frac{1}{2}, \mu, \nu, -\mu, -\nu, \mu + \nu, -\mu - \nu$.

References

Babbitt, D.G., Varadarajan, V.S. (1983) Formal reduction theory of meromorphic differential equations: a group theoretic view, Pacific J.Math. **109**(1983), 1-80.

Barnes (1907), The asymptotic expansion of integral functions defined by generalized hypergeometric series, Proc.London Math.Soc (2)**5**(1907) 59-116.

Beukers, F.,Brownawell, W.D., Heckman,G. (1988) Siegel normality, Annals of Math. **127** (1988) 279-308.

Beukers, F., Heckman, G. (1989) Monodromy for the hypergeometric function ${}_nF_{n-1}$, Inv.Math. **95** (1989) 325-354.

Beukers, F., Peters, C.A.M. (1984) A family of K3-surfaces and $\zeta(3)$, J.reine angew. Math. **351** 42-54.

Deligne, P. (1987) Catégories tannakiennes, in *Grothendieck Festschrift*, part. II, Birkhäuser 1991.

Duval, A., Mitschi, C. (1989) Matrices de Stokes et groupe de Galois des équations hypergéometriques confluentes généralisées, Pacific J. Math. **138** (1989), 25-56.

Erdélyi, A. et al. (1953) *Higher transcendental functions* (Bateman manuscript project) Vol II,Ch.7 McGraw-Hill, New York 1953.

Fano, G. (1900) Über lineare homogene Differentialgleichungen, Math.Annalen **53** 493-590.

Freudenthal, H., de Vries, H. (1969) *Linear Lie groups*, Academic Press New York 1969.

Gray, J. (1986) *Linear differential equations and group theory from Riemann to Poincaré* Birkhäuser, Boston 1986.

Hille, E. (1976) *Ordinary differential equations in the complex domain* Wiley, New York 1976.

Humphreys, J.E. (1972) *Introduction to Lie algebras and representation theory* Springer Verlag, New York 1972.

Humphreys, J. E. (1975) *Linear algebraic groups* Springer Verlag, New York 1975.

Ince, E.L. (1926) *Ordinary differential equations* Dover Publication (reprinted) New York 1956.

Kaplansky, I. (1957) *Differential algebras* Hermann, Paris 1957

Katz, N.M. (1970) Nilpotent connections and the monodromy theorem: Applications of a result of Turrittin, Publ.Math IHES **39** 175-232.

Katz, N.M. (1987) On the calculation of some differential Galois groups, Inv.Math. **87** (1987) 13-61.

Katz, N.M. (1990) *Exponential sums and differential equations* Ann. of Math. Study **124** Princeton University Press, Princeton 1990

Katz, N.M., Pink,R. (1987) A note on pseudo-CM representations and differential Galois groups, Duke Math.J. **54** (1987) 57-65.

Klein, F. (1933) *Vorlesungen über die hypergeometrische Funktion* Springer Verlag, New York 1933.

Kolchin, E.R.1 (1948) Algebraic matrix groups and the Picard-Vessiot theory of homogeneous linear ordinary differential equations, Annals of Math. **49** (1948) 1-42.
Kolchin, E.R.2 (1948) Existence theorems connected with the Picard- Vessiot theory of homogeneous linear ordinary differential equations, Bull. Amer. Math. Soc. **54** (1948) 927-932.
Kolchin, E.R. (1973) *Differential algebra and algebraic groups*, Academic Press, New York 1973.
Kolchin, E.R. (1985) *Differential algebraic groups*, Academic Press, New York 1985.
Kostant, B. (1958) A characterisation of the classical groups, Duke Math. J. **25** (1958) 107-123.
Levelt, A.H.M. (1961) *Hypergeometric functions*,thesis University of Amsterdam 1961.
Levelt, A.H.M. (1975) Jordan decomposition for a class of singular differential operators, Arkiv Math. **13** (1975) 1-27.
Levelt, A.H.M. (1990) Differential Galois theory and tensor products, Indag. Math. **1** (new series) (1990) 439-449.
Martinet, J. Ramis, J.P. (1989) Théorie de Galois différentielle in *Computer Algebra and Differential Equations*, ed. E.Tournier, Academic Press, New York 1989.
Meijer, C.S. (1946) On the G-function I, Indag.Math. **8** (1946) 124-134.
Picard, E. (1898) *Traité d'analyse* Vol III, Chapter 17, Gauthier-Villars, Paris 1898. Reprinted as *Analogies entre la théorie des équations différentielles linéaires et la théorie des équations algébriques*, Gauthier-Villars Paris, 1936.
Pommaret, J.F. (1983) *Differential Galois theory*, Gordon and Breach New York, 1983.
Poole, E.G.C. *Introduction to the theory of linear differential equations* Oxford 1936, reprinted in Dover Publ.
Ritt,J.F. (1932) *Differential equations from the algebraic standpoint* AMS Coll. Publ. XIV, New York 1932
Schwartz , H.A. (1873) Über diejenigen Fälle in welchen die Gaussische hypergeometrische Reihe einer algebraische Funktion ihres vierten Elementes darstellt, Crelle J. **75** (1873) 292-335
Sibuya, Y. (1977) Stokes phenomena, Bull.American Math.Soc. **83** (1977), 1075-1077.
Singer, M. (1989) An outline of differential Galois theory in *Computer Algebra and Differential Equations*, ed. E.Tournier, Academic Press, New York 1989.
Springer, T.A.(1981) *Linear algebraic groups*, Progress in math.9, Birkhäuser, Boston 1986.
Turrittin, H.L. (1955) Convergent solutions of ordinary linear homogeneous differential equations in the neighbourhood of an irregular singular point, Acta Math. **93** (1955) 27-66.
Vessiot, E. (1892) Sur les intégrations des équations différentielles linéaires, Ann. Sci. École Norm. Sup.(3)**9** (1892), 197-280.
Wasow, W. (1965) *Asymptotic expansions for ordinary differential equations* Wiley, New York 1965.

Chapter 9

p-adic Numbers and Ultrametricity

by Gilles Christol

A. Ultrametric Spaces

The aim of this first part is to define classical ultrametric sets and especially p-adic numbers. This will be achieved through a general procedure described in Section 2. In Section 1 are given general definitions about ultrametric sets. These definitions may well be meaningless for the unwarned reader. Thus we suggest beginning at 2.1.2, reading Section 1 as and when needed. The ring $\mathbb{Z}_p$ of p-adic numbers is introduced in Section 3.2 and the ring $\widehat{\mathbb{Z}}$, a generalization of $\mathbb{Z}_p$ without any privileged prime, is introduced in Section 3.3. The ring $\widehat{\mathbb{Z}}$ has already been used by physicists, as explained in 3.4.

1. Ultrametric Distances

1.1 Definitions

1.1.1. A distance d on a set E is said to be *ultrametric* if it satisfies the usual conditions :

$$d(a,b) \geq 0 \ ; \ d(a,b) = 0 \iff a = b \ ; \ d(a,b) = d(b,a) \ ,$$

and if the triangle inequality is replaced by the stronger one :

$$d(a,c) \leq \max\left[d(a,b), d(b,c)\right] \ , \tag{1.1}$$

called the *ultrametric inequality.*

1.1.2. As a consequence of inequality (1.1), it is easily seen that :

$$d(a,b) < d(b,c) \ \Rightarrow \ d(a,c) = d(b,c) \ ,$$

If we think geometrically, this means that 'all triangles are isosceles'. An amusing consequence is that any point in a disk is its center. For let

$$D_a(r) = \{x; d(a,x) \leq r\}$$

$$D_a(r^-) = \{x; d(a,x) < r\} \quad .$$

Then for each b in $D_a(r)$, $D_b(r) = D_a(r)$. Moreover, if the 'circumference' $\{x; d(a,x) = r\}$ is non empty, it contains every 'open disk' $D_b(r^-)$ about any one of its points b , and is bigger (often much bigger) than the 'interior' $D_a(r^-)$.

Remarks.

* The set E is endowed with its metric space topology, and topological properties such as completeness of E will refer to that topology.

* The notions of 'open disk', 'interior', 'circumference' are set in quotation marks because they are *not meant in the topological sense*. In fact any disk $D_a(r)$ or $D_a(r^-)$ is both open and closed. By definition of the metric topology, each point of E has a fundamental system of neighbourhoods which consists of disks which are both open and closed. A topological space with this property is said to be *totally disconnected*.

1.2 Absolute Value and Valuations

1.2.1. If the set E is a commutative group (with composition denoted additively), it is natural to ask the distance to be compatible with addition :

$$d(a+c, b+c) = d(a,b) \quad .$$

Then the distance is entirely determined by the numbers $d(a,0) = d(0,a)$. Another amusing consequence of ultrametricity is that a series converges in an ultrametric group as soon as its general term goes to zero.

1.2.2. If the set E is a commutative ring, compatibility is expressed by :

$$d(a,b) = |a-b| \quad ,$$

where $|.|$ is an *ultrametric absolute value*, i.e., satisfies :

$$\left\{ \begin{array}{ll} |a| \geq 0 \ ; & |a| = 0 \iff a = 0 \ ; \\ |a+b| \leq \max(|a|,|b|) \ ; & |ab| = |a||b| \ . \end{array} \right. \tag{1.2}$$

A ring endowed with an ultrametric absolute value is called a *valued ring*. A ring can be valued if and only if it has no zero divisor, indeed on any ring without zero divisor there is a trivial valuation defined by :

$$\forall (a \neq 0) \quad , |a| = 1 \ .$$

Then a valued ring has a fraction field to which the absolute value can be extended by :

$$|a/b| = |a|/|b| \quad .$$

So valued rings are the subrings of valued fields. Such fields are often called *non-Archimedean* by reference to the Archimedean property of $\mathbb{R}$ endowed with the classical absolute value (or $\mathbb{C}$ endowed with the norm,...) that one can reach all points from zero with steps of any given length.

1.2.3. If K is a valued field, the set $\{|a| \; ; a \in K\}$ is a multiplicative subgroup of $\mathbb{R}^+$. In many circumstances it is more convenient to use 'additive' notations [1]. For that purpose we choose some number $\alpha > 1$, and we define :

$$v(a) = \begin{cases} -\log_\alpha(|a|) & \text{if } a \neq 0 \\ \infty & \text{if } a = 0 \end{cases} .$$

Translating ultrametric properties, one gets :

$$\begin{aligned} v(a) = \infty &\iff a = 0 \; ; \\ v(a+b) &\geq \inf(v(a), v(b)) \; ; \\ v(ab) &= v(a) + v(b) \; . \end{aligned} \tag{1.3}$$

A map from a field K to $\mathbb{R} \cup \{\infty\}$ that satisfies (1.3) is called a *valuation*. Given a valuation v, one gets back an ultrametric distance by

$$d(a,b) = \alpha^{-v(a-b)} \; .$$

Clearly the distances constructed in this way with various α's are all equivalent (i. e. define the same topology).

2. A General Procedure for Obtaining Ultrametric Spaces

2.1 Construction

2.1.1. We will give a general description of a large class of ultrametric spaces that contains, among others, complete *discretely* valued rings (i. e., valued rings with valuations in $\mathbb{N}$). We think that the view it gives of ultrametric spaces is more realistic than the geometric representation. Examples of 3) illustrate the following 'theory'.

2.1.2. Let us examine an object under various magnifications. For each magnification the object seems to be composed of a set (finite or not) of points. For instance under the 'zero' magnification there is only one point, namely the object itself. Now successively increasing the magnification reveals each point to be a cluster of smaller points, each of which can in its turn be split into subpoints and so on. Our purpose is to describe the object as it appears under an infinite magnification. We construct a mathematical model of the situation by giving an infinite sequence E_n ($n = 0, 1, \ldots$) of sets connected by maps φ_n from E_{n+1} onto E_n. The situation can also be represented by a tree whose vertices

[1] For the sake of simplicity we define v only in the case of absolute values, but the additive notation is useful generally as will be seen in 2.

are the elements of $\bigcup E_n$ and whose (directed) edges connect $a_{n+1} \in E_{n+1}$ to $\varphi_n(a_{n+1})$. In fig.1 we represent the case where each point of E_n is the image of exactly two elements of E_{n+1}.

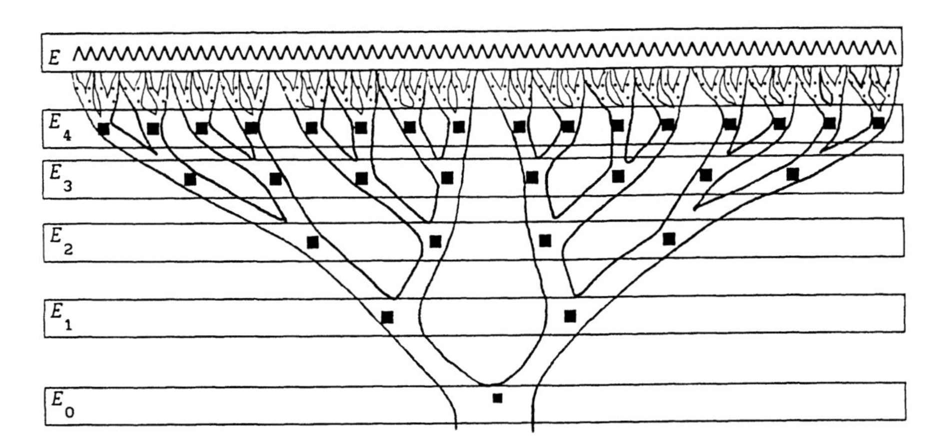

Fig. 1.

Now to represent our object, we consider the set E of the 'leaves of the tree' (or paths of the tree starting from E_0 and of infinite length) i.e., more precisely, the set of sequences $a = \{a_n\}$ such that :

$$a_n \in E_n \ ; \ \varphi_n(a_{n+1}) = a_n \ . \tag{2.1}$$

Classically one uses the notation :

$$E = \varprojlim E_n$$

and calls E the *inverse limit* of the E_n .

2.1.3. Let us verify that the set E is equipped with an ultrametric topology. For that purpose let us define first a 'natural' distance on the tree just defined: let $\{d_n\}$ be any strictly decreasing sequence of real numbers going to zero (for instance $d_n = \alpha^n$ for any $\alpha < 1$), and let us agree that any edge connecting a vertex in E_{n+1} to a vertex in E_n has a length of $\frac{1}{2}\ (d_n - d_{n+1})$. Then any infinite directed path ending in E_n has a length of $\frac{1}{2}\ d_n$ and any path in the tree has finite length. Now there is an obvious distance on E , namely the length of the (shortest) path going from one 'leaf' to an other by passing through the tree. This definition can be easily expressed in mathematically correct terms. Let a and b be two distinct elements of E, i.e. two sequences $\{a_n\}$ and $\{b_n\}$ satisfying (2.1) (hence $a_n = b_n$ implies $a_i = b_i$ for $i \leq n$), let :

$$v(a,b) = \ \sup\{n\, ; \ a_n = b_n\}$$

(this is the level where the 'common ancestor' of a and b lies). Then the formula:

$$d(a,b) = d_{v(a,b)}$$

defines an ultrametric distance on E. Moreover the topological space E so obtained is independent of the sequence $\{d_n\}$ and *complete*. Indeed if $a(m)$ is a Cauchy sequence in E, then, for m large enough, the $a(m)$ stay in the same disk (as E is ultrametric, to go outside a disk one needs to make a step larger than its radius) i.e. the sequence $a(m)_n$ is ultimately constant. Thus, if a is defined such that $a_n = \lim a(m)_n$ for all n, one sees easily that $a = \lim a(m)$. The space E is a *compact* space if and only if the sets E_n are finite. Indeed if E is compact, the 'open' disks of radius d_n are in one to one correspondence with the elements on E_n, but they are a disjoint covering of E. Thus E_n must be finite. Conversely, if the E_n are finite, for any sequence $a(m)$ in E, one can find a in E such that for all n one has $a(m)_n = a_n$ for an infinite number of m. Thus a is a limit point of the sequence and E is compact. If v is a valuation, we have to choose $d_n = \alpha^n$ for some $\alpha < 1$ for the distance to become an absolute value.

2.2 Further Properties

2.2.1. If the E_n are (commutative) groups and the φ_n are group homomorphisms then E itself is a group with the rule of composition :

$$\{a_n\} + \{b_n\} = \{a_n + b_n\}$$

with which the distance is compatible in the sense of 1.2.1.

2.2.2. In the same way, if the E_n are rings and the φ_n ring homomorphisms, then E becomes a ring for the product :

$$\{a_n\}\{b_n\} = \{a_n b_n\} \quad .$$

But there is no reason for this product to be compatible with the distance in the sense of 1.2.2, i.e., for :

$$v(a) = v(a,0) = \inf\{n; a_n \neq 0\}$$

to satisfy $v(ab) = v(a)v(b)$. For instance, if the E_n have no zero divisors one only gets :

$$v(ab) \geq \sup(v(a), v(b)) \ ,$$

which is enough to ensure the continuity of the product.

2.2.3. In many examples it is possible, for each vertex, to select one edge among those abutting in it. In other words, E_n can be 'lifted' in E_{n+1} , i.e., there exist applications ψ_n from E_n to E_{n+1} such that $\varphi_n \circ \psi_n =$ Identity. If this situation occurs, when describing E it is more concise to skip at each level the information already known from the previous one. More precisely elements of E can be given in the following way. Let $E_{[n]}$ be the quotient group $E_{n+1}/\psi_n(E_n)$ (isomorphic to the kernel of φ_n), then the element $\{a_n\}$ of E is entirely determined by the sequence $a_{[0]}, \ldots, a_{[n]}, \ldots$, where :

$$a_{[0]} = a_1 \text{ and } a_{[n]} \text{ is the image of } a_{n+1} \text{ in } E_{n+1}/\psi(E_n) \ .$$

This representation can be viewed as a kind of generalized decimal expansion. Moreover for each n there is a 'canonical lifting' Φ from E_n to E where $\Phi(a_n)$ is given by the infinite path going through a_n and made of the chosen edges.

2.2.4. An analogous situation occurs under the following circumstances. Let A be a ring and I_n a sequence of ideals of A such that :

$$I_{n+1} \subset I_n \quad \text{and} \quad \bigcap I_n = \{0\} \ .$$

Let E_n be the ring A/I_n. Reduction modulo I_n gives a homomorphism φ_n from E_{n+1} to E_n, and thus a distance on the ring $E = \varprojlim E_n$ as shown in 2.1. There is a canonical embedding of A in E : each a in A is attached to the sequence $\{a \bmod I_n\}$. Then A is a dense topological subset of E. In other words E is the completion of A for the topology constructed from the I_n. The v defined in 2.2.2 is actually a valuation when $I_n = I^n$ for some ideal $I \neq A$ (such that $\bigcap I^n = 0$).

3. Basic Examples

3.1 Polynomials and Power Series

3.1.1. Let k be a field and let E_n be the additive group of polynomials with coefficients in k of degree less than or equal to n. The natural projection from E_{n+1} to E_n, namely 'forget the term of degree $n+1$ ', enables one to construct a group E along the line of 2.1 (use 2.2.1 for the group structure).

3.1.2. The description of this set is more concrete if we use the trick of 2.2.3 (indeed E_n is a subset of E_{n+1}) : it appears to be the set $k[[x]]$ of 'formal power series over k ' i. e. the set of power series :

$$a = \sum_{n=0}^{\infty} a_{[n]} x^n \ , \ a_{[n]} \in k$$

without any condition concerning the $a_{[n]}$ (they are formal because, for instance if $k = \mathbb{C}$, most of them are nowhere convergent). The distance between two power series a and b is $\alpha^{v(a-b)}$ $(\alpha < 1)$ where $v(a-b)$ is the valuation of $a-b$, namely the smallest integer such that the coefficient of the term of degree n in $a-b$ is not zero.

3.1.3. Now there is a classical product on $k[[x]]$ given by :

$$ab_{[n]} = \sum_{k=0}^{n} a_{[k]} b_{[n-k]}$$

which is an extension of the product of polynomials. For that law, the valuation just defined satisfies Condition (1.3) and is actually a valuation ! This valuation is called the $x-$*adic valuation*. This designation recalls that we are in a situation where 2.2.4 applies, taking for A the ring $k[x]$ of all polynomials with coefficients in k and for I the ideal $xk[x]$ (then $I_n = I^n$ is the ideal $x^n k[x]$). In that description, E_n appears as the set of polynomials modulo x^n and $k[[x]]$ as the completion of $k[x]$ with respect to the $x-$adic valuation (a polynomial is a power series whose coefficients are eventually zero !).

3.1.4. As $k[[x]]$ is a valued ring it has a fraction field, namely the field $k((x))$ of formal power series over k , whose elements are :

$$a = \sum_{n=v(a)}^{\infty} a_{[n]} x^n \ , \quad a_{[n]} \in k \ , \quad a_{[v(a)]} \neq 0$$

where $v(a)$ belongs to $\mathbb{Z}$ and is, obviously, the valuation of a . The field $k(x)$ of rational fractions over k is the fraction field of the ring of polynomials which is then contained in $k((x))$. The x-adic valuation of a rational fraction f is given by writing it in the form :

$$x^{v(f)} P(x)/Q(x) \ ; \ P(0) \neq 0 \ ; \ Q(0) \neq 0$$

(which enables us to expand it in Laurent series 'near 0'). Actually there are many other valuations defined over $k(x)$ namely one for each irreducible polynomial of $k[x]$. For instance, if k is algebraically closed, the valuations of $k(x)$ are in one to one correspondence with the points of the 'projective line' over k , i. e. $k \cup \{\infty\}$: the valuation associated with the point ξ is given by the expansion of the rational fraction in Laurent series near ξ (i. e. in $k((x-\xi))$ for finite ξ and in $k((1/x))$ if $\xi = \infty$).

3.2 p-adic Numbers

3.2.1. Let p be a prime number and let $E_n = \mathbb{Z}/p^n\mathbb{Z}$ be the ring of 'integers modulo p^n '. The projection[2] from E_{n+1} to E_n given by the remainder when dividing by p^n enables one to construct a ring E along the lines of 2.1 (use 2.2.1 and 2.2.2 for the ring structure). Figure 2 shows the construction when $p = 2$. This ring is denoted by $\mathbb{Z}_p$ and called *ring of p-adic integers.*

3.2.2. Let us now describe $\mathbb{Z}_p$ more concretely. Using $[0, 1, \ldots, p^n - 1]$ as a set of representatives, E_n appears as a subset of E_{n+1}. Even if this injection is not a ring homomorphism (for instance $1 + (p^n - 1)$ is 0 in E_n but not in E_{n+1}), it allows us to apply the trick of 2.2.3. The ring $\mathbb{Z}_p$ appears as the set of *Hensel series* :

$$a = \sum_{n=0}^{\infty} a_{[n]} \, p^n \ , \quad a_{[n]} \in [0, \ldots, p-1] \ .$$

With these notations, the (p-*adic*) *valuation* of a p-adic integer a is the smallest integer n such that $a_{[n]} \neq 0$. It is not difficult to prove that this is actually a

[2] There is another natural projection given by the quotient when dividing by p, but this second one is not a ring homomorphism and will not concern us.

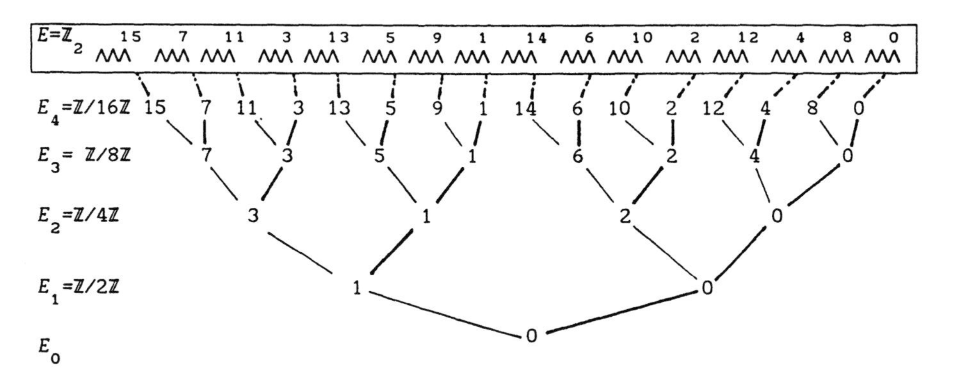

Fig. 2. The 'tree' of 2-adic numbers

valuation. Classically, the number α used to define the corresponding absolute value is chosen to be $\frac{1}{p}$ so that the (*p-adic*) *distance* between to p-adic integers a and b is $p^{-v(a-b)}$.

Remarks.

* An infinite sequence $a_{[n]}$ of digits in $[0,\ldots,p-1]$ can represent an element of three distinct additive groups :
  1) a real number $a_{[0]}, a_{[1]} \cdots a_{[n]} \cdots$ of $\mathbb{R}/p\mathbb{Z}$,
  2) a power series $a_{[0]} + a_{[1]}x + \cdots + a_{[n]}x^n + \cdots$ of $\mathbb{Z}/p\mathbb{Z}[[x]]$,
  3) a p-adic integer $a_{[0]} + a_{[1]}p + \cdots + a_{[n]}p^n + \cdots$ of $\mathbb{Z}_p$.
  Each of these groups is characterized by the way we carry out additions. For power series we behave like dunces and forget to carry, for real numbers, as everybody knows, we carry to the left and for p-adic integers we carry to the right. This fact explains why small perturbations can change every digit in the real case but not in the other two cases, where you cannot disturb digits lying before the one you change.
* We use $[0, 1, \ldots, p^n - 1]$ as a set of representatives, but other choices are available, for instance $[(1-p)/2, \ldots, (p-1)/2]$ when $p \neq 2$. Each choice leads to a distinct representation of $\mathbb{Z}_p$ by mean of a Hensel expansion. The differences appear only in explicit calculations.

3.2.3. In the above description, a positive integer m belongs to E_n as soon as $p^n > m$. Then, using 2.2.3, it can be viewed as an element of $\mathbb{Z}_p$. Its (finite) Hensel series is its expansion in base p . Its p-adic valuation is easily seen to be the power of the prime p that occurs in its prime power decomposition : for instance the 3-adic valuation of $2250 = 2.3^2.5^3$ is 2. In that way, $\mathbb{Z}_p$ becomes the completion of the set $\mathbb{N}$ of natural integers endowed with the p-adic distance.

As an exercise, let the reader verify that negative integers correspond to Hensel series whose coefficients are ultimately equal to $(p-1)$: for instance, in $\mathbb{Z}_3$ one has :

$$-5 = 1 + 1.3 + 2.3^2 + 2.3^3 + 2.3^4 + \cdots \quad .$$

3.2.4. As $\mathbb{Z}_p$ is a valued ring it has a fraction field, namely the field $\mathbb{Q}_p$ of *p−adic numbers*. Using Hensel representations, elements of $\mathbb{Q}_p$ are

$$a = \sum_{n=v(a)}^{\infty} a_{[n]}\, p^n \ , \ a_{[n]} \in [0, \ldots, p-1] \ , \ a_{[v(a)]} \neq 0 \ ,$$

where $v(a)$ belongs to $\mathbb{Z}$ and is, obviously, the (p-adic) valuation of a. The field $\mathbb{Q}$ of rational numbers is contained in $\mathbb{Q}_p$, the rational numbers appearing as the eventually periodic Hensel series. For instance, in $\mathbb{Q}_3$ one has :

$$\frac{1}{24} = \frac{2}{3} + \frac{2+3}{1-9} = 2.\frac{1}{3} + 2 + 3 + 2.3^2 + 3^3 + \cdots \quad .$$

The p-adic valuation $v(r)$ of a rational number r is given by writing it in the form :

$$p^{v(r)}\ \frac{m}{d} \quad ; \ m \text{ and } d \text{ prime to } p.$$

3.2.5. For each prime p we have defined a p-adic valuation (resp. absolute value) over $\mathbb{Q}$. We will denote it by v_p (resp. $|.|_p$) when confusion may occur. Besides the classical absolute value (often denoted by $|.|_\infty$ by analogy with Example 3.1) they are essentially the only ones, more precisely any non trivial valuation over $\mathbb{Q}$ must be equivalent to one of them. Moreover they are connected by the *product formula* :

$$|r|_\infty \prod_{p \text{ prime}} |r|_p = 1 \quad .$$

3.3 The Ring $\widehat{\mathbb{Z}}$ and Adelic Numbers

3.3.1. As it is difficult to believe that any single prime plays a special role in nature, we generalize the construction of Example 3.2 without any privileged prime. The starting point is the family of finite sets $E_n = \mathbb{Z}/n\mathbb{Z}$ of 'integers modulo n'. For any integer k there is a projection from E_{kn} to E_n given by the remainder when dividing by n. These projections enable us to construct a ring, denoted by $\widehat{\mathbb{Z}}$, along a method generalizing that of 2.1. However things

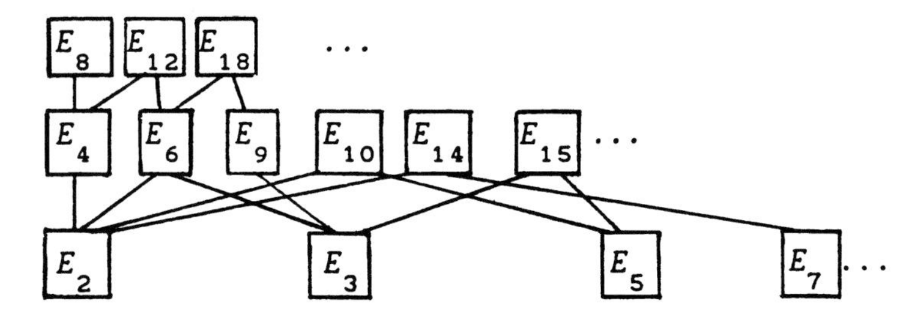

Fig. 3.

are more complicated because the sets E_n are no longer stacked but ordered in a more intricate way shown in fig.3.

Notice that going through two different paths from E_n to E_m gives the same projection. At the price of introducing an arbitrary choice, we will work on a sub-stack of the array and then recover the 'classical' construction 2.1.

3.3.2. A sequence of integers $\{u_n\}$ is said to go *multiplicatively to infinity* if, for any given integer k, there exists an integer N such that, for $n > N$, u_n becomes a multiple of k. For instance the sequence $n!$ goes multiplicatively to infinity. We choose once and for all a sequence $\{u_n\}$ among the numerous ones that satisfy the two conditions :

$$\begin{array}{ll} (3.3.2.1) & \left\{ \begin{array}{l} \{u_n\} \text{ goes multiplicatively to infinity} \\ u_0 = 1 \text{ and for each } n \text{ , } u_n \text{ divides } u_{n+1} \text{ .} \end{array} \right. \\ (3.3.2.2) & \end{array}$$

3.3.3. Let us change notations slightly and denote by E_n the set $\mathbb{Z}/u_n\mathbb{Z}$. The condition (3.3.2.2) implies that we can apply the construction of 2.1 and the algebraic remarks of 2.2 to the sequence of rings E_n and obtain a ring $E = \widehat{\mathbb{Z}}$. Choosing for instance the sequence $\{1, 2, 6, 12, 60, \ldots\}$ for u_n, we get the 'tree' of Fig.4.

Then the condition (3.3.2.1) says that distinct integers will be eventually separated in the construction, and hence that $\mathbb{Z}$ is a subring of $\widehat{\mathbb{Z}}$. The topology induced on $\mathbb{Z}$ is described by the following remark : *a sequence of integers converges to* 0 *in* $\widehat{\mathbb{Z}}$ *if and only if it goes multiplicatively to infinity* (notice that this is independent of the chosen sequence $\{u_n\}$. Actually we have a special case of 2.2.4 with $A = \mathbb{Z}$ and $I_n = u_n\mathbb{Z}$. Hence $\mathbb{Z}$ is dense in $\widehat{\mathbb{Z}}$ and $\widehat{\mathbb{Z}}$ appears as the set of 'limits' of sequences $\{s_n\}$ of integers such that $s_n - s_m$ becomes a multiple of any fixed integer when n and m are large enough.

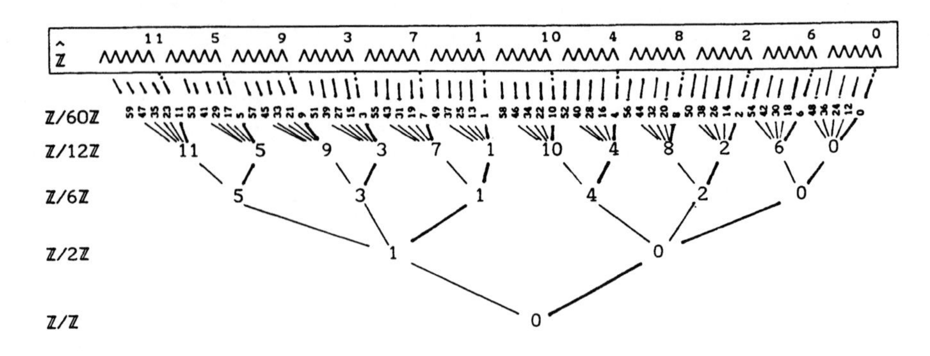

Fig. 4. A $\widehat{\mathbb{Z}}$-tree

Remark. As there are always new primes available, one cannot have $I_{n+m} \supset I_n.I_m$ for all integers n and m , hence the v defined in 2.2.2 is not a valuation, and $\widehat{\mathbb{Z}}$ is not a valued ring but only an ultrametric compact and complete ring.

3.3.4. Let n be an integer, from its prime decomposition :

$$n = \prod_{p \text{ prime}} p^{v_p(n)}$$

we get the decomposition (this is the Chinese Remainder Theorem) :

$$\mathbb{Z}/n\mathbb{Z} = \prod_{p \text{ prime}} \mathbb{Z}/p^{v_p(n)}\mathbb{Z}$$

(actually there is only a finite number of primes in the product, namely those dividing n). We can apply this decomposition for each u_n . For any prime p , the condition 3.3.2.2 says that the sequence $v_p(u_n)$ increases and the condition 3.1.2.1 that it goes to infinity. Then the decomposition gives in the limit :

$$\widehat{\mathbb{Z}} = \prod_{p \text{ prime}} \mathbb{Z}_p \ .$$

Here $\widehat{\mathbb{Z}}$ is a compact set, as an (infinite) product of compact sets. This allows us to write an element a of $\widehat{\mathbb{Z}}$ as a family $(a_2, a_3, a_5, \ldots, a_p, \ldots)$ of p-adic numbers. The addition and multiplication laws act component by component. As an exercise the reader can recover the a_n of the construction 2.1 from the a_p just defined.

3.3.5. The ring $\widehat{\mathbb{Z}}$ has many zero divisors, as is easily seen from 3.3.4 (hence it cannot be valued !). Therefore it has no fraction field. However a kind of 'fraction ring' can be defined, namely the ring of *finite adeles* :

$$\mathbb{A}_f = \{(a_2, a_3, \ldots, a_p, \ldots) ;\\ a_p \in \mathbb{Q}_p,\ a_p \in \mathbb{Z}_p \text{ for all but a finite number of } p\text{'s}\} \ .$$

The *adeles* $\mathbb{A}$ are constructed in the same way but with one more component a_∞ lying in $\mathbb{R}$, so that all completions of $\mathbb{Q}$ are considered. The group of *(finite) ideles* is the multiplicative subgroup of invertible elements of (finite) adeles. They are the (finite) adeles with non zero components and all but a finite number of components of (p-adic) valuation 0 (equivalently of absolute value 1). Given an idele, the number :

$$|a_\infty|_\infty \prod_{p \text{ prime}} |a_p|_p$$

is called its volume. Any rational number a is identified with the (finite) adele all components of which are equal to a (if a is an integer we get the element of $\widehat{\mathbb{Z}}$ already obtained in 3.3.3.). If $a \neq 0$ we obtain an idele and the product formula asserts that it has a unit volume. It is worth pointing out that $\mathbb{Q}$ is a discrete subgroup of $\mathbb{A}$ for addition, as $\mathbb{Z}$ is a discrete subgroup of $\mathbb{R}$ (indeed two rational numbers have distance 1 from each other for all but a finite number of p-adic distances). Moreover the analogy goes further and the quotient in both cases is compact (using the Chinese Remainder Theorem one proves $\mathbb{A}/\mathbb{Q} = \mathbb{R}/\mathbb{Z} \prod \mathbb{Z}_p$).

3.4 Parisi's Matrices

3.4.1. Let us recall the definition of matrices introduced by Parisi as a 'replica symmetry breaking' in the Sherrington-Kirkpatrick mean field model of a spin glass (for details see Rammal, Toulouse, and Virasoro 1986). Let $1 = m_N \leq \cdots \leq m_0 = n$ be integers such that m_i is a multiple of m_{i+1} and let $Q_i (0 \leq i \leq N-1)$ be a sequence of real numbers. From these data, for $0 \leq i \leq N$, one can construct a $(n/m_i) \times (n/m_i)$-matrix $\mathbf{Q}^{(i)}$ following the two recursion rules :

a) $\mathbf{Q}^{(0)} = [0]$,

b) The matrix $\mathbf{Q}^{(i+1)}$ is constructed by substituting for the entry $\mathbf{Q}^{(i)}_{a,b}$ of $\mathbf{Q}^{(i)}$ a $(m_i/m_{i+1}) \times (m_i/m_{i+1})$-matrix $\mathbf{P}^{(i),a,b}$ defined by :

$$\mathbf{P}^{(i),a,b}_{c,d} = \begin{cases} \mathbf{Q}^{(i)}_{a,b} & \text{if } a \neq b \\ Q_i & \text{if } a = b \text{ and } c \neq d \\ 0 & \text{if } a = b \text{ and } c = d \end{cases}$$

At the end of the process we get a so called *Parisi matrix* $\mathbf{Q} = \mathbf{Q}^{(N)}$.

3.4.2. We verify now that, by means of a slight change of notation, a Parisi matrix, obtained from a strictly decreasing sequence Q_i of positive real numbers, is nothing else than the table of mutual distances 'at level n' of $\widehat{\mathbb{Z}}$ endowed with one of the distances defined in 3.3.3. Thus it is not surprising that calculations

	11	5	9	3	7	1	10	4	8	2	6	0
11	0	β	α	α	α	α	1	1	1	1	1	1
5	β	0	α	α	α	α	1	1	1	1	1	1
9	α	α	0	β	α	α	1	1	1	1	1	1
3	α	α	β	0	α	α	1	1	1	1	1	1
7	α	α	α	α	0	β	1	1	1	1	1	1
1	α	α	α	α	β	0	1	1	1	1	1	1
10	1	1	1	1	1	1	0	β	α	α	α	α
4	1	1	1	1	1	1	β	0	α	α	α	α
8	1	1	1	1	1	1	α	α	0	β	α	α
2	1	1	1	1	1	1	α	α	β	0	α	α
6	1	1	1	1	1	1	α	α	α	α	0	β
0	1	1	1	1	1	1	α	α	α	α	β	0

Fig. 5.

based on such matrices lead to an ultrametric space. More precisely choose a sequence $\{u_n\}$ of integers satisfying (3.3.2.1) and (3.3.2.2) and a strictly decreasing sequence $\{d_n\}$ of real numbers going to zero such that, for $0 \leq i \leq N$:

$$u_i = m_{N-i} \; ; \; d_i = Q_{N-i}$$

and construct the associated distance on $\widehat{\mathbb{Z}}$ following 3.3.2 and 2.1.3. Then if we order the subset $\{0, 1, \ldots, n-1\}$ of $\widehat{\mathbb{Z}}$ according to the remainder when dividing by the u_i's , the table of mutual distances is exactly the Parisi matrix. We think an example is better than a complicated explanation. Figure 5 gives the Parisi matrix for the sequences $\{1, 2, 6, 12, \ldots\}$ and $\{1, \alpha, \beta, \ldots\}$. We must point out that Parisi matrices are constructed usually from increasing sequences Q_i and that the associated distance is in fact given by $1 - Q_i$. When working in this last situation, it would be more convenient to have diagonal entries equal to 1 instead of 0.

B. Complete Non-Archimedean Valued Fields

4. General Properties

4.1 Definitions

4.1.1. Let K be a non-Archimedean valued field i. e. (see 1.2.2) a field endowed with an absolute value $|.|$ satisfying (1.2). The set :

$$\mathcal{O}_K = \{a \in K; \; |a| \leq 1\}$$

is a valued ring (ultrametricity implies the stability with respect to addition). It is called the *valuation ring of* K. The set :

$$\mathcal{M}_K = \{a \in K;\ |a| < 1\}$$

is an ideal of $\mathcal{O}_K$. In fact this is *the* maximal ideal because every element u of $\mathcal{O}_K$ not in $\mathcal{M}_K$ is invertible. Hence the set :

$$k = \mathcal{O}_K/\mathcal{M}_K$$

is a field called the *residue field of* K. For instance for $k((x))$, the valuation ring is $k[[x]]$ and the residue field is k. For $\mathbb{Q}_p$ the valuation ring is $\mathbb{Z}_p$ and the residue field is $\mathbb{Z}/p\mathbb{Z}$, i. e. *the* field with p elements, also denoted by $\mathbb{F}_p$. We will use the following usual notation :

If $a \in \mathcal{O}_K$, its image in k is denoted by $\overline{a}$.

For instance if a is an integer in $\mathbb{Z}_p$ then $\overline{a}$ is its residue (mod p) in $\mathbb{Z}/p\mathbb{Z}$.

4.1.2. A field is said to be of characteristic zero if it contains $\mathbb{Z}$. Otherwise its characteristic is the smallest integer p for which one has $p.1 = 0$. This integer must be prime. Denote by $\mathrm{ch}(K)$ the characteristic of K. Examining the characteristics of a non-Archimedean valued field and of its residue field leads to three possibilities :

1) $\mathrm{ch}(K) = \mathrm{ch}(k) = 0$,
2) $\mathrm{ch}(K) = \mathrm{ch}(k) = p$,
3) $\mathrm{ch}(K) = 0$; $\mathrm{ch}(k) = p$.

Let us assume moreover that the field K is complete (for the topology defined by its valuation); then :

* In case of *equal characteristics* (cases 1 or 2), the field K contains subfields isomorphic to $k((x))$. More explicitly, for any a in the maximal ideal, there is an embedding i, given by $i(x) = a$, of $k((x))$ into K. The image of i is denoted by $k((a))$.
* In case of *unequal characteristics* (case 3), the field K contains a subfield isomorphic to $\mathbb{Q}_p$.

4.2 Looking for Squares in $\mathbb{Q}_p$

Our next aim is to solve algebraic equations in complete non-Archimedean valued fields. Let us begin by the simplest case, namely second degree equations in the field of p-adic numbers. An elementary calculation reduces the question to solving the equation $x^2 = a$. The existence of solution(s) in the field $\mathbb{Q}_p$ will be studied in three steps :

4.2.1. As the valuation of any element of $\mathbb{Q}_p$ is an integer, for a to be a square its valuation has to be an *even* integer (the trivial case $a = 0$ being omitted). For instance p is not a square in $\mathbb{Q}_p$.

4.2.2. Let us suppose $v(a)$ to be even, i.e., $a = p^{2n}u$, $|u| = 1$. Then it is enough to solve $x^2 = u$. Taking residues, we get $\overline{x}^2 = \overline{u}$ (in $\mathbb{F}_p$) and find a second condition for a to be a square in $\mathbb{Q}_p$, namely that $\overline{u}$ has to be a square in $\mathbb{F}_p$. For instance neither 2 nor -1 are squares in $\mathbb{Q}_3$ because 2 is not a square in $\mathbb{F}_3$.

Remark. Squares in $\mathbb{F}_p$ can be found directly by squaring all numbers in $\mathbb{F}_p$ or better by using the well known reciprocity law of Gauss, which we recall for the sake of completeness. Let p be a prime number and n an integer prime with p. The *Legendre symbol* $\left(\frac{n}{p}\right)$ is 1 (resp. -1) if n is (resp. is not) a square in $\mathbb{F}_p$. The Legendre symbol can be easily computed by means of the following rules where p and q are distinct odd prime numbers :

$$\left(\frac{n+p}{p}\right) = \left(\frac{n}{p}\right) , \left(\frac{n.m}{p}\right) = \left(\frac{n}{p}\right)\left(\frac{m}{p}\right)$$

$$\left(\frac{-1}{p}\right) = (-1)^{(p-1)/2} , \left(\frac{2}{p}\right) = (-1)^{(p^2-1)/8}$$

$$\left(\frac{q}{p}\right)\left(\frac{p}{q}\right) = (-1)^{(p-1)(q-1)/4} \quad \text{(reciprocity law)} .$$

4.2.3. Now $\overline{u}$ is a square in $\mathbb{F}_p$ if and only if there exists α in $\mathbb{Z}_p$ such that $u - \alpha^2$ belongs to $p\mathbb{Z}_p$. For $\overline{u} \neq 0$, α is invertible in $\mathbb{Z}_p$ and the map :

$$\ell(x) = \frac{1}{2\alpha}\left(\frac{u-\alpha^2}{p} - px^2\right)$$

is a contraction of $\mathbb{Z}_p$. Its unique fixed point v satisfies :

$$u = (\alpha + pv)^2$$

and gives a square root of u (starting with $-\alpha$ one gets $-v$ and the other square root). We sum up our discussion by the following equivalence :

$$\exists b \in \mathbb{Q}_p : a = b^2 \iff a = 0 \text{ or } \begin{cases} a = p^{2n}u \, , \, |u| = 1 \\ \overline{u} = \alpha^2 \text{ (in } \mathbb{F}_p) \, . \end{cases}$$

Remark. In $\mathbb{R}$ there is only one condition to be satisfied for x to be a square, namely $x \geq 0$, but in $\mathbb{Q}_p$ there are two conditions to be satisfied. Roughly speaking one half of real numbers are squares (compared to one quarter for p-adic numbers).

4.2.4. Example. As $-1 = 2 \times 2 \pmod 5$ the polynomial $x^2 + 1$ has a root a in $\mathbb{Z}_5$ such that $\overline{a} = 2$. It is a good exercise to do explicit calculations. Supposing the root to be $2 + 5b$, one finds $b = 1 + 5(b + b^2)$, which enables us to find the Hensel series recursively :

$$a = 2 + 5 + 2.5^2 + 5^3 + 3.5^4 + 4.5^5 + 2.5^6 + 3.5^7 + \cdots \quad .$$

4.3 Hensel's Lemma

Basically, Hensel's lemma gives a generalization to any monic polynomial P with coefficients in $\mathbb{Z}_p$ of the principle explained in 4.2.3 : it roughly says that for each root α of $\overline{P}$ (the polynomial obtained by taking residues coefficient-wise) there exists in $\mathbb{Z}_p$ one root a of P such that $\overline{a} = \alpha$. It represents an even more basic tool for solving algebraic equations in a *complete* ultrametric field K since any polynomial P whose coefficients lie in K can be broken into a product of polynomials each of which has all its roots with the same valuation.

4.3.1. *Hensel's Lemma.* Let K be a complete ultrametric field and let $P = \sum a_i x^i$ be a polynomial whose coefficients a_i lie in $\mathcal{O}_K$. Suppose there exist two polynomials q and r whose coefficients lie in the residue field k, respectively prime, and such that $\overline{P} = \sum \overline{a}_i x^i = q\,r$. Then there exist two polynomials Q and R whose coefficients lie in $\mathcal{O}_K$ and satisfying : $P = Q\,R$, $\overline{Q} = q$, $\overline{R} = r$, $\deg(Q) = \deg(q)$. The proof of this lemma is based on Newton's algorithm.

Remarks.

* The primality condition means that q and r have no common factors; it is essential as shown by the polynomial $P = x^2 + p$ of $\mathbb{Z}_p[x]$ which is irreducible but verifies $\overline{P} = x\,x$.

* There are several other Hensel lemmas (which deal with analytic functions, differential operators,... instead of polynomials); each of them states that, under some mild condition (here primality), any break at the residue field level is actually a 'total' break.

4.3.2. When the polynomial q is of degree one, Hensel's lemma becomes : Let K be a complete ultrametric field and let P be a polynomial whose coefficients lie in $\mathcal{O}_K$. Let α be a simple root of $\overline{P}$ (i. e. $\overline{P}(\alpha) = 0$ but $\overline{P}'(\alpha) \neq 0$). Then there exists a root a of P in $\mathcal{O}_K$ such that $\overline{a} = \alpha$.

4.3.3 Example. One can apply the method given in the proof of Hensel's lemma (i. e. Newton's method) to find roots of the polynomial $x^2 + 1$. The root lying 'near 2' appears to be the limit of the sequence :

$$u_0 = 2, u_{n+1} = u_n - (u_n^2 + 1)/2u_n \quad .$$

In this particular case, explicit calculations are more involved if this general method is used instead of the method of 4.2.3.

4.3.4 Newton's Polygon. With each polynomial :

$$P = \sum a_n\, x^n$$

(more generally with each power or Laurent series) one associates the set Δ of points Δ_n with coordinates $(n, v(a_n))$. Let x be in K. A direct calculation shows that the straight line $L_n(x)$ with slope $v(x)$ which goes through the point

Δ_n intersects the y-axis at the point $(0, v(a_n x^n))$. By the ultrametric inequality (1.3) one knows that the point $(0, v(P(x)))$ is above one line $L_n(x)$ at least. Moreover, the ultrametric inequality is actually an equality when only one term of the sum has the smallest valuation. In other words, *if there is only one point of Δ on the lowest line $L_n(x)$ then this line intersects the y-axis at the point $(0, v(P(x)))$.*

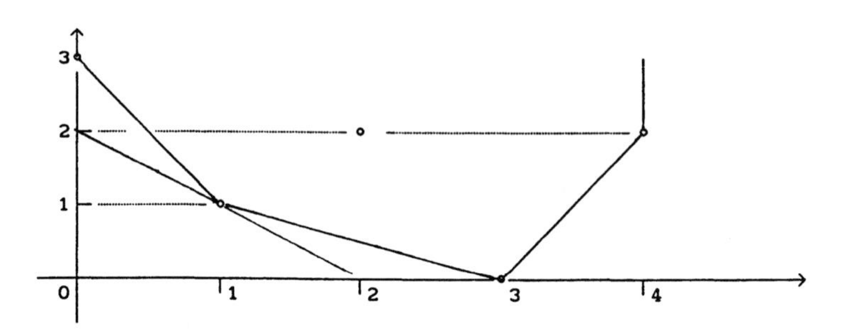

Fig. 6. Newton polygon for the polynomial $P(x) = 25x^4 + 2x^3 + 25x^2 + 5x + 250$, and graphic computation of the valuation of $P(5)$.

Conversely, if x is a root of P, as $v(P(x)) = \infty$, the previous situation cannot happen and the lowest line must encounter two points of Δ at least. Hence this lowest line must be an edge of the convex hull $\mathcal{N}(P)$ of the set $\Delta \cup \{\infty\}$ where ∞ means a point at infinity in the $0y$-direction. By removing vertical edges from the boundary of $\mathcal{N}(P)$ one gets a polygonal line *called the Newton polygon of P*. Thus we proved the slopes of the edges of the Newton polygon to be the (possible) valuations of the roots of the polynomial. Actually one can go further by applying Hensel's lemma. Let us choose an edge of the Newton polygon of P. Let λ denote its slope and m its length. Then there exist polynomials Q and R such that :

1) $P = QR$.
2) The polynomial Q is of degree m and has a Newton polygon consisting of a single edge of slope λ,
3) The slopes of the edges of the Newton polygon of R are distinct from λ.

From this one obtains a basic result : the irreducible polynomials of $K(x)$ have Newton polygons consisting of a single edge. Thus their roots in any non-Archimedean valued field containing K (and whose valuation extends that of K) have same valuation, namely the slope of the edge of the Newton polygon.

5. Extensions

5.1 Finite Extensions

5.1.1. Let K be a complete non-Archimedean valued field and let L be a finite (algebraic) extension of K, say of degree n. Then there exists a unique absolute value on L which extends the absolute value of K. Let α belong to L and let $P(x) = x^d + \cdots + a_0$ be the minimal monic polynomial of $K[x]$ such that $P(\alpha) = 0$. Then the valuation of α is given by the slope of the unique edge of the Newton polygon of P and one must have :

$$|\alpha| = |a_0|^{1/d} \text{ or } v(\alpha) = v(a_0)/d \quad . \tag{5.1}$$

It remains to verify that this actually gives an absolute value on L.

Remark. For any α in L , the degree d of $K(\alpha)$ over K divides the degree n of L over K (n/d is the degree of L over $K(\alpha)$).

5.1.2. Let us suppose now that K is *discretely valued* i.e. that the set $v(K)$ of valuations of its elements is a discrete subgroup of $\mathbb{R}$ (for instance $\mathbb{Q}_p$ is discretely valued because $v(\mathbb{Q}_p) = \mathbb{Z}$) and that L/K is a finite extension of degree n. By construction the set $v(L)$ is an additive subgroup of $\frac{1}{n}\, v(K)$ and hence must be $\frac{1}{e}\, v(K)$ for some integer e. This integer is called the *ramification index* of L/K. On the other hand, the residue field $\overline{L}$ appears to be a finite extension of the residue field $\overline{K}$. Let f be its degree. Then one can prove that $n = ef$.

5.1.3. Let now L be an extension of $\mathbb{Q}_p$ of finite degree n and ramification index e. Let us choose an element π of L such that $v(\pi) = 1/e$. As $\overline{L}$ is an extension of $\mathbb{F}_p$, say of degree f, it must be the finite field with p^f elements. For each element α of $\overline{L}$ choose an *antecedent* $\underline{\alpha}$ such that $\overline{\underline{\alpha}} = \alpha$ and denote by U the set of all antecedents (it has exactly p^f elements). It is easy to prove that each element a of L has a unique expansion :

$$a = \sum_{ev(a)\leq i}^{\infty} a_i \pi^i \ ; \ a_i \in U$$

which looks very much like a Hensel expansion.

5.2 Algebraically Closed Extensions. Construction of $\mathbb{C}_p$

5.2.1. Let K be any field. Its *algebraic closure* K^{alg} is, roughly speaking, the set of all roots of polynomials in $K[x]$. It is also the smallest field containing K which is *algebraically closed* i.e. in which any polynomial is the product of polynomials of degree one.

5.2.2. When K is a complete non-Archimedean field, there exists a unique absolute value over K^{alg} , given by Formula (5.1), which extends the absolute value of K. Hence the field K^{alg} is a non-Archimedean field but unfortunately it is not complete in general. However a deep result of Krasner says that *the completion* of K^{alg} is an algebraically closed complete non-Archimedean field.

5.2.3. The completion of the algebraic closure of $\mathbb{Q}_p$ is 'classically' denoted by $\mathbb{C}_p$ because this field plays the role of $\mathbb{C}$ in many respects. The set of valuations $v(\mathbb{C}_p)$ is $\mathbb{Q}$ so $\mathbb{C}_p$ is not discretely valued. The residue field of $\mathbb{C}_p$ is the algebraic closure of the field $\mathbb{F}_p$. Unlike $\mathbb{Q}_p$, or more generally any complete discretely valued field, there are, up to isomorphism, several *immediate extensions* of $\mathbb{C}_p$, i.e. fields with the same set of valuations and residue field as $\mathbb{C}_p$. This phenomenon will be made clear in the next paragraph.

5.3 Spherically Complete Extensions

5.3.1. A curious properties of $\mathbb{C}_p$ is that it contains decreasing sequences of disks with void intersection, i.e. disks :

$$D_n = D(a_n, r_n^-) = \{x \in \mathbb{C}_p \ ; 0 \leq |x - a_n| < r_n)$$

such that $D_{n+1} \subseteq D_n$ but $\cap D_n = \emptyset$. As $\mathbb{C}_p$ is a complete field, when this situation occurs the sequence r_n is strictly decreasing but does not go to 0 (and so the situation cannot occur in $\mathbb{Q}_p$ where the r_n belong to $p^{\mathbb{Z}}$!).

5.3.2. A metric space is said to be *spherically complete* if any decreasing sequence of disks in it has a non-empty intersection. Spherically complete spaces are complete (apply the definition to sequences of disks with radii going to zero) but the converse is not true. However one can prove the following results:

* Any complete and *discretely valued* non-Archimedean field is spherically complete (easy),
* Any algebraically closed non-Archimedean field which is *maximally complete* i.e. with no immediate extension (strict extension with same set of valuations and residue field) is spherically complete (not very difficult),
* Any algebraically closed non-Archimedean field is contained in a maximally complete field. The latter is algebraically closed hence spherically complete (very technical : the key point is to prove that the family of immediate extensions is actually a set !).

These results show the existence of a field, denoted by Ω_p, which contains $\mathbb{C}_p$ (and $\mathbb{Q}_p$) and is both algebraically complete and spherically complete. This field is useful under certain circumstances, but is difficult to visualize because it is so big.

5.4 Ultrametric Banach Spaces

5.4.1. A vector space E over the non-Archimedean valued field K is said to be ultrametric if it is endowed with a norm $\|.\|$ satisfying the classical properties but where the triangular inequality is replaced by the ultrametric one :

$$\|v + w\| \leq \text{Max}(\|v\|, \|w\|) \ .$$

A vector space is said to be a Banach space if it is complete. Even though the proofs may be different, ultrametric Banach spaces have most of the properties of classical Banach spaces. For instance if E is a *finite dimensional* K-vector space and K is a *complete* non-Archimedean valued field, then all ultrametric norms over E are equivalent.

5.4.2. However there are some noticeable exceptions. The most famous one is the Hahn-Banach theorem[3], which is false for non spherically complete fields. Indeed let L be a spherically complete field and let K be a complete but not spherically complete sub-field of L; then there is no K-linear map bounded by 1 on L which extends the identity map of K .

5.4.3. Let E be a K-Banach space. The family (e_i) is said to be a *normal basis* of E if for all x in E there exist x_i in K such that :

$$x = \sum_{i \in I} x_i \, e_i \ ; \ \|x\| = \sup_{i \in I} |x_i| \ .$$

The existence and construction of normal bases is a fundamental issue when studying Banach spaces. As an example let $\mathcal{C}$ be the set of continuous functions from $\mathbb{Z}_p$ to $\mathbb{C}_p$. This is a $\mathbb{C}_p$-Banach space for the norm :

$$\|f\| = \sup_{x \in \mathbb{Z}_p} |f(x)|$$

of uniform convergence and the family (e_i) given by :

$$e_i(x) = x(x-1)\cdots(x-i+1)/i!$$

is proved to be a normal basis for it. More precisely if f belongs to $\mathcal{C}$ one has :

$$f = \sum_{i=0}^{\infty} a_i \, e_i \ ; \ a_n = \sum_{i=0}^{n} (-1)^{n-i} \, e_i(n) \, f(i) \ .$$

Conversely this formula (Newton's interpolation) enables to decide whether a function from $\mathbb{N}$ to $\mathbb{C}_p$ can be continued into a function of $\mathcal{C}$: its interpolation coefficients a_i, defined by the above formula, must go to zero. For instance the function :

$$n \longrightarrow a^n = \sum_{i=0}^{\infty} (a-1)^i \, e_i(n)$$

is the restriction to $\mathbb{N}$ of a function of $\mathcal{C}$ if and only if $|a-1| < 1$. In other words the number a^x can be defined for a in $1 + p\mathbb{Z}_p$ and x in $\mathbb{Z}_p$.

[3] Any linear form defined over some sub-vector space F of E and bounded by M on F can be extended to a linear form on E bounded by M.

6. Analytic Functions

6.1 Generalities

6.1.1. Let us choose once for all a complete algebraically closed non-Archimedean field K, for instance $K = \mathbb{C}_p$. A power series :

$$f(x) = \sum_{n=0}^{\infty} a_n (x - \alpha)^n$$

converges if and only if its general term goes to zero i.e. it converges in the 'open' disk :

$$D(\alpha, r^-) \; ; \; r = 1/\limsup \sqrt[n]{|a_n|}$$

as in the complex case. Moreover the power series converges for x such that $|x - \alpha| = r$ if and only if :

$$\lim |a_n| \, r^n = 0$$

in which case it converges in the 'closed' disk $D(a, r)$.

6.1.2. A function f defined in some ('open' or 'closed') disk $D(\alpha, r)$ of K is said to be *analytic* if it is the sum of a power series converging in that disk. Now because of the ultrametric nature of the distance in K, any point β in $D(\alpha, r)$ is a center and f can be expanded in a power series around β. This power series converges in the disk $D(\beta, r) = D(\alpha, r)$.

6.1.3. A function defined over a union of disks and analytic in each of them is called *locally analytic*. Again by ultrametricity, two disks are either concentric [4] or disjoint. Hence if a function is defined on two non concentric disks and analytic on each one of them, there is no connection between the values of f in the two disks (just as in the complex case for functions holomorphic on a set with two connected components). As a consequence of this, one cannot construct analytic continuation by using the same trick as in the complex case. However theories of analytic continuation do exist (see 6.2 for instance) but are of a quite different kind.

6.2 Basic Examples

To illustrate general properties we give basic examples of analytic functions over $\mathbb{C}_p$.

6.2.1 The Exponential. Let $[r]$ denote the integer part of the real r. A simple calculation shows that :

$$v(n!) = [n/p] + [n/p^2] + \cdots + [n/p^h] + \cdots \quad .$$

From this one deduces that the function :

[4] In the sense that there exists a common center.

$$\exp(x) = \sum_{n=0}^{\infty} x^n/n!$$

is only defined in a small disk namely the disk $D(0, \rho^-)$ with :

$$\rho = |p|^{1/(p-1)} < 1 \ .$$

Moreover for any x in that disk one has :

$$|\exp(x) - 1| \le \rho < 1 \ .$$

Hence :

$$|\exp(x)| = 1 \ .$$

In other words the exponential function is *bounded* (by 1) on its disk of convergence and has no zero in it. More generally a bounded analytic function whose coefficients lie in a discretely valued field has only a finite number of zeros in its disk of convergence (proof by means of the Newton polygon). However non-bounded functions can have an infinite number of zeros as will be shown for the logarithm function. Moreover, bounded function with coefficients in a non discretely valued field can also have an infinite number of zeros, for instance the function : $\sum a_n\, x^n \ ; \ a_n^n = p \ , \ a_n \in \mathbb{C}_p$.

6.2.2 The Logarithm. The logarithm function is defined by :

$$\log(x) = -\sum_{n=1}^{\infty} \frac{(-1)^n}{n} \ (x-1)^n \ .$$

It converges on the disk $D(1, 1^-)$ and for x and y in that disk one has

$$\log(xy) = \log(x) + \log(y) \ .$$

In particular, if ζ is a k-th root of unity in $D(1, 1^-)$ one has :

$$\log(\zeta) = \frac{1}{k} \log(\zeta^k) = \frac{1}{k} \log(1) = 0 \ .$$

To find zeros of the logarithm function we have to figure out which k-th roots of unity lie in $D(1, 1^-)$. If ξ is a k-th root of unity, $\xi - 1$ is a root of the polynomial :

$$P_k(x) = (1+x)^k - 1 \ .$$

Let $k = p^h d$ with $(d, p) = 1$. One finds :

$$P_k(x) = P_{p^h}(x)\, Q(x)$$
$$Q(x) = [1 + (1+x)^{p^h} + ... + (1+x)^{(d-1)p^h}] \ .$$

As $Q(0) = d$ is prime to p, by looking at the Newton polygon of Q one sees that no root of Q can lie in $D(0, 1^-)$. Therefore only the p^h-th roots of unity can lie in the disk $D(1, 1^-)$. Conversely, by looking at the Newton polygon of the polynomial $P_{p^h}(x)$ it is easy to show that the p^h-th roots of unity lie in the disk $D(1,1^-)$ and thus are zeros of the logarithm. Thus we constructed an infinite set of zeros of this function. It remains to prove that we have obtained all zeros in this manner. For that purpose, we consider the Newton polygon of the logarithm function constructed in Fig.7.

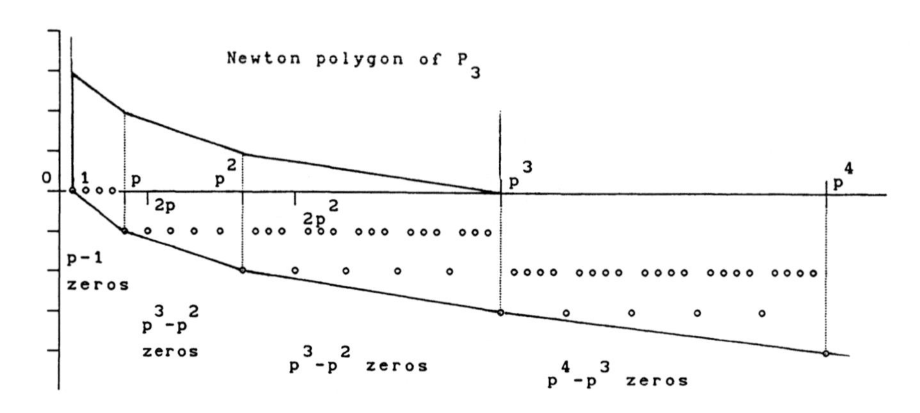

Fig. 7. The Newton polygon of the logarithm function

The link between the Newton polygons of the logarithm function and of the polynomials P_h is contained in the useful formula :

$$\log(x) = \lim_{h\to\infty} [1 - x^{p^h}]/p^h \quad \text{for } |x| < 1 \ .$$

As for polynomials, the slopes of the edges of the Newton polygon of the logarithm function give the valuations of the zeros. Putting all the information together one finally obtains : For each integers $h \geq 0$ the logarithm function has exactly $p^h - p^{h-1}$ zeros such that :

$$v(\zeta - 1) = 1/(p^h - p^{h-1}) \quad \text{i.e. } |\zeta - 1| = |p|^{p^{1-h}/(p-1)} = \rho^{1/p^{h-1}}$$

which are the roots of unity :

$$\zeta^{p^h} = 1 \ ; \quad \zeta^{p^{h-1}} \neq 1 \ .$$

This exhausts the set of zeros distinct from 1 . If $|x| < \rho$, so that $\exp(x)$ does exist, one has :

$$\log(\exp(x)) = x \ .$$

In the opposite direction :

$$\exp(\log(x)) = x$$

holds when $|\log(x)| < \rho$ i. e. when $|x-1| < \rho$, where ρ is the radius of convergence of the exponential function.

6.2.3 The Artin-Hasse Exponential. The analytic function :

$$\ell(x) = x + x^p/p + x^{p^2}/p^2 + \cdots + x^{p^h}/p^h + \cdots$$

is defined in the disk $D(0,1^-)$ and looks (p-adically) very much like the logarithm $\log(1-x)$, for the two functions have the same Newton polygon. For instance the zeros of the two functions have the same valuations. It is also easy to check on the Newton polygon that :

$$|\ell(x)| \leq |x| \;\text{ if } |x| \leq \rho$$

where ρ is the radius of convergence of the exponential function.

One defines the Artin-Hasse exponential by taking the formal expansion :

$$A(x) = \exp(\ell(x)) = \sum_{n=0}^{\infty} e_n x^n \quad . \tag{6.1}$$

A priori this power series exists for $|x| < \rho$, but in fact $|e_n| \leq 1$, and hence $A(X)$ converges for $|x| < 1$ (compare with the function $x = \exp(\log(x))$ which exists a priori only on $D(1,\rho^-)$ but in fact everywhere).

Proof. One finds by formal calculations :

$$A(x)^p/A(x^p) = \exp(px) = 1 + pg(x) ,$$

where the coefficients of g are p-adic integers. On the other hand, as for any power series in $\mathbb{Q}[[x]]$, there exist b_n (in $\mathbb{Q}$) such that :

$$A(x) = \prod_{n=1}^{\infty}(1 - b_n x^n)$$

thus :

$$A(x)^p/A(x^p) = \prod_{n=1}^{\infty}(1 - b_n x^n)^p/(1 - b_n x^{np}) = \prod_{n=1}^{\infty}(1 + p\, b_n x^n + \cdots)$$

by induction on n one deduces that the b_n are p-adic integers as well as the e_n. □

As the e_n are integers and $e_1 = 1$, one finds, for $|x| < 1$:

$$|A(x) - 1 - x| \leq |x^2| \quad \text{hence} \quad |A(x) - 1| = |x| \quad . \tag{6.2}$$

Now let us choose one number ξ in $\mathbb{C}_p$ such that :

$$\ell(\xi) = 0 \; ; \; |\xi| = \rho$$

(by considering the Newton polygon of ℓ one sees that there exist exactly $p-1$ numbers with these properties). To compute $A(\xi)$ one must be very cautious: using (6.1) carelessly gives $A(\xi) = 1$ which contradicts (6.2). Indeed formula (6.1) is true only formally, hence for $|x| < \rho$. So it cannot be applied for ξ. But one actually has for any x in $D(0, \rho)$:

$$A(x)^p = \exp(p\,\ell(x))$$

because now $|p\,\ell(x)| \leq |p||x| < \rho$. Thus $A(\xi)$ is a p-th root of unity. Moreover by (6.2) one knows that it is *the* p-th root of unity that lies in $D(1+\xi, \rho^2)$. Thus the Artin-Hasse exponential gives an analytic representation of the p-th roots of unity.

6.2.4. In practice and to make calculations easier one works with a truncated function $\ell(x)$ keeping only the first two terms. We leave it as an exercise to check the following properties : let π be a root of the polynomial $X^{p-1} + p = 0$ ('Dwork's π'). The function :

$$E_\pi(x) = \exp(\pi x - \pi x^p) = \sum_{n=0}^{\infty} f_n\, x^n$$

(Dwork's exponential) which a priori converges in $D(0, 1^-)$, is actually defined in $D(0, p^{(p-1)/p^2})$. Moreover the value $E(1)$ is the p-th root of unity which lies in the disk $D(1+\pi, \rho^2)$.

6.3 Analytic Elements

We only give a short overview of the most classical theory of analytic continuation. It is due to Krasner. Let Δ be a subset of $\mathbb{C}_p$ and let :

$$H_0(\Delta) = \{f \in \mathbb{C}_p(x) \ \text{ without any pole in } \Delta\} \ .$$

We define the 'norm of uniform convergence' by the formula :

$$\|f\| = \sup_{x \in \Delta} |f(x)|$$

and consider the completion $H(\Delta)$ of $H_0(\Delta)$ for this norm (obviously this is a Banach space). The functions in $H(\Delta)$ are called *analytic elements* in Δ. When Δ is not too complicated, for instance if it is the complement in a disk of a finite union of smaller disks, $H(\Delta)$ has good properties, for instance the zeros of analytic elements are isolated. Hence on such Δ there is an analytic continuation theory. Namely if two analytic elements take same values on a small disk, then they are equal everywhere.

C. Integration

There are two entirely distinct theories of integration on ultrametric groups depending upon the field on which the functions to be integrated are defined. When dealing with $\mathbb{C}$-valued functions, this is a particular chapter in the general theory of integration over groups, but dealing with $\mathbb{C}_p$-valued functions leads to entirely new phenomena.

7. $\mathbb{C}$-valued Integration

7.1 Integration over a Profinite Group

Let G_n be finite Abelian groups and let :

$$G = \varprojlim G_n$$

be their inverse limit for homomorphisms from G_{n+1} onto G_n (see 2.1.2 and 2.2.1). In this situation G is said to be a *profinite group*. It is then a compact group. Hence there is a (unique) Haar measure μ on it such that :

$$\int_G d\mu(x) = 1 \quad .$$

The Haar measure on G can be easily calculated. Let a be in G_n and let :

$$D_n(a) = \{x \in G\, ; \ x_n = a\}$$

(recall that elements x in G are sequences $\{x_n \in G_n\}$). As Haar measures are invariant by translation one finds :

$$\mu(D_n(a)) = 1/\#(G_n) \quad ,$$

where $\#(E)$ denotes the number of elements in E. Now, let f be a continuous function from G to $\mathbb{C}$, and for each a in G_n let us choose an antecedent $\underline{a}$ in G such that $a = \underline{a}_n$. One finds :

$$\int_G f(x)\, d\mu(x) = \lim_{n\to\infty} \frac{1}{\#(G_n)} \sum_{a\in G_n} f(\underline{a}) \quad . \tag{7.1}$$

7.2 Integration over $\mathbb{Z}_p$

7.2.1. By construction $\mathbb{Z}_p$ is a profinite group. Now for any number x of $\mathbb{Z}_p$ the absolute value $|x|$ is in $\mathbb{R}$ so that, for s in $\mathbb{R}^+$, the function $x \longrightarrow |x|^s$ is a continuous function. As an example we will integrate it over $\mathbb{Z}_p$. Taking $\{0, 1, \ldots, p^n - 1\}$ as set of representatives for $\mathbb{Z}/p^n\mathbb{Z}$, equation (7.1) becomes :

$$\int_{\mathbb{Z}_p} |x|^s \, d\mu(x) = \lim_{n\to\infty} \frac{1}{p^n} \sum_{a=0}^{p^n-1} |a|^s \quad .$$

Grouping together numbers with the same absolute value, one readily computes (remember $|p| = 1/p$) :

$$\begin{aligned}\int_{\mathbb{Z}_p} |x|^s \, d\mu(x) &= \lim_{n\to\infty} \frac{1}{p^n} \left[(p^n - p^{n-1}) 1^s + (p^{n-1} - p^{n-2}) |p|^s + \cdots \right. \\ &\qquad \left. \cdots + (p-1) |p|^{s(n-1)} \right] \\ &= \lim_{n\to\infty} \frac{(p-1)}{p} (1 + p^{-s-1} + \cdots + p^{(-s-1)(n-1)}) \\ &= (1-p^{-1})/(1-p^{-1-s}) \quad .\end{aligned}$$

7.2.2. The above computation can be generalized in the following way. Let $f(x_1, \ldots, x_r)$ be a polynomial in $\mathbb{Z}_p[x_1, \ldots, x_r]$ and let :

$$N_n = \#\{(x_1, \ldots, x_r) \in (\mathbb{Z}/p^n\mathbb{Z})^r \ ; \ f(x_1, \ldots, x_r) \equiv 0 \ (\mathrm{mod}\ p^n)\}$$

be the number of solutions 'approximate at order n' of f [5]. For instance in the case $r = 1$ and $f(x) = x$, one has $N_n = 1$ for all n. One defines the so-called *Poincaré series* :

$$P(T) = \sum_{n=0}^{\infty} N_n T^n$$

and one computes, for s in $\mathbb{R}^+$:

$$\int_{\mathbb{Z}_p^r} |f(x_1, \ldots, x_r)|^s \, d\mu(x_1) \cdots d\mu(x_r) = p^s - (p^s - 1) P(p^{-r-s}) \quad .$$

This formula enabled Igusa to prove $P(T)$ to be a rational fraction.

7.3 Integration over $\widehat{\mathbb{Z}}$ and the Replica Trick

Let $\{u_n\}$ be a sequence satisfying conditions 3.3.2 and let g be a continuous function from $\widehat{\mathbb{Z}}$ to $\mathbb{C}$. Formula 7.1 gives :

$$\begin{aligned}\int_{\widehat{\mathbb{Z}}} g(z) \, d\mu(z) &= \lim_{n\to\infty} \frac{1}{u_n} \sum_{a=0}^{u_n-1} g(a) \\ &= \lim_{n \times\to\infty} \frac{1}{n} \sum_{a=0}^{n-1} g(a) \quad ,\end{aligned}$$

[5] caution : an exact solution of $f(x_1, ..., x_r) = 0$ gives an approximate solution but the converse is false.

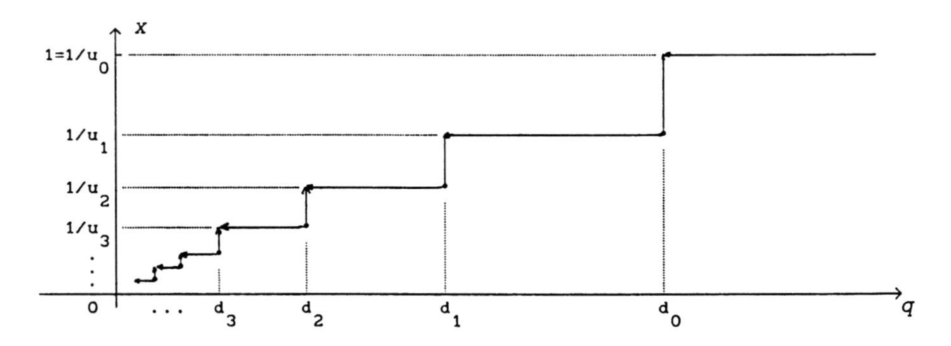

Fig. 8.

where $n \times\!\!\rightarrow \infty$ means that n goes multiplicatively to infinity. Now suppose a distance d has been defined on $\widehat{\mathbb{Z}}$ by means of a decreasing sequence $\{d_n\}$ as in 2.1.3 and 3.4.2. Using the corresponding Parisi matrices one computes for any function f from $\mathbb{R}$ to $\mathbb{C}$:

$$
\begin{aligned}
\int_{\widehat{\mathbb{Z}}} f(d(z,0))\, d\mu(z) &= \lim_{n\to\infty} \frac{1}{u_n} \sum_{a=0}^{u_n-1} f(d(a,0)) \\
&= \lim_{n\to\infty} \frac{1}{u_n(u_n-1)} \sum_{a=0}^{u_n-1} \sum_{b=0}^{u_n-1} f(\mathbf{Q}_{a,b}^{(n)}) \ .
\end{aligned} \tag{7.2}
$$

On the other hand the integral can be computed by a classical method. Namely let us define a function x from $\mathbb{R}^+$ to $[0,1]$ by :

$$x(q) = \mu\{z \in \widehat{\mathbb{Z}}\,;\ d(z,0) < q\}$$

and let q be the 'inverse' function from $[0,1]$ to $\mathbb{R}$:

$$q(x) = d_n \quad \text{when} \quad \frac{1}{u_{n+1}} \le x < \frac{1}{u_n}$$

shown in Fig.8. Then one finds easily :

$$
\begin{aligned}
\int_{\widehat{\mathbb{Z}}} f(d(z,0))\, d\mu(z) &= \int_0^\infty f(q)\, dx(q) \\
&= \int_0^1 f(q(x))\, dx \ .
\end{aligned} \tag{7.3}
$$

Formulas (7.2) and (7.3) recall those used when dealing with the replica trick. The main distinction is that in our formulation, although the sequence u_n goes to zero in $\widehat{\mathbb{Z}}$, it has to be considered as an $\mathbb{R}$-valued sequence in formula (7.2) and thus goes to infinity.

8. $\mathbb{C}_p$-valued Integration

8.1 Measures on $\mathbb{Z}_p$

8.1.1. A 'p-adic' measure μ on $\mathbb{Z}_p$ is a *bounded* finitely additive map from finite unions of disks of $\mathbb{Z}_p$ into $\mathbb{C}_p$. Then, if f is a continuous function from $\mathbb{Z}_p$ to $\mathbb{C}_p$, one defines the integral of f by the formula :

$$\int_{\mathbb{Z}_p} f(x)\, d\mu(x) = \lim_{n\to\infty} \sum_{a=0}^{p^n-1} f(a)\ \mu(a+p^n\mathbb{Z}_p) \ . \tag{8.1}$$

It is not difficult to prove that the limit does exist, but the boundedness of μ is essential in the proof.

8.1.2. Let us suppose that there exists an p-adic measure μ invariant by translation. The measure of a disk would depend only upon its radius. So, normalizing by $\mu(\mathbb{Z}_p)=1$, we should get, for any a in $\mathbb{Z}_p$:

$$\mu(a+p^n\mathbb{Z}_p) = 1/p^n \ .$$

As $|1/p^n|$ goes to infinity with n, the 'measure' μ cannot be bounded. Hence *there is no p-adic Haar measure !* However let us suppose the Haar measure does exist and compute its moments by means of its characteristic function. For any z in $\mathbb{C}_p$ such that $|z| < \rho = |p|^{1/(p-1)}$ one obtains :

$$\begin{aligned}
\sum_{k=0}^{\infty} \frac{z^k}{k!} \int_{\mathbb{Z}_p} x^k\, d\mu(x) &= \int_{\mathbb{Z}_p} e^{xz}\, d\mu(x) \\
&= \lim_{n\to\infty} \frac{1}{p^n} \sum_{a=0}^{p^n-1} e^{az} = \lim_{n\to\infty} \frac{e^{p^n z}-1}{p^n(e^z-1)} \\
&= \frac{z}{e^z-1} = \sum_{k=1}^{\infty} \frac{B_k}{k!}\, z^k \\
&= \sum_{k=1}^{\infty} -k\,\zeta(1-k)\ \frac{z^k}{k!} \ ,
\end{aligned}$$

where the B_k are Bernoulli numbers and ζ is the famous Riemann zeta function. The aim of the next paragraph is to give a correct version of this computation, namely to express values of the zeta function at negative integers by means of an integral over $\mathbb{Z}_p$.

8.2 A p-adic Zeta Function

8.2.1. As $\mathbb{C}$ and $\mathbb{C}_p$ are algebraically closed they both contain exactly $p-1$ $(p-1)$-th roots of unity. Moreover, by Fermat's theorem the polynomial $x^{p-1}-1$ has $p-1$ distinct roots in $\mathbb{F}_p$ (namely $1,2,\ldots,p-1$) hence, according to Hensel's lemma, it has $p-1$ roots in $\mathbb{Z}_p$, one in each disk $D(i,|p|)$ $(1\le i<p)$. So the

$(p-1)$-th roots of unity of $\mathbb{C}_p$ actually lie in $\mathbb{Z}_p$. Let us choose a one to one mapping between $(p-1)$-th roots of unity in $\mathbb{C}$ and $(p-1)$-th roots of unity in $\mathbb{C}_p$ which preserves multiplication (group isomorphism). For instance for $p=5$ there are two such isomorphisms, one of which is given by :

$$1 \longrightarrow 1, \quad -1 \longrightarrow -1, \quad i \longrightarrow 2+\cdots, \quad -i \longrightarrow 3+\cdots \ .$$

The other is obtained by exchanging i and $-i$.

8.2.2. We will suppose now that $p>2$. Let α be a $(p-1)$-th root of unity in $\mathbb{C}_p$ distinct from 1, and let μ_α be the p-adic measure defined by

$$\mu_\alpha(\mathbb{Z}_p)=1\,,\ \mu_\alpha(a+p^n\mathbb{Z}_p)=\alpha^a \quad \text{for } n\geq 1$$

As $\alpha^p=\alpha$, the measure of $a+p^n\mathbb{Z}_p$ does not depend of the (integer) a as center. Moreover the formula :

$$\sum_{b=0}^{p-1}\mu(a+p^n b+p^{n+1}\mathbb{Z}_p)=\sum_{b=0}^{p-1}\alpha^{a+p^n b}=\alpha^a\ \frac{\alpha^{p^{n+1}}-1}{\alpha^{p^n}-1}=\alpha^a$$

guarantees the (finite) additivity of μ_α.

8.2.3. Let us compute the moments of μ_α :

$$\begin{aligned}\sum_{k=0}^{\infty}\frac{z^k}{k!}\int_{\mathbb{Z}_p}x^k\,d\mu_\alpha(x)&=\int_{\mathbb{Z}_p}e^{xz}\,d\mu_\alpha(x)\\ &=\lim_{n\to\infty}\sum_{a=0}^{p^n-1}e^{az}\,\alpha^a=\lim_{n\to\infty}\frac{e^{p^n z}\,\alpha^{p^n}-1}{e^z\alpha-1}\\ &=\frac{\alpha-1}{e^z\alpha-1}=(1-\alpha)\sum_{k=0}^{\infty}L(-k,\alpha)\ \frac{z^k}{k!}\ .\end{aligned}$$

The numbers $L(-k,\alpha)$ defined by the last equality lie in $\mathbb{C}_p$, but they can be expressed as numbers in $\mathbb{Q}[\alpha]$. Thus by means of the isomorphism in 8.2.1 they can be viewed as elements of $\mathbb{C}$. Now this last equality becomes the Taylor expansion of a well known holomorphic function. A classical formula asserts that the $L(-k,\alpha)$'s are the values, at negative integers, of a holomorphic function defined for $\mathrm{Re}(s)>1$ by :

$$L(s,\alpha)=\sum_{n=1}^{\infty}\frac{\alpha^n}{n^s}$$

8.2.4. We readily compute (for $\mathrm{Re}(s)>1$, hence for all s by analytic continuation) :

$$\sum_{\alpha^{p-1}=1}L(s,\alpha)=\sum_{n=1}^{\infty}\frac{p-1}{[(p-1)n]^s}=(p-1)^{1-s}\zeta(s)\ .$$

Hence it is natural to define a 'pseudo Haar measure' on $\mathbb{Z}_p$ by :

$$\mu = \sum_{\substack{\alpha^{p-1}=1 \\ \alpha\neq 1}} \frac{1}{1-\alpha} \mu_\alpha \ .$$

Putting everything together and noticing that $L(s,1) = \zeta(s)$ one finally obtains for every integer $k \geq 0$:

$$\int_{\mathbb{Z}_p} x^k \, d\mu(x) = \sum_{\substack{\alpha^{p-1}=1 \\ \alpha\neq 1}} L(-k,\alpha) = [(p-1)^{1+k} - 1]\zeta(-k) \ .$$

As $\zeta(-k) = -B_k/(k-1)$ these numbers lie in $\mathbb{Q}$, hence in $\mathbb{C}_p$. For instance, for $k = 0$, one has $\mu(\mathbb{Z}_p) = (p-2)/2$.

8.2.5. Let $\mathbb{Z}_p^* = \mathbb{Z}_p - p\mathbb{Z}_p$. As a immediate consequence of the definition one has :

$$\mu(px) = \mu(x)$$

so that $\mu(\mathbb{Z}_p^*) = 0$. More generally :

$$\int_{\mathbb{Z}_p^*} x^k \, d\mu(x) = (1-p^k) \int_{\mathbb{Z}_p} x^k \, d\mu(x) = [(p-1)^{1+k} - 1](1-p^k)\zeta(-k) \ .$$

8.2.6. On the other hand, the formula :

$$a^p = (b + a - b)^p = b^p + p(a-b)[\cdots] + (a-b)^p$$

shows that for any a in $\mathbb{Z}_p$ and $n \geq 1$:

$$|a-b| \leq |p|^n \ \Rightarrow \ |a^p - b^p| \leq |p|^{n+1} \ .$$

But, for any a in $\mathbb{Z}_p^*$ one has :

$$|a^{p-1} - 1| \leq |p|$$

hence for any integer $n \geq 0$:

$$|a^{(p-1)p^n} - 1| \leq |p|^n \ .$$

Roughly speaking this means that if $k - k$' is divisible by $(p-1)p^n$ for n large enough then a^k and $a^{k'}$ are close in $\mathbb{Z}_p$. In other words the function $k \longrightarrow a^k$ can be continued so as to obtain a continuous function from

$$\varprojlim \mathbb{Z}/(p-1)p^n\mathbb{Z} = \mathbb{Z}/(p-1)\mathbb{Z} \times \mathbb{Z}_p$$

to $\mathbb{Z}_p$.

8.2.7. Now using 8.2.6 it is not difficult to prove that :

$$\zeta_p(s) = \frac{1}{(p-1)^{1-s}-1} \int_{\mathbb{Z}_p^*} x^{-s}\, d\mu(x)$$

is a function from $\mathbb{Z}/(p-1)\mathbb{Z} \times \mathbb{Z}_p$ to $\mathbb{Z}_p$, defined and continuous outside 1. Moreover from 8.2.5 one has, for all integers $k \geq 0$:

$$\zeta_p(-k) = (1-p^k)\zeta(-k) \ .$$

Remarks.

* The function ζ_p was discovered by Kubota and Leopoldt.
* The factor $(1-p^{-s})$ is exactly the p-th factor in the Euler expansion of ζ.
* The continuity of ζ_p gives many congruences between values at negative integers, hence between Bernoulli numbers. These congruences where already known as Kummer's congruences.
* The formula $a^s = \exp[\log(a)s]$, true for a in $1+p\mathbb{Z}_p$ and $|s| \leq 1$, shows the function ζ_p to be (the restriction of) an analytic function on each component $\mathbb{Z}_p$ of $\mathbb{Z}/(p-1)\mathbb{Z} \times \mathbb{Z}_p$.

8.2.8. As a typical example of p-adic computation we show how to compute the 'residue' of ζ_p near 1. Clearly we only have to compute :

$$\int_{\mathbb{Z}_p^*} x^{-1}\, d\mu(x) = \sum_{\substack{\alpha^{p-1}=1 \\ \alpha \neq 1}} \frac{1}{1-\alpha} \lim_{n\to\infty} \sum_{\substack{a=1 \\ (a,p)=1}}^{p^n-1} \frac{\alpha^a}{a} \ .$$

For $|x| < 1$ one checks easily :

$$\lim_{n\to\infty} \sum_{\substack{a=1 \\ (a,p)=1}}^{p^n-1} \frac{x^a}{a} = \sum_{\substack{a=1 \\ (a,p)=1}}^{\infty} \frac{x^a}{a}$$

$$= \log(1-x) - \frac{1}{p}\log(1-x^p) = \frac{1}{p}\log\left(\frac{(1-x)^p}{1-x^p}\right) \ .$$

Let :

$$\Delta = \{x \in \mathbb{C}_p \ ; \ |x| \leq 1 \, , \ |x-1| = 1\} = D(0,1) - D(1,1^-) \ .$$

One has :

$$\sum_{\substack{a=1 \\ (a,p)=1}}^{p^{n+1}-1} \frac{x^a}{a} = \sum_{\substack{a=1 \\ (a,p)=1}}^{p^n-1} \frac{x^a}{a} \sum_{b=0}^{p-1} \frac{a}{a+p^n b} x^{p^n b} \ .$$

But on Δ the sequence

$$\sum_{b=1}^{p-1} x^{p^n b} = \frac{x^{p^{n+1}} - x^{p^n}}{x^{p^n}-1}$$

converges uniformly to zero. From this one easily deduces that the sequence

$$\sum_{\substack{a=1 \\ (a,p)=1}}^{p^n-1} \frac{x^a}{a}$$

uniformly converges in Δ, i.e., that its limit is an analytic element in Δ . On the other hand there exists a polynomial P in $\mathbb{Z}[x]$ such that :

$$\frac{(1-x)^p}{1-x^p} = 1 - p\frac{P(x)}{1-x^p}$$

and for x in Δ one has :

$$\left|\frac{P(x)}{1-x^p}\right| \leq 1 \ .$$

Hence the function :

$$\log\left(\frac{(1-x)^p}{1-x^p}\right) = \lim_{n\to\infty} \sum_{k=1}^{n} p^n \left(\frac{P(x)}{1-x^p}\right)^n$$

is also an analytic element in Δ . As we know the two analytic elements agree for $|x| < 1$, by Krasner's theorem they agree everywhere in Δ , for instance at the $(p-1)$-th roots of unity (distinct from 1). Finally we get the formula :

$$\int_{\mathbb{Z}_p^*} x^{-1}\, d\mu(x) = \sum_{\substack{\alpha^{p-1}=1 \\ \alpha\neq 1}} \frac{1}{1-\alpha} \log((1-\alpha)^{p-1}) \ .$$

Here the $(p-1)$-th power must be left in the argument for the logarithm to be defined.

8.3 The p-adic Gamma Function

To define a p-adic gamma function we use an entirely distinct viewpoint. The aim is to find a p-adic analogue to functions like $\Gamma(a) = \int x^{a-1}e^x\, dx$ i.e. defined by integrating a differential depending upon a parameter. The basic remark is that when integrating on a *closed* path exact differentials give zero. So, following Dwork-Boyarski, we will act as if the path 0—∞ were closed somewhere.

8.3.1. Let π be a root of the polynomial $X^{p-1}+p$ (cf. 6.2.4) and let $H^\dagger$ be the space of *overconvergent* analytic functions, i.e. functions which are analytic in a disk $D(0,r)$ of $\mathbb{C}_p$ for some $r > 1$ (depending upon the function). For each 'parameter' a in $\mathbb{Z}_p$ let :

$$\Omega_a = x^a e^{\pi x} H^\dagger \ .$$

If $f = x^a e^{\pi x} g$ lies in Ω_a then $xf' = x^a e^{\pi x}(ag + \pi x g + x g')$ also lies in Ω_a so it makes sense to define :

$$W_a = \Omega_a / (x \frac{d}{dx}\Omega_a) \ .$$

Taking $g = x^k$ $(k \geq 0)$ in the above computation one finds :

$$x^a\ e^{\pi x}\ x^{k+1} \simeq -\frac{(a+k)}{\pi}\ x^a\ e^{\pi x}\ x^k \tag{8.2}$$

where $\simeq$ means that the difference of the two sides is in $x\,\frac{d}{dx}\,\Omega_a$. One deduces immediately that W_a is one dimensional over $\mathbb{C}_p$ with $x^a\ e^{\pi x}$ as a basis.

8.3.2. Define an operator, called inverse Frobenius, which acts on analytic functions by the following formula :

$$\psi(\sum a_n x^n) = \sum a_{np} x^n \quad .$$

One checks that this operator preserves $H^\dagger$, and the formula

$$\psi(xf') = p\,x\,[\psi(f)]' \tag{8.3}$$

asserts that it preserves derivatives. To explain how it acts on Ω_a we need to introduce the unique number b of $\mathbb{Z}_p$ which satisfies :

$$a = p\,b - r\ ;\ r \in \mathbb{Z}\ ,\ 0 \leq r < p \quad .$$

With these notations one finds :

$$\psi(x^a\ e^{\pi x}\ g) = x^b\ e^{\pi x}\ \psi(x^{a-pb}\ e^{\pi(x-x^p)}\ g) \quad .$$

As explained in 6.2.4 the function $e^{\pi(x-x^p)}$ is overconvergent so that the function $\psi(x^a\ e^{\pi x}\ g)$ belongs to Ω_b (as $r < p$, terms containing a negative power of x disappear when applying ψ). Formula (8.3) asserts that the action of ψ is compatible with taking quotients. So one obtains an operator, denoted α , from W_a to W_b. By definition the p-adic gamma function is the 1×1 matrix of the operator α , namely one has :

$$\alpha(x^a\ e^{\pi x}) = \pi^r\ \Gamma_p(a)\ x^b\ e^{\pi x} \quad . \tag{8.4}$$

8.3.3. In W_0 , using (8.2) one gets $x^k\ e^{\pi x} \simeq \text{Cst} \cdot x\,e^{\pi x} \simeq 0$. Hence for $a = 0$ one has $b = r = 0$ and one computes in W_0 :

$$\Gamma_p(0)\ e^{\pi x} \simeq e^{\pi x}\ \psi(e^{\pi(x-x^p)}) \simeq e^{\pi x} \quad .$$

If $|a| = 1$, p does not divide a and one has $r > 0$, $a+1 = pb - (r-1)$. Using the definition of Γ_p, (8.2) and (8.3) one computes in W_{a+1} :

$$\begin{aligned} \Gamma_p(a+1)\ x^b\ e^{\pi x} &\simeq \pi^{-r+1}\ \psi(x^{a+1} e^{\pi x}) \\ &\simeq \pi^{-r+1}\ \psi(-\frac{a}{\pi}\ x^a\ e^{\pi x}) \\ &\simeq -a\,\Gamma_p(a)\ x^b\ e^{\pi x} \quad . \end{aligned}$$

If $|a| < 1$, one has $a = pb$, $r = 0$ and $a+1 = p(b+1) - (p-1)$. One computes:

$$\Gamma_p(a+1)\, x^{b+1}\, e^{\pi x} \simeq \pi^{-p+1}\, \psi(x^{a+1} e^{\pi x}) \simeq -\frac{1}{p}\,\psi(-\frac{a}{\pi}\, x^a\, e^{\pi x})$$
$$\simeq \frac{b}{\pi}\,\Gamma_p(a)\, x^b e^{\pi x} \simeq -\Gamma_p(a+1)\, x^{b+1}\, e^{\pi x} \quad .$$

Summarizing one obtains :

$$\Gamma_p(0) = 1\ ,\ \Gamma_p(a+1)/\Gamma_p(a) = \begin{cases} -1 & \text{if } |a| < 1 \\ -a & \text{if } |a| = 1 \ . \end{cases}$$

This was the original Morita definition for the p-adic gamma function. It can be used to compute its values at positive integers, namely :

$$\Gamma_p(k) = (-1)^k \prod_{\substack{i=1 \\ p \nmid i}}^{k} i$$

which reminds us of the values of the classical Γ function.

8.3.4. Let

$$e^{\pi(x-x^p)} = \sum_{n=0}^{\infty} c_n x^n$$

one can prove that $v(c_n) \geq n(p-1)/p^2$. Using (8.2) once more one readily computes, for $r \in \mathbb{Z}\ ,\ 0 \leq r < p-1\ ,\ b \in \mathbb{Z}_p$:

$$\begin{aligned} \pi^r\ \Gamma_p(pb-r)\ x^b\ e^{\pi x} &\simeq x^b\, e^{\pi x}\, \psi(x^{-r} e^{\pi(x-x^p)}) \\ &= x^b\, e^{\pi x}\, \psi(\sum_{n=0}^{\infty} c_n x^{n-r}) \\ &= x^b\, e^{\pi x} \sum_{n=0}^{\infty} c_{np+r} x^n \\ &\simeq x^b\, e^{\pi x} \sum_{n=0}^{\infty} c_{np+r}\, (-\pi)^{-n}\, (b+n-1)(b+n-2)\ldots(b) \end{aligned}$$

Thus $\Gamma_p(pb-r) = h_r(b)$ where the function h_r is defined by :

$$h_r(x) = \pi^{-r} \sum_{n=0}^{\infty} c_{np+r}\, (-\pi)^{-n}\, (x+n-1)(x+n-2)\cdots(x) \quad .$$

It is not very difficult to check that this formula defines a function h_r which is analytic in the disk $D(0, |p|^{e}\ ^{-})$ where :

$$e = p^{-1} + (p-1)^{-1} - 1 < 0 \quad .$$

Hence the p-adic Γ_p function is, in each disk $r + p\mathbb{Z}_p$, the restriction of an analytic function.

References

1 Elementary books to go further.

About algebraic properties (Section 5.1) :

Borevitch Z. I., Chafarevitch I. R. : *Number Theory*, Academic Press (1966) (French translation: Gauthier-Villars 1967).

About analytic properties (Section 6) :

Amice Y. : *Les nombres p-adiques* Collection Sup. P.U.F. (1975).

About p-adic measures, gamma function,... (Section 7, 8) :

Koblitz N. : *p-adic Analysis : A short Course on Recent Work* , London Math. Soc. Lecture Notes 46 (1980).

About connections with physics :

Rammal R., Toulouse G., Virasoro M. : *Ultrametricity for physicists* , Rev. Mod. Phys. 58 (1986) 765.

Each of these books contains numerous references. Some exercises at an elementary level can be found in :

Parent D. P. : *Exercices de théorie des nombres* Gauthier Villars (1978) (English translation: Springer (1984), Japanese translation (1987))

2 More specialised references.

On valuations :

Schilling O. : *The theory of valuations* Math Survey IV (1950).

On Banach spaces :

Monna A. F. : Rapport sur la théorie des espaces linéaires topologiques sur un corps valué non-archimédien.

Gruson L., Van Der Put M. : Banach spaces.

both in *Table ronde d'analyse non archimédienne 1972 Paris*, Bull. Soc. Math. France Mémoire 39-40 (1974) p 255-278 and 55-100.

With an introduction to rigid analytic geometry, very complete and self contained, but rather difficult for non specialists :

Bosh S. , Güntzer U., Remmert R. : *Non-Archimedean Analysis* Grundlehren 261 Springer (1984).

Chapter 10

Introduction to Lattice Geometry

by Marjorie Senechal

1. Introduction

Lattices arise in many areas of number theory and physics. In this volume, for example, they appear in chapters on algebraic numbers, elliptic functions, and quasicrystals. One reason for their ubiquity is that the lattice is the basic framework for periodic structures, both algebraic and geometric. One reason for their importance is that many nonperiodic structures can be described as irrational sections of lattices in higher dimensions.

(1.1) Definition. *A lattice Λ is the group T, isomorphic to $\mathbb{Z}^n$, of vector sums*

$$\{\sum_{i=1}^{n} \alpha_i \mathbf{a}_i, \alpha_i \in \mathbb{Z}\},$$

where $\mathbf{a}_1, \mathbf{a}_2, \ldots, \mathbf{a}_n$ are n linearly independent vectors in $\mathbb{R}^n$. (We also use the word lattice for the orbit of any point of $\mathbb{R}^n$ under the action of T.)

Thus by their very definition, lattices are geometric as well as algebraic entities (Figure 1). Their geometry is studied in at least three contexts.

(*i.*) The metric properties of a lattice are most easily studied by means of positive definite quadratic forms in n variables: indeed, these properties are encoded in them. Let A be the $n \times n$ matrix whose columns are n linearly independent vectors $\mathbf{a}_1, \ldots \mathbf{a}_n$. Then the columns are a basis for a lattice Λ; A is said to be a *generator matrix* for Λ. The product $A^T A = M$ is a symmetric positive definite $n \times n$ matrix with entries $m_{ij} = (\mathbf{a}_i, \mathbf{a}_j)$. This matrix is a *Gram matrix*, or 'metric tensor' for Λ; by means of it, we can recover Λ up to congruence. M is also the matrix associated with the quadratic form $\mathbf{x}^T M \mathbf{x}$, where $\mathbf{x} \in \mathbb{R}^n$ is a column vector. Thus, alternatively, a basis being chosen, we can identify lattices with their quadratic forms.

We will use the notation $N(\mathbf{x}) = (\mathbf{x}, \mathbf{x})$ for the *norm* or squared length $|\mathbf{x}|^2$ of $\mathbf{x}$, computed in the usual way with respect to an orthonormal basis. Because a lattice is a discrete point set, it has a minimal positive norm which we will denote by r. Thus $\forall \mathbf{x} \in \Lambda$, $\mathbf{x} \neq \mathbf{0}$, $N(\mathbf{x}) \geq r$. In particular, if $\mathbf{x} \in \Lambda$ is written in terms of the basis A, then $\mathbf{x}^T M \mathbf{x} = N(\mathbf{x})$ and $(\mathbf{x}, \mathbf{y}) = \mathbf{x}^T M \mathbf{y}$. The *volume* of the lattice, $V(\Lambda)$, is $|\det A|$; it is invariant under change of basis since $|\det AX| = |\det A|$ when $X \in \mathrm{GL}(n, \mathbb{Z})$. The determinant of the lattice is the determinant of its Gram matrix; hence $\det \Lambda = V(\Lambda)^2$.

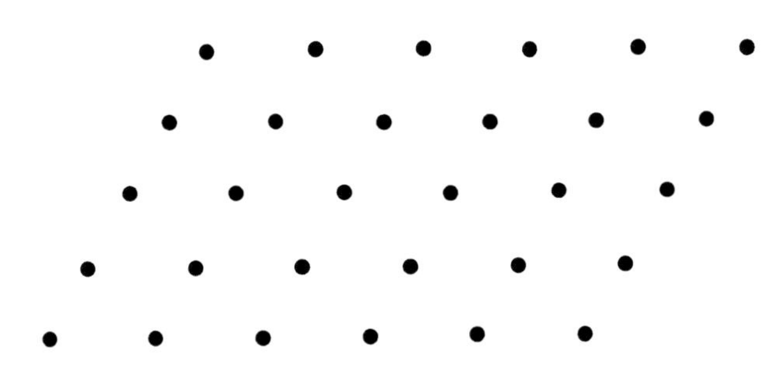

Fig. 1. A lattice in $\mathbb{R}^2$.

The study of lattice symmetry is the study of finite subgroups of $\mathrm{GL}(n, \mathbb{Z})$. The symmetry group of a lattice is the semidirect product of its translation subgroup T and the isotropy subgroup $E(\Lambda)$ of the vector $\mathbf{0} \in \Lambda$. Clearly, $E(\Lambda)$ is a finite subgroup of $O(n)$. Since T is invariant under $E(\Lambda)$, $E(\Lambda)$ has a faithful representation in $\mathrm{Aut}(T) = \mathrm{GL}(n, \mathbb{Z})$. Conversely, any finite subgroup $W \subset \mathrm{GL}(n, \mathbb{Z})$ is a subgroup of the isotropy group of $\mathbf{0}$ for some lattice (a simple construction for a lattice invariant under W is given in (Conway and Sloane, 1988)).

A group of isometries in $\mathbb{R}^n$ which has a fixed point is called a *point group.* Since the right regular representation of any finite group can be realized by permutation matrices, every finite point group has a representation in $\mathrm{GL}(m, \mathbb{Z})$ for some m; m need not be equal to n. Determining the relation between m and n is a problem that goes back at least to Minkowski (Minkowski, 1887); we will discuss it in detail in Section 4.

(*ii.*) There is a close relation between lattices and tilings of $\mathbb{R}^n$. Indeed, lattice theory has evolved side by side with tiling theory: historically, their development has been greatly stimulated by theoretical crystallography. Since ancient times, natural philosophers have attempted to explain the structure of matter as combinations of a very few basic geometric units. As far as crystals are concerned, these efforts first began to achieve success in the early 19th century, when Haüy proposed a model envisioning crystals as stacks of building blocks. Haüy's theory accounted for many of the known morphological properties of crystals, and helped to establish the idea that a crystal is a modular, periodic structure. Since periodicity could be described more simply and more

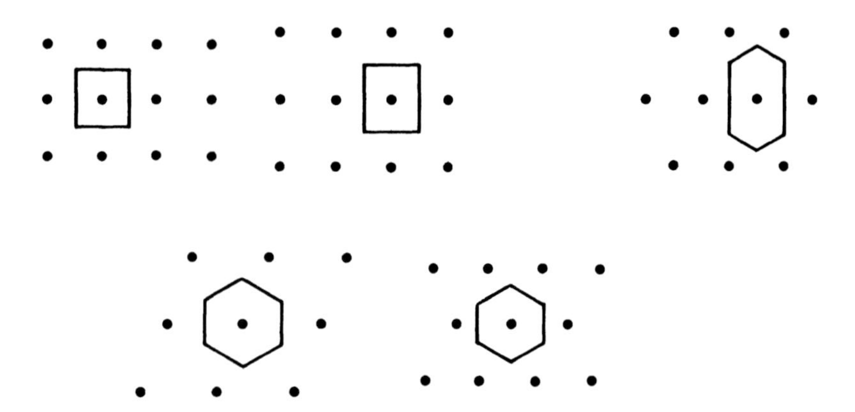

Fig. 2. Dirichlet domains for plane lattices.

generally by lattices than by Haüy's bricks, lattices were soon the preferred way to describe crystals. Today, lattices and tilings in $\mathbb{R}^n$ are considered to be alternative models. For example, the lattice nodes can be taken to be the vertices, or the centers, of tilings of $\mathbb{R}^n$ by congruent parallel polytopes.

Tilings help us to visualize the geometric properties of lattices. In crystallography, the tiles are usually chosen to be copies of the box-like polytope whose vertices are the endpoints of the 2^n vectors $\epsilon_1 \mathbf{a}_1 + \ldots + \epsilon_n \mathbf{a}_n$ of Λ for which $\epsilon_i = 0$ or 1. This polytope is called a *unit cell*; its volume is $V(\Lambda)$. It has $2n$ pairwise parallel $(n-1)$-dimensional faces, and lattice points only at its vertices. There are infinitely many choices of unit cell, corresponding to the infinitely many choices of lattice basis. The unit cell is convenient for classification purposes, but it does not display the symmetry of Λ. There is however an alternative tile, the *Voronoï polytope* of Λ, which does display the symmetry of Λ and also gives a great deal of additional information about its geometry.

The history of Voronoï polytopes begins with early work on the classification of binary quadratic forms. Lagrange had defined a positive definite binary quadratic form $ax^2+2bxy+cy^2$ to be 'reduced' (the distinguished representative of its equivalence class under unimodular transformations) if $0 \leq 2b \leq a$ and $2b \leq c$; such a form exists in every class. For example, the equivalent quadratic forms associated with the two bases $\{(1,0), (0,1)\}$ and $\{(2,1), (1,1)\}$ for the square lattice are x^2+y^2 and $5x^2+6xy+2y^2$; the former is reduced but the latter is not. Dirichlet noted that Lagrange's conditions are always satisfied when the basis vectors $\mathbf{a}_1, \mathbf{a}_2$ for the corresponding lattice are chosen to have minimal lengths and non-negative scalar product. He also pointed out that the perpendicular bisectors of $\pm\mathbf{a}_1, \pm\mathbf{a}_2$ and $\pm(\mathbf{a}_1 - \mathbf{a}_2)$ define a convex polygon which is the region of the plane nearer to 0 than to any other point of Λ (Figure 2). This polygon is a rectangle or a hexagon, according as $(\mathbf{a}_1, \mathbf{a}_2)$ is or is not equal to 0.

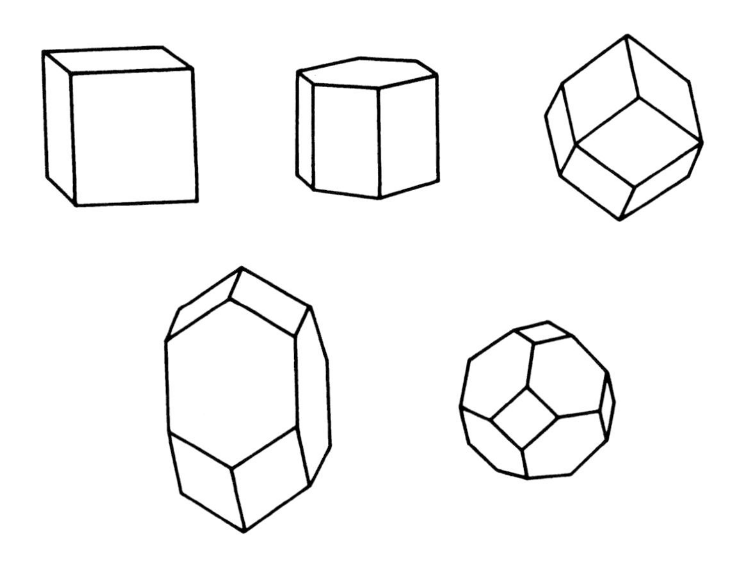

Fig. 3. Fedorov's five parallelohedra.

Dirichlet's polygon is the famous *Dirichlet domain*, a concept which has been redisovered or reintroduced in many different contexts and consequently is known by various names, including 'Wigner-Seitz cells', and 'Brillouin zones'. By 1885, Fedorov had enumerated the 5 combinatorial types of *parallelohedra*, polyhedra which fill $\mathbb{R}^3$ in parallel position (Fedorov, 1885); they are shown in Figure 3.

Early in this century, Voronoï extended the concept to higher dimensional lattices (Voronoï, 1908); thus we attach his name to them, but in honor of Dirichlet's contribution we denote them by the letter D. More recently, Delone and his colleagues discovered the 52 combinatorial types of Voronoï polytopes in $\mathbb{R}^4$ (Delone, 1929; Shtogrin, 1973). The number of types in $\mathbb{R}^5$ is not known; Engel estimates it to be about 75,000 (Engel, private communication). The Voronoï polytope is a remarkably useful tool for investigating lattice geometry. But although the construction is well known, it is not used to the extent that it could be. One goal of this article is to demonstrate its potential.

(*iii.*) Lattices are classified in different ways for different purposes. Most of these classification schemes can be described as quotient and submanifolds of $\mathrm{GL}(n,\mathbb{R})$, with induced topologies (see, e.g., (Schwarzenberger, 1980) and (Michel and Mozryzmas, 1989)).

Two matrices A_1 and A_2 are generator matrices for the same lattice if and only if $A_2 = A_1X$, where $X \in \mathrm{GL}(n,\mathbb{Z})$. Thus there is a one-one correspondence between lattices and left cosets of $\mathrm{GL}(n,\mathbb{Z})$ in $\mathrm{GL}(n,\mathbb{R})$. Two lattices are congruent if for each generator matrix A of Λ_1 there is a generator matrix B of Λ_2 such that $B = OA$, where $O \in O(n)$. Thus to identify congruent lattices we take for the space of lattices the double quotient

$$L(n) = O(n)\backslash \mathrm{GL}(n,\mathbb{R})/\mathrm{GL}(n,\mathbb{Z}).$$

Here $V\backslash U$ and U/V are the sets of right and left cosets of V in U, respectively.

Crystallographers classify lattices in various ways appropriate for the study of crystallographic groups (Michel and Mozryzmas, 1989). The key to this classification is the partition of $L(n)$ into strata under the action of different groups. (A group action $G \to \mathrm{Aut}(M)$ partitions a mathematical structure M into orbits. The sets of orbits whose isotropy groups are conjugate under G are called *strata.*) Under the action of $O(n)$ on $L(n)$, the strata are called *crystal systems*. To each stratum there corresponds an $O(n)$-conjugation class of isotropy groups. The strata obtained by the action of $\mathrm{GL}(n,\mathbb{Z})$ on $L(n)$ are called the *Bravais classes* of lattices. The number of Bravais classes is always larger than the number of crystal systems (for $n = 3$, the numbers are 14 and 7, respectively). However, Jordan proved that the number of Bravais classes is finite in any dimension.

In number theoretic contexts, where lattices are often identified with quadratic forms, the Gram matrices are a principal tool. M_1 and M_2 are Gram matrices for the same lattice if there is an $X \in \mathrm{GL}(n,\mathbb{Z})$ such that $M_2 = X^T M_1 X$; congruent lattices have the same Gram matrices. The positive definite symmetric $n \times n$ matrices or, equivalently, positive definite quadratic forms, constitute an open convex cone $P(n) \subset \mathbb{R}^{n(n+1)/2}$. We will denote the set of equivalence classes $\{X^T M X\}$ by $Q(n)$. The commutative diagram below summarizes the relations among these various classifications.

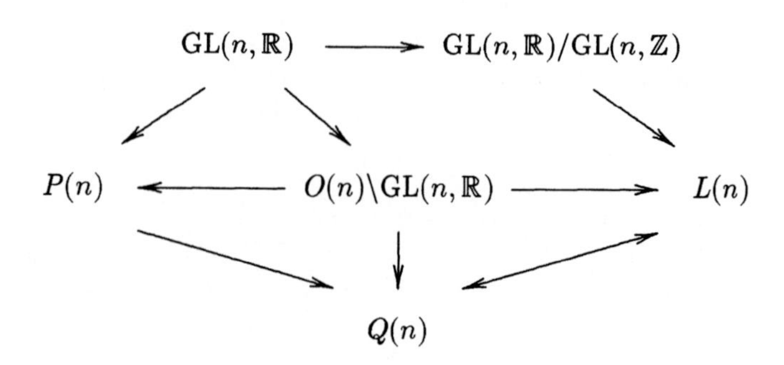

Lattices can also be classified by their Voronoï polytopes, which in turn can be classified by their symmetry (crystal system or Bravais class) or by their combinatorial structure. The partition of $L(n)$ according to combinatorial type does not coincide with its stratifications by symmetry. A single combinatorial structure can often be realized with different symmetries; the Voronoï polytope is very sensitive to parameter changes. Also, different combinatorial structures can have the same symmetry.

A concise history of mathematical crystallography and the role that Hilbert's 18th problem has played in its development can be found in (Senechal, 1985). Important background reading, in addition to the works already cited, includes the fundamental work of Coxeter (Coxeter, 1940, 1973, 1985, 1988), (Grünbaum and Shephard, 1987), the definitive work on tilings and patterns in the plane, and (Gruber and Lekkerkerker, 1987), a comprehensive, recently updated, survey of geometry of numbers, the field created by Minkowski. A recent encyclopedic book and supplementary series of papers by Conway and Sloane (Conway and Sloane, 1988) is an important summary of much of what is known today about lattices in (relatively) low dimensions. The present article draws on all of these sources but includes much that is new. This material, based on joint work with Michel and Engel, will appear in more complete form in our monograph (Engel, Michel, and Senechal, 1992?).

2. Voronoï Polytopes

The construction of a Voronoï polytope of a lattice point is a straightforward generalization of the construction of the Dirichlet domain. We simply join $\mathbf{0} \in \Lambda$ to each of the other lattice points and construct the perpendicular bisectors of these line segments. The Voronoï polytope $D(\mathbf{0})$ is the smallest convex region about $\mathbf{0}$ bounded by these hyperplanes. (Although in principle this construction calls for an infinite number of operations, it can be shown that a small number suffices.) The *facets* of the polytope are its $(n-1)$-dimensional faces; the lattice vectors whose bisectors are facets will be called *facet vectors*. Since every lattice point is a center of symmetry for Λ, and so is the midpoint between any pair of lattice points, $D(\mathbf{0})$ and its facets are centrosymmetric.

An equivalent description of $D(\mathbf{O})$ is

$$D(\mathbf{O}) = \{\mathbf{x} \in \mathbb{R}^n | \ \ |(\mathbf{x},\mathbf{f})| \leq N(\mathbf{f})/2\}$$

for each facet vector $\mathbf{f}$.

Carrying out the Voronoï construction for each lattice point, we obtain a *Voronoï tiling* of $\mathbb{R}^n$ by convex, parallel polytopes which share whole facets. Since the tiles are all congruent we can speak of 'the' Voronoï polytope of Λ. We will usually denote it simply by D; sometimes it will be necessary to use $D(\Lambda)$.

Voronoï polytopes belong to the class of polytopes known as *parallelotopes*, polytopes which tile $\mathbb{R}^n$ when placed facet-to-facet in parallel position. Venkov showed that parallelotopes are completely characterized by three properties: they are centro-symmetric, their facets are centrosymmetric, and each closed sequence of facets linked by parallel $(n-2)$-dimensional faces is of length 4 or length 6 (Venkov, 1954); the last condition derives from the fact that plane 'parallelogons' are quadrilaterals or hexagons. A long-standing conjecture asserts that every parallelotope is an affine image of a Voronoï parallelotope; it has been established only for $n \leq 4$.

Since each Voronoï polytope of a lattice Λ contains exactly one lattice point, its volume is equal to that of a unit cell, and thus it is an alternative tile which can be used to represent Λ. Since, by its construction, the Voronoï polytope is invariant under $E(\Lambda)$, it displays the symmetry of the lattice in a much more visual form than does the generator matrix A or the Gram matrix M. Moreover, since each facet of the polytope lies in the bisector of a lattice vector, the polytope provides us with the details of the distribution of lattice points. The number $p(m)$ of lattice points of Λ at squared distance m^2 from the origin is given by the theta series $\sum_{\mathbf{x}\in\Lambda} q^{(\mathbf{x},\mathbf{x})} = \sum_{m=0}^{\infty} p(m)q^{m^2}$, but the series does not tell us how the points are distributed in each concentric shell. Indeed, there are several examples of pairs of distinct lattices with the same theta series (Serre, 1970). But these lattices have different Voronoï polytopes, as we show below.

Voronoï characterized the facet vectors of Λ: they are the vectors that are 'relatively short'. Let $m\Lambda = \{m\mathbf{x}, \mathbf{x} \in \Lambda\}$, for m a positive integer. We say that $\mathbf{x}$ and $\mathbf{y}$ are congruent modulo $m\Lambda$, or belong to the same coset modulo $m\Lambda$, if $\mathbf{x} - \mathbf{y} \in m\Lambda$.

(2.1) Theorem. *Let* $\mathbf{x} \in \Lambda$, $\mathbf{x} \neq \mathbf{0}$. *Then* $\mathbf{x}$ *is a facet vector if and only if* $N(\mathbf{x}) < N(\mathbf{y})$ *for all* $\mathbf{y} \equiv \mathbf{x} \pmod{2\Lambda}$, $\mathbf{y} \neq \pm\mathbf{x}$.

Voronoï's proof can be given a simple geometric interpretation.

Proof. Let $\mathbf{x}$ be a facet vector and $\mathbf{y} \equiv \mathbf{x}$ modulo 2Λ. Then $\mathbf{x} - \mathbf{y} \in 2\Lambda$ so $\mathbf{p} = (\mathbf{x}-\mathbf{y})/2 \in \Lambda$. Suppose $N(\mathbf{y}) \leq N(\mathbf{x})$. Then the perpendicular bisector of $\mathbf{p}$ cuts $\mathbf{x}$ at or below its midpoint. But then the bisector of $\mathbf{x}$ does not contribute a facet to D. Conversely, suppose that $\mathbf{x}$ is the (unique) shortest vector in its coset and let $\mathbf{p}$ be any lattice vector. Let $\mathbf{y} = 2\mathbf{p} - \mathbf{x}$. Then $\mathbf{y} \equiv \mathbf{x}$ modulo 2Λ. By the above reasoning, the perpendicular bisector of $\mathbf{p}$ does not cut off the midpoint of $\mathbf{x}$. Thus $\mathbf{x}$ is a facet vector (Figure 4.) Since D is centrosymmetric, if $\mathbf{x}$ is a facet vector so is $-\mathbf{x}$. □

This simple theorem has many important corollaries, including the following.

(2.2) Corollary. (Minkowski) *The number of facet vectors is at least* $2n$ *and at most* $2(2^n - 1)$.

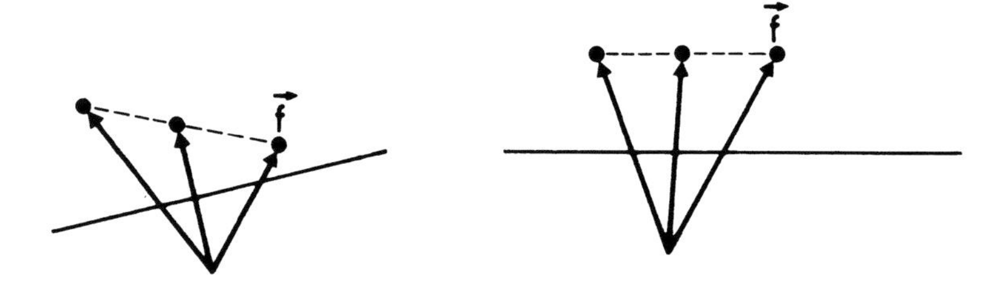

Fig. 4. A vector f is a facet vector if and only if it is the shortest vector in its coset modulo 2Λ.

Proof. Since any centrosymmetric n-dimensional polytope has at least $2n$ facets, the smallest possible number of facet vectors is $2n$. There are 2^n cosets of 2Λ in Λ; we can label them with the n-tuples $(\epsilon_1, \ldots, \epsilon_n)$, where $\epsilon_i = 0$ or 1. Λ itself is represented by $(0, \ldots, 0)$. Thus there are $2^n - 1$ cosets corresponding to possible facet vectors. We double this number in order to include $-\mathbf{f}$ for each $\mathbf{f}$. □

The minimum is attained in every dimension by the integer (hypercubic) lattice I_n and all lattices which are orthogonal direct sums of 1-dimensional lattices. The maximum is also attained in every dimension: indeed, almost all lattices have $2(2^n - 1)$ facet vectors (see Sec. 3).

(2.3) Corollary. *Every finite subgroup of* $\mathrm{GL}(n, \mathbb{Z})$ *has an orbit of size* S *such that* $S \leq 2(2^n - 1)$.

Proof. The set of facet vectors is invariant under the action of the point group of Λ. □

(2.4) Corollary. *If* $\mathbf{f}$ *is a facet vector, then it is the shortest vector in its coset* (mod $m\Lambda$) *for every* $m \geq 2$. *Thus linearly independent facet vectors belong to different cosets modulo* $m\Lambda$.

Proof. Let f be a facet vector and $\mathbf{x} = \mathbf{f} + m\mathbf{v}$ for some $\mathbf{v} \in \Lambda$. It follows from Theorem 2.1 that if $N(\mathbf{f}) \geq N(\mathbf{x})$ then m is odd. For every vector $\mathbf{y} = \mathbf{x} + \alpha m\mathbf{v}$, where $0 < \alpha < 1$, we have $N(\mathbf{y}) < N(\mathbf{f})$ (Figure 5). But

$$N(\mathbf{f}-(m-1)\mathbf{v}) = N(\mathbf{x}+\mathbf{v}) > N(f)$$

because $m-1$ is even, a contradiction. □

(2.5) Corollary. *Every vector of norm $< 2r$ is a facet vector.*

Proof. We need to show that two such vectors cannot belong to the same coset modulo 2Λ. Let $\mathbf{x}, \mathbf{y} \in \Lambda$, $\mathbf{y} \neq \mathbf{x}$, and suppose $N(\mathbf{x})$ and $N(\mathbf{y})$ are both $< 2r$. Then if $\mathbf{x} \pm \mathbf{y} \in 2\Lambda$, we would have $N(\mathbf{x} \pm \mathbf{y}) \geq 4r$, so

$$0 \leq N(\mathbf{x} \pm \mathbf{y}) - 4r = N(\mathbf{x}) + N(\mathbf{y}) - 4r \pm 2(\mathbf{x}, \mathbf{y})$$

which is impossible. □

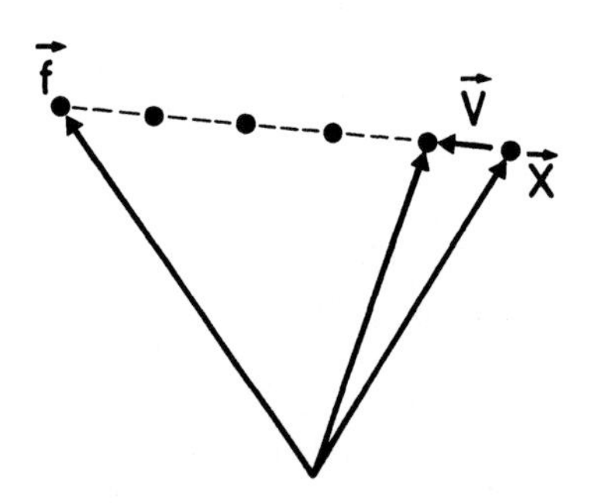

Fig. 5. Facet vectors belong to different cosets modulo m.

(2.6) Corollary. *The vectors of minimal norm r are always facet vectors.*

The positive definite integral quadratic forms of determinant $d \leq 25$ in low dimensions are completely classified in (Conway and Sloane, 1988b). Here we present a few illustrative examples of some important lattices and their Voronoï polytopes.

a) *The integer lattice I_n.* Up to similarity, this is the unique lattice with an orthonormal basis. It is an orthogonal direct sum of one-dimensional lattices, and its Voronoï polytope is the n-dimensional hypercube.

b) *Root lattices.* The lattices whose point groups are $\mathbb{Z}$-irreducible and are generated by reflections have been studied the most intensively, partly because the tools for studying them are available, and partly because they arise in other 'natural' ways – for example, they can be constructed by stacking lattices of equal spheres, and they play an important role in the theory of Lie algebras and groups. These lattices are called *root lattices.* The root lattices are described in detail in (Conway and Sloane, 1988a), (Coxeter, 1973) and (Engel, Michel, and Senechal, 1992?). For our purposes, the following brief remarks will suffice. There are two infinite families of root lattices, usually denoted A_n and D_n, and one finite family, E_n, $n = 6, 7, 8$. All can be described in terms of the integer lattice.

The A_n lattice , $n \geq 1$, is conveniently defined to be an n-dimensional sublattice of I_{n+1}; it is the set of $(x_1, \ldots, x_{n+1}) \in I_{n+1}$ such that $(x_1, \ldots, x_{n+1}) \cdot (1, \ldots, 1) = 0$. Thus its shortest vectors have the form $(1, -1, 0, \ldots, 0)$, and there are $n(n+1)$ of them. Obviously $r = 2$.

The D_n lattice is the n-dimensional 'checkerboard' lattice; its points are the $(x_1, \ldots, x_n) \in I_n$ with $\sum x_i \in 2\mathbb{Z}$. Its vectors of norm r, which again is equal to 2, are those with two nonzero coordinates, which may be ± 1; thus there are $2n(n-1)$ vectors of minimal norm. $D_2 = I_2$ and $D_3 = A_3$, so this family is assumed to begin with $n = 4$.

The E_n family is somewhat harder to visualize. The Voronoï polytope of D_8 has two congruent orbits of 128 vertices whose squared distance from the origin is $r = 2$; each of these orbits, extended by all of the symmetries of D_8, is a copy of D_8. If we add one of them to D_8 we obtain the lattice called E_8; it has $240 = 112 + 128$ vectors of minimal norm. The other two members of the E_n family can be obtained as 7 and 6 dimensional sublattices of E_8.

The Voronoï polytope of a root lattice is *isohedral*: its facets are equivalent under $E(\Lambda)$. The isotropy groups are generated by reflections and hence can be described by Coxeter diagrams; the nodes can be interpreted as the vertices of a simplex which is a fundamental region of the corresponding infinite group (Figure 6).

One can show that the Voronoï polytope of a root lattice is the union of the fundamental simplexes obtained by reflection in the hyperplanes containing the walls of the simplex which meet at $\mathbf{0}$. The walls opposite $\mathbf{0}$ comprise the facets of the domain. Thus all the facets are congruent and the facet vectors are those of minimal norm.

c) *Duals of root lattices.* Every lattice has a *dual lattice* Λ^*, where

$$\Lambda^* = \{\mathbf{y} \in \mathbb{R}^n | (\mathbf{y}, \mathbf{x}) \in \mathbb{Z}, \forall \mathbf{x} \in \Lambda\}.$$

The dual lattice is a useful construction in number theory as well as in many branches of crystallography, including the interpretation of diffraction patterns.

Obviously, $\Lambda^{**} = \Lambda$. The isotropy subgroups $E(\Lambda)$ and $E(\Lambda^*)$ are conjugate in $\mathrm{GL}(n, \mathbb{R})$ but not necessarily in $\mathrm{GL}(n, \mathbb{Z})$. If $\Lambda \subseteq \Lambda^*$, then Λ is said to be *integral*; note that in this case all of the entries of the Gram matrix are integers. (In particular, the root lattices are integral.) If $\Lambda = \Lambda^*$, then Λ is *self-dual*. For integral Λ, $[\Lambda^* : \Lambda] = \det \Lambda$. Thus Λ is self-dual if and only if $\det \Lambda = 1$.

In (Engel, Michel, and Senechal, 1992?) we prove:

(2.7) Theorem. *For the root lattices, vectors of the dual lattice are located only at vertices of the fundamental simplexes, though not all vertices need be dual*

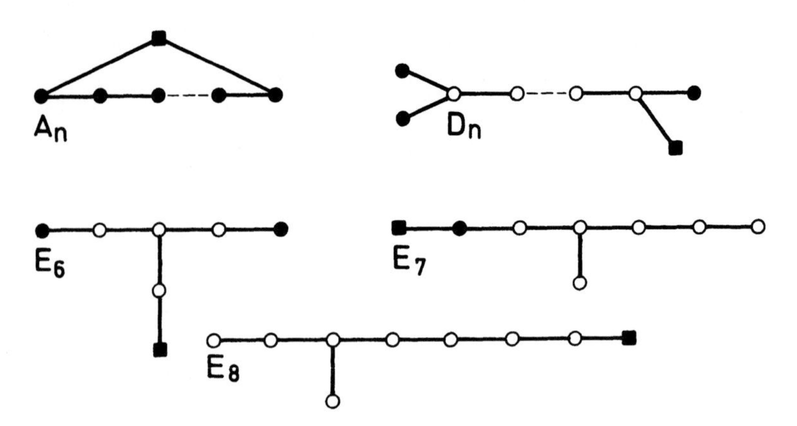

Fig. 6. Coxeter diagrams for the root lattices. Each node represents a vertex of a fundamental region. Solid nodes are elements of the dual lattice, as are the square marked nodes, which represent $(0,\ldots,0)$.

lattice vectors. If Λ is a root lattice, the vertices of a given simplex which are elements of Λ^* constitute a complete set of coset representatives of Λ in Λ^*, and different coset representatives belong to different cosets mod $2\Lambda^*$.

It follows that we can obtain the Voronoï polytope of Λ^* by adjoining the vertex vectors which belong to Λ^* to Λ and bisecting them and their images under $E(\Lambda)$. This amounts to truncating $D(\Lambda)$ at those vertices. Notice (Fig. 6) that every vertex of the fundamental simplex of A_n belongs to A_n^*; it follows that A_n^* has the maximum number of facet vectors, $2(2^n - 1)$.

The vectors of Λ^* are orthogonal to lattice planes of Λ and conversely, so the lattice hyperplanes of one are parallel to the facets of the Voronoï polytope of the other. (Alas, $D(\Lambda)$ and $D(\Lambda^*)$ are not dual to one another in the usual geometric sense.)

d) *Intermediate lattices.* A lattice Λ' such that $\Lambda \subseteq \Lambda' \subseteq \Lambda^*$ is said to be an *intermediate* lattice. To construct the intermediate lattices Λ', we add some but not all of the coset representatives of Λ in Λ^* to Λ. It follows from the Proposition above that these coset representatives are facet vectors for $D(\Lambda')$. Thus $D(\Lambda')$ is also obtained by suitable truncation of $D(\Lambda)$.

e) *Orthogonal direct sums.* Let $\bar{\Lambda}$ be a direct sum of sublattices Λ_i, $i = 1,\ldots,m$ of dimensions n_i. Then dim $\bar{\Lambda} = \sum n_i = n$. We embed the lattices Λ_i in $\mathbb{R}^n$ in the usual way and denote then by $\bar{\Lambda}_i$. In (Engel, Michel, and Senechal, 1992?) we show that $x \in \mathbb{R}^n$ is in $D(\bar{\Lambda})$ if and only if it is in the intersection of the cylinders $D_i = D(\Lambda_i) \oplus \mathbb{R}^{n-n_i}$. Thus $D(\bar{\Lambda})$ is a product polytope $\prod D_i$. We see immediately that D_{16} and $E_8 \oplus E_8$, which have the same theta series, have different Voronoï polytopes. It is not difficult to prove that $\bar{\Lambda}^*$ is the direct sum of the Λ_i^*. Thus when the Λ_i are root lattices it is straightforward to construct the Voronoï polytopes of $\bar{\Lambda}^*$ and the intermediate lattices $\bar{\Lambda}'$.

The root lattices, together with their duals, intermediate lattices and direct sums, are present in some form in many if not most of the lattices that occur in number theory. For example, they are the building blocks for the construction of lattices by 'gluing' (see, e.g., (Conway and Sloane, 1988a)).

3. The Sets S, F, and C

For a deeper discussion of lattice geometry, we find it helpful to introduce the following sets:

short vectors: $S = \{\mathbf{x} \in \Lambda | N(\mathbf{x}) = r\}$,

facet vectors: $F = \{\mathbf{x} \in \Lambda |\ \mathbf{x} \text{ is a facet vector}\}$,

corona vectors: $C = \{\mathbf{x} \in \Lambda | D(\mathbf{0}) \cap D(\mathbf{x}) \neq \emptyset\}$.

(3.1) Proposition. $S \subseteq F \subseteq C$.

Proof. We have already seen that $S \subseteq F$. Since the definition of a facet vector $\mathbf{f}$ implies that $D(\mathbf{0}) \cap D(\mathbf{f})$ is a common facet, $F \subseteq C$. □

(3.2) Proposition. *If* $\mathbf{x} \in C$, *then* $N(\mathbf{x}) \leq N(\mathbf{y})$ *for all* $\mathbf{y} \in \mathbf{x} + 2\Lambda$, *and* $N(\mathbf{x}) < N(\mathbf{y})$ *for all* $\mathbf{y} \in \mathbf{x} + 3\Lambda$.

Proof. This is yet another corollary of Theorem 2.1. □

The 'if' cannot be replaced by 'iff'.

(3.3) Corollary. $|C| \leq 3^n - 1$.

This maximum is attained in every dimension (for example, by the lattices I_n).

It follows from the discussion in Section 2 that if Λ is a root lattice or a direct sum of root lattices and integer lattices, then $S = F$. We will now show that the converse is also true. The key lemma is due to Witt (Witt, 1941):

(3.4) Lemma. *A finite set of lattice vectors which is irreducible (i.e., cannot be decomposed into an orthogonal direct sum) and invariant under reflection in the hyperplanes orthogonal to each of them belongs to one of the lattice families* A_n, D_n *or* E_n.

I_n is of course invariant under such reflections but is excluded from this list because it is the orthogonal direct sum of n identical one-dimensional lattices.

Witt's proof is based on the fact that the hyperplanes divide space into a finite number of regions which are permuted by the reflections. It follows that the permutations generate a Coxeter group and the configuration of hyperplanes can be represented by a Coxeter diagram. Choosing those diagrams whose groups can be extended to infinite ones, the lemma follows.

From Witt's lemma we derive the following remarkably simple characterization of the root lattices.

(3.5) Theorem. *Let Λ be irreducible. Then $S = F$ if and only if Λ is a root lattice.*

Proof. We must show that if $S = F$ then Λ is a root lattice. Let $\mathbf{s}, \mathbf{t} \in S$. Reflection in the hyperplane orthogonal to $\mathbf{s}$ maps $\mathbf{t}$ to $\mathbf{t} - 2\mathbf{s}(\mathbf{s}, \mathbf{t})/r$. Corollary 2.4 implies that Λ has no vectors whose norms lie properly between r and $2r$, so $N(\mathbf{s} - \mathbf{t}) = (\mathbf{s}, \mathbf{s}) + (\mathbf{t}, \mathbf{t}) - 2(\mathbf{s}, \mathbf{t}) = 2(r - (\mathbf{s}, \mathbf{t})) = r$ or $2r$ or is $> 2r$. In the first case, $(\mathbf{s}, \mathbf{t}) = r/2$, in the second $(\mathbf{s}, \mathbf{t}) = 0$, and in the third $N(\mathbf{s} + \mathbf{t}) = r$ and hence $(\mathbf{s}, \mathbf{t}) = -r/2$. In each case the image of $\mathbf{t}$ under reflection is in Λ and has norm r. Thus S is invariant under reflection in this hyperplane, and the conclusion follows from Witt's theorem. □

Since the isotropy group of an orthogonal direct sum of congruent lattices is the wreath product of the isotropy groups of these lattices, we have immediately:

(3.6) Corollary. *The Voronoï polytope of Λ is isohedral if and only if Λ is a root lattice or a direct sum of identical root lattices.*

Indeed, the theorem proves more: the Voronoï polytope is *inscribable* (that is, its facets are tangent to an inscribed sphere) if and only if Λ is I_n or a root lattice or a direct sum of root lattices and a properly scaled I_k.

Although we will probably never know the number of different Voronoï polytopes in $\mathbb{R}^n$ for $n > 5$, we do know exactly how many of them are isohedral!

(3.7) Corollary. *The number of isohedral Voronoï parallelotopes in $\mathbb{R}^n$ is*

$$I(n) = d(n) + e(n) + f(n)$$

where $d(n)$ is the number of divisors of n (including 1 and n), $e(n)$ is the number of divisors of n which are greater than 3 (again including n itself), and $f(n) = 0,1,2$ or 3 according as none, one, two or three of the numbers 6,7, 8 divides n.

Proof. If $k|n$, then we can also form the direct sum of n/k copies of A_k (I_n is the sum of n copies of A_1). Thus the number of isohedral polytopes which are direct sums of lattices of the A type is $d(n)$. The same is true for copies of D

lattices, but here $k \geq 4$. Finally, $f(n)$ counts the number of direct sums that we can build with lattices of the E type. □

(3.8) Corollary. *If Λ is not a root lattice or an appropriate direct sum of root lattices, then Λ has vectors of norm $> r$ and $< 2r$.*

It follows from Corollary 2.2 and Proposition 3.2 that $F = C$ if and only if $|F| = 2(2^n - 1)$. The case $F = C$ is very different from the case $S = F$. For example, almost all plane lattices satisfy the condition $F = C$. For, when $F = C$, $D(\Lambda)$ has six edges. $D(\Lambda)$ is *not* a hexagon only if Λ has generators $\mathbf{a}_1$ and $\mathbf{a}_2$ which satisfy the very special condition $(\mathbf{a}_1, \mathbf{a}_2) = 0$.

Lattices for which exactly $n + 1$ parallelotopes meet at each vertex of the Voronoï tiling of $\mathbb{R}^n$ were called *primitive* by Voronoï (Voronoï, 1908); primitive lattices have the property $F = C$. Voronoï showed (Voronoï, 1909) that under the classification of primitive lattices by the combinatorial type of their Voronoï polytope, $P(n)$ is partitioned into a finite number of open sets of dimension $n(n+1)/2$. Thus the set of lattices in $P(n)$ for which $F = C$ contains an open dense set. We conclude that

(3.9) Theorem. *The condition $F = C$ is generic.*

Engel has shown that beginning with $n = 4$, there are also nonprimitive parallelotopes for which $F = C$ (Engel, 1986).

4. Sphere Packings

If S generates F (and thus Λ), then any lattice point can be reached from any other by a path of vectors of minimal length. Such a lattice can be constructed by packing equal spheres: if we place spheres of radius $\frac{\sqrt{r}}{2}$ at each lattice point, we can go from any sphere to any other by an unbroken sequence of center-to-center links of length $\sqrt{r}$. There are many lattices which are generated by S but for which $S \neq F$. For example, a very small perturbation of a root lattice can produce a lattice with this property.

This leads us to the very interesting topic of lattice (and nonlattice) sphere packings, and the problem of finding packings with maximal densities. The *density* δ of a sphere-packing lattice is defined to be the ratio of the volume of a unit sphere to the volume of its Voronoï polytope. In a sphere-packing lattice, the number of spheres touching each sphere is $|S|$. This number is sometimes called the *contact number* or 'kissing number'.

Various more or less successful techniques have been developed for finding the densest lattice packings in a given dimension. For example, Conway and Sloane have shown that lattice packings of high density can be constructed by 'lamination' (Conway and Sloane, 1982).

The process is straightforward. It follows from their construction that a vertex of a Voronoï cell of any n-dimensional lattice Λ_n is equidistant from the center of the cells that share that vertex. It is thus the center of a spherical 'hole' in the lattice. Since the lattice is invariant under its translations, the holes are arranged in a finite number of lattices congruent to Λ_n. Suppose that Λ_n is a sphere-packing lattice. We 'inflate' the spheres to $n+1$-dimensional spheres of the same radius, and then create an $(n+1)$-dimensional lattice Λ_{n+1} by stacking copies of the inflated Λ_n one above the other, with the spheres of one layer placed above the centers of a lattice of holes in the layer below.

To obtain dense packings, we start with $n = 1$ and at each stage of the lamination process choose a lattice of 'deep' holes – holes of maximal radius. In fact, we obtain the densest possible lattice packings in dimensions one through eight in this way, and very dense packings in other dimensions. The extremely dense twenty-four dimensional Leech lattice appears in this series.

Evidently, the deeper the hole the larger the number of cells meeting at its center (and thus the larger the value of $|S|$ in Λ_{n+1}), and the denser the lattice we obtain by stacking above it. (It seems that neither of these statements has been ever proved). This suggests that δ varies directly with $|S|$. Unfortunately, the situation is more complicated, even in three dimensional space. For example, there are lattices with minimal $|S|$ (that is, 6) whose density is arbitrarily close to that of the densest lattice, in which $|S| = 12$. Thus a lattice with $|S| = 6$ can be denser than a lattice with $|S| = 8$. It appears that the relation between δ and $|S|$ has never been completely clarified.

There is, however, a theorem of Voronoï (Voronoï, 1907) which does link them. We will state it without proof. First we must define three more important classes of lattices.

1. A lattice for which δ is a local maximum is said to be *extreme*; if δ is a global maximum the lattice is *absolutely extreme*. For $n = 2, 3, \ldots, 8$, the absolutely extreme lattices are the root lattices A_2, A_3, D_4, D_5, E_6, E_7, E_8.

2. A *perfect* lattice is one whose Gram matrix is completely determined by its minimal vectors. ('Perfect' is not synonymous with 'S generates F'.)

3. A *eutactic* lattice is one whose minimal vectors are parallel to the vectors of a eutactic star. (A eutactic star is a set $\pm a_1, \ldots, \pm a_s$ of vectors in $\mathbb{R}^n$ with the property that there exists a $\mu > 0$ such that every $x \in \mathbb{R}^n$ satisfies $(x, x) = \mu \sum (a_k, x)^2$.)

The extreme property is related to the density δ; the perfect property is related to the contact number $|S|$, and as Coxeter notes (Coxeter, 1973), eutaxy means 'good arrangement, orderly disposition'. Voronoï proved:

(4.1) Theorem. *A lattice is extreme if and only if it is perfect and eutactic.*

For further details, consult (Conway and Sloane, 1988b) and (Coxeter, 1951).

5. Integral Representations and the Crystallographic Restriction

Haüy noted that his building-block theory of crystal structure implied that no crystal can have icosahedral symmetry. More generally, for each n there are finite subgroups of $O(n)$ which cannot be represented in $\mathrm{GL}(n, \mathbb{Z})$. Which subgroups of $O(n)$ *do* have such representations?

(5.1) Lemma. (Minkowski) *Let $B \in \mathrm{GL}(n, \mathbb{Z})$ be of finite order. If B is congruent to the identity matrix I modulo m, where m is an integer greater than 2, then $B = I$.*

We will give a very simple proof using Voronoï polytopes.

Proof. Let Λ be a lattice invariant under B. If $B = I + mA$, for some $A \in \mathrm{GL}(n, \mathbb{Z})$, then $B(\mathbf{x} + m\mathbf{v}) = \mathbf{x} + m(A\mathbf{x} + \mathbf{v} + mA\mathbf{v})$ for every $\mathbf{x}$, $\mathbf{v}$ in Λ. Thus B fixes the cosets of $m\Lambda$. Corollary 2.4 states that, for any lattice, the elements of F belong to different cosets of $m\Lambda$. Thus since B is an isometry it fixes the shortest members of these cosets. Since there are n independent facet vectors, $B = I$. □

(5.2) Theorem. (Minkowski) *Every homomorphism $\mu : E \to \mathrm{GL}(n, \mathbb{Z}_q)$, where E is a finite subgroup of $\mathrm{GL}(n, \mathbb{Z})$ and q is an odd prime, is an injection.*

Proof. It follows from the Lemma that Ker $\mu = \{I\}$. □

A partial answer to our question is,

(5.3) Corollary. *If m is the order of a finite subgroup $W \subset \mathrm{GL}(n, \mathbb{Z})$, then for each odd prime q, m divides $\kappa(n, q) = 2q^{n(n-1)/2} \prod_{k=2}^{n}(q^k - 1)$.*

Proof. For each odd prime q, the map $\psi : W \to \mathrm{GL}(n, \mathbb{Z}_q)$ is a homomorphism. Thus W has a faithful representation in $\mathrm{GL}(n, \mathbb{Z}_q)$, which is a finite group of order $\kappa(n, q)$ (the formula is due to Galois). □

The question then is, which integers m divide $\kappa(n, q)$? Let $\lambda(m)$ be the 'universal exponent of m', defined as follows: $\lambda(1) = 1$, $\lambda(2^\alpha) = \phi(2^\alpha)$ if $\alpha = 1$ or 2, and $\phi(2^\alpha)/2$ if $\alpha > 2$, $\lambda(p^\alpha) = \phi(p^\alpha)$ for p an odd prime, and $\lambda(2^\alpha p_1^{\alpha_1} \dots p_r^{\alpha_r}) =< \lambda(2^\alpha), \lambda(p_1^{\alpha_1}), \dots, \lambda(p_r^{\alpha_r}) >$. Here $\phi(d)$, Euler's function, is the number of integers less than and relatively prime to d, and $< a, b >$ is the least common multiple of a and b.

(5.4) Corollary. *The condition $\lambda(m) \leq n$ is sufficient to ensure that m divides $\kappa(n, q)$ for every q which does not divide m.*

Proof. Let m have the prime-power decomposition $2^\alpha p_1^{\alpha_1} \dots p_r^{\alpha_r}$. $\lambda(m)$ is the smallest positive integer t such that $q^t \equiv 1 \pmod{m}$ for every prime q which does not divide m (see, e.g., (LeVecque, 1956)). By Corollary 5.3, for each odd prime q that does not divide m, m divides $\prod_{k=2}^{n}(q^k - 1)$. □

Neither Corollary 5.3 nor a sharper condition also due to Minkowski guarantees the existence of a subgroup of $\mathrm{GL}(n, \mathbb{Z})$ of order m. However, in the case that W is a cyclic group, we can get a necessary and sufficient condition. This result is known as the *crystallographic restriction.*

As we have seen, the Voronoï polytopes for lattices in $\mathbb{R}^2$ are either quadrilaterals or hexagons. Thus the order of a single element of $\mathrm{GL}(2, \mathbb{Z})$ can be only 1, 2, 3, 4 or 6. It is not difficult to see that this 'restriction' holds in $\mathbb{R}^3$ as well.

This is a special case of the following condition for arbitrary n. Let $g(m)$ be the least value of n for which an element of order m appears in $\mathrm{GL}(n, \mathbb{Z})$.

(5.5) Lemma. *If the order of $B \in \mathrm{GL}(n, \mathbb{Z})$ is a power of a single prime, $m = p^\alpha$, then $g(m) = \phi(p^\alpha)$.*

Proof. B is an isometry, so its eigenvalues are m^{th} roots of unity. One of them, say ζ, must be a primitive m^{th} root, because the order of B is the maximum of the orders of the eigenvalues. The roots of the minimal polynomial of ζ are precisely the primitive m^{th} roots of unity, and since this polynomial divides the characteristic equation of B, all of the primitive m^{th} roots must be eigenvalues. The number of primitive roots is $\phi(p^\alpha)$, so $n \geq \phi(p^\alpha)$. In fact, the integral lattice in the cyclotomic field $R(\zeta)$ has dimension $\phi(p^\alpha)$, so this must be the lowest dimension n in which an element of order p^α appears in $\mathrm{GL}(n, \mathbb{Z})$. □

Suppose now that m is a product of distinct primes.

(5.6) Theorem. *Let $B \in \mathrm{GL}(n, \mathbb{Z})$ have finite order $m = p_1^{\alpha_1} \ldots p_r^{\alpha_r}$. Then*

$$g(m) = \sum_{p_i^{\alpha_i} \neq 2} \phi(p_i^{\alpha_i}).$$

Proof. If m is odd, then $g(2m) = g(m)$ because $-B \in \mathrm{GL}(n, \mathbb{Z})$. Thus we assume that if $2^\alpha | m$, then $\alpha > 1$. Clearly $g(m) \leq \sum \phi(p_i^{\alpha_i})$, because an element of that order can be constructed as a direct sum of elements of $\mathrm{GL}(\phi(p_i^{\alpha_i}, \mathbb{Z})$. But also $g(m) \geq \sum_\phi (p_i^{\alpha_i})$: by the reasoning used in the case of a single prime power, the set of eigenvalues of B must include at least one of the $p_i^{\alpha_i th}$ primitive roots of unity, $i = 1, \ldots, r$, and hence it contains all of them. □.

As Pleasants points out (Pleasants, 1985), proofs in the literature which mistakenly assert that $\phi(m) = g(m)$ implicitly assume that for *any* m, a primitive m^{th} root of unity must be an eigenvalue of B. Hiller, whose notation we have followed here, gives a table of admissible orders m for $n \leq 23$ (Hiller, 1985).

We have seen that every finite subgroup $W \subset \mathrm{GL}(n, \mathbb{Z})$ is isomorphic to a finite subgroup $W' \subset \mathrm{GL}(n, \mathbb{Z}_3)$. Obviously the entries in the matrices of W' can be taken to be 1, 0 and -1. In some cases this is also true for the group W. For example, looking in the International Tables for X-Ray Crystallography, we find that *all* the matrices for the 14 Bravais lattices in $\mathbb{R}^3$ have entries 0, 1 and -1. Evidently, for each lattice in $\mathbb{R}^3$ a basis can be chosen so that its point group E can be represented in this simple form. This raises the interesting question of whether such a basis can be found for each lattice in every dimension.

The vectors sums $\sum_{i=1}^{n} \alpha_i \mathbf{a}_i$, where $\alpha_i \in \{0, 1, -1\}$ form a box which is a unit cell for 2Λ centered at the origin. The vectors which appear as the jth-columns of the matrices representing E constitute the orbit of the basis vector $\mathbf{a}_j$. The matrices will have all their entries 0, 1 and -1 if and only if the orbits of all the basis vectors lie in this box.

At first sight, determining whether such a basis exists appears to be an (impossibly difficult) exercise in group conjugation. However, a geometric approach has proved to be much more successful. Obviously, the shorter the basis vectors and the more symmetrical the box, the closer the basis will be to having this property. Since the Voronoï polytope $D(\Lambda)$ is invariant under E, and since facet vectors are 'relatively short', perhaps the basis we are seeking can be found in the set F.

Although F generates Λ, it is not obvious that a lattice basis can be extracted from this set. This is in sharp distinction to a vector space, in which every spanning set contains a basis. Voronoï asserted, but did not prove, that

(5.7) Theorem. *Every lattice contains a basis of facet vectors.*

We believe this theorem to be true, and have proved it for large classes of lattices (Engel, Michel, and Senechal, 1992?).

Plesken has determined the maximal subgroups of $\mathrm{GL}(n, \mathbb{Z})$ for $n \leq 10$ (Plesken, 1977, 1980). By finding and examining bases of facet vectors for these lattices we have been able to prove

(5.8) Theorem. *For $n \leq 7$, every finite subgroup of* $\mathrm{GL}(n, \mathbb{Z})$ *can be represented by matrices with entries 0, 1 and -1.*

The proof of the theorem (Engel, Michel, and Senechal, 1992?) is quite complicated, involving both algebraic and geometric arguments and extensive use of the computer.

In dimensions greater than 7, '$0, 1, -1$' bases can be found for many families of lattices, including all the root lattices A_n, B_n and D_n, but in general such a basis does not exist. The first exception occurs in $\mathbb{R}^8$. There are $0, 1, -1$ bases for E_6 and E_7 (this is guaranteed by Theorem 5.4) but *not*, as J.H. Conway first pointed out (private communication), for E_8. One way to see why not is study the corresponding Voronoï tiling. E_8 is a sphere packing lattice which can be constructed by stacking layers of spheres in congruent E_7 configurations

(Conway and Sloane, 1988a). Replacing the spheres of E_8 by Voronoï polytopes, we see that the Voronoï tiling can be divided into layers -1, 0, $1, \ldots,$ the tile centers in each layer constituting a lattice of E_7 type. We find that $D(\mathbf{0})$ meets 56 others in layer -1, 126 in layer 0, and 56 in layer 1. But $2 \times 56 + 126 = 138$, while $D(E_8)$ has 240 facets. Thus each Voronoï polytope must share one facet with a polytope in layer -2 and one with a polytope in layer 2. F cannot be contained in our box no matter what basis we use to construct it. It follows that we cannot represent E_8 by matrices with only 0, 1 and -1.

More generally, E_8 is one of a series of intermediate lattices of type $D_n^+ = D_n \cup (\frac{1}{2}, \ldots, \frac{1}{2}) + D_n$ (defined only for even n). The Voronoï polytope of D_n^+ is obtained from that of D_n by truncating the latter at half of the vertices of type $(\frac{1}{2}, \ldots, \frac{1}{2})$; thus these vertices are facet vectors for D_n^+. It can be shown that one of them must appear in every basis of facet vectors. Since their norm is $n/4$, these vertices recede from the center as n increases. The 'optimal' matrix representation for the point group of D_n^+ will always have at least one entry no less than $\geq [n/4]$.

Acknowledgements It is a pleasure to thank P.J. Cassidy, R. Connolly, and W. Whitely for stimulating discussions, and H. S. M. Coxeter, P. Engel, L. Michel, J. J. Seidel and a referee for their comments on various drafts of this manuscript.

References

Conway, J.H. and Sloane, N.J.A., (1982), Laminated Lattices, Annals of Mathematics **116** 593 - 620.

Conway, J. H., and Sloane, N. J. A., (1988a) Sphere Packings, Lattices and Groups, Springer-Verlag, NY, 1988.

Conway, J. H. and Sloane, N. J. A. (1988b) Low-dimensional lattices. I. Quadratic forms of small determinant., Proc. R. Soc. Lond. A **418** (1988) 17-41. II. Subgroups of GL(n, Z). *ibid*, **419** (1988) 29-68. III. Perfect forms, *ibid* **418** (1988) 43-80. IV. The mass formula, *ibid*, **419** (1988) 259-286.

Coxeter, H.S.M. (1940), Regular and Semiregular Polytopes I, Mathematische Zeitschrift **46**, 380-407.

Coxeter, H.S.M. (1951), Extreme Forms, Canadian Journal of Mathematics **3** 391 - 441.

Coxeter, H.S.M. (1973), Regular Polytopes, Dover, N.Y., 1973.

Coxeter, H.S.M. (1985), Regular and Semiregular Polytopes II, Mathematische Zeitschrift **188**, (1985) 559-591 (1985)

Coxeter, H.S.M. (1988), Regular and Semiregular Polytopes III, Mathematische Zeitschrift **200** 3-45.

Delone (Delaunay), B.N. (1929), Sur la partition regulière de l'espace à 4 dimensions, Isv. Akad. Nauk. S.S.S.R., Otdel. Fiz. Mat. Nauk 79-110, 145-164.

Engel, P. (1986), Geometric Crystallography, Reidel, Dordrecht.

Engel, P., Michel, L. and Senechal, M. (1992?), Lattice Geometry, in preparation.

Fedorov, E. S. (1885), Nachala Ucheniya o Figurakh, Notices of the Imperial Petersburg Mineralogical Society, 2nd series **21** 1-279.

Gruber, P. and Lekkerkerker, G. (1987), Geometry of Numbers, Elsevier/North Holland, Amsterdam.

Grünbaum, B. and Shephard, G.C. (1987) Tilings and Patterns, Freeman, NY.

Hiller, H. (1985), The crystallographic restriction in higher dimensions, Acta Crystallographica, **A41** 541-544.
LeVecque, J. (1956), Introduction to Number Theory, Volume I, Addison-Wesley, Reading.
Michel, L. and Mozryzmas, J. (1989), Les concepts fondamentaux de la cristallographie, CR Acad. Sci. Paris, **308**, Serie II, p. 151-158.
Minkowski, H. (1887) Zur Theorie der positiven quadratischen Formen, J. reine angew. Math., **101** 196-202.
Pleasants, P.A.B. (1985), Quasicrystallography: some interesting new patterns, Elementary and Analytic Theory of Numbers, Banach Center Publications **17** PWN - Polish Scientific Publishers, Warsaw.
Plesken, W. (1977), On maximal finite subgroups of GL(n, Z), Mathematics of Computation, Parts I, II: **31** (1977)536-551, 552-573.
Plesken, W. (1980), On maximal finite subgroups of GL(n, Z), Mathematics of Computation, parts III, IV, V: **34** (1980) 245-258, 259-275, 277-301.
Schwarzenberger, R. L. E. (1980), N-dimensional Crystallography, Pitman, London.
Senechal, M. (1985) Introduction to Mathematical Crystallography, in I.H.E.S. Workshop on Mathematical Crystallography, IHES preprint P/85/47.
Serre, J. P. (1970), Cours d'Arithmetique, Presses Universitaires de France, Paris.
Shtogrin, M.I., (1973), Regular Dirichlet - VoronoïPartitions, Proceedings of the Steklov Institute of Mathematics No. 123, Moscow.
Venkov, B.A. (1954), On a class of Euclidean polytopes, Vestnik Leningrad Univ., Ser. Mat. Fiz. Him., **9** 234-265.
Voronoï, G. (1907), Sur quelques propriétés des formes quadratiques positives parfaites, J. reine angew. Math. **133** 97 - 198.
Voronoï, G. (1908), Recherches sur les paralléloèdres primitifs, J. reine angew. Math. **134** 198-287.
Voronoï, G. (1909) Part II of the previous reference, *ibid*, **136** 67 - 181.
Witt, E. (1941), Spiegelungsgruppen und Aufsählung halbeinfacher Liescher Ringe, Abhandl. Math. Sem. Hamburg **A4** 189 - 322.

Chapter 11

A Short Introduction to Quasicrystallography

by André Katz

1. Introduction

Since the discovery (in the fall of 1984) of icosahedral quasicrystals, their structure determination has motivated a great deal of work. However, these efforts have been up to now rather unsuccessful and no systematic approach to this question is known.

In this paper, we explain why this question is difficult and develop the basic problems related to the description of the microscopic (atomic) structure of quasicrystals. We focus on the 'quasiperiodic' point of view, and we emphasize the topological problems which arise when one goes from (point) atoms in the periodic case to 'atomic surfaces' of non-zero dimension in the quasiperiodic case. Although the paper is not written in a mathematically rigorous way, we hope that it will prove legible by mathematicians, and will stimulate the communication between mathematicians and physicists.

After a brief review of classical crystallography, we give the necessary information about the experimental and phenomenological aspects of quasicrystals. Then we set up the general framework for the description of quasiperiodic sets of points, as sections of a periodic lattice of 'atomic surfaces' embedded in a higher dimensional space, in the way introduced several years ago by Janner and Janssen in their theory of incommensurate crystals (Janner and Janssen (1977)). We explain in particular how to compute the Fourier transform of these structures, which is of primary importance for crystallography. The next topic is the study of the notion of symmetry for such structures, which is not exactly the same for ordinary crystals and for quasicrystals (for the sake of clarity we restrict the discussion to the icosahedral case). As an important example, we develop in the following Section the description of the Penrose-like tilings, which are the best known quasiperiodic systems, and we explain how one can systematically study their local properties, such as the frequencies of the tiles or the classification of local patterns.

In the following Section, we examine the special properties of the Penrose-like tilings which should be preserved in any realistic model of quasicrystals, in connection with the physically important problem of the *propagation of order*. As explained in the last Section, this study results in natural topological constraints on the possible atomic surfaces, which on one hand allow us to distinguish between quasicrystals and other kinds of quasiperiodic systems such as modulated crystals, and on the other to recover to a certain extent the classical notion of integral stoechiometric coefficients.

2. Phenomenology of Quasicrystals

Let us first recall the fundamentals of experimental crystallography: the main tool used for the study of the microscopic structure of solid state materials is the diffraction of X-rays, electrons or neutrons. X-rays and electrons interact mainly with the electrons of the material, while neutrons interact with the nuclei. Consider for instance the diffraction of X-rays, and let $\rho(\mathbf{r})$ be the electron density, where $\mathbf{r} \in \mathbb{R}^3$. Then the output of the diffraction experiments is essentially the squared modulus $|\widehat{\rho}(\mathbf{k})|^2$ of the Fourier transform $\widehat{\rho}(\mathbf{k})$ of $\rho(\mathbf{r})$.

For an infinite and perfect periodic crystal, the function $\rho(\mathbf{r})$ is periodic and $|\widehat{\rho}(\mathbf{k})|^2$ is an array of Dirac peaks (called Bragg peaks by physicists). One measures the coefficients of these Bragg peaks (the so-called *intensities*) and the task of crystallography is to reconstruct $\rho(\mathbf{r})$ or at least the atomic positions from these diffraction data. Observe that this is not a simple task since the phase of $\widehat{\rho}(\mathbf{k})$ is not known: only the squared modulus $|\widehat{\rho}(\mathbf{k})|^2$ is measured. In order to find the structure, crystallographers have to obtain and use all possible kinds of extra information, such as the density and stoechiometry of the compound under study and its chemical properties, the comparison with already known structures, and so on.

One of the most important features of a crystalline structure is its *symmetry group*. Let us recall that in classical crystallography one usually defines three symmetry groups: the *space group* is the largest subgroup of the displacement group (the Euclidean group of isometries of the affine space $\mathbb{R}^3$) which leaves the crystal invariant. The intersection of the space group with the translation group $\mathbb{R}^3$ is the *translation group* of the crystal (which is an invariant subgoup of the space group), and the quotient group of the space group by the translation group is the *point group*, a finite subgroup of the orthogonal group $\mathcal{O}(3)$. Now, it is known since the 19th century that only a finite number of subgroups of $\mathcal{O}(3)$ may occur as point groups of a crystalline structure. In particular, five-fold axes are forbidden and a crystal cannot exhibit icosahedral symmetry. This is why the discovery of quasicrystals was so surprising.

Quasicrystals were first discovered in ultra-quenched alloys of aluminium and manganese, prepared by the same method used for metallic glasses: the melted alloy is dropped onto a spinning wheel of copper, which is at room temperature. This produces a thin ribbon of alloy, which is cooled (quenched) at very high rates (up to several million Kelvins per second), depending on the experimental conditions.

As is well known, if the quenching rate is high enough, there is no time for the crystallization process to take place, and one gets an amorphous material (a metallic glass). These metallic glasses are characterized in diffraction experiments by the absence of well-defined Bragg peaks, which are the signature of long range positional order. Roughly speaking, a metallic glass is simply a liquid which has been made infinitely viscous by the cooling.

For lower quenching rates, one gets a microcrystalline structure, which is characterized by the presence of Bragg peaks in diffraction experiments, in a way compatible with the classical laws of crystallography.

In the special case of *Al–Mn* alloys, and for intermediate cooling rates, Shechtman, Blech, Gratias and Cahn (Shechtman et al. 1984) observed a new phase, which has been named icosahedral quasicrystal. This phase appears in small grains of a few microns size embedded in an aluminium matrix.

The striking feature of this phase is that its electron diffraction patterns exhibit both sharp Bragg peaks and icosahedral symmetry. As mentioned above, such a situation is forbidden by the laws of classical crystallography. Since this initial discovery, icosahedral quasicrystals have been observed in several compounds, mainly ternary aluminium-based alloys. Up to now, the best quasicrystalline samples (with single quasicrystals as large as a few millimeters) have been obtained in *Al–Cu–Fe* alloys. On the other hand, other kinds of 'two-dimensional' quasicrystals have been obtained: they are characterized by a periodic arrangement of atoms along one direction and a non-periodic arrangement in the planes orthogonal to that direction, which appears as a symmetry axis of order eight, ten or twelve, all of these cases being forbidden by the laws of classical crystallography. For general informations about quasicrystals, the reader is referred to the following books: International Workshop on Aperiodic Crystals (Les Houches)(1986) and The Physics of Quasicrystals (1987).

Soon after it was recognized that these extraordinary electron diffraction patterns were not artifacts, three different approaches were devised to interpret them: the so-called *icosahedral glass* theory, the theory of *multiple twinning* and the theory of *quasiperiodic structures*. This last point of view will be developed in the remaining of this paper. Let us say a word about the two first:

Following the point of view of icosahedral glasses, the quasicrystals are to be seen as 'random' packings of icosahedrally symmetric clusters of atoms. The randomness in these packings is not so easy to characterize: the idea is that during the growth of the solid phase from the melt, whole icosahedral clusters stick on the solid in a way compatible with steric hindrance, and when several sites are available for a cluster, then the choice is made at random. It can be shown, in particular with numerical simulations, that such a growth process may result in a partial positional ordering, which manifests itself through the appearance of peaks in the diffraction pattern (this is the 'Hendricks-Teller' effect), and the distribution of these peaks presents the icosahedral symmetry. However, these peaks are not infinitely thin like ideal Bragg peaks, and this broadening reflects the 'intrinsic disorder' of the structure which is referred to in the name icosahedral *glass*.

The theory of multiple twinning adopts quite the opposite point of view: the 'quasicrystallite' are considered to be 'twins' of a normal (cubic) crystal with a large unit cell, i.e., pieces of crystals glued together in an icosahedrally symmetric way around a common center. The diffraction pattern of such a multiple twin is the superimposition of the diffraction patterns of the different parts, and thus will present the icosahedral symmetry. However, since crystals cannot present the icosahedral symmetry, one cannot achieve an exact superimposition of the diffraction peaks coming from the components of the twin. Thus in this theory the icosahedral symmetry of the diffraction pattern is due to the approximate superimposition of a set of peaks and if the crystallites are small enough, the peak widths are large enough for this set of peaks to appear as an unique peak centered on an exact icosahedral site.

It should be emphasized that at the time when these theories were proposed, they were compatible with the available experimental data: both of them require rather large diffraction peaks (the first one because of intrinsic disorder and the second for the splitting of the peaks to be hidden). However, the progressive improvement of 'quality' and size of the quasicrystals prepared by chemists led to narrower and narrower diffraction peaks. In particular, the *Al–Cu–Fe* system produces very good single quasicrystals with peak widths far smaller than those predicted by the icosahedral glass model and which would require unrealistically large unit cells for a multiple twinning model to work. One the other hand, the theory of quasiperiodic structures constitutes a framework large enough to allow the description of both some randomness of the structure (through a 'roughening' of the cut, see below) and the transition to periodic crystals with large unit cells which are in a precise sense 'close' to a quasiperiodic system (the so-called *approximating* structures).

3. Description of Quasiperiodic Sets of Points

Let us first recall the definition of quasiperiodicity (see for instance Besicovitch (1932)): a function (of d real variables) on an affine (d-dimensional) space $\mathbf{E}$ is said to be *quasiperiodic* if it is the restriction to $\mathbf{E}$ (embedded as an affine subspace) of a periodic function of n real variables defined in a higher dimensional space $\mathbb{R}^n$. Of course, if the embedded space $\mathbf{E}$ (which will be referred to as the 'cut') is rationally oriented with respect to the lattice of periods of the periodic function (i.e., if $\mathbf{E}$ is parallel to a lattice subspace) then the restriction of this function to $\mathbf{E}$ is also periodic. But if the orientation of the cut is irrational, that is, if the vector subspace parallel to the cut contains no point of the lattice besides the origin, then the restriction is not a periodic function. Observe that from an analytical point of view, the restriction to $\mathbf{E}$ can be considered as the density of the measure on $\mathbf{E}$ defined by the (multiplicative) product of the periodic function in $\mathbb{R}^n$ by the Lebesgue measure carried by $\mathbf{E}$. This product is well defined as soon as the function is smooth.

3.1 The Atomic Surfaces

To describe a quasiperiodic set of points rather than a quasiperiodic function, the natural generalization is to attach to each of these points a Dirac delta and to consider them as the restriction μ of a periodic measure instead of the restriction of a periodic (smooth) function. Let Λ be the lattice of periods of this periodic measure. Then this measure $\widetilde{\eta}$ is the inverse image of a measure η defined on the torus $\mathbb{T}^n = \mathbb{R}^n/\Lambda$ through the canonical projection $\mathbb{R}^n \mapsto \mathbb{T}^n$. The carrier σ of η is to be a compact submanifold in $\mathbb{T}^n$ with a dimension $n-d$ equal to the codimension of the cut $\mathbf{E}$, and σ will be in the general case a manifold with boundary. For physical reasons to be made clear below, it is sufficient to consider measures with a continuous density along their carriers. The manifold σ is called the 'atomic surface' and its canonical lift in $\mathbb{R}^n$ is the periodic set Σ of atomic surfaces. They do not correspond exactly to the atomic surfaces defined by Janner and Janssen: as will be made clear, the latter (which could be called 'large atomic surfaces') appear as the unions of our submanifolds over whole sublattices of the high-dimensional lattice, and they have no boundary.

To determine a structure in quasicrystallography amounts to construct an appropriate lattice Λ in the relevant high-dimensional space $\mathbb{R}^n$, and then to construct the atomic surface σ in the torus $\mathbb{R}^n/\Lambda$. Observe that this framework is a natural generalization of classical crystallography: in that case too, the first step is to find the space group of the crystal, and the second is to determine the atomic positions in the unit cell, that is in the 3-dimensional torus. We can say that this second step is to determine the embedding of a 0-dimensional compact manifold (a finite set of points...) in $\mathbb{T}^3$ just as it is for quasicrystals to determine the embedding of the atomic surface in $\mathbb{T}^n$.

Now, the main difficulty of quasicrystallography becomes clear: loosely speaking, the position of a manifold in $\mathbb{T}^n$ depends on 'infinitely more' parameters than the position of a finite set of points in $\mathbb{T}^3$. On the other hand, we get from diffraction experiments essentially the same amount of data: a finite set of numbers, measured of course with a finite accuracy. It seems that the only way to solve this difficulty is to find enough a priori constraints on the atomic surface to make its position in $\mathbb{T}^n$ depend on finitely many real parameters only, and in this paper we report on some progress in this direction.

Let us return to the description of our framework. From an analytical point of view, the set of intersections between the cut and the periodic set Σ of atomic surfaces carries the multiplicative product μ between the measure $\widetilde{\eta}$ defined on the atomic surfaces and the Lebesgue measure on the cut $\mathbf{E}$. With our smoothness assumptions, this product is well defined as soon as the cut is transverse to the atomic surfaces, and we shall only consider situations in which this condition is generic with respect to the translations of the cut. The transversality condition means that the cut does not intersect any atomic surface on its boundary or on a point where the tangent space to the atomic surface has a non-zero intersection with the cut. Observe that asking for the genericity of this property is not a strong restriction and simply means, in a first approach, that the boundary is not a too complicated set, such as a fractal set.

To sum up, the definition of a quasiperiodic set of points involves: first, a periodic lattice of atomic surfaces, and second, a plane cut through this lattice. Observe that for a given *direction* of the cut, one defines at once infinitely many different structures. In fact, for a generic atomic surface, two cuts yield isometric structures if and only if they are mapped on each other by a translation belonging to the lattice, up to a translation parallel to the cut. These translations may be described as all the vectors in $\mathbb{R}^n$ which project on the subspace orthogonal to the direction of the cut onto the projection of the lattice, in such a way that the different structures are classified by the quotient of this subspace by the projection of the lattice. Since this last set is countable, we see that by shifting the cut we generate an uncountable infinity of different (non-isometric) structures.

However, from a crystallographic point of view, all these structures are indistinguishable from each other, for they all yield the same intensities in their diffraction patterns, as we shall now compute.

3.2 Fourier Transform

Let us first remark that the Fourier transform $\widehat{\mu}$ of the measure μ on the cut obtained as the restriction to the cut of the measure $\widetilde{\eta}$ carried by the atomic surfaces, is well defined in physically relevant situations. In fact, we have in any atomic system a minimal distance between any two points, so that if η is bounded we see immediately that the measure μ is a slowly growing generalized function and thus admits a Fourier transform.

To compute it, one has only to reproduce in Fourier (reciprocal) space the Fourier image of the construction of this measure: in direct space, we take the measure η on the atomic surface attached to one fundamental domain of the lattice Λ (we identify here the chosen fundamental domain with the torus $\mathbb{T}^n$). Then we construct the corresponding measure $\widetilde{\eta}$ on the lattice Σ of atomic surfaces by taking the convolution product of η with the measure defined by one Dirac delta at each vertex of the lattice Λ, and finally we take the multiplicative product by the Lebesgue measure carried by the cut, and consider the result as a measure on the cut.

Accordingly, we have to consider the Fourier transform $\widehat{\eta}$ of η in Fourier space, to take its multiplicative product by the Fourier image of the lattice of Dirac deltas (which is the reciprocal lattice Λ^* of Dirac deltas) and finally to take the convolution product of the result by the Fourier image of the Lebesgue measure carried by the cut (Duneau and Katz (1985), Kalugin et al. (1985), Elser (1986)).

The two first steps are always well defined in our context: since the atomic surface attached to one vertex of the n-dimensional lattice is compact, the Fourier transform $\widehat{\eta}$ exists and is a smooth (analytic) function, so that the multiplicative product by the reciprocal lattice exists and simply results in attaching to each Dirac delta in this lattice Λ^*, a coefficient which is the value of this smooth function at the same point.

To develop the last step, let us first consider the case where the cut is identical to $\mathbf{E}^{\|}$, the vector subspace of $\mathbb{R}^n$ defining the direction of the cut $\mathbf{E}$, and let $\mathbf{E}^{\perp}$ be the orthogonal subspace. Then the Lebesgue measure $d\xi$ carried by the cut can be written as the tensor product of the Lebesgue measure $dx_{\|}$ on $\mathbf{E}^{\|}$ with the Dirac delta at the origin in $\mathbf{E}^{\perp}$:

$$d\xi = dx_{\|} \otimes \delta_{0\in\mathbf{E}^{\perp}}$$

so that:

$$d\xi \cdot \widetilde{\eta} = d\xi \cdot (\eta * \Lambda) = \mu \otimes \delta_{0\in\mathbf{E}^{\perp}}$$

The Fourier transform is the tensor product of the Fourier transforms of the factors: we get the Lebesgue measure carried in the reciprocal space by the subspace $\mathbf{E}^{\perp *}$ dual to $\mathbf{E}^{\perp}$:

$$\widehat{d\xi} = \delta_{0\in\mathbf{E}^{\|*}} \otimes dx_{\perp}$$

The convolution product of this measure by the lattice $\widehat{\eta} \cdot \Lambda^*$ of weighted Dirac deltas simply amounts to placing at each vertex of Λ^* a copy of $\mathbf{E}^{\perp}$, bearing the Lebesgue measure weighted by a coefficient which is the value of $\widehat{\eta}$ at this vertex. The result has the form of a tensor product between the Lebesgue measure in $\mathbf{E}^{\perp *}$ and a generalized function $\widehat{\mu}$ in the orthogonal subspace $\mathbf{E}^{\|*}$:

$$\widehat{d\xi} * (\widehat{\eta} \cdot \Lambda^*) = \widehat{\mu} \otimes dx_{\perp}$$

and this generalized function $\widehat{\mu}$ is the Fourier transform of the measure μ defined on the cut.

Observe that this last step simply corresponds to the projection of the lattice $\widehat{\eta} \cdot \Lambda^*$ of weighted Dirac deltas on $\mathbf{E}^{\|*}$ along $\mathbf{E}^{\perp *}$.

To deal with the case where the cut $\mathbf{E}$ is no longer identical to the subspace $\mathbf{E}^{\|}$, let us denote $\mathbf{x}_{\perp}$ the vector in $\mathbf{E}^{\perp}$ such that $\mathbf{E} = \mathbf{E}^{\perp} + \mathbf{x}_{\perp}$ and observe that we get the same result by shifting the cut through the vector $\mathbf{x}_{\perp}$ and shifting the lattice Σ of atomic surfaces through $-\mathbf{x}_{\perp}$, keeping $\mathbf{E}^{\|}$ as the cut. Such a shift of the atomic surfaces amounts to multiplying the function $\widehat{\eta}(\mathbf{k})$ by the phase factor $e^{-i\mathbf{k}\cdot\mathbf{x}_{\perp}}$, whose 'sampling' at each $\mathbf{k} \in \Lambda^*$ yields a different phase factor for each of the Dirac deltas of $e^{-i\mathbf{k}\cdot\mathbf{x}_{\perp}} \cdot \widehat{\eta} \cdot \Lambda^*$.

Let us make some comments about this computation:

The Fourier transform $\widehat{\mu}$ of the structure is the sum of a countable family of weighted Dirac deltas, carried by the $\mathbb{Z}$–module projection of Λ^* on $\mathbf{E}^{\|*}$, and this justifies the reference to the *quasiperiodicity* of this structure. Observe that for a 'completely irrational' orientation of the cut with respect to the lattice Λ, this $\mathbb{Z}$–module is dense in $\mathbf{E}^{\|*}$.

However, the family of coefficients is not in general locally summable (although it is locally ℓ^2 as soon as the atomic surface σ is a compact manifold with a smooth enough boundary). For this reason, the very existence of the sum may depend on the phases of the coefficients, and this restriction is the 'Fourier image' of the transversality condition stated in the direct space: the Fourier transform of the structure is well defined only when the structure itself is well defined. As mentioned above, this requires a transversality condition between the cut and the lattice Σ of atomic surfaces, which may break down for special positions of the cut (for instance if Σ has boundaries). In the Fourier space, these special positions correspond to special combinations of phase factors such that the (locally infinite) sum of weighted Dirac deltas does not exist as a generalized function.

On the other hand, since a shift of the cut results only in a change of the phase factor for each of the coefficients of the Dirac deltas, we see that such a shift does not alter the *intensities*, which are the squared modulus of the coefficients of the peaks (this is true only if the projection of Λ^* on $\mathbf{E}^{\|*}$ is one to one, but when one starts from experiments, it is always possible, and in fact natural, to construct the high-dimensional lattice Λ so that this property is verified). Since the diffraction experiments record only the intensities, we conclude that the 'large' infinity of different structures corresponding to the allowed parallel cuts are indistinguishable from the experimental (diffraction) point of view.

Finally, observe that the carrier of the Fourier transform being a dense subset of the Fourier space $\mathbf{E}^{\|*}$ raises specific problems for the interpretation of experimental data. The fact that it is actually possible to 'index' the diffraction peaks unambiguously (that is, to find for each peak the vertex of Λ^* of which this peak is the projection on $\mathbf{E}^{\|*}$) is linked with the following important (experimental) property: there is a strong 'hierarchical' structure for the diffraction patterns, in the sense that the set of peaks with an intensity greater than any strictly positive threshold is always a discrete set. Although the exact significance of this fact for the geometry of the atomic surface is not completely understood, we shall see on the example of the Penrose tilings that it seems to correspond to smoothness features of the atomic surface σ.

In order to show how the previous geometric framework may be directly relevant for the structural study of quasicrystals, we shall now present the results of the high dimensional generalization of the classical Patterson analysis.

3.3 Patterson Analysis

Let us first recall the hypothesis underlying classical crystallography: because of finite instrumental resolution, finite size effects, inelastic scattering and so on, the diffraction measurements do not result in infinitely thin peaks which could directly be considered as Dirac deltas. The crystallographic interpretation of these data requires some hypothesis, the main of which being that *the sample is 'sufficiently well' described as a finite piece of a periodic structure.* Then it is easy to show that the best approximation to the diffraction pattern which would result from such an ideal structure is provided by the assignment of *the integrated intensity of each diffraction peak* as the coefficient of a Dirac delta *located on the mass-center of each peak.* Of course, one has to verify that the positions assigned in this way in the Fourier space coincide 'sufficiently well' with some allowed reciprocal lattice (this is nothing but the first of the numerous verifications of the consistency of their work which are made by crystallographers).

Considering the coefficients resulting from this integration to be the set of squared modulus $|\widehat{\rho}(\mathbf{k})|^2$ of the Fourier coefficients of the periodic diffracting function $\rho(\mathbf{x})$, one can then compute their back Fourier transform. One obtains in this way a function proportional to $(\rho * \check{\rho})(\mathbf{x})$, (where $\check{\rho}$ denotes as usual the function $\mathbf{x} \mapsto \rho(-\mathbf{x})$) which describes the set of interatomic distances in the crystalline sample: more precisely, the function $(\rho * \check{\rho})(\mathbf{x})$ (the Patterson function) takes large values for the translations $\mathbf{x}$ which superimpose a large part of the crystal on itself. This is by definition the case for the periods of the crystal, so that the Patterson function is periodic with at least the periodicity of the crystal itself. But if there are several atoms in the unit cell of the crystal, then the Patterson function takes also large values for the vectors $\mathbf{x}$ which map an atom on a different atom of the cell. Thus it yields useful information about the relative atomic positions inside the crystalline cell.

Observe however that the maxima of the Patterson function are not sharp, for two main reasons. The first one is that the diffracting function ρ is in general smooth (it is in general the electronic density, which is spread over all the atomic volume). The second is that only a finite number of intensities enter the computation of the back Fourier transform (those which are large enough to allow an accurate measurement). This is why the interatomic distances deduced from the Patterson analysis are not in general accurate enough to lead directly to the atomic structure of the crystal.

Let us now turn to (icosahedral) quasicrystals. As already mentioned, the most prominent feature of their diffraction patterns is that they present sharp Bragg peaks and icosahedral symmetry. Since the icosahedral symmetry is not compatible with any lattice symmetry, it is clearly impossible to index these peaks in a 3-dimensional (reciprocal) lattice. But it is natural to try and index the peak positions in a $\mathbb{Z}$–module of rank greater than 3, invariant under the icosahedral group. This happens to be possible with a $\mathbb{Z}$–module of rank 6 and one is naturally led to develop the same analysis as for an ordinary crystal, i. e., to compute the integrated intensity for each peak and consider it as the squared modulus of a Fourier coefficient of an ideal quasiperiodic structure (Gratias et al. (1988)).

Following in the reverse way the computation of the Fourier transform of a quasiperiodic function, it is then natural to embed the 3-dimensional reciprocal space in $\mathbb{R}^6$ as the previous $\mathbf{E}^{\|*}$, and to choose in $\mathbb{R}^6$ a lattice Λ^* projecting in a one to one way onto the 'experimental' $\mathbb{Z}$–module. Then one can lift in a unique way the diffraction data on the lattice Λ^* and consider them as the Fourier series of a periodic function in $\mathbb{R}^6$, which is the generalization of the Patterson function. Some 2-dimensional sections of this function computed for

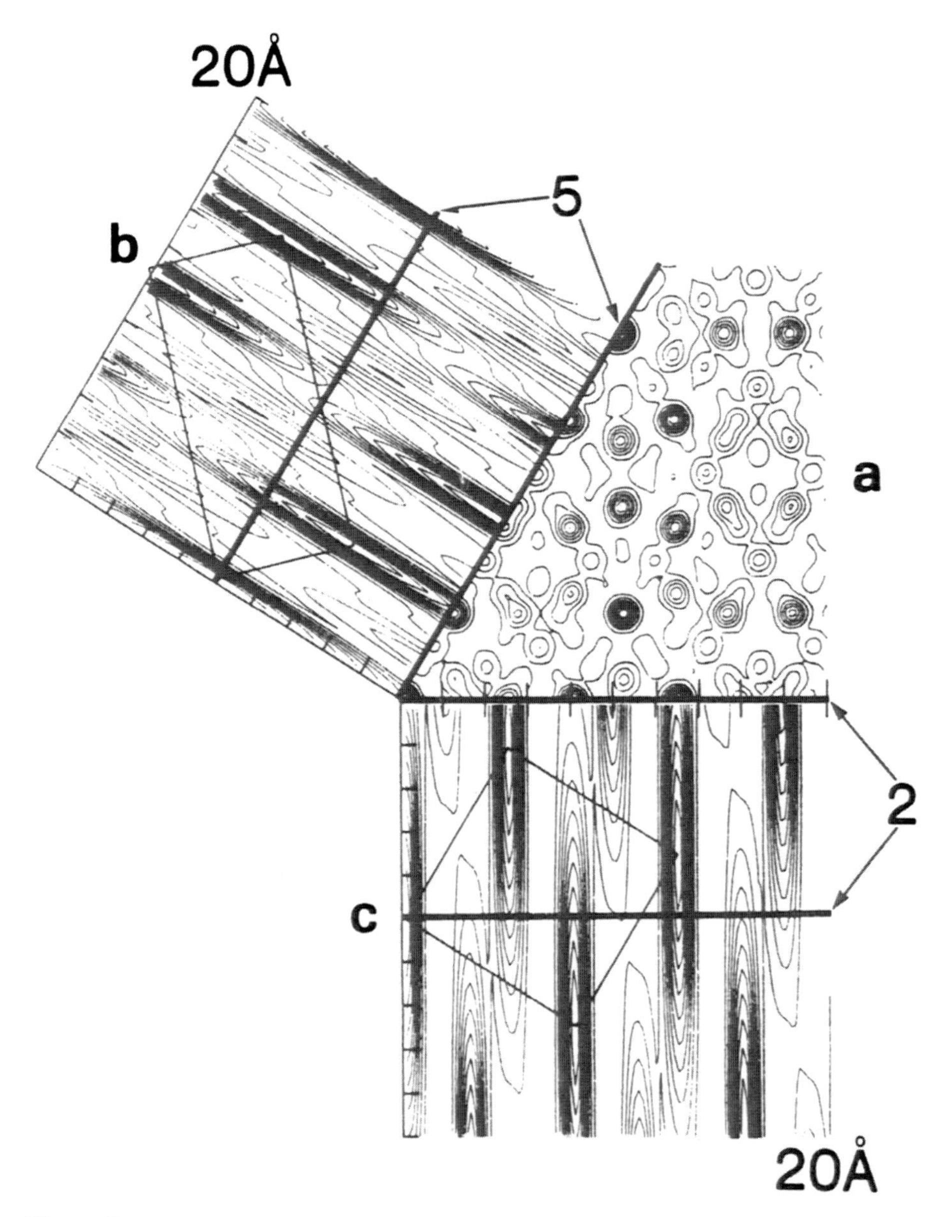

Fig. 1. The 6-dimensional Patterson function of the quasicrystal *Al–Mn–Si* computed from X-ray diffraction data and showing (a) the trace on a plane of the physical space spanned by a five-fold and a two-fold axis, (b) the trace on a five-fold plane of $\mathbb{R}^6$ and (c) the trace on a two-fold plane of $\mathbb{R}^6$ (courtesy of D. Gratias)

the quasicrystal *Al–Mn–Si* are depicted on Fig. 1. Let us make some comments about its specific features.

Since we have been dealing with atomic surfaces, it is worthwhile to explain the relation of the atomic surface and the diffracting density ρ. Let us first consider the case of a periodic crystal. For each atomic species a, one defines its atomic form factor F_a and the corresponding partial density is the sum of all the atomic form factors centered on all the vertices of the sublattice corresponding to the atomic species. Finally the function ρ is the sum of the previous partial densities over the different atomic species.

In other words, we consider for each atomic species the convolution product of the atomic form factor by the sublattice of Dirac deltas describing the positions of the corresponding atoms, and then sum over the different chemical species.

For quasicrystals, we have to consider a (higher-dimensional) lattice Σ_a of atomic surfaces as the generalization of the periodic lattice of atomic positions for the atomic species a, and the measure $\widetilde{\eta}_a$ instead of the corresponding lattice of Dirac deltas. To express the 'spreading' of the atomic surface *along the 3-dimensional direction* $\mathbf{E}^{\|}$ *only*, we have to construct the measure carried by $\mathbf{E}^{\|}$ with the density F_a with respect to the Lebesgue measure on $\mathbf{E}^{\|}$, and to compute the convolution product of $\widetilde{\eta}_a$ by this last measure. Let ρ' be the sum of these convolution products over the set of atomic species: it is clear that ρ' is a periodic function (not necessary smooth in the $\mathbf{E}^{\perp}$ direction) which yields upon restriction to the cut $\mathbf{E}$ the diffracting density ρ.

Now, the 6-dimensional Patterson function depicted on Fig. 1 is essentially proportional to $\rho' * \check{\rho}'$. As one can see, this function takes large values along two 'sheets', parallel to $\mathbf{E}^{\perp}$ and centered on the vertices and the body centers of the 6-dimensional simple cubic lattice. Moreover, the width of this function, measured along $\mathbf{E}^{\|}$, is qualitatively the same as the width of the Patterson function of a periodic crystal, i. e., is compatible with the sizes of the atoms of the material.

These two main features of the 6-dimensional Patterson function strongly suggest that the atomic surfaces corresponding to this quasicrystal contain pieces not far from being parallel to $\mathbf{E}^{\perp}$. In fact, if the atomic surfaces were curved or 'modulated' along $\mathbf{E}^{\|}$, this modulation would reflect in $\rho' * \check{\rho}'$ through an additional blurring which is not observed, or at least may be neglected in first approximation

Finally, one can draw a main qualitative conclusion from this Patterson analysis: it is that the system of atomic surfaces does not look so complicated. It seems that this system is made of only a small number of (3-dimensional) sheets parallel to $\mathbf{E}^{\perp}$ (much like the atomic surfaces of the Penrose tilings, see Section 5), so that the structural problem seems to reduce to the study of their shapes in $\mathbf{E}^{\perp}$. However, this study is very difficult, mainly because, as mentioned above, these shapes depend *a priori* on infinitely many parameters.

4. Symmetry

We will now turn to the study of the symmetry properties of quasicrystals. Since there is no general agreement among specialists on these questions, our discussion on this topic will be more speculative than the rest of this paper. For the sake of simplicity, we shall consider in this Section only icosahedral quasicrystals.

To begin with, let us remark that the definition of these symmetry properties is not obvious, and that it is not possible to give a straightforward generalization of the definitions available in classical crystallography.

In fact, one defines in classical crystallography the *space group* as the largest subgroup of the group of Euclidean motions which leaves the structure invariant. But for quasiperiodic structures, this group is in general reduced to the identity, so that this definition does not make sense and we have to 'weaken' in some sense our notion of symmetry.

Observe that this situation has heavy consequences for the 'solid state physics' of quasicrystals. The point is that the existence of a translation group for crystals is the main tool used in the theory of thermal vibrations (phonons), the theory of electronic states and so on. These classical theories do not admit any simple generalization for quasiperiodic media, and in fact very little is known on these important questions for quasicrystals. In that sense, the 'weakening' of the notion of symmetry for quasicrystals is not formal: the main theoretical consequences of the ordinary notion of symmetry are actually lost.

Now, the natural starting point is again the diffraction pattern. One can associate two groups with it. The first one is obviously the subgroup of the rotation group $\mathcal{O}(3)$ which leaves this pattern invariant: it is the icosahedral group Y_h, from which are named the icosahedral quasicrystals. The second is the symmetry group of the $\mathbb{Z}$–module which carries the diffraction pattern. This last group contains the $\mathbb{Z}$–module itself as its translation subgroup, and the quotient by this translation group contains, besides the icosahedral group, the so-called *self-similarity* transformations.

4.1 The Icosahedral 'Point Group'

In order to establish the physical meaning of the icosahedral symmetry in this context, let us refer once again to the case of periodic crystals.

From a formal point of view, observe first that if the translation group L of a crystal is invariant under the action of the point group P, then the reciprocal lattice L^* is invariant through the dual action of P. But the diffraction pattern is always invariant through the inversion with respect to the origin, simply because this operation changes the sign of the phases in the Fourier transform and these phases are lost in the diffraction pattern. Thus we see that for a periodic crystal the relationship between the point group and the symmetry group of the diffraction pattern is the following: if the point group contains the inversion with respect to the origin, then these two groups are identical; and if the point group does not contain the inversion, then the symmetry group of the diffraction pattern is a two fold extension of the point group. For the quasicrystals, this means that we cannot at this level make any distinction between the group Y_h, which contains the inversion and is of order 120, and the group Y, which does not contain the inversion and is of order 60.

From a physical, or phenomenological point of view, one may consider that the relevant symmetry properties of a crystal depend on the problem under consideration, and more precisely on the natural length scales of the problem. For instance, if we study the interaction of the crystal with X-rays (whose wave length is of the order of magnitude of the interatomic distances in the crystal), then the relevant group to consider is the space group, since the X photons 'see' the inhomogeneity of the material on the atomic scale. But if we consider the propagation of light in a transparent crystal, then the relevant group is the point group: in fact, the wave length of light is of the order of a micron, and at this scale the crystal is homogeneous, although its anisotropy may still have effects on the light propagation.

Since the symmetry group of the diffraction pattern does not embody more information than this anisotropy, we may — cautiously — conclude that, as a phenomenological symmetry group, the icosahedral group of quasicrystals describe the anisotropy of the material on length scales large with respect to the atomic scale.

4.2 The Invariance Group of the $\mathbb{Z}$–module

Let us recall more precisely the main steps of the strategy used in classical crystallography to determine a structure: one first identifies the point group (up to the inversion with respect to the origin) by looking at the symmetry of the diffraction pattern. Then one finds the space group of the structure by a closer examination of the diffraction pattern: for a given point symmetry, one has to choose among a finite list of possible space groups, which correspond to different reciprocal lattices, and the reciprocal lattice is closely related to the carrier of the diffraction pattern (up to the inversion ...). When the space group (and thus the unit cell of the crystal) is known, it remains to find the location of the atoms inside the unit cell. One considers that the job is finished when one has found a model such that the simulated diffraction computed from the model coincides 'sufficiently well' with the experimental diffraction data.

It is important to observe that the fact that the symmetries of the diffraction data actually reflect symmetry properties of the crystal itself does not play an essential role in this strategy, so that the same steps (or at least the first two) can be worked out for quasicrystals essentially in the same way, although, as explained above, there is no (non trivial) invariance group of isometries in the real space for quasiperiodic structures.

In fact, one can classify the $\mathbb{Z}$–modules invariant in $\mathbb{R}^3$ under the action of the icosahedral group using the same systematic group-theoretical methods which are used to classify the crystallographic lattices. Since there is already an extensive literature available on this subject, we shall not enter into much details, and give only the main results (see Janssen (1986), Cartier (1987), Martinais (1987 and 1988) and Levitov and Rhyner (1988)).

The first point is that the rank of such invariant $\mathbb{Z}$–modules is 6 or a multiple of 6. However, those $\mathbb{Z}$–modules which are of rank $6n$ with $n > 1$ appear as the sum of $\mathbb{Z}$–modules of rank 6. Thus we have only to classify the $\mathbb{Z}$–modules of rank 6. Observe that such a situation does not appear in classical crystallography where one classifies invariant lattices, i. e., discrete $\mathbb{Z}$–modules of rank 3, because the sum of two different lattices is a $\mathbb{Z}$–module of rank > 3 and is no longer a discrete lattice.

The second point is that there are only 3 different kinds (in the sense of Bravais lattices) of icosahedrally invariant $\mathbb{Z}$–modules of rank 6. They are referred to as the primitive P, the face-centered F and the body-centered I. Let us now explain where these names come from.

Observe that a $\mathbb{Z}$–module of rank 6 in $\mathbb{R}^3$ can be considered as the projection in $\mathbb{R}^3$ (embedded in $\mathbb{R}^6$) of a 6-dimensional lattice $\mathbb{Z}^6$. The choice of this 6-dimensional lattice is not unique, but it turns out that it is possible to choose a cubic one. In particular, the 'simple cubic' P comes from the following construction:

Consider in $\mathbb{R}^3$ a regular icosahedron and the twelve vectors joining the center to the vertices. Choose six vectors among the twelve, no two of them being opposite. Then there exists an unique embedding of $\mathbb{R}^3$ as a vector subspace of $\mathbb{R}^6$ such that the canonical orthonormal basis of $\mathbb{R}^6$ projects orthogonally on these six vectors of the icosahedron. Now consider in $\mathbb{R}^6$ the action of the icosahedral group which permutes the six basis vectors just as the ordinary action of the icosahedral group permutes the six chosen vectors in $\mathbb{R}^3$. Observe that since the icosahedral group acts on the basis vectors by (signed) permutations, this action leaves invariant the whole lattice generated by the basis, i. e., the simple cubic lattice Λ^* made of the points whose six coordinates are integers.

The 6-dimensional action of the icosahedral group just defined is not irreducible, and upon reduction we find the two non-equivalent 3-dimensional irreducible representations of the group. They operate on two 3-dimensional invariant subspaces, one of them being of course the image of our embedding of $\mathbb{R}^3$, and the other the orthogonal subspace. Following our notations and in order to emphasize that the initial $\mathbb{Z}$–module under study is the carrier of the diffraction pattern of a quasicrystal, it is natural to denote these two invariant subspaces respectively $\mathbf{E}^{\|*}$ and $\mathbf{E}^{\perp *}$. Since the 6-dimensional action of the icosahedral group is orthogonal and since the simple cubic group is self-dual (it is its own reciprocal lattice), we find in the 'direct' 6-dimensional space exactly the same geometry: the simple cubic lattice Λ and two orthogonal invariant subspaces, which are identified with our previous $\mathbf{E}^{\|}$ and $\mathbf{E}^{\perp}$.

It is easy to compute the two projectors $\pi^{\|}$ and $\pi^{\perp}$ onto $\mathbf{E}^{\|}$ and $\mathbf{E}^{\perp}$ respectively. One finds:

$$\pi^{\|} = \frac{1}{2\sqrt{5}} \begin{pmatrix} \sqrt{5} & 1 & -1 & -1 & 1 & 1 \\ 1 & \sqrt{5} & 1 & -1 & -1 & 1 \\ -1 & 1 & \sqrt{5} & 1 & -1 & 1 \\ -1 & -1 & 1 & \sqrt{5} & 1 & 1 \\ 1 & -1 & -1 & 1 & \sqrt{5} & 1 \\ 1 & 1 & 1 & 1 & 1 & \sqrt{5} \end{pmatrix}$$

and

$$\pi^{\perp} = \frac{1}{2\sqrt{5}} \begin{pmatrix} \sqrt{5} & -1 & 1 & 1 & -1 & -1 \\ -1 & \sqrt{5} & -1 & 1 & 1 & -1 \\ 1 & -1 & \sqrt{5} & -1 & 1 & -1 \\ 1 & 1 & -1 & \sqrt{5} & -1 & -1 \\ -1 & 1 & 1 & -1 & \sqrt{5} & -1 \\ -1 & -1 & -1 & -1 & -1 & \sqrt{5} \end{pmatrix}$$

Observe that although the choice of the cubic lattice is a handy one, it has no physical significance: we are interested only in the icosahedral group, which is a (very) small subgroup of the hyperoctahedral group which leaves invariant the cubic lattice. As a straightforward consequence of Shur's lemma, it is clear that one can make any dilatation or contraction along $\mathbf{E}^{\perp}$, keeping the (physically relevant) icosahedral symmetry while breaking the (irrelevant) hypercubic one. In physical words, one may say that, since no actual operation can relate the 'real' space $\mathbf{E}^{\|}$ and the 'fictitious' space $\mathbf{E}^{\perp}$, one can choose arbitrarily different unit lengths in each of them.

We have thus found a first icosahedrally symmetric $\mathbb{Z}$–module, which is the projection along $\pi^{\|}$ of the simple cubic group (called P for primitive) of $\mathbb{R}^6$. It now easy to find two more. In fact, since the 6-dimensional action of the icosahedral group implies only signed permutations, it follows that two sublattices of the simple cubic lattice are separately invariant under this action: the sublattice (of index 2) of vertices such that the sum of their six coordinates is an even number and the sublattice (of index 32) of vertices such that their six coordinates are simultaneously even or odd. As in the 3-dimensional case, each of these three invariant lattices is a sublattice of each of the two others. For instance, the simple cubic is a sublattice (of index 2) of the last one, 'scaled' by a factor one half, i. e., the lattice of points in $\mathbb{R}^6$ with coordinates simultaneously integers or half-integers. This shows that this last lattice is simply the body-centered cubic denoted I. In the same manner, the simple cubic is shown to be a sublattice (of index 32) of the remaining one, which is in this way identified with the cubic lattice with a vertex added at the center of the even-dimensional faces (i. e., the facets of dimension 2 and 4 and the bodies (dimension 6) are centered). By analogy with the 3-dimensional case, this last lattice is simply referred to as the face-centered F lattice. Finally, one can easily show that (again as in the 3-dimensional case) the simple cubic P is self-dual and that the two lattices F and I are the reciprocal lattice of each other.

One can show — but this is not trivial, see for instance Martinais (1988) — that although there are two more invariant sublattices of the simple cubic lattice, the three *projections* on $\mathbf{E}^{\|}$ of the three lattices P, F and I (or equivalently, the three *projections* on $\mathbf{E}^{\|*}$ of P^*, F^* and I^*, which are of more concern as long as we are dealing with diffraction data) are *the only three icosahedrally invariant $\mathbb{Z}$–modules in $\mathbb{R}^3$*.

This classification plays the same role for icosahedral quasicrystals that the classification of the three cubic Bravais lattices for the cubic crystals. One can go further and classify completely the space groups, but we shall not develop on this point, referring to the above-mentioned literature for more details.

We can now return to the analysis of diffraction data: as a consequence of the fact that the three $\mathbb{Z}$–modules P^*, F^* and I^* are submodules of P^*, it is always possible to index the diffraction pattern in a simple cubic lattice with an adequate edge length. Then, if the quasicrystal under study actually corresponds to the primitive P^*, one sees non-vanishing peaks on each of the possible sites. More precisely, since one observes in real experiments only a finite number of peaks, one verifies that the simple cubic is the smallest $\mathbb{Z}$–module in which fall all the observed diffraction peaks. On the contrary, if the relevant $\mathbb{Z}$–module for the given quasicrystal is F^* or I^*, some peaks belonging to P^* are systematically missing. Thus one can decide what is the relevant $\mathbb{Z}$–module, and up to now two of them have been observed: the primitive P^* in most of the quasicrystals (in particular *Al–Mn–Si*) and the face-centered F^* in *Al–Cu–Fe*.

Finally, one sees that the two first steps of the strategy sketched above are easily worked out for quasicrystals: one is able to find the relevant $\mathbb{Z}$–module together with its scale (see below), to lift it as a (reciprocal) lattice in $\mathbb{R}^6$, and finally to construct the corresponding lattice in the direct space. The last step is far more difficult: it consists in the construction of the atomic surfaces in the high-dimensional unit cell thus determined. No systematic approach to this step is known, but before coming to this question let us discuss a specific feature of our $\mathbb{Z}$–modules, which has direct implications for the choice of the length scale for the high-dimensional lattice but may be of broader physical significance.

4.3 Self-similarity

Besides the icosahedral and the translational symmetries, the $\mathbb{Z}$–module which carries the diffraction pattern possesses another symmetry: it is invariant through a group of homotheties.

The simplest way to find it is to consider the 6-dimensional lattice which projects onto this $\mathbb{Z}$–module, and let us choose the simple cubic case for the sake of clarity. This lattice is invariant under the group $Gl(6, \mathbb{Z})$. On the other hand, Shur's lemma entails that the automorphisms of $\mathbb{R}^6$ which commute with the icosahedral group are of the form $\lambda\pi^{\|} + \lambda'\pi^{\perp}$, for any real numbers λ and λ'. Thus the question is whether there exist matrices of this last form with integer entries, and one easily shows that they do exist and are all the powers of the following:

$$M = \begin{pmatrix} 2 & 1 & -1 & -1 & 1 & 1 \\ 1 & 2 & 1 & -1 & -1 & 1 \\ -1 & 1 & 2 & 1 & -1 & 1 \\ -1 & -1 & 1 & 2 & 1 & 1 \\ 1 & -1 & -1 & 1 & 2 & 1 \\ 1 & 1 & 1 & 1 & 1 & 2 \end{pmatrix}$$

which is such that:

$$M = \tau^3 \cdot \pi^{\parallel} - \tau^{-3} \cdot \pi^{\perp}$$

where $\tau = (\sqrt{5}+1)/2$ is the golden ratio.

Thus one sees that our $\mathbb{Z}$–module is invariant under the homotheties whose ratio is a power of $\tau^3 = \sqrt{5}+2$. As a consequence, the indexation scheme of the diffraction pattern must contain not only the orientation of the chosen basis vectors, but also their length, measured (in the reciprocal space), in nm^{-1}.

One can show in the same way that the two other $\mathbb{Z}$–modules F^* and I^* are invariant through the dilatation of ratio τ (and its powers). Let us remark that this fact allows an easy distinction between the primitive $\mathbb{Z}$–module P^* and, say, the face-centered F^*: for the last one, one finds for each peak along a five-fold axis, another peak τ times farther. For the primitive P^*, some of these peaks are missing and one has to go τ^3 times farther to find the next one.

One the other hand, one can wonder if these special self-similarity properties, which depend on the 'point symmetry' and are not a general feature of quasiperiodic structures, are or not relevant for the physics of quasicrystals. Observe once again that this symmetry of the $\mathbb{Z}$–module does not reflect any real symmetry of the structure in the direct space: in fact, it is clear that a packing of atoms cannot be preserved by a group of homotheties. However, there are several instances in physics where a symmetry, although 'broken' in real systems, plays an important role in their understanding. Very little is known about the self-similarity properties of quasicrystals, but we shall show in the special case of Penrose-like tilings what kind of real-space features they correspond to.

5. Penrose Tilings and Related Structures

We shall now turn to the description of the best known class of quasiperiodic systems, which are the Penrose and 'Penrose-like' tilings. In our context, we define a tiling as a partition of the space into connected tiles which belong to a *finite* set of shapes. The 'Penrose-like' tilings are named after Roger Penrose, who discovered a strikingly simple non-periodic tiling of the plane with five-fold symmetry (Penrose (1977)). Although his construction relied mainly on self-similarity properties and the quasiperiodicity of his tilings was recognized only later by crystallographers and solid state physicists, these tilings are in fact the simplest non-trivial quasiperiodic set of points that one can imagine.

Besides providing the reader with explicit examples of quasiperiodic structures, our goal in this Section is to analyze some properties of these tilings, which are not generic for quasiperiodic structures but are useful for the understanding of quasicrystals. On one hand, quasicrystals are not at all generic quasiperiodic systems, as we have already seen with Patterson functions, and share many properties with Penrose-like tilings. One the other hand, it is worthwhile to emphasize here that the relationship between quasiperiodic sets of points and quasiperiodic tilings is not at all the same that in the periodic case. As is well known, one can attach a periodic tiling to any periodic distribution of points, in such a way that the theories of both objects are essentially the same. This relation, which accounts for the relevance of periodic tilings in classical crystallography, is no longer true in the quasiperiodic case, and for instance the tiles of which are made the tilings we are to deal with have a priori nothing to do with 'unit cells' for quasicrystals.

5.1 A Low Dimensional Example

Let us start with the general definition of what we call tilings of the Penrose type (Katz and Duneau (1986)): to construct a d-dimensional tiling, consider in $\mathbb{R}^n$ the simple cubic lattice $\mathbb{Z}^n$ generated by the canonical orthonormal basis of $\mathbb{R}^n$. Then choose any d-dimensional subspace $\mathbf{E}^{\|}$ in $\mathbb{R}^n$, and denote $\mathbf{E}^{\perp}$ the orthogonal subspace. Let γ_n be the unit hypercube spanned by the basis of $\mathbb{R}^n$. We define the atomic surface σ in the n-dimensional torus $\mathbb{T}^n = \mathbb{R}^n/\mathbb{Z}^n$ by projecting orthogonally γ_n on $\mathbf{E}^{\perp}$ and then projecting this polyhedron in $\mathbb{T}^n$ along the canonical projection $\mathbb{R}^n \mapsto \mathbb{T}^n$. The atomic surface σ constructed in this way is a $(n-d)$-dimensional manifold with boundary in $\mathbb{T}^n$, and the corresponding lattice Σ of atomic surfaces in $\mathbb{R}^n$ is made of copies of the $(n-d)$-dimensional polyhedron $\pi^{\perp}(\gamma_n)$, each copy σ_ξ being embedded in an affine subspace parallel to $\mathbf{E}^{\perp}$ and attached to the vertex ξ of $\mathbb{Z}^n$.

The vertices of our tilings are the intersections of Σ with any d-dimensional plane cut $\mathbf{E}$ parallel to $\mathbf{E}^{\|}$, and which is everywhere transverse to Σ, i. e., which does not intersect any of the boundaries $\partial\sigma_\xi$ for $\xi \in \mathbb{Z}^n$. Observe that since the cut $\mathbf{E}$ is of dimension d and the boundaries $\partial\sigma_\xi$ are (piece-wise linear) manifolds of dimension $n-d-1$, their non-intersection is a generic property in $\mathbb{R}^n$, which means that almost all choices of $\mathbf{E}$ will work.

Of course, we are mainly interested in cases where the orientation of $\mathbf{E}^{\|}$ is irrational with respect to the lattice $\mathbb{Z}^n$: if this orientation is rational, that is, if $\mathbf{E}^{\|}$ goes through d independent lattice points besides the origin, then these d vectors span a sublattice of $\mathbb{Z}^n$ which is a symmetry group for the tilings, so that we get periodic structures. Notice that there are intermediate possibilities between such periodic cases and the 'completely irrational' cases where $\mathbf{E}^{\|}$ intersects the lattice on the origin only. Such situations appear when one wants to describe the original Penrose tilings while sticking strictly to the previous definition, and also in the description of some quasicrystals which are periodic along one direction and quasiperiodic in the orthogonal planes.

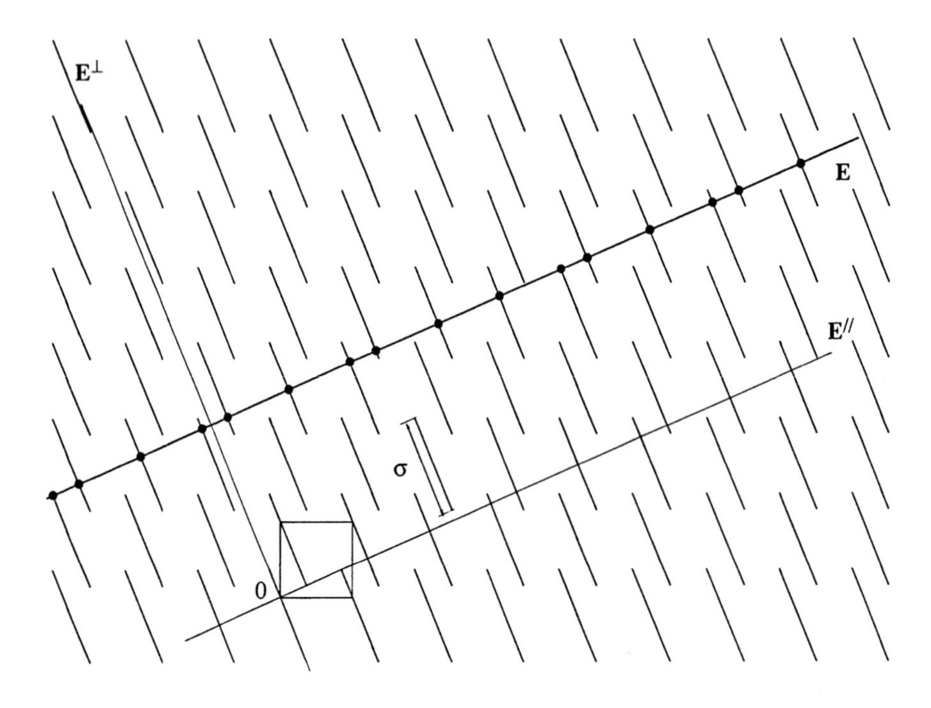

Fig. 2. The construction of a quasiperiodic tiling in the lowest dimensions.

The simplest example of this construction is obtained with $n = 2$ and $d = 1$, and is depicted on Fig. 2. For an irrationally oriented $\mathbf{E}^{\parallel}$, we get a quasiperiodic tiling of the cut $\mathbf{E}$ by means of two segments, which are the projections of the two edges (horizontal and vertical) of the unit square γ_2. Although this construction may look rather trivial, it deserves attention because the most important features of this class of tilings already appear in this simple case and may be discussed in a dimension-independent way.

5.2 The Oblique Tiling

To prove that we actually get a tiling by means of the projections of the edges of the square, the best way is to construct the so-called *oblique tiling*. The idea is the following: consider any tiling of $\mathbb{R}^n$ and any plane cut through this tiling. Each time the cut is generic, that is, intersects transversely the boundaries of the tiles, the traces of the tiles on the cut make up a covering of the cut, without overlapping nor hole. But this covering is not a tiling in general, since there is no reason for the traces of the tiles to belong to a finite set of shapes. For instance, consider a cut $\mathbf{E}$ with an irrational slope through the standard square tiling of the plane: since there is no minimal distance between the vertices of the

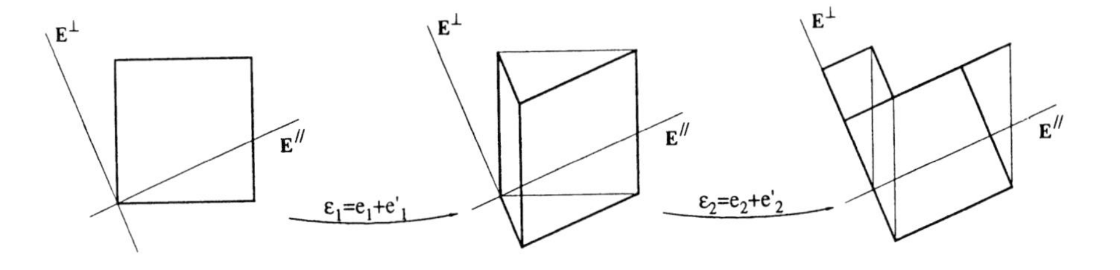

Fig. 3. Construction of the tiles of the oblique tiling.

tiling and the cut, there is no minimal length for the segments of the induced covering of the cut, and this entails that there are infinitely many different lengths in this covering, which therefore is not a tiling.

However, it is possible to adapt the shape of the tiles of a periodic tiling of $\mathbb{R}^n$ to the orientation of the cut, in order to obtain only a finite number of shapes in the generic cuts: the trick is to make the boundaries of the tiles parallel to either the direction of the cut $\mathbf{E}^{\|}$ or to the orthogonal subspace $\mathbf{E}^{\perp}$.

For our low dimensions case, the construction of this oblique tiling is the following: let us start with the unit square, spanned by the canonical basis $(\varepsilon_1, \varepsilon_2)$ of $\mathbb{R}^2$. We define $e_i = \pi^{\|}(\varepsilon_i)$ and $e'_i = \pi^{\perp}(\varepsilon_i) \quad (i = 1,2)$, in such a way that $\varepsilon_i = e_i + e'_i$. Observe that $\varepsilon_1 \wedge \varepsilon_2 = e_1 \wedge \varepsilon_2 + e'_1 \wedge \varepsilon_2 = 1$ (the area of the unit square), and that the two wedge products $e_1 \wedge \varepsilon_2$ and $e'_1 \wedge \varepsilon_2$ are both positive.

This means that the two parallelograms spanned by $\{e_1, \varepsilon_2\}$ and $\{e'_1, \varepsilon_2\}$ do not overlap, so that the union $\{e'_1, \varepsilon_2\} \cup (e'_1 + \{e_1, \varepsilon_2\})$ is still a unit cell for the lattice $\mathbb{Z}^2$, as depicted on the second part of Fig. 3. Now, let us proceed to the same decomposition for the vector ε_2 and each parallelogram: we get a new unit cell of $\mathbb{Z}^2$ made of four subcells spanned by $\{e'_1, e_2\}$, $\{e'_1, e'_2\}$, $\{e_1, e_2\}$ and $\{e_1, e'_2\}$. But the two subcells spanned by $\{e'_1, e'_2\}$ and $\{e_1, e_2\}$ are flat and we can omit them, so that we obtain finally only two subcells whose union is a fundamental domain of $\mathbb{Z}^2$ (last part of Fig. 3). The corresponding tiling of the plane (Fig. 4) is the oblique tiling.

Observe that whatever the order of the decomposition, the resulting tiling is the same. Since each tile is the product of the projection of a basis vector on $\mathbf{E}^{\|}$ by the projection of the other on $\mathbf{E}^{\perp}$, it is clear that any cut $\mathbf{E}$ parallel to $\mathbf{E}^{\|}$ which does not intersect the lattice inherits a tiling by means of the two projections e_1 and e_2, which is our quasiperiodic tiling: in fact, the pieces of the boundaries of the tiles of the oblique tiling which are parallel to $\mathbf{E}^{\perp}$ are by their very construction identical to the lattice Σ of atomic surfaces.

Due to its recursive character, the same argument works in any dimension n. Since we double the number of subcells each time we operate the decompo-

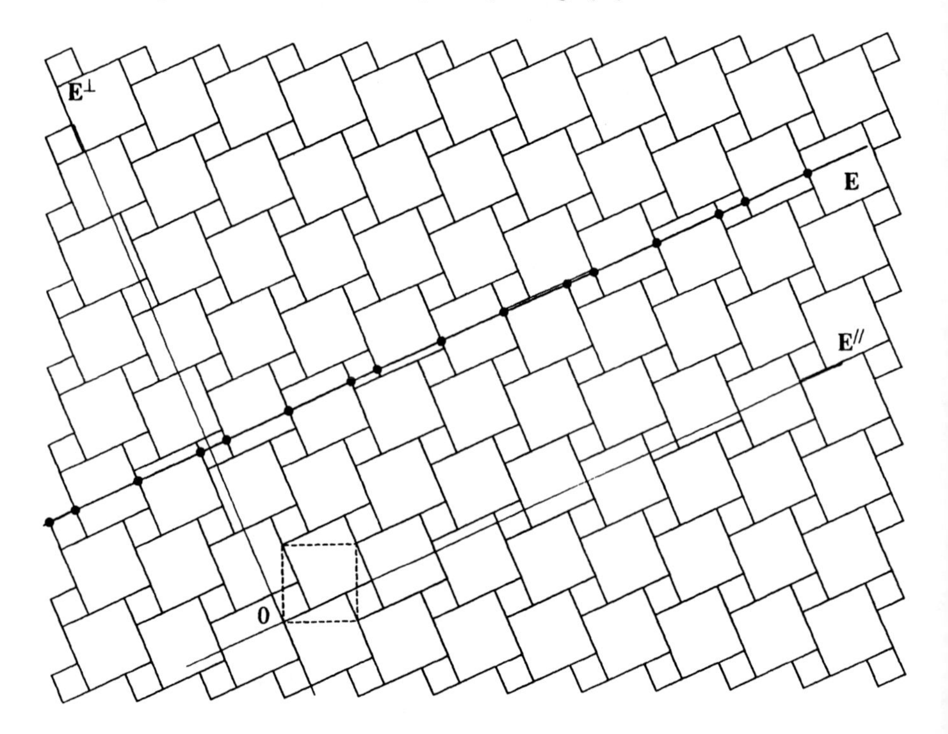

Fig. 4. The oblique tiling of the plane associated with the quasiperiodic tiling of the line.

sition $\varepsilon_i = e_i + e'_i$, we end with 2^n subcells. But only those which are spanned by d projections e_i of basis vectors on $\mathbf{E}^{\|}$ and $(n-d)$ projections e'_i on $\mathbf{E}^{\perp}$ have a non-zero volume and their number is $\binom{n}{d}$. As in the low dimensions case, one easily verifies that the traces of these subcells on $\mathbf{E}^{\perp}$, which are parallelohedra spanned by $(n-d)$ projections e'_i, exactly cover the atomic surface $\sigma_0 = \pi^{\perp}(\gamma_n)$ attached to the origin, so that our construction yields in the general case a tiling of the d-dimensional cut $\mathbf{E}$ by means of the projections of the $\binom{n}{d}$ d-dimensional facets of the hypercube γ_n.

Since the oblique tiling is periodic, it projects in the n-dimensional torus on a partition of the torus, which presents the following special property: each cell in this partition is the sum (in the sense of the sum of subsets in the Abelian group $\mathbb{T}^n$) of a subset contained in the projection of $\mathbf{E}^{\|}$ and of a subset contained in the projection of $\mathbf{E}^{\perp}$. The first one is the projection in $\mathbb{T}^n$ of a type of tile of the quasiperiodic tiling, while the second, as we shall see below, is the projection in $\mathbb{T}^n$ of the corresponding 'existence domain' of this type of tile. Conversely, given any finite partition of $\mathbb{T}^n$ with this special property, we can lift it in $\mathbb{R}^n$ to obtain a periodic tiling, which will give a quasiperiodic tiling upon restriction to any cut parallel to $\mathbf{E}^{\|}$ which is everywhere transverse to the boundaries of the tiles of this periodic tiling of $\mathbb{R}^n$. Thus we conclude that quasiperiodic tilings are in one to one correspondence with this special kind of partitions of the torus, which have been called 'Penrose partitions' by V. I. Arnol'd (Arnol'd (1988)).

5.3 Fourier Transform

The simple geometry of the atomic surface associated to the Penrose-like tilings permits the explicit computation of their Fourier transform. Let us work out the low dimensional case, and let μ be the measure defined by attaching a Dirac delta to each vertex of the tiling defined by the cut $\mathbf{E}^{\|}$ (we choose this special cut in order to avoid the additional phase factor associated with the translation of the cut along $\mathbf{E}^{\perp}$).

The simplicity of this peculiar case stems from the fact that the measure η carried by a single atomic surface (say, the component σ_0 attached to the origin of $\mathbb{R}^2$) is the tensor product of a Dirac delta at the origin of $\mathbf{E}^{\|}$ by a measure in $\mathbf{E}^{\perp}$. And since we want all the vertices in the tiling to have the same weight, this last measure has a density (with respect to the Lebesgue measure of $\mathbf{E}^{\perp}$) which is simply the characteristic function $\mathbb{1}_{\sigma_0}$ of the segment $\sigma_0 = \pi^{\perp}(\gamma_2)$, up to an irrelevant multiplicative constant. Thus:

$$\eta = \mathbb{1}_{\sigma_0} \otimes \delta_{0 \in \mathbf{E}^{\perp}}$$

so that its Fourier transform is the tensor product of the corresponding Fourier transforms and may be written:

$$\widehat{\eta}(\kappa) = \frac{\sin(\pi^{\perp *}(\kappa))}{\pi^{\perp *}(\kappa)}$$

The Fourier transform $\widehat{\widetilde{\eta}}$ of the periodic measure $\widetilde{\eta}$ carried by the whole lattice Σ of atomic surfaces is simply the 'sampling' of this function by the reciprocal lattice Λ^* of the square lattice Λ, which is itself a square lattice:

$$\widehat{\widetilde{\eta}} = \sum_{\kappa \in \mathbb{Z}^2} \frac{\sin(\pi^{\perp *}(\kappa))}{\pi^{\perp *}(\kappa)} \delta_{\kappa}$$

and finally we obtain the Fourier transform $\widehat{\mu}$ by a projection on $\mathbf{E}^{\| *}$:

$$\widehat{\mu} = \sum_{\kappa \in \mathbb{Z}^2} \frac{\sin(\pi^{\perp *}(\kappa))}{\pi^{\perp *}(\kappa)} \delta_{\pi^{\| *}(\kappa)}$$

If the orientation of $\mathbf{E}^{\|}$ is irrational, then the carrier $\pi^{\| *}(\Lambda^*)$ is a dense $\mathbb{Z}$–module, as we already know. We have developed this simple calculation mainly to illustrate the two following properties:

First, the 'hierarchical' structure of the Fourier transform. The coefficient of each Dirac delta in $\widehat{\mu}$ is the value of $\widehat{\eta}$ on the lattice vertex which projects on this point, and since the function $\widehat{\eta}$ is decreasing in the direction of $\mathbf{E}^{\perp *}$, this coefficient will be small if the vertex is far away from $\mathbf{E}^{\parallel *}$. It is clear that the set of Dirac deltas with a modulus of their coefficients greater than any strictly positive threshold come from a strip of *finite* width parallel to $\mathbf{E}^{\parallel *}$, so that this set is discrete. However, the decreasing rate of $\widehat{\eta}$ is not simple to characterize for a generic atomic surface; the 'trend' is that if the atomic surface has a lot of holes, for instance, then the decrease of $\widehat{\eta}$ will be slower and the hierarchical structure of $\widehat{\mu}$ less marked.

Second, the summability properties of the family of coefficients. Consider a point $\mathbf{k} \in \mathbf{E}^{\parallel *}$ and the affine line parallel to $\mathbf{E}^{\perp *}$ going through $\mathbf{k}$. The set of Dirac deltas contributing to $\widehat{\mu}$ in a small ball centered on $\mathbf{k}$ come from a strip which is the product of the affine line by the ball. Due to the irrational orientation of $\mathbf{E}^{\perp *}$, the vertices of Λ^* falling in this strip are evenly (and even quasiperiodically) distributed around the line, in such a way that the summability properties of the family of coefficients of this set of Dirac deltas is the same that the integrability properties of the restriction of $\widehat{\eta}$ to the line: namely those of the function $(\sin(k))/k$. Thus we see that the family of coefficients is locally ℓ^2 but not locally ℓ^1.

5.4 Local Isomorphism

Returning to direct space, let us now discuss a very important property of the Penrose like tilings, which seems (experimentally) to be shared by the icosahedral quasicrystals. It is known as the *local isomorphism property*, and it is two-fold: the first part is that *any finite packing of tiles which appears in a given tiling, appears infinitely many times in the same tiling with a well-defined frequency.* The second part asserts that *any finite packing of tiles which appears in a given tiling, appears in any tiling defined by a cut* $\mathbf{E}$ *parallel to the same* $\mathbf{E}^{\parallel}$.

As will be made clear below, the first part is true for any Penrose-like tiling, while the second is true only when the $\mathbb{Z}$–module $\pi^{\perp}(\Lambda)$ is dense in $\mathbf{E}^{\perp}$, i. e., in 'completely irrational' cases.

To explain these properties, let us return to the low dimensional model of Fig. 2. We want to show that any finite patch of tiles which appears somewhere in the tiling in repeated infinitely many times. Let $\{\sigma_{\xi_i}\}$ $(\xi_i \in \mathbb{Z}^2)$ be the finite family of atomic surfaces whose intersections with the cut $\mathbf{E}$ define the vertices of our patch of tiles and consider the intersection $\beta = \bigcap_i \pi^{\perp}(\sigma_{\xi_i})$ of the projections on $\mathbf{E}^{\perp}$ of the atomic surfaces σ_{ξ_i}. The important point is that since the cut intersects each atomic surface σ_{ξ_i} in its interior, the *finite* intersection β has a non empty interior in $\mathbf{E}^{\perp}$.

Consider the set of translations $\zeta \in \mathbb{Z}^2$ such that $\mathbf{E} - \zeta$ intersects β. This set is infinite, because it contains 0 and the projection $\pi^{\perp}(\mathbb{Z}^2)$ is not discrete (and is in fact dense) as soon as $\mathbf{E}^{\parallel}$ is not a lattice subspace (in which case we have a periodic tiling and we have a whole sublattice of ζ's leaving the cut invariant). In any case, we see that for each ζ in this set, the cut $\mathbf{E}$ intersects the family $\{\sigma_{\xi_i+\zeta}\}$, which corresponds to a copy of our initial patch of tiles, shifted by $z = \pi^{\parallel}(\zeta)$, and this proves the first part of the local isomorphism property.

Observe that the family of translations z thus defined is in fact a quasiperiodic set and has a well-defined density in $\mathbf{E}^{\parallel}$. In fact, it is obtained through the same canonical construction, using β instead of σ as the prototypic atomic surface.

For the second part, consider two tilings defined by the two parallel cuts $\mathbf{E}$ and $\mathbf{E}'$. We want to see whether any patch of tiles which appears in $\mathbf{E}$ appears also in $\mathbf{E}'$. The argument runs as in the previous case, but we have now to define the set of translations ζ such that $\mathbf{E}' - \zeta$ intersects β. Then it is no longer always true that this set is not empty if $\mathbf{E}' \neq \mathbf{E}$. For the low dimensional example, it is clear that $\pi^{\perp}(\mathbb{Z}^2)$ is dense if $\mathbf{E}^{\parallel}$ is not rationally oriented, so that we find infinitely many ζ's and the second part of the local isomorphism property holds. If $\mathbf{E}^{\parallel}$ is a lattice subspace, then $\pi^{\perp}(\mathbb{Z}^2)$ is a discrete lattice in $\mathbf{E}^{\perp}$, and if the initial patch of tiles is large enough, then β is a segment so small that for an arbitrary choice of $\mathbf{E}'$, there may be no ζ at all.

There are interesting questions on the relation between the size of the finite patch under consideration and the frequency of its repetitions: it is clear that this frequency, which is the density of the translations z, is measured by the length of the intersection β, so that the question is equivalent to the following: how fast does the intersection β shrinks when the length of the patch grows?

For this low dimensional example, the answer is given by the classical theory of Diophantine approximation in number theory, and depends on number theoretical properties of the slope. If this slope is an algebraic number it can be shown, as a consequence of Roth's theorem, that for almost all patches the mean distance between two successive copies of the patch is proportional to its length: this corresponds to the fact that algebraic numbers are not very well approximated by rational numbers. On the contrary, if the slope is very well approximated by rational numbers, for instance if it is a Liouville number, then the mean distance between copies of a patch can grow faster than any power of the length of that patch.

In higher dimensions, the proof of the local isomorphism property is exactly the same: the intersection β is no longer a line segment, but a $(n-d)$-dimensional polyhedron embedded in $\mathbf{E}^{\perp}$. The only new feature is that there are more possibilities than just the periodic and completely irrational cases, and one is led to define the so-called 'local isomorphism classes' to classify the tilings in the intermediate cases where $\pi^{\perp}(\Lambda)$ is neither dense nor discrete. However, we shall not develop this point since it seems irrelevant as far as quasicrystals are concerned: recall that our starting point is the $\mathbb{Z}$–module carrying the diffraction pattern, and that the natural choice of the high-dimensional lattice has dimension equal to the rank of the $\mathbb{Z}$–module, in such a way that we are always in the 'completely irrational' case. Then the second part of the local isomorphism property holds: in other words, there is only one local isomorphism class.

To summarize, the local isomorphism property follows from two features of the construction: the first one is the orientation $\mathbf{E}^{\|}$ of the cut, which should be completely irrational for both parts of the property to hold. The second, which is the most important, is that the prototypic atomic surface σ is *contained in a $(n-d)$-dimensional plane* (which is of course transverse to $\mathbf{E}^{\|}$, but is not necessarily parallel to $\mathbf{E}^{\perp}$). This last condition is required for each local pattern (i. e., the restriction to a *bounded* subset $U \subset \mathbf{E}$ of the cut) to be *locally constant* with respect to the translations of the cut along $\mathbf{E}^{\perp}$. In fact, consider the set Σ_U of the atomic surfaces σ_ξ intersecting $\mathbf{E}$ on U. Since the projections (parallel to $\mathbf{E}^{\|}$) of Σ_U on $\mathbf{E}^{\perp}$ (or on any $(n-d)$-dimensional subspace transverse to $\mathbf{E}^{\|}$) has a non empty intersection $\beta = \bigcap_{\sigma_\xi \in \Sigma_U} \pi^{\perp}(\sigma_\xi)$, the cut will still intersect the same Σ_U if we shift the cut, as long as $\mathbf{E}$ still intersects β. And if the atomic surfaces are parallel, the relative positions of the points $\sigma_\xi \cap \mathbf{E}$ will remain constant under such allowed shifts.

The local constancy of the local patterns is a 'rigidity' feature which depends only on the atomic surfaces being embedded in planes. This is by their very definition the case of the Penrose-like tilings, but it also occurs (at least in the first approximation) for the icosahedral quasicrystals, as shown by the Patterson analysis (for quasicrystals, the atomic surface is made of several 'sheets', but it is clear that the previous argument remains valid in that case). Such 'rigid' structures share with tilings the following property: for any given finite radius r, there exist only a finite number of different atomic configurations within any ball of radius r. Of course, this number grows with r (it remains bounded for a periodic crystal), but the situation is nevertheless quite different for quasicrystals and for displacive modulated crystals, which are quasiperiodic structures described by 'curved' atomic surfaces, and in which the mutual distances between atoms changes continuously upon a shift of the cut.

5.4.1 Pseudo-group of Translations. The local isomorphism property can also be described by a pseudo-group of translations for these 'rigid' quasiperiodic structures. In our context, a set of translations forms a pseudo-group if each translation in this set maps a finite fraction of the vertices onto vertices of the structure. The subset of vertices which are mapped to other vertices constitutes the domain of the translation, and the product (composition) of two translations is defined only on the intersection (if not empty) of the image of the first one and the domain of the second.

To see how such a pseudo-group arises in our constructions, consider a quasiperiodic system associated with a prototypic atomic surface σ_0 (attached to the origin of $\mathbb{R}^n$) embedded in $\mathbf{E}^{\perp}$. Then consider any $\zeta \in \mathbb{Z}^n$ such that the intersection $\beta_\zeta = \sigma_0 \cap (\sigma_0 + \pi^{\perp}(\zeta))$ is not empty. It is clear that for any $\mathbf{E}$ the cut $\mathbf{E} - \zeta$ intersects all the atomic surfaces which are intersected by $\mathbf{E}$ on the 'subsurfaces' $\beta_\zeta + \mathbb{Z}^n$, so that the translation $z = \pi^{\|}(\zeta) \in \mathbf{E}^{\|}$ maps a subset of the structure defined by $\mathbf{E}$ onto a subset of the structure defined by $\mathbf{E} - \zeta$. The domain of the translation z is the set of vertices of the first structure which correspond to atomic surfaces intersected by $\mathbf{E}$ on the subsurface, and they form a finite fraction of the whole structure measured by the ratio of the volume of β_ζ and the volume of the whole atomic surface σ. We see that the domain of a translation z is a large fraction of the whole structure when z is the projection $\pi^{\|}(\zeta)$ of a $\zeta \in \mathbb{Z}^n$ such that the other projection $\pi^{\perp}(\zeta)$ is small, and this requires z to be large. Finally, the pseudo-group of translations is very simple to describe: it is the projection on $\mathbf{E}^{\|}$ of the subset of $\mathbb{Z}^n$ which projects on $\mathbf{E}^{\perp}$ on translations small enough to map the atomic surface σ_0 onto an image which intersects σ_0. Unfortunately, it is not known whether the existence of such a structure, which seems to embody all the 'translational symmetry' of quasicrystals, has or not consequences for the propagation problems (for phonons and electrons) in these systems.

5.4.2 Classification of Local Patterns. Let us now show that the same tools allow the classification of local patterns in the context of 'rigid' structures, and specifically the classification of the possible packings of tiles around a common vertex in the case of the Penrose-like tilings (Katz (1989 a)).

Let us consider the general case of a d-dimensional Penrose-like tiling, defined by a cut $\mathbf{E}$ in $\mathbb{R}^n$ and the projection $\sigma_0 = \pi^{\perp}(\gamma_n)$ (where γ_n is the unit hypercube of $\mathbb{R}^n$) as the prototypic atomic surface attached to the origin. Let us first determine all the vertices of the tiling which are the origins of a given edge $e_i = \pi^{\|}(\varepsilon_i)$. In order to yield such an edge attached to a vertex $x = \mathbf{E} \cap \sigma_\xi$, the cut $\mathbf{E}$ must intersect also the atomic surface $\sigma_{\xi+\varepsilon_i}$, and this requires x to fall in $\sigma_\xi \cap (\sigma_{\xi+\varepsilon_i} - e_i)$. But since $\pi^{\perp}(e_i) = 0$, this last intersection is simply the projection of the $(n-1)$-dimensional facet $\gamma_{n-1}^{\hat{\imath}}$ of γ_n which does not contain ε_i.

We shall call this projection $\pi^{\perp}(\gamma_{n-1}^{\hat{\imath}})$, which is a polytope inscribed in the atomic surface, the *existence domain* of the edge: each atomic surface which is intersected by the cut on this existence domain yields in the tiling a vertex which is the origin of an edge e_i. Observe that we could as well consider the existence domain of the extremities of the same edge: it is of course the other facet $\pi^{\perp}(\gamma_{n-1}^{\hat{\imath}} + \varepsilon_i)$.

One can easily iterate this argument d times. For instance, a vertex x is the origin of a two-dimensional facet spanned by $\{e_i, e_j\}$ if and only if both x and $x + e_j$ are the origin of an edge e_i. The corresponding existence domain is easily seen to be the projection of the $(n-2)$-dimensional facet $\gamma_{n-2}^{\hat{\imath}\hat{\jmath}}$ of γ_n which does not contain either ε_i nor ε_j. And finally, the existence domains of the tiles spanned by d vectors $\{e_i, e_j, \ldots, e_k\}$ are the subpolyhedra of the atomic surface spanned by the $(n-d)$ remaining vectors $\{e'_l, e'_m, \ldots, e'_n\}$. Observe that each of these polyhedra is present 2^d times in the atomic surface, according to the vertex considered as the origin of the corresponding tile. Moreover, the relative abundance of each kind of tile is proportional to the volume of its existence domain.

By considering the superposition of all the existence domains of tiles, one gets a cellular decomposition of the atomic surface, in which each cell corresponds to the set of vertices attached to a family of tiles such that the given cell is the intersection of their existence domain. One can achieve in this way the complete classification of the 'vertex neighborhoods' of the tiling, i. e., the possible arrangements of tiles around a common vertex. Of course, the explicit computations become difficult quickly as the dimensions n or d increase, although there is nothing more to compute than the intersection of a family of $(n-d)$-dimensional polyhedra.

More generally, one can develop the same considerations for any 'rigid' quasiperiodic structure associated with a system of atomic surfaces embedded in parallel subspaces: each local pattern made of p points which appears in the structure admits an existence domain which is the intersection of p suitably shifted copies of the atomic surface, and the frequency of each particular pattern is proportional to the volume of its existence domain. In other words, in the 'rigid' cases, one can systematically study the finite configurations which appears in the quasiperiodic structures, as soon as the corresponding atomic surfaces are given.

5.5 Self-Similarity

Up to now, we have considered any irrational orientation for the pair of subspaces $(\mathbf{E}^{\|}, \mathbf{E}^{\perp})$. Let us now specialize to the case where there exist lattice-preserving linear transformations which commute with the two projections $(\pi^{\|}, \pi^{\perp})$. Such transformations are elements M of $Gl(n, \mathbb{Z})$ which preserve the subspaces $\mathbf{E}^{\|}$ and $\mathbf{E}^{\perp}$.

Given a tiling associated to the canonical atomic surface $\pi^{\perp}(\gamma_n)$, one can construct a new atomic surface as $\pi^{\perp}M(\gamma_n)$, by taking the projection of the image through M of the unit hypercube. Since M preserves $\mathbf{E}^{\|}$ and $\mathbf{E}^{\perp}$, it transforms any cut $\mathbf{E}$ into a parallel cut $M(\mathbf{E})$ which carries the image under the restriction of M to $\mathbf{E}^{\|}$ of the tiling carried by $\mathbf{E}$.

Of special interest is the case where M operates on $\mathbf{E}^{\|}$ and $\mathbf{E}^{\perp}$ by homotheties, since in this case the image of the tiling under M is a tiling of the same type, but at a different scale. This happens in particular when there is an invariance point group of the high-dimensional lattice, such that $\mathbf{E}^{\|}$ and $\mathbf{E}^{\perp}$ are the only two invariant subspaces, carrying irreducible non-equivalent representations of the invariance group. As we have already seen, such a situation occurs for icosahedral symmetry, but in order to give an explicit example it is worthwhile to consider a two-dimensional case. The simplest choice is the octagonal tilings (which were first introduced by R. Ammann, see Grünbaum and Shephard (1987)).

5.5.1 The Octagonal Tilings. These tilings are obtained in a straightforward way as Penrose-like tilings with $n = 4$, $d = 2$, and the orientation of the pair $(\mathbf{E}^{\|}, \mathbf{E}^{\perp})$ prescribed by the following symmetry considerations, which are a paraphrase of the icosahedral case:

Consider a regular octagon in the Euclidean plane and choose four of its vertices, no two of them being opposite. Consider the four vectors joining the center of the octagon to these vertices. There exists an unique embedding of the plane in $\mathbb{R}^4$ such that the canonical orthonormal basis of $\mathbb{R}^4$ projects orthogonally on our four vectors. Now consider the symmetry group of the octagon. Since it permutes the vertices of the octagon, we can define a 4-dimensional action of this group by the condition that it permutes the basis vectors in the same way as the vertices of the octagon, and since this action involves only signed permutations, the lattice $\mathbb{Z}^4$ spanned by the basis is preserved by this action. Then we decompose it into irreducible representations and we find two of them, one carried by our embedded plane, which is identified with $\mathbf{E}^{\|}$, and the other by the orthogonal plane, identified with $\mathbf{E}^{\perp}$.

It is easy to see that the prototypic atomic surface $\pi^{\perp}(\gamma_4)$ is an octagon, and that the six 2-dimensional facets of γ_4 fall under $\pi^{\|}$ on two orbits of tiles: two squares (with orientations differing by a $\pi/4$ rotation), and four rhombs with an acute angle of $\pi/4$, again mapped on each other by rotations which are multiples of $\pi/4$. A sample of an octagonal tiling is shown on Fig. 5.

If we choose our initial vertices in the octagon such that the angles between consecutively numbered vectors $e_i = \pi^{\|}(\varepsilon_i)$, $(i = 1, \ldots, 4)$ are equal to $\pi/4$, then the angle between the corresponding projections $e'_i = \pi^{\perp}(\varepsilon_i)$ are $3\pi/4$, so that one has equalities of the type: $e_1 + e_2 + e_3 = (\sqrt{2}+1)e_2$ while $e'_1 + e'_2 + e'_3 = -(\sqrt{2}-1)e'_2$. This suggests that we study the matrix:

$$M = \begin{pmatrix} 1 & 1 & 0 & -1 \\ 1 & 1 & 1 & 0 \\ 0 & 1 & 1 & 1 \\ -1 & 0 & 1 & 1 \end{pmatrix}$$

M is easily seen to have all the required properties: its determinant is 1, so that it belongs to $Gl(4, \mathbb{Z})$, and it commutes with the action of the octagonal group, so that it reduces on $\mathbf{E}^{\|}$ to a dilatation of ratio $(\sqrt{2}+1)$ and on $\mathbf{E}^{\perp}$ to a contraction of ratio $-(\sqrt{2}-1)$.

Now, our general argument shows that if we replace our original atomic surface by an octagon $(\sqrt{2}-1)$ times smaller, then we will find in any cut an octagonal tiling scaled by a factor $(\sqrt{2}+1)$. In particular, if we consider both atomic surfaces: our original one containing the smaller one, we see that we can 'extract' from any tiling the vertices of a larger one, by discarding all the vertices which correspond to intersections of the cut with the large octagon, falling outside the small one. In the cut, one can describe this operation as the regrouping of clusters of tiles to form larger tiles, and this is called a *deflation*. Since the matrix M is invertible, this process may be done in the reverse way: it is possible to 'dissect' the tiles of a given tiling, in order to obtain a tiling of the same type, but with an edge length shortened by a factor $(\sqrt{2}-1)$. This is called an *inflation*, because it enlarges the number of the tiles.

Observe that for these considerations we are not interested in comparing the tiling carried by $\mathbf{E}$ and $M(\mathbf{E})$, because the 'absolute' position of the cut is in general difficult to assess (due to the local isomorphism property) unless the

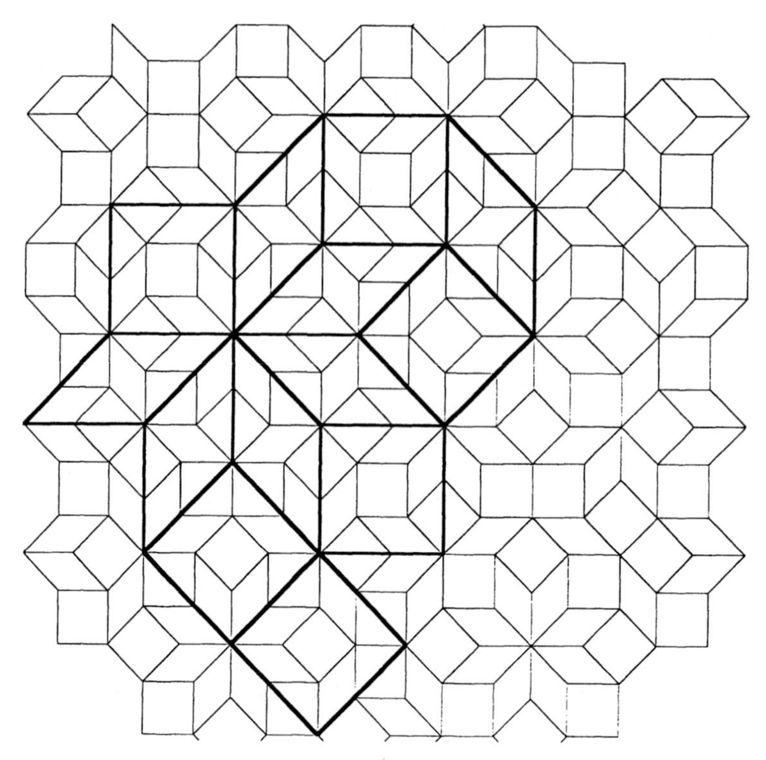

Fig. 5. A sample of the octagonal tiling, showing some deflated tiles.

tiling has special (global) symmetry properties. On the contrary, we are interested in comparing two tilings carried by the same cut. This entails that the position of the small atomic surface inside the large one is irrelevant: whatever this position, the discarding process explained above will lead to a 'deflated' tiling. We conclude that in the octagonal case the inflation/deflation constructions are not uniquely defined (although it is possible to devised decorations of the tiles such that these constructions may be made unique for the decorated tiles).

This is in sharp contrast with the original Penrose tilings. As is well known, the uniqueness of the inflation/deflation operations has played a key role as well in the constructions of Roger Penrose as in the early study of these tilings by R. Penrose and J. Conway (see Gardner (1977)). Without entering into much details, let us say a word about the special case of these original Penrose tilings. One possible approach is the following: these tilings may be constructed from a 4-dimensional space equipped with a lattice Λ which is not the simple cubic, but is invariant under the group $\mathbb{Z}_5$ (this lattice is generated by any four of the five vectors whose convex hull is the symmetric 4-dimensional simplex, and the action of $\mathbb{Z}_5$ permutes these five vectors). In this context, the atomic surface which yields the Penrose tilings is made of four 'sheets' (connected components) for each vertex of the lattice, and these components are pentagons of two sizes. Finally, the deflation matrix M maps the two large pentagons *exactly* onto the small ones (and the small ones inside the large ones) in such a way that the deflated set of atomic surfaces sit in only one position inside the original one, and this entails the uniqueness of the deflation in this case. In fact, the deflation process consists in keeping *all* the vertices corresponding to the original small pentagons (which are identical to the deflated large pentagons), and discarding most of the vertices corresponding to the original large pentagons (of which the deflated small pentagons are only a small subset).

Returning to the general case, let us stress that the properties which are referred to as the 'self-similarity' of the quasiperiodic tilings or of the quasicrystals are rather heterogeneous: on one hand, there is the actual and well-defined self-similarity of the $\mathbb{Z}$–modules, connected with symmetry-preserving automorphisms of the high-dimensional lattices Λ. One the other hand, the notion of self-similarity can have only a weaker meaning for the quasiperiodic structures themselves, since the previous symmetry is broken as soon as one introduces the atomic surfaces (which of course cannot be invariant through a group of homotheties)

However, some breakings may be weaker than others: in the case of 'self-similar' Penrose or Penrose-like tilings, the remaining 'self-similarity' is that, roughly speaking, it is possible to build clusters of tiles which yield new larger tiles of the same shape, arranged in the same type of tilings. There is an obvious analogy in the case of quasicrystals: one can wonder if some clusters of atoms are packed in the quasicrystal in the same way as single atoms, and study the implications of this kind of hypothesis for the geometry of the atomic surfaces. Very little is known at the present time on these questions.

6. Propagation of Order

6.1 The Main Problem of Quasicrystallography

Besides the local isomorphism property, there are other features of the Penrose-like tilings which may help to understand the structure of quasicrystals, and which are linked with the major problem of the *propagation of order* in these materials. In fact, structure determination is only one of the problems raised by the discovery of quasicrystals. The main one is to understand the simple possibility of non periodic long range order. In a perfect quasiperiodic structure the positions of any two atoms are correlated, whatever the distance between them, while the number of different neighborhoods of atoms grows with the size of the neighborhood.

Moreover, it is reasonable to consider that in a metallic alloy the effective interatomic forces are essentially short range, so that the actual position of a given atom is controlled by the positions of atoms which are at distances much larger than the range of their interactions. We are thus led to imagine that the ordering propagates in the material, and we have to understand how such propagation can occur.

It seems that the problem splits into two sub-problems. The first one deals with the existence of local constraints which could force the quasiperiodic order (for Penrose-like tilings, this is the question of 'matching rules'). The second sub-problem concerns the possibility of atomic rearrangements, which could allow for the relaxation of defects or of strain during the growth of the quasicrystal, and would thus permit the long range manifestation of a locally-governed propensity for icosahedral symmetry. It turns out that these two requirements: the existence of local constraints forcing the quasiperiodic order and the possibility of low energy rearrangements of atoms, lead to severe constraints on the atomic surfaces, as we shall now explain.

6.2 Matching Rules

Matching rules are known for the original Penrose tilings, for the octagonal tilings and for the 3-dimensional Penrose-like tilings with icosahedral symmetry (Katz (1988)). They consist in decorations of the tiles, such that any tiling in which the decorations 'match' (in a sense which depends on the decoration) for each pair of adjacent tiles is a quasiperiodic tiling. Although there is not yet a complete theory for the matching rules, it is possible to give the following general ideas about their existence and construction.

For the sake of clarity, let us consider the octagonal tilings. Given any tiling of the plane by means of the two squares and the four rhombs which appear in the octagonal tilings, it is quite clear that it is always possible to lift this tiling in $\mathbb{R}^4$, i. e., to find a cut in $\mathbb{R}^4$ such that the vertices of the given tiling are the projections on $\mathbf{E}^{\parallel}$ of the atomic surfaces intersected by this cut. As we have seen, the tiling is a quasiperiodic octagonal tiling if and only if this cut may be chosen as a plane $\mathbf{E}$ parallel to $\mathbf{E}^{\parallel}$. Thus in our framework, the matching rules correspond to a set of local constraints on the cut such that any cut fulfilling these constraints is homotopic to a plane parallel to $\mathbf{E}^{\parallel}$.

It is essential to observe that such local constraints cannot exist for a generic orientation of $\mathbf{E}^{\parallel}$. Loosely speaking, in order for such local rules to exist, we must be able to recognize *locally* the orientation of $\mathbf{E}^{\parallel}$. Let us explain in the general case (tilings defined by a d-dimensional cut through $\mathbb{R}^n$ equipped with a lattice Λ) which kind of special orientation of $\mathbf{E}^{\parallel}$ may give rise to matching rules for the corresponding Penrose-like tilings.

Since we are dealing with the orientations of d-dimensional planes in $\mathbb{R}^n$, it is natural to consider the Grassmannian manifold G^n_d of these planes. Recall that, given a flag of embedded vector subspaces $\{F^1 \subset F^2 \subset \cdots \subset F^{n-1} \subset \mathbb{R}^n\}$, one can define a stratification of G^n_d by looking at the transversality of the d-dimensional planes to the components of the flag. Let us define a stratification of the same kind which will allow us to characterize the orientation of the d-dimensional planes with respect to the lattice Λ. Consider in $\mathbb{R}^n$ the ball of radius r centered at the origin and recall that a lattice p-plane is a p-dimensional plane in $\mathbb{R}^n$ which goes through p affinely independent vertices of Λ besides the origin. Now, consider the (finite) family of lattice planes defined by vertices of Λ falling in the previous ball, and the associated stratification of G_d^n obtained by looking at the transversality properties of the d-dimensional planes with respect to this family of lattice planes. It is clear that if we enlarge the radius r, we obtain a finer stratification.

It turns out — although we do not have a rigorous proof of it — that a necessary condition for the existence of matching rules for the Penrose-like tilings corresponding to an orientation $\mathbf{E}^{\parallel}$ for the cut, is that the direction $\mathbf{E}^{\parallel} \subset G_d^n$ falls on a stratum of dimension zero of such a stratification, and the smallest r for which this occurs measures the 'range' of the matching rules.

In other words, a necessary condition for the existence of matching rules is that the direction $\mathbf{E}^{\parallel}$ is simultaneously non-transverse to a family of lattice planes, large enough to completely determine this direction. Let us give some examples.

First, the low dimensional case of Fig. 2: in that trivial case, the Grassmannian G_1^2 is a circle, the lattice planes are lines and $\mathbf{E}^{\parallel}$ is either transverse to a lattice line or identical to it, so that the stratification of the circle is made of arc segments bounded by rational directions p/q such that the point $(p,q) \in \mathbb{Z}^2$ falls inside the disk of radius r. It follows from our condition that matching rules exist for these 1-dimensional tilings only if they are periodic.

Second, the octagonal case. Referring to the notations of Sect. 5.5, let us exhibit the family of lattice planes which are non-transverse to $\mathbf{E}^{\parallel}$: as already mentioned, the sum $(e_1 + e_3)$ is parallel to e_2 in $\mathbf{E}^{\parallel}$ and the sum $(e_1' + e_3')$ is parallel to e_2' in $\mathbf{E}^{\perp}$. It follows that the lattice plane spanned by ε_2 and $(\varepsilon_1 + \varepsilon_3)$ intersects $\mathbf{E}^{\parallel}$ along a line (which is the same as its projection) and thus is not transverse to $\mathbf{E}^{\parallel}$. By application of the octagonal group, we find four planes of this kind. To find another family of four planes, observe that (e_1+e_4) is parallel to (e_2+e_3) in $\mathbf{E}^{\parallel}$ and $(e_1'+e_4')$ is parallel to $(e_2'+e_3')$ in $\mathbf{E}^{\perp}$. It follows as previously that the lattice plane spanned by $(\varepsilon_1+\varepsilon_4)$ and $(\varepsilon_2+\varepsilon_3)$ is not transverse to $\mathbf{E}^{\parallel}$, and the octagonal group again provides us with four planes of this kind. It is easy to see that the non-transversality to these eight planes completely determine the direction $\mathbf{E}^{\parallel}$. In fact, the non-transversality to each quadruplet of planes defines a 1-dimensional stratum in the Grassmannian G_2^4, each of which is made of planes invariant under a realization of the symmetry group of the square. These two strata intersect at a point which presents the full octagonal symmetry.

One can develop similar considerations for the icosahedral case. One finds an orbit of 15 4-dimensional lattice planes which are non-transverse to $\mathbf{E}^{\parallel}$ and such that this property enforces the icosahedral symmetry of $\mathbf{E}^{\parallel}$. It may be possible that a set of 3-dimensional lattice planes is sufficient.

Let us make two comments: first, the existence of a point symmetry group seems not to be required for our necessary condition to hold. But it helps, since we get a whole orbit of non-transverse lattice planes as soon as we have found one. Second, in the octagonal and icosahedral cases where there exist self-similarity transformations, one can easily verify that each of the non-transverse lattice planes is invariant under the self-similarity operations. This could indicate some relationship between the existence of such a family of non-transverse lattice planes and the existence of self-similarity properties.

Finally, we must emphasize that even for this first step (the characterization of 'good' orientations for $\mathbf{E}^{\|}$), the situation is far from being thoroughly investigated. In particular, all of the cases for which we know matching rules possess both a 'large' point group and self-similarity transformations. It would be very interesting to have an example (if it exists) of a system possessing matching rules, but with less symmetry.

The next step is to show how one can actually construct matching rules for Penrose-like tilings corresponding to these 'good' orientations of $\mathbf{E}^{\|}$. To explain the main idea, let us first return to the low dimensional model depicted on Fig. 2. It is clear that with an arbitrary 'wavy' cut, which is only required to be a section of the projection $\pi^{\|}$ (i. e., to project in a one to one way on $\mathbf{E}^{\|}$), we can select by intersection any set of atomic surfaces, even surfaces which do not build the two tiles of the quasiperiodic tilings obtained with a straight cut parallel to $\mathbf{E}^{\|}$. Now consider the oblique tiling of Fig. 4, and add the following condition on the cut: it is required not to cross the boundaries of the oblique tiles which are parallel to $\mathbf{E}^{\|}$.

One immediately sees that this restricted class of cuts selects only whole tiles: the cut is 'channeled' between the boundaries of the oblique tiles in such a way that it is forced to go out of any oblique tile through an atomic surface which is the extremity of a tile of the quasiperiodic tiling. However, this channeling is only local: each time the cut goes out from a large oblique tile, it may either enter another large tile or a small one, and no global ordering is forced by this new requirement.

Nevertheless, these considerations suggest that we can try to enforce more long range order in the tiling by enlarging the 'forbidden set' parallel to $\mathbf{E}^{\|}$ that the cut is required not to cross. For the low dimensional case, it is clear that there are only two qualitatively different situations. Either the orientation of $\mathbf{E}^{\|}$ is rational, and if we enlarge the forbidden set sufficiently, then these line segments will connect to form whole straight lines which give rise to a global channeling of the cut corresponding of course to periodic tilings, or the orientation of $\mathbf{E}^{\|}$ is irrational, in which case, whatever the finite enlargement of the forbidden set, no topological change will occur and no global channeling will appear. Thus we recover the fact that the kind of matching rules we are dealing with exist in the low dimensional case only for periodic tilings.

In higher dimension the geometry is far trickier, and we shall give here only a short account of the construction. Observe first that the boundaries of the oblique tiles fall into two subsets: those which are the product of a tile of the quasiperiodic tiling and the boundary of its existence domain, and those which are the product of the existence domain of a tile and the boundary of this tile. If we take the first subset as the forbidden set, then it is again true that any cut which does not cross this forbidden set selects whole tiles (this is the local channeling) but no global channeling is forced at this stage.

To understand how a *finite* enlargement of the forbidden set can result in global constraints on the cut, we have to look more closely at these boundaries 'parallel to $\mathbf{E}^{\|}$' of the oblique tiles. For this, consider their traces in the fibers of $\pi^{\perp}$. As we already know, a generic fiber does not intersect this forbidden set and is a well-behaved cut carrying a quasiperiodic tiling. Now, the key point is that non-generic fibers intersect the forbidden sets attached to *infinitely many* vertices of Λ. This is the consequence of two independent features of the construction: the non-transversality property of $\mathbf{E}^{\|}$, and the special choice of the atomic surfaces, which is bounded in $\mathbf{E}^{\perp}$ by the traces on $\mathbf{E}^{\perp}$ of lattice planes non-transverse to $\mathbf{E}^{\|}$. The net result is that one finds in these non-generic fibers of $\pi^{\perp}$ a (quasiperiodic) distribution of traces of the forbidden set, in such a way that a finite enlargement of this set will connect these traces and yield an infinite forbidden set, just as in the low dimensional periodic case.

However, the codimension of the forbidden set depends on the dimension of the tiling: for the low dimensional model, the total space is $\mathbb{R}^2$ and the forbidden set is 1-dimensional, so that when the pieces of the forbidden set are glued together (in the periodic case), the plane is disconnected in a family of parallel strips and any cut confined in one of these strips yields a periodic tiling. Consider now the octagonal case: the total space is $\mathbb{R}^4$, and one can show that the forbidden set (suitably enlarged to give matching rules) admits as a deformation retract a family of 2-dimensional lattice affine subspaces, which of course do not disconnect $\mathbb{R}^4$. Observe however that the requirement for a 2-dimensional cut not to cross a family of 2-dimensional planes in $\mathbb{R}^4$ is a strong constraint, which happens to be sufficient to force the cut to be homotopic to a plane parallel to $\mathbf{E}^{\|}$. The proof of this main conclusion is not simple and will not be developed here.

Let us now turn to the consequences of this (partial) theory of matching rules for tilings, to understand the propagation of order in quasicrystals. The notion of 'matching rules' for packings of atoms must clearly be defined in the following way (Levitov (1988)): we shall say that a quasiperiodic structure admits local rules if there exists a finite family of finite configurations, such that any structure which induces around each vertex a configuration belonging to this family, is quasiperiodic. Of course, such a definition assumes the structure to be 'rigid' in the sense that the corresponding atomic surfaces are embedded in parallel planes, since otherwise we have infinitely many possible configurations around each point.

It seems clear that the previous discussion on the possible orientations of $\mathbf{E}^{\|}$ remains valid for these more general structures, so that it is not impossible for icosahedral quasicrystals to admit local rules. More interestingly, the discussion on matching rules strongly suggests — we do not claim that we have any proof of this assertion — that in order for local rules to exist for a quasicrystal, *the atomic surfaces must be polyhedra bounded by the traces of lattice planes non-transverse to* $\mathbf{E}^{\|}$, and to get *short range* local rules, i. e., for the configurations in the family of 'local models' to be as small as possible, we have to look for lattice planes with the smaller possible Miller indices: namely, the best candidates are the 15 symmetry planes of the icosahedral group.

To conclude, let us emphasize that the existence of local rules may be not required for real quasicrystals. It is possible that their stability is controlled by long range interactions, such as some version of the Hume-Rothery effect. All that we can say is that, if we admit that local rules exist for quasicrystals, then we get severe constraints on the possible atomic surfaces, which are likely to be polyhedra bounded by symmetry planes. Of course, we hope that such constraints will help us to find the structure — but this is just a hope.

7. Topological Properties of Atomic Surfaces

Let us now turn to our last topic, which has still to do with the propagation of order, but from a quite different point of view. Recall that quasicrystals grow from the melt, and that in a (perfect) quasicrystal the position of any atom is correlated with the positions of atoms which are far away. Even if the energetics of local configurations make it possible to have local rules as discussed in the previous Section, the fact that quasicrystals actually grow without too many defects may require a mechanism which allows the rearrangement of atoms in the solid phase in order to achieve the long range ordering.

Thus one can imagine the growth of a quasicrystal as a two step process. Atoms in the melt stick on the boundary of the growing quasicrystal at positions governed by short range interactions and which are not necessarily compatible with the quasiperiodic order, but which nevertheless could perhaps be described by a 'wavy' cut in our framework. Then, as new layers of atoms stick to the solid phase, the propensity to form a quasicrystal manifests itself through a *bulk reconstruction*, a process through which atoms jump to their final positions and which geometrically corresponds to the progressive flattening of the cut behind the 'quasicrystallization front'. We require that the deformation of the cut and its progressive flattening should be energetically possible at (or below) the quasicrystallization temperature, i.e., should be a low activation energy process.

7.1 The Closeness Condition

To understand how this last condition yields topological constraints on the atomic surfaces, let us return once more to the low dimensional model of Fig. 2, and examine what happens when one moves the cut (say, upward). Each times the cut reaches the upper tip of an atomic surface, it begins at the same time to intersect a neighboring atomic surface, because the projection on $\mathbf{E}^{\perp}$ of the upper tip of any atomic surface coincide with the projection of the lower tip of a neighboring one. This depends of course on the choice of the atomic surface (the length of the segment in this low-dimensional model) and can be interpreted as the 'jump' of a vertex from a site to a neighboring one.

This is the only elementary change in the tiling upon a shift of the cut but, since the orientation of the cut is irrational, such elementary changes occur infinitely many times for any finite translation of the cut, resulting in an uncountable infinity of different tilings.

Observe that from a physical point of view these 'jumps' should be low activation energy processes, for two reasons:
First, they correspond to a symmetry operation of the nearest neighbor configuration, so that if the effective range of the interatomic interaction is short, the energies of the two configurations related by the jump should be close to each other (they may differ only by the second and further shells, whose contributions to the energy is small).
Second, the energy barrier between the two equilibria of the jumping atom should be low, simply for geometrical reasons: the distance between the two sites is small and during the jump the atom does not come too close to its neighbors.

Finally, we conclude that for this special choice of atomic surface, (and in fact for any Penrose-like tiling), the cut should be 'glissile', i. e., should allow the bulk reconstruction.

Conversely, let us examine what happens if we alter the length of the atomic surface. Suppose for instance that we decrease its length a little, and let us move the cut upward. When the cut reaches the upper end point of an atomic surface and leaves it, the corresponding atom has no neighboring site to jump to. The prescription is that there should no longer exist an atom on this site, but of course atoms cannot vanish and we have to see it as a point defect in the structure (an insertion).

Similarly, when the cut reaches the lower end point of an atomic surface, resulting in a new intersection, no atom can magically appear and we find a vacancy at this site. Such defects are of relatively high energy and the system will resist producing them: the end points of the atomic surfaces will 'anchor' the cut and the flattening of the cut will not occur. We see that the cut will be 'glissile' only when the appearance of an inserted atom coincide with the appearance of a neighboring vacancy, so that the inserted atom can fall in the vacancy and both defects cancel: this is the previous 'jump'.

To summarize this discussion, observe that point defects appear each time the cut crosses the boundary of an atomic surface, in such a way that the requirement that such defects do not exist corresponds to the atomic surface having no boundary. In fact, one sees immediately that we do not change our quasiperiodic structure if we 'complete' the atomic surfaces by the addition of the segments parallel to $\mathbf{E}^{\|}$ which connect the lower end-point of any atomic surface to the upper end-point of its neighbor, and which are the 'jump trajectories': these new segments are generically not intersected by the cut just because they are parallel to it.

We can now formulate our topological constraint in the general case: in order to allow for low activation energy rearrangements of the structure, the atomic surface σ embedded in the high-dimensional torus should be a close manifold without boundary. Observe that the Penrose-like tilings are of this type: it is always possible to 'complete' the atomic surface, which is the projection $\pi^{\perp}(\gamma_n)$ wound in the torus $\mathbb{T}^n$, by gluing on its boundary some new pieces which are nowhere transverse to the direction of $\mathbf{E}^{\parallel}$, and thus are generically not intersected by the cut. Let us describe this construction for the octagonal case.

Rather than carrying out the construction in the torus $\mathbb{T}^4$, let us construct it periodically in $\mathbb{R}^4$. Consider an octagonal atomic surface and one of the edges, say e'_1, of its boundary. Observe that this edge is mapped on the opposite side of the octagon by the translation $(e'_2 + e'_3 + e'_4)$ (or the opposite) so that the translation $(e_2 + e_3 + e_4)$ maps this edge e'_1 on the boundary of another atomic surface (because the sum $(e'_2 + e'_3 + e'_4) + (e_2 + e_3 + e_4) = (\varepsilon_2 + \varepsilon_3 + \varepsilon_4)$ is a lattice vector). Thus we can make the boundaries of atomic surfaces vanish by gluing on each edge e'_i of the boundary a rectangle spanned by e'_i and $(e_j + e_k + e_l)$, where (i, j, k, l) is a permutation of $(1, \dots, 4)$. Of course, this first step introduces new boundaries, which are the segments $(e_j + e_k + e_l)$, but it is easy to see that these new boundaries enclose a periodic set of octagons contained in planes parallel to $\mathbf{E}^{\parallel}$, so that we can in a second step add these octagons and we are left with no boundaries at all.

The completed atomic surface is the projection of this construction in the torus $\mathbb{T}^4$, and is a manifold without boundary which is easy to identify: by deforming the manifold so as to shrink the images of the rectangles and of the additional octagon, we see that the topology of this manifold is the same as what we obtain by gluing directly each edge of the atomic surface on the parallel opposite edge, and it is well known that this operation yields a two handled torus. Finally, we conclude that the completed atomic surface of the octagonal tiling is a two handled torus embedded in the 4-dimensional torus. It is clear that the construction depends only on the geometry of $\pi^{\perp}(\gamma_n)$ and may be carried in $n - d$ steps in the general case of Penrose-like tilings.

7.1.1 The Definition of the Measure η. It may seem curious that we can modify (complete) the atomic surface without changing the structure, and in particular without changing its Fourier transform, whose computation relies on the atomic surface σ. The reason is that the measure η which is involved in this computation is completely determined by the geometry of the atomic surface, through the condition that each point in the structure appears with the same weight.

As a first example, consider our low dimensional model and suppose that we tilt all the atomic surfaces without changing the length of their projection on $\mathbf{E}^{\perp}$. Since the length of the atomic surface is then multiplied by the inverse of the sine of the angle between $\mathbf{E}^{\perp}$ and the atomic surface, we see that in order not to change the weight of the vertices in the cut, we have to multiply by the same sine the (constant) density of the measure η carried by the atomic surface. This remark leads to the following construction:

Let us deal with a structure of dimension d defined by a d-dimensional cut in $\mathbb{R}^n$. Consider $\mathbb{R}^n$ the canonical volume form ω. Since it is constant, it projects on a well defined volume form still denoted ω on the torus $\mathbb{T}^n$. Similarly, consider d constant vector fields defined by an orthonormal basis of $\mathbf{E}^{\|}$ in $\mathbb{R}^n$. They project in $\mathbb{T}^n$ on d constant vector fields $(v_1, v_2, \ldots, v_d)$, which span at each point the tangent space to the leaf of the foliation defined by $\mathbf{E}^{\|}$. Now, take the inner product of ω by the wedge product of these d vector fields: we get a constant $(n-d)$-differential form $\omega(v_1 \wedge v_2 \wedge \ldots \wedge v_d)$ on $\mathbb{T}^n$, which is null by definition on any $(n-d)$-dimensional vector subspace tangent to $\mathbb{T}^n$ and non transverse to the foliation $\mathbf{E}^{\|}$. Finally, the density of the measure η on the atomic surface σ is the restriction of $\omega(v_1 \wedge v_2 \wedge \ldots \wedge v_d)$ to σ, i.e., the inverse image of $\omega(v_1 \wedge v_2 \wedge \ldots \wedge v_d)$ with respect to the embedding of σ in $\mathbb{T}^n$.

With this definition, it is now clear that η is zero on the new pieces we have added to the original atomic surface, since they are non-transverse to $\mathbf{E}^{\|}$, so that nothing is changed neither in the structure nor in its Fourier transform.

Moreover, observe that the measure η must be of *constant sign* (which we can choose positive) in order for the structure to make sense physically: a negative value for η on an intersection would correspond to an 'antiatom'. This corresponds to an additional constraint on the embedding of σ (which thus must be orientable) in $\mathbb{T}^n$. Namely, at each transverse intersection point between σ and the cut $\mathbf{E}$ wound in $\mathbb{T}^n$, the relative orientation of their tangent space must be the same.

Finally, the measure η must be positive or zero along σ, and the points where $\eta = 0$ appear as a 'fold locus' for the atomic surface σ. From a topological point of view, it is interesting to discuss the stability of this 'fold locus', i.e., whether it is possible to remove it by a small deformation of σ. This is not always possible, and may correspond to the distinction to be made between quasicrystals and modulated crystals, as we shall now see.

7.2 Quasicrystals versus Modulated Crystals

Let us recall that one calls (displacive) modulated crystals a kind of structure in which atoms lie at positions which are close to the sites of a periodic lattice, from which they are shifted by a periodic modulation which is incommensurate with respect to the lattice parameters. There also exist structures which can be described as interpenetrating modulated crystals. They are called 'intergrowths'. All these structures are clearly quasiperiodic and can be described within the previous framework, which was in fact introduced by Janner and Janssen precisely to describe them, several years before the discovery of quasicrystals. By their very definition, the modulated structures can be demodulated, i.e., one can decrease the modulation function down to zero and recover a periodic crystal. This entails that the corresponding atomic surface is a m-dimensional torus (where m is the number of independent modulations), in the case of a modulated crystals, or a set of similar tori in the case of intergrowth. Conversely, let us show that each time the atomic surface is (up to a small deformation) everywhere transverse to the foliation defined by $\mathbf{E}^{\|}$, then it corresponds to a modulated crystal or an intergrowth:

Suppose that we are given in the high dimensional torus $\mathbb{T}^n$ such an atomic surface everywhere transverse to the foliation defined by the projection of $\mathbf{E}^{\|}$ on $\mathbb{T}^n$. Let us lift the atomic surface in $\mathbb{R}^n$ and call a 'large atomic surface' any connected component of the lift. It is easy to show that any such large atomic surface projects in a one to one way onto $\mathbf{E}^{\perp}$: consider any cut parallel to $\mathbf{E}^{\|}$ and an intersection point with any large atomic surface; since the cut is everywhere transverse to the large atomic surface, one can follow the intersection when one shifts the cut along any path drawn in $\mathbf{E}^{\perp}$. This shows that the projection of the large atomic surface on $\mathbf{E}^{\perp}$ is a covering map. But since $\mathbf{E}^{\perp}$ is a contractible set and the large atomic surface is connected, this covering is a homeomorphism. It is then immediate that each large atomic surface is invariant under a subgroup of rank $\dim(\mathbf{E}^{\perp})$ of the translation group of $\mathbb{R}^n$, and thus projects in $\mathbb{T}^n$ onto a torus of dimension $\dim(\mathbf{E}^{\perp})$.

Of course, one may find in $\mathbb{T}^n$ either one torus or several tori as the atomic surface, and these two cases correspond respectively to modulated crystals and to intergrowth structures. Thus it is attractive to reserve (under the closeness hypothesis) the word 'quasicrystal' to the cases where the atomic surface presents a stable fold locus, in such a way that we find in $\mathbb{R}^n$ a small number of 'large atomic surfaces' and not a whole lattice of them. This is the case for the Penrose-like tilings (as soon as we are not in too low dimensions) and for instance it is easy to see that there is only one connected component in $\mathbb{R}^4$ above the two handled torus which defines the octagonal tilings.

Such a topological situation causes a strange behaviour of the system under vibrations or fluctuations of the cut. For modulated crystals, a vibration of the cut simply results in vibrations of atoms (the so-called phason modes), but for a quasicrystal, the effects of such a vibration are quite different:

In fact, consider an atomic surface with a (topologically stable) fold locus in $\mathbb{T}^n$, lift it in $\mathbb{R}^n$ and consider a connected component of this lift. Now, choose an intersection point between this component and the cut, and try and follow this intersection when moving the cut along a path defined in $\mathbf{E}^{\perp}$. It is clear that it is possible to lift such a path if and only if it does not cross the projection on $\mathbf{E}^{\perp}$ of the fold locus, because at these points the inverse image of the tangent vector to the path is not defined. Thus we see that the projection is a covering map only in the complement of the projection of the fold locus, which is in general dense in $\mathbf{E}^{\perp}$ and is to be considered as a ramification locus for this covering. In particular, if we consider a closed path contained in this complement we see that after the transport of the cut along this path, we recover the same quasicrystal (since it is defined by the initial cut) but that some atoms have been permuted (in fact, a finite fraction which depends on the area enclosed by the loop in $\mathbf{E}^{\perp}$). This striking feature of quasicrystals, first mentioned in Frenkel et al. (1986), shows that for quasicrystals, the atomic displacements induced by a vibration of the cut are not only vibrations of atoms around their equilibrium positions. They necessarily involve displacements of atoms from one equilibrium position to another, and could result in special mass-diffusion kinetics. However, no experimental results have been up to now reported in this direction.

7.3 The Homology Class of the Atomic Surface

As a last consequence of the closeness condition, let us show how this hypothesis allows us to recover to a certain extent the classical notion of integral stoechiometric coefficients (Kalugin (1989), Katz (1989 b)).

For the sake of simplicity, we shall restrict the following discussion to the case of 2-dimensional octagonal quasicrystals. Observe first that, as an immediate consequence of its closeness, a compact atomic surface without boundary embedded in $\mathbb{T}^4$ is homologically a cycle and thus defines a homology class. Although its homology class does not provide much information about an embedded manifold, it is worthwhile considering it because it is directly related to the density (or more precisely the atomic concentration) of the quasicrystal.

The simplest way to handle such homology classes is to make use of the intersection index. Recall that on the n-th homology group of any compact manifold of even dimension $2n$, there is a natural pairing which is the homological version of the Poincaré duality usually defined in cohomology. Concretely, the pairing between two classes (their intersection index) is defined as follows: consider any representatives of the two classes; it is always possible to choose them so that they are transverse and thus intersect in a finite number of points. Then the choice of an orientation allows us to assign a sign ± 1 to each of these points and the intersection index between the two classes is simply the sum of these signs. As is well known, the intersection index depends only on the homology classes and is a non-degenerate pairing (a scalar product). In other words, an embedded manifold is homologous to zero if and only if its intersection indices with the elements of a basis of the n-th homology group are all zero.

For the octagonal quasicrystals, $n = 2$ and the $2n$-dimensional manifold is $\mathbb{T}^4$. Using the four canonical angles θ_i as coordinates on $\mathbb{T}^4$, a convenient basis of $H_2(\mathbb{T}^4)$ is provided by the classes of the six 2-dimensional tori $\mathbb{T}_{ij}$ defined by the equations $\theta_k = \theta_l = 0$, where (i, j, k, l) is a permutation of $(1, \ldots, 4)$. Thus the homology class of an atomic surface σ is characterized by the 6 integers $(\sigma \mid \mathbb{T}_{ij})$. This is only a very crude characterization of σ, since quite different σ's may define the same homology class. Nevertheless, this is sufficient to compute the atomic concentration of the quasicrystal, as we will now show.

Let us first notice that our basis is invariant (up to a sign) under the action of the octagonal group, and that there are only two orbits under this action, associated with the two types of tiles (the squares and the rhombs) of which the octagonal tilings are made. It is clear that since σ is itself invariant, the intersection indices of σ with two tori related by a symmetry operation have to be equal, so that on account of symmetry the number of independent indices reduces to 2.

To compute the atomic concentration, let us consider a large volume V in a plane cut parallel to $\mathbf{E}^{\parallel}$ and identified with the physical space. Let N be the number of intersection points in V between the cut and the set of atomic surfaces. Then the atomic concentration is equal to the ratio N/V and we are looking for an expression of this ratio in terms of intersection indices. For this, let us build on the cut the octagonal tiling corresponding to the given simple cubic lattice of $\mathbb{R}^4$ and consider the projection of a tile in $\mathbb{T}^4$. The key point is to observe that this tile wound round $\mathbb{T}^4$ along the projection of $\mathbf{E}^{\parallel}$ can be completed to yield a manifold isotopic to one of the 2-dimensional tori $\mathbb{T}_{ij}$ by gluing on its boundary a set of cells which are not transverse to $\mathbf{E}^{\perp}$: the construction is exactly the same as for the closing of the atomic surface, with the exchange of $\mathbf{E}^{\perp}$ and $\mathbf{E}^{\parallel}$.

For instance, let us consider the square spanned by e_1 and e_3. Let us glue the square $\{e_1, e_3'\}$ along the edge e_1, the square $\{e_1', e_3\}$ along the edge e_3 and the square $\{e_1', e_3'\}$ attached to the origin of the tile and glued along the two edges e_1' and e_3' of the newly added squares. Then one verifies easily that the free edges of this construction are mapped on each other either by $e_1 + e_1' = \varepsilon_1$ or by $e_3 + e_3' = \varepsilon_3$ so that this construction closes on an object homotopic to $\mathbb{T}_{13}$ upon projection in $\mathbb{T}^4$.

This shows that the mean number of atoms falling in each direction of tile is equal to the intersection index of the atomic surface with the corresponding torus $\mathbb{T}_{ij}$. If the atomic surface σ is actually closed, then this mean number is an integer and we recover in some sense the usual notion of integral stoechiometric coefficients. On the other hand, we cannot measure this mean number of atom per tile for each kind of tile: we can only compute the mean over the set of the 6 directions of tiles, using the density of the quasicrystal to assess the atomic concentration and the diffraction data to assess the unit length in $\mathbb{R}^4$ or the length of the edges of the corresponding tiles.

Let v_s and v_r be the areas of the square and of the rhomb, and n_s and n_r their respective numbers (for each of the orientations) inside the volume V. As already mentioned, the intersection indices $(\sigma \mid \mathbb{T}_{ij})$ depend only on the type (square or rhomb) of the tile associated with the torus $\mathbb{T}_{ij}$. Let i_s and i_r denote these two numbers. We have:

$$V = 2n_s\, v_s + 4n_r\, v_r \quad \text{and} \quad N = 2n_s\, i_s + 4n_r\, i_r \quad \text{so that :}$$

$$N/V = (n_s\, i_s + 2n_r\, i_r)/(n_s\, v_s + 2n_r\, v_r)$$

As one can easily see, the existence domain of a square is a square and the existence domain of a rhomb is a rhomb, so that if V is large enough:

$$n_s \approx \sqrt{2}\, n_r \quad \text{while} \quad v_s = \sqrt{2}\, v_r$$

in such a way that:

$$N/V = 1/(2v_s)(i_s + \sqrt{2}\, i_r)$$

Finally, we see that the measurable quantity $2v_s\, N/V$ should be of the form $i_s + \sqrt{2}\, i_r$. Of course, one can trivially remark that these linear combinations are dense in $\mathbb{R}$, but we are expecting reasonably small intersection indices.

The same computations can be developed for the icosahedral quasicrystals. In this case, one computes from the experimental data a number which should fall in the modulus $\mathbb{Z}(\tau)$ if the closeness hypothesis is verified. It happens that one can fit these numbers very well in the two cases of *Al–Mn–Si* and *Al–Cu–Fe* for which the density has been measured with sufficient accuracy. This yields additional constraints to be satisfied by the atomic surfaces.

However, it is necessary to stress that, as for the existence of local rules, the closeness hypothesis is not a physically undisputable requirement, but merely a plausible assumption which may help to find a structural model.

As the reader may have observed, the research of the atomic structure of quasicrystals is a difficult task and is far from its achievement, even if it is not completely hopeless.

References

Arnol'd, V. I. (1988) Physica D **33** (1988) 21

Besicovitch, A. S. (1932) Almost Periodic Functions. Cambridge University Press 1932

Cartier, P. (1987) C. R. Acad. Sci. Paris, **304** série II (1987) 789

Duneau, M. and Katz, A. (1985) Phys. Rev. Lett. **54** (1985) 2688

Elser, V. (1986) Acta Cryst. A **42** (1986)

Frenkel, D. M., Henley, C. L. and Siggia, E. D. (1986) Phys. Rev. B **34** (1986) 3649

Gardner, M. (1977) Sci. Amer. **236** 110

Gratias, D., Cahn, J. W., and Mozer, B. (1988) Phys. Rev. B **38** (1988) 1643

Grünbaum, B. and Shephard, G. C. (1987) Tilings and Patterns. W. H. Freeman 1987

Janner, A. and Janssen, T. (1977) Phys. Rev. B **15** (1977) 649

Janssen, T. (1986) Acta Cryst. A **42** (1986) 261

Kalugin, P. A., Kitayev, A. Yu. and Levitov, L. S. (1985) J. Physique Lett. **46** (1985) 601

Kalugin, P. A. (1989) Europhys. Lett., **9** (1989) 545

Katz, A. and Duneau, M. (1986) J. Phys. France **47** (1986) 181

Katz, A. (1988) Commun. Math. Phys. **118** (1988) 263

Katz, A. (1989 a) in Jarić, M. (ed.) Introduction to the Mathematics of Quasicrystals (Aperiodicity and Order vol. 2) Academic Press 1989

Katz, A. (1989 b) in Jarić, M. and Lundqvist, S. (eds.) Quasicrystals — Adriatico Anniversary Research Conference. World Scientific, Singapore 1990

Les Houches — International Workshop on Aperiodic Crystals. Gratias, D. and Michel, L. (eds.) J. Phys. France **47** C3 1986

Levitov, L. S. (1988) Commun. Math. Phys. **119** (1988) 627

Levitov, L. S. and Rhyner, J. (1988) J. Phys. France **49** (1988) 1835

Martinais, D. (1987) C. R. Acad. Sci. Paris, **305** série I (1987) 509

Martinais, D. (1988) Classification explicite de certains groupes cristallographiques (Thèse de Doctorat) Paris 1988

Penrose, R. (1979) Mathematical Intelligencer **2** (1979) 32

Shechtman, D., Blech, I., Gratias, D., and Cahn, J. W. (1984) Phys. Rev. Lett. **53** (1984) 1951

The Physics of Quasicrystals. Steinhardt, P. and Ostlund, S. (eds.) World Scientific, Singapore 1987

Chapter 12

Gap Labelling Theorems for Schrödinger Operators

by Jean Bellissard

Table of Contents

1. Introduction

Quantum Mechanics was born in 1900, when Planck [PL00] investigated the laws of black body radiation. He found the correct formula for the power spectrum in terms of the light frequency. Einstein's interpretation in 1905 by means of energy quanta [EI05] was confirmed by his interpretation of the photoelectric effect. However it took years before Quantum Mechanics became such a solid body of knowledge that it could not be avoided by any reasonable physicist. There is no doubt today that it is a fundamental theory of matter, and that it has changed daily life through new technology, in a way which has never been known before in human history.

My purpose in this introduction, is not to develop some philosophy about human society, but rather to reinterpret some basic facts in Quantum Mechanics, in the light of new sophisticated mathematical techniques which I have proposed to use a few years ago to get a gap labelling for quantum systems of aperiodic media.

Aperiodic media have been the focus of attention during the last twenty years in Solid State Physics: disorder in metals or semiconductors, charge density waves in quasi one dimensional organic conductors, superlattices, structure of glasses, quasicrystals, high temperature superconductors. They provide new materials with interesting unusual physical properties which are to be used in modern technology sooner or later. While periodic media are now well understood through Bloch theory, giving rise to band spectrum for electrons, the correct mathematical framework for aperiodic material is not completely developed yet. One proposal that I have attempted to give during the last few years is that Non Commutative Geometry and Topology, from the point of view developed by A. Connes, is the most accurate candidate for it. It is accurate both from a fundamental point of view, as I will try to show in this introduction, but also from a very practical point of view, through its efficiency in computing real things in real experiments.

In the present work I will explain only one piece in this game, namely how to obtain accurately the gap labelling for complicated band spectra, by computing K-groups of observable algebras appearing as natural objects associated to electrons motion.

However the reader must know that other pieces have been already worked at, such as the existence and quantization of plateaus for the Quantum Hall conductance [BE88a, BE88b, XI88, NB90], semiclassical calculations to explain the behavior of Bloch electrons in a uniform magnetic field [BR90] (a subject of interest in the field of high temperature superconductors), a semiclassical Birkhoff expansion and Nekhoroshev's type estimates in Quantum Mechanics [BV90], and electronic properties of scale invariant homogeneous media such as fractals or quasicrystals [SB90]. It seems also to be the correct tool to investigate rigorously properties of quantum systems which are classically chaotic, a field in which no rigorous result has been proved yet, even though physicists have accumulated an enormous body of knowledge [BE90b].

My main motivation in pushing toward Non Commutative techniques, comes from the fact that in many instances, physicists are using wave functions which are not well adapted in most problems encountered in modern Quantum Physics, essentially because they are defined up to a very troublesome phase factor. But if we are to abandon wave functions, we must explain how quantum interferences enter into the game. For indeed, each typical quantum phenomenon is due to quantum interferences: gaps in electron spectra, localization in disordered or aperiodic media, level repulsion in quantum chaos, phase quantization in Aharonov-Bohm effect, flux quantization in superconductors or in dirty metals, quantization of Hall conductance in the Quantum Hall effect, etc.

It is the purpose of this introduction to give a hint in this direction.

Acknowledgements. I wish to thank J.B. Bost for explaining the index theorem for Tœplitz operators and the Pimsner-Voiculescu exact sequence. I thank also all my collaborators without whom almost none of the results here could have been found. I thank also A. Barelli for reading the manuscript.

1.1 Waves

The key idea in Einstein's explanation of black body radiation was that everything behaved as if the energy exchanged between matter and light were quantized as an integer multiple of a small unit. This small unit, called a quantum, is given by the famous Planck formula

$$E = h\nu \,, \tag{1.1.1}$$

where ν is the light frequency, and h a constant, Planck's constant.

In 1905, in the paper for which he was awarded the Nobel prize [EI05], Einstein used this result to explain the photoelectric effect. He claimed that Planck's quantum was actually the energy of a particle, the particle of light that he called 'photon' with momentum $p = h\sigma$, where σ was the number of waves per unit length. With such a simple idea, namely that light was actually made of particles, he could explain why there was no electric current as long as the light frequency was smaller than a threshold frequency ν_0, and why that current was proportional to the difference $\nu - \nu_0$, for $\nu > \nu_0$. Thanks to that theory, Planck's constant could be measured, and it appeared that it had the same value as the one given by black body radiation. Such an amazing result was an indication that Planck's formula had a universal meaning, namely that indeed light was made of photons as explained by Einstein.

In 1911, Lord Rutherford reported upon a series of experiments he had performed on α-ray diffraction patterns produced by various targets [RU11]. He concluded that atoms were not such simple objects as their name 'atom' could suggest. According to his picture, they were made of a heavy nucleus, supporting all the mass, around which electrons would move like planets around the sun. The attracting forces however, were Coulomb forces, namely purely electrostatic, in such a way that the nucleus had a positive charge $+ne$, if n was the number of electrons gravitating around, and e the electron charge.

It was immediately realized that such a structure was unstable from a purely classical point of view. For an accelerated charged particle must radiate some electromagnetic field, ending in a permanent loss of energy, which could be gained only in the collapse of the electrons on to the nucleus. Such a simple minded argument is questioned nowadays [GB82], because it does not take into account the possibility of regaining energy from the electromagnetic field floating around. But even with this, a classical treatment of atoms would be much too complicated, and would not explain the universality of Planck's constant, resulting in "the inadequacy of classical mechanics and electromagnetism to explain the inherent stability of atoms" [BO83].

By the time, a very elegant solution to this paradox was proposed a year later by N. Bohr in 1913 [BO13, BO14, BO15a, BO15b, BO18]. It started with the remark that Planck's constant had the dimension of an action. So that he was led to postulate that the electron classical phase space orbits γ giving rise to a stable motion, were such that their classical action was quantized according to the rule

$$\int_{\gamma} p \cdot dq = nh \; , n \in \mathbb{Z} \; . \tag{1.1.2}$$

For an electron on a circular orbit in a hydrogen atom, this simple rule together with the usual rules of classical mechanics, give quantization of energy levels according to $E_n = -E_0/n^2 (n = 1, 2, \ldots)$, $E_0 = 2\pi^2 k^2 m e^4 / h^2 \approx 13.6\,\text{eV}$.

In addition, whenever an electron jumps from the orbit n to the orbit m, the change in energy will be compensated by emitting or absorbing a photon of frequency $\nu_{n,m}$, given by Planck's law. It leads immediately to the famous Balmer formula

$$\nu_{n,m} = R \left| \frac{1}{n^2} - \frac{1}{m^2} \right| \; , \qquad R = \frac{E_0}{h} = \frac{2\pi^2 k^2 m e^4}{h^3} \; , \tag{1.1.3}$$

which was known from the mid-nineteenth century by Balmer for $n = 2$, and then generalized by Rydberg for any pair n, m. It was used as a phenomenological formula to explain the ray spectrum emitted by hydrogen atoms. Amazingly enough the value of Rydberg's constant given by (1.1.3), agreed with the experimental to one part in 10^4! It could not be just a coincidence.

In the years following the First World War, more and more physicists were involved in the problem of understanding properties of atoms. The Stern-Gerlach experiments in 1922 [SG22] gave support to the idea of stationary states, while the Compton effect discovered in 1924 [CO23] confirmed the views of Einstein about photons.

As elegant and simple as it was, this argument did not explain yet why quantization of the electron action was required to get stability. And this is precisely the point where Quantum Mechanics had to be developed.

The key idea came in 1925 and was exposed in L. de Broglie's thesis [BR25]. Until Einstein's paper in 1905, and since the beginning of the 19th century, light was obviously a wave phenomenon. Interferences and diffraction had been successfully interpreted by physicists like Young, Fresnel, through Huygens's principle. By the 1860's, Maxwell had shown that they were actually electromagnetic waves of very short wavelength. However, Einstein had reintroduced also very successfully, an old concept, quite popular in the 17th century, namely that light was made of corpuscules.

L. de Broglie also pointed out that matter was obviously known to be made of particles. Chemists from the 19th century had used atoms and molecules to explain chemical reactions, and Mendeleev had also successfully classified elementary atoms. Moreover, several works on brownian motion in the first decade of the 20th century, gave some reality to the existence of atoms and molecules and the Avogadro number could be measured. By the end of the 19th century, J. J. Thompson had found that electric current was created by particles that he called electrons. Radioactivity gave also new particles like α particles. At last, Rutherford had confirmed the intricate relationships between these particles to constitute atoms.

There was clearly a dissymmetry between the Einstein treatment of radiations and the way particles like electrons were looked at. L. de Broglie claimed that the wave-corpuscule duality should be a universal principle: to each particle is associated a wave and vice-versa. This idea was soon convincingly confirmed by electron interference phenomena, a very widely accepted fact nowadays.

To give support to that idea, one has to go back to the very definition of a wave. The simplest example is provided by plane waves represented by a function proportional to

$$\psi(t, \mathbf{x}) = \mathrm{e}^{\mathbf{i}(\omega t - \mathbf{kx})} , \tag{1.1.4}$$

where ω is the pulsation namely $\omega/2\pi$ is the number of waves per unit time, while $\mathbf{k}$ is the wave vector, with direction equal to the direction of the wave and $|\mathbf{k}|/2\pi$ is the number of waves per unit length. The phase is then constant on the planes $\mathbf{kx} = \omega t + \text{const.}$ resulting in a phase velocity $\mathbf{v}_\phi = \mathbf{k}\omega/|\mathbf{k}|^2$.

Now one can argue that plane waves are only an idealization of real ones even for free particles. Actually, a real wave is never pure, it is usually what one called at that time a 'wave packet', namely a superposition of plane waves in the form

$$\psi(t, \mathbf{x}) = \int d^3\mathbf{k} \mathrm{e}^{\mathbf{i}(\omega_{\mathbf{k}} t - \mathbf{kx})} f(\mathbf{k}) , \tag{1.1.5}$$

with a $\mathbf{k}$-dependent pulsation (dispersion law). If f is a regular function decreasing rapidly at infinity in $\mathbf{k}$, so does $\psi(t, \mathbf{x})$, as a function of $\mathbf{x}$ at each time t. Its maximal value is reached at points $\mathbf{x}$'s for which the phase factor in the integral is stationary, namely

$$t\nabla_{\mathbf{k}}\omega = \mathbf{x} + \text{const.} , \tag{1.1.6}$$

resulting in a 'group velocity' $\mathbf{v}_g = \nabla_{\mathbf{k}}\omega$ if $\nabla_{\mathbf{k}}$ is the gradient with respect to $\mathbf{k}$. It is then natural to assume that the particle associated to this wave has a velocity given by the group velocity, and that its energy E is related to the pulsation ω through Planck's formula namely $E = \hbar\omega$ ($\hbar = h/2\pi$). We then remark that in classical mechanics, the energy E is represented by the Hamiltonian function $H(\mathbf{q},\mathbf{p})$ (where now $\mathbf{q}$ represents the position of the classical particle). Moreover the Hamilton equations of motion give for the velocity

$$\mathbf{v} = d\mathbf{q}/dt = \nabla_{\mathbf{p}}H \,. \tag{1.1.7}$$

The similarity between (1.1.6) for the group velocity and (1.1.7) for the particle velocity is striking and led de Broglie to supplement Planck's formula $E = h\nu = \hbar\omega$ by a similar expression for the momentum, namely

$$\mathbf{p} = \hbar\mathbf{k} \qquad \text{(de Broglie's formula)}\,, \tag{1.1.8}$$

extending the Einstein formula for the photon, to every free particle.

However the previous argument does not apply to particles submitted to forces, like for instance to electrons in the hydrogen atom. Clearly, free wave packets cannot represent such a particle. But we may extend the argument in the following way: on a very short distance $d\mathbf{x}$, and during a short amount of time dt, the wave can be approximated by plane waves, so that its phase factor increases by the amount

$$d\phi = \omega dt - \mathbf{k}d\mathbf{x} = \{Hdt - \mathbf{p}d\mathbf{q}\}/\hbar \,. \tag{1.1.9}$$

Let us consider first a particle in a stationary state, namely such that its associated wave is monochromatic, which means that ω is fixed. Then $H(\mathbf{q},\mathbf{p}) = E =$ const. If γ represents its classical orbits in phase space, the total variation of the phase along this orbit will be

$$\Delta\phi = -\int_\gamma \mathbf{p}d\mathbf{q}/\hbar \,. \tag{1.1.10}$$

For closed orbits, the corresponding wave functions must be single-valued, resulting in the relation $\Delta\phi = 2\pi n\,(n \in \mathbb{Z})$ which is nothing but Bohr's quantization condition

$$\int_\gamma \mathbf{p}d\mathbf{q} = nh \,, \qquad (n \in \mathbb{Z}) \,. \tag{1.1.11}$$

Therefore stationary states are only stationary waves, giving a coherent scheme for their stability: *stability comes from constructive interferences for the wave associated to the particle.* Actually a real wave will be a superposition of pure waves. Going back to (1.1.9), and remarking that $Hdt - \mathbf{p}d\mathbf{q} = -L(\mathbf{q}, d\mathbf{q}/dt)dt$ where L is the Lagrangian, a pure wave will have the form

$$\psi(t,\mathbf{x}) = \exp\left(-\frac{\mathbf{i}}{\hbar}\int_{t_0}^{t} ds L(\mathbf{q}(s),\mathbf{q}'(s))\right) , \quad \mathbf{q}(t_0) = \mathbf{x}_0 , \mathbf{q}(t) = \mathbf{x} , \tag{1.1.12}$$

along a trajectory $s \to q(s)$ (here $\mathbf{q}' = d\mathbf{q}/ds$), which may or may not be a solution of the Hamilton equations. The importance of the Lagrangian in building up phases was emphasized in 1936 by Dirac. In his 1942 thesis [FE48], R.P. Feynman started from this expression to give an integral representation of the wave function in the formal form:

$$\psi(t,\mathbf{x}) = \int \prod_{s=0}^{s=t} D\mathbf{q}(s) \exp\left(-\frac{\mathbf{i}}{\hbar}\int_{t_0}^{t} ds L(\mathbf{q}(s),\mathbf{q}'(s))\right) \psi(0,\mathbf{q}(0)) , \tag{1.1.13}$$

where one integrates over the set of all trajectories such that $\mathbf{q}(t) = \mathbf{x}$. This formal expression can be seen as a superposition of pure waves, like a wave packet is a superposition of plane waves.

Again, we can argue that the main contribution to this integral comes from trajectories which produce stationary phase factors. By the Maupertuis principle these trajectories are precisely the classical orbits of the corresponding classical system.

For a single particle in a potential, represented by a Hamiltonian $H = \mathbf{p}^2/2m + V(\mathbf{q})$, Feynman showed that the wave function in (1.1.13) is a solution of Schrödinger equation

$$\mathbf{i}\hbar\frac{\partial\psi}{\partial t} = -\frac{\hbar^2}{2m}\Delta\psi + V\psi , \tag{1.1.14}$$

and conversely, he showed that every solution of Schrödinger equation can be written as the path integral (1.1.13), establishing a complete formal equivalence between the two formalisms. Notice that Schrödinger introduced this equation in 1926 [SR26] on the basis of a variational principle instead and the success was so high that this logical scheme was forgotten for a long time.

E. Nelson [NE64] proved in 1964 that one can give a mathematical rigorous meaning of (1.1.13) for most values of the mass m, whereas Albeverio et al. [AH76] developed a formalism based upon Fresnel integrals, which gives a mathematical status to (1.1.13) for potentials V which are Fourier transforms of positive measures. This result has been improved recently by Fujiwara [FU80, FU90]. Still, the mathematical status of the Feynman path integral is rather unclear. It will require quite a lot of improvements to make it a useful mathematical tool.

Even though (1.1.13) is quite formal, it has been used over and over, especially in Quantum Field Theory. Still today it is one of the most powerful tools for intuition, in dealing with proper definitions and properties of models both in Particle and Solid State Physics.

1.2 Particles

In the previous Section, we described the usual way of representing waves, leading to Schrödinger equation. However, it is to be noted that from a historical viewpoint, waves appeared later in the 1926 paper by E. Schrödinger [SR26], only a few months after the work of Heisenberg, Born and Jordan on what was called at the time 'Matrix Mechanics' [BH25].

The description of point-like particles had been the purpose of Classical Mechanics, which reached a very sophisticated level by the middle of the nineteenth century, with the works of Lagrange, Liouville, Hamilton, Jacobi. They had introduced a new function for the mechanical energy, the so-called 'Hamiltonian', giving rise to a symmetric treatment of momentum and position through the so-called 'phase space'. Moreover, the symplectic structure of the equations of motion was conserved by a special family of changes of coordinates called 'canonical transformations'. Instead of computing the solutions by various techniques, like perturbation theory, Jacobi proposed a new method: to compute a canonical change of coordinates transforming the equations of motion into a family of trivial ones. Liouville had introduced also the notion of action-angle variables [AR78], which had the property of being a universal choice of coordinates, at least locally in phase space, and one could define properly the notion of completely integrable system.

In the beginning of the twentieth century, this part of classical mechanics was taught in universities as the most sophisticated piece of knowledge.

No surprise then that Bohr pointed out the importance of action integrals in dealing with quantization. Soon enough after his 1913 seminal paper, A. Sommerfeld [SO15] had extended his quantization condition to get the two other quantum numbers for a complete description of stationary states of the hydrogen atom. In a 1917 paper [EI17], Einstein generalized the method to non-separable Hamiltonians, by means of Jacobi's method: through various canonical transformations, one expresses the classical Hamiltonian in terms of action variables only and then one replaces each action variable J_k by $n_k\hbar$, where n_k is an integer. In this way it became possible to treat more complicated problems like the emission spectrum of atoms and molecules [SR82, JR82, MA85]. The method was improved later on by Brillouin [BR26] and Keller [KE58] and is known nowadays as the EBK quantization scheme.

However, the method did not succeed at the time, in explaining details of atomic spectra: the case of the helium atom was especially emphasized in the beginning of the twenties. The exclusion principle partly responsible for the discrepancy was discovered only in 1925 by W. Pauli. Moreover, the Stern and Gerlach experiment in 1922 [SG22], coming after many difficulties in interpreting the Zeeman multiplets, showed the limits of the method in that this quantization condition was unable to explain why angular momenta in atoms required to introduce half-integers as quantum numbers. Sure enough there was something wrong about the old theory of quanta.

During the year 1925, W. Heisenberg using a method of 'systematic guessing' [VW67], was 'fabricating quantum mechanics' [HE25, HE85]. His first motivation was to compute the line intensities in the hydrogen spectrum. He used a generalization of Fourier analysis which proved quite hard and forced him to consider the simpler problem of a harmonic oscillator. That was enough to produce new rules, soon recognized as the rules of matrix multiplication. Immediately after his paper was published, M. Born conjectured that the basic equation in this theory was the 'commutation rule', and that was confirmed some days later by a calculation of his pupil P. Jordan who showed that within the Heisenberg rules, the canonical commutation relations $[q,p] = \mathbf{i}\hbar$ were correct [BJ25]. This led to the extraordinary 'three men paper' published in November 1925 by M. Born, W. Heisenberg and P. Jordan [BH25], in which the basic principles of quantum mechanics are settled in a way which requires no change in the light of later improvements. Moreover several difficulties were solved at once like formulæ for perturbation theory, the 'anomalous' Zeeman effect (quantization rules for the angular momentum), and they also recovered the Planck formula for black body radiation through using the statistical approach of P. Debye. A few weeks after, W. Pauli gave a treatment of the hydrogen atom, using this new mechanics [PA26].

The first step in Heisenberg's intuition is related to the fact that only observable quantities must enter in building up a theory. One of the main problem in the old theory comes from the fact that one is dealing with classical orbits which are obviously meaningless (this will be made more precise in 1927 [HE27, JA74], by means of the uncertainty principle). However, there was surely a notion of 'stationary states' which could be observed through spectral lines, and interpreted as energy levels. To compute line intensities, it was necessary to consider transitions from one stationary state labelled by n to another one labelled by $(n-l)$. In dealing with a time dependent quantity $x(t)$, there is a frequency $\nu(n, n-l) = \omega(n, n-l)/2\pi$ associated to such a transition: this is the frequency of the light emitted by the atom. The observed rule for transition is given by

$$(1.2.1) \qquad \omega(n, n-l) + \omega(n-l, n-l-l') = \omega(n, n-l-l')\,,$$

which permits to write the transition frequency as

$$(1.2.2) \qquad \omega(n, n-l) = \{W(n) - W(n-l)\}/\hbar\,,$$

where $W(n)$ is called a spectral term; according to Planck's formula, it is an energy (defined up to an additive constant term).

The second step is Bohr's correspondence principle according to which quantum mechanics must agree with classical laws for large quantum numbers n. Thus for fixed l's and large n's (2.2) gives

$$(1.2.3) \qquad \omega(n, n-l) = l \cdot \omega(n) = l/\hbar \partial W/\partial n\,,$$

where $1/\hbar\partial W/\partial n$ can be understood as $\partial W/\partial J$ if $J = n\hbar$ is the classical action integral, in complete analogy with the classical calculation of particle frequencies.

The emission of electromagnetic waves is then classically governed by the laws of electrodynamics. For instance, the electric field at a distance $\mathbf{r}$ from the emitting electron is given by

$$\mathcal{E} \approx e/r^3\{\mathbf{r} \times (\mathbf{r} \times \mathbf{v}')\}\ , \tag{1.2.4}$$

where $\mathbf{v}' = d^2\mathbf{r}/dt^2$ is the electron acceleration. It implies that classically a quantity $x(t)$, like the electron position or its velocity, can be expanded in Fourier transform, in terms involving all possible transition frequencies. The quantum assumption made by Heisenberg is that the correspondence should be as follows

$$\begin{aligned} &\text{in quantum theory } \ x(t) = \sum a(n, n-l)e^{i\omega(n,n-l)t}\ , \\ &\text{in classical theory } \ x(t) = \sum a_l(n)e^{i\omega(n)\cdot l\cdot t}\ . \end{aligned} \tag{1.2.5}$$

He then addressed the question of how to compute quantities like $x(t)y(t)$, whenever x and y are two observable quantities. Using (1.2.1) and (1.2.4) he then found immediately

$$(x\cdot y)(n, n-l) = \sum_{l'} x(n, n-l')y(n-l', n-l)\ . \tag{1.2.6}$$

This law is nothing but matrix multiplication as M. Born realized soon after [VW67].

Differentiating $x(t)$ with respect to time, one gets the so-called Heisenberg equations of motion which are, in matrix language

$$dx/dt = (xW - Wx)/i\hbar\ , \tag{1.2.7}$$

where W is the diagonal matrix given by (1.2.2).

The third step is now to express the Bohr-Sommerfeld-Einstein quantization condition (1.1.2). In order to do so, let us consider a one dimensional particle, and let us write the action as

$$nh = \int_\gamma pdq = \int dtmv^2(t) = 2\pi m \sum_l |q_l(n)|^2 l^2\omega(n)\ , \tag{1.2.8}$$

If one formally differentiates both sides with respect to n (!), one gets

$$\hbar = h/2\pi = m\sum_l l\cdot \partial/\partial n\{|q_l(n)|^2 l\omega(n)\}\ , \tag{1.2.9}$$

with its quantum equivalent [HE25]

$$\hbar = 2m\sum_{l>0}\{|q(n, n+l)|^2\omega(n, n+l) - |q(n, n-l)|^2\omega(n, n-l)\}\ . \tag{1.2.10}$$

In the matrix language, this is nothing but the diagonal part of the canonical commutation relation $\langle n|(q \cdot p - p \cdot q)|n\rangle = \mathbf{i}\hbar$, where $p = mdq/dt$ is the momentum. Using the Heisenberg equation of motion, Jordan could prove that the commutator $qp - pq$ is actually diagonal in such a way that

$$q \cdot p - p \cdot q = \mathbf{i}\hbar \ . \tag{1.2.11}$$

In his first 1925 paper, Heisenberg used this new mechanics to compute the spectrum of an anharmonic oscillator. He actually made a very important remark: in order to get the spectrum of a quantized harmonic oscillator, he imposed the existence of a 'ground state', namely he demanded that the energy be bounded from below. This can be expressed in modern language by asking the observable algebra to contain the notion of 'positive elements'.

1.3 Why is the Set of Observables a C^*-Algebra?

The Heisenberg arguments have justified the non commutative approach to quantum mechanics. The correspondence principle can be sharpened now in the following way:

(i) classical observable quantities are functions on the phase space. Quantum mechanically, they are replaced by elements of a non commutative algebra $\mathcal{A}$ over the complex numbers.

(ii) the observable algebra must admit an involution $A \rightarrow A^*$ such that $(A+B)^* = A^* + B^*$, $(AB)^* = B^*A^*$, (if $\lambda \in \mathbb{C}$, λ^* is the complex conjugate of λ). It expresses the fact that we need real numbers to measure physical quantities.

(iii) a measurement process is described by states, namely linear forms τ: $\mathcal{A} \rightarrow \mathbb{C}$, such that $\tau(A^*A) \geq 0$ for any A in $\mathcal{A}$ and $\tau(\mathbf{1}) = 1$.

The positivity property (iii) is actually crucial for quantization. Indeed one can exhibit examples of algebras, generated by elements $q = q^*$ and $p = p^*$, satisfying (1.2.11), for which the Hamiltonian $H = p^2 + q^2$ corresponding to a harmonic oscillator, would not admit quantized energy levels. In its first 1925 paper, Heisenberg found that the frequencies in this model satisfied $W(n) = \hbar\omega(n+\text{const.})$. To fix the arbitrary constant, he insists in having a lowest energy level, namely a ground state, which gives the famous $n + 1/2$.

If one insists that the observable algebra be entirely defined by measurement processes, one may use the Gelfand-Naimark-Segal construction [SA71, BR79, PE79, TA79] from any state τ, to get a Hilbert space $\mathcal{H}_\tau$, and a unit vector ζ_τ, for which any observable A is represented by an operator $\pi_\tau(A)$ such that $\pi_\tau(A^*) = \pi_\tau(A)^*$ and $\tau(A) = \langle\zeta_\tau|\pi_\tau(A)\zeta_\tau\rangle$.

Practical calculations will not be possible without permitting to take limits of sequences of observables. Otherwise it would be like ignoring real numbers and working only with rationals. The main problem is that there are many ways of defining a topology on such an algebra. The measurement process provides us with a way of defining a natural topology.

For technical simplicity, one can restrict oneself to the set of observables giving rise for each state, to bounded operators in the GNS representation, in such a way that if we set

$$\|A\| = \sup_\tau |\tau(A)| \ , \tag{1.3.1}$$

we get a norm on $\mathcal{A}$ satisfying

$$\|A^*A\| = \|A\|^2 \ . \tag{1.3.2}$$

Then one can include in $\mathcal{A}$ all possible elements obtained by a limiting procedure: this means that we ask $\mathcal{A}$ to be complete. Such an algebra is called a C^*-algebra:

Definition. 1) A *C^*-algebra* is an algebra over the complex field with an involution satisfying (ii), and a norm satisfying (1.3.2), for which it is complete.

2) A **-homomorphism* from the C^*-algebra $\mathcal{A}$ to the C^*-algebra $\mathcal{B}$ is a linear mapping $\alpha : \mathcal{A} \to \mathcal{B}$ such that $\alpha(AB) = \alpha(A)\alpha(B)$ and $\alpha(A^*) = \alpha(A)^*$ for any $A, B \in \mathcal{A}$. It is a *-isomorphism if it is invertible, and a *-automorphism of $\mathcal{A}$ whenever it is an isomorphism from $\mathcal{A}$ to $\mathcal{A}$.

3) A *one-parameter group of *-automorphisms* is a family $\{\alpha_t; t \in \mathbb{R}\}$ of *-automorphisms of $\mathcal{A}$ such that $\alpha_{s+t} = \alpha_s \circ \alpha_t$ for any $s, t \in \mathbb{R}$; it is point-wise norm continuous whenever for any $A \in \mathcal{A}$ the mapping $t \in \mathbb{R} \to \alpha_t(A) \in \mathcal{A}$ is continuous in norm.

4) A **-derivation* on a C^*-algebra $\mathcal{A}$, is a linear map δ defined on a dense subalgebra $\mathcal{D}(\delta)$ in $\mathcal{A}$, and such that $\delta(AB) = \delta(A)B + A\delta(B)$, $\delta(A^*) = \delta(A)^*$ for any $A, B \in \mathcal{D}(\delta)$.

5) A *-derivation δ generates a one-parameter group of point-wise norm continuous *-automorphisms $\{\alpha_t; t \in \mathbb{R}\}$ if and only if for $A \in \mathcal{D}(\delta), \delta(A) = d\alpha_t(A)/dt$ at $t = 0$, namely if one can write $\alpha_t = \exp\{t\delta\}$.

As we see, C^*-algebras emerge as very natural objects from Heisenberg's construction. Moreover, the norm which has been constructed here has another canonical property: eq. (1.3.2) implies that the square of the norm of A is nothing but the spectral radius of A^*A; hence the topology given by a C^*-norm comes entirely from the algebraic structure (algebra and positivity)! In particular every *-homomorphism is automatically norm-continuous namely $\|\alpha(A)\| \leq \|A\|, A \in \mathcal{A}$, showing that the algebraic structure is sufficient to define the topology.

The restriction to bounded operators is not actually essential, in that 'good observables' can be computed through bounded operators, by means of resolvents (Green's functions), or any kind of functional calculus. However, as pointed out by von Neumann, unbounded observables raise a very difficult technical problem, due to the domain of definition which may lead to the impossibility of computing bounded functions. This is why he defined self-adjoint operators [FA75]. They are precisely those unbounded symmetric operators for which the Schrödinger equation $\mathbf{i}\partial\psi/\partial t = H\psi$ admits a unique solution for all time with a given initial data. Then the functional calculus and spectral theory exist for self-adjoint operators and permit to reduce their study to the case of bounded operators.

The correspondence principle is now supplemented by requiring that the classical symplectic structure should survive quantum mechanically. By analogy with the previous calculation, this is done by defining Poisson's brackets and canonical variables as follows

(iv) Poisson's brackets are given by [DI26]

$$\{A, B\} = [A, B]/\mathbf{i}\hbar\,, \qquad [A, B] = AB - BA\,. \tag{1.3.3}$$

Then one sees that the map $\mathcal{L}_H : A \to \{H, A\}$ satisfies all axioms of a derivation namely it is linear and

$$\mathcal{L}_H(AB) = \mathcal{L}_H(A)B + A\mathcal{L}_H(B)\,, \qquad \mathcal{L}_H(A)^* = \mathcal{L}_{H^*}(A^*)\,. \tag{1.3.4}$$

Quantum equations of motion will then be nothing but Hamilton-Jacobi's ones namely

$$dA/dt = \mathcal{L}_H(A)\,. \tag{1.3.5}$$

If $H = H^*$ is unbounded $\mathcal{L}_H$ is an unbounded derivation, which requires to define properly its domain of definition [BR79] in such a way that solutions of (1.3.5) be given by $A(t) = \exp\{t\mathcal{L}_H\}(A)$. More generally, any canonical transformation is generated by products of operators on $\mathcal{A}$ of the form $\exp\{\mathcal{L}_H\}$ with $H = H^*$. Going back to (1.3.3), one sees that they correspond formally to unitary transformations of the form $A \to SAS^{-1}$. In this way perturbation theory will be quite simple to develop in a completely similar way with classical mechanics [BV90].

(v) the observable algebra $\mathcal{A}$ can be constructed by means of a family of bounded functions of two families of elements $Q_k = Q_k^*$ and $P_k = P_k^*$ ($k = 1, \cdots, D$) satisfying the canonical commutation relations

$$\{Q_k, Q_l\} = 0 = \{P_k, P_l\}\,, \{Q_k, P_l\} = \delta_{kl}\mathbf{1}\,. \tag{1.3.6}$$

As we will see there exist many different non isomorphic such algebras, depending upon the problem we want to investigate. The simplest one is generated by smooth fast decreasing functions of the P's and the Q's, through Weyl's quantization formula, and gives rise to the algebra $\mathcal{K}$ of compact operators, which is used in ordinary Quantum Mechanics. However, when dealing with periodic or aperiodic media we will get different algebras, giving rise to various kinds of spectra.

The connection with waves and Schrödinger's point of view was done in 1926 by Schrödinger and Pauli [JA74]. In modern language, wave functions provide a 'representation' of the observable algebra. For ordinary Quantum Mechanics, Weyl's theorem [RSII] shows that there is a unique (up to unitary equivalence) such representation, namely the Hilbert space is $L^2(\mathbb{R}^D)$ and Q_k

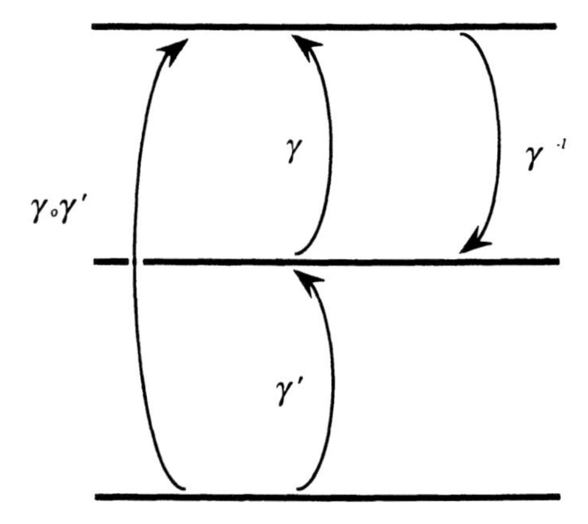

Fig. 1. Groupoid as transitions between stationary states

is the operator of multiplication by the k-th coordinate x_k, while P_k is the differential operator $-\mathbf{i}\hbar\partial/\partial x_k$. Therefore for a $3D$ particle in a potential $H = \mathbf{P}^2/2m + V(\mathbf{Q})$ is nothing but the Schrödinger operator (see eq. (1.1.14))

$$H = -\hbar^2/2m\Delta + V(\mathbf{x})\,. \tag{1.3.7}$$

The three men paper [BH25] has been so cleanly written that most of the textbooks in quantum mechanics have forgotten the original intuition of Heisenberg and usually introduce the operator aspect as a consequence of the Schrödinger equation, after having defined the notion of wave functions, of states as vectors in a Hilbert space, and then coming to the point where partial differential operators can be algebraically interpreted.

Actually, the original approach by Heisenberg contains much more structure than a first look may show. A lot of progress were made by mathematicians during the seventies to use this construction in building various C^*-algebras. The breakthrough went with the notion of a groupoid [HA78a, HA78b, RE80, CO79], which is nothing but the abstract generalization of the notion of transition between stationary states as defined by Bohr and Heisenberg.

A groupoid **G** is a family of two sets $\{G^{(0)}, G\}$ where $G^{(0)}$ is called the basis and generalizes the set of stationary states. An element γ of G is called an 'arrow' and represents a transition between two elements of $G^{(0)}$ its source $s(\gamma)$, and its range $r(\gamma)$. Giving two arrows γ and γ' such that $r(\gamma') = s(\gamma)$, one can define a new one $\gamma\circ\gamma'$ (see Fig. 1) with $r(\gamma\circ\gamma') = r(\gamma)$ and $s(\gamma\circ\gamma') = s(\gamma')$. Actually we demand that an element x of the basis be considered as a special arrow with $r(x) = s(x) = x$, $x\circ\gamma = \gamma$ for every arrow γ such that $r(\gamma) = x$, and $\gamma\circ x = \gamma$ for every arrow γ such that $s(\gamma) = x$; in other words, x is a unit. At last we demand that each arrow $\gamma \in G$ admits an inverse γ^{-1} with $r(\gamma^{-1}) = s(\gamma), s(\gamma^{-1}) = r(\gamma)$ and $\gamma^{-1}\circ\gamma = s(\gamma), \gamma\circ\gamma^{-1} = r(\gamma)$. Examples of groupoids are known:

1) if the basis has only one element, then **G** is a group.

2) On the other extreme, given a set X, a groupoid $\mathbf{G}(X)$ is defined with X as a basis and $G = X \times X$. Then $r(x,y) = x$, $s(x,y) = y$, and x is identified with the arrow (x,x); we define the product by $(x,y) \circ (y,z) = (x,z)$, and the inverse by $(x,y)^{-1} = (y,x)$.

3) Let now M be a compact space, and let Γ be a locally compact group. We assume that Γ acts on M through a group of homeomorphisms $g \in \Gamma \to f_g$ such that $f_{gg'} = f_g \circ f_{g'}$. Then we get a groupoid by letting M be the basis, and $G = M \times \Gamma$. Then $r(x,g) = x$, $s(x,g) = f_g^{-1}(x)$, $(x,g) \circ (f_g^{-1}(x), g') = (x, gg')$, $(x,g)^{-1} = (f_g^{-1}(x), g^{-1})$, and x will be identified with $(x,1)$ where 1 is the unit of Γ.

One can easily define the notion of measurable, topological, differentiable groupoid, by asking $G^{(0)}$ and G to be measurable spaces, topological spaces, or manifolds, and demanding that the maps defining the range, the source, the product and the inverse be measurable, continuous or smooth (see [CO79], for more details). In particular, foliations on a smooth compact manifold are described by differentiable groupoids.

Let $\mathbf{G}$ be a locally compact groupoid. We say that $\mathbf{G}$ is discrete whenever for every x in the basis, the set $G^{(x)} = \{\gamma \in G; r(\gamma) = x\}$ is discrete. In this case we consider the topological vector space $\mathcal{C}_c(G)$ of complex continuous functions with compact support on G. It becomes a *-algebra if we define the product and the adjoint as follows

$$ff'(\gamma) = \sum_{\gamma' \in G^{(x)}} f(\gamma') f'(\gamma'^{-1} \circ \gamma)\,, \qquad f^*(\gamma) = f(\gamma^{-1})^*\,. \tag{1.3.8}$$

If $\mathbf{G}$ is given by the example 2) with $X = \{1, 2, \cdots, n\}$, a continuous function with compact support on G is nothing but a matrix $f = ((f(i,j)))$, and the product is nothing but the usual matrix multiplication, whereas the * is nothing but the usual hermitian conjugate.

For every x in the basis of $\mathbf{G}$, we then define the Hilbert space $H_x = l^2(G^{(x)})$. A representation π_x of $\mathcal{C}_c(G)$ in this Hilbert space is provided by setting

$$(\pi_x(f)\psi)(\gamma) = \sum_{\gamma' \in G^{(x)}} f(\gamma^{-1} \circ \gamma')\psi(\gamma')\,. \tag{1.3.9}$$

One can check that $\pi_x(ff') = \pi_x(f)\pi_x(f')$, and that $\pi_x(f)^* = \pi_x(f^*)$ for every f, f' in $\mathcal{C}_c(G)$. Therefore we get a C^*-norm by setting

$$\|f\| = \sup_{x \in G^{(0)}} \|\pi_x(f)\|\,, \tag{1.3.10}$$

By completing $\mathcal{C}_c(G)$ with respect to this norm, we get a C^*-algebra $C^*(\mathbf{G})$. The generalization of this construction to every locally compact groupoid can be found in [RE80].

1.4 The Index Theorem Versus Wave-Particle Duality

In this Section we want to illustrate the wave-particle duality in a mathematical problem namely the Index Theorem for Tœplitz operators.

Let f be a complex continuous function on the real line $\mathbb{R}$, periodic of period 2π, which never vanishes. Then it is quite elementary to show that it can be written as

$$f(x) = \exp\{\mathbf{i}nx + \phi(x)\}\ , \tag{1.4.1}$$

where ϕ is a complex continuous function on the real line $\mathbb{R}$, periodic of period 2π, and n is an integer. Both n and ϕ are uniquely defined through (1.4.1). Indeed the integer n can be computed through a Cauchy formula:

$$n = \frac{1}{2\mathbf{i}\pi}\int_0^{2\pi} dx\, \frac{f'(x)}{f(x)} \tag{1.4.2}$$

showing that it can be seen as the winding number of the closed path $\gamma = f(\mathbb{T})$ in the complex plane given by the image of the torus $\mathbb{T} = \mathbb{R}/\mathbb{Z}$ by f. Obviously, this winding number does not change under a continuous deformation of this path, namely, n is a homotopic invariant. Actually, f is homotopic to the function $e_n : x \to \exp\{\mathbf{i}nx\}$, simply by letting ϕ go to zero uniformly in (1.4.1). So, since a continuous function never vanishes if and only if it has a continuous inverse, the winding number n classifies the homotopy classes of invertible functions in the algebras $\mathcal{C}(\mathbb{T})$.

We then remark that n is defined by the requirement that the phase of f changes by a multiple of 2π whenever x varies through one period. This is therefore a condition very similar to the Bohr quantization rule (1.1.2). Moreover we see here that it is expressed as the evaluation of the closed 1-form $\eta = dz/2\mathbf{i}\pi z$ over the path $\gamma = f(\mathbb{T})$ defined by f

$$n = \int_\gamma \eta\ . \tag{1.4.3}$$

Since η is closed, it defines a de Rham cohomology class $[\eta]$ in the first cohomology group $H^1(\mathbb{C}\backslash\{0\}, \mathbb{C})$ over the pointed complex plane $\mathbb{C}\backslash\{0\}$. By Cauchy's formula the evaluation of $[\eta]$ on any closed path is an integer so $[\eta] \in H^1(\mathbb{C}\backslash\{0\}, \mathbb{Z})$.

On the other hand, γ is a closed path, and therefore it defines a homology class $[\gamma] = f^*[\mathbb{T}]$ in $H_1(\mathbb{C}\backslash\{0\}, \mathbb{Z})$. By the de Rham duality one can therefore write n in the following abstract way

$$n = \langle [\eta] | f^*[\mathbb{T}] \rangle \tag{1.4.4}$$

exhibiting the homological content of the 'wave' aspect of Bohr's quantization rule.

Let us now consider the Hardy space $\mathcal{H}(D)$ namely the space of holomorphic functions on the unit disc $D = \{z \in \mathbb{C}; |z| < 1\}$ with square integrable boundary value on the unit circle $S_1 = \{z \in \mathbb{C}; |z| = 1\} \approx \mathbb{T}$. $\mathcal{H}(D)$ is a closed subspace of the Hilbert space $L^2(\mathbb{T})$ namely the subspace of function having a Fourier series vanishing on negative frequencies. We will denote by P the orthogonal projection onto $\mathcal{H}(D)$.

For $f \in C(\mathbb{T})$, we will denote by $T(f)$ the restriction to $\mathcal{H}(D)$ of the operator of multiplication by f. It is called the Tœplitz operator associated to f. The main property of $T(f)$ is the following

Proposition 1.4.1. *The Tœplitz operator $T(f)$ is Fredholm. More precisely, the operators $T(f)T(f^{-1}) - \mathbf{1}$ and $T(f^{-1})T(f) - \mathbf{1}$ are compact.*

Proof. First of all, it is easy to check that $\|T(f)\| \leq \|f\|_\infty$ where $\|f\|_\infty = \sup_{x \in \mathbb{T}}|f(x)|$. Therefore, by Stone-Weierstrass theorem, we can approximate f by a sequence of trigonometric polynomials. Since the set of compact operators is norm closed it is enough to prove the theorem for trigonometric polynomials.

We can write $T(f) = PfP$ where f denotes here the operator of multiplication by f in $L^2(\mathbb{T})$. Thus $T(f)T(f^{-1}) - \mathbf{1} = P[f, P]f^{-1}P$. It is enough to show that $[f, P]$ is compact. By linear combination, it is sufficient to consider the case for which $f(x) = \exp\{\mathbf{i}nx\}$. An elementary calculation shows that it is a finite rank operator. □

It is well known [RSIV, CF87] that a Fredholm operator T admits finite dimensional kernel and cokernel. The index is then defined as follows:

$$\text{Ind}(T) = \dim(\text{Ker}(T)) - \dim(\text{Coker}(T)) \,. \tag{1.4.5}$$

It satisfies the following properties:

(i) Ind is norm continuous (homotopy invariance),

(ii) If K is any compact operator $\text{Ind}(T + K) = \text{Ind}(T)$,

(iii) $\text{Ind}(T^*) = -\text{Ind}(T)$.

The main result about Tœplitz operators is the following theorem

Index Theorem. *The index of the Tœplitz operator $T(f)$ is equal to the winding number of f namely:*

$$\text{Ind}(T(f)) = -\langle [\eta] | f^*[\mathbb{T}] \rangle \tag{1.4.6}$$

where $\eta = dz/2\mathbf{i}\pi z$.

Proof. By the homotopy invariance of both sides, and thanks to the formula (1.4.1), it is sufficient to prove this formula for $f(x) = e_n(x) = \exp\{\mathbf{i}nx\}$. The formula (1.4.4) shows that the right-hand side is nothing but n. Let us consider $g \in \mathcal{H}(D)$. Then if $n \geq 0$, $T(e_n)g(z) = z^n g(z)$ showing that $\text{Ker}(T(f)) = \{0\}$. If $n < 0$ $T(e_n)g(z) = \text{Reg}(g(z)/z^{|n|})$ where $\text{Reg}(h)$ denotes the regular part of the meromorphic function h; thus $T(e_n)g = 0$ if and only if g is a polynomial of degree $|n| - 1$, namely $\dim\{\text{Coker}(T(f))\} = |n| = -n$. Since $\text{Coker}(T) = \text{Ker}(T^*)$, and $T(e_n)^* = T(e_{-n})$ we get immediately the result. □

The Index Formula (1.4.6) is interesting in that, while the right hand side expresses in some way a property of 'waves', related to the necessity that the phase of f varies by a multiple of 2π as x varies on one period, the left hand side gives an expression of the winding number in term of an operator, which may be interpreted as the analog of the 'particle' interpretation in quantum mechanics.

1.5 The Sturm-Liouville Theory on an Interval

Let us consider now a Schrödinger equation on a finite interval, namely we seek for solutions of the following ordinary differential equation

$$\psi''(x) + (E - V(x))\psi(x) = 0 \qquad x \in (-L, L) \tag{1.5.1}$$

where we assume that V is a continuous non negative function on $(-L, L)$.

Let us consider first the point of view of waves. It is well known [AR74] that given x_0 in $(-L, L)$ and $A, B \in \mathbb{C}$, there is a unique solution such that $\psi(x_0) = A$ and $\psi'(x_0) = B$. This solution is of class $\mathcal{C}^2$. It is real whenever E, A and B are real. In particular for every value of E, we can find unique solutions $\psi_\pm$ such that

$$\psi_\pm(\pm L) = 0\,, \qquad \pm\psi'_\pm(\pm L) = -1\,, \tag{1.5.2}$$

They are real for real E's, and whenever $E \leq 0$, these two solutions are positive and convex everywhere on $(-L, L)$. Moreover they are analytic entire functions of E in the complex plane.

In what follows we will restrict ourselves to the solution ψ_- which will be denoted by ψ_E to specify the value of the energy. The case of ψ_+ can be treated in a similar way. The main result is contained in the following theorem

Theorem 1. *(i) The set of E's for which $\psi_E(L) = 0$ is an infinite discrete sequence of positive numbers $0 < E_0 < E_1 < \cdots < E_{n-1} < \cdots$ converging to ∞.*

(ii) If $E_{n-1} < E < E_n$, the solution ψ_E has exactly n simple, isolated zeroes $-L < x_1(E) < x_2(E) < \cdots < x_n(E) < L$ such that $dx_k(E)/dE < 0$ and $x_n(E)$ converges to L as E tends to E_{n-1} from above.

Proof. Let us define the Wronskian of two functions ψ, ϕ by $W_x(\phi, \psi) = \phi(x)\psi'(x) - \phi'(x)\psi(x)$. Differentiating and using the Schrödinger equation, we get the following identities

$$W_x(\psi_E, \psi_E^*) = 2\mathbf{i}\mathrm{Im}(E) \int_{[-L,x]} d\zeta |\psi_E(\zeta)|^2\,, \tag{1.5.3a}$$

$$W_x(\partial_E\psi_E, \psi_E) = \int_{[-L,x]} d\zeta \psi_E(\zeta)^2\,. \tag{1.5.3b}$$

The first identity applied to $x = L$ shows that if $\mathrm{Im}(E) \neq 0, \psi_E(L) \neq 0$. The second one shows that if $\psi_E(L) = 0$, then $\psi'_E(L) \neq 0$ (by the uniqueness theorem), and therefore $\partial_E \psi_E(L) \neq 0$. Thus any E such that $\psi_E(L) = 0$ is necessarily real and isolated. Since ψ_E is positive for $E \leq 0$, each such value is automatically positive. So we get a discrete set $0 < E_0 < E_1 < \cdots < E_{n-1} < \cdots$ corresponding to solutions of $\psi_E(L) = 0$. We will prove below that it is actually infinite and unbounded.

On the other hand, if for some x in $[-L, L]$ $\psi_E(x) = 0$, $\psi'_E(x)$ cannot vanish for otherwise, by the uniqueness theorem, we would have $\psi_E = 0$, contradicting the fact that $\psi'_E(-L) = 1$. Therefore, each zero of ψ_E is simple and isolated. Let $x(E)$ be such a zero. Differentiating the identity $\psi_E(x(E)) = 0$ with respect to E, and using (1.5.3b), we get (since for E real, ψ_E is real)

$$dx(E)/dE = -\{\psi'_E(x(E))\}^{-2} \int_{[-L,x(E)]} d\zeta \psi_E(\zeta)^2 < 0 \,. \tag{1.5.4}$$

The only way for a new zero to hold as E increases, is to appear at the extremity L of the interval namely for $E = E_n$ for some n. Otherwise, there would be a value $E' \neq E_n$ for all n's and a point x' in $(-L, L)$ such that for ε small enough and $E' - \varepsilon < E < E'$, ψ_E admits $-L < x_1(E) < x_2(E) < \cdots < x_k(E) < L$ as zeroes for some k, and for $E' < E < E' + \varepsilon$ there would be a new zero $x'(E)$ between say $x_j(E)$ and $x_{j+1}(E)$, converging to x' as E converges to E' from above. Since the function ψ_E is smooth in E, it converges point-wise to ψ'_E as $E \to E'$, and in particular $(-)^j \psi_{E'}(x) \geq 0$ for $x_j(E) < x < x_{j+1}(E)$. Thus x' would be a double zero of $\psi_{E'}$ which is impossible. Thus as E increases, one and only one zero appears every time E takes values in the set of E_n's.

Let us now introduce the real function $\theta_E(x)$ defined by $\tan(\theta_E(x)/2) = \sqrt{E}\psi_E(x)/\psi'_E(x)$ which may also be defined as the unique solution of

$$d\theta_E(x)/dx = 2\sqrt{E} - 2V(x)/\sqrt{E}\sin^2(\theta_E(x)/2)\,, \qquad \theta_E(-L) = 0\,. \tag{1.5.5}$$

Since $V \geq 0$ we get $\sqrt{E}/\pi - \langle V \rangle/\pi\sqrt{E} \leq \theta_E(L)/4\pi L \leq \sqrt{E}/\pi$, with $\langle V \rangle = (1/2L)\int_{[-L,L]} dx V(x)$. On the other hand, $\theta_E(L)/2\pi = n$ for $E = E_{n-1}$ as one can easily see from the definition, so we get

$$n\pi/2L \leq E_{n-1}^{1/2} \leq n\pi/4L + \{n^2\pi^2/16L^2 + \langle V \rangle\}^{1/2}\,, \tag{1.5.6}$$

showing that the sequence $\{E_n; n \in N\}$ is infinite and unbounded. □

Let us now consider the vector $\Psi_E(x) = (\psi_E(x), \psi'_E(x)/\sqrt{E}) \in \mathbb{R}^2$. Thanks to the uniqueness theorem, it never vanishes for $-L \leq x \leq L$. Thus it defines a unique line $\Delta_E(x)$ in the projective space $\mathbb{PR}(1)$. We will identify $\mathbb{PR}(1)$ with the unit circle through the stereographic projection so that $\Delta_E(x)$ is parametrized by the angle $\theta_E(x) (\mathrm{mod}\, 2\pi)$ defined above. Therefore denoting by η the closed 1-form $\eta = d\theta/2\pi$ on the unit circle, the previous theorem shows that

$$E_{n-1} < E < E_n \; , \qquad n = [\langle \eta | \Delta_E(-L, L) \rangle] \; , \tag{1.5.7}$$

where $[x]$ represents the integer part of x.

Let us now consider the operator point of view. Let H be the self adjoint operator defined by $H = H_0 + V$ where H_0 is defined as $-\partial^2/\partial x^2$ with Dirichlet boundary conditions on $[-L, L]$. By standard arguments, H_0 is a positive operator with a compact resolvent and therefore it has a discrete unbounded spectrum on the real positive axis. Since $V \geq 0$ the same is true for H. E is an eigenvalue of H if and only if the corresponding solutions $\psi_\pm$ both vanish on $\pm L$, in which case they are proportional. Thus the spectrum of H coincides with the family $\{E_n;\, n \in \mathbb{N}\}$ defined previously.

Let now P_E be the eigenprojection $\chi\{H \leq E\}$ of H onto the energies smaller than or equal to E. Then we get the following gap labelling theorem

Theorem 2 (The First Gap Labelling Theorem). *Let H be the self adjoint operator on $L^2(-L, L)$ defined by $H = -\partial^2/\partial x^2 + V$ with Dirichlet boundary conditions. We denote by $0 < E_0 < E_1 < \cdots < E_{n-1} \cdots$ the eigenvalues of H. For $E \in \mathbb{R}$, let P_E be the eigenprojection of H on energies smaller than or equal to E.*

Let also ψ_E be the unique solution of $-\psi_E''(x) + V(x)\psi_E(x) = E\psi_E(x)$ such that $\psi_E(-L) = 0$, $\psi_E'(-L) = 1$. Let $\Delta_E(x)$ be the line defined by the vector $(\psi_E(x), \psi_E'(x)/\sqrt{E})$ in $\mathbb{R}^2$, and let η be the canonical closed 1-form on $\mathbb{PR}(1)$.

Then if $E_{n-1} \leq E < E_n$ we get

$$\mathrm{Tr}(P_E) = [\langle \eta | \Delta_E(-L, L) \rangle] = n \; . \tag{1.5.8}$$

Remark. Here as in the index formula, we get a formula with two sides: the r.h.s. represents the operator, namely the particle point of view, while the l.h.s. represents the wave function point of view.

2. Homogeneous Media

2.1 Breaking the Translation Symmetry

This Section will be devoted to the description of the formalism required to treat more complicated problems arising in Physics, and especially in the physical properties of solids. The main tool in Solid State Physics is the Bloch theory valid for periodic crystals [MA76]. It gives rise to the theory of bands. Much effort has been devoted during the fifties and the sixties to the explicit calculation of bands and Bloch waves in real crystals. Considering the great number of crystal symmetries, one can imagine how difficult it was to exhaust all possible cases. The next class of problems solid state physicists were interested in, was the transport properties of these materials: electric conduction (metals, semiconductors, insulators, superconductors), thermal properties (phonons, heat capacity, diffusion constant). This requires the use of the Green-Kubo theory, which is not yet completely justified because of its conceptual difficulty, but which can be considered as a satisfactory and widely accepted phenomenological theory. Bloch's theory and Green-Kubo's theory are basic tools in dealing with the subject as long as we are not dealing with the many-body problem.

However, most materials are actually aperiodic. First of all, even though the periodic case is an excellent approximation, there is usually quite a lot of defects in real crystals. Moreover, temperature produces migration of atoms in a solid, leading to some randomness in the distribution of forces acting on the electrons. Roughness of interfaces may also produce random forces in $2D$ devices. However, defects and thermal fluctuations, as long as they are small, can be treated as a first order perturbation of band theory.

But aperiodicity may be produced on large scale for physical reasons. The oldest example is probably the effect of a uniform magnetic field on electronic properties of a crystal. It has been the focus of attention since the very early days of Solid State Physics with the works of Landau [LA30] and Peierls [PE33] in the thirties devoted to the electron diamagnetism of metals. As a matter of fact, a uniform magnetic field breaks the translation symmetry of the Bloch waves because it produces a non translation invariant phase factor. On the other hand, the response of a solid to a uniform magnetic field is one of the most useful tools for experiments to get information on the microscopic properties. The reason is that a magnetic field breaks the time reversal symmetry, and will enable to separate various effects. For instance the Hall effect permits to measure the sign and the density of charge carriers in a conductor. Moreover, de Haas-van Alfen oscillations of the magnetoresistance give precise informations on the shape of the Fermi surface. Still the calculation of the electronic energy spectrum in this case required the contribution of hundreds of the best physicists during the last forty years. One of the most spectacular results is probably the calculation by Hofstadter in 1976 [HO76, WI84, SO85] of the energy spectrum for a $2D$ electron in a square crystal submitted to a perpendicular magnetic field: it has a fractal structure which is still under study now by mathematicians [BS82b, EL82b, HS87, GH89, BR90].

Other materials are intrinsically aperiodic. The charge density wave in a one-dimensional chain submitted to a Peierls instability, is modulated at a frequency determined by the Fermi quasimomentum, giving rise to a quasiperiodic effective potential for electrons. Quasicrystals discovered in 1984 [SB84], have their atoms located on a quasiperiodic lattice. Amorphous materials, have their atoms located on a very aperiodic lattice, which nevertheless may be generated by a deterministic geometry [MS83]. All these states of matter are actually stable (or may be strongly metastable).

So there is a need for a mathematical framework liable to describe these situations, and to permit the calculation of physical quantities of interest such as the electronic spectrum, the density of states, thermodynamical quantities (e.g. the heat capacity), the transport coefficients. The main difficulty is that translation invariance is broken, in such a way that there is no Bloch decomposition of the problem.

The main tool we develop here is the description of a Non Commutative Manifold, namely the Brillouin zone of an aperiodic medium, in order to replace Bloch theory for aperiodic media!

2.2 Non-Commutative Geometry

An ordinary locally compact manifold M is usually described as a set with a topology which makes it a locally compact space. In addition, one usually defines a family of charts, namely mappings from open sets in $\mathbb{R}^n$ into M satisfying compatibility and smoothness conditions [BO67a]. It allows to define the space of $\mathcal{C}^\infty$ functions on M with compact support; it is a commutative Frechet *-algebra. The space $\mathcal{C}_0(M)$ of continuous functions vanishing to zero at infinity on M is the completion of the space $\mathcal{C}^\infty(M)$ under the sup-norm. It is a commutative C^*-algebra. Then smooth real vector fields with compact support are nothing but a family of *-derivations of this algebra generating a one-parameter group of point-wise norm continuous *-automorphisms.

Conversely, the manifold structure of M can be recovered from the data of $\mathcal{C}_0^\infty(M)$, namely from $\mathcal{C}_0(M)$ and the family of all smooth vector fields with compact support, the dense subalgebra given by the common domain of all polynomials in these vector fields being nothing but $\mathcal{C}_0^\infty(M)$.

Gelfand's theorem [SA71, BO67b] asserts that every commutative C^*-algebra $\mathcal{A}$ is the space of continuous functions vanishing to zero at infinity on a locally compact space M. In this latter case, M is constructed as the set of characters of $\mathcal{A}$ namely the set of *-homomorphisms from $\mathcal{A}$ to $\mathbb{C}$, and the identification between points x of M and a character χ of A is given through $\chi(f) = f(x)$ for all $f \in \mathcal{A}$.

A non-commutative locally compact space will be defined by analogy with the commutative case as the data of a non commutative C^*-algebra $\mathcal{A}$, which will represent the space of continuous functions vanishing to zero at infinity on a virtual object which will be the non commutative space itself [CO90]. The fact of the matter is that only functions on a non commutative space are defined, not the space itself at this level of generality. The game consists in expressing every geometrical property of an ordinary manifold M as an algebraic property on $\mathcal{C}_0(M)$, and in extending this last property as a definition in the non-commutative case. In this way, M becomes a smooth manifold whenever we have defined on $\mathcal{A}$ a family of *-derivations generating one-parameter groups of point-wise norm continuous *-automorphisms. In much the same way, a vector bundle $\mathcal{E}$ is defined through the data of its smooth sections, which is nothing but a module (in the algebraic sense) over the algebra $\mathcal{A}$. One can then associate to each derivation δ a connection ∇ on this bundle by means of a linear map $\nabla : \mathcal{E} \to \mathcal{E}$ such that $\nabla(f\eta) = \delta(f)\eta + f\nabla(\eta)$ for $f \in \mathcal{A}, \eta \in \mathcal{E}$. Such a connection is not unique: it is defined up to a module homomorphism.

In much the same way one can integrate functions. This can be done in the simplest cases by means of a trace on $\mathcal{A}$, namely a densely defined linear map τ on $\mathcal{A}$ such that $\tau(A^*A) = \tau(AA^*) \geq 0$ for any $A \in \mathcal{A}$. This is a natural generalization of the integral. However, there are C^*-algebras on which no non trivial trace exists, but as far as we will be concerned in this paper, all C^*-algebras we will consider will have a faithful trace. A trace is faithful if for each non zero element $A \in \mathcal{A}, \tau(AA^*) > 0$.

More generally, one can define a differential algebra $\Omega(\mathcal{A})$ as the linear space generated by symbols of the form $A_0 dA_1 \cdots dA_n, A_i \in \mathcal{A}$. This algebra becomes a *-algebra if we impose the relations $d(AB) = (dA)B + A(dB)$ and $d(A^*) = (dA)^*$. Moreover it is graded over $\mathbb{Z}$ if we define the degree of $A_0 dA_1 \cdots dA_n$ as n. The differential d is extended to $\Omega(\mathcal{A})$ by linearity and by imposing $d^2 = 0$. Then, a closed current is a trace on this graded algebra, namely a linear map $\tau : \Omega(\mathcal{A}) \to \mathbb{C}$ such that

$$\tau(\eta\eta') = (-1)^{\deg(\eta)\deg(\eta')}\tau(\eta'\eta), \quad \tau(d\eta) = 0, \qquad \text{for all } \eta, \eta' \in \Omega(\mathcal{A}) .$$

The definition and the study of cyclic and de Rham cohomology defined on closed currents in this non commutative context, have been the subject of A. Connes's work and we will refer the reader to [CO82, CO83] for a complete description.

2.3 Periodic Media: Bloch Theory and the Brillouin Zone

Let us consider first a Schrödinger operator on $L^2(\mathbb{R}^n)$ of the form

$$H = P^2/2m + V ,$$

where V is a continuous and periodic function on $\mathbb{R}^n$ with a lattice of periods Γ. It means that H commutes with the operators $T(a)$ of translation by $a \in \Gamma$. Γ is a Bravais lattice namely a discrete subgroup of $\mathbb{R}^n$ such that the linear space it generates is equal to $\mathbb{R}^n$. We will set $\mathbf{V} = \mathbb{R}^n/\Gamma$; $\mathbf{V}$ can be represented by means of the unit cell of the lattice, which is a fundamental domain for Γ (the Voronoi cell).

The Bloch theory consists in diagonalizing simultaneously H and the $T(a)$'s. To do so, we introduce the reciprocal lattice Γ^* as the set of b's in $\mathbb{R}^n$ such that the scalar product $\langle b|a\rangle$ is an integer multiple of 2π for any $a \in \Gamma$. The Brillouin zone will be defined as the quotient $\mathbf{B} = \mathbb{R}^n/\Gamma^*$ and is isomorphic to a n-dimensional torus. Notice that this definition of the Brillouin zone is not exactly the same as the definition given by Solid State physicists [MA76].

Now, the generalized eigenfunctions of the $T(a)$'s are Bloch waves, namely functions $\psi_k(x)$ on $\mathbb{R}^n \times \mathbb{R}^n$ such that

$$\psi_{k+b}(x + a) = e^{i\langle k|a\rangle}\psi_k(x) . \tag{2.3.1}$$

Here k is defined modulo Γ^* namely, $k \to \psi_k(x)$ can be seen as a function over $\mathbf{B}$. Moreover, for every pair $\psi_k(x), \phi_k(x)$ of Bloch waves, the map $x \in \mathbb{R}^n \to \psi_k(x)^* \phi_k(x)$ is Γ-periodic and can therefore be seen as a function over $\mathbf{V}$.

In order to get a mathematically clean picture we introduce the Hilbert space $\mathcal{W}$ built as the space of ψ's satisfying (2.3.1), and such that

(2.3.2) $$\|\psi\|^2 = \int_{\mathbf{V}\times\mathbf{B}} d^n x d^n k |\psi_k(x)|^2 < \infty$$

$\mathcal{W}$ is actually isomorphic to $\mathcal{H} = L^2(\mathbb{R}^n)$ and a unitary transformation from $\mathcal{H}$ to $\mathcal{W}$ is defined by the following 'Wannier transform':

(2.3.3a) $$Wf(x,k) = \sum_{a \in \Gamma} f(x-a) e^{\mathrm{i}\langle k|a\rangle}, \quad f \in \mathcal{H},$$

(2.3.3b) $$W^*\psi(x) = (1/|\mathbf{B}|) \int_{\mathbf{B}} d^n k \psi_k(x), \quad \psi \in \mathcal{W},$$

which satisfies $WW^* = \mathbf{1}_{\mathcal{W}}, W^*W = \mathbf{1}_{\mathcal{H}}$. In particular the norms are conserved.

We then introduce for each $k \in \mathbf{B}$ the Hilbert space $\mathcal{H}_k$ of functions u on $\mathbb{R}^n$, such that

(2.3.4) $$u(x+a) = e^{\mathrm{i}\langle k|a\rangle} u(x), \qquad \|u\|^2 = \int_{\mathbf{V}} d^n x |u(x)|^2 < \infty.$$

Then an element ψ of $\mathcal{W}$ defines a square integrable section $k \to \psi_k \in \mathcal{H}_k$ and conversely, each such section defines an element of $\mathcal{W}$, namely $\mathcal{W}$ can be identified with the direct integral [DI69]

(2.3.5) $$\mathcal{W} = \int_{\mathbf{B}}^{\oplus} d^n k \mathcal{H}_k.$$

Then both $WT(a)W^*$ and WHW^* leave each fiber $\mathcal{H}_k$ invariant. $WT(a)W^*$ is nothing but the operator of multiplication by $e^{\mathrm{i}\langle k|a\rangle}$. Hence the C^*-algebra the $T(a)$'s generate, is nothing but the algebra $\mathcal{C}(\mathbf{B})$ of continuous functions over $\mathbf{B}$ acting on $\mathcal{W}$ by multiplication.

On the other hand, it is standard to check that the self-adjoint operator H is transformed by W into the family of partial differential operators

(2.3.6) $$H_k = -(\hbar^2/2m) \sum_{i \in [1,n]} \partial^2/\partial x_i^2 + V(x),$$

with domain $\mathcal{D}(H_k)$ given by the space of elements u's in $\mathcal{H}_k$ such that $\partial^2 u/\partial x_i^2 \in \mathcal{H}_k$. It is a standard result [RSII] that the H_k's have compact resolvents: this is because for $V=0$, the spectrum of $\mathcal{H}_k$ is the discrete set of eigenvalues $\{(\hbar^2/2m)(k+b)^2; b \in \Gamma^*\}$, showing that its resolvent is indeed compact, whereas V is a bounded operator.

We can then identify $\mathcal{H}_k$ with $l^2(\Gamma^*)$ by means of the choice of the orthonormal basis $\{u_b(k); b \in \Gamma^*\}$, where

$$u_b(k) : x \in \mathbb{R}^n \to e^{i\langle k+b|x\rangle} \in \mathbb{C}\,, \tag{2.3.7}$$

whereas the resolvent $(z\mathbf{1}-H_k)^{-1}$ becomes a compact operator $R_k(z)$ on $l^2(\Gamma^*)$ the matrix of which being $R_k(z)_{b,b'} = \langle u_b(k)|(z\mathbf{1}-H_k)^{-1}u_{b'}(k)\rangle$. If $L(b)$ is the translation by $b \in \Gamma^*$ in $l^2(\Gamma^*)$, we actually get

$$L(b)R_k(z)L(b)^{-1} = R_{k+b}(z)\,, \tag{2.3.8}$$

and the map $k \in \mathbb{R}^n \to R_k(z) \in \mathcal{K}(l^2(\Gamma^*))$ is norm continuous (here $\mathcal{K}(l^2(\Gamma^*))$ denotes the C^*-algebra of compact operators on $l^2(\Gamma^*)$). In other words, $R(z)$ defines a continuous map from $\mathbb{R}^n$ to the C^*-algebra $\mathcal{K}$ of compact operators, satisfying the covariance condition (2.3.8). Here we must notice that given two Hilbert spaces with a countable basis $\mathcal{H}$ and $\mathcal{H}'$, the C^*-algebras $\mathcal{K}(\mathcal{H})$ and $\mathcal{K}(\mathcal{H}')$ are actually isomorphic, and this allows us to denote them by $\mathcal{K}$ without referring to the Hilbert spaces on which they act.

A better representation will be given by the data of a family $\{\psi_j \in \mathcal{W}; j \in \mathbb{N}^n\}$, such that $\psi_j(k) : k \in \mathbf{B} \to \psi_j(.;k) \in L^2_{\text{loc}}(\mathbb{R}^n)$ is continuous, and

$$\int_V d^n x \psi_j(x,k)^* \psi_{j'}(x,k) = \delta_{j,j'}\,. \tag{2.3.9}$$

It is tedious but easy to prove that such a family exists and can be constructed in such a way as to be $\mathcal{C}^\infty$ with respect to k. Clearly, $\{\psi_j(k); j \in \mathbb{N}^n\}$ gives an orthonormal basis of $\mathcal{H}_k$.

Let $b \in \Gamma^* \to n(b) \in \mathbb{N}^n$ be a bijection. We then denote by $S(k)$ the unitary operator from $l^2(\Gamma^*)$ into $l^2(\mathbb{N}^n)$ defined by

$$S(k) = \sum_b |\psi_{n(b)}(k)\rangle\langle u_b(k)|\,. \tag{2.3.10}$$

It satisfies $S(k)L(b) = S(k-b)$ for $b \in \Gamma^*$. Then $R'_k(z) = S(k)R_k(z)S(k)^*$ is norm continuous and is Γ^*-periodic in k, and defines therefore a continuous mapping from $\mathbf{B}$ into $\mathcal{K}$ namely an element of the C^*-algebra $\mathcal{C}(\mathbf{B}) \otimes \mathcal{K}$.

The main result of this Section is the following

Theorem 3. *The C^*-algebra generated by the family $\{T(x)R(z)T(x)^{-1}; x \in \mathbb{R}^n\}$ of translated of the resolvent of H is isomorphic to $\mathcal{C}(\mathbf{B}) \otimes \mathcal{K}$.*

This result is actually the intuitive key to understand the point of view developed later on. For indeed, the algebra $\mathcal{C}(\mathbf{B}) \otimes \mathcal{K}$ can be identified with the algebra of the Brillouin zone.

Even though $\mathcal{C}(\mathbf{B}) \otimes \mathcal{K}$ is already non commutative, its non commutative part comes from $\mathcal{K}$ which represents possible degeneracies. More precisely, given a complex vector bundle E over $\mathbf{B}$, we will denote by $\mathcal{E}$ the space of continuous sections of E: $\mathcal{E}$ is then a module over $\mathcal{C}(\mathbf{B})$ for the point-wise multiplication.

However, each section can be seen locally as a continuous map from $\mathbf{B}$ into $\mathbb{C}^N$ for some N, and therefore we can multiply it point-wise with a $N \times N$-matrix valued continuous function over $\mathbf{B}$, namely by an element of $\mathcal{C}(\mathbf{B}) \otimes M_N(\mathbb{C})$ which is canonically imbedded in $\mathcal{C}(\mathbf{B}) \otimes \mathcal{K}$. Such an operator is a module homomorphism of $\mathcal{E}$, and we can see that each module homomorphism is of this type. Moreover, the Swann-Serre theorem [RI82] asserts that for every vector bundle E there is N big enough and a projection P in $\mathcal{C}(\mathbf{B}) \otimes M_N(\mathbb{C})$, such that its module of sections $\mathcal{E}$ is isomorphic to the module $P\mathcal{C}(\mathbf{B}) \otimes \mathbb{C}^N = \{f : k \in \mathbf{B} \to C^N; Pf = f\}$; conversely, each projection in $\mathcal{C}(\mathbf{B}) \otimes \mathcal{K}$ gives rise to such a vector bundle. Hence $\mathcal{C}(\mathbf{B}) \otimes \mathcal{K}$ contains not only the construction of $\mathbf{B}$ itself but also that of every vector bundle over $\mathbf{B}$.

This way of reasoning applies as well to every locally compact space M provided $\mathcal{C}_0(M) \otimes \mathcal{K}$ replaces $\mathcal{C}(\mathbf{B}) \otimes \mathcal{K}$. $\mathcal{C}_0(M) \otimes \mathcal{K}$ will be called the algebra of M; in particular, $\mathcal{K}$ represents the algebra of a point.

We say that a C^*-algebra $\mathcal{A}$ is stable whenever it is isomorphic to $\mathcal{A} \otimes \mathcal{K}$. It is a standard result that $\mathcal{K}$ is stable [BL86]. Therefore for any C^*-algebra $\mathcal{A}$, $\mathcal{A} \otimes \mathcal{K}$ is always stable, and is called the stabilized of $\mathcal{A}$. Two C^*-algebras $\mathcal{A}$ and $\mathcal{B}$ are stably isomorphic whenever $\mathcal{A} \otimes \mathcal{K}$ is isomorphic to $\mathcal{B} \otimes \mathcal{K}$.

The Theorem 3 shows that if H is a periodic Schrödinger operator, the algebra of the Brillouin zone is nothing but the algebra $C^*(H)$ generated by the family of all translated of the resolvent of the Hamiltonian H. But the construction of $C^*(H)$ does not require the periodicity of the potential V, and can be done for an aperiodic potential as well. Therefore, $C^*(H)$ will be called the 'Non Commutative Brillouin zone' or the 'Brillouin zone' of H.

2.4 Homogeneous Schrödinger Operators [BE86]

We see that two ingredients are necessary to define a Non-Commutative Brillouin zone, namely the energy operator and the action of the translation group on H. Actually, this is not so surprising if we deal with a homogeneous medium. For indeed, a homogeneous medium with infinite volume looks translation invariant at a macroscopic scale, even though translation invariance may be microscopically broken. For this reason, there is no natural choice of an origin in space. In particular if H is a Hamiltonian describing one particle in this medium, we can choose to replace it by any of its translated $H_a = T(a)HT(a)^{-1}$ $(a \in \mathbb{R}^n)$ and the physics will be the same. This choice is entirely arbitrary, so that the smallest possible set of observables must contain at least the full family $\{H_a; a \in \mathbb{R}^n\}$.

Actually H is not a bounded operator in general, so that calculations are made easier if we consider its resolvent instead. On the other hand as has already been argued in Section 1.3, the set of observables must be a C^*-algebra. So we must deal with the algebra $C^*(H)$ anyway. Let us define precisely what we mean by 'homogeneity' of the medium described by H.

Definition. Let $\mathcal{H}$ be a Hilbert space with a countable basis. Let $\mathbf{G}$ be a locally compact group (for instance $\mathbb{R}^n$ or $\mathbb{Z}^n$). Let U be a unitary projective representation of $\mathbf{G}$ namely for each $a \in \mathbf{G}$, there is a unitary operator $U(a)$ acting on $\mathcal{H}$ such that the family $U = \{U(a); a \in \mathbf{G}\}$ satisfies the following properties:

(i) $U(a)U(b) = U(a+b)e^{i\phi(a,b)}$ for all $a, b \in \mathbf{G}$, where $\phi(a,b)$ is some phase factor.

(ii) For each $\psi \in \mathcal{H}$, the map $a \in \mathbf{G} \to U(a)\psi \in \mathcal{H}$ is continuous.

Then a self adjoint operator H on $\mathcal{H}$ is *homogeneous* with respect to $\mathbf{G}$ if the family $S = \{R_a(z) = U(a)(z\mathbf{1} - H)^{-1}U(a)^{-1}; a \in \mathbf{G}\}$ admits a compact strong closure.

Remark. A sequence A_n of bounded operators on $\mathcal{H}$ converges strongly to the bounded operator A if for every $\psi \in \mathcal{H}$, the sequence $\{A_n\psi\}$ of vectors in $\mathcal{H}$ converges in norm to $A\psi$. So the set S has a strong compact closure if given $\varepsilon > 0$, and a finite set $\psi_1, \cdots, \psi_N$ of vectors in $\mathcal{H}$, there is a finite set $a_1, \cdots, a_n$ in $\mathbf{G}$ such that for every a in $\mathbf{G}$ and every $1 \leq j \leq N$ there is $1 \leq i \leq n$ such that $\|(R_a(z) - R_{a_i}(z))\psi_j\| \leq \varepsilon$. In other words, the full family of translated of $R(z)$ being well approximated on vectors by a finite number of them, repeats itself infinitely many times up to infinity.

The virtue of this definition comes from the construction of the 'hull'. Indeed let z be in the resolvent set $\rho(H)$ of H, and let H be homogeneous. Then let $\Omega(z)$ be the strong closure of the family $\{R_a(z) = U(a)(z\mathbf{1} - H)^{-1}U(a)^{-1}; a \in \mathbf{G}\}$. It is therefore a compact space, which is metrizable since the Hilbert space $\mathcal{H}$ has a countable basis. Moreover, it is endowed with an action of the group $\mathbf{G}$ by means of the representation U. This action defines a group of homeomorphisms of $\Omega(z)$.

We first remark that if z' is another point in $\rho(H)$ the spaces $\Omega(z)$ and $\Omega(z')$ are actually homeomorphic. For indeed, we have for $a \in \mathbf{G}$

$$\begin{aligned} R_a(z') &= \{\mathbf{1} + (z' - z)R_a(z)\}^{-1} R_a(z)\,, \\ \{\mathbf{1} + (z' - z)R_a(z)\}^{-1} &= \{\mathbf{1} - (z' - z)R_a(z')\}\,, \end{aligned} \tag{2.4.1}$$

so that $\{R_{a_i}(z); i \geq 0\}$ converges strongly to R if and only if $\{R_{a_i}(z'); i \geq 0\}$ converges strongly to some R' and the map $R \in \Omega(z) \to R' \in \Omega(z')$ is an homeomorphism. Identifying them gives rise to an abstract compact metrizable space Ω endowed with an action of $\mathbf{G}$ by a group of homeomorphisms. If $\omega \in \Omega$ and $a \in \mathbf{G}$ we will denote by $T^a.\omega$ the result of the action of a on ω, and by $R_\omega(z)$ the representative of ω in $\Omega(z)$. Then one gets

$$U(a)R_\omega(z)U(a)^* = R_{T^a.\omega}(z)\,, \tag{2.4.2}$$

$$R_\omega(z') - R_\omega(z) = (z - z')R_\omega(z')R_\omega(z) = (z - z')R_\omega(z)R_\omega(z')\,. \tag{2.4.3}$$

In addition, $z \to R_\omega(z)$ is norm-holomorphic in $\rho(H)$ for every $\omega \in \Omega$, and $\omega \to R_\omega(z)$ is strongly continuous.

Definition. Let H be a homogeneous operator on the Hilbert space $\mathcal{H}$ with respect to the representation U of the locally compact group $\mathbf{G}$. Then the *hull* of H is the dynamical system $(\Omega, \mathbf{G}, U)$ where Ω is the compact space given by the strong closure of the family $\{R_a(z) = U(a)(z\mathbf{1} - H)^{-1}U(a)^{-1}; a \in \mathbf{G}\}$, and $\mathbf{G}$ acts on Ω through U.

The equation (2.4.3) is not sufficient in general to insure that $R_\omega(z)$ is the resolvent of some self-adjoint operator H_ω, for indeed one may have $R_\omega(z) = 0$ if no additional assumption is demanded. A sufficient condition is that H be given by $H_0 + V$ where H_0 is self-adjoint and $\mathbf{G}$-invariant, whereas V is relatively bounded with respect to H_0, i.e. $\|(z - H_0)^{-1}V\| < \infty$, and $\lim_{|z|\to\infty}\|(z - H_0)^{-1}V\| = 0$. Then,

$$R_\omega(z) = \{\mathbf{1} - (z - H_0)^{-1}V_\omega\}^{-1}(z - H_0)^{-1}$$

where $(z - H_0)^{-1}V_\omega$ is defined as the strong limit of $(z - H_0)^{-1}V_{a_i}$, which obviously exists. So $R_\omega(z)$ is the resolvent of $H_0 + V_\omega$.

In the case of Schrödinger operator, the situation becomes simpler. Let us consider the case of a particle in $\mathbb{R}^n$ submitted to a bounded potential V and a uniform magnetic field with vector potential A, such that

$$\partial_\mu A_\nu - \partial_\nu A_\mu = B_{\mu,\nu} = \text{const.} \tag{2.4.4}$$

The Schrödinger operator is given by

$$H = (1/2m) \sum_{\mu\in[1,n]} (P_\mu - eA_\mu)^2 + V = H_0 + V\,. \tag{2.4.5}$$

The unperturbed part H_0 is actually translation invariant provided one uses magnetic translations [ZA64] defined by (if $a \in \mathbb{R}^n$, $\psi \in L^2(\mathbb{R}^n)$)

$$U(a)\psi(x) = \exp\left\{(\mathrm{i}e/\hbar)\int_{[x-a,x]} dx'^\mu A_\mu(x')\right\}\psi(x-a)\,. \tag{2.4.6}$$

It is easy to check that the $U(a)$'s give a projective representation of the translation group. The main result in this case is given by

Theorem 4. *Let H be given by* (2.4.4) *and* (2.4.5) *with V a measurable essentially bounded function over $\mathbb{R}^n$. Let $\mathcal{B}_s\{L^2(\mathbb{R}^n)\}$ represent the space of bounded linear operators on $L^2(\mathbb{R}^n)$ with the strong topology, let $L^\infty_{\mathbb{R}}(\mathbb{R}^n)$ be the space of measurable essentially bounded real functions over $\mathbb{R}^n$ with the weak topology of $L^1(\mathbb{R}^n)$, and let z be a complex number with non zero imaginary part. Then, the map $V \in L^\infty_{\mathbb{R}}(\mathbb{R}^n) \to \{z - H_0 - V\}^{-1} \in \mathcal{B}_s\{L^2(\mathbb{R}^n)\}$ is continuous.*

The proof of this theorem can be found in [NB90, Appendix]. As a consequence we get

Corollary 2.4.1. *Let H be given by* (2.4.4) *and* (2.4.5) *with V a real, measurable, essentially bounded function over $\mathbb{R}^n$. Then H is homogeneous with respect to the representation U (Eq. (2.4.6)) of the translation group.*

Proof. Indeed, any ball in $L^\infty_{\mathbb{R}}(\mathbb{R}^n)$ is compact for the weak topology of $L^1(\mathbb{R}^n)$. Moreover, $U(a)VU(a)^* = V_a$ where $V_a(x) = V(x-a)$ almost surely. So V_a belongs to the ball $\{V' \in L^\infty_{\mathbb{R}}(\mathbb{R}^n); \|V'\| \le \|V\|\}$, and the weak closure of the family $\{V_a; a \in \mathbb{R}^n\}$ in $L^\infty_{\mathbb{R}}(\mathbb{R}^n)$ is compact. Thanks to Theorem 4, it follows that the strong closure of the family $\{(z - H_0 - V_a)^{-1}; a \in \mathbb{R}^n\}$ is the direct image of a compact set by a continuous function, and is therefore compact. □

Another consequence of this result is given by the following characterization of the hull

Corollary 2.4.2. *Let H be as in Corollary 2.4.1. Then the hull of H is homeomorphic to the hull of V namely the weak closure of the family $\Omega = \{V_a; a \in \mathbb{R}^n\}$ in $L^\infty_{\mathbb{R}}(\mathbb{R}^n)$. Moreover, there is a Borelian function v on Ω such that $V_\omega(x) = v(T^{-x}\omega)$ for almost every $x \in \mathbb{R}^n$ and all $\omega \in \Omega$. If in addition V is uniformly continuous and bounded, v is continuous.*

Proof. Let ρ_k be non negative functions on $\mathbb{R}^n$ such that $\int_{\mathbb{R}^n} d^n x \rho_k(x) = 1$ and that for each $\delta > 0$ $\lim_{k\to\infty} \int_{|x|>\delta} d^n x \rho_k(x) = 0$. V_ω is an element of $L^\infty_{\mathbb{R}}(\mathbb{R}^n)$. Let $v_k(\omega)$ be defined by $v_k(\omega) = \int_{\mathbb{R}^n} d^n x V_\omega(x)\rho_k(x)$. By definition of the hull, this is a sequence of continuous functions on Ω uniformly bounded by $\|V\|_\infty$. We set $v(\omega) = \lim_{k\to\infty} v_k(\omega)$ if the limit exists. This is a Borelian function because if I is a closed interval in $\mathbb{R}$, the set $\sum(I) = \{\omega \in \Omega; v(\omega) \in I\}$ is given as $\sum(I) = \cap_{n\ge1} \cup_{k\ge1} \cap_{p\ge k} \{\omega \in \Omega;\ v_p(\omega) \in I^{(1/n)}\}$, where $I^{(\varepsilon)}$ is the set of points in $\mathbb{R}$ within the distance ε from I. Since v_k is continuous, $\sum(I)$ is a Borel set. If now $F \in L^1(\mathbb{R}^n)$, one gets $\int_{\mathbb{R}^n} d^n x v(T^{-x}\omega)F(x) = \lim_{k\to\infty} \int_{\mathbb{R}^{2n}} d^n x d^n y V_{T^{-x}\omega}(y)\rho_k(y)F(x)$, which by the covariance property is nothing but $\lim_{k\to\infty} \int_{\mathbb{R}^{2n}} d^n x d^n y V_\omega(x)\rho_k(y)F(x-y)$. Since the convolution $\lim_{k\to\infty} \int_{\mathbb{R}^n} d^n y \rho_k(y)F(x-y)$ converges to F in $L^1(\mathbb{R}^n)$, it follows that $v(T^{-x}\omega) = V_\omega(x)$ for almost all x's, and all $\omega \in \Omega$.

Let V be uniformly continuous and bounded on $\mathbb{R}^n$. Then V_ω exists as an element of $L^\infty_{\mathbb{R}}(\mathbb{R}^n)$. We claim that $(v_k)_{k\ge0}$ is a Cauchy sequence for the uniform topology. For indeed, by definition of the hull, for each $\omega \in \Omega$, there is a sequence $\{a_i\}$ in $\mathbb{R}^n$, such that $\int_{\mathbb{R}^n} d^n x V_\omega(x)F(x) = \lim_{i\to\infty} \int_{\mathbb{R}^n} d^n x V(x-a_i)F(x)$, for every $F \in L^1(\mathbb{R}^n)$. In particular, we get $|v_k(\omega) - v_{k'}(\omega)| \le \lim_{i\to\infty} \int_{\mathbb{R}^2} d^n x d^n y |V(x-a_i) - V(y-a_i)|\rho_k(x)\rho_{k'}(y)$. Since V is uniformly continuous, given $\varepsilon > 0$, there is $\delta > 0$ such that if $|x-y| < \delta \Rightarrow |V(x) - V(y)| < \varepsilon/2$. Therefore, using $|x-y| > \delta \Rightarrow |x| > \delta/2$ or $|y| > \delta/2$, we get whenever $k < k'$

$$\int_{\mathbb{R}^{2n}} d^n x d^n y |V(x-a_i) - V(y-a_i)|\rho_k(x)\rho_{k'}(y)$$
$$\le \varepsilon/2 + 2\|V\|_\infty \left(\int_{|x|>\delta/2} d^n x \rho_k(x) + \int_{|x|>\delta/2} d^n x \rho_{k'}(x) \right).$$

Choosing N big enough, for $k, k' > N$ the right hand side is dominated by ε, proving the claim. Therefore, the sequence v_k converges uniformly to a continuous function v. □

The situation is very similar and actually technically simpler for a discrete Schrödinger operator on a lattice. Let us consider a Bravais lattice, that will be identified with $\mathbb{Z}^n$, and let us consider the Hilbert space $l^2(\mathbb{Z}^n)$ of square summable sequences indexed by n-uples $x = \{x_1, \cdots, x_n\} \in \mathbb{Z}^n$. Let us consider the operator H acting on $l^2(\mathbb{Z}^n)$ as follows

$$H\psi(x) = \sum_{x' \in \mathbb{Z}^n} H(x,x')\psi(x') , \qquad \psi \in l^2(\mathbb{Z}^n) , \tag{2.4.7}$$

where the sequence $\{H(x,x'); x' \in \mathbb{Z}^n\}$ satisfies $|H(x,x')| \leq f(x-x')$ with $\sum_{a\in\mathbb{Z}^n} f(a) < \infty$. Therefore H is bounded, and there is no need to consider resolvents anymore. Then we get the following theorem, where we let now U be the unitary representation of the translation group $\mathbb{Z}^n$ given by

$$U(a)\psi(x) = \psi(x-a) , \qquad \psi \in l^2(\in \mathbb{Z}^n) . \tag{2.4.8}$$

Theorem 5. *Let H be given by (2.4.7). Then H is homogeneous with respect to the representation U (Eq. (2.4.8)) of the translation group.*

Moreover, if Ω is the hull of H, there is a continuous function vanishing at infinity h on $\Omega \times \mathbb{Z}^n$, such that $H_\omega(x,x') = h(T^{-x}\omega, x'-x)$, for every pair $(x,x') \in \mathbb{Z}^n$ and $\omega \in \Omega$.

Proof. Let $D(a)$ be the disc in the complex plane centered at zero with radius $f(a)$. Let Ω_0 be the product space $\prod_{(x,x')\in\mathbb{Z}^n\times\mathbb{Z}^n} D(x-x')$. By Tychonov's theorem, it is a compact space for the product topology. The sequence $\underline{H}(a) = \{H(x-a,x'-a); (x,x') \in \mathbb{Z}^n \times \mathbb{Z}^n\}$ belongs to Ω_0 for every $a \in \mathbb{Z}^n$. Therefore the closure Ω of the family $\{\underline{H}(a); a \in \mathbb{Z}^n\}$ is compact. For $\omega \in \Omega$ let $H_\omega(x,x')$ be the (x,x') component of ω. For the product topology the projections on components being continuous, the map $\omega \in \Omega \to H_\omega(x,x') \in \mathbb{C}$ is continuous for each pair $(x,x') \in \mathbb{Z}^n \times \mathbb{Z}^n$. Moreover we easily get $|H_\omega(x,x')| \leq f(x-x')$ for all pairs $(x,x') \in \mathbb{Z}^n \times \mathbb{Z}^n$. Let now H_ω be the operator defined by (2.4.7) with $\underline{H}$ replaced by $\underline{H}_\omega$. Its norm is bounded by $\|H_\omega\| \leq \sum_{a\in\mathbb{Z}^n} f(a)$, and the map $\omega \in \Omega \to H_\omega$ is weakly continuous, and therefore strongly continuous. In particular, the family $\{H_\omega; \omega \in \Omega\}$ is strongly compact. We also remark that if ω is the limit point of the sequence $\underline{H}(a_i)$ in Ω_0 then H_ω is the strong limit of the sequence $U(a_i)HU(a_i)^*$. At last, let T^a be the action of the translation by a in Ω_0 defined by $T^a\underline{H}' = \{H'(x-a,x'-a); (x,x') \in \mathbb{Z}^n \times \mathbb{Z}^n\}$ whenever $\underline{H}' \in \Omega_0$ then one has

$$U(a)H_\omega U(a)^* = H_{T^a_\omega} \qquad a \in \mathbb{Z}^n, \omega \in \Omega . \tag{2.4.9}$$

So that H is indeed homogeneous, and its hull is precisely Ω. Moreover, (2.4.9) implies that $H_\omega(x-a,x'-a) = H_{T^a_\omega}(x,x')$, namely, $H_\omega(x,x') = H_{T^{-x}\omega}(0,x'-x)$. Thus $h(\omega,x) = H_\omega(0,x)$. □

Some examples of Hamiltonians for which the hull can be explicitly computed will be especially studied in Section 3.

2.5 The Non-Commutative Brillouin Zone

Let us consider now a topological compact space Ω with a $\mathbb{R}^n$-action by a group $\{T^a; a \in \mathbb{R}^n\}$ of homeomorphisms. Given a uniform magnetic field $B = \{B_{\mu,\nu}\}$ we can associate to this dynamical system a C^*-algebra $C^*(\Omega \times \mathbb{R}^n, B)$ defined as follows. We first consider the topological vector space $\mathcal{C}_K(\Omega \times \mathbb{R}^n)$ of continuous functions with compact support on $\Omega \times \mathbb{R}^n$. It is endowed with the following structure of *-algebra

$$\begin{aligned} fg(\omega, x) &= \int_{\mathbb{R}^n} d^n y f(\omega, y) g(T^{-y}\omega, x-y) e^{i\pi(e/h)B.x\wedge y} , \\ f^*(\omega, x) &= f(T^{-x}\omega, -x)^* , \end{aligned} \tag{2.5.1}$$

where $B.x\wedge y = B_{\mu,\nu} x_\mu y_\nu$, $f, g \in \mathcal{C}_K(\Omega \times \mathbb{R}^n)$, and $\omega \in \Omega$, $x \in \mathbb{R}^n$. Here e is the electric charge of the particle and $h = 2\pi\hbar$ is Planck's constant. Let us remark here that this construction is very similar to the one given for the algebra of a groupoid (see Section 1.3). Here the groupoid is the set $\Gamma = \Omega \times \mathbb{R}^n$, its basis is $\Gamma^{(0)} = \Omega$, and the laws are given by: $r(\omega, x) = \omega, s(\omega, x) = T^{-x}\omega, (\omega, x) = (\omega, y) \circ (T^{-y}\omega, x-y), (\omega, x)^{-1} = (T^{-x}\omega, -x)$.

This *-algebra is represented on $L^2(\mathbb{R}^n)$ by the family of representations $\{\pi_\omega; \omega \in \Omega\}$ given by

$$\pi_\omega(f)\psi(x) = \int_{\mathbb{R}^n} d^n y f(T^{-x}\omega, y-x) e^{i\pi(e/h)B.y\wedge x} \psi(y) , \quad \psi \in L^2(\mathbb{R}^n) \tag{2.5.2}$$

namely, π_ω is linear and $\pi_\omega(fg) = \pi_\omega(f)\pi_\omega(g)$ and $\pi_\omega(f)^* = \pi_\omega(f^*)$. In addition, $\pi_\omega(f)$ is a bounded operator, for $\|\pi_\omega(f)\| \leq \|f\|_{\infty,1}$ where

$$\|f\|_{\infty,1} = \mathrm{Max}\{\sup_{\omega\in\Omega} \int_{\mathbb{R}^n} d^n y |f(\omega, y)|, \sup_{\omega\in\Omega} \int_{\mathbb{R}^n} d^n y |f^*(\omega, y)|\} . \tag{2.5.3}$$

It is a norm such that $\|fg\|_{\infty,1} \leq \|f\|_{\infty,1}, \|f\|_{\infty,1} = \|f^*\|_{\infty,1}$, and we will denote by $L^{\infty,1}(\Omega \times \mathbb{R}^n; B)$ the completion of $\mathcal{C}_K(\Omega \times \mathbb{R}^n)$ under this norm. We then remark that these representations are related by the covariance condition:

$$U(a)\pi_\omega(f)U(a)^{-1} = \pi_{T^a\omega}(f) . \tag{2.5.4}$$

Now we set

$$\|f\| = \sup_{\omega\in\Omega} \|\pi_\omega(f)\| , \tag{2.5.5}$$

which defines a C^*-norm on $L^{\infty,1}(\Omega \times \mathbb{R}^n; B)$, and permits to define $C^*(\Omega \times \mathbb{R}^n, B)$ as the completion of $\mathcal{C}_K(\Omega \times \mathbb{R}^n)$ or of $L^{\infty,1}(\Omega \times \mathbb{R}^n; B)$ under this norm.

The main result of this Section is summarized in the:

Theorem 6. *Let A_μ be the vector potential of a uniform magnetic field B, and let V be in $L^\infty(\mathbb{R}^n)$. Let H be the operator*

$$H = (1/2m) \sum_{\mu \in [1,n]} (P_\mu - eA_\mu)^2 + V = H_B + V$$

and we denote by Ω its hull. Then for each z in the resolvent set of H, and for every $x \in \mathbb{R}^n$ there is an element $r(z;x) \in C^*(\Omega \times \mathbb{R}^n, B)$, such that for each $\omega \in \Omega, \pi_\omega(r(z;x)) = (z - H_{T^{-x}\omega})^{-1}$.

Proof. Since the magnetic field B is uniform, it defines a real antisymmetric $N \times N$ matrix. By Cartan's theorem, there is an orthonormal basis in $\mathbb{R}^n\{e_1, e_2, \cdots, e_{2l-1}, e_{2l}, e_{2l+1}, \cdots, e_n\}$ and positive real numbers $B_j (1 \leq j \leq l \leq n/2)$ such that

$$\begin{aligned} Be_{2j-1} &= B_j e_{2j} \,, \\ Be_{2j} &= -B_j e_{2j-1} \,. \end{aligned} \quad \text{if } 1 \leq j \leq l, \qquad Be_m = 0, \quad \text{if } m \geq 2l+1 \,, \tag{2.5.6}$$

Let A_μ be the vector potential in the symmetric gauge, namely, $A_\mu = -(1/2)B_{\mu\nu}x_\nu$. Then, the operators $K_\mu = -\mathbf{i}\partial_\mu - 2\pi e/hA_\mu$ obey the following commutation relations

$$\begin{aligned} &[K_{2j-1}, K_{2j}] = \mathbf{i}2\pi e/hB_j \,, \qquad 1 \leq j \leq l \,, \\ &[K_\mu, K_\nu] = 0 \text{ otherwise} \,. \end{aligned} \tag{2.5.7}$$

Therefore at $V = 0$, the Schrödinger operator H_B will satisfy

$$\begin{aligned} \exp\{-tH_B\} = &\prod_{j \in [1,l]} \exp\left\{-t(h^2/8\pi m)(K_{2j-1}^2 + K_{2j}^2)\right\} \\ &\prod_{m \in [2j+1,n]} \exp\left\{-t(h^2/8\pi m)K_m^2\right\} \,. \end{aligned} \tag{2.5.8}$$

In particular, it acts on $\psi \in L^2(\mathbb{R}^n)$ through

$$\exp\{-tH_B\}(x) = \int_{\mathbb{R}^n} d^n y f_B(x-y;t) e^{\mathbf{i}\pi(e/h)B.y\wedge x}\psi(y) \tag{2.5.9}$$

where $f_B(x;t)) = \prod_{j \in [1,l]} f_2(\mathbf{x}_j; t; B_j) \prod_{m \in [2j+1,n]} f_1(x_m; t)$, with the following notations

(i) if $x = (x_1, \cdots, x_n)$, then $\mathbf{x}_j = (x_{2j-1}, x_{2j}) \in \mathbb{R}^2$ for $1 \leq j \leq l$,
(ii) $f_1(x;t) = (t.h/2\pi m)^{-1}\exp\{-2\pi m x^2/t.h^2\}$ if $x \in \mathbb{R}$,
(iii) $f_2(\mathbf{x};t;B) = (eB/h)(2\sinh(t.eBh/4\pi m))^{-1}$
$\exp\{-\pi eB\mathbf{x}^2/(2h.\tanh(t.eBh/4\pi m))\}$.

In particular, the kernel $f_B(x;t)$ is smooth and fast decreasing in x. In view of (2.5.2) and (2.5.3), we see that $f_B(t) \in L^{\infty,1}(\Omega \times \mathbb{R}^n; B)$ and that $\pi_\omega(f_B(t)) = \exp\{-tH_B\}$ whatever Ω and $\omega \in \Omega$.

Now by Dyson's expansion, we get

$$e^{-t(H_B+V)} = e^{-tH_B} + \sum_{n=0}^{\infty}(-)^n \int_0^t ds_1 \int_0^{s_1} ds_2 \ldots \int_0^{s_{n-1}} ds_n e^{-(t-s_1)H_B}$$
$$V e^{-(s_1-s_2)H_B} \ldots V e^{-s_n H_B} .$$

Thus it is enough to show that $V(t,s) = \exp\{-tH_B\}V \exp\{-sH_B\}$ has the form $\pi_\omega(g)$ for some $g \in L^{\infty,1}(\Omega\times\mathbb{R}^n; B)$. For indeed the Dyson series converges in norm since V is bounded. Thanks to (2.5.8), we get $V(t,s) = \pi_\omega(v(t,s))$ with

$$\text{(2.5.10)} \qquad v(t,s)(\omega,x) = \int_{\mathbb{R}^n} d^n y f_B(y;t) V_\omega(y) f_B(x-y;s) e^{\mathbf{i}\pi(e/h)B.y\wedge x} .$$

Now we recall that V_ω is the weak limit in $L^\infty(\mathbb{R}^n)$ of a sequence of the form $V(.-a_i)$ as $i\to\infty$. Since f_B is smooth and fast decreasing, it follows that the left hand side of (2.5.10) is continuous in both ω and x, and that its $L^{\infty,1}(\Omega\times\mathbb{R}^n;B)$-norm is estimated by

$$\text{(2.5.11)} \qquad \|v(t,s)\|_{\infty,1} \le \|V\|_\infty \|f_B(t)\|_{\infty,1}\|f_B(s)\|_{\infty,1} \le \text{const.}\,\|V\|_\infty ,$$

uniformly in s,t. Therefore the Dyson expansion converges in $L^{\infty,1}(\Omega\times\mathbb{R}^n;B)$, and $\exp\{-t(H_B+V_\omega)\} = \pi_\omega(f_{B,V}(t))$ for some $f_{B,V}(t) \in L^{\infty,1}(\Omega\times\mathbb{R}^n;B)$. It follows from (2.5.5) that if $V(x) \ge V_-$ almost surely, we have

$$\text{(2.5.12)} \qquad \|f_{B,V}(t)\| \le e^{-tV_-} .$$

In particular, since $(z-H_\omega)^{-1} = \int_{[0,\infty]} dt\exp\{-t(H_B+V_\omega-z)\}$ provided $\mathrm{Re}(z) < V_-$, the theorem is proved. □

Problem. When is the C^*algebra $\mathcal{A}$ generated by the Hamiltonian and the family of its translated, identical to $C^*(\Omega\times\mathbb{R}^n, B)$? In most examples, one can prove the equality. But is there a general condition on the potential implying it?

2.6 Non-Commutative Calculus

In the previous Section we constructed the non commutative Brillouin zone, which is identified with the C^*-algebra $\mathcal{A} = C^*(\Omega\times\mathbb{R}^n, B)$. We now want to show that it is indeed a manifold. In order to do so we will first describe the integration theory and then we will define a differential structure.

We have already seen in Section 2.2 that one can indeed integrate functions by means of a trace on $\mathcal{A}$. Let $\mathbf{P}$ be now a probability measure on Ω, invariant and ergodic under the action of $\mathbb{R}^n$. We associate to $\mathbf{P}$ a normalized trace τ on $\mathcal{A} = C^*(\Omega\times\mathbb{R}^n, B)$ as follows:

$$\text{(2.6.1)} \qquad \tau(f) = \int_\Omega \mathbf{P}(d\omega) f(\omega,0) , \qquad f \in \mathcal{C}_K(\Omega\times\mathbb{R}^n) .$$

One can easily check that this formula defines a trace on the dense subalgebra $\mathcal{C}_K(\Omega \times \mathbb{R}^n)$. If $p \geq 1$, we denote by $L^p(\mathcal{A}, \tau)$ the completion of $\mathcal{C}_K(\Omega \times \mathbb{R}^n)$ under the norm:

$$\|f\|_{L^p} = \tau(\{ff^*\}^{p/2}) . \tag{2.6.2}$$

In particular, one can check that the space $L^2(\mathcal{A}, \tau)$ is a Hilbert space (GNS construction) identical to $L^2(\Omega \times \mathbb{R}^n)$. The map $\phi \in L^2(\mathcal{A}, \tau) \to f\phi \in L^2(\mathcal{A}, \tau)$ defines a representation π_{GNS} of $\mathcal{A}$. The weak closure $\mathcal{W} = \pi_{\mathrm{GNS}}(\mathcal{A})''$, denoted by $L^\infty(\mathcal{A}, \tau)$, is a Von Neumann algebra. By construction the trace τ extends to a trace on this algebra. We remark that if H is a self-adjoint element of $\mathcal{A}$, its eigenprojections are in general elements of the Von Neumann algebra $\mathcal{W}$.

The main result is provided by the following theorem

Theorem 7. *Let f be an element of $\mathcal{C}_K(\Omega \times \mathbb{R}^n)$. Then its trace can be obtained as the trace per unit volume of the operator $\pi_\omega(f)$, namely for* **P***-almost all ω's*

$$\tau(f) = \lim_{\Lambda \uparrow \mathbb{R}^n} 1/|\Lambda| \mathrm{Tr}_\Lambda(\pi_\omega(f)) , \tag{2.6.3}$$

where Tr_Λ is the restriction of the usual trace in $L^2(\mathbb{R}^n)$ onto $L^2(\Lambda)$.

Remark. In the limit $\Lambda \uparrow \mathbb{R}^n$ the subsets Λ are measurable, their union adds up to $\mathbb{R}^n$, and $\lim_{\Lambda \uparrow \mathbb{R}^n} |\Lambda \Delta (\Lambda + a)|/|\Lambda| = 0$, where Δ represents the symmetric difference (Følner sequence [GR69]) .

Proof. Since f belongs to $\mathcal{C}_K(\Omega \times \mathbb{R}^n)$, the operator $\pi_\omega(f)$ admits a smooth kernel given by $F_\omega(x, y) = f(T^{-x}\omega, y - x)e^{\mathbf{i}B.x\wedge y}$ (see Eq. (2.5.2)). By Fredholm theory [SI79], the trace of this operator restricted to Λ is given by:

$$\mathrm{Tr}_\Lambda(\pi_\omega(f)) = \int_\Lambda d^n x F_\omega(x, x) = \int_\Lambda d^n x f(T^{-x}\omega, 0)) .$$

By Birkhoff's ergodic theorem, we get for **P**-almost all ω's

$$\lim_{\Lambda \uparrow \mathbb{R}^n} 1/|\Lambda| \mathrm{Tr}_\Lambda(\pi_\omega(f)) = \lim_{\Lambda \uparrow \mathbb{R}^n} 1/|\Lambda| \int_\Lambda d^n x f(T^{-x}\omega, 0) = \int_\Lambda \mathbf{P}(d\omega) f(\omega, 0) ,$$

which proves the result. □

To define the differential structure, we denote by ∂_μ the linear map from $\mathcal{C}_K(\Omega \times \mathbb{R}^n)$ into $\mathcal{C}_K(\Omega \times \mathbb{R}^n)$ defined by:

$$\partial_\mu f(\omega, x) = \mathbf{i} x_\mu f(\omega, x) . \tag{2.6.4}$$

It is quite easy to check that it is a family of mutually commuting *-derivations (see Definition in Section 1.3), which generates an n-parameter group of *-automorphisms namely:

$$\rho_\varsigma(f)(\omega, x) = e^{\mathbf{i}\langle \varsigma, x \rangle} f(\omega, x) . \tag{2.6.5}$$

This group extends by continuity to the C^*-algebra $\mathcal{A}$, and therefore the ∂'s are generators. It is an elementary calculation to show that

$$\pi_\omega(\partial_\mu f) = \mathbf{i}[X_\mu, \pi_\omega(f)] \ , \tag{2.6.6}$$

where X_μ is the position operator namely the multiplication by x_μ in $L^2(\mathbb{R}^n)$.

In the periodic case (see Section 2.3), let $k \in \mathbb{R}^n \to f_k$ be the compact operator valued map associated with f in the Wannier representation (Eq. (2.3.6)), such that $L(b) f_k L(b)^{-1} = f_{k+b}$ for $b \in \Gamma^*$. Then it is easy to check that $\{\mathbf{i}[X_\mu, f]\}_k = \partial f_k / \partial k_\mu$, showing that ∂_μ is nothing but the derivation with respect to k in the Brillouin zone. In such a way that the differential structure defined by the ∂'s is a natural generalization of the differential structure on the Brillouin zone in the non commutative case.

3. The Lattice Case: Physical Examples

In Section 2 we have introduced the Non-Commutative Brillouin zone of an electron in a potential submitted to a uniform magnetic field. The relation between the Schrödinger operator in $\mathbb{R}^n$ and the lattice problem has been discussed over and over again and is known in textbooks as the 'tight binding representation'. Recently, several mathematical studies have been proposed to justify such a representation [BE88b, GM90]. One can see this representation in many ways: the usual one consists in starting from Bloch's theory for the periodic crystal. Then one reduces the study to the bands lying in the vicinity of the Fermi level, since only electrons with energy close to the Fermi energy up to thermal fluctuations, contribute to physical effects like electric conductivity. In the roughest approximation, only one band contributes, but in general a finite number of them has to be considered. We then describe the corresponding Hilbert subspace selected in this way by using a basis of Wannier functions. Then adding either a weak magnetic field, or impurities in the crystal, or even the Coulomb potential between electrons results in getting a lattice operator, with short range interaction, where the lattice sites actually label the Wannier basis. In the case of a weak magnetic field, we obtain the main approximation for the effective Hamiltonian by means of the so-called 'Peierls substitution', namely if $E(k)$ is the energy band function as a function of the quasimomentum $k \in \mathbf{B}$, we simply replace k by the operator $K = (P - eA)/\hbar$, where $P = -\mathbf{i}\hbar\nabla$ is the momentum operator, and A is the magnetic vector potential. Since $E(k)$ is periodic in k, it can be expanded in Fourier series, it is enough to define by means of a Weyl quantization, the operators $\exp\{\mathbf{i}\langle K|b\rangle\}$ where $b \in \Gamma^*$. These operators are called 'magnetic translations' [ZA64]. In Section 3.1, we will focus on this question.

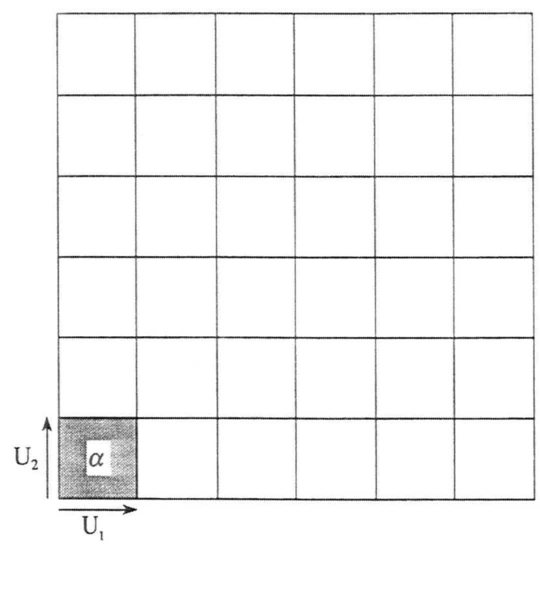

Fig. 2. The square lattice in two dimensions. U_1 and U_2 are the magnetic translations, ϕ is the flux through the unit cell, $\phi_0 = h/e$ is the flux quantum, and $\alpha = \phi/\phi_0$.

3.1 Two-Dimensional Lattice Electrons in a Uniform Magnetic Field

We consider a 2D lattice, that will be identified with $\mathbb{Z}^2$. Examples of such lattices with a symmetry group are the square lattice, the triangular lattice, the honeycomb lattice (see Fig. 2). We can also consider lattices like the rectangular one or even a rhombic lattice. We assume that this lattice is imbedded in a plane of the real 3D space, and is submitted to a perpendicular uniform magnetic field B. The Hilbert space of states is identified with $l^2(\mathbb{Z}^2)$, and the most important class of operators acting on it is the set of magnetic translations denoted by $W(m)$, $m \in \mathbb{Z}^2$, acting as follows:

$$W(m)\psi(x) = e^{i\pi\alpha x\wedge m}\psi(x-m), \quad x \in \mathbb{Z}^2, \psi \in l^2(\mathbb{Z}^2)\,, \tag{3.1.1}$$

where $\alpha = \phi/\phi_0$, $\phi_0 = h/e$ is the quantum of flux, whereas ϕ is the magnetic flux through a unit cell. They are unitaries and satisfy the following commutation rule:

$$W(m)W(m') = e^{i\pi\alpha m\wedge m'}W(m+m')\,. \tag{3.1.2}$$

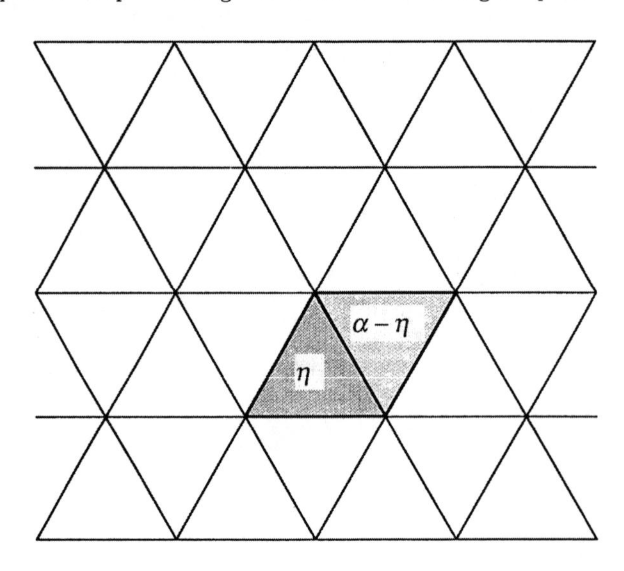

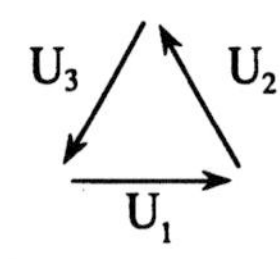

Fig. 3. The triangular lattice with the three magnetic translations and the two different normalized fluxes η and $\alpha-\eta$.

If we set $U_1 = W(1,0)$ and $U_2 = W(0,1)$, we get $W(m) = U_1^{m_1} U_2^{m_2} e^{i\pi\alpha m_1 m_2}$ whereas they obey the following commutation rule:

$$U_1 U_2 = e^{2i\pi\alpha} U_2 U_1 \ . \tag{3.1.3}$$

It turns out that all models constructed so far to represent a lattice electron in a magnetic field, are given by a Hamiltonian belonging to the C^*-algebra $\mathcal{A}_\alpha$ generated by U_1 and U_2. The simplest example is provided by the Harper model: if the energy band of the electron is given by $E(k) = 2t\{\cos a_1 k_1 + \cos a_2 k_2\}$ (square lattice) the Peierls substitution gives

$$H_S = t\{U_1 + U_1^* + U_2 + U_2^*\} \ . \tag{3.1.4}$$

After the gauge transformation $\psi(m_1, m_2) = e^{-i\pi\alpha m_1 m_2}\phi(m_1, m_2 - 1)$ the corresponding Schrödinger equation $H\psi = E\psi$ leads to the so-called 'Harper equation' [HA55]

$$\begin{aligned} &\phi(m_1+1, m_2) + \phi(m_1-1, m_2) + e^{-2i\pi\alpha m_1}\phi(m_1, m_2+1) \\ &\quad + e^{2i\pi\alpha m_1}\phi(m_1, m_2-1) = E\phi(m_1, m_2) \ . \end{aligned} \tag{3.1.5}$$

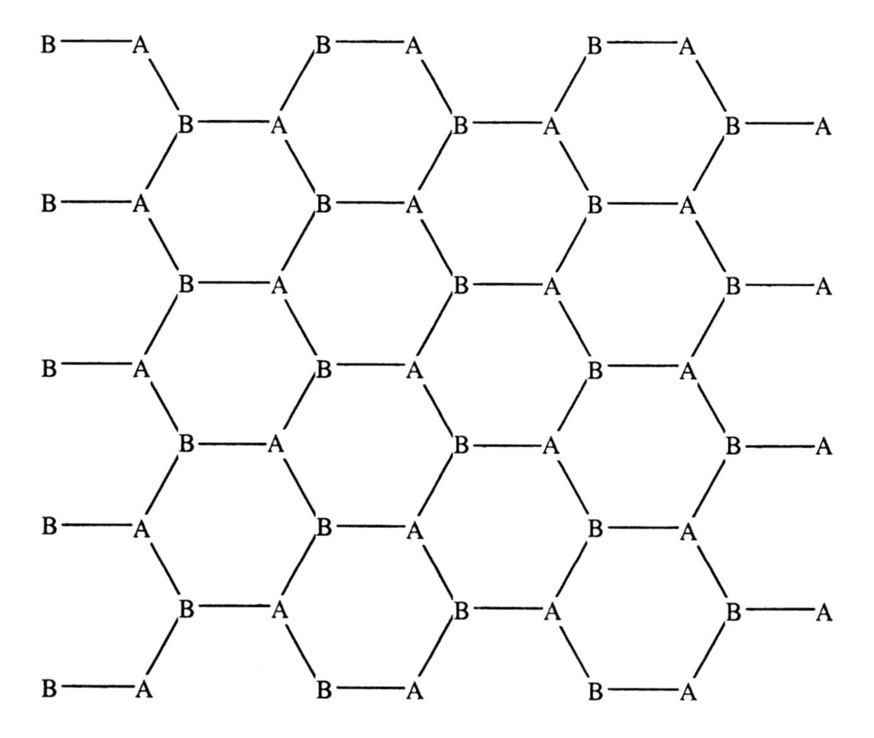

Fig. 4. The honeycomb lattice with its two sublattices.

In much the same way one can describe the nearest neighbour model for a triangular or hexagonal lattice as follows: one introduces the unitary operator U_3 defined by

$$U_1 U_2 U_3 = e^{2\mathrm{i}\pi\eta}\mathbf{1}\ . \tag{3.1.6}$$

These three operators will represent the translation in the three directions of the triangular lattice, in which we suppose that the flux through a triangle with vertex up is given by η (in units of the flux quantum), whereas the flux through triangles with vertex down is $\alpha - \eta$ (see Fig. 3). The nearest neighbour Hamiltonian will simply be

$$H_T = t\{U_1 + U_1^* + U_2 + U_2^* + U_3 + U_3^*\}\ . \tag{3.1.7}$$

The honeycomb lattice can be decomposed into two sublattices Γ_A and Γ_B (see Fig. 4), so that we can decompose the Hilbert space into the direct sum of the two subspaces $\mathcal{H}_I = l^2(\Gamma_I), I = A, B$, and the Hamiltonian can be written in matrix form as

$$H_H = \begin{vmatrix} 0 & U_1 + U_2 + U_3 \\ U_1^* + U_2^* + U_3^* & 0 \end{vmatrix}\ , \tag{3.1.8}$$

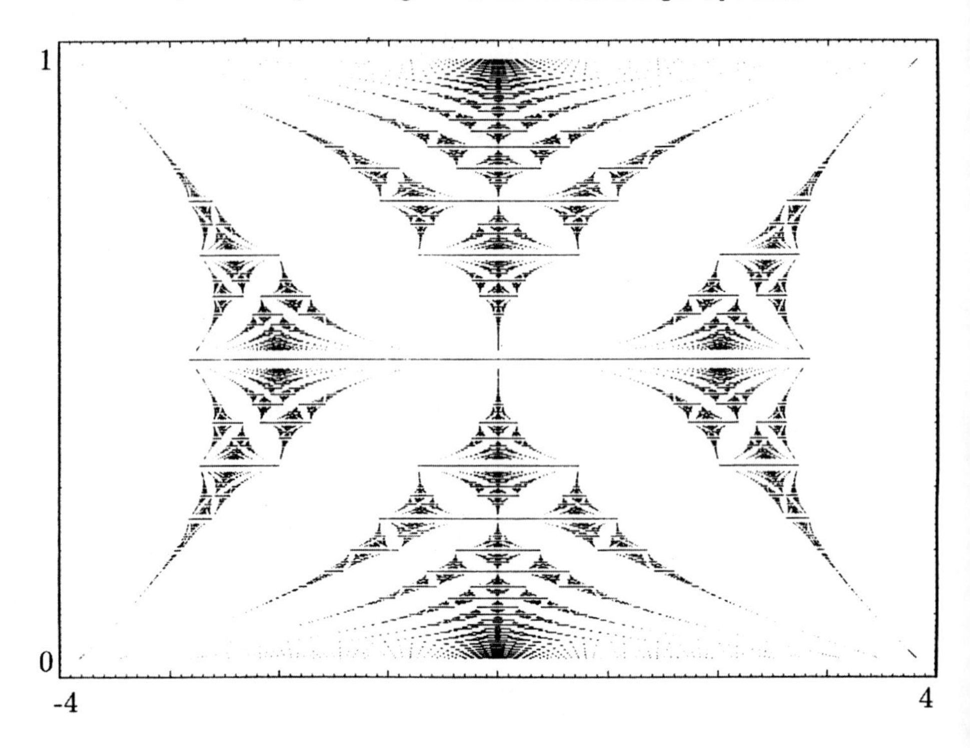

Fig. 5. The Hofstadter spectrum, namely the spectrum of $H = U_1 + U_1^* + U_2 + U_2^*$ as a function of α [HO76].

Fig. 5. The Hofstadter spectrum, namely the spectrum of $H = U_1 + U_1^* + U_2 + U_2^*$ as a function of α [HO76].

where the U's refer to an underlying triangular lattice in which α has to be replaced by $\alpha/3$ and η by $\alpha/6$.

More generally one can easily take into account interactions with any sites by adding to the Hamiltonian other monomials in the U's. One can also easily describe a model in which the unit cell contains different sites, occupied by different species of ions by introducing a decomposition of the Hilbert space similar to the one of the honeycomb lattice.

The non-commutative manifold associated to $\mathcal{A}_\alpha$ is a non-commutative torus. For indeed we have two non commuting unitary generators. This algebra was introduced and studied by M. Rieffel [RI81] in 1978, and has been called the 'rotation algebra'. For indeed, it can also be constructed as the crossed product of the algebra of continuous functions on the unit circle, by the action of $\mathbb{Z}$ given by the rotation by α on the circle. Then U_1 is nothing but the map $x \in \mathbb{T} \to e^{ix} \in \mathbb{C}$, whereas U_2 is the rotation by α.

The trace on $\mathcal{A}_\alpha$ is defined by analogy with the Haar measure on $\mathbb{T}^2$ by means of the Fourier expansion namely:

$$\tau(W(m)) = \delta_{m,0} , \qquad m \in \mathbb{Z}^2 . \tag{3.1.9}$$

Again this trace is also equal to the trace per unit volume in the representation of $\mathcal{A}_\alpha$ defined by Eq. (3.1.1) on $l^2(\mathbb{Z}^2)$.

On the other hand the differential structure is defined by the two commuting derivations ∂_1 and ∂_2 given by:

$$\partial_\mu U_\nu = \mathbf{i}\delta_{\mu,\nu} U_\nu . \tag{3.1.10}$$

In much the same way, if $X_\mu(\mu = 1, 2)$ represents the (discrete) position operator on $l^2(\mathbb{Z}^2)$, we also get:

$$\partial_\mu f = \mathbf{i}[X_\mu, f] . \tag{3.1.11}$$

Let us mention the work of A. Connes and M. Rieffel [CR87, RI90] which classifies the fiber bundles on this non commutative torus, and also the moduli space of connections.

One of the remarkable facts about the previous models, and also about most smooth self adjoint elements of this C^*-algebra (namely at least of class $\mathcal{C}^k$ for $k > 2$), is that their spectrum is a Cantor set provided α is irrational. One sees in Fig. 5 above the famous Hofstadter spectrum representing the spectrum of the Harper Hamiltonian as a function of α. One sees a remarkable fractal structure which has been investigated by many physicists [HO76, CW78, CW79, CW81, WI84, RA85, SO85, BK90] and mathematicians [BS82b, EL82b, HS87, GH89, BR90], without having been completely understood quantitatively so far. We refer the reader to [BE89] for a review.

3.2 Quasicrystals

In 1984, in a famous paper, Shechtman et al. [SB84] announced the discovery of a new type of crystalline phase in rapidly cooled alloys of Aluminium and Manganese, for which the diffraction pattern was point like but with a fivefold symmetry. This is forbidden by theorems on crystalline groups in 3 dimensions. However, if one breaks the translation invariance it is possible to get such a diffraction pattern provided the lattice of sites where the ions lie is quasiperiodic.

This discovery created an enormous amount of interest. Actually aperiodic tilings of the space had already been studied since the end of the seventies. The first example was provided by Penrose in 1979 [PN79] who gave a tiling of the plane with two kinds of tiles, having a fivefold symmetry. This was followed by several works concerning a systematic construction of such tilings in particular by Conway [GA77], de Bruijn [BR81]. Mackay [MK82] and then Mosseri and Sadoc [MS83, NS85] gave arguments for the relevance of such tilings in solid state physics. A 3-dimensional generalization of the Penrose tiling was theoretically described by Kramer and Neri [KN84]. Experimentally, the diffraction pattern created by such tilings was observed empirically by Mackay,

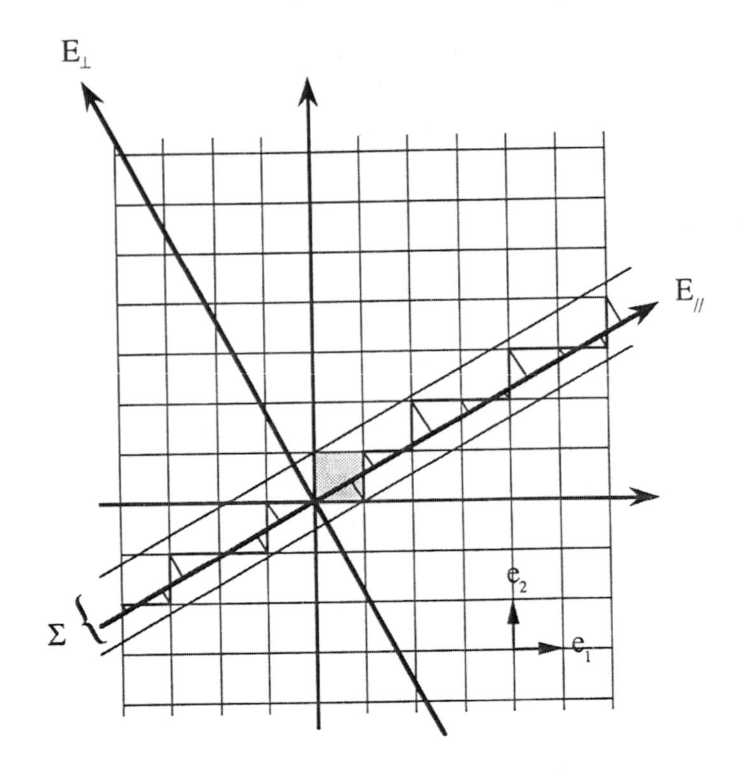

Fig. 6. The cut and projection method for a one dimensional quasicrystal.

whereas Levine and Steinhardt [LS84] suggested to consider quasiperiodicity as a basis to get a quasicrystal with icosahedral symmetry.

In this Section we will rather use the projection method introduced independently by Elser [EL85] and Duneau-Katz [DK85] (see Fig. 6). In this method, a quasicrystal is seen as the projection of a lattice $\mathbb{Z}^\nu$ of higher dimension contained in $\mathbb{R}^\nu$, on a subspace $E \approx \mathbb{R}^n$ irrationally oriented with respect to the canonical basis defining $\mathbb{Z}^\nu$ which will be identified with the physical space. We then call $E_\perp$ the subspace perpendicular to E. We will denote by Π and $\Pi_\perp$ the orthogonal projections onto E and $E_\perp$. Let $\varepsilon(\mu)(1 \leq \mu \leq \nu)$ be the unit vectors of the canonical basis of $\mathbb{R}^\nu$. We set $\varepsilon(\mu + j\nu) = (-)^j \varepsilon(\mu)$, for $j \in \mathbb{Z}$. Then we set $e(\mu) = \Pi(\varepsilon(\mu))$ and $e'(\mu) = \Pi_\perp(\varepsilon(\mu))$. Let now Δ be the (half-closed) unit cube $[0,1]^\nu$ of $\mathbb{R}^\nu, \zeta \in \mathbb{R}^\nu$, and let $\sum(\zeta)$ be the strip obtained by translating $\Delta + \zeta$ along E, namely $\sum(\zeta) = \Delta + E + \zeta$. In the sequel we shall drop the reference to ζ. We denote by $\mathcal{S} = \sum \cap \mathbb{R}^\nu$ the set of lattice points contained in the strip $\sum$. The quasilattice $\mathcal{L}$ will be obtained as the orthogonal projection of $\mathcal{S}$ on E (see Fig. 7). It is a discrete subset of E, and there is a one-to-one correspondence between $\mathcal{L}$ and $\mathcal{S}$ [DK85]. The 'acceptance zone' is the intersection $\Omega = \sum \cap E_{\perp}$ of $\sum$ with the perpendicular space (see Fig. 8).

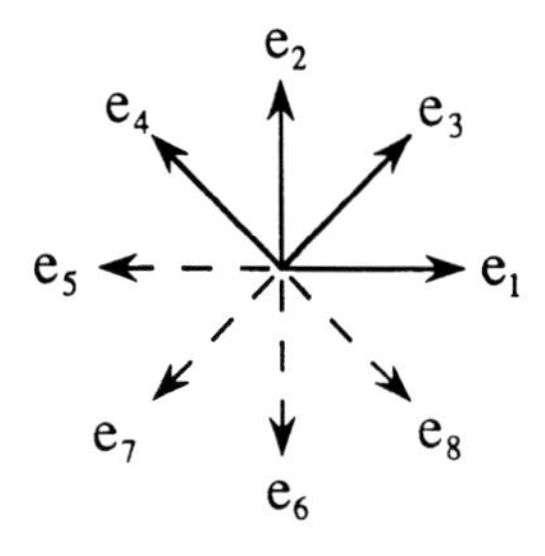

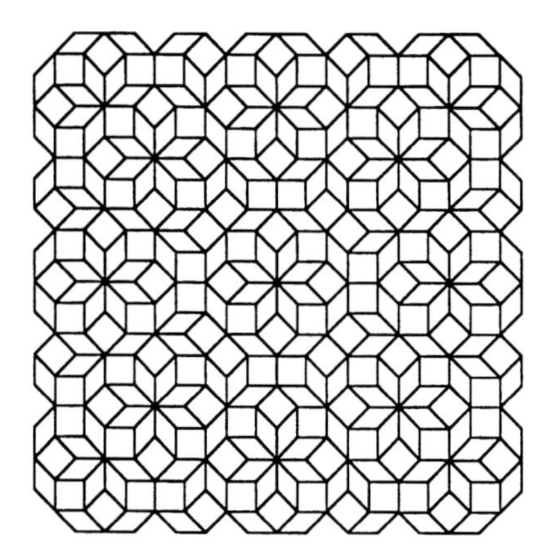

Fig. 7. The octagonal lattice, obtained as a projection of $\mathbb{Z}^4$ on $\mathbb{R}^2$ in such a way as to conserve the 8-fold symmetry. The projection of the canonical basis of $\mathbb{Z}^4$ on $\mathbb{R}^2$ is shown above.

It is a polyhedron in a $(\nu - n)$-dimensional space. The projection of $\sum$ on Ω is discrete if E is rationally oriented and dense if it is irrationally oriented. If ν is high enough one can also get intermediate situations where this projection is dense in a union of polyhedra of lower dimension. A substrip $\sum'$ of $\sum$ is a subset of $\sum$ invariant by translation along E. Such a substrip will be identified with its projection $\Omega' = \Pi_{\perp}\{\sum'\}$ in the acceptance zone. Ω' will be called the 'acceptance zone' of $\sum'$.

To describe the quantum mechanics of a particle on $\mathcal{L}$, we first introduce the large Hilbert space $\mathcal{K} = l^2(\mathbb{Z}^{\nu})$ and the physical Hilbert space $\mathcal{H} = l^2(\mathcal{L})$. $\mathcal{H}$ can be seen as a subspace of $\mathcal{K}$ if we identify $\mathcal{L}$ with $\mathcal{S}$. Denoting by χ_{Ω}

the characteristic function of $\mathcal{S}$ in $\mathbb{Z}^\nu$ we can identify the orthogonal projection onto $\mathcal{H}$ in $\mathcal{K}$ with the operator of multiplication by χ_Ω. More generally, if $\sum'$ is a substrip of $\sum$, the orthogonal projection in $\mathcal{K}$ onto $l^2(\sum' \cap \mathbb{Z}^\nu)$ will be denoted by $\chi_{\Omega'}$, if Ω' is the acceptance zone of $\sum'$.

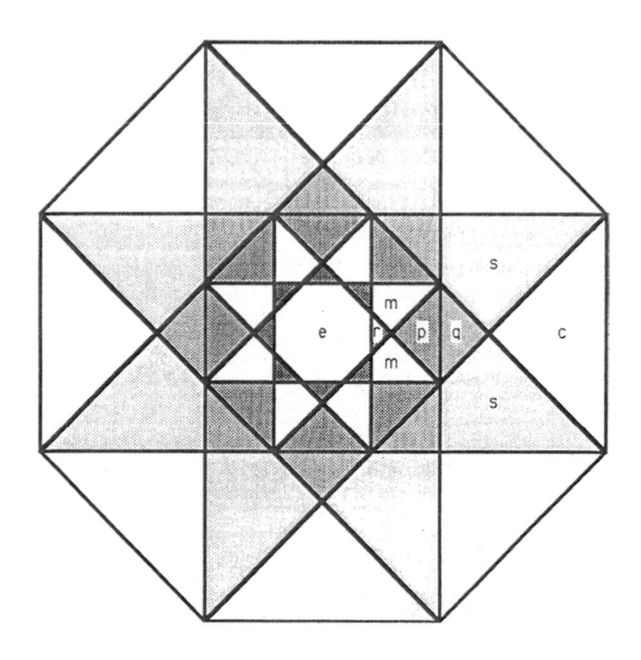

Fig. 8. The acceptance zone of the octagonal lattice. The various subsets correspond to the projections of points in the octagonal lattice having a given local environment [SB90].

A free particle on the lattice $\mathbb{Z}^\nu$ is usually described by the algebra generated by the translation operators $T_\mu (1 \leq \mu \leq \nu)$. It is isomorphic to the algebra $\mathcal{C}(\mathbb{T}^\nu)$ of continuous functions on the ν-dimensional torus, namely the Brillouin zone in this case. For convenience, we will set $T_{\mu+\nu} = T_\mu^* = T_\mu^{-1}$, and $T_{\mu+2\nu} = T_\mu$. By analogy, we shall consider the algebra $\mathcal{Q}(\nu, E)$ generated by the restrictions $S_\mu = \chi_\Omega T_\mu \chi_\Omega$ of the translation operator to $\mathcal{H}$. For instance, the Laplacian $\Delta_\mathcal{L}$ on the lattice $\mathcal{L}$ is nothing but the Hamiltonian given by

$$\Delta_\mathcal{L} = \sum_{\mu=1}^{2\nu} (S_\mu - \mathbf{1}) \,. \tag{3.2.1}$$

The operators S_μ are no longer unitary, but they are partial isometries, and they do not commute anymore. Thus $\mathcal{Q}(\nu, E)$ describes the set of continuous functions on a non commutative compact space (the identity belongs to $\mathcal{Q}(\nu, E)$).The commutation rules are given as follows:

$$S_\mu S_\mu^* = \chi_{\Omega \cap \{\Omega + e'(\mu)\}}\,, \qquad S_\mu^* S_\mu = \chi_{\Omega \cap \{\Omega - e'(\mu)\}}\,. \tag{3.2.2}$$

Actually we get $S_\mu = \chi_{\Omega \cap \{\Omega + e'(\mu)\}} T_\mu \chi_{\Omega \cap \{\Omega - e'(\mu)\}}$. More generally, if $x \in \mathbb{Z}^\nu$, $T(x)$ will denote the translation by x in $\mathcal{K}$. Then we set $S(x) = \chi_\Omega T(x) \chi_\Omega$, in particular $S_\mu = S(\varepsilon(\mu))$. It follows immediately that:

$$S(\varepsilon(\mu_1))S(\varepsilon(\mu_2))\ldots S(\varepsilon(\mu_N)) = \chi_{\Omega(\underline{\mu})} T(\varepsilon(\mu_1) + \varepsilon(\mu_2) + \ldots + \varepsilon(\mu_N)) \tag{3.2.3}$$

where $\underline{\mu} = (\mu_1, \mu_2, \ldots, \mu_N)$, and

$$\Omega(\underline{\mu}) = \Omega \cap \{\Omega + e'(\mu_1)\} \cap \ldots \cap \{\Omega + e'(\mu_1) + \ldots + e'(\mu_N)\}\,. \tag{3.2.4}$$

This latter set is nothing but the acceptance zone of the set of points x in $\sum$ such that the path $\gamma(x) = (x_0, \ldots, x_N)$ is entirely included in $\sum$, whenever $x_0 = x$, $x_k = x_{k-1} + \varepsilon(\mu_k)$. In particular, any element f of the algebra $\mathcal{Q}(\nu, E)$ can be approximated in norm, by a sequence of 'trigonometric polynomials' namely finite sums of the form

$$f = \sum_{\underline{\mu}} c(\underline{\mu}) \chi_{\Omega(\underline{\mu})} T(\varepsilon(\mu_1) + \varepsilon(\mu_2) + \ldots + \varepsilon(\mu_N))\,, \tag{3.2.5}$$

where the $c(\underline{\mu})$'s are complex coefficients.

For $\underline{\mu} = (\mu_1, \mu_2, \ldots, \mu_N)$ such that $1 \le \mu_i \le 2\nu$, we set $x(\underline{\mu}) = \varepsilon(\mu_1) + \varepsilon(\mu_2) + \ldots + \varepsilon(\mu_N)$, and for $a \in \mathbb{Z}^\nu$, we denote by $I(a)$ the set of $\underline{\mu}$'s such that $x(\underline{\mu}) = a$. A linear form on $\mathcal{Q}(\nu, E)$ is then defined by linearity as follows:

$$\tau\{\chi_{\Omega_{\underline{\mu}}} T(x(\underline{\mu}))\} = \delta_{x(\underline{\mu}),0} |\Omega(\underline{\mu})|\,, \tag{3.2.6}$$

where $|.|$ denotes the Lebesgue measure in $E_\perp$. Since the Lebesgue measure is translation invariant, one can easily check that if f, g are two trigonometric polynomials, $\tau(fg) = \tau(gf)$ and that $\tau(ff^*) > 0$ as soon as f is non zero in $\mathcal{Q}(\nu, E)$. At last, $\tau(\mathbf{1}) = 1$ and therefore τ extends uniquely to a normalized trace on $\mathcal{Q}(\nu, E)$.

As in the previous Section, one defines the differential structure on $\mathcal{Q}(\nu, E)$ by means of the group of *-automorphisms η_a ($a \in \mathbb{T}^\nu$) defined by linearity by:

$$\eta_a\{\chi_{\Omega(\underline{\mu})} S(x(\underline{\mu}))\} = e^{i\langle a | x(\underline{\mu})\rangle} \chi_{\Omega(\underline{\mu})} S(x(\underline{\mu}))\,, \tag{3.2.7}$$

and the derivation ∂_μ is given by:

$$\partial_\mu f = d\eta_a(f)/da_\mu|_{a=0}\,. \tag{3.2.8}$$

A special case concerns the 1D quasicrystals. Then $\nu = 2$, and E is a line in $\mathbb{R}^2$ (see Fig. 6). Let ω be the slope of E and we set $\alpha = \omega/(\omega + 1)$. It is sufficient to choose ζ of the form $\zeta = (0, n_0 + \eta)$ with $n_0 \in \mathbb{Z}$, and $-1 \le \eta < \omega$. The lattice $\mathcal{L}$ is then a sublattice of E, and it is in one-to-one correspondence with $\mathbb{Z}$. More precisely, if $x = (m,n) \in \mathcal{S}$, one set, $l(x) = m + n - n_0$. The map $x \in \mathcal{S} \rightarrow l(x) \in \mathbb{Z}$, is one-to-one, with inverse given by $n = n_0 + 1 + [l\alpha - \theta]$, $m = l - 1 - [l\alpha - \theta]$, provided $\theta = \alpha - (1-\alpha)\eta \in [0,1]$ (here, $[r]$ denotes the integer part of r). In particular, $\mathcal{H}$ is naturally isomorphic to $l^2(\mathbb{Z})$.

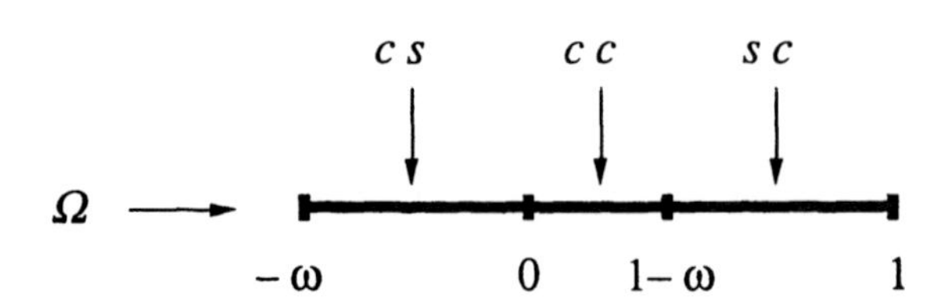

Fig. 9. The acceptance zone for the one dimensional quasilattice shown in Fig. 6. The subintervals correspond to the projections of the lattice points with indicated environment, c stands for horizontal bonds whereas s stands for the vertical ones.

Then setting $T = S_1 + S_2$, T is unitary and is nothing but the translation by one in $\mathcal{H}$. On the other hand, we can easily check that $\chi_\alpha = \chi_{\Omega \cap \{\Omega - e'(2)\}} = S_2^* S_2$ (see Fig. 9) is identified with the operator $\chi_{\alpha,\theta}$ of multiplication by $\chi_{[1-\alpha,1)}(l\alpha - \theta)$ in $\mathcal{H}$ where $\chi_{[1-\alpha,1)}$ is the characteristic function of $[1-\alpha, 1)$ in the unit circle. Hence, the algebra $\mathcal{Q}\alpha$ is generated by T and χ_α. By the previous remark, the Abelian C^*-algebra generated by the family $\{\chi_{\alpha,n} = T^n \chi_\alpha T^{-n}\}$ is naturally isomorphic to the algebra $\mathcal{C}_\alpha$ generated by the characteristic functions $\chi_{[1+(n-1)\alpha, 1+n\alpha)}$ of the interval $[1 + (n-1)\alpha, 1 + n\alpha)$ of the unit circle. Thus $\mathcal{Q}_\alpha = \mathcal{Q}(2, E)$ appears as the cross product of $\mathcal{C}_\alpha$ by the rotation by α on the unit circle. This algebra was already studied by Cuntz. In particular, when α is irrational, $\mathcal{C}_\alpha$ contains all continuous functions on the unit circle, and therefore, $\mathcal{Q}_\alpha$ contains the irrational rotation algebra $\mathcal{A}_\alpha$.

It will be convenient to define a 'universal' C^*-algebra formally given by $\mathcal{Q} = \cup_{\alpha \in [0,1]} \mathcal{Q}_\alpha$. In order to do so, let χ_n be the map on $\mathbb{T}^2$ defined by $\chi_n(\alpha, x) = \chi_{[1+(n-1)\alpha, 1+n\alpha)}(x)$. We shall denote by $\mathcal{C}$ the Abelian C^*-algebra generated by the χ_n's. If f is the diffeomorphism of $\mathbb{T}^2$ defined by $f(\alpha, x) = (\alpha, x - \alpha)$, then $\mathcal{Q}$ appears as the cross product of $\mathcal{C}$ by f. For an irrational α, the set $\mathcal{I}_\alpha$ of elements of $\mathcal{Q}$ vanishing at α is a closed two-sided ideal. Then, $\mathcal{Q}_\alpha$ can be identified with the quotient $\mathcal{Q}/\mathcal{I}_\alpha$. The quotient map will be denoted by ρ_α [BI90].

One model has been the focus of attention in this latter case, namely the 'Kohmoto model' [KK83, OP83, KO84, OK85, CA86, KL86, LP86, SU87, BI89, LE89, SU89, BI90], the Hamiltonian of which acting on $l^2(\mathbb{Z})$ as follows:

(3.2.9) $$H_{(\alpha,\theta)} = T + T^* + V\chi_{(\alpha,\theta)} = S_1 + S_2 + S_1^* + S_2^* + V S_2^* S_2 \ ,$$

where $V \geq 0$ is a coupling constant. It describes a 1D Schrödinger operator with a potential taking two values in a quasiperiodic way with quasiperiods 1 and α. This operator and operators related to it have been also proposed to describe the behavior of 1D electrons in a Charge Density Wave (CDW) once the linear chain is submitted to a Peierls instability. The energy spectrum of the Kohmoto Hamiltonian has been computed first by Ostlund and Kim [OK85]. It has a beautiful fractal structure which can be investigated by means of a renormalization group analysis [KK83, OP83, KO84, SO85, KL86, CA86, LP86, SU87, BE89, BI89, LE89, SU89]. More precisely one has

Proposition 3.2.1 [BI90]. *Let H be a self adjoint element in Q. Then the gap boundaries of the spectrum of $\rho_\alpha(H)$ are continuous with respect to α at any irrational number.*

Proposition 3.2.2 [BI89]. *For any number α in $[0,1]$, the spectrum $\sum(\alpha)$ of the operator $H_{(\alpha,\theta)}$ acting on $l^2(\mathbb{Z})$ and given by Eq. (3.2.9), is independent of $\theta \in [0,1)$.*

Proposition 3.2.3 [BI89]. *For any irrational number α in $[0,1]$, and any θ in [0,1), the operator $H_{(\alpha,\theta)}$ acting on $l^2(\mathbb{Z})$ and given by Eq. (3.2.9), has a Cantor spectrum of zero Lebesgue measure, and its spectral measure is singular continuous.* (See Fig. 10)

In dimension $D > 1$, the character of the spectrum is not known rigorously yet. However, several theoretical works have been performed in 2D and they indicate that two regimes can be obtained, depending upon the values of the Fourier coefficients of the Hamiltonian under consideration: an 'insulator' regime, for which the spectrum looks like a Cantor set, and a 'metallic' regime for which the spectrum has very few gaps and a continuous (but fractal) density of states.

The oldest works [see for reviews BE89 and SO87, KS86, ON86, TF86, HK87], consist in restricting the lattice to a finite box Λ, and to diagonalize numerically by brute force the finite dimensional matrix obtained by restricting the Hamiltonian to Λ. The main difficulty with this method comes from surface states. For small boxes (with about 2000 sites), it is estimated that surface states may contribute up to 20 % of the spectrum. It is then necessary to eliminate the surface state contribution from the spectrum to get a good approximation of the infinite volume limit. To solve this difficulty one may approximate the lattice $\mathcal{L}$ by a periodic lattice with large period, and use Bloch's theory to compute the spectrum numerically. This can be done by replacing the space E by a 'rational' one close to it [MO88, JD88].

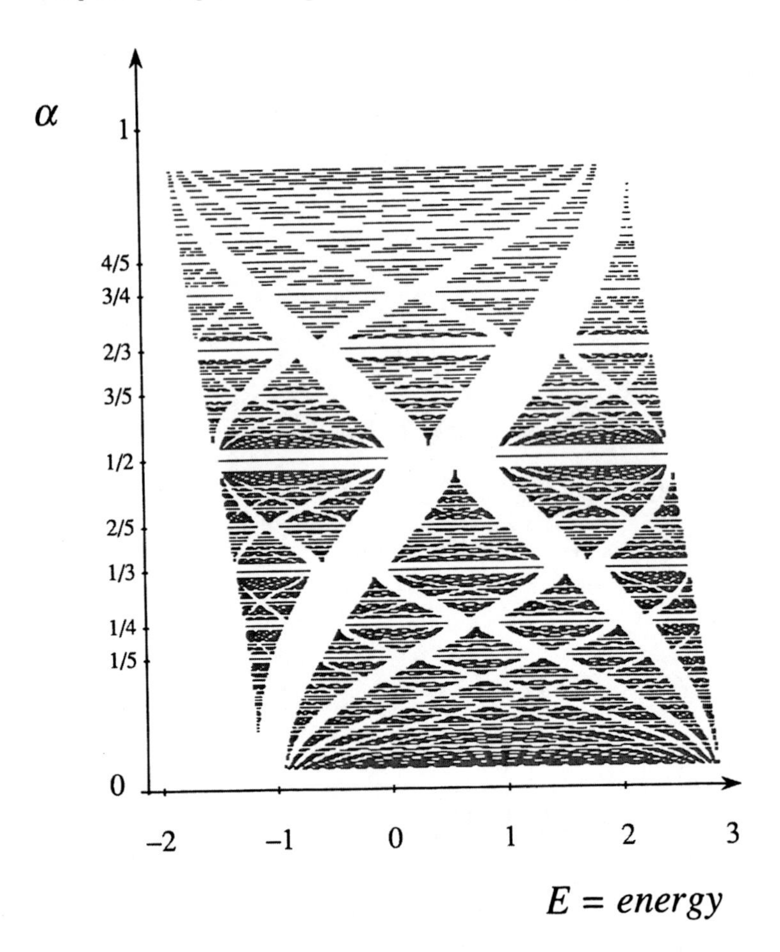

Fig. 10. The spectrum of the Kohmoto Hamiltonian (eq. 3.2.9) as a function of α [OK85].

More recently a solvable model called the 'labyrinth model', has been found by C. Sire [SI89], using a Cartesian product of two 1D chains. It exhibits a 'metal-insulator' transition. In the insulating regime this Hamiltonian admits a Cantor spectrum of zero Lebesgue measure.

At last a Renormalization Group approach has been proposed to compute numerically the spectrum in the infinite volume limit [SB90]. Even though it is in principle rigorous, the practical calculation on the computer requires approximations. The cases of Penrose and octagonal lattices in 2D with nearest neighbour interactions have been performed, and also give a Cantor spectrum in the insulating regime [BS 91][1].

[1] See also the note added before the references for the corrected edition (May 1995).

3.3 Superlattices and Automatic Sequences

Superlattices are made of two species of doped semiconductors piled in on top of each other to produce a 1D chain of quantum wells. A qualitative model to describe electronic wave functions is given by a 1D discrete Schrödinger operator of the form

$$H\psi(n) = \psi(n+1) + \psi(n-1) + \lambda V(n)\psi(n)\ , \psi \in l^2(\mathbb{Z})\,, \tag{3.3.1}$$

where $V(n)$ represents the effect of each quantum well, and λ is a positive parameter which will play the role of a coupling constant. An interesting situation consists in building up the superlattice by means of a deterministic rule. The simplest one is given by a periodic array, in which we alternate the two species in a periodic way. But in general the rule will be aperiodic. One widely studied example is the Fibonacci sequence: given two letters a and b, one substitutes the word ab to a, and the word a to b. Starting from a one generates an infinite sequence $abaababaabaababa\ldots$ in which the frequency of a's is given by the golden mean $(\sqrt{5}-1)/2$. Another example (the oldest one actually) of such a substitution is the Thue-Morse sequence [TU06, MO21, QU87]. It is obtained through the substitution $a \to ab, b \to ba$, and gives $abbabaabbaababba\ldots$ Also if $S_2(n)$ is the sum of the digits of n in a dyadic expansion, then the sequence $(u(n))_{n\geq 0}$ defined by setting $u(n) = a$ if $S_2(n)$ is even, and $u(n) = b$ if $S_2(n)$ is odd, is the Thue-Morse sequence again.

To define a substitution, we start with the data of a finite set A called an 'alphabet'. Elements of A are called 'letters'. Elements of the Cartesian product A^k are called 'words of length k'. The disjoint union $A^* = \cup_{k\geq 1} A^k$, is the set of words. The length of a word w is denoted by $|w|$. A substitution is a map ζ from A to A^*, which associates to each letter $a \in A$ a word $\zeta(a)$. We extend ζ into a map from A^* into A^* by concatenation namely $\zeta(a_0 a_1 \ldots a_n) = \zeta(a_0)\zeta(a_1)\ldots\zeta(a_n)$. In particular one can define the iterates ζ^n of ζ. We shall assume the following [see QU87 p. 89]:

(S1) $\lim_{n\to\infty} |\zeta^n(a)| = +\infty$ for every $a \in A$.
(S2) there is a letter $0 \in A$ such that $\zeta(0)$ begins with 0.
(S3) for every $a \in A$, there is $k \geq 0$ such that $\zeta^k(a)$ contains 0.

The first condition insures the existence of an infinite sequence. The second is actually always fulfilled [QU87 p. 88], so that the 'substitution sequence' $u = \lim_{n\to\infty} \zeta^n(0)$ exists in the infinite product space $A^{\mathbb{N}}$ such that $\zeta(u) = u$. Let T be the shift in $A^{\mathbb{N}}$, namely $(Tw)_k = w_{k+1}$, and let Ω be the compact subspace of $A^{\mathbb{N}}$ given as the closure of the family $O(u) = \{T^n u;\ n \geq 0\}$. Then (Ω, T) is a topological dynamical system, and the condition (S3) insures that it is minimal, namely that every orbit is dense [QU87, Theorem V.2]. By P. Michel's theorem [QU87, Theorem V.13], it is therefore uniquely ergodic namely it admits a unique ergodic invariant probability measure that will be denoted by $\mathbf{P}$.

In much the same way, defining $u'_{-k} = u'_{k-1} = u_{k-1}$, for $k \geq 1$, one gets a two-sided sequence in $A^{\mathbb{Z}}$. Ω' is the closure of the orbit of u' under the two-sided shift T' defined similarly on $A^{\mathbb{Z}}$. It is then easy to check that the dynamical system (Ω', T') is still uniquely ergodic. Moreover, the unique invariant probability measure $\mathbf{P}'$ is uniquely determined by $\mathbf{P}$. Thus each non empty open set of Ω' has a positive measure, implying that (Ω', T') is minimal (see Proposition IV.5 in [QU87]).

A sequence $\{V(n);\, n \in \mathbb{Z}\}$ of real numbers is a 'substitution potential' whenever there is a substitution fulfilling conditions (S1–3) and a mapping $\mathcal{V}$ from A to $\mathbb{R}$ such that

$$V(n) = \mathcal{V}(u_n) = V(-n-1)\,, \qquad \text{for } n \geq 0\,. \tag{3.3.2}$$

It will be convenient to extend $\mathcal{V}$ to Ω' by setting $v(w) = \mathcal{V}(w_0)$, for $w \in \Omega'$. Then for $w \in \Omega$, we define the potential V_w by:

$$V_w(n) \equiv v(T^n w)\,, \Rightarrow V = V_u\,. \tag{3.3.3}$$

Following the method described in Section 2, we deduce that the Hamiltonian given in Eq. (3.3.1) belongs to the C^*-Algebra $C^*(\Omega', T')$ which is the crossed-product of the space $\mathcal{C}(\Omega')$ by the action of $\mathbb{Z}$ defined by T'. An element F in this algebra is described by a kernel $F(w, n)$, which is for each $n \in \mathbb{Z}$, a continuous function of w on Ω'. For each $w \in \Omega'$ one gets a representation $(\pi_w, \mathcal{H}_w)$ of this algebra, such that $\mathcal{H}_w = l^2(\mathbb{Z})$, and if U denotes the translation by 1 in $l^2(\mathbb{Z})$, $f \in \mathcal{C}(\Omega')$:

$$\pi_w(T') = U^* \qquad \pi_w(f) = \text{Multiplication by } f(T'^n w)\,. \tag{3.3.4}$$

This is equivalent to saying that the matrix elements of $\pi_w(F)$ are:

$$\langle n|\pi_w(F)|m\rangle = F(T'^n w, m-n)\,. \tag{3.3.5}$$

We immediately get the 'covariance property' $U^{-1}\pi_w(F)U = \pi_{T'w}(F)$. Since (Ω', T') is minimal, it follows that the spectrum of $\pi_w(F)$ (as a set) does not depend on $w \in \Omega'$.

A trace τ on this algebra is defined as the trace per unit volume. Using Birkhoff's theorem [HA56], and Eq. (3.3.5), it can easily be shown that it is given as follows:

$$\tau(F) \equiv \lim_{N\to\infty} N^{-1} \sum_{0\leq n<N} \langle n|\pi_w(F)|n\rangle = \int_{\Omega'} \mathbf{P}'(dw) F(w, 0)\,. \tag{3.3.6}$$

The unique ergodicity of P' implies that this limit is uniform in w. In much the same way as in the previous Sections, a differential structure is defined by one derivation ∂ such that:

$$\partial F(w, n) \equiv \mathbf{i} n F(w, n)\,, \quad \Rightarrow \pi_w(\partial F) = -\mathbf{i}[X, \pi_w(F)]\,, \tag{3.3.7}$$

where X is the position operator namely $X\psi(n) = n\psi(n)$ for $\psi \in l^2(\mathbb{Z}), n \in \mathbb{Z}$.

Very few results have been obtained yet on the spectral properties of the Hamiltonian H above. The Fibonacci sequence is actually a special case of a 1D quasicrystal, and gives rise to Kohmoto's model.

J.M. Luck [LU90], gave a series of theoretical arguments to show that several non almost-periodic examples of substitutions give rise to a Cantor spectrum. The width of the gaps follows a scaling behaviour in λ as $\lambda \to 0$. This has been rigorously proved in two cases [AA86, AP88, BE90a, DP90, BB90]: the Thue-Morse (given by $a \to ab$, $b \to ba$) and the period-doubling sequence (given by $a \to ab$, $b \to aa$). More precisely one has[2]:

Proposition 3.3.1. *Let H be given by* (3.3.1) *above, where V is defined either by the Thue-Morse or by the period doubling sequence. Then H has a Cantor spectrum of zero Lebesgue measure. Its spectral measure is singular continuous. Moreover as $\lambda \approx 0$, the gap widths W obey to the following asymptotics:*

(i) for the Thue-Morse sequence: $W \approx \text{const.}\, \lambda^{\sigma} b(\lambda)$, *with* $\sigma = Ln4/Ln3$, *and b is a continuous function bounded away from zero and such that* $b(\lambda) = b(\lambda/3)$;

(ii) for the period doubling sequence, we get two families of gaps with

- *either* $W \approx \text{const.}\, \lambda$,
- *or* $W \approx \text{const.} \exp\{3Ln2/\lambda\}\lambda^{Ln2}$.

4. K-Theory and Gap Labelling

This Section is devoted to the exposition of the general theory of gap labelling. The operator side of the formula is given by a natural countable group associated to each algebra, the group K_0 [AT67, KA78, CU82, BL86] which serves to label gaps of an aperiodic Hamiltonian in a stable way under perturbations. In Section 4.1, we introduce the Integrated Density of States (IDS) [CF87], using the trace per unit volume, to count the number of states below some energy level. In Sections 4.2 and 4.3, we will give the general properties of the K_0-group, generalizing in this way the homotopy invariant aspect of the index of a Fredholm operator. The wave aspect of the gap labelling will be investigated in the next Sections. Most of the content of this Section can be found in [BE86], already. We will postpone the study of examples to Section 5.

4.1 Integrated Density of States and Shubin's Formula

The operators we have studied up to now model the motion of a single particle in a solid with infinitely many fixed ions producing the potential. This is the 'one-body' approximation. However in a solid with infinitely many ions, there is an infinite number of electrons. This is not a problem as long as we can neglect the electron-electron interaction, and also the deformation of the ionic

[2] See also the note added before the references for the corrected edition (May 1995).

potential by the electrons themselves (electron-phonon interaction). For then the total energy is simply the sum of the individual single electron energies. If we take into account the Pauli exclusion principle, electrons being fermions, the ground state energy of the electron gas will simply be given by adding up all filled individual energy states, each of them being occupied by one electron only. The main technical difficulty comes from having to distribute an infinite number of fermions on a continuum of energy levels. The key fact about it comes from the homogeneity of the crystal we are investigating: electrons in far apart areas tend to ignore each other; moreover, spectral properties will tend to be almost translation invariant, at least if we compare large regions. Therefore it will be enough to consider the number of available energy levels per unit volume. This is how the integrated density of states IDS is defined.

Let $\mathcal{H}$ be the Hilbert space $L^2(\mathbb{R}^n)$, and let us consider the Schrödinger operator given by (2.4.5) $H = (1/2m)\sum_{\mu\in[1,n]}(P_\mu - eA_\mu)^2 + V = H_B + V$, where A is the vector potential associated to a uniform magnetic field B, and V is a real measurable essentially bounded function on $\mathbb{R}^n$. For any rectangular box Λ, we denote by H_Λ the restriction of H to Λ with some boundary conditions (for instance, Dirichlet or periodic boundary conditions). Then H_Λ has a discrete spectrum bounded from below. Let $N_\Lambda(E)$ be the number of eigenvalues of H_Λ smaller than or equal to E. Because of the homogeneity condition, if we translate Λ by ζ, we expect $N_{\Lambda+\zeta}(E) = N_\Lambda(E) + o(|\Lambda|)$ (where $|.|$ denotes the Lebesgue measure), and if Λ and Λ' are two large non intersecting boxes, $N_{\Lambda\cup\Lambda'}(E) = N_\Lambda(E) + N_{\Lambda'}(E) + o(|\Lambda\cup\Lambda'|)$. In other words, we expect $N_\Lambda(E)$ to increase with Λ as $|\Lambda|$. So we define the IDS as:

$$\mathcal{N}(E) = \lim_{\Lambda\uparrow\mathbb{R}^n} N_\Lambda(E)/|\Lambda| \,. \tag{4.1.1}$$

Here the limit $\Lambda\uparrow\mathbb{R}^n$ is understood in the following sense: we consider a 'Følner sequence' [GR69] namely a sequence Λ_k of bounded open sets, with piecewise smooth boundary, with union $\mathbb{R}^n$, and such that $|\Lambda_k\Delta(\Lambda_k + a)|/|\Lambda_k|$ converges to zero as $k\to\infty$ (Δ is the symmetric difference of sets).

The very same definition can apply to the case of a Hamiltonian on a lattice (Section 2.4, Eq. (2.4.7)) and we can also extend this definition to the case where the lattice is a homogeneous sublattice of $\mathbb{R}^n$. We will leave to the reader the obvious extension to this case.

One remarks that $N_\Lambda(E)$ is also the trace of the eigenprojection $\chi(H_\Lambda \le E)$ onto eigenstates of H_Λ with energy less than or equal to E. So that the IDS appears as

$$\mathcal{N}(E) = \lim_{\Lambda\uparrow\mathbb{R}^n} \mathrm{Tr}\{\chi(H_\Lambda \le E)\}/|\Lambda| \,. \tag{4.1.2}$$

The first rigorous work on the IDS goes back to Benderskii-Pastur [BP70], who proved the existence of the limit for a one-dimensional Schrödinger operator on a lattice with a random potential. Then the existence and smoothness properties of the derivative $d\mathcal{N}$ as a Stieljes-Lebesgue measure, were proved by different methods with an increasing degree of generality for the Schrödinger operator with random potential by Pastur [PA73], Nakao [NA77], Kirsch and Martinelli [KM82]. The algebraic approach goes back to the work of Shubin [SH79] inspired by the Index theory of Coburn, Moyer and Singer [CM73] on uniformly elliptic operators with almost periodic coefficients. The extension to more general coefficients is elementary and has been given in the discrete case by the author in [BE86], and for the continuum case it is given below [BL85].

The formula (4.1.2) is very reminiscent to the formula defining the trace per unit volume of the eigenprojector $\chi(H \leq E)$ of the infinite volume limit. Notice that this projector does not belong in general to the C^*-algebra $\mathcal{A}$ of H but to the Von Neumann algebra $L^\infty(\mathcal{A}, \tau)$ associated to the trace per unit volume by means of the GNS construction.

Definition : Shubin's Formula.

We say that H obeys *Shubin's formula* whenever the IDS satisfies:

(4.1.3) $$\mathcal{N}(E) = \tau\{\chi(H \leq E)\} ,$$

where τ is the trace per unit volume in the C^*-algebra of H.

To establish Shubin's formula, we have to compare $\mathrm{Tr}\{\chi(H_\Lambda \leq E)\}/|\Lambda|$ with $\mathrm{Tr}\{\chi_\Lambda \chi(H \leq E)\}/|\Lambda|$ as $\Lambda \uparrow \mathbb{R}^D$, where χ_Λ is the characteristic function of Λ , and to show that they are equal. The first result in this respect was provided by Shubin [SH79] whenever H is a uniformly elliptic partial differential operator with almost periodic coefficients.

Actually, Shubin's proof can be extended to a more general situation. Let us formulate the result. Let Ω be a compact space with a $\mathbb{R}^D$-action by a group $\{T^a;\ a \in \mathbb{R}^D\}$ of homeomorphisms. We assume that there is $\omega_0 \in \Omega$ the orbit of which being dense in Ω. Let $U^\infty(\Omega, \mathbb{R}^n)$ be the space of smooth functions f on $\mathbb{R}^n$, with bounded derivatives of every order, such that there exists a continuous function F on Ω for which $f(x) = F(T^{-x}\omega_0)$. We set $f_\omega(x) = F(T^{-x}\omega)$. Let H_ω be a uniformly elliptic formally self-adjoint operator of the form:

(4.1.4) $$H_\omega = \sum_{|\alpha| \leq m} h_\omega^{(\alpha)}(x) D^\alpha ,$$

where $h^{(\alpha)} \in U^\infty(\Omega, \mathbb{R}^n)$, $\alpha = (\alpha_1, \cdots, \alpha_n) \in \mathbb{N}^n$ is a multi-index, $|\alpha| = \alpha_1 + \cdots + \alpha_n$, whereas $D^\alpha = \prod_{1 \leq \mu \leq n} \{-\mathrm{i}\partial_\mu\}^{\alpha_\mu}$. By uniformly elliptic we mean that the principal symbol $h_\omega^{(m)}$ satisfies:

(4.1.5) $$h_\omega^{(m)}(x, \zeta) = \sum_{|\alpha| = m} h_\omega^{(\alpha)}(x) \zeta^\alpha \geq \varepsilon |\zeta|^m , \quad \varepsilon > 0\,; x, \zeta \in \mathbb{R}^n .$$

By formally self-adjoint, we mean that for any pair ϕ, ψ of functions in the Schwartz space $\mathcal{S}(\mathbb{R}^n)$ we have:

(4.1.6) $$\langle \phi | H_\omega \psi \rangle = \langle H_\omega \phi | \psi \rangle .$$

Then it is well-known that H_ω admits a unique self-adjoint extension, still denoted by H_ω and that $\mathcal{S}(\mathbb{R}^n)$ is a core [SH79, HÖ85]. Moreover, it satisfies the covariance condition

$$U(a)H_\omega U(a)^* = H_{T^a\omega} \,. \tag{4.1.7}$$

Now for any Λ in the Følner sequence defined above, we consider the operator $H_{\omega\Lambda}$ defined by (4.1.4), with domain $\mathcal{D}(\Lambda)$ given by the space of functions ψ in the Sobolev space $\mathcal{H}^m(\Lambda)$ satisfying the boundary conditions $B_j\psi|_{\partial\Lambda} = 0\,(1 \le j \le m/2)$; we assume the operators B_j to have order not exceeding $m-1$, and to be self-adjoint. It is known that $H_{\omega,\Lambda}$ is self-adjoint and has a discrete spectrum, bounded from below. Therefore $N_\Lambda(E)$ exists and is finite. The main result of Shubin [SH79, Theorem 2.1] can be rephrased as follows:

Theorem 8. *If H_ω is a uniformly elliptic self-adjoint operator with coefficients in $U^\infty(\Omega, \mathbb{R}^n)$, then:*

(i) there is $r(z) \in C^*(\Omega \times \mathbb{R}^n, B = 0)$ such that $\pi_\omega(r(z)) = \{z - H_\omega\}^{-1}$, for every complex z in the resolvent set of H_ω;

(ii) Shubin's formula holds.

Sketch of the proof. Let us only indicate the punch line of the proof. First of all, standard results show that the elementary solution of the heat equation $\partial u/\partial t = -H_\omega u$ admits a regular kernel $G_\omega(t; x, y)$ which depends smoothly upon x and y, and satisfies for any multi-indices α, β, γ the following type of estimate:

$$|\partial_x^\alpha \partial_y^\beta \partial_t^\gamma G_\omega(t; x, y)| \le C t^{-(n+|\alpha|+|\beta|+m\gamma)/m} \exp\{-\varepsilon(|x-y|^m/t)^{1/m}\} \,, \tag{4.1.8}$$

where $0 < t < T$, and the constant $C > 0$ depends only on α, β, γ and T. The covariance condition shows that if we set $G(t; \omega, x) = G_\omega(t; 0, x)$, then $G_\omega(t; x, y) = G(t; T^{-x}\omega, y - x)$. Thus $G(t; \omega, x)$ defines for each $t > 0$, an element $g(t)$ of $L^{\infty,1}(\Omega \times \mathbb{R}^n; B = 0)$, showing that $e^{-tH_\omega} = \pi_\omega(g(t))$. It follows from the covariance condition, the strong continuity of $\pi_\omega(g(t))$ with respect to ω, and the density of the orbit of ω_0 in Ω, that the spectrum of H_ω is contained in the spectrum of H. The resolvent is given by $(H_\omega - z)^{-1} = \int_{[0,\infty]} \mathrm{d}t \exp\{-t(H_\omega - z)\}$ for $\mathrm{Re}(z)$ sufficiently negative, showing that $r(z)$ exists.

On the other hand, $\mathcal{N}(E)$ is a non decreasing non negative function of E. Its derivative exists as a Stieljes-Lebesgue measure. It is thus sufficient to show that its Laplace transform $N(t) = \int \mathcal{N}(dE)e^{-Et}$ is equal to $\tau\{\exp(-tH)\}$. If we call $f_{\omega,\Lambda}$ the kernel of the operator $\exp\{-tH_\omega\} - \exp\{-tH_{\omega,\Lambda}\}$, one has the following estimate:

$$|\partial_x^\alpha \partial_t^\gamma f_{\omega,\Lambda}(t; x, y)| \le C t^{-(n+|\alpha|+m\gamma)/m} \exp\{-\varepsilon[(|x-y| + \mathrm{dist}(y, \partial\Lambda))^m/t]^{1/m}\} \,, \tag{4.1.9}$$

showing that $\lim_{\Lambda\uparrow\mathbb{R}^n} \mathrm{Tr}\{f_{\omega,\Lambda}\}/|\Lambda| = 0$, and therefore the Shubin formula holds. □

In the discrete case the situation is much simpler, from a technical point of view. Let G be a (non necessarily Abelian) countable discrete amenable group [GR69]. By amenable we mean that there is a Følner sequence, namely an increasing sequence $(\Lambda_n)_{n\geq 0}$ of finite subsets of G, with union equal to G, and such that $\lim_{n\to\infty} |\Lambda_n \Delta(a^{-1}\Lambda_n)|/|\Lambda_n| = 0$. In practice, G will be $\mathbb{Z}^n$, but it may perfectly contain non Abelian parts, corresponding to discrete symmetries. Let Ω be a compact space endowed with a G-action by homeomorphisms, and let $\mathbf{P}$ be a G-invariant ergodic probability measure. The set $\Gamma = \Omega \times G$ is a topological groupoid if we set $r(\omega, x) = \omega, s(\omega, x) = x^{-1}\omega, (\omega, y) \circ (y^{-1}\omega, y^{-1}x) = (\omega, x), (\omega, x)^{-1} = (x^{-1}\omega, x^{-1})$. Let $\Gamma^{(2)}$ denote the set of pairs (γ_1, γ_2) with $r(\gamma_2) = s(\gamma_1)$. A unitary cocycle is a mapping $\delta : \Gamma^{(2)} \to S_1$, where S_1 is the unit circle, such that $\delta(\gamma_1,\gamma_2)\delta(\gamma_1 \circ \gamma_2, \gamma_3) = \delta(\gamma_1, \gamma_2 \circ \gamma_3)\delta(\gamma_2, \gamma_3)$. We can then construct the C^*-algebra $C^*(\Omega, G, \delta)$ by means of the method of Section 2.5 [RE80], where now G replaces $\mathbb{R}^n$, the Haar measure on G (here the discrete sum) replacing Lebesgue's measure, and δ replacing $e^{i\pi(e/h)B.x\wedge y}$. We also get a family π_ω of representations on $l^2(G)$ in a similar way. Then [BE86 Appendix]:

Theorem 9. *Let H be a self-adjoint element of $C^*(\Omega, G, \delta)$. Then the Shubin formula holds for H.*

In what follows, the spectrum of H is understood as the spectrum in the C^*-algebra, namely

$$\mathrm{Sp}(H) = \bigcup_{\omega\in\Omega} \sigma\{\pi_\omega(H)\} \ . \tag{4.1.10}$$

The main properties of the IDS are listed below. From Shubin's formula it is easy to get:

Proposition 4.1.1. *Let H be a homogeneous self-adjoint Hamiltonian, and let $\mathcal{A}$ be the C^*-algebra it generates by translation. Let τ be a translation invariant trace, for which H obeys to Shubin's formula.*

Then its IDS *is a non negative, non decreasing function of E, which is constant on each gap of* Sp(H).

The IDS also depends upon the choice of a translation invariant ergodic probability measure $\mathbf{P}$ on the hull of the Hamiltonian. We say that a trace τ on a C^*-algebra $\mathcal{A}$ is faithful if any f in $L^1(\mathcal{A}, \tau) \cap \mathcal{A}$ with $\tau(f^*f) = 0$ necessarily vanishes. Then

Proposition 4.1.2. *Let H be as in Proposition* 4.1.1. *If τ is faithful, the spectrum of H coincides with the set of points $E \in \mathbb{R}$ in the vicinity of which the IDS is not constant.*

Proof. Let E be a point in the spectrum of H (the spectrum is the set of $E \in \mathbb{R}$ for which $(E\mathbf{1}-H)$ has no inverse in $\mathcal{A}$). Then for any $\delta > 0$ the eigenprojection $\chi(E - \delta < H \leq E + \delta)$ is a non zero positive element of the Von Neumann algebra $L^\infty(\mathcal{A}, \tau)$. Since τ is faithful, we have

$$\mathcal{N}(E+\delta) - \mathcal{N}(E-\delta) = \tau(\chi(E-\delta < H \leq E+\delta)) > 0\,,$$

proving the result. □

Remark. Suppose we consider the operator $H = -\Delta + V$ on $\mathbb{R}^n$ where V decays at infinity. Then the hull of H is the one-point compactification of $\mathbb{R}^n$. The only translation invariant ergodic probability measure on the hull $\Omega = \mathbb{R}^n \cup \{\infty\}$ is the Dirac measure at ∞. Therefore, the trace per unit volume cannot be faithful. In particular the IDS does not take into account the discrete spectrum of H, since the density of such eigenvalues is zero.

Another example of such phenomena is provided by strong limits when $\alpha \to p/q$ of the Kohmoto model (see Section 3.2), which may exhibit simple isolated eigenvalues in the gaps of the spectrum of the periodic model [BI90].

Proposition 4.1.3. *With the assumption of Proposition* 4.1.1, *any discontinuity point of the IDS is an eigenvalue of H with an infinite multiplicity.*

Proof. Let E be such a discontinuity point. It means that there is $\varepsilon > 0$, such that for every $\delta > 0$, $\mathcal{N}(E+\delta) - \mathcal{N}(E-\delta) = \tau(\chi(E-\delta < H \leq E+\delta)) \geq \varepsilon$. Taking the limit $\delta \to 0$, we get a non zero eigenprojection $\chi(H = E)$ corresponding to the eigenvalue E. Since $\tau(\chi(H = E)) \geq \varepsilon > 0$, it follows that the multiplicity per unit volume is bigger than or equal to ε, and therefore the multiplicity of E is infinite. □

Remark. Many examples of physical models have given such eigenvalues. For instance, the discrete Laplace operator on a Sierpinski lattice [RA84] has a pure-point spectrum containing an infinite family of isolated eigenvalues of infinite multiplicity, the other eigenvalues being limit points and having infinite multiplicity as well. So that the IDS has no continuity point on the spectrum. Another example was provided by Kohmoto et al.[KS86], namely the discrete Laplace operator on a Penrose lattice in 2D, for which $E = 0$ is an isolated eigenvalue with a non zero multiplicity per unit area.

Actually one has the following result proved by Craig and Simon [CS83] and in a very elegant way, by Delyon and Souillard [DS84]:

Proposition 4.1.4. *Let $H = \Delta + V$ be self adjoint, bounded and homogeneous on $\mathbb{Z}^n$. Then its IDS is continuous.*

Sketch of the proof. Taking Λ a finite parallelepipedic box, the eigenvalues of H_Λ have multiplicity not bigger than $O(|\partial\Lambda|)$. This is due to the special geometry of the hypercubic lattice $\mathbb{Z}^n$. Therefore the multiplicity per unit volume of any eigenvalue must vanish. □

Remark. 1) Such a general argument is not available yet in the continuum. However with reasonable assumptions on the potential one may expect such a result.

2) Craig and Simon have actually proved that the IDS is Log-Hölder continuous [CS83].

3) The extension of this result to Hamiltonians with short range interactions should be possible.

4.2 Gap Labelling and the Group K_0

In the previous Section, we have introduced the IDS, which is constant on each gap of the spectrum. A label of a spectral gap $\{g\}$ is given by the value $\mathcal{N}(g)$ taken by the IDS on this gap. This is a real number which has been recognized to be very rigid under perturbations of the Hamiltonian. In this Section we will label gaps in a different way, which will explain such a rigidity.

For indeed, one can associate to $\{g\}$ the eigenprojection $P(g)$ of the Hamiltonian H, on the interval $(-\infty, E]$ of energies where E is any point in $\{g\}$. Clearly, $P(g)$ does not depend upon the choice of E in $\{g\}$. Moreover, if f is any smooth function on $\mathbb{R}$ such that $0 \leq f \leq 1$, $f(E) = 1$ for $E \leq \inf\{g\}$, and $f(E) = 0$ for $E \geq \sup\{g\}$, then $P(g) = f(H)$. In particular, $P(g)$ belongs to the C^*-algebra $\mathcal{A}$ of H . However, the data of $P(g)$ contained not only an information about the spectral gap $\{g\}$, but also about the nature of the spectral measure of H. Changing H by a unitary transformation, or more generally by an algebraic automorphism will not change the spectrum as a set. So that it is sufficient to label $\{g\}$ by means of the equivalence class of $P(g)$ under unitary transformations.

This is a highly non trivial fact, for the set of equivalence classes of projections of $\mathcal{A}$ is usually a rather small set. To illustrate this claim, let $\mathcal{K}$ be the algebra of compact operators on a separable Hilbert space. Then a projection is compact if and only if it is finite dimensional. Moreover, two such projections are unitarily equivalent if and only if they have the same dimension. So, up to unitary equivalence, the set of projections in $\mathcal{K}$ is nothing but the set $\mathbb{N}$ of integers.

Actually we will use the Von Neumann definition of equivalence [PE79], namely:

Definition Two projections P and Q of a C^*-algebra $\mathcal{A}$ are *equivalent* if there is $U \in \mathcal{A}$ such that $UU^* = P$ and $U^*U = Q$. We will then write $P \approx Q$.

We will now give a few lemmas.

Lemma 4.2.1 [PE79]. *Let P and Q be two projections in a C^*-algebra $\mathcal{A}$ such that $\|P-Q\| < 1$. Then they are equivalent. In particular any norm continuous path $t \in \mathbb{R} \to P(t) \in \mathrm{Proj}(\mathcal{A})$ is made of mutually equivalent projections.*

Proof. We consider $F = PQ$ in $\mathcal{A}$. We claim that its square FF^* is invertible in the subalgebra $P\mathcal{A}P$. Actually, since P is the unit in $P\mathcal{A}P$, we get $\|FF^* - P\| \leq \|Q - P\| < 1$. Therefore the series $\sum_{n\geq 0} c_n(P - FF^*)^n$ converges absolutely in norm provides $\limsup_{n\to\infty} |c_n|^{1/n} \leq 1$. In particular $(FF^*)^{-1/2} = (P - (P - FF^*))^{-1/2}$ can be constructed as the sum of such a norm convergent series in $P\mathcal{A}P$. If we now set $U = (FF^*)^{-1/2}F$ then we get an element in $\mathcal{A}$ which satisfies $UU^* = P$. On the other hand $U^*U = F^*\{FF^*\}^{-1}F$ is the sum of the series $\sum_{n\geq 0} F^*(P - FF^*)^n F = \sum_{n\geq 0}(Q - F^*F)^n F^*F = Q$. Thus $P \approx Q$. □

Let $\mathcal{A}$ be a separable C^*-algebra. We will denote by $\mathcal{P}(\mathcal{A})$ the set of equivalence classes of projections in $\mathcal{A}$, and by $[P]$ the equivalence class of P. The next important property is

Lemma 4.2.2 [PE79]. *Let $\mathcal{A}$ be a separable C^*-algebra. The set $\mathcal{P}(\mathcal{A})$ of equivalence classes of projections in $\mathcal{A}$ is countable.*

Proof. Since $\mathcal{A}$ is separable, there is a countable set $(A_n)_{n\geq 0}$ in $\mathcal{A}$ which is norm dense in the unit ball. Given P a projection, and $\varepsilon < 1/2$, there is an $n \in \mathbb{N}$ such that $\|P - A_n\| \leq \varepsilon$. Replacing A_n by $(A_n + A_n^*)/2$ does not change this estimate, so that we can assume A_n to be self-adjoint. Thus its spectrum is contained into the union of two closed real intervals of width ε, the first centered at zero, the other at one. If the eigenprojection corresponding to the latter piece of the spectrum of A_n is denoted by P_n, we get $\|P_n - A_n\| \leq \varepsilon$, showing that $\|P - P_n\| \leq 1$, namely $P \approx P_n$. Therefore the set of equivalence classes of projections contains at most a countable family of elements. □

Two projections P and Q in $\mathcal{A}$ are orthogonal whenever $PQ = QP = 0$. In this latter case, the operator sum $P + Q$ is a new projection called the direct sum of P and Q, and it is denoted by $P \oplus Q$. We then get:

Lemma 4.2.3. *Let P and Q be two orthogonal projections in the C^*-algebra $\mathcal{A}$. Then the equivalence class of their direct sum depends only upon the equivalence classes of P and of Q . In particular, a sum is defined on the set $\mathcal{E}$ of pairs $([P],[Q])$ in $\mathcal{P}(\mathcal{A})$ such that there are $P' \approx P$ and $Q' \approx Q$ with $P'Q' = Q'P' = 0$, by $[P]+[Q] \equiv [P' \oplus Q']$. This composition law is commutative and associative.*

Proof. Suppose that there are projections $P' \approx P$ and $Q' \approx Q$ such that $P'Q' = Q'P' = 0$. This means that there are U and V in $\mathcal{A}$ such that $UU^* = P$, $U^*U = P'$, $VV^* = Q$, $V^*V = Q'$. One immediately checks that $(\mathbf{1} - P)U = 0$, and similar relations with U^*, V, V^*. It implies in particular that $VU^* = UV^* = V^*U = U^*V = 0$. Thus if we set $W = U + V$, we get $WW^* = P \oplus Q$ and $W^*W = P' \oplus Q'$. Hence, $P \oplus Q \approx P' \oplus Q'$. So that the sum is defined on the set $\mathcal{E}$ of pairs $([P],[Q])$ in $\mathcal{P}(\mathcal{A})$ such that there are $P' \approx P$ and $Q' \approx Q$ with $P'Q' = Q'P' = 0$, by $[P] + [Q] \equiv [P' \oplus Q']$. Since two orthogonal projections commute, the sum is commutative. Because the direct sum comes from the operator sum it is associative as well. □

The main problem is that $\mathcal{E}$ can be rather small, and our first task is to enlarge this definition in order to define the sum everywhere. The main idea consists in replacing the algebra $\mathcal{A}$ by the stabilized algebra $\mathcal{A}\otimes\mathcal{K}$. We have seen that such an algebra is naturally associated to a non commutative manifold in that it gives the set of multipliers of sections on any fiber bundle. We will imbed $\mathcal{A}$ into $\mathcal{A}\otimes\mathcal{K}$ by identifying $A \in \mathcal{A}$ with the matrix $\{i(A)\}_{m,n} = A\delta_{m,0}\delta_{n,0}$, if $\mathcal{K}$ is identified with the algebra generated by finite dimensional matrices indexed by $\mathbb{N}$. Actually we get:

Lemma 4.2.4. *Given any pair P and Q of projections in the C^*-algebra $\mathcal{A}\otimes\mathcal{K}$, there is always a pair P', Q' of mutually orthogonal projections in $\mathcal{A}\otimes\mathcal{K}$ such that $P' \approx P$ and $Q' \approx Q$. In particular the sum $[P]+[Q] \equiv [P' \oplus Q']$ is always defined.*

Proof. By definition any element in $\mathcal{A}\otimes\mathcal{K}$ can be approximated in norm by a finite matrix with elements in $\mathcal{A}$, namely by an element of $\mathcal{A}\otimes M_N(C)$ for some $N \in \mathbb{N}$. In particular if P is a projection in $A \otimes \mathcal{K}$, given $\varepsilon < 1/2$, one can find $A = A^*$ in $\mathcal{A}\otimes M_N(C)$ such that $\|P - A\| \leq \varepsilon$. Thus the spectrum of A is contained in the union of two disks of radius ε one centered at the origin, the other one centered at $z = 1$. The spectral projection P' corresponding to the latter one satisfies $\|P' - A\| \leq \varepsilon$, and $P' \in \mathcal{A}\otimes M_N(C)$. We get that $\|P - P'\| < 1$, namely $P \approx P'$. Thus one can always suppose that P is in $\mathcal{A}\otimes M_N(C)$. If now Q is another projection in $\mathcal{A}\otimes M_N(C)$, Q is equivalent to the matrix Q' in $\mathcal{A}\otimes M_{2N}(C)$ with $\{Q'\}_{m,n} = 0$ for $m,n \notin [N, 2N-1]$ and $\{Q'\}_{N+m,N+n} = \{Q\}_{m,n}$ if $m,n \in [0, N-1]$. In particular, $PQ' = Q'P = 0$. Therefore in $\mathcal{A}\otimes\mathcal{K}$ one can always add two equivalence classes. □

In other words, if $\mathcal{A}$ is a stable algebra, the set $\mathcal{P}(\mathcal{A})$ of equivalence classes of projections is an Abelian monoid with neutral element given by the class of the zero projection. If $\mathcal{A}$ is not stable, we will still denote by $\mathcal{P}(\mathcal{A})$ the set of equivalence classes of projections of the stabilized algebra $\mathcal{A}\otimes\mathcal{K}$.

There is a standard way to construct a group from such a monoid [CU82], generalizing the construction of $\mathbb{Z}$ from $\mathbb{N}$. We consider on the set of pairs $([P],[Q]) \in \mathcal{P}(\mathcal{A}) \times \mathcal{P}(\mathcal{A})$, the equivalence relation

$$(4.2.1)\quad ([P],[Q])\mathcal{R}([P'],[Q']) \Leftrightarrow \exists [R] \in \mathcal{P}(\mathcal{A})\ [P]+[Q']+[R] = [P']+[Q]+[R]$$

Roughly speaking it means that $[P]-[Q] = [P']-[Q']$. We get

Theorem 10. *(i) For a separable C^*-algebra $\mathcal{A}$, the set $\mathrm{K}_0(\mathcal{A}) = \mathcal{P}(\mathcal{A}\otimes\mathcal{K}) \times \mathcal{P}(\mathcal{A}\otimes\mathcal{K})/\mathcal{R}$ is countable and has a natural structure of Abelian group.*

(ii) Any real valued map ϕ defined on the set of projections of $\mathcal{A}$ such that $P \approx Q \Rightarrow \phi(P) = \phi(Q)$, and $\phi(P \oplus Q) = \phi(P) + \phi(Q)$, defines canonically a group homomorphism ϕ^* from $\mathrm{K}_0(\mathcal{A})$ into $\mathbb{R}$.

(iii) Any trace τ on $\mathcal{A}$ defines in a unique way a group homomorphism τ^* such that if P is a projection on $\mathcal{A}, \tau(P) = \tau^*([P])$ where $[P]$ is the class of P in $\mathrm{K}_0(\mathcal{A})$.

Proof. (i) Since $\mathcal{A} \otimes \mathcal{K}$ is separable the set $\mathcal{P}(\mathcal{A})$ is countable and so is $\mathrm{K}_0(\mathcal{A})$. Moreover, the equivalence relation $\mathcal{R}$ is compatible with the addition in $\mathcal{P}(\mathcal{A})$, namely if $([P_1],[Q_1])\mathcal{R}([P_1'],[Q_1'])$ and $([P_2],[Q_2])\mathcal{R}([P_2'],[Q_2'])$ then from (4.2.1) there is an $[R]$ such that

$$[P_1] + [P_2] + [Q_1'] + [Q_2'] + [R] = [P_1'] + [P_2'] + [Q_1] + [Q_2] + [R] ,$$

namely $([P_1] + [P_2], [Q_1] + [Q_2])\mathcal{R}([P_1'] + [P_2'], [Q_1'] + [Q_2'])$, showing that the addition on $\mathcal{P}(\mathcal{A}) \times \mathcal{P}(\mathcal{A})$ defines on the quotient space a structure of Abelian monoid. The neutral element is the class of $[0] \equiv ([0],[0])$, namely the set of $([R],[R])$'s with $[R]$ in $\mathcal{P}(\mathcal{A})$. Thus for any element $([P],[Q])$ in $\mathcal{P}(\mathcal{A}) \times \mathcal{P}(\mathcal{A})$ the sum $([P],[Q]) + ([Q],[P])$ is equivalent to $[0]$, showing that in the quotient space any element has an opposite element: so it is a group. Clearly $\mathcal{P}(\mathcal{A})$ is imbedded in this group through the map $[P] \longrightarrow ([P],[0])$ and the quotient map, giving a homomorphism of monoid.

From now on, we will identify $[P]$ with its image in $\mathrm{K}_0(\mathcal{A})$. The assertions (ii) and (iii) are direct consequences of the definition of $\mathrm{K}_0(\mathcal{A})$. □

Theorem 11 (The Abstract Gap Labelling Theorem). *Let H be a homogeneous self-adjoint operator satisfying Shubin's formula (4.1.3). Let $\mathcal{A}$ be the C^*-algebra it generates together with the translation group (its non commutative Brillouin zone). Let $\mathrm{Sp}(H)$ be its spectrum in $\mathcal{A}$. Then for any gap $\{g\}$ of $\mathrm{Sp}(H)$,*

(i) the value of the IDS of H on $\{g\}$ belongs to the countable set of real numbers $[0, \tau(\mathbf{1})] \cap \tau^*(\mathrm{K}_0(\mathcal{A}))$.

(ii) the equivalence class $n(g) = [P(g)] \in \mathrm{K}_0(\mathcal{A})$, gives a labelling invariant under small perturbations of the Hamiltonian H within $\mathcal{A}$.

(iii) If S is a subset of $\mathbb{R}$ which is closed and open in $\mathrm{Sp}(H)$, then $n(S) = [P_S] \in \mathrm{K}_0(\mathcal{A})$ where P_S is the eigenprojection of H corresponding to S, is a labelling for each such part of the spectrum.

As a consequence, we get the stability of the gap labelling namely the property (ii) above which is immediate from the Lemma 4.2.1. Let us make this point more precise:

Proposition 4.2.5 (Stability and Sum Rules for the Labelling). *Let $t \in \mathbb{R} \to H(t)$ be a continuous family (in the norm-resolvent sense) of self-adjoint operators with resolvent in a separable C^*-algebra $\mathcal{A}$.*

(i) invariance: *the gap edges of H are continuous and the labelling of a gap $\{g(t)\}$ is independent of t as long as the gap does not close.*

(ii) sum rule: *suppose that for $t \in [t_0, t_1]$, the spectrum of $H(t)$ contains an open-closed subset $S(t)$ such that $S(t_0) = S_+ \cup S_-$ and $S(t_1) = S'_+ \cup S'_-$ where $S_\pm$ and $S'_\pm$ are open-closed in $\mathrm{Sp}(H(t_0))$, $\mathrm{Sp}(H(t_1))$; then $n(S_+) + n(S_-) = n(S'_+) + n(S'_-)$.*

4.3 Properties of the K-Groups

In Section 4.2, we have defined the group $\mathrm{K}_0(\mathcal{A})$ in an abstract way, and we have shown how it can be used to label the gaps or the open-closed subsets of the spectrum of a homogeneous Hamiltonian. Obviously we will need to compute more explicitly the group $\mathrm{K}_0(\mathcal{A})$ and its image under the trace per unit volume in order to get a labelling of the gaps. In this Section let us give some general rules for such a computation, which we will use in the last Section to get the result in various classes of examples. The proofs of these properties are too long to be reproduced here, so that we will invite the reader to look into the references if he wants to have the proofs.

Let us first define the group $\mathrm{K}_1(\mathcal{A})$. Let $\mathcal{A}$ be a separable C^*-algebra with a unit. $\mathrm{GL}_n(\mathcal{A})$ is the group of invertible elements of the algebra $M_n(\mathcal{A})$. One can consider $\mathrm{GL}_n(\mathcal{A})$ as the subgroup of $\mathrm{GL}_{n+1}(\mathcal{A})$ made of matrices X with coefficients in $\mathcal{A}$, such that $X_{n+1,m} = X_{m,n+1} = \delta_{n+1,m}$. $\mathrm{GL}(\mathcal{A})$ is then the inductive limit of this family, namely the norm closure of their union. Let $\mathrm{GL}(\mathcal{A})_0$ be the connected component of the identity in $\mathrm{GL}(\mathcal{A})$. One set:

$$\mathrm{K}_1(\mathcal{A}) = \pi_0(\mathrm{GL}(\mathcal{A})) = \mathrm{GL}(\mathcal{A})/\mathrm{GL}(\mathcal{A})_0 \,. \tag{4.3.1}$$

Here $\pi_0(M)$ denotes the set of connected components of the topological space M. If A has no unit, one sets $\mathrm{K}_1(A) = \mathrm{K}_1(\mathcal{A}^\sim)$.

So $\mathrm{K}_1(\mathcal{A})$ is an Abelian group. For indeed the path $B(t)$ defined by

$$B(t) = \begin{bmatrix} \cos t & \sin t \\ -\sin t & \cos t \end{bmatrix} \begin{bmatrix} B & 0 \\ 0 & \mathbf{1} \end{bmatrix} \begin{bmatrix} \cos t & -\sin t \\ \sin t & \cos t \end{bmatrix}$$

permits to connect the following matrices in $\mathrm{GL}(\mathcal{A})$

$$\begin{bmatrix} AB & 0 \\ 0 & \mathbf{1} \end{bmatrix} \approx \begin{bmatrix} A & 0 \\ 0 & B \end{bmatrix} \approx \begin{bmatrix} BA & 0 \\ 0 & \mathbf{1} \end{bmatrix} .$$

Since $\mathcal{A}$ is separable, it is easy to check that $\mathrm{K}_1(\mathcal{A})$ is countable.

The main general properties of $\mathrm{K}_0(\mathcal{A})$ and $\mathrm{K}_1(\mathcal{A})$ are summarized in the following theorem

Theorem 12. *Let $\mathcal{A}, \mathcal{B}$ be separable C^*-algebras*

*(i) If $f : \mathcal{A} \to \mathcal{B}$ is a *-homomorphism, then f defines in a unique way groups homomorphisms f_* from $\mathrm{K}_i(\mathcal{A})$ into $\mathrm{K}_i(\mathcal{B})$ $(i = 0, 1)$. They satisfy* $\mathrm{id}_* = \mathrm{id}$, *and* $(f \circ g)_* = f_* \circ g_*$.

(ii) $\mathrm{K}_i(\mathcal{A} \oplus \mathcal{B})$ *is isomorphic to* $\mathrm{K}_i(\mathcal{A}) \oplus \mathrm{K}_i(\mathcal{B})$.

(iii) If $\mathcal{A}$ is the inductive limit of a sequence $(\mathcal{A}_n)_{n \geq 0}$ of C^*-algebras then $\mathrm{K}_i(\mathcal{A})$ *is the inductive limit of the groups* $\mathrm{K}_i(\mathcal{A}_n)$.

(iv) If $\mathcal{C}_0(\mathbb{R})$ is the space of continuous functions on $\mathbb{R}$ vanishing at infinity, $\mathrm{K}_i(\mathcal{A})$ *is isomorphic to* $\mathrm{K}_{i+1}(\mathcal{C}_0(\mathbb{R}) \otimes \mathcal{A})$ *(where $i + 2 \equiv i$).*

(v) $\mathrm{K}_0(\mathcal{A})$ *is isomorphic to the group* $\pi_1(\mathrm{GL}(\mathcal{A}))$ *of homotopy classes of closed paths in* $\mathrm{GL}(\mathcal{A})$. *If τ is a trace on $\mathcal{A}$ and if $t \in [0,1] \to U(t)$ is a closed path in* $\mathrm{GL}(\mathcal{A})$ *we get* [CO81]:

$$\tau_*([U]) = 1/2\mathbf{i}\pi \int_{[0,1]} \mathrm{d}t \tau(U(t)^{-1}U'(t)) \tag{4.3.2}$$

*(vi) If $\phi : \mathcal{J} \to \mathcal{A}$, and $\psi : \mathcal{A} \to \mathcal{B}$ are *-homomorphisms such that the sequence $0 \to \mathcal{J} \to \mathcal{A} \to \mathcal{B} \to 0$ is exact, then the following six-terms sequence is exact* [CU82, BL86]:

$$\begin{array}{ccccc}
\mathrm{K}_0(\mathcal{J}) & \xrightarrow{\phi^*} & \mathrm{K}_0(\mathcal{A}) & \xrightarrow{\psi^*} & \mathrm{K}_0(\mathcal{B}) \\
\mathrm{Ind} \uparrow & & & & \downarrow \mathrm{Exp} \\
\mathrm{K}_1(\mathcal{B}) & \xleftarrow[\psi^*]{} & \mathrm{K}_1(\mathcal{A}) & \xleftarrow[\phi^*]{} & \mathrm{K}_1(\mathcal{J})
\end{array}$$

In the previous theorem, Ind and Exp are the 'connection' automorphisms defined as follows assuming that $\mathcal{A}$ has a unit. Let P be a projection in $\mathcal{B} \otimes \mathcal{K}$, and let A be a self-adjoint element of $\mathcal{A} \otimes \mathcal{K}$ such that $\psi \otimes \mathrm{id}(A) = P$. One gets:

$$\psi \otimes \mathrm{id}(e^{2\mathbf{i}\pi A}) = e^{2\mathbf{i}\pi P} = \mathbf{1} \, .$$

So that $B = e^{2\mathbf{i}\pi A} \in (\mathcal{J} \otimes \mathcal{K})^{\sim}$ and is unitary in $(\mathcal{J} \otimes \mathcal{K})^{\sim}$. The class of B gives an element of $\mathrm{K}_1(\mathcal{J})$ which is by definition $\mathrm{Exp}([P])$. One can check that this definition makes sense. In much the same way, let now U be an element of $(\mathcal{B} \otimes \mathcal{K}^{\sim})$ which we may assume without loss of generality to be the image by $\psi \otimes \mathrm{id}$ of a partial isometry W in $(\mathcal{A} \otimes \mathcal{K})^{\sim}$. Then $\mathrm{Ind}([U])$ is the class of $[WW^*] - [W^*W]$ in $\mathrm{K}_0(\mathcal{J})$. One can also check that this definition makes sense.

In order to compute these groups in practice we will also need two other kinds of results. The first one is a theorem by Pimsner and Voiculescu [PV80a], and concerns $\mathbb{Z}$-actions on a C^*-algebra.

Theorem 13. *Let $\mathcal{A}$ be a separable C^*-algebra, and α be a *-automorphism of $\mathcal{A}$. The crossed product of $\mathcal{A}$ by $\mathbb{Z}$ via α is the C^*-algebra generated by $\mathcal{A}$ and a unitary U such that $UAU^{-1} = \alpha(A)$ for all A's in $\mathcal{A}$. Then there exists a six-terms exact sequence of the form:*

$$\begin{array}{ccccc}
\mathrm{K}_0(\mathcal{A}) & \xrightarrow{\mathrm{id}-\alpha_*} & \mathrm{K}_0(\mathcal{A}) & \xrightarrow{j_*} & \mathrm{K}_0(\mathcal{A}\rtimes_\alpha \mathbb{Z}) \\
\mathrm{Ind}\uparrow & & & & \downarrow \mathrm{Exp} \\
\mathrm{K}_1(\mathcal{A}\rtimes_\alpha \mathbb{Z}) & \xleftarrow[j_*]{} & \mathrm{K}_1(\mathcal{A}) & \xleftarrow[\mathrm{id}-\alpha_*]{} & \mathrm{K}_1(\mathcal{A})
\end{array}$$

where j is the canonical injection of $\mathcal{A}$ into the crossed product $\mathcal{B} = \mathcal{A} \rtimes_\alpha \mathbb{Z}$.

The connection homomorphisms in Theorem 13, are defined from the following exact sequence (called the Toeplitz extension of $\mathcal{A}$): let S be a non unitary isometry such that $S^*S = 1$, $SS^* = \mathbf{1} - P$, with $P \neq 0$. The C^*-algebra $C^*(S)$ defined by S does not depend upon the choice of S [CO67]. Choosing U in $\mathcal{B}$ such that $\alpha(A) = U^*AU$, one gets a homomorphism ψ from $\mathcal{A} \otimes \mathcal{K}$ into the subalgebra $\mathcal{T}$ of $\mathcal{B} \otimes C^*(S)$ generated by $\mathcal{U} \otimes S$ and $\mathcal{A} \otimes \mathbf{1}$ by the formula:

$$\psi(A \otimes e_{i,j}) = (U \otimes S)^i A \otimes P(U \otimes S)^{*j} ,$$

where $e_{i,j}$ is the matrix with all its elements equal to zero but for the element (i,j), equal to one. Then the image of ψ is the ideal $\mathcal{J}$ generated by $\mathbf{1} \otimes P$ in $\mathcal{T}$ and the quotient algebra $\mathcal{T}/\mathcal{J}$ is isomorphic to $\mathcal{B}$. In other words the sequence $0 \to \mathcal{A} \otimes \mathcal{K} \to \mathcal{T} \to \mathcal{B} \to 0$ is exact.

We will use this exact sequence to compute the gap labelling of the 1D Schrödinger operator with a potential given by an automatic sequence.

At last, let us indicate the Connes analog of Thom's isomorphism [CO81]

Theorem 14. *Let $\mathcal{A}$ be a separable C^*-algebra, and α be a one parameter group of *-automorphisms of $\mathcal{A}$. Then there is a natural isomorphism between $\mathrm{K}_i(\mathcal{A} \rtimes_\alpha \mathbb{R})$ and $\mathrm{K}_i(\mathcal{A} \otimes \mathcal{C}_0(\mathbb{R})) \approx \mathrm{K}_{i+1}(\mathcal{A})$, for $i \in \mathbb{Z}/2$.*

These rules allow us to compute the K-groups for inductive limits, ideals, quotients, extensions, tensor product by $M_n, \mathcal{K}, \mathcal{C}_0(\mathbb{R})$, and by crossed products by $\mathbb{Z}$ or $\mathbb{R}$. It is more than enough for the class of C^*-algebras we have developed previously in this paper to compute it.

5. Gap Labelling Theorems for 1D Discrete Hamiltonians

5.1 Completely Disconnected Hull

Let Ω be a compact metrizable space. In this Section we will assume that Ω is completely disconnected. Let T be a homeomorphism of Ω. We will assume that T is 'topologically transitive' namely that it admits at least one dense orbit in Ω. At last μ will denote a T-invariant ergodic probability measure on Ω.

We will consider 1D discrete Schrödinger operators of the form:

$$(5.1.1)\quad H_\omega\psi(n) = \psi(n+1)+\psi(n-1)+v(T^{-n}\omega)\psi(n)\,, \qquad \omega \in \Omega, \psi \in l^2(\mathbb{Z})\,,$$

where v is a continuous function on Ω. More generally, we will consider bounded self-adjoint operators acting on $l^2(\mathbb{Z})$ as follows:

$$(5.1.2)\qquad H_\omega\psi(n) = \sum_{m\in\mathbb{Z}} v(T^{-n}\omega; m-n)\psi(m)\,, \quad \omega \in \Omega, \psi \in l^2(\mathbb{Z})\,,$$

where each map $v_j : \omega \in \Omega \to v(\omega; j) \in C$ is continuous on Ω.

The oldest examples investigated in the literature are the case of a 1D Schrödinger operator with random potentials of Bernouilli type [CF87] or with a limit periodic potential [MO81, AS81]. On the lattice they give rise to a completely disconnected hull indeed. The case of 1D quasicrystals falls into this class (see Section 3.2). More recently potentials with a hierarchical structure had been investigated [KC88, LM88, KL89]. They also belong to such a class, together with potentials given by a substitution (Section 3.3). More generally every operator like the one given by (5.1.1) in which the coefficients take finitely many values, is of this type.

Recall that the C^*-algebra generated by the family of translated of H is the crossed product $C^*(\Omega,T) = \mathcal{C}(\Omega) \rtimes_T \mathbb{Z}$ of the algebra $\mathcal{C}(\Omega)$ of continuous functions on Ω, by the $\mathbb{Z}$-action defined by T. The probability μ defines a trace τ on it, which coincides with a trace per unit length.

Thanks to results of Section 4, the gap labelling will be given by the K_0-group of $C^*(\Omega,T)$, whereas the values of the IDS on the gaps will belong to the image of this group by the trace τ. The main tool to compute the K_0-group is provided by the Pimsner-Voiculescu exact sequence (cf. Section 4.3) where $\mathcal{A} = \mathcal{C}(\Omega)$ and $\alpha = T_*$ is the automorphism induced by T. We need first the following lemmas proved in [see BB91]:

Lemma 5.1.1. *Let Ω be a completely disconnected compact metrizable space. Then* $\mathrm{K}_1(\mathcal{C}(\Omega)) = \{0\}$.

Lemma 5.1.2. *Let Ω be a completely disconnected compact metrizable space. Then* $\mathrm{K}_0(\mathcal{C}(\Omega))$ *is isomorphic to the group $\mathcal{C}(\Omega,\mathbb{Z})$ of integer valued continuous functions on Ω.*

The main result of this Section will be the following

Theorem 15. *Let Ω be a completely disconnected compact metrizable space, and T a homeomorphism on Ω. Then:*

(i) If T is topologically transitive $\mathrm{K}_1(\mathcal{C}(\Omega) \rtimes_T \mathbb{Z})$ is isomorphic to $\mathbb{Z}$.

(ii) $\mathrm{K}_0(\mathcal{C}(\Omega) \rtimes_T \mathbb{Z})$ is isomorphic to the quotient $\mathcal{C}(\Omega, \mathbb{Z})/\mathcal{E}_D$, where $\mathcal{E}_D$ is the subgroup $\mathcal{E}_D = \{f \in \mathcal{C}(\Omega, \mathbb{Z}); \exists g \in \mathcal{C}(\Omega, \mathbb{Z}), f = g - g \circ T^{-1}\}$.

(iii) Let μ be a T-invariant ergodic probability measure on Ω, and τ be the corresponding trace on $\mathcal{C}(\Omega) \rtimes_T \mathbb{Z}$. Then, the image of $\mathrm{K}_0(\mathcal{C}(\Omega) \rtimes_T \mathbb{Z})$ by τ is equal to the countable subgroup $\mu(\mathcal{C}(\Omega, \mathbb{Z}))$ of $\mathbb{R}$.

Proof. Thank to Lemmas 5.1.1 and 5.1.2, and using the Pimsner-Voiculescu exact sequence for $\mathcal{A} = \mathcal{C}(\Omega), \alpha = T_*$ the automorphism defined by T, we get the following exact sequence

$$(5.1.3)\quad 0 \to \mathrm{K}_1(\mathcal{C}(\Omega) \rtimes_T \mathbb{Z}) \to \mathcal{C}(\Omega, \mathbb{Z}) \xrightarrow{\mathrm{id}-T_*} \mathcal{C}(\Omega, \mathbb{Z}) \to \mathrm{K}_0(\mathcal{C}(\Omega) \rtimes_T \mathbb{Z}) \to 0\,,$$

where the middle arrow is given by $\mathrm{id} - T_*$. Since the sequence is exact, it follows that $\mathrm{K}_1(\mathcal{C}(\Omega) \rtimes_T \mathbb{Z})$ is isomorphic to the kernel of $\mathrm{id} - T_*$. If T is topologically transitive, the kernel of $\mathrm{id} - T_*$ is the set of constant functions in $\mathcal{C}(\Omega, \mathbb{Z})$ namely $\mathbb{Z}$ and (i) is proved.

For the same reason, since the sequence is exact, the fourth arrow i_* is surjective. Thus $\mathrm{K}_0(\mathcal{C}(\Omega) \rtimes_T \mathbb{Z}) \approx \mathcal{C}(\Omega, \mathbb{Z})/\mathrm{Ker}(i_*)$, which proves (ii) because $\mathrm{Ker}(i_*) = \mathrm{Im}\{\mathrm{id} - T_*\} = \mathcal{E}_D$.

If i is the canonical injection from $\mathcal{C}(\Omega)$ into $\mathcal{C}(\Omega) \rtimes_T \mathbb{Z}$, it follows from the definition of the trace given by μ that $\tau \circ i = \mu$. By functoriality $\tau_* \circ i_* = \mu_*$ on the corresponding K_0-groups. One can show that μ_* is nothing but μ acting on $\mathcal{C}(\Omega, \mathbb{Z})$ (see the proof of the Lemma 5.1.2 in [BB91]), and i_* is surjective, (iii) holds. □

5.2 Potentials Taking Finitely Many Values

Let us now consider a Hamiltonian on $l^2(\mathbb{Z})$ of the form given by:

$$(5.2.1)\qquad H\psi(n) = \sum_{|m| \leq N} t_m(n)\psi(n+m)\,, \psi \in l^2(\mathbb{Z})\,,$$

where the coefficients take finitely many values. Then the hull Ω can be constructed as to be the closure of an orbit of T in $A^{\mathbb{Z}}$ where A is a finite set (the possible values of the coefficients), and T the two-sided shift on $A^{\mathbb{Z}}$. It is therefore completely disconnected. By a 'letter' we mean a point in the finite set A, while a 'word' will be a finite sequence $w = (a_0 a_1 \cdots a_n)$ of letters; $n = |w|$ is the length of w. Let μ be a T-invariant ergodic probability measure on Ω.

Given a sequence $\omega = (\omega(n))_{n \in \mathbb{Z}} \in A^{\mathbb{Z}}$, the occurrence number $L_\omega(w)$ of a word w in ω is defined (whenever it exists) by:

(5.2.2)

$$L_\omega(w) = \lim_{L\to\infty} 1/(2L+1)\#\{n \in [-L,L]; (\omega(n+1),\cdots,\omega(n+N)) = w\}\,.$$

By Birkhoff's ergodic theorem, this limit exists for μ-almost all $\omega \in \Omega$; it is independent of ω and coincides with $\mu(\chi_w)$ where χ_w is the characteristic function of the cylinder set $\Omega_w = \{\omega \in \Omega; (\omega(0),\cdots,\omega(N-1)) = w\}$. A direct consequence of Theorem 15 is given by:

Proposition 5.2.1. *The values of the IDS of H on the spectral gaps are linear combinations with integer coefficients of the occurrence numbers of any possible word of A.*

Ex 1: Bernouilli Process

Corollary 5.2.2. *If $\Omega = \{-1,+1\}^{\mathbb{Z}}$, T is the two-sided shift, and $\mu = \otimes_{n\in\mathbb{Z}}\mu_p$ the Bernouilli measure where $\mu_p\{-1\} = p, \mu_p\{+1\} = 1-p$, the possible values of the IDS on gaps of H given by (5.1.2), are linear combinations with integer coefficients of the numbers $p^m(1-p)^n$.*

Remark. 1) In this last corollary, T is not topologically transitive on Ω. Nevertheless (ii) and (iii) of Theorem 15 still hold.

2) If H is given by (5.1.1), where $V(n) = v(T^{-n}\omega)$ takes values $\pm V$ with probabilities $p, (1-p)$ respectively, we cannot expect more than one gap. However for models with longer range interaction of the type (5.1.2), or another function v in $\mathcal{C}(\Omega)$, one may expect more than one gap, even though their number may be finite.

Ex 2: Induced Process

Another example of systems described by such a construction is given as follows: let $(M,S,\mathbf{P})$ be a dynamical system, namely M is a compact metrizable space, S is a homeomorphism of M, and $\mathbf{P}$ is a S-invariant ergodic probability measure on M. Let now Λ_i $(1 \le i \le N)$ be a finite family of $\mathbf{P}$-measurable subsets of M. Let $\mathcal{A}_0$ be the *-algebra generated by the functions $\{\chi_i \circ S^n;\ 1 \le i \le N,\ n \in \mathbb{N}\}$, where χ_i is the characteristic function of Λ_i. Without loss of generality we can assume that the Λ_i's form a partition of M. Let $\mathcal{A}$ be the norm closure of $\mathcal{A}_0$ in $L^\infty(M,\mathbf{P})$. $\mathcal{A}$ is an Abelian C^*-algebra, so that there is a compact space Ω, (its spectrum namely the set of characters of $\mathcal{A}$), such that $\mathcal{A}$ is isomorphic to $\mathcal{C}(\Omega)$. The map $g \in \mathcal{A}_0 \to g \circ S \in \mathcal{A}_0$ induces a *-automorphism of $\mathcal{A}$, which in turn gives rise to a homeomorphism also denoted by S, of Ω. The map $g \in \mathcal{A}_0 \to \int d\mathbf{P}(\omega) g(\omega) \in \mathbb{C}$ extends to a trace on $\mathcal{A}$ which in turn defines a S invariant ergodic probability measure on Ω. Thus we get a new dynamical system $(\Omega, S, \mathbf{P})$, induced by the family $\{\Lambda_i; 1 \le i \le N\}$.

The partition $\mathcal{P} = \{\Lambda_i; 1 \leq i \leq N\}$ is called 'generating' if there is a compatible metric on M for which, for any $\varepsilon > 0$, there are $m \leq n$ such that the partition $\mathcal{P}_{m,n} = S^m\mathcal{P} \wedge S^{m+1}\mathcal{P} \wedge \cdots \wedge S^n\mathcal{P}$ contains only atoms of diameter less than or equal to ε.

Application of the Theorem 15 gives [BB91]:

Proposition 5.2.3 . *(i) The compact space Ω induced by the partition $\mathcal{P} = \{\Lambda_i; 1 \leq i \leq N\}$ is completely disconnected.*

(ii) If $\mathcal{P}$ is generating, $\mathcal{C}(M)$ is a closed subalgebra of $\mathcal{C}(\Omega)$. If in addition (M, S) is uniquely ergodic, then (Ω, S) is also uniquely ergodic, and the unique ergodic invariant measures on M and Ω agree.

(iii) The IDS of a self adjoint element of $\mathcal{C}(\Omega) \rtimes_S \mathbb{Z}$ takes values in the $\mathbb{Z}$-module generated by the numbers $\mathbf{P}(\Lambda_{i(0)} \cap S\Lambda_{i(1)} \cap \cdots S^k\Lambda_{i(k)})$ where $1 \leq i(n) \leq N$ for all n's.

Among the examples which have been investigated in the literature, let us mention the following:

Ex 3: The Kohmoto model [KK83, OP83, KO84, OK85, KL86, LP86, CA87, SU87, WI89, BI89, LE89, SU89, BI90]

The algebra is generated by the characteristic function $\chi_{[1-\alpha,1)}$ of the interval $\Lambda_0 = [1-\alpha, 1)$ on the torus $M = \mathbb{T}$, with $S = R_\alpha$ is the rotation by $\alpha \in (0,1)\backslash\mathbb{Q}$ and $\mathbf{P} = \lambda$ is the normalized Lebesgue measure. Then $(\mathbb{T}, R_\alpha)$ is uniquely ergodic. The Kohmoto model is given by:

(5.2.3)

$$H_x\psi(n) = \psi(n+1) + \psi(n-1) + V\chi_{[1-\alpha,1)}(n\alpha - x)\psi(n)\ , x \in \mathbb{T}\ , \psi \in l^2(\mathbb{Z})\ .$$

Hamiltonians on a 1D quasicrystal also belong to the same algebra. Applying the Proposition 5.2.3 we get:

Proposition 5.2.4. *The IDS of the Kohmoto model, or of Hamiltonians in the same algebra, takes values in the set $(\mathbb{Z} + \mathbb{Z}\alpha) \cap [0,1]$. If we denote by Ω the corresponding hull, the K_0-group of the algebra $\mathcal{C}(\Omega) \rtimes_\alpha \mathbb{Z}$ defined above, is isomorphic to $\mathbb{Z}^2$. Its image under the trace defined by the Lebesgue measure on Ω is the dense subgroup $\mathbb{Z} + \mathbb{Z}\alpha$ of $\mathbb{R}$.*

Ex 4: The B-S Model [BS82a]:

It is given by the same dynamical system $(\mathbb{T}, R_\alpha, \lambda)$, but now $\Lambda_0 = [1-\beta, 1)$, where β is an irrational number rationally independent of α. One possible example of Hamiltonian is given by:

(5.2.4)

$$H_x\psi(n) = \psi(n+1) + \psi(n-1) + V\chi_{[1-\beta,1)}(n\alpha - x)\psi(n)\ , x \in \mathbb{T}\ , \psi \in l^2(\mathbb{Z})\ .$$

More generally, one can introduce the sets $\Lambda_i = [1-\beta_i, 1]$ $(1 \leq i \leq N)$, where the numbers $1, \alpha, \beta_i$ are all rationally independent. Then the same argument leads to a gap labelling given by the $\mathbb{Z}$-module generated by $1, \alpha$ and the β_i 's.

5.3 Some Almost Periodic Hamiltonians

1) Limit-Periodic Case. Let us consider the Hamiltonian H given by (5.1.2), where the coefficients are limit periodic sequences on $\mathbb{Z}$. We recall that a sequence $\mathbf{V} = (V(n))_{n\in\mathbb{Z}}$ is limit periodic [BO47] if it is the uniform limit of a family of periodic sequences. It implies that there is a sequence $(q_i)_{i\in\mathbb{N}}$ of integers, such that $q_0 = 1$, $q_{i+1}/q_i = a_i \in \mathbb{N}\backslash\{0,1\}$, and for each $i \geq 0$, a q_i-periodic sequence $\mathbf{V}_i$ on $\mathbb{Z}$, such that $\sup_{n\in\mathbb{Z}}|\mathbf{V}(n) - \mathbf{V}_i(n)| \to 0$ as $i \to \infty$.

Such models have been investigated in [MO81, AS81, BB82, KC88, LM88, KL89].

The hull of such a sequence can then be constructed as follows: as a set $\Omega = \Pi_{i\in\mathbb{N}}\{0,1,\cdots,a_i-1\}$. If $\omega', \omega'' \in \Omega$, the sum $\omega' + \omega'' = \omega$ is defined as follows: ω_0 is the unique integer in $\{0,1,\cdots,a_0-1\}$ such that $\omega_0' + \omega_0'' = \omega_0 + a_0 r_1$, where r_1 takes values 0 or 1. By recursion if $r_i \in \{0,1\}$ is defined, then ω_i is the unique integer in $\{0,1,\cdots,a_i-1\}$ such that $\omega_i' + \omega_i'' + r_i = \omega_i + a_i r_{i+1}$, where now r_{i+1} takes values 0 or 1. One can check that this sum is associative and commutative, that the sequence 0 with all coordinates equal to 0 is a neutral element, and that every element $\omega \in \Omega$ has an opposite ω' defined by $\omega_0' = a_0 - \omega_0, w_i' = a_i - 1 - \omega_i\,(i \geq 1)$. Thus Ω is a compact Abelian group.

We then denote by ε the element $(\delta_{i,0})_{i\in\mathbb{N}}$ and we check that $T\omega = \omega + \varepsilon$ is a homeomorphism of Ω, whereas the orbit of 0 is dense. Actually, if $n \in \mathbb{N}$, one can decompose n in a unique way in the form $n = \nu_0 + \nu_1 q_1 + \cdots + \nu_L q_L$ with $0 \leq \nu_i < a_i$ and we check that $T^n 0 = n\varepsilon = (\nu_0, \nu_1 \cdots, \nu_L, 0, \cdots, 0, \cdots)$, namely $n \in \mathbb{N} \to T^n 0 \in \Omega$ extends in a unique way to a group homomorphism with a dense image.

Any character χ of Ω is associated to a unique rational number (mod 1) of the form k/q_L by $\chi(\omega) = \Pi_{i\leq L} e^{2i\pi k\omega_i q_i/q_L}$. Thus the dual group of Ω can be identified with the subgroup of the torus $\mathbb{T}$ given by numbers of the form k/q_L; it is isomorphic to the inductive limit $\Omega^* = \mathrm{Lim}_{L\in N}\mathbb{Z}/q_L\mathbb{Z}$, where the injection of $\mathbb{Z}/q_L\mathbb{Z}$ into $\mathbb{Z}/q_{L+1}\mathbb{Z}$ is given by the multiplication by a_{L+1}. The 'frequency module' is the $\mathbb{Z}$-module generated by the k/q_L's in $\mathbb{Q}$. In the sequel we will denote by $\mathbb{Z}[a_i; i \in \mathbb{N}]$ the group Ω so constructed. Clearly it is completely disconnected. So we can summarize it as follows:

Proposition 5.3.1. *A bounded sequence $(V(n))_{n\in\mathbb{Z}}$ is limit periodic if and only if there is a sequence $(a_i)_{i\in\mathbb{N}}$ of integers bigger than 1, and a continuous function v on the compact group $\mathbb{Z}[a_i; i \in \mathbb{N}]$ such that $V(n) = v(n\varepsilon)$.*

T is uniquely ergodic, since any T-invariant measure must coincide with the Haar measure μ. Since Ω is a product space, and one can show that the Haar measure is the product measure $\mu = \otimes_{i\in\mathbb{N}}\mu_i$ where μ_i is the uniform measure on $\{0, 1, \cdots, a_i - 1\}$, namely the measure which affects the weight $1/a_i$ to each point. So as a corollary of Theorem 15 we get:

Theorem 16. *Let H be given by (5.1.2) where the coefficients are limit periodic sequences. Then the IDS on the gaps of H takes values in the frequency module of the hull.*

2) Harper's Model and related ones [HA55, HO76, RI81, BE88b]. Let us now consider the models described in Section 3.1. The algebra is now $\mathcal{A}_\alpha = C^*(U, V)$ where U and V are two unitary operators such that $UV = e^{2i\pi\alpha}VU$. Let us remark that this algebra is isomorphic to the crossed product $\mathcal{C}(\mathbb{T}) \rtimes_\alpha \mathbb{Z}$ where $\mathbb{Z}$ acts on $\mathbb{T}$ through the rotation R_α by α. For indeed, $\mathcal{C}(\mathbb{T})$ is the C^*-algebra generated by one unitary operator V, namely the function $V : x \in \mathbb{T} \to e^{2i\pi x} \in \mathbb{C}$. Thus $V \circ R_\alpha = UVU^{-1}$, showing that U is the generator of the rotation in the crossed product. We get the following gap labelling in that case [RI81, PV80b]:

Theorem 17. *Let α be an irrational number in $(0, 1)$. The K_0-group of the algebra $\mathcal{A}_\alpha = C^*(U, V) \approx \mathcal{C}(\mathbb{T}) \rtimes_\alpha \mathbb{Z}$ defined above, is isomorphic to $\mathbb{Z}^2$. Its image under the trace defined by the Lebesgue measure on $\mathbb{T}$ is the dense subgroup $\mathbb{Z} + \mathbb{Z}\alpha$ of $\mathbb{R}$.*

Proof. Let α be irrational now, and let $\mathcal{B}_\alpha = \mathcal{C}(\Omega) \rtimes_\alpha \mathbb{Z}$ be the algebra corresponding to the Kohmoto model (see Section 5.2). Since the characteristic functions of the intervals $[n\alpha, m\alpha]$ generate $\mathcal{C}(\Omega)$, and since these intervals can be as small as we want, it follows that $\mathcal{C}(\Omega)$ contains $\mathcal{C}(\mathbb{T})$ as a closed subalgebra. Thus $\mathcal{A}_\alpha$ is contained as a closed subalgebra of $\mathcal{B}_\alpha$. In particular, by functoriality, $\mathrm{K}_0(\mathcal{A}_\alpha)$ is a subgroup of $\mathrm{K}_0(\mathcal{B}_\alpha)$. Since $\mathrm{K}_0(\mathcal{B}_\alpha) \approx \mathbb{Z}^2$, it is enough to show that the generators of $\mathrm{K}_0(\mathcal{B}_\alpha)$ can be taken in $\mathrm{K}_0(\mathcal{A}_\alpha)$. M. Rieffel [RI81] found one projection given by:

$$P_R = fU + g + U^{-1}f \;. \tag{5.3.1}$$

Here f and g are continuous functions on $\mathbb{T}$ constructed as follows: given $0 < \varepsilon < \alpha, 1 - \alpha$, $g(x) = 0$ if $\alpha \le x \le 1 - \varepsilon$, $g(x) = 1$ if $0 \le x \le \alpha - \varepsilon$, $0 \le g(x + \alpha) = 1 - g(x) \le 1$ if $1 - \varepsilon \le x \le 1$; then $f(x) = \{g(x) - g(x)^2\}^{1/2}$ if $1 - \varepsilon \le x \le 1$, and $f(x) = 0$ otherwise on $[0, 1]$. We continue them by periodicity on $\mathbb{R}$. We can even choose f and g in $\mathcal{C}^\infty(\mathbb{T})$. Then P_R is a projection in $\mathcal{A}_\alpha$. Now, let S be the element of $\mathcal{B}_\alpha$ given by $S = aU + b$, where $a(x) = g(x + \alpha)^{1/2}\chi_{[1-a,1)}$, $b(x) = g(x)^{1/2}\chi_{[1-\alpha,1)}$. We check that both $a, b \in \mathcal{C}(\Omega)$, and that $SS^* = \chi_{[1-\alpha,1)}$ whereas $S^*S = P_R$. Thus $\chi_{[1-\alpha,1)} \approx P_R$ in $\mathcal{B}_\alpha$, which shows that the two generators of $\mathrm{K}_0(\mathcal{B}_\alpha)$, namely $[\mathbf{1}]$ and $[\chi_{[1-\alpha,1)}]$ can be taken in $\mathcal{A}_\alpha$. Thus $\mathrm{K}_0(\mathcal{B}_\alpha) = \mathrm{K}_0(\mathcal{A}_\alpha)$. □

3) Denjoy's Diffeomorphism of the Circle [HE79]. We consider now the dynamical system $(\mathbb{T}, S)$, where S is an orientation preserving homeomorphism. One can lift S as an increasing function from $\mathbb{R}$ to $\mathbb{R}$, also denoted by S, such that $S(x+1) = S(x) + 1$. By Poincaré's theorem [HE79], the rotation number $\alpha = \lim_{n\to\infty}\{S^n(x) - x\}/n$ is well defined modulo 1. We will assume that it is irrational. Then, S is uniquely ergodic, and has no periodic orbits. If μ is the corresponding invariant measure, we set $h(x) = \int_{[0,x]} d\mu$ to get $h(x+1) = h(x)+1$ and $h \circ S = R_\alpha \circ h$, showing that S and the rotation R_α are semi-conjugate. The support M of μ is always the unique minimal S-invariant closed subset of $\mathbb{T}$. Denjoy's theorem [HE79] asserts that if S has a bounded-variation first derivative, then $M = \mathbb{T}$ and h is a homeomorphism. In this latter case, h induces an isomorphism between the C^*-algebras $\mathcal{C}(\mathbb{T}) \rtimes_\alpha \mathbb{Z}$ and $\mathcal{C}(\mathbb{T}) \rtimes_S \mathbb{Z}$.

Examples with $M \neq \mathbb{T}$ have been constructed by Denjoy [HE79]. Then, M is a nowhere dense set without isolated points, so it is completely disconnected. Its complement is a countable union of intervals $(a_{j,n}, b_{j,n})$ (the 'gaps' of M) where $n \in \mathbb{N}$, $j \in J$ (J is a countable set), $a_{j,n} < b_{j,n}$, and $S^n(a_{j,0}) = a_{j,n}, S^n(b_{j,0}) = b_{j,n}$. h is constant on each gap, and we denote by $\theta_{j,n}$ its value on $(a_{j,n}, b_{j,n})$. It follows that $h(S(x)) = h(x) + \alpha \pmod 1$, namely $\theta_{j,n} = \{\theta_{j,0} + n\alpha\}$ ($\{x\}$ denotes the fractional part of x). From Theorem 15, we get [PS86]:

Proposition 5.3.2. *If S is a homeomorphism of $\mathbb{T}$, with irrational rotation number and minimal set $M \neq \mathbb{T}$, the values of the IDS of a self adjoint element of $\mathcal{C}(\mathbb{T}) \rtimes_S \mathbb{Z}$ or $\mathcal{C}(M) \rtimes_S \mathbb{Z}$ on spectral gaps belong to the $\mathbb{Z}$-module generated by the numbers $\{\theta_{j,0} - \theta_{j',0} + n\alpha\}$, where $j, j' \in J$.*

Remark. Every point x outside M is non-wandering, and $\lim_{n\to\pm\infty} \mathrm{dist}(S^n x, M) = 0$. Let a be the closest point of M from x. If v is a continuous function on $\mathbb{T}$, the potential $V_x(n) = v(S^{-n}x)$ differs from V_a by a sequence converging to zero as $|n| \to \infty$. In this way, the corresponding Schrödinger operator $H_x = \Delta + V_x$ differs from $H_a = \Delta + V_a$ by a potential converging to zero at infinity, namely a *localized impurity*. It will give rise to isolated eigenvalues of multiplicity one which cannot be seen in the IDS, since their multiplicity per unit length is zero. For example if v is supported by the gap $(a_{j,0}, b_{j,0})$, V_x is a continuous family of rank one operators vanishing on gaps $(a_{j',n}, b_{j',n})$ with $j' \neq j$. However, as x varies in $\mathbb{T}$, this eigenvalue also varies continuously, giving rise to a band in $\mathrm{Sp}(H) = \cup_{x\in\mathbb{T}} \mathrm{Sp}(H_x)$.

5.4 Automatic Sequences: Potentials Given by a Substitution

Let us consider now the situation where Ω is generated by a substitution (see Section 3.3). Namely let A be a finite set (the 'alphabet') with L elements, and let $A^* = \cup_{k\leq 1} A^k$ be the set of words. A substitution is a map ζ from A to A^*, which can be extended to A^* by concatenation. We denote by $M_{b,a}(\zeta)$ the occurrence number of the letter b in the word $\zeta(a)$: it gives an $L \times L$ matrix $M(\zeta)$ with integer coefficients and called the 'occurrence matrix'. Then if ζ and η are two substitutions we get

$$M(\zeta)M(\eta) = M(\zeta\eta)\,. \tag{5.4.1}$$

By Perron-Frobenius's theorem, $M(\zeta)$ has an eigenvalue θ of highest module which is positive with corresponding eigenvector V having non negative coordinates. If in addition $M(\zeta)$ is primitive, namely if there is an integer n such that $M(\zeta)^n$ has positive coefficients (a condition fulfilled whenever S3 is satisfied), this eigenvalue is simple and V has positive coordinates. We will normalize V by the condition $\sum_a V_a = 1$.

We now assume that the substitution ζ satisfies the hypotheses S1-S3 of Section 3.3. In particular there is a letter 0 such that the word $\zeta(0)$ begins with 0. So that there is an infinite word, or an infinite sequence of letters $u = \lim_{n\to\infty}\zeta^n(0)$. The occurrence number of a letter a in this sequence is then given by:

$$L(a) = \lim_{n\to\infty} M(\zeta^n)_{a,0} \left\{ \sum_b M(\zeta^n)_{b,0} \right\}^{-1} = V_a\,. \tag{5.4.2}$$

The limit is reached as a consequence of the spectral theorem. In particular, if Ω is the two-sided hull of u, and if T is the two-sided shift, any T-invariant ergodic measure μ on Ω satisfies $\mu(\chi_a) = V_a$, where χ_a is the characteristic function of the set of doubly infinite sequences w of letters such that $w(0) = a$.

More generally, let A_N be the set of all words of length N in the sequence u. The substitution ζ induces on A_N a substitution ζ_N for any $N \geq 1$ as follows: A_N is now considered as the set of letters of a new alphabet. If $w \in A_N$ begins with a assume that $\zeta(w) = a_0 a_1 \cdots a_n$ while the length of $\zeta(a)$ is m. Then we set:

$$\zeta_N(w) = (a_0 a_1 \cdots a_{N-1})(a_1 a_2 \cdots a_N) \cdots (a_{m-1} a_m \cdots a_{m+N-2})\,. \tag{5.4.3}$$

So defined, $|\zeta_N(w)| = |\zeta(a)|$ and we will extend it to A_N^* by concatenation. In much the same way, we get an occurrence matrix denoted by $M_N(\zeta)$, again satisfying (5.4.1), with integer coefficients. In particular, if the axioms S1–S3 are satisfied, $M_N(\zeta)$ is primitive and admits the same highest eigenvalues θ as $M(\zeta)$ [QU87, Proposition V.15], with a corresponding eigenvector $V^{(N)}$ having positive coordinates and normalized according to $\sum_w V_w^{(N)} = 1$. Now we remark that if $u = 0u_1 \cdots u_{N-1} \cdots$ one has $\zeta_N^n(0u_1 \cdots u_{N-1}) = (0u_1 \cdots u_{N-1})(u_1 \cdots u_N) \cdots$ in such a way that the occurrence number of a word w in u is nothing but the limit:

$$L(w) = \lim_{n\to\infty} M_N(\zeta^n)_{w,0^{(N)}} \left\{ \sum_v M_N(\zeta^n)_{v,0^{(N)}} \right\}^{-1} = V_w^{(N)}\,, \tag{5.4.4}$$

where $0^{(N)} = 0u_1 \cdots u_{N-1}$. In particular any T-invariant ergodic measure μ on Ω satisfies $\mu(\chi_w) = V_w^{(N)}$, where χ_w is the characteristic function of the set of doubly infinite sequences of letters such that $v(0)v(1)\cdots v(N) = w$. This implies that Ω is uniquely ergodic [QU87 Theorem V.13].

Our main result for substitutions is the following:

Theorem 18. *Let H be given by Eq. (5.1.2) where the coefficients are determined by a substitution ζ on a finite alphabet A, which satisfies the hypothesis S1–S3. Then the values of the IDS of H on the spectral gaps (contained in $[0,1]$) belong to the $\mathbb{Z}[\theta^{-1}]$-module generated by the coordinates of the normalized positive eigenvectors V and $V^{(2)}$ of the occurrence matrices $M(\zeta)$ and $M_2(\zeta)$, where θ is the common highest eigenvalue of each of them.*

Proof. For p large enough, namely if $\theta^p >$ const. N, ζ_{N^p} is entirely determined on $w \in A_N$ by the knowledge of the first two letters of w. In the sequel, let p assume such a condition. To compute μ it is thus sufficient to compute the positive eigenvectors of all the $M_N(\zeta)$'s. Actually there is a remarkable property of these matrices namely $M_2(\zeta)$ will suffice to get it. To see this one defines $\pi_{N,2}$ as the map from A_N into A_2 which gives the restriction to the first two letters extended to the set of corresponding words by concatenation. One also defines $\tau_{2,N,p}$ as the map from A_2 into A_N^* given by:

(5.4.5)

$$\tau_{2,N,p}(w) = (a_0 a_1 \dots a_{N-1})(a_1 a_2 \dots a_N) \dots (a_{|\zeta^p(a)|-1} a_{|\zeta^p(a)|} \cdots a_{|\zeta^p(a)|+N-2}) ,$$

where $\zeta^p(w) = a_0 a_1 \dots a_n$ and w begins by a. Then one immediately obtains:

$$\tau_{2,N,p} \circ \pi_{N,2} = \zeta_N^p, \quad \pi_{N,2} \circ \tau_{2,N,p} = \zeta_2^p, \quad \zeta_N \circ_{2,N,p} = \tau_{2,N,p} \circ \zeta_2 . \tag{5.4.6}$$

If we denote by $M_{2,N,p}$ the occurrence matrix associated to $\tau_{2,N,p}$, we get :

$$M_{2,N,p} M_2 = M_N M_{2,N,p} . \tag{5.4.7}$$

In particular M_2 and M_N have the same non zero eigenvalues and $V_N = M_{2,N,p}(V^{(2)})$ is a positive eigenvector of M_N associated to the highest eigenvalue θ. We just need to normalize V_N to get the occurrence number of any word in u. This is done by remarking that $\sum_w \{M_{2,N,p}\}_{w,v} = |\tau_{2,N,p}(v)| = |\zeta^p(v_0)| = |\zeta_2^p(v)| = \sum_{v'} \{M_2^p\}_{v',v}$ if v begins by the letter v_0. Thus, $\sum_v \{M_{2,N,p} V^{(2)}\}_v = \sum_v \{M_2^p V^{(2)}\}_v = \theta^p \sum_v V_v^{(2)} = \theta^p$.

Consequently, the occurrence number of a word w in u is therefore given by the coordinates of $V^{(N)} = V_N / \sum_w \{V_N\}_w = V_N \theta^{-p}$, which, since $M_{2,N,p}$ has integer coefficients, belongs to the set of linear combinations with integer coefficients of the coordinates of $V^{(2)}$ divided by some power of θ. Since $M_{2,N,p}$ has integer coefficients, this set is the $\mathbb{Z}[\theta^{-1}]$-module generated by the coordinates of $V^{(2)}$, where $\mathbb{Z}[X]$ is the set of polynomials in X with integer coefficients. □

To illustrate this result, let us treat a few examples.

1) The **Fibonacci Sequence** is given by an alphabet with two letters $A = \{0,1\}$, and the substitution is $\zeta(0) = 01$, $\zeta(1) = 0$, which obeys S1-S3. The alphabet A_2 contains only the words $\{00, 01, 10\}$, giving $\zeta_2(00) = (01)(10)$, $\zeta_2(01) = (01)(10)$, $\zeta_2(10) = (00)$, and:

$$M(\zeta) = \begin{bmatrix} 1 & 1 \\ 1 & 0 \end{bmatrix}, \qquad M_2(\zeta) = \begin{bmatrix} 0 & 0 & 1 \\ 1 & 1 & 0 \\ 1 & 1 & 0 \end{bmatrix}, \tag{5.4.8}$$

and the highest eigenvalue is given by $\theta = (\sqrt{5}+1)/2$, namely the inverse of the golden mean $\sigma = (\sqrt{5}-1)/2$. The corresponding eigenvectors V and V_2 are given by:

$$V = \begin{bmatrix} \sigma \\ 1-\sigma \end{bmatrix}, \qquad V_2 = \begin{bmatrix} 2\sigma - 1 \\ 1-\sigma \\ 1-\sigma \end{bmatrix}. \tag{5.4.9}$$

Thus, since $\sigma^2 = 1 - \sigma$, the IDS on the gaps takes values of the form $m + n\sigma$ where $m, n \in \mathbb{Z}$.

2) The **Thue-Morse** sequence is again made with two letters with $\zeta(0) = 01$, $\zeta(1) = 10$. This sequence also obeys S1–S3. The alphabet A_2 contains the four words $\{00, 01, 10, 11\}$ and we get $\zeta_2(00) = (01)(10)$, $\zeta_2(01) = (01)(11)$, $\zeta_2(10) = (10)(00)$, $\zeta_2(11) = (10)(01)$. The two matrices $M(\zeta)$ and $M_2(\zeta)$ are given by:

$$M(\zeta) = \begin{bmatrix} 1 & 1 \\ 1 & 1 \end{bmatrix}, \qquad M_2(\zeta) = \begin{bmatrix} 0 & 0 & 1 & 0 \\ 1 & 1 & 0 & 1 \\ 1 & 0 & 1 & 1 \\ 0 & 1 & 0 & 0 \end{bmatrix}, \tag{5.4.10}$$

and the highest eigenvalue is given by $\theta = 2$. The corresponding eigenvectors are:

$$V = \begin{bmatrix} 1/2 \\ 1/2 \end{bmatrix}, \qquad V_2 = \begin{bmatrix} 1/6 \\ 1/3 \\ 1/3 \\ 1/6 \end{bmatrix}. \tag{5.4.11}$$

Thus the IDS in gaps will take values in the set $1/3\mathbb{Z}[1/2] \cap [0,1]$, namely the set of numbers of the form $k/(3 \cdot 2^N)$ where $k \in N$ and $N \in \mathbb{N}$. We then remark that for the Hamiltonian $H = -\Delta + V$, where Δ is the discrete Laplacian on $\mathbb{Z}$, whereas $V(n)$ takes values V_0 or V_1 according to whether $u_n = 0$ or 1, it has been shown [BE90a] that all gaps corresponding to $k = 3j+1$ or $3j+2$ ($j \in N$) are indeed open, whereas the others are closed due to a special symmetry of the Thue-Morse potential. However a generic perturbation of V in $\mathcal{C}(\Omega)$ will open these gaps too.

3) The **period-doubling** sequence [BB90] is also defined with two letters by $\zeta(0) = 01, \zeta(1) = 00$. This sequence also obeys S1–S3. We get $A_2 = \{00, 01, 10\}$

and $\zeta_2(00) = (01)(10), \zeta_2(01) = (01)(10), \zeta_2(10) = (00)(00)$. The two matrices $M(\zeta)$ and $M_2(\zeta)$ are given by:

$$(5.4.12) \qquad M(\zeta) = \begin{bmatrix} 1 & 2 \\ 1 & 0 \end{bmatrix}, \qquad M_2(\zeta) \begin{bmatrix} 0 & 0 & 2 \\ 1 & 1 & 0 \\ 1 & 1 & 0 \end{bmatrix},$$

with the highest eigenvalue $\theta = 2$, and eigenvectors $V = (2/3, 1/3)$, whereas $V_2 = (1/3, 1/3, 1/3)$. So again the IDS takes values on gaps in the set of $k/(3 \cdot 2^N)$ where $k \in \mathbb{N}$ and $N \in \mathbb{N}$. If $H = -\Delta + V$, where Δ, is the discrete Laplacian on $\mathbb{Z}$, whereas $V(n)$ takes values V_0 or V_1 according to whether $u_n = 0$ or 1, it has been shown [BB90] that indeed all gaps are open.

4) The **Rudin-Shapiro** sequence $(r_n)_{n \geq 0}$ [RU59, SH51, QU87] is defined recursively by $r_0 = 1, r_{2n} = r_n, r_{2n+1} = (-1)^n r_n$. It is actually given by $r_n = (-1)^{f(n)}$ where $f(n)$ is the number of 11 in the dyadic representation of n. It can also be defined through the substitution involving 4 letters [CK80] given by $\zeta(0) = 02, \zeta(1) = 32, \zeta(2) = 01, \zeta(3) = 31$. If $u = \lim_{n \to \infty} \zeta^n(0)$, and if τ is the map from A to $\{-1, +1\}$ given by $\tau(0) = \tau(2) = 1, \tau(1) = \tau(3) = -1$, then $\tau(u_n) = r_n$. The alphabet A_2 contains the eight words $\{01, 02, 10, 13, 20, 23, 31, 32\}$ and we get $\zeta_2(01) = (02)(23), \zeta_2(02) = (02)(20), \zeta_2(10) = (32)(20), \zeta_2(13) = (32)(23), \zeta_2(20) = (01)(10), \zeta_2(23) = (01)(13), \zeta_2(31) = (31)(13), \zeta_2(32) = (31)(10)$. The two matrices $M(\zeta)$ and $M_2(\zeta)$ are given by Eq. (5.4.13a and b) below. We get $\theta = 2$, and $V = (1/4, 1/4, 1/4, 1/4)$, whereas $V_2 = (1/8, 1/8, 1/8, 1/8, 1/8, 1/8, 1/8, 1/8)$. Thus the values of the IDS on gaps are of the form $k2^{-N}$ where $k \in \mathbb{N}$ and $N \in \mathbb{N}$. The structure of the gaps in that case is quite involved [LU90] and no rigorous result has been proved yet in this case.

$$(5.4.13a) \qquad M(\zeta) = \begin{bmatrix} 1 & 0 & 1 & 0 \\ 0 & 0 & 1 & 1 \\ 1 & 1 & 0 & 0 \\ 0 & 1 & 0 & 1 \end{bmatrix},$$

$$(5.4.13b) \qquad M_2(\zeta) = \begin{bmatrix} 0 & 0 & 0 & 0 & 1 & 1 & 0 & 0 \\ 1 & 1 & 0 & 0 & 0 & 0 & 0 & 0 \\ 0 & 0 & 0 & 0 & 1 & 0 & 0 & 1 \\ 0 & 0 & 0 & 0 & 0 & 1 & 1 & 0 \\ 0 & 1 & 1 & 0 & 0 & 0 & 0 & 0 \\ 1 & 0 & 0 & 1 & 0 & 0 & 0 & 0 \\ 0 & 0 & 0 & 0 & 0 & 0 & 1 & 1 \\ 0 & 0 & 1 & 1 & 0 & 0 & 0 & 0 \end{bmatrix}.$$

6. Gap Labelling Theorems

Whereas in Section 5 we investigated 1D discrete Schrödinger operators it remains to compute the gap labels in a more general situation, namely either for discrete Hamiltonians in higher dimension, or for Hamiltonians given by homogeneous pseudodifferential operators on $\mathbb{R}^n$.

It turns out that this latter case can be solved using geometrical techniques coming from the study of smooth foliations on manifolds and developed by A. Connes [CO82]. On the other hand, the discrete case which corresponds in the physicist's language to the so-called "tight-binding representation" [BE86], can be solved by associating a continuous system, its suspension, the C^*-algebra of which (i.e. the Non-Commutative Brillouin zone) being 'Morita equivalent' to the algebra of the discrete case. As a result, their K-groups are the same and the gap labelling of the discrete version can be computed through Connes formulæ from the continuous one.

Specializing to one dimensional situations, we get a much more precise result in the case of ODE's thanks to the approach proposed by R. Johnson [JM82, JO83, JO86]. It gives a generalization of the Sturm-Liouville gap labelling theorem (see Section 1.5) for homogeneous 1D systems. Very recently [JO90] he proposed an extension of this theory to odd dimensions for Schrödinger operators.

6.1 Connes Formulæ for Group Actions

In Section 2 we have shown that the Non-Commutative Brillouin zone for a homogeneous Hamiltonian H acting on $L^2(\mathbb{R}^n)$, is given by the C^*-algebra $\mathcal{A} = C(\Omega) \rtimes_T \mathbb{R}^n$, where Ω is a compact metrizable space and T is a continuous action of the translation group $\mathbb{R}^n$ on Ω. Let $\mathbf{P}$ be a T-invariant ergodic probability measure on Ω. Then we get a trace $\tau_{\mathbf{P}}$ on $\mathcal{C}(\Omega) \rtimes_T \mathbb{R}^n$ equal for $\mathbf{P}$-almost all $\omega \in \Omega$ to the trace per unit volume of the representative H_ω of H. If a uniform magnetic field is present, the Non-Commutative Brillouin zone is $\mathcal{A} = C^*(\Omega \times \mathbb{R}^n, B)$, described in Section 2.5.

If H is bounded below, Shubin's formula asserts that the IDS on spectral gaps of H is equal to $\tau(\chi_{\leq E}(H))$ where $\chi_{\leq E}(H)$ is the eigenprojection of H corresponding to energies smaller than E. Since E is in a spectral gap, the continuous functional calculus implies that $\chi_{\leq E}(H)$ is a projection in the C^*-algebra $\mathcal{A}$ and therefore the IDS takes values in the countable subgroup $\tau_*(\mathrm{K}_0(\mathcal{A}))$. Thus as in Section 5 we want to get rules for calculating this group. The first important tool is Theorem 14 of A. Connes [CO81] on crossed products, which relates the K-groups of the Non-Commutative Brillouin zone to the topology of the space Ω. This space represents the lack of translation invariance of H, namely the amount of disorder in the system described by H. An important consequence is the following

Proposition 6.1.1. *We get:*

(*i*) $\mathrm{K}_0(C^*(\Omega \times \mathbb{R}^n, B)) \approx \mathrm{K}_0(\mathcal{C}(\Omega))$ whenever n is even,

(*ii*) $\mathrm{K}_0(C^*(\Omega \times \mathbb{R}^n, B)) \approx \mathrm{K}_1(\mathcal{C}(\Omega))$ whenever n is odd.

Sketch of the proof. The main remark about $C^*(\Omega \times \mathbb{R}^n, B)$ is that it can be written as a double crossed product [XI88] as follows $C^*(\Omega \times \mathbb{R}^n, B) \approx ((\mathcal{C}(\Omega) \rtimes_{T'} \mathbb{R}^{l+s}) \rtimes_\beta \mathbb{R}^l)$, that is to be described below. Theorem 14 used twice, we then get the result.

Now the double crossed product structure comes from the way the magnetic field acts (see Section 2.5, Eq. (2.5.6)): there is a decomposition of $x \in \mathbb{R}^n$ into $x = (x_+ \oplus x_- \oplus x_0) \in \mathcal{E}_+ \oplus \mathcal{E}_- \oplus \mathcal{E}_0$ such that $\dim(E_\pm) = l$, $\dim(\mathcal{E}_0) = s = n - 2l$, and $(Bx)_+ = \mathbf{b}x_-$, $(Bx)_- = -\mathbf{b}x_0$, $(Bx)_0 = 0$, where $\mathbf{b}$ is an $l \times l$ real symmetric matrix. Then if f is a continuous function with compact support on $\mathbb{R}^n$, we define $\underline{f}$ by $\underline{f}(\omega, x) = f(\omega, x)e^{\mathrm{i}\pi(e/h)\langle x_+|\mathbf{b}x_-\rangle}$. The map $f \to \underline{f}$ defines a *-isomorphism between $C^*(\Omega \times \mathbb{R}^n, B)$ and an algebra $\mathcal{A}'$ on which the structure is

$$
\begin{aligned}
\underline{f}_1\underline{f}_2(\omega, x) &= \int_{\mathbb{R}^n} d^n y \underline{f}_1(\omega, x)\underline{f}_2(T^{-y}\omega, x - y)e^{\mathrm{i}\pi(e/h)\langle x_+ - y_+|\mathbf{b}y\rangle} , \\
\underline{f}^*(\omega, x) &= \underline{f}(T^{-x}\omega, -x)e^{\mathrm{i}2\pi(e/h)\langle x_+|\mathbf{b}x_-\rangle} .
\end{aligned} \tag{6.1.1}
$$

If we define on the crossed product $\mathcal{C}(\Omega) \rtimes_{T'} \mathcal{E}_+ \oplus \mathcal{E}_0$ (where T' is the restriction of T to $\mathcal{E}_+ \oplus \mathcal{E}_0$), a group of *-automorphisms by $\beta_{x''}(g)(\omega, x') = g(T'^{-x''}\omega, x')e^{\mathrm{i}\pi(e/h)\langle x'|\mathbf{b}x''\rangle}$ for $x' \in \mathcal{E}_+ \oplus \mathcal{E}_0, x'' \in \mathcal{E}_-$, it is then tedious but elementary to check that $(\mathcal{C}(\Omega) \rtimes_{T'} \mathcal{E}_+ \oplus \mathcal{E}_0) \rtimes_\beta \mathcal{E}_-$ is isomorphic to $\mathcal{A}'$. □

This result shows that there is certainly a difference in the treatment of odd or even dimensions. In the odd case, we associate to the projection $\chi_{\leq E}(H)$ an invertible matrix valued map on Ω, whereas in the even case, one has to find a fiber bundle over Ω, or equivalently a projection valued map on Ω. As we will see this problem is not yet solved. A solution has been found by R. Johnson in 1D for Schrödinger operators, and he has recently found an important step to treat the odd dimensional case.

However, there is another general result by A. Connes which permits to compute the group $\tau_*(\mathrm{K}_0(\mathcal{C}(\Omega) \rtimes_T \mathbb{R}^n))$ in many relevant situations. Let us assume that Ω is a smooth manifold, and that the action of $\mathbb{R}^n$ is smooth namely that T^a is a diffeomorphism for $a \in \mathbb{R}^n$. Therefore, one gets smooth vector fields $X_1, X_2, \cdots, X_n$ on Ω formally defined by

$$
X_\mu(\omega) = \left.\frac{\partial T^a\omega}{\partial a_\mu}\right|_{a=0} , \qquad 1 \leq \mu \leq n . \tag{6.1.2}
$$

Now a smooth differential form η of degree n on Ω, will give a volume element on each n-dimensional subspace of the tangent space $T_\omega\Omega$ at $\omega \in \Omega$. Evaluating it on the subspace tangent to the orbits of $\mathbb{R}^n$ gives rise to the smooth function $\langle \eta | X_1 \wedge X_2 \wedge \cdots \wedge X_n \rangle$ on Ω. The averaged value of this function

(6.1.3)
$$\langle C|\eta\rangle = \int_{\Omega} \mathbf{P}(d\omega)\langle\eta|X_1 \wedge X_2 \wedge \cdots \wedge X_n\rangle$$

defines a de Rham current of degree n called the 'Ruelle-Sullivan current' of the smooth dynamical system $(\Omega, T, \mathbf{P})$. Its main properties are the following:

(i) C is *closed* namely $\langle C|d\theta\rangle = 0$ for any $(n-1)$-form θ

(ii) C is *positive* namely for any n-form η positive along the orbits of $\mathbb{R}^n$, $\langle C|\eta\rangle \geq 0$ (η is positive along the orbits of $\mathbb{R}^n$ whenever $\langle\eta|X_1 \wedge X_2 \wedge \cdots \wedge X_n\rangle \geq 0$).

One can show [CO82] that any such current is automatically of the form given by Eq. (6.1.3). Since it is closed, it defines a class [C] in the de Rham homology group $\mathrm{H}_n(\Omega, \mathbb{R})$, and therefore, its evaluation $\langle C|\eta\rangle$ on a *closed* form η can also be written in terms of the duality between the homology and the cohomology as $\langle[C]|[\eta]\rangle$ where $[\eta]$ is the cohomology class of η. Recall that a closed n-form η has integer coefficients whenever its evaluation on any n-cycle is an integer. The set of such forms defines a discrete countable subgroup $\mathrm{H}^n(\Omega, \mathbb{Z})$ of the n-th cohomology group.

The next theorem by A. Connes is the main result concerning the gap labelling in the continuous case [CO82]:

Theorem 19. *If $\mathbb{R}^n$ acts freely on Ω by means of diffeomorphisms, the countable subgroup $\tau_*(\mathrm{K}_0(\mathcal{C}(\Omega) \rtimes_T \mathbb{R}^n))$ of $\mathbb{R}$ coincides with the group $\langle[C]|\mathrm{H}^n(\Omega, \mathbb{Z})\rangle$ obtained by evaluating the Ruelle-Sullivan current C on the n-th cohomology group with integer coefficients.*

In practice, it will require the calculation of a set of closed n-forms with integer coefficients generating $\mathrm{H}^n(\Omega, \mathbb{Z})$. The evaluation of C on it is then purely computational.

Let us now relate the Proposition 6.1.1 to Theorem 19 by showing how the trace acts on $\mathrm{K}_{i+n}(\mathcal{C}(\Omega))$ through the isomorphism with $\mathrm{K}_0(\mathcal{C}(\Omega) \rtimes_T \mathbb{R}^n)$. In the case $n = 1$, $\mathrm{K}_0(\mathcal{C}(\Omega) \rtimes_T \mathbb{R})$ is isomorphic to $\mathrm{K}_1(\mathcal{C}(\Omega))$. A typical element of the latter may be generated by a smooth map $\omega \in \Omega \to U(\omega) \in \mathrm{GL}_N(\mathbb{C})$. Then we get a closed 1-form by considering $\eta_U = (1/2\mathrm{i}\pi)\mathrm{Tr}(U^{-1}dU)$. It represents the differential of the logarithm of $\mathrm{Det}(U)$ (up to the normalization factor). It is well known that the variation of $\mathrm{LogDet}(U)$ on any closed path in Ω is an integer multiple of $2\mathrm{i}\pi$, which implies that η has integer coefficients. Actually any element in $\mathrm{H}^1(\Omega, \mathbb{Z})$ can be obtained in this way and therefore it is enough to consider numbers of the form $\langle C|\eta_U\rangle$. It is simple to check that if U is homotopic to U', η_U and $\eta_{U'}$ are also homotopic, and therefore they admit the same equivalence class in $\mathrm{H}_1(\Omega, \mathbb{Z})$. Thus $U \to \eta_U$ defines a map η_* from $\mathrm{K}_1(\mathcal{C}(\Omega))$ into $\mathrm{H}^1(\Omega, \mathbb{Z})$ Moreover, it is elementary to check that $\eta_{UV} = \eta_U + \eta_V$, showing that η_* is actually a group homomorphism. Using Connes's Theorem 19 and Birkhoff's ergodic theorem, all possible gap labels are given by numbers (where $U_\omega(a) = U(T^{-a}\omega)$)

$$\begin{aligned}\langle [C] | \eta_*[U] \rangle &= \int_\Omega \mathbf{P}(d\omega) \langle (1/2\mathbf{i}\pi) \mathrm{Tr}(U^{-1} dU) | X \rangle = \\ &= \lim_{L \to \infty} (1/2\mathbf{i}\pi)(1/2L) \int_{-L}^{+L} da\, \mathrm{Tr}(U_\omega(a)^{-1} dU_\omega(a)/da)\,,\end{aligned} \tag{6.1.4}$$

which is nothing but the average of the variation of the phase of $\mathrm{Det}(U)$ along $\mathbf{P}$-almost every orbit. We will see later on how to associate canonically to each gap of a 1D Schrödinger equation a map $\omega \in \Omega \to U(\omega) \in \mathbb{C}\backslash\{0\}$.

We can generalize this construction to the odd dimensional case by considering the n-form $\eta_U^{(n)} = c_n \mathrm{Tr}((U^{-1}dU)^n)$ instead, where c_n is a suitable normalization factor insuring that $\eta_U^{(n)}$ has integer coefficients.

If now $n = 2$, a typical element of $\mathrm{K}_0(\mathcal{C}(\Omega))$ is given by a smooth map $\omega \in \Omega \to P(\omega) \in M_N(\mathbb{C})$ such that $P(\omega)^2 = P(\omega)$. Such a map defines a fiber bundle over Ω by taking the image of the projection $P(\omega)$ as the fiber above ω. A closed differential 2-form is then given by the trace of the curvature of this bundle, namely the second Chern class $\theta_P = (1/2\mathbf{i}\pi)\mathrm{Tr}(PdPdP)$. That θ_P has integer coefficients is a classical result about Chern classes. The map $P \to \theta_P$ defines also a group homomorphism θ_* between $\mathrm{K}_0(\mathcal{C}(\Omega))$ and $\mathrm{H}^2(\Omega, \mathbb{Z})$. It is surjective also (but not injective!), so that gap labels are provided by numbers $\langle [C] | \theta_*[P] \rangle$ for all possible P's. Remark that using again Birkhoff's ergodic theorem and setting $P_\omega(a) = P(T^{-a}\omega)$,

$$\langle [C] | \theta_*[P] \rangle = \lim_{L \to \infty} (1/2\mathbf{i}\pi)(1/2L) \int_{[-L,L]} da \mathrm{Tr}(P_\omega[\partial P_\omega/\partial a_1, \partial P_\omega/\partial a_2])\,, \tag{6.1.5}$$

which is the averaged Chern class of the bundle over Ω defined by P along $\mathbf{P}$-almost every orbit. The main problem is that nobody yet has been able to associate explicitly such a bundle to a given spectral gap of the original Hamiltonian H.

The generalization for $n = 2p$ consists in replacing θ_P by the higher Chern classes, namely $\theta_P^{(n)} = c_n \mathrm{Tr}(P(dPdP)^{n/2})$ where c_n is a suitable normalization constant to make sure that we have an n-form with integer coefficients.

Let us finish by remarking that formulae (6.1.4) and (6.1.5) do not require the smoothness of Ω. We conjecture that they are still true whenever Ω is a compact space, owing to the fact that it is always possible to regularize any continuous function over Ω along the orbits of $\mathbb{R}^n$ by means of a convolution with a smooth function on $\mathbb{R}^n$.

6.2 Gap Labelling Theorems for Quasiperiodic Hamiltonians on $\mathbb{R}^n$

Thanks to Theorem 19, we are able to compute possible gap labels in full generality, at least if the disorder space Ω is a manifold on which $\mathbb{R}^n$ acts freely by diffeomorphisms. By a free action, we mean that if there are $\omega \in \Omega$, $a, b \in \mathbb{R}^n$ such that $T^a\omega = T^b\omega$ then $a = b$.

Our first important result concerns the quasi periodic potentials. Recall that V is a quasi periodic function on $\mathbb{R}^n$ whenever there exists $\nu > n$, and a continuous function $\mathcal{V}$ on $\mathbb{R}^\nu$ periodic of period 1 in each variable, such that if [BO47]

$$V_\omega(x_1, \cdots, x_n) = \mathcal{V}(\omega_1 - \sum_j \alpha_{1j}x_j, \cdots, \omega_\nu - \sum_j \alpha_{\nu j}x_j) , \tag{6.2.1}$$

then $V_{\omega=0} = V$. Moreover the rectangular $\nu \times n$ matrix $\alpha = ((\alpha_{\mu j}))$ is called the 'frequency matrix'. It is 'irrational' if the subspace $\Delta_\alpha = \{\alpha x; x \in \mathbb{R}^n\}$ intersects the lattice $\mathbb{Z}^\nu$ at $\{0\}$ only. The hull of such a function is then the torus $\Omega = \mathbb{T}^\nu$ and $\mathbb{R}^n$ acts on it through the translation $T^x\omega = \omega + \alpha x$. This action is always smooth (it is actually analytic), and it is free if and only if α is irrational.

There is a unique T-invariant ergodic probability measure on $\mathbb{T}^\nu$ namely the Lebesgue measure $\mathbf{P}(d\omega) = d^\nu\omega$. The action of $\mathbb{R}^n$ is defined by the n constant vector fields $X_i = (\alpha_{\mu j})_{\mu\in[1,\nu]}$. The calculation of the n-th cohomology group of $\mathbb{T}^\nu$ is actually quite easy. The generators are the n-forms $d\omega_{i(1)} \wedge \cdots \wedge d\omega_{i(n)}$ where $1 \leq i(1) < i(2) < \cdots < i(n) \leq \nu$. By definition, they have integer coefficients so that $\mathrm{H}_n(\mathbb{T}^\nu, \mathbb{Z}) \approx \mathbb{Z}^N$ with $N = \nu!/(\nu - n)!n!$. Therefore we immediately get [BL85]

Proposition 6.2.1. *Let H be a pseudo-differential operator on $\mathbb{R}^n$ bounded from below with quasiperiodic coefficients, the frequency matrix α of which being irrational. Then its* IDS *on spectral gaps takes values in the dense subgroup $\sum_\beta n_\beta \det(\beta)$ where the sum runs over the set of square submatrices β of maximal rank of α.*

6.3 Johnson's Approach for Schrödinger Operators

Let us consider now the one-dimensional Schrödinger operator acting on $L^2(\mathbb{R})$ by

$$H_\omega\psi(x) = -d^2\psi/dx^2 + v(T^{-x}\omega)\psi(x) = E\psi(x) , \tag{6.3.1}$$

where Ω is a compact metrizable space, $\omega \in \Omega$, T is a group action of $\mathbb{R}$ by homeomorphisms and v is a continuous real function on Ω. This equation can be written as a first order differential system:

$$d\Psi/dx = M(T^{-x}\omega)\Psi(x) , \tag{6.3.2}$$

with

$$\Psi(x) = \begin{bmatrix} \psi(x) \\ \frac{\psi'(x)}{\sqrt{E}} \end{bmatrix} , \qquad M(\omega) = \begin{bmatrix} 0 & \sqrt{E} \\ -\sqrt{E} + \frac{v(\omega)}{\sqrt{E}} & 0 \end{bmatrix} , \tag{6.3.3}$$

The full spectrum of H is the union $\mathrm{sp}(H) = \cup_{\omega\in\Omega}\mathrm{Sp}(H_\omega)$. Remark that if $\omega \in \Omega$ has a dense orbit, then $\mathrm{sp}(H) = \mathrm{Sp}(H_\omega)$. We will also denote by $\mathbf{P}$ a T-invariant ergodic probability measure on Ω.

Proposition 6.3.1. *If E belongs to a spectral gap of the operator $H_\omega = -d^2/dx^2 + v(T^{-x}\omega)$ defined by Eq. (6.3.1), up to normalization, there is a unique real solution Ψ_+ (resp. Ψ_-) of Eq. (6.3.2) converging to zero at $+\infty$ (resp. $-\infty$). This solution belongs to $C^2(\mathbb{R})$, depends continuously on $\omega \in \Omega$ and decays exponentially fast at infinity. Moreover these two solutions give linearly independent vectors of $\mathbb{R}^2$ at every $x \in \mathbb{R}$.*

Sketch of a proof. Uniqueness: given two solutions Ψ_1 and Ψ_2 of (6.3.2) their Wronskian is constant for M is traceless. Thus if they both converge to zero at $+\infty$ their Wronskian vanishes identically and they must be equal up to a constant.

Existence: let us consider the Green function $G_E(\omega; x, y)$ defined by

$$G_E(\omega; x, y) = \langle \delta_x | (H_\omega - E)^{-1} \delta_y \rangle\ , \tag{6.3.4}$$

where δ_x is the Dirac measure at x. It is easy to check that $u_x = (\mathbf{1} + p^2)^{-1/2}\delta_x$ belongs to $L^2(\mathbb{R})$ if $p = -\mathrm{i}d/dx$. Then u_x is Hölder continuous of exponent $\alpha < 1/2$ and bounded with respect to x. Moreover, $E_0 \leq -(1 + \|v\|) \Rightarrow H_\omega - E_0 \geq \mathbf{1} + p^2$. Using the resolvent equation, it follows that $(\mathbf{1}+p^2)^{1/2}(H_\omega - z)^{-1}(\mathbf{1} + p^2)^{1/2}$ is a family of bounded operators, strongly continuous with respect to $\omega \in \Omega$, and norm-analytic with respect to z in the resolvent set of H_ω. Thus Eq. (6.3.4) defines for every $R > 0$, a bounded continuous function of the variables $(E, \omega, x, y) \in \mathbb{C}\backslash \mathrm{sp}(H) \times \Omega \times \mathbb{R}^2$, analytic in E, continuous in ω, Hölder continuous of exponent $\alpha < 1/2$ in (x, y).

On the other hand, in the sense of distributions, one gets

$$(\partial^2/\partial x^2 + v(T^{-x}\omega) - E)\, G_E(\omega; x, y) = \delta(x - y)\ . \tag{6.3.5}$$

In particular, if $x > x_0$, the map $\psi : x \in \mathbb{R} \to G_E(\omega; x, x_0)$ is a solution of (6.3.1), which, by uniqueness, can be continued in a unique way as a solution of (6.3.1) on the full line. Standard results on ODE's show that this solution is automatically in $C^2(\mathbb{R})$.

To show that ψ decays exponentially fast at $+\infty$, let $W(a)$ be the unitary operator on $L^2(\mathbb{R})$ of multiplication by $e^{\mathrm{i}ax}$. Then $H_\omega(a) = W(a)H_\omega W(a)^* = H_\omega + 2ap + a^2$ and

$$\langle \delta_x | (H_\omega(a) - E)^{-1} \delta_y \rangle = e^{\mathrm{i}a(x-y)} G_E(\omega; x, y)\ . \tag{6.3.6}$$

Since p is H_ω-bounded, it follows that $H_\omega(a)$ is analytic with respect to a in the norm-resolvent sense. In particular, if $E \notin \mathrm{Sp}(H_\omega)$ there is $\rho_E > 0$ such that if $a \in \mathbb{C}$ and $|a| \leq \rho_E$, one has $\|(\mathbf{1}+p^2)^{1/2}(H_\omega(a) - E)^{-1}(\mathbf{1}+p^2)^{1/2}\| \leq C_E < \infty$. Thus, using (6.3.6) we get

$$|G_E(\omega; x, y)| \leq C_E e^{-\rho_E |x-y|} \,. \tag{6.3.7}$$

Hence ψ is exponentially decaying at $+\infty$ and defines the solution Ψ_+ of the proposition. A similar argument holds for Ψ_-. □

Let us now consider the trivial bundle $\mathcal{E}_0 = \Omega \times \mathbb{R}^2$. Using the previous proposition, one can find for each $\omega \in \Omega$ a vector $\Phi_\pm(\omega)$, unique up to a normalization factor, such that the unique solution of (6.3.2) with initial condition $\Phi_\pm(\omega)$ at $x = 0$, decays exponentially fast at $\pm\infty$. Moreover, we know that these vectors vary continuously with respect to ω, and are linearly independent. This gives a splitting $\mathcal{E}_0 = \mathcal{E}_+ \oplus \mathcal{E}_-$ into the Whitney sum of two line bundles. Denoting by $\Phi_\pm(\omega, x)$ the solution of (6.3.2) with initial condition $\Phi_\pm(\omega)$ at $x = 0$, we remark that $\Phi_\pm(T^{-a}\omega, x - a)$ satisfies also the equation (6.3.2) and also converges to zero at $\pm\infty$. By the uniqueness theorem, it follows that $\Phi_\pm(\omega, x) \in \mathcal{E}_+(T^{-x}\omega)$. On the other hand, as x varies, $\Phi_\pm(\omega, x)$ never vanishes otherwise it would be identically zero, and therefore it rotates around the origin in $\mathbb{R}^2$. Let us parametrize $\Phi_\pm(\omega, x)$ by means of the angle $\theta_\pm(T^{-x}\omega)$ with the first axis. If x varies between $-L$ and $+L$, the total variation of this angle can be written as $\Delta_L\theta_\pm = \int_{-L}^{+L} d/dx(\theta_\pm(T^{-x}(\omega))$ and $\Delta_L\theta_\pm/\pi$ differs from the number of zeroes of $\psi_\pm$ in the interval $[-L, +L]$ by at most 2. Sturm-Liouville's theory (see Section 1.5) shows that this number of zeroes is equal to the number of eigenvalues smaller than E of the Hamiltonian H_L given by the restriction of H_ω to $[-L, +L]$ with suitable boundary conditions. Thus as $L \to \infty$, $\Delta_L\theta_\pm/2L\pi$ converges $\mathbf{P}$-almost surely to the IDS $\mathcal{N}(E)$. Using Birkhoff's ergodic theorem again, one gets

$$\mathcal{N}(E) = 1/\pi \int_\Omega d\mathbf{P}(\omega) d/dx(\theta_\pm(T^{-x}\omega))|_{x=0} \,. \tag{6.3.8}$$

If Ω were a manifold, and if the action of $\mathbb{R}$ were smooth, denoting by X the vector field defined by the flow, we could write

$$d/dx(\theta_\pm(T^{-x}\omega))|_{x=0} = \langle d\theta_\pm | X\rangle_\omega \,. \tag{6.3.9}$$

Moreover if we identify $\mathbb{R}^2$ with the complex plane $\mathbb{C}$, $d\theta_\pm$ is homologous to $-\mathbf{i}\Phi_\pm^{-1} d\Phi_\pm$. Adopting these notations even if Ω is not a manifold, we get for the value of the IDS on the previous gap, a Connes formula with an explicit 1-form, namely the rotation angle of a solution vanishing at infinity [JM82, JO83, JO86, BE86]:

Theorem 20. *If $H_\omega = -d^2/dx^2 + v(T^{-x}\omega)$ acts on $L^2(\mathbb{R})$, where $v \in \mathcal{C}(\Omega)$, and $(\Omega, \mathbb{R}, \mathbf{P})$ is a topological dynamical system with $\mathbf{P}$ an invariant ergodic probability, the* IDS *on gaps is given by*

$$\mathcal{N}(E) = \tau(\chi_{\leq E}(H)) = 1/\mathbf{i}\pi \int_\Omega d\mathbf{P} \langle \Phi_\pm^{-1} d\Phi_\pm | X\rangle$$

where $\Phi_\pm = \Psi_\pm + \mathbf{i}\Psi'_\pm/\sqrt{E}$ *and* $\Psi_\pm$ *is the solution of the Schrödinger equation* $H_\omega\psi = E\psi$ *vanishing at* $\pm\infty$.

Remark. We get therefore a result compatible with the form found in Section 6.1. For indeed here the maps $\omega \in \Omega \to \Phi_\pm(\omega) \in \mathbb{C}$ have inverses and they define elements in $\mathrm{GL}(\mathcal{C}(\Omega))$.

The previous construction has been extended recently by R. Johnson [JO90] to the case of a Schrödinger operator of the type given by Eq. (6.3.1) on $\mathbb{R}^n$, for n odd. Let D be a bounded open domain with $\mathcal{C}^\infty$ boundary ∂D given as the zero set of a $\mathcal{C}^\infty$ function from $\mathbb{R}^n$ into $\mathbb{R}$, having no critical point on ∂D. Let $V(x)$ be a bounded continuous function on the closure of D. We consider the Schrödinger operator $H_D = -\Delta + V$ with Dirichlet boundary conditions on ∂D, and we assume that the potential V is such that H_D has a simple spectrum given by the eigenvalues $E_1 < E_2 < \cdots < E_i < \cdots$. If $(\psi_i)_{i\geq 1}$ are the corresponding eigenfunctions, there is $y \in D$ such that $\psi_i(y) \neq 0$ for every $i \geq 1$. Without loss of generality, one can choose the origin of the coordinates in such a way that $y = 0$.

As in (6.3.4) let $G_D(E; x, y)$ be the Green function of H_D at the energy E, and we set $g(x, E) = G_D(E; x, 0)$. Let ζ be the map from $D \times \mathbb{R}$ to the projective space $\mathbb{RP}(n)$ (the manifold of lines in $\mathbb{R}^{n+1}$) given by

$$(6.3.10) \qquad \zeta(x, E) = [g(x, E), \partial_1 g(x, E), \cdots, \partial_n g(x, E)] \ ,$$

where $\partial_i = \partial/\partial x_i$, and $[\mathbf{u}]$ denotes the line through the vector $\mathbf{u}$. This map is well defined at every 'non singular point', namely points where the vector in brackets does not vanish. Whenever g does not vanish, this line is nothing but the line $[1, \nabla_x \mathrm{Log}(g)]$.

Since n is odd, the space $\mathbb{RP}(n)$ is an orientable manifold. So let 'vol' be its volume form normalized in such a way that

$$(6.3.11) \qquad \int_{\mathbb{RP}(n)} \mathrm{vol} = \Omega_n/2 \ , \qquad \Omega_n = \text{volume of the } n\text{-sphere } S^n \ .$$

The 'oscillation' of g is the integral of the pull-back $\zeta^*\mathrm{vol}$, which can be viewed as a rigorous definition of the differential of $\mathrm{Log}(g)$. More precisely, in order to avoid the singular points, we fix the interval $I = [E_0, E]$ where $E_0 < E$ .Let $\sum$ be the union of the hypersurfaces $\partial D \times I$ and $D \times \{E\}$, and we set:

$$(6.3.12) \qquad O(D; E) = \int_{\sum} \zeta^*\mathrm{vol} - \sum_{j\in[1,M]} \mathrm{Ind}(s_j) \ ,$$

where $s_1, \cdots, s_M$ are the singular points of g (see [JO90] for a precise definition of them), and $\mathrm{Ind}(s)$ is the index of the lift of the map ζ to S^n at the point s. The main result of Johnson is the following:

Proposition 6.3.2. *With the previous notation, if n is odd, and if E is not an eigenvalue of H_D, the oscillation number satisfies the formula:*

$$O(D;E) = -\Omega_n N_D(E) \tag{6.3.13}$$

where $N_D(E)$ is the number of eigenvalues of H_D in the interval $[E_0, E]$.

Let now assume that Ω is a topological compact metrizable space, endowed with an action of $\mathbb{R}^n$ by a group of homeomorphisms. **P** will denote an invariant ergodic probability measure on ω. Let v be a continuous function on Ω, and for $\omega \in \Omega$ let V_ω be the potential on $\mathbb{R}^n$ given by $V_\omega(x) = v(T^{-x}\omega)$. Let $(D_m)_{m\geq 1}$ be a Følner sequence (see Section 4.1) of bounded open domains in $\mathbb{R}^n$ satisfying the same conditions as D above, and covering $\mathbb{R}^n$. Then the mean oscillation number is defined as

$$O^{\sim}(E) = \lim_{n\to\infty} |D_n|^{-1} O(D_n;E)\,, \tag{6.3.14}$$

and by using the Proposition 6.3.2 it follows that the IDS is given by

$$\mathcal{N}(E) = -\Omega_n^{-1} O^{\sim}(E)\,. \tag{6.3.15}$$

We will conclude this Section by addressing the following question:

Problem. Prove that the mean oscillation number can be written in the form $\langle C|\eta\rangle$, where C is the Ruelle-Sullivan current and η a closed n-form related to ζ^*vol above.

6.4 Strong Morita Equivalence and Tight-Binding Approximation

Our last Section will concern the relation between Hamiltonians on the continuum and on a discrete lattice. In 1D this is related to the so-called 'Poincaré' section, and its converse the so-called 'suspension' construction. Let $(\Omega, T, \mathbb{R})$ be a dynamical system, namely Ω is a compact metrizable space endowed with an action of $\mathbb{R}$ by a group of homeomorphisms $(T^s)_{s\in\mathbb{R}}$. By a 'smooth transversal' (or a Poincaré section), we mean a compact subspace N such that for every $\omega \in \Omega$, the orbit of ω meets N and the set $\mathcal{L}(\omega) = \{s \in \mathbb{R}; T^{-s}\omega \in N\}$ is discrete, non empty and depends continuously on $\omega \in \Omega$. We then define on N the 'first return map' ϕ as follows: if $\zeta \in N$, let $t(\zeta)$ be the lowest positive real number t such that $T^t\zeta \in N$, and we set $\phi(\zeta) = T^{t(\zeta)}\zeta$. By hypothesis on N the map $\zeta \in N \to t(\zeta) \in (0,\infty)$ is continuous (in particular there is $t_- > 0$ such that $t(\zeta) \geq t_-$ for all $\zeta \in N$). Then (N, ϕ) gives a $\mathbb{Z}$-action on N.

Conversely, let (N, ϕ) be a $\mathbb{Z}$-action on a compact metrizable space N, and let $\zeta \in N \to t(\zeta) \in (0,\infty)$ be a continuous function. Let us consider on the space $N \times \mathbb{R}$, the map $\Phi : (\zeta, t) \to (\phi(\zeta), t - t(\zeta))$. The suspension of N is then the space $SN = N \times \mathbb{R}/\Phi$, namely the compact topological space obtained by identifying (ζ, t) with $\Phi(\zeta, t)$. We then consider the flow $T^s : (\zeta, t) \to (\zeta, t + s)$ on $N \times \mathbb{R}$. Since T and Φ commute, it follows that T defines an $\mathbb{R}$-action on SN. Then N can be identified with the transversal $N \times \{0\}$ in SN, and one can easily check that the first return map coincides with ϕ. If now N is a smooth transversal of the dynamical system $(\Omega, T, \mathbb{R})$ the map $h : (\zeta, t) \in N \times \mathbb{R} \to T^t\zeta \in \Omega$ defines a homeomorphism of SN onto Ω which intertwins the corresponding $\mathbb{R}$-actions.

The main question is to find the relation between the C^*-algebras $\mathcal{B} = C(\Omega) \rtimes_T \mathbb{R}$ and $\mathcal{C} = C(N) \rtimes_\phi \mathbb{Z}$. These two algebras are actually 'strongly Morita equivalent' [RI82, AMS81]. In particular they are 'stably isomorphic' namely $\mathcal{B} \otimes \mathcal{K}$ and $\mathcal{C} \otimes \mathcal{K}$ are isomorphic, implying that they have the same K-groups. In this Section we wish to develop this theory to apply to our problem.

More generally, let $(\Omega, T, \mathbb{R}^n)$ be a dynamical system in n-dimensions. A smooth transversal is defined as before where $\mathbb{R}^n$ replaces $\mathbb{R}$. Then we get immediately:

Lemma 6.4.1. *If N is a smooth transversal of $(\Omega, T, \mathbb{R}^n)$ there is $R > 0$ such that if $\omega \in \Omega$, $s, t \in \mathcal{L}(\omega)$, then $|s - t| \geq R$. Moreover, $s \in \mathcal{L}(\omega)$ if and only if $(s - t) \in \mathcal{L}(T^{-t}\omega)$.*

Let now Γ_N be the set of pairs (ζ, s) where $\zeta \in N$ and $s \in \mathcal{L}(\zeta)$. Since $\mathcal{L}(\zeta)$ is discrete, Γ_N is endowed with an obvious topology which makes it a locally compact space. We define by $\mathcal{C}_0$ the space $\mathcal{C}_c(\Gamma_N)$ of continuous functions on Γ_N with compact support endowed with the structure of *-algebra given by:

$$(6.4.1)\quad c_1 c_2(\zeta, s) = \sum_{t \in \mathcal{L}(\zeta)} c_1(\zeta, t) c_2(T^{-t}\zeta, s - t)\,, \qquad c^*(\zeta, t) = c(T^{-t}\zeta, -t)^*\,.$$

For $\zeta \in N$, we then denote by π_ζ the *-representation of $\mathcal{C}_0$ on $l^2(\mathcal{L}(\zeta))$ given by:

$$(6.4.2)\qquad \pi_\zeta(c)\psi(s) = \sum_{t \in \mathcal{L}(\zeta)} c(T^{-s}\zeta, t - s)\psi(t)\,.$$

Then a C^*-norm is given by

$$(6.4.3)\qquad \|c\| = \sup_{\zeta \in N} \|\pi_\zeta(c)\|$$

We will denote by $\mathcal{C} = C^*(N)$ the completion of $\mathcal{C}_0$ under this norm. We remark that $\mathcal{C}_0$ has a unit namely $\mathbf{1}(\zeta, s) = \delta_{s,0}$. We will denote by $\mathcal{B}_0$ the dense subalgebra $\mathcal{C}_c(\Omega \times \mathbb{R}^n)$ of $\mathcal{B} = \mathcal{C}(\Omega) \times_T \mathbb{R}^n$.

Remark. If $\zeta \in N$, the set $\mathcal{L}(\zeta)$ is a deformed lattice in $\mathbb{R}^n$. The C^*-algebra $C^*(N)$ is then exactly the algebra containing all Hamiltonians on this lattice. It is even possible to define properly the notion of covariance [CO79, CO82, BE86]. In [BE86], a connection between the Schrödinger operator in $\mathbb{R}^n$ with a potential given by a function on Ω, and operators on the lattices $\mathcal{L}(\zeta)$ has been described justifying the so-called 'tight-binding' representation in Solid State Physics.

Now let χ_0 be the space $\mathcal{C}_c(N \times \mathbb{R}^n)$. We define on it a structure of $\mathcal{B}_0 - \mathcal{C}_0$-bimodule as follows (where $b \in \mathcal{B}_0, c \in \mathcal{C}_0, k \in \chi_0$):

$$
\begin{aligned}
kc(\zeta, s) &= \sum_{t \in \mathcal{L}(\zeta)} k(T^{-t}\zeta, s-t)c(T^{-t}\zeta, -t) , \\
bk(\zeta, s) &= \int_{\mathbb{R}^n} d^n t b(T^s \zeta, t) k(\zeta, s-t) .
\end{aligned} \tag{6.4.4}
$$

Moreover, we define a 'hermitian' structure [RI82] by means of scalar products with values in the algebras $\mathcal{B}_0$ and $\mathcal{C}_0$, namely if $k, k' \in \chi_0$:

$$
\begin{aligned}
\langle k|k' \rangle_{\mathcal{B}}(\omega, s) &= \sum_{t \in \mathcal{L}(\omega)} k(T^{-t}\omega, t) k'(T^{-t}\omega, t-s)^* , \\
\langle k|k' \rangle_{\mathcal{C}}(\zeta, s) &= \int_{\mathbb{R}^n} d^n t k(\zeta, t)^* k'(T^{-t}\zeta, s-t) .
\end{aligned} \tag{6.4.5}
$$

then we get the following properties which will be left to the reader [RI82]:

Lemma 6.4.2. *The hermitian bimodule structure defined by* (6.4.4) *and* (6.4.5) *satisfies the following identities (here $b, b' \in \mathcal{B}_0, c, c' \in \mathcal{C}_0, k, k', k'' \in \chi_0$)*

(i) $bk \in \chi_0, \quad kc \in \chi_0$.

(ii) $b(b'k) = (bb')k, \quad (kc)c' = k(cc'), \quad (bk)c = b(kc)$

(iii) $\langle k|k' \rangle_{\mathcal{B}} \in \mathcal{B}_0, \quad \langle k|k' \rangle_{\mathcal{C}} \in \mathcal{C}_0$.

(iv) $\langle bk|k' \rangle_{\mathcal{B}} = b\langle k|k' \rangle_{\mathcal{B}} \quad \langle k|k'c \rangle_{\mathcal{C}} = \langle k|k' \rangle_{\mathcal{C}} c$.

(v) $\langle k|k' \rangle_{\mathcal{B}}^* = \langle k'|k \rangle_{\mathcal{B}}, \quad \langle k|k' \rangle_{\mathcal{C}}^* = \langle k'|k \rangle_{\mathcal{C}}$.

(vi) $\langle k|k \rangle_{\mathcal{B}} \geq 0$ *and* $\langle k|k \rangle_{\mathcal{B}} = 0 \Rightarrow k = 0$ *(same property with* $\mathcal{C}$*).*

(vii) $\langle kc|k' \rangle_{\mathcal{B}} = \langle k|k'c^* \rangle_{\mathcal{B}} \quad \langle k|bk' \rangle_{\mathcal{C}} = \langle b^*k|k' \rangle_{\mathcal{C}}$

(viii) $k\langle k'|k'' \rangle_{\mathcal{C}} = \langle k|k' \rangle_{\mathcal{B}} k''$.

(ix) $\langle kc|kc \rangle_{\mathcal{B}} \leq \|c\|^2 \langle k|k \rangle_{\mathcal{B}}, \langle bk|bk \rangle_{\mathcal{C}} \leq \|b\|^2 \langle k|k \rangle_{\mathcal{C}}$

(x) $\langle \chi_0|\chi_0 \rangle_{\mathcal{B}}$ *is dense in* $\mathcal{B}_0$, $\langle \chi_0|\chi_0 \rangle_{\mathcal{C}} = \mathcal{C}_0$.

(xi) There is $u \in \chi_0$ *such that* $\mathbf{1}_{\mathcal{C}} = \langle u|u \rangle_{\mathcal{C}}$.

We simply want to prove *(xi)* for it will be useful in practice. Let g be a continuous function on $\mathbb{R}^n$ with compact support contained in the open ball of radius $R/2$, and such that $\int_{\mathbb{R}^n} d^n t |g(t)|^2 = 1$. Then $u(\zeta, s) \equiv g(s)$ satisfies *(xi)*, in particular u is not unique. Then a norm is defined on χ_0 by setting:

$$
\|k\| = \|\langle k|k \rangle_{\mathcal{B}}\|^{1/2} = \|\langle k|k \rangle_{\mathcal{C}}\|^{1/2} . \tag{6.4.6}
$$

We will denote by χ the $\mathcal{C} - \mathcal{B}$-bimodule obtained by completing χ_0 under this norm. It also satisfies all the properties of the Lemma 6.4.2.

Now let $\mathbf{P}$ be an invariant ergodic probability measure on Ω. Then it induces on N a probability measure μ with the property that whenever $s : \zeta \in N \to s(\zeta) \in \mathbb{R}^n$ is a continuous function such that $s(\zeta) \in \mathcal{L}(\zeta)$, then μ is invariant by the map $\phi_s(\zeta) = T^{s(\zeta)}\zeta$. Moreover any Borelian set F in N which is ϕ_s-invariant for every s satisfies $\mu(F) = 0$ or 1. Then we get a trace on each of the algebras $\mathcal{B}_0$ and $\mathcal{C}_0$ via:

$$\tau_{\mathbf{P}}(b) = \int_{\Omega} \mathbf{P}(d\omega) b(\omega, 0) , \qquad \tau_\mu(c) = \int_N \mu(d\zeta) c(\zeta, 0) . \tag{6.4.7}$$

We then get easily:

Lemma 6.4.3. *The traces $\tau_{\mathbf{P}}$ and τ_μ satisfy $\tau_{\mathbf{P}}(\langle k|k'\rangle_{\mathcal{B}}) = \tau_\mu(\langle k'|k\rangle_{\mathcal{C}})$, for every k, k' in χ_0.*

Now let $c \in \mathcal{C}_0$, and $u \in \chi_0$ such that $\mathbf{1}_{\mathcal{C}} = \langle u|u\rangle_{\mathcal{C}}$. Then we set $\rho(c) = \langle uc|u\rangle_{\mathcal{B}}$. It is easy to check that $\rho(c) \in \mathcal{B}_0$, and that ρ is a *-endomorphism, namely it satisfies $\rho(cc') = \rho(c)\rho(c')$ and $\rho(c)^* = \rho(c^*)$. Thus ρ extends as a *-endomorphism from $\mathcal{C}$ into $\mathcal{B}$. Moreover, $\rho(\mathbf{1}_{\mathcal{C}})$ is a projection in $\mathcal{B}_0$, and for every projection P in $\mathcal{C}_0$, $\rho(P)$ is a projection in $\mathcal{B}_0$.

Conversely, given any projection Q in $\mathcal{B}_0$, using the property (x), we can find two finite families $\underline{k} = (k_i)_{1\leq i\leq I}$ and $\underline{k}' = (k'_i)_{1\leq i \leq I}$ in χ_0 such that $\|Q - \sum_{i\in[1,I]} \langle k_i|k'_i\rangle_{\mathcal{B}}\| \leq \varepsilon < 1/2$. It is actually possible to choose $\underline{k}' = \underline{k}$ if we accept to replace χ_0 by χ. Then by an argument similar to the one used in the proof of Lemma 4.2.2, one gets a projection $Q' \approx Q$ in the form $Q' = \sum_{i=1}^{I} \langle k_i|k_i\rangle_{\mathcal{B}}$ for some $\underline{k} \in \chi^I$. If we now set $P' = ((\langle k_i|k_j\rangle_{\mathcal{C}}))_{i,j}$, we get a projection in the matrix algebra $M_I(\mathcal{C})$.

More generally through the replacement of $\mathcal{B}_0$ by $M_L(\mathcal{B}_0)$, of $\mathcal{C}_0$ by $M_N(\mathcal{C}_0)$, and of χ_0 by $M_{L\times N}(\chi_0)$, we get in an obvious way a $M_L(\mathcal{B}_0)$–$M_N(\mathcal{C}_0)$ hermitian bimodule, and any projection in $M_L(\mathcal{B}_0)$ will give rise in the same manner, to a projection in $M_{N'}(C)$, for some N'. This is the basic argument leading to [BR77, RI82]:

Theorem 21. *(i) The C^*-algebras $\mathcal{B} = \mathcal{C}(\Omega) \rtimes_T \mathbb{R}^n$ and $\mathcal{C} = C^*(N)$ are stably isomorphic, namely $\mathcal{B} \otimes \mathcal{K}$ and $\mathcal{C} \otimes \mathcal{K}$ are isomorphic (not in a canonical way).*

(ii) The K_0-groups of $\mathcal{B}$ and $\mathcal{C}$ are isomorphic.

(iii) Their images by the traces coincide namely $\tau_{\mathbf{P}}(\mathrm{K}_0(\mathcal{B})) = \tau_\mu(\mathrm{K}_0(\mathcal{C}))$.

Remark. The property (iii) is a direct consequence of the Lemma 6.4.3.

As a consequence of Theorem 21, the computation of the gap labelling for a discrete system is equivalent to that of its suspension. This gives immediately some results in practice. First of all let us consider a Hamiltonian on $\mathbb{Z}^n$ with quasiperiodic coefficients. It means that it has the form:

$$H_\xi \psi(m) = \sum_{m' \in \mathbb{Z}^n} h(\xi - \alpha m, m' - m)\psi(m') , \qquad \psi \in l^2(\mathbb{Z}^n) , \tag{6.4.8}$$

where $h \in \mathcal{C}(\mathbb{T}^\nu) \rtimes_\alpha \mathbb{Z}^n$, and α is a $\nu \times n$ matrix with rationally independent real column vectors, acting on the n-dimensional torus $\mathbb{T}^\nu$ by translation namely $T^m \xi = \xi - \alpha m$.

The suspension of the dynamical system $(\mathbb{T}^\nu, \alpha, \mathbb{Z}^n)$ is actually given by $(\mathbb{T}^{\nu+n}, \beta, \mathbb{R}^n)$, where the matrix β is the $(\nu+n) \times n$ real matrix given by $\beta = [\alpha, \mathbf{1}_n]$ obtained by gluing together the $\nu \times n$ matrix α and the $n \times n$ identity matrix $\mathbf{1}_n$. Using the Proposition 6.2.1, we get the following result [BE81, EL82a, DS83, BL85].

Proposition 6.4.4. *Let H be as in Eq. (6.4.8). Then the* IDS *on spectral gaps takes values in the $\mathbb{Z}$-module in $\mathbb{R}$ generated by* 1 *and all the minors of the matrix α.*

In particular if $n = 1$, this module is generated by 1 and the components of the line α.

Another consequence was given by A. Connes, and concerns the absence of gaps in the spectrum.

Proposition 6.4.5. *Let Ω be a manifold such that the first cohomology group* $\mathrm{H}^1(\Omega, \mathbb{Z}) = 0$. *If ϕ is a minimal diffeomorphism, the algebra $\mathcal{C}(\Omega) \rtimes_\phi \mathbb{Z}$ is simple, has a unit and no non trivial projection.*

In particular, if $(H_\omega)_{\omega \in \Omega}$ is a covariant family of self adjoint bounded operators on $l^2(\mathbb{Z})$ of the form:

$$H_\omega \psi(m) = \sum_{m' \in \mathbb{Z}} h(\phi^{-m}(\omega), m' - m)\psi(m') \,, \qquad \psi \in l^2(\mathbb{Z}^n) \,,$$

with $h \in \mathcal{C}(\Omega) \rtimes_\phi \mathbb{Z}$ then for every $\omega \in \Omega$, the spectrum of H_ω is connected.

Sketch of the proof [CO81]. First of all, the minimality of ϕ implies the simplicity of $\mathcal{C}(\Omega) \rtimes_\phi \mathbb{Z}$ [SA79].

Let μ be a ϕ-invariant probability measure on Ω. Let $S\Omega$ be the suspension corresponding to the constant first return time $t(\omega) = 1$, then the measure $\mathbf{P} = \mu(d\omega)dt$ is invariant and ergodic on $S\Omega$. Using Theorem 21, the image of the K_0-group of $\mathcal{C}(\Omega) \rtimes_\phi \mathbb{Z}$ by the trace τ_μ is identical with the image of the K_0-group of its suspension $\mathcal{C}(S\Omega) \rtimes_T \mathbb{R}$ by the induced trace τ_P. By Theorem 19, this last set is given by $\langle [C] | H^1(S\Omega, \mathbb{Z}) \rangle$, where C is the Ruelle-Sullivan current induced by $\mathbf{P}$. The flow on $S\Omega$ is then generated by the vector field $\partial/\partial t$. Since Ω is connected (ϕ is minimal), and $H^1(\Omega, \mathbb{Z}) = 0$, it follows that $H^1(S\Omega, \mathbb{Z}) = \mathbb{Z}$ with the 1-form dt as generator. Then one clearly has $\langle [C] | [dt] \rangle = 1$, for P is a probability, and we get $\tau_{\mathbf{P}}(\mathrm{K}_0(\mathcal{C}(S\Omega) \rtimes_T \mathbb{R})) = \tau_\mu(\mathrm{K}_0(\mathcal{C}(\Omega) \rtimes_\phi \mathbb{Z})) = \mathbb{Z}$. Now let P be a projection in $\mathcal{C}(\Omega) \rtimes_\phi \mathbb{Z}$, it follows that its trace is an integer, and therefore it must be 0 or 1, because $0 \leq P \leq \mathbf{1} \Rightarrow 0 \leq \tau_\mu(P) \leq 1$. It is elementary to check that since ϕ is minimal, the support of any invariant measure on Ω is Ω itself, implying that the trace τ_μ is faithful, and therefore $P = 0$ or $\mathbf{1}$.

In particular, since ϕ is minimal, and since $TH_\omega T^{-1} = H_{\phi(\omega)}$, the spectrum of H_ω is independent of $\omega \in \Omega$. If it had a gap $G = (a,b)$ then the eigenprojection $\chi_{(-\infty,E]}(H_\omega)$, with $E \in G$, would define a non trivial projection (i.e. different from 0 or 1) in the C^*-algebra $\mathcal{C}(\Omega) \rtimes_\phi \mathbb{Z}$ leading to a contradiction.

Notes added to the second print (May 1995). 1) End of Section 3.2: in a series of papers the spectrum of the octagonal lattice has been performed numerically using the periodic approximants [PS92, SP93, SG94, SI94, PJ95]. There is an 'insulating regime' with Cantor spectrum. The Lebesgue measure of the spectrum may not necessarily vanish. There is also a 'metallic regime' with no spectral gaps. In this regime there is level repulsion and anomalous diffusion. The spectrum is likely to be singular continuous. 2) End of Section 3.3: for more recent results see [BG93, JS94, HK94].

References

[AA86] Axel, F. Allouche, J.P., Kléman M., Mendès-France, M. Peyrière J.: Vibrational modes in a one-dimensional quasi-alloy, the Morse case. J. Phys. C3, **47** (1986) 181–187.

[AH76] Albeverio, S., Høegh-Krohn, R.: Mathematical theory of Feynman path integral. Lecture Notes in Mathematics, vol. 523. Springer, Berlin Heidelberg New York 1976.

[AP88] Axel, F., Peyrière, J.: C.R. Acad. Sci. Paris, série II **306** (1988) 179–182.

[AR74] Arnold, V.I.: Equations différentielles ordinaires. Ed. Mir, Moscow 1974.

[AR78] Arnold, V.I.: Méthodes mathématiques de la mécanique classique. Ed. Mir, Moscow 1978.

[AS81] Avron, J., Simon, B.: Almost periodic Schrödinger operators, I. Limit Periodic Potentials. Commun. Math. Phys. **82** (1981) 101–120.

[AT67] Atiyah M.: K-theory. Benjamin, New York Amsterdam 1967.

[BB82] Bellissard, J., Bessis, D., Moussa, P.: Chaotic states of almost periodic Schrödinger operators. Phys. Rev. Lett. **49** (1982) 701–704.

[BB90] Bellissard, J., Bovier A., Ghez, J.M.: Spectral properties of a tight binding hamiltonian with period doubling potential. Commun. Math. Phys. **135** (1991) 379–399.

[BB91] Bellissard, J., Bovier, A., Ghez, J.M.: Gap labelling theorems for 1D discrete Schrödinger operators, CPT preprint Marseille (May 1991), submitted to Rev. Math. Phys.

[BE81] Bellissard, J.: Schrödinger operators with an almost periodic potential. In: Mathematical problems in theoretical physics (R. Schrader and R. Seiler, eds.). Lecture Notes in Physics, vol. 153. Springer, Berlin Heidelberg New York 1982, pp. 356–359.

[BE86] Bellissard, J.: K-Theory of C^*-algebras in solid state physics. In: Statistical mechanics and field theory, mathematical aspects (T.C. Dorlas, M.N. Hugenholtz, M. Winnink, eds.) Lecture Notes in Physics, vol. 257. Springer, Berlin Heidelberg New York 1986, pp. 99–156.

[BE88a] Bellissard, J.: Ordinary quantum Hall effect and non-commutative cohomology. In: Localization in disordered systems (W. Weller and P. Ziesche, eds.). Teubner, Leipzig 1988.

[BE88b] Bellissard, J.: C^*-algebras in solid state physics. In: Operator algebras and applications (D.E. Evans and M. Takesaki, eds.). Cambridge Univ. Press, Cambridge 1988.

[BE89] Bellissard, J.: Almost periodicity in solid state physics and C^*-algebras. In: The Harald Bohr centenary (C. Berg, F. Fuglede, eds.) Royal Danish Acad. Sciences, MfM 42:3 (1989) 35–75.

[BE90a] Bellissard, J.: Spectral properties of Schrödinger's operator with a Thue-Morse potential. In: Number theory and physics (J.M. Luck, P. Moussa, M. Waldschmidt, eds.) Springer Proc. in Phys. **47** (1990).

[BE90b] Bellissard, J.: Etats localisés du rotateur pulsé. In: Journées Equations aux dérivées partielles. St Jean-de-Monts, SMF, 28 Mai-1er Juin 1990, Ecole Polytechnique, Palaiseau (1990). pp. XIII, 1–14.
[BG77] Brown, L.G., Green, P., Rieffel, M.A.: Stable isomorphisms and strong Morita equivalence. Pacific J. Math. **71** (1977) 349–363.
[BG93] Bovier, A., Ghez, J.-M.: Spectral properties of one-dimensional Schrödinger operators with potential generated by substitutions, Commun. Math. Phys, **158**, (1993), 45–66; Erratum in Commun. Math. Phys. (1994).
[BH25] Born, M., Heisenberg, W., Jordan, P.: Zur Quantenmechanik II. Z. Phys. **35** (1925) 557–615 (see [HE85, VW67]).
[BI89] Bellissard, J., Iochum, B., Scoppola, E., Testard, D.: Spectral properties of one dimensional quasicrystals. Commun. Math. Phys. **125** (1989) 527–543.
[BI90] Bellissard, J., Iochum, B., Testard, D.: Continuity properties of electronic spectrum of 1D quasicrystals. Commun. Math. Phys. **141** (1991) 353–380.
[BJ25] Born, M., Jordan P.: Zur Quantenmechanik. Z. Phys. **34** (1925) 858–888 (see [VW67]).
[BK90] Bellissard, J., Kreft, C., Seiler, R.: Analysis of the spectrum of a particle on a triangular lattice with two magnetic fluxes by an algebraic and numerical method. J. Phys. A **24** (1991) 2329–2353.
[BL85] Bellissard, J., Lima R., Testard, D.: Almost periodic Schrödinger operators. In: Mathematics + Physics, Lectures on recent results, vol. 1 (L. Streit, ed.). World Science Publ., Singapore, Philadelphia 1985, pp. 1–64.
[BL86] Blackadar, B.: *K*-Theory for operator algebras. MSRI publications **5**. Springer, Berlin Heidelberg New York 1986.
[BO13] Bohr, N.: On the constitution of atoms and molecules. Phil. Mag. **26** (1913) pp. 1, 476, 857.
[BO14] Bohr, N.: On the effect of electric and magnetic fields on spectral lines. Phil. Mag. **27** (1914) 506.
[BO15a] Bohr, N.: On the series spectrum of hydrogen and the structure of the atom. Phil. Mag. **29** (1915) 332.
[BO15b] Bohr, N.: On the quantum theory of radiation and the structure of the atom. Phil. Mag. **30** (1915) 394.
[BO18] Bohr, N.: On the quantum theory of Line-spectra. D. Kgl. Danske Viden. Selsk. Skrifter, Naturvidensk. og Mathem. Afd. 8. Rœkke, IV.1 (1918) 1–36. (Reproduced in [VW67]).
[BO47] Bohr, H.: Almost periodic functions. Chelsea Publ. Co., New York 1947.
[BO67a] Bourbaki, N.: Variétés différentielles et analytiques. Fasc. de résultats. Hermann, Paris 1967.
[BO67b] Bourbaki, N.: Théories spectrales. Hermann, Paris 1967.
[BO83] Bohr N.: Discussion with Einstein on epistemological problems in atomic physics (see p. 9). In: Quantum Theory and Measurement (J.A. Wheeler and W.H. Zurek, eds.). Princeton Series in Physics, Princeton, NJ (1983), pp. 9–49.
[BP70] Benderskii, M., Pastur, L.: On the spectrum of the one dimensional Schrödinger equation with a random potential. Math. Sborn. USSR **11** (1970) 245.
[BR25] Broglie, L. de: Nature **112** (1923) 540. Thèse (1924). Ann. Physique [10] (1925) 2.
[BR26] Brillouin, L.: J. Phys. Radium **7** (1926) 353–368.
[BR77] Brown, L.G.: Stable isomorphisms of hereditary subalgebras of C^*-algebras. Pacific J. Math. **71** (1977) 335–348.
[BR79] Bratteli, O., Robinson, D.W.: Operator algebras and quantum statistical mechanics. Springer, Berlin Heidelberg New York 1979.
[BR81] Bruijn, N.J. de: Sequences of zeroes and ones generated by special production rules. Kon. Neder. Akad. Wetensch. Proc. A **84** (1981) 27–37.
[BR90] Bellissard, J., Rammal, R.: An algebraic semiclassical approach to Bloch electrons in a magnetic field. J. Phys. France **51** (1990) 1803–1830.
[BS82a] Bellissard J., Scoppola E.: The density of states for an almost periodic Schrödinger operator and the frequency module: a counter-example. Commun. Math. Phys. **85** (1982) 301–308.

[BS82b] Bellissard, J., Simon, B.: Cantor spectrum for the almost Mathieu equation. J. Funct. Anal.**48** (1982) 408–419.

[BS91] Benza, V.G., Sire C.: Band spectrum for the octagonal qyasicrystal: gaps, finite measure, and chaos, CPT preprint Marseille (May 1991), submitted to Phys. Rev. Letters.

[BV90] Bellissard, J., Vittot, M.: Heisenberg's picture and non-commutative geometry of the semi classical limit in quantum mechanics. Ann. Inst. Henri Poincaré **52** (1990), 175–235.

[CA86] Casdagli, M.: Symbolic dynamics for the renormalization map of a quasiperiodic Schrödinger equation. Commun. Math. Phys. **107** (1986) 295–318.

[CF87] Cycon, H.L., Froese, R.G., Kirsch, W., Simon, B.: Schrödinger Operators. Text and Monographs in Physics. Springer, Berlin Heidelberg New York (1987).

[CK80] Christol, G., Kamae, T., Mendès-France, M., Rauzy, G. Suites algébriques, automates et substitutions. Bull. SMF **108** (1980) 401–419.

[CM73] Coburn, L.A., Moyer, R.D., Singer, I.D.: C^*-algebras of almost periodic pseudo-differential operators. Acta Math. **139** (1973) 279–307.

[CO23] Compton, A.H.: Proc. Nat. Acad. Sci. **9** (1923) 359.

[CO67] Coburn, L.A.: Bull. Amer. Math. Soc. **73** (1967) 722.

[CO79] Connes, A.: Sur la théorie non commutative de l'intégration. Lecture Notes in Mathematics, vol. **725** Springer, Berlin Heidelberg New York 1979.

[CO81] Connes, A.: An analogue of the Thom isomorphism for crossed products of a C^*-algebra by an action of $\mathbb{R}$. Adv. Math. **39** (1981) 31–55.

[CO82] Connes, A.: A survey of foliation and operator algebras. In: Operator algebras and applications. Proc. Symposia Pure Math. **38** (1982), Part I, 521–628.

[CO83] Connes, A.: Non-commutative differential geometry, Part I. The Chern character in K-homology, Part II, de Rham homology and non-commutative algebra. (Preprint IHES 1983) Publ. Math. **62** (1986) 44–144.

[CO90] Connes, A.: Géométrie non commutative. InterEditions, Paris (1990).

[CR87] Connes, A., Rieffel, M.: Yang-Mills for non-commutative two tori. Contemp. Math. **62** (1987) 237–266.

[CS83] Craig, W., Simon, B.: Log Hölder continuity of the integrated density of states for stochastic Jacobi matrices. Commun. Math. Phys. **90** (1983) 207–218.

[CU82] Cuntz, J.: The internal structure of simple C^*-algebras. In: Operator algebras and applications. Proc. Symposia Pure Math. **38** (1982) Part I, 85–115.

[CW78] Claro, F.H., Wannier, W.H.: Closure of bands for Bloch electrons in a magnetic field. Phys. Status Sol. B **88** (1978) K147–151.

[CW79] Claro, F.H., Wannier, W.H.: Magnetic subband structure of electrons on a hexagonal lattice. Phys. Rev. B **19** (1979) 6068–6074.

[CW81] Claro, F.H., Wannier, W.H.: Spectrum of tight binding electrons on a square lattice with magnetic field. Phys. Status Sol. B **104** (1981) K31–34.

[DI26] Dirac, P.A.M.: The fundamental equations of quantum mechanics. Proc. Roy. Soc. London A **109** (1926) 642–653.

[DI69] Dixmier, J.: Les algèbres d'opérateurs dans l'espace hilbertien (2nd Printing). Gauthier-Villars, Paris (1969).

[DK85] Duneau, M., Katz, A.: Quasiperiodic patterns. Phys. Rev. Lett. **54** (1985) 2688–2691; Quasiperiodic patterns and icosahedral symmetry. J. de Phys. **47** (1986) 181-196.

[DP90] Delyon, F., Peyrière, J.: Recurrence of the eigenstates of a Schrödinger operator with automatic potential. J. Stat. Phys. **64** (1991) 363–368.

[DS83] Delyon, F., Souillard, B.: The rotation number for finite difference operators and its properties. Commun. Math. Phys. **89** (1983) 415.

[DS84] Delyon, F., Souillard, B.: Remark on the continuity of the density of states of ergodic finite difference operators. Commun. Math. Phys. **94** (1984) 289–291.

[EI05] Einstein, A.: Über einen die Erzeugung und Verwandlung des Lichtes betreffenden heuristischen Gesichtspunkt. Ann. Phys. **17** (1905) 132.

[EI17] Einstein, A.: Zum Quantensatz von Sommerfeld und Epstein. Verh. der D. Physikal. Ges. **19** (1917) 82–92.

[EL82a] Elliott, G.: On the K-theory of the C^*-algebra generated by a projective representation of $\mathbb{Z}^n$. In: Operator algebras and applications. Proc. Symposia Pure Math. **38** (1982) Part I, 17–180 and in: Operator algebras and group representations. G. Arsene Ed., Pitman, London 1983.
[EL82b] Elliott, G.: Gaps in the spectrum of an almost periodic Schrödinger operator. C.R. Math. Ref. Acad. Sci. Canada **4** (1982) 255–259.
[EL85] Elser, V.: Indexing problem in quasicrystal diffraction. Phys. Rev. B **32** (1985) 4892–4898; the diffraction pattern of projected structures. Acta Crystallogr. A **42** (1986) 36–43.
[FA75] Faris, W.G.: Self-adjoint operators. Lecture Notes in Math. **433** (1975).
[FE48] Feynman, R.P.: Space-time approach to non-relativistic quantum mechanics. Rev. Mod. Phys. **20** (1948) 367–385.
[FU80] Fujiwara, D.: Remarks on convergence of some Feynman path integrals. Duke Math. J. **47** (1980) 559–600.
[FU90] Fujiwara, D.: The Feynman integral as an improper integral over Sobolev space. In: Journées Equations aux dérivées partielles. St Jean-de-Monts, SMF, 28 Mai-1er Juin 1990, pp. XIV, 1–15, Ecole Polytechnique, Palaiseau 1990.
[GA77] Gardner, M.: Sci. Am. **236** no. 1 (1977) 110.
[GB82] Galgani, L., Benettin, G.: Planck's formula for classical oscillators with stochastic thresholds. Lett. al Nuovo Cim. **35** (1982) 93–96.
[GH89] Guillement, J.P., Helffer, B., Treton, P.: Walk inside Hofstadter's butterfly. J. de Phys. **50** (1990) 2019–2058.
[GM90] Gerard, C., Martinez, A., Sjöstrand, J.: A mathematical approach to the effective Hamiltonian in perturbed periodic problems. Preprint Univ. Paris-Sud 1990, pp. 90–14.
[GR69] Greenleaf, F.: Invariant means on topological groups. Van Nostrand Reinhold, New York 1969.
[HA55] Harper, P.G.: Single band motion of conduction electrons in a uniform magnetic field. Proc. Phys. Soc. London A **68** (1955) 874.
[HA56] Halmos, P.R.: Lectures on ergodic theory. Chelsea Publ. Co., New York 1956.
[HA78a] Hahn, P.: Haar measure for measure groupoids. Trans. Amer. Math. Soc. **242**, no. 519 (1978) 1–33.
[HA78b] Hahn, P.: The regular representation of measure groupoids. Trans. Amer. Math. Soc. **242** n° 519 (1978) 35–72.
[HE25] Heisenberg, W.: Über quantentheoretische Umdeutung kinematischer und mechanischer Beziehungen. Z. Phys. **33** (1925) 878–893 (see [HE85, VW67]).
[HE27] Heisenberg, W.: Über den anschaulichen Inhalt der quantentheoretischen Kinematik und Mechanik. Z. Phys.**43** (1927) 172–198 (see [HE85]).
[HE79] Herman, M.R.: Sur la conjugaison différentiable des difféomorphismes du cercle à des rotations. Pub. Math. IHES **49** (1979) 5–234.
[HE85] Heisenberg, W.: Gesammelte Werke, Collected Works (W.Blum, H.P.Dürr and H.Rechenberg, eds.).(Springer, Berlin Heidelberg New York 1985.
[HK87] Hatakeyama, T., Kamimura, H.: Electronic properties of a Penrose tiling lattice in a magnetic field. Solid State Comm. **62** (1987) 79–83.
[HK94] Hof, A., Knill, O., Simon, B.: Singular continuous spectrum for palindromic Schrödinger operators, preprint Caltech, (1994).
[HO76] Hofstadter, D.R.: Energy levels and wave functions of Bloch electrons in a rational or irrational magnetic field. Phys. Rev. B **14** (1976) 2239.
[HÖ85] Hörmander, L.: The analysis of linear partial differential operators III. Grundlehren der mathematischen Wissenschaften, vol. 274. Springer, Berlin Heidelberg New York 1985.
[HS87] Helffer, B., Sjöstrand, J.: Analyse semi-classique pour l'équation de Harper I, II, III., in: Mémoires de la SMF (1990).
[JA74] Jammer, M.: The philosophy of Quantum Mechanics. John Wiley and Sons, New York London Sydney Toronto 1974.
[JD88] Janot, C., Dubois, J.M. (eds.): Quasicrystalline materials. Grenoble 21–25 March 1988, World Scientific Pub., Singapore 1988.

[JM82] Johnson, R., Moser, J.: The rotation number for almost periodic potentials. Commun. Math. Phys. **84** (1982) 403–438.

[JO83] Johnson, R.: A review of recent works on almost periodic differential and difference operators. Acta Appl. Math. **1** (1983) 241–261.

[JO86] Johnson, R.A.: Exponential dichotomy, rotation number, and linear differential equations. J. Differential Equations **61** (1986) 54–78.

[JO90] Johnson, R.: Oscillation theory for the odd-dimensional Schrödinger operator. (Preprint IMA) Univ. of Minnesota, Minneapolis (1990).

[JR82] Jaffe, C., Reinhard, W.P.: Uniform semiclassical quantization of regular and chaotic classical dynamics on the Hénon-Heiles surface. J. Chem. Phys. **77** (1982) 5191–5203.

[JS94] Jitomirskaya, S., Simon, B.: Operators with singular continuous spectrum: III. Almost periodic Schrödinger operators, Comm. Math. Phys., **165**, 201–205, (1994).

[KA78] Karoubi, M.: K-theory. An introduction. Grundlehren der mathematischen Wissenschaften, vol. 226. Springer, Berlin Heidelberg New York 1978.

[KC88] Keirstead, W.P., Cecatto, H.A., Huberman, B.A.: Vibrational properties of hierarchical systems. J. Stat. Phys. **53** (1988) 733–757.

[KE58] Keller, J.B.: Corrected Bohr-Sommerfeld quantum conditions for nonseparable systems. Ann. Phys. **4** (1958) 180–188.

[KK83] Kadanoff, L.P., Kohmoto, M., Tang, C.: Localization problem in one dimension: mapping and escape. Phys. Rev. Lett. **50** (1983) 1870–1872.

[KL86] Kalugin, P.A., Kitaev, A.Yu., Levitov, S: Electron spectrum of a one dimensional quasi-crystal. Sov. Phys. JETP **64** (1986) 410–415.

[KM82] Kirsch, W., Martinelli, F.: On the density of states of Schrödinger operators with a random potential. J. Phys. A **15** (1982) 2139–2156.

[KN84] Kramer, P., Neri, R.: On periodic and non periodic space fillings of E^n obtained by projections. Acta Crystallogr. A **40** (1984) 580–587.

[KO84] Kohmoto, M., Oono, Y.: Cantor spectrum for an almost periodic Schrödinger operator and a dynamical map. Phys. Lett. **102 A** (1984) 145–148.

[KS86] Kohmoto, M., Sutherland, B.: Electronic states on a Penrose lattice. Phys. Rev. Lett. **56** (1986) 2740–2743.

[KS89] Kunz, H., Livi, R., Suto, A.: Cantor spectrum and singular continuity for a hierarchical hamiltonian. Commun. Math. Phys. **122** (1989) 643–679.

[LA30] Landau, L.: Diamagnetismus der Metalle. Z. Phys. **64** (1930) 629–637.

[LE89] Levitov, L.S.: Renormalization group for a quasiperiodic Schrödinger operator. J. de Phys. **50** (1989) 707–716.

[LM88] Livi, R., Maritan, A., Ruffo, S.:The spectrum of a 1-D hierarchical model. J. Stat. Phys. **52** (1988) 595–608.

[LP86] Luck, J.M., Petritis, D.: Phonon spectra in one dimensional quasicrystals. J. Stat. Phys. **42** (1986) 289–310.

[LS84] Levine, D., Steinhardt: Quasicrystals: a new class of ordered structure. Phys. Rev. Lett. **53** (1984) 2477–2480.

[LU90] Luck, J.M.: Cantor spectra and scaling of gap widths in deterministic aperiodic systems. Phys. Rev. B **39** (1989) 5834–5849.

[MA76] Mermin, D., Ashcroft: Solid state physics. Saunders, Philadelphia Tokyo 1976.

[MA85] Marcus, R.A.: Aspect of intramolecular dynamics in Chemistry. In: Chaotic behaviour in Quantum Systems. (G. Casati, ed.) NATO ASI B120, Plenum, New York 1985.

[MK82] Mackay, A.L.: Physica **114 A** (1982) 566–613; Sov. Phys. Crystallogr. **26** (1981) 5.

[MO21] Morse, Recurrent geodesics on a surface of negative curvature. Trans. Amer. Math. Soc. **22** (1921) 84–100.

[MO81] Moser, J.: An example of a Schrödinger equation with an almost periodic potential and nowhere dense spectrum. Commun. Math. Helv. **56** (1981) 198.

[MO88] Mosseri, R., Oguey, C., Duneau, M.: A new approach to quasicrystal approximants. In: Ref. [JD88].

[MS83] Mosseri, R., Sadoc, J.F.: In: Structure of non crystalline materials 1982. Taylor and Francis London 1983.

[NA77] Nakao, S.: On the spectral distribution for the Schrödinger operator with a random potential. Japan J. Math. **3** (1977) 111.
[NB90] Nakamura, S., Bellissard, J.: Low energy bands do not contribute to the quantum Hall effect. Commun. Math. Phys. **131** (1990) 283–305.
[NE64] Nelson, E.: Feynman integrals and the Schrödinger equation. J. Math. Phys. **5** (1964) 332–343.
[NS85] Nelson, D., Sachdev, S.: Statistical mechanics of pentagonal and icosahedral order in dense liquids. Phys. Rev. B **32** (1985) 1480–1502.
[OK85] Ostlund, S., Kim, S.: Renormalization of quasiperiodic mappings. Physica Scripta **9** (1985) 193–198.
[ON86] Odagaki, T., Nguyen, D.: Electronic and vibrational spectra of two- dimensional quasicrystals. Phys. Rev. B **33** (1986) 2184–2190.
[OP83] Ostlund, S., Pandit, R., Rand, D., Schellnhuber, H.J., Siggia, E.D.: One dimensional Schrödinger equation with an almost periodic potential. Phys. Rev. Lett. **50** (1983) 1873–1877.
[PA26] Pauli, W.: Über das Wasserstoffspektrum vom Standpunkt der neuen Quantenmechanik. Z. Phys. **36** (1926) 336–363 (see [VW67]).
[PA73] Pastur, L.: Spectra of random selfadjoint operators. Usp. Mat. Nauk. **28** (1973) 3.
[PE33] Peierls, R.: Zur Theorie des Diamagnetismus von Leitungelektronen. Z. Phys. **80** (1933) 763–791.
[PE79] Pedersen, G.: C^*-algebras and their automorphism groups. Academic Press, New York 1979.
[PJ95] Piéchon, F., Jagannathan, A.: Energy-level statistics of electrons in a two-dimensional quasicrystal, Phys. Rev. **B51** (1995), 179–184.
[PL00] Planck, M.: Zur Theorie des Gesetzes der Energieverteilung im Normalspektrum. Verh. der D. Physikal. Ges. **2** (1900) 237.
[PN79] Penrose, R.: Bull. Inst. Math. **10** (1974) 266; The Mathematical Intelligencer **2** (1979) 22–37.
[PS86] Putnam, I., Schmidt, K., Skau, C.: C^*-algebras associated with Denjoy diffeomorphisms of the circle. J. of Operator Theory **16** (1986) 99–126.
[PS92] Passaro, B., Sire, C. , Benza, V.G.: Chaos and diffusion in 2D quasicrystals, Phys. Rev. B **46** (1992) 13751–13755
[PV80a] Pimsner, M., Voiculescu, D.: Exact sequences for K-groups and Ext groups of certain cross-product C^*-algebras. J. Operator Theory. **4** (1980) 93–118.
[PV80b] Pimsner, M., Voiculescu, D.: Imbedding the irrational rotation C^*-algebra into an AF algebra. J. Operator Theory **4** (1980) 201–211.
[QU87] Queffelec, M.: Substitution dynamical systems-Spectral analysis. Lecture Notes in Math. Springer, Berlin Heidelberg New York 1987, vol. **1294**.
[RA84] Rammal, R.: Spectrum of harmonic excitations on fractals. J. de Phys. **45** (1984) 191–206.
[RA85] Rammal, R.: Landau level spectrum of Bloch electrons on a honeycomb lattice. J. de Phys. **46** (1985) 1345–1354.
[RE80] Renault, J.: A groupoid approach to C^*-algebras. Lecture Notes in Math. **793** (1980).
[RI81] Rieffel, M.A.: C^*-algebras associated with irrational rotations. Pac. J. Math. **95** (2), (1981) 415–419.
[RI82] Rieffel, M.A.: Morita equivalence for operator algebras. In: Operator algebras and applications. Proc. Symposia Pure Math. **38**, Part I (1982) 285–298.
[RI90] Rieffel, M.A.: Critical points of Yang-Mills for noncommutative two-tori. J. Diff. Geom. **31** (1990) 535–546.
[RSII] Reed, M., Simon, B.: Methods of Modern Mathematical Physics, II. Fourier Analysis, Self-Adjointness. Academic Press, London 1975.
[RSIV] Reed, M., Simon, B.: Methods of modern mathematical physics, IV. Analysis of Operators. Academic Press, London 1978.
[RU11] Rutherford, E.: The scattering of α and β particles by matter and the structure of the atom. Phil. Mag. **21** (1911) 669.
[RU59] Rudin, W.: Some theorems on Fourier coefficients. Proc. AMS **10** (1959) 855–859.

[SA71] Sakai, S.: C^*-algebras and W^*-algebras. Ergebnisse der Mathematik und ihrer Grenzgebiete, Bd. 60. Springer, Berlin Heidelberg New York 1971.

[SA79] Sauvageot, J.L.: Idéaux primitifs induits dans les produits croisés. J. Funct. Anal. **32** (1979) 381–392.

[SB84] Shechtman, D., Blech, I., Gratias, D., Cahn, J.V.: Metallic phase with long range orientational order and no translational symmetry. Phys. Rev. Lett. **53** (1984) 1951–1953.

[SB90] Sire, C., Bellissard, J.: Renormalization group for the octagonal quasi-periodic tiling. Europhys. Lett. **11** (1990) 439–443.

[SG22] Stern, O., Gerlach, W.: Z. Phys. **9** (1922) 349.

[SG94] Sire, C., Gratias, D.: Introduction to the physics of quasicrystals, NATO School "Phase transformations in alloys", p. 127, Plenum Press NY (1994);

[SH79] Shubin, M.: The spectral theory and the index of elliptic operators with almost periodic coefficients. Russ. Math. Surv. **34** (1979) 109–157.

[SI79] Simon, B.: Trace ideals and their applications. London Math. Soc. Lecture Notes **35** Cambridge Univ. Press 1979.

[SI89] Sire, C.: Electronic spectrum of a 2D quasi-crystal related to the octagonal quasi-periodic tiling. Europhys. Lett. **10** (1990) 483–488.

[SI94] Sire, C.: Properties of quasiperiodic Hamiltonians : spectra, wave functions and transport, "Winter School on Quasicrystals", Les Éditions de Physique, Paris (1994).

[SO15] Sommerfeld, A.: Ber. Akad. München (1915) pp. 425, 459 (see [VW67] p. 95).

[SO85] Sokoloff, J.B.: Unusual band structure, wave functions and electrical conductance in crystals with incommensurate periodic potentials. Phys. Reports **126** (1985) 189–244.

[SO87] Steinhardt, P.J., Ostlund, S.: The physics of quasicrystals. World Scientific Pub., Singapore 1987.

[SP93] Sire, C., Passaro, B., Benza, V.: Electronic properties of 2d quasicrystals. Level spacing distribution and diffusion, J. Non Crystalline Sol. **153** &**154** (1993) 420–425.

[SR26] Schrödinger, E.: Quantisierung als Eigenwertproblem. Ann. Phys. **79** (1926) 361-376.

[SR82] Shirts, R.B., Reinhardt, W.P.: Approximate constants of motion for classically chaotic vibrational dynamics: Vague Tori, semiclassical quantization, and classical intramolecular energy flow. J. Chem. Phys. **77** (1982) 5204–5217.

[SU87] Sütó, A.: The spectrum of a quasi-periodic Schrödinger operator. Commun. Math. Phys. **111** (1987) 409–415.

[SU89] Sütó, A.: Singular continuous spectrum on a Cantor set of zero Lebesgue measure for the Fibonacci Hamiltonian. J. Stat. Phys. **56** (1989) 525–531.

[TA79] Takesaki, M.: Theory of operator algebras. Springer, Berlin Heidelberg New York 1979.

[TF86] Tsunetsugu, H., Fujiwara, T., Ueda, K., Tokihiro, T.: Eigenstates in a 2-dimensional Penrose tiling. J. Phys. Soc. Japan **55** (1986) 1420–1423.

[TU06] Thue, A.: Über die gegenseitige Lage gleicher Teile gewisser Zeichenreichen. Videnskabsselskabets Skrifter I Mat. nat., Christiania (1906).

[VW67] Waerden, B.L. van der: Sources of quantum mechanics. North-Holland Pub. Co., Amsterdam 1967.

[WI84] Wilkinson, M.: Critical properties of electron eigenstates in incommensurate systems. Proc. Roy. Soc. London A **391** (1984) 305–350.

[WI89] Wijnands, F.: Energy spectra for one-dimensional quasiperiodic potentials: bandwidth, scaling, mapping and relation with local isomorphism. J. Phys. A **22** (1989) 3267–3282.

[XI88] Xia, J.: Geometric invariants of the quantum Hall effect. Commun. Math. Phys. **119** (1988) 29–50.

[ZA64] Zak, J.: Magnetic translation groups. Phys. Rev. A **134** (1964) 1602–1607; Magnetic translation groups II: Irreducible representations. Phys. Rev. A **134** (1964) 1607-1611.

Chapter 13

Circle Maps: Irrationally Winding

by Predrag Cvitanović

1. Introduction

In these lectures we shall discuss circle maps as an example of a physically interesting chaotic dynamical system with rich number-theoretic structure. Circle maps arise in physics in a variety of contexts. One setting is the classical *Hamiltonian mechanics*; a typical island of stability in a Hamiltonian 2-d map is an infinite sequence of concentric KAM tori and chaotic regions. In the crudest approximation, the radius can here be treated as an external parameter Ω, and the angular motion can be modelled by a map periodic in the angular variable (Shenker 1982; Shenker and Kadanoff 1982). In *holomorphic dynamics* circle maps arise from the winding of the complex phase factors as one moves around the Mandelbrot cacti (Cvitanović and Myrheim 1983 and 1989). In the context of *dissipative* dynamical systems one of the most common and experimentally well explored routes to chaos is the two-frequency mode-locking route. Interaction of pairs of frequencies is of deep theoretical interest due to the generality of this phenomenon; as the energy input into a dissipative dynamical system (for example, a Couette flow) is increased, typically first one and then two of intrinsic modes of the system are excited. After two Hopf bifurcations (a fixed point with inward spiraling stability has become unstable and outward spirals to a limit cycle) a system lives on a two-torus. Such systems tend to mode-lock: the system adjusts its internal frequencies slightly so that they fall in step and minimize the internal dissipation. In such case the ratio of the two frequencies is a rational number. An irrational frequency ratio corresponds to a quasiperiodic motion - a curve that never quite repeats itself. If the mode-locked states overlap, chaos sets in. For a nice discussion of physical applications of circle maps, see the references (Jensen et al. 1983 and 1984; Bak et al. 1985). Typical examples are dynamical systems such as the Duffing oscillator and models of the Josephson junction, which possess a natural frequency ω_1 and are in addition driven by an external frequency ω_2. Periodicity is in this case imposed by the driving frequency, and the dissipation confines the system to a low dimensional attractor; as the ratio ω_1/ω_2 is varied, the system sweeps through infinitely many mode-locked states. The likelihood that a mode-locking occurs depends on the strength of the coupling of the internal and the external frequencies.

By losing all of the 'island-within-island' structure of real systems, circle map models skirt the problems of determining the symbolic dynamics for a realistic Hamiltonian system, but they do retain some of the essential features of such systems, such as the golden mean renormalization (Greene 1979; Shenker and Kadanoff 1982) and non-hyperbolicity in form of sequences of cycles accumulating toward the borders of stability. In particular, in such systems there are orbits that stay 'glued' arbitrarily close to stable regions for arbitrarily long times. As this is a generic phenomenon in physically interesting dynamical systems, such as the Hamiltonian systems with coexisting elliptic islands of stability and hyperbolic homoclinic webs, development of good computational techniques is here of utmost practical importance.

We shall start by briefly summarizing the results of the 'local' renormalization theory for transitions from quasiperiodicity to chaos. In experimental tests of this theory one adjusts the external frequency to make the frequency ratio as far as possible from being mode-locked. This is most readily attained by tuning the ratio to the 'golden mean' $(\sqrt{5}-1)/2$. The choice of the golden mean is dictated by number theory: the golden mean is the irrational number for which it is hardest to give good rational approximants. As experimental measurements have limited accuracy, physicists usually do not expect that such number-theoretic subtleties as how irrational a number is should be of any physical interest. However, in the dynamical systems theory to chaos the starting point is the enumeration of asymptotic motions of a dynamical system, and through this enumeration number theory enters and comes to play a central role.

Number theory comes in full strength in the 'global' theory of circle maps, the study of universal properties of the entire irrational winding set – the main topic of these lectures. We shall concentrate here on the example of a global property of the irrational winding set discovered by Jensen, Bak, and Bohr (Jensen et al. 1983 and 1984; Bak et al. 1985): the set of irrational windings for critical circle maps with cubic inflection has the Hausdorff dimension $D_H = 0.870\ldots$, and the numerical work indicates that this dimension is *universal.* The universality (or even existence) of this dimension has not yet been rigorously established. We shall offer here a rather pretty explanation (Cvitanović, Gunaratne and Vinson 1990) of this universality in form of the explicit formula (39) which expresses this Hausdorff dimension as an average over the Shenker universal scaling numbers (Shenker 1982; Feigenbaum, Kadanoff and Shenker 1982; Ostlund et al. 1983). The renormalization theory of critical circle maps demands at present rather tedious numerical computations, and our intuition is much facilitated by approximating circle maps by number-theoretic models. The model that we shall use here to illustrate the basic concepts might at first glance appear trivial, but we find it very instructive, as much that is obscured for the critical maps by numerical work is here readily number-theoretically accessible. Indicative of the depth of mathematics lurking behind physicists' conjectures is the fact that the properties that one would like to establish about the renormalization theory of critical circle maps might turn out to be related to number-theoretic abysses such as the Riemann conjecture, already in the context of the 'trivial' models.

The literature on circle maps is overwhelming, ranging from pristine Bourbakese (Herman 1979; Yoccoz 1984) to palpitating chicken hearts (Glass et al. 1983), and attempting a comprehensive survey would be a hopeless undertaking. The choice of topics covered here is of necessity only a fragment of what is known about the diffeomorphisms of the circle.

2. Mode Locking

The Poincaré section of a dynamical system evolving on a two-torus is topologically a circle. A convenient way to study such systems is to neglect the radial variation of the Poincaré section, and model the angular variable by a map of a circle onto itself. Both quantitatively and qualitatively this behavior is often well described (Arnold 1961, 1965 and 1983) by 1-dimensional circle maps $x \to x' = f(x)$, $f(x+1) = f(x) + 1$ restricted to the circle, such as the *sine map*

$$x_{n+1} = x_n + \Omega - \frac{k}{2\pi}\sin(2\pi x_n) \qquad \text{mod } 1 \ . \tag{1}$$

$f(x)$ is assumed to be continuous, have a continuous first derivative, and a continuous second derivative at the inflection point. For the generic, physically relevant case (the only one considered here) the inflection is cubic. Here k parametrizes the strength of the mode-mode interaction, and Ω parametrizes the ω_1/ω_2 frequency ratio. For $k = 0$, the map is a simple rotation (the *shift map*)

$$x_{n+1} = x_n + \Omega \qquad \text{mod } 1 \ , \tag{2}$$

and Ω is the winding number

$$W(k, \Omega) = \lim_{n\to\infty} x_n/n \ . \tag{3}$$

If the map is monotonically increasing ($k < 1$ in (1)), it is called *subcritical*. For subcritical maps much of the asymptotic behavior is given by the trivial (shift map) scalings (Herman 1979; Yoccoz 1984). For invertible maps and rational winding numbers $W = P/Q$ the asymptotic iterates of the map converge to a unique Q-cycle attractor

$$f^Q(x_i) = x_i + P, \quad i = 0, 1, 2, \cdots, Q-1 \ .$$

For any rational winding number, there is a finite interval of parameter values for which the iterates of the circle map are attracted to the P/Q cycle. This interval is called the P/Q *mode-locked* (or *stability*) interval, and its width is given by

$$\Delta_{P/Q} = Q^{-2\mu_{P/Q}} = \Omega^{right}_{P/Q} - \Omega^{left}_{P/Q} \,. \tag{4}$$

Parametrizing mode lockings by the exponent μ rather than the width Δ will be convenient for description of the distribution of the mode-locking widths, as the exponents μ turn out to be of bounded variation. The stability of the P/Q cycle is defined as

$$\Lambda_{P/Q} = \frac{\partial x_Q}{\partial x_0} = f'(x_0)f'(x_1)\cdots f'(x_{Q-1})$$

For a stable cycle $|\Lambda|$ lies between 0 (the superstable value, the 'center' of the stability interval) and 1 (the $\Omega^{right}_{P/Q}$, $\Omega^{left}_{P/Q}$ ends of the stability interval (4)).

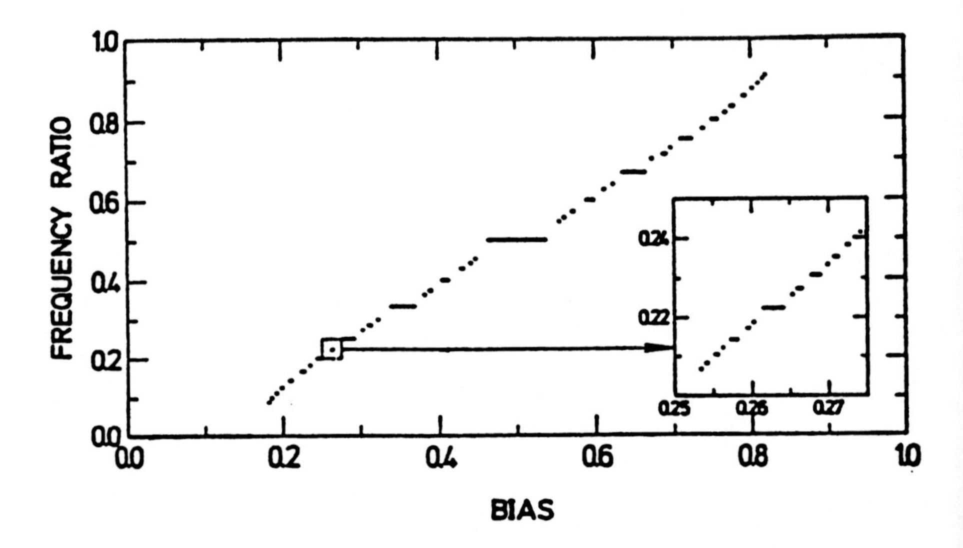

Fig. 1. The critical circle map ($k = 1$ in (1)) devil's staircase (Jensen et al. 1983 and 1984; Bak et al. 1985): the winding number W as function of the parameter Ω.

For the shift map, the stability intervals are shrunk to points. As Ω is varied from 0 to 1, the iterates of a circle map either mode-lock, with the winding number given by a rational number $P/Q \in (0,1)$, or do not mode-lock, in which case the winding number is irrational. A plot of the winding number W as a function of the shift parameter Ω is a convenient visualization of the mode-locking structure of circle maps. It yields a monotonic 'devil's staircase' of fig. 1 whose self-similar structure we are to unravel.

Circle maps with zero slope at the inflection point x_c

$$f'(x_c) = 0, f''(x_c) = 0$$

($k = 1$, $x_c = 0$ in (1)) are called *critical*: they delineate the borderline of chaos in this scenario. As the non-linearity parameter k increases, the mode-locked intervals become wider, and for the critical circle maps ($k = 1$) they fill out the whole interval. For numerical evidence see the references (Jensen et al. 1983 and 1984; Bak et al. 1985: Lanford 1985); a proof that the set of irrational windings is of zero measure is given by Swiatek (Swiatek 1988). A critical map has a superstable P/Q cycle for any rational P/Q, as the stability of any cycle that includes the inflection point equals zero. If the map is non-invertible ($k > 1$), it is called supercritical; the bifurcation structure of this regime is extremely rich and beyond the scope of these (and most other such) lectures.

For physicists the interesting case is the critical case; the shift map is 'easy' number theory (Farey rationals, continued fractions) which one uses as a guide to organization of the non-trivial critical case. In particular, the problem of organizing subcritical mode lockings reduces to the problem of organizing rationals on the unit interval. The self-similar structure of the devil's staircase suggests a systematic way of separating the mode lockings into hierarchies of levels. The set of rationals P/Q clearly possesses rich number-theoretic structure, which we shall utilize here to formulate three different partitionings of rationals:

1. Farey series
2. Continued fractions of fixed length
3. Farey tree levels

3. Farey Series Partitioning

Intuitively, the longer the cycle, the finer the tuning of the parameter Ω required to attain it; given finite time and resolution, we expect to be able to resolve cycles up to some maximal length Q. This is the physical motivation for partitioning mode lockings into sets of cycle length up to Q (Artuso, Cvitanović and Kenny 1989; Cvitanović 1987a). In number theory such set of rationals is called a *Farey series* (Hardy and Wright 1938).

(1) Definition. *The Farey series $\mathcal{F}_Q$ of order Q is the monotonically increasing sequence of all irreducible rationals between 0 and 1 whose denominators do not exceed Q. Thus P_i/Q_i belongs to $\mathcal{F}_Q$ if $0 < P_i \leq Q_i \leq Q$ and $(P_i|Q_i) = 1$.*

For example

$$\mathcal{F}_5 = \left\{ \frac{1}{5}, \frac{1}{4}, \frac{1}{3}, \frac{2}{5}, \frac{1}{2}, \frac{3}{5}, \frac{2}{3}, \frac{3}{4}, \frac{4}{5}, \frac{1}{1} \right\}$$

A Farey sequence can be generated by observing that if P_{i-1}/Q_{i-1} and P_i/Q_i are consecutive terms of $\mathcal{F}_Q$, then

$$P_i Q_{i-1} - P_{i-1} Q_i = 1 .$$

The number of terms in the Farey series F_Q is given by

$$\Phi(Q) = \sum_{n=1}^{Q} \phi(n) = \frac{3Q^2}{\pi^2} + O(Q \ln Q). \tag{5}$$

Here the Euler function $\phi(Q)$ is the number of integers not exceeding and relatively prime to Q. For example, $\phi(1) = 1$, $\phi(2) = 1$, $\phi(3) = 2, \ldots, \phi(12) = 4, \phi(13) = 12, \ldots$ As $\phi(Q)$ is a highly irregular function of Q, the asymptotic limits are not approached smoothly: incrementing Q by 1 increases $\Phi(Q)$ by anything from 2 to Q terms. We refer to this fact as the 'Euler noise'.

The Euler noise poses a serious obstacle for numerical calculations with the Farey series partitionings; it blocks smooth extrapolations to $Q \to \infty$ limits from finite Q data. While this in practice renders inaccurate most Farey-sequence partitioned averages, the finite Q Hausdorff dimension estimates exhibit (for reasons that we do not understand) surprising numerical stability, and the Farey series partitioning actually yields the *best* numerical value of the Hausdorff dimension (30) of any methods used so far; for example (Artuso, Cvitanović and Kenny 1989; Cvitanović 1987a), the sine map (1) estimate based on $240 \leq Q \leq 250$ Farey series partitions yields $D_H = .87012 \pm .00001$. The quoted error refers to the variation of D_H over this range of Q; as the computation is not asymptotic, such numerical stability can underestimate the actual error by a large factor.

4. Continued Fraction Partitioning

From a number-theorist's point of view, the *continued fraction partitioning* of the unit interval is the most venerable organization of rationals, preferred already by Gauss. The continued fraction partitioning is obtained by deleting successively mode-locked intervals (points in the case of the shift map) corresponding to continued fractions of increasing length. The first level is obtained by deleting $\Delta_{[1]}, \Delta_{[2]}, \cdots, \Delta_{[a_1]}, \cdots$ mode-lockings; their complement are the *covering* intervals $\ell_1, \ell_2, \ldots, \ell_{a_1}, \ldots$ which contain all windings, rational and irrational, whose continued fraction expansion starts with $[a_1, \ldots]$ and is of length at least 2. The second level is obtained by deleting $\Delta_{[1,2]}, \Delta_{[1,3]}, \Delta_{[2,2]}, \Delta_{[2,3]}, \cdots, \Delta_{[n,m]}, \cdots$ and so on, as illustrated in fig. 2

(2) Definition. *The n-th level continued fraction partition $S_n = \{a_1 a_2 \cdots a_n\}$ is the monotonically increasing sequence of all rationals P_i/Q_i between 0 and 1 whose continued fraction expansion is of length n:*

$$\frac{P_i}{Q_i} = [a_1, a_2, \cdots, a_n] = \cfrac{1}{a_1 + \cfrac{1}{a_2 + \ldots \cfrac{1}{a_n}}}$$

The object of interest, the set of the irrational winding numbers, is in this partitioning labelled by $\mathcal{S}_\infty = \{a_1 a_2 a_3 \cdots\}$, $a_k \in Z^+$, i.e., the set of winding numbers with infinite continued fraction expansions. The continued fraction labelling is particularly appealing in the present context because of the close connection of the Gauss shift to the renormalization transformation R, discussed below. The Gauss shift (see for instance Billingsley (Billingsley 1965))

$$T(x) = \begin{cases} \frac{1}{x} - \left[\frac{1}{x}\right] & x \neq 0 \\ 0\,, & x = 0 \end{cases} \tag{6}$$

($[\cdots]$ denotes the integer part) acts as a shift on the continued fraction representation of numbers on the unit interval

$$x = [a_1, a_2, a_3, \ldots] \rightarrow T(x) = [a_2, a_3, \ldots]\,, \tag{7}$$

and maps 'daughter' intervals $\ell_{a_1 a_2 a_3 \ldots}$ into the 'mother' interval $\ell_{a_2 a_3 \ldots}$.

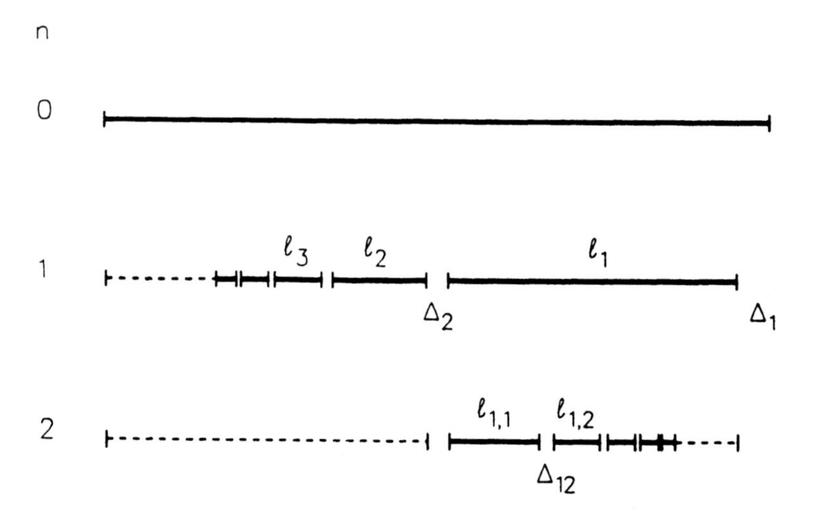

Fig. 2. Continued fraction partitioning of the irrational winding set (Artuso,Aurell and Cvitanović 1990b). At level n=1 all mode locking intervals $\Delta_{[a]}$ with winding numbers 1/1, 1/2, 1/3, ..., $1/a$, ... are deleted, and the cover consists of the complement intervals l_a. At level n=2 the mode locking intervals $\Delta_{[a,2]}$, $\Delta_{[a,3]}, \ldots$ are deleted from each cover l_a, and so on.

However natural the continued fractions partitioning might seem to a number theorist, it is problematic for an experimentalist, as it requires measuring infinity of mode-lockings even at the first step of the partitioning. This problem can be overcome both numerically and experimentally by some understanding of the asymptotics of mode-lockings with large continued fraction entries (Artuso et al. 1990b; Cvitanović, Gunaratne and Vinson 1990). Alternatively, a finite partition can be generated by the partitioning scheme to be described next.

5. Farey Tree Partitioning

The *Farey tree partitioning* is a systematic bisection of rationals: it is based on the observation that roughly halfways between any two large stability intervals (such as 1/2 and 1/3) in the devil's staircase of fig. .1 there is the next largest stability interval (such as 2/5). The winding number of this interval is given by the Farey mediant $(P+P')/(Q+Q')$ of the parent mode-lockings P/Q and P'/Q' (Hardy and Wright 1938). This kind of cycle 'gluing' is rather general and by no means restricted to circle maps; it can be attained whenever it is possible to arrange that the Qth iterate deviation caused by shifting a parameter from the correct value for the Q-cycle is exactly compensated by the Q'th iterate deviation from closing the Q'-cycle; in this way the two near cycles can be glued together into an exact cycle of length $Q+Q'$. The Farey tree is obtained by starting with the ends of the unit interval written as 0/1 and 1/1, and then recursively bisecting intervals by means of Farey mediants. This kind of hierarchy of rationals is rather new, and, as far as we are aware, not previously studied by number theorists. It is appealing both from the experimental and from the the golden-mean renormalization (Feigenbaum 1987a) point of view, but it has a serious drawback of lumping together mode-locking intervals of wildly different sizes on the same level of the Farey tree. The Farey tree partitioning was introduced in the references (Williams and Browne 1947; MacKay 1982; Cvitanović and Myrheim 1983 and 1989; Cvitanović, Shraiman and Söderberg 1985) and its thermodynamics is discussed in details by Feigenbaum.

(3) Definition. *The n-th Farey tree level T_n is the monotonically increasing sequence of those continued fractions $[a_1, a_2, \ldots, a_k]$ whose entries $a_i \geq 1$, $i = 1, 2, \ldots, k-1$, $a_k \geq 2$, add up to $\sum_{i=1}^{k} a_i = n+2$.*

For example

$$T_2 = \{[4], [2,2], [1,1,2], [1,3]\} = \left(\frac{1}{4}, \frac{1}{5}, \frac{3}{5}, \frac{3}{4}\right).$$

The number of terms in T_n is 2^n. Each rational in T_{n-1} has two 'daughters' in T_n, given by

$$[\cdots, a] \longrightarrow \{\ [\cdots, a-1, 2]\ ,\ [\cdots, a+1]\ \}$$

Iteration of this rule places all rationals on a binary tree, labelling each by a unique binary label (Cvitanović, Shraiman and Söderberg 1985). The transcription from the binary Farey labels to the continued fraction labels follows from the mother-daughter relation above; each block $1\cdots 0$ ('1' followed by $a-1$ zeros) corresponds to entry $[\cdots, a, \cdots]$ in the continued fraction label. The Farey tree has a variety of interesting symmetries (such as 'flipping heads and tails' relations obtained by reversing the order of the continued-fraction entries) with as yet unexploited implications for the renormalization theory: some of these are discussed in reference (Cvitanović, Shraiman and Söderberg 1985).

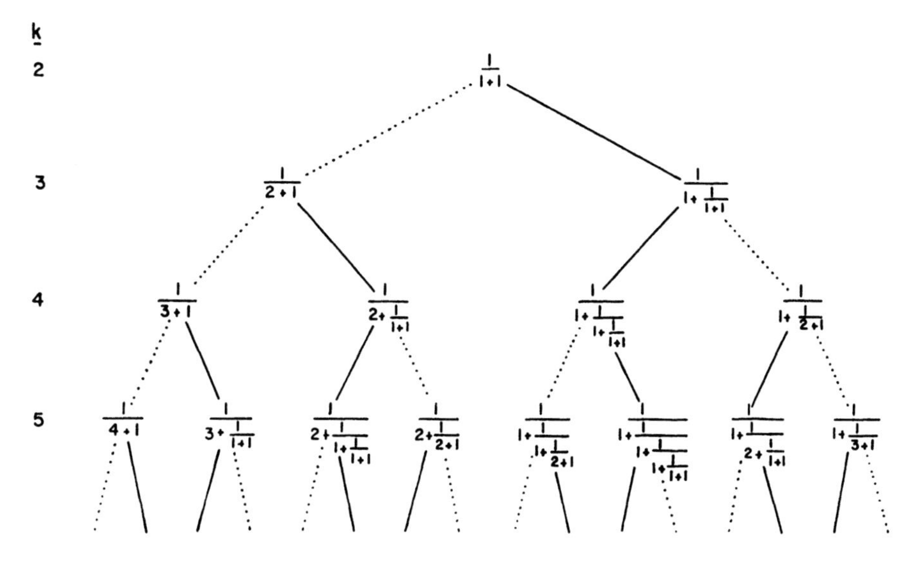

Fig. 3. The Farey tree in the continued fraction representation (from Cvitanović and Myrheim 1983 and 1989)

The smallest and the largest denominator in T_n are respectively given by

$$[n-2] = \frac{1}{n-2}, \quad [1,1,\ldots,1,2] = \frac{F_{n+2}}{F_{n+1}} \propto \rho \,, \tag{8}$$

where the Fibonacci numbers F_n are defined by $F_{n+1} = F_n + F_{n-1}$; $F_0 = 0$, $F_1 = 1$, and ρ is the golden mean ratio

$$\rho = \frac{1+\sqrt{5}}{2} = 1.61803\ldots \tag{9}$$

Note the enormous spread in the cycle lengths on the same level of the Farey tree: $n \leq Q \leq \rho^n$. The cycles whose length grows only as a power of the Farey tree level will cause strong non-hyperbolic effects in the evaluation of various averages.

The Farey tree rationals can be generated by backward iterates of 1/2 by the Farey presentation function (Feigenbaum 1987a and 1988a):

$$\begin{aligned} f_0(x) &= x/(1-x) \qquad 0 \le x < 1/2 \\ f_1(x) &= (1-x)/x \qquad 1/2 < x \le 1 \,. \end{aligned} \tag{10}$$

(the utility of the presentation function is discussed at length by Feigenbaum in the above references). The Gauss shift (6) corresponds to replacing the binary Farey presentation function branch f_0 in (10) by an infinity of branches

$$\begin{aligned} f_a(x) &= f_1 \circ f_0^{(a-1)}(x) = \frac{1}{x} - a, \qquad \frac{1}{a-1} < x \le \frac{1}{a}, \\ f_{ab\cdots c}(x) &= f_c \circ \cdots \circ f_b \circ f_a(x) \,. \end{aligned} \tag{11}$$

A rational $x = [a_1, a_2, \ldots, a_k]$ is 'annihilated' by the kth iterate of the Gauss shift, $f_{a_1 a_2 \cdots a_k}(x) = 0$. The above maps look innocent enough, but note that what is being partitioned is not the dynamical space, but the parameter space. The flow described by (10) and by its non-trivial circle-map generalizations will turn out to be a *renormalization group* flow in the function space of dynamical systems, not an ordinary flow in the phase space of a particular dynamical system.

Having defined the three partitioning schemes, we now briefly summarize the results of the circle-map renormalization theory.

6. Local Theory: 'Golden Mean' Renormalization

Possible trajectories of a dynamical system are of three qualitatively distinct types: they are either asymptotically unstable (positive Lyapunov exponent), asymptotically marginal (vanishing Lyapunov) or asymptotically stable (negative Lyapunov). The asymptotically stable orbits can be treated by the traditional integrable system methods. The asymptotically unstable orbits build up chaos, and can be dealt with using the machinery of the hyperbolic, 'Axiom A' dynamical systems theory (Ruelle 1978). Here we shall concentrate on the third class of orbits, the asymptotically marginal ones. I call them the 'border of order'; they lie between order and chaos, and remain on that border to all times.

The way to pinpoint a point on the border of order is to recursively adjust the parameters so that at the recurrence times $t = n_1, n_2, n_3, \cdots$ the trajectory passes through a region of contraction sufficiently strong to compensate for the accumulated expansion of the preceding n_i steps, but not so strong as to force the trajectory into a stable attracting orbit. The *renormalization operation R* implements this procedure by recursively magnifying the neighbourhood of a point on the border in the dynamical space (by rescaling by a factor α), in the parameter space (by shifting the parameter origin onto the border and rescaling by a factor δ), and by replacing the initial map f by the nth iterate f^n restricted to the magnified neighbourhood

$$f_p(x) \to Rf_p(x) = \alpha f^n_{p/\delta}(x/\alpha)$$

There are by now many examples of such renormalizations in which the new function, framed in a smaller box, is a rescaling of the original function, i. e. the fix-point function of the renormalization operator R. The best known is the period doubling renormalization, with the recurrence times $n_i = 2^i$. The simplest circle map example is the golden mean renormalization (Shenker 1982), with recurrence times $n_i = F_i$ given by the Fibonacci numbers (8). Intuitively, in this context a metric self-similarity arises because iterates of critical maps are themselves critical, i. e. they also have cubic inflection points with vanishing derivatives.

The renormalization operator appropriate to circle maps (Feigenbaum, Kadanoff and Shenker 1982; Ostlund et al. 1983) acts as a generalization of the Gauss shift (11); it maps a circle map (represented as a pair of functions (g, f), see fig. 4), of winding number $[a, b, c, \ldots]$ into a rescaled map of winding number $[b, c, \ldots]$:

$$R_a \begin{pmatrix} g \\ f \end{pmatrix} = \begin{pmatrix} \alpha g^{a-1} \circ f \circ \alpha^{-1} \\ \alpha g^{a-1} \circ f \circ g \circ \alpha^{-1} \end{pmatrix}, \tag{12}$$

Acting on a map with winding number $[a, a, a, \ldots]$, R_a returns a map with the same winding number $[a, a, \ldots]$, so the fixed point of R_a has a quadratic irrational winding number $W = [a, a, a, \ldots]$. This fixed point has a single expanding eigenvalue δ_a. Similarly, the renormalization transformation $R_{a_p} \ldots R_{a_2} R_{a_1} \equiv R_{a_1 a_2 \ldots a_p}$ has a fixed point of winding number $W_p = [a_1, a_2, \ldots, a_p, a_1, a_2, \ldots]$, with a single expanding eigenvalue δ_p (Feigenbaum, Kadanoff and Shenker 1982; Ostlund et al. 1983). A computer assisted proof for the golden mean winding number has been carried out by Mestel (Mestel 1985; we assume that there is a single expanding eigenvalue for any periodic renormalization).

For short repeating blocks, δ can be estimated numerically by comparing successive continued fraction approximants to W. Consider the P_r/Q_r rational approximation to a quadratic irrational winding number W_p whose continued fraction expansion consists of r repeats of a block p. Let Ω_r be the parameter for which the map (1) has a superstable cycle of rotation number $P_r/Q_r = [p, p, \ldots, p]$. The δ_p can then be estimated by extrapolating from (Shenker 1982)

$$\Omega_r - \Omega_{r+1} \propto \delta_p^{-r}. \tag{13}$$

What this means is that the 'devil's staircase' of fig.4 is self-similar under magnification by factor δ_p around any quadratic irrational W_p.

The fundamental result of the renormalization theory (and the reason why all this is so interesting) is that the ratios of successive P_r/Q_r mode-locked

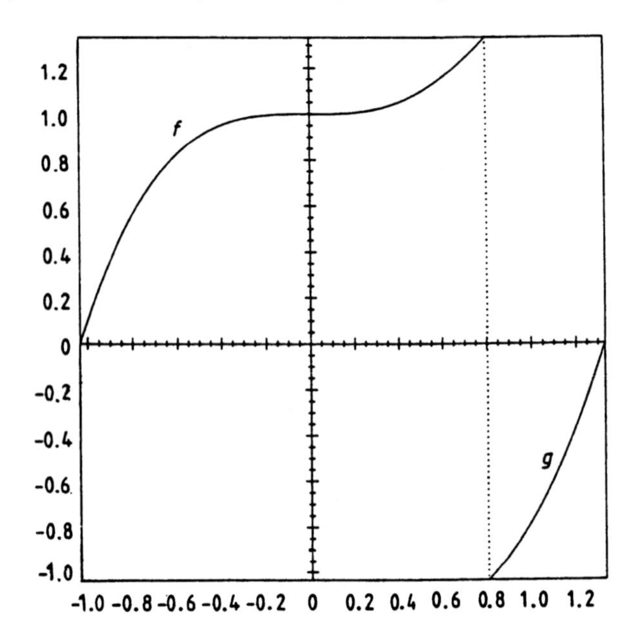

Fig. 4. The golden-mean winding number fixed-point function pair (f, g) for critical circle maps with cubic inflection point. The symbolic dynamics dictates a unique framing such that the functions (f, g) are defined on intervals $(\bar{x} \leq x \leq \bar{x}/\alpha, \bar{x}/\alpha \leq x \leq \bar{x}\alpha)$, $\bar{x} = f^{-1}(0)$: in this framing, the circle map (f, g) has continuous derivatives across the f–g junctions (from Feigenbaum 1988b).

intervals converge to *universal* limits. The simplest example of (13) is the sequence of Fibonacci number continued fraction approximants to the golden mean winding number $W = [1, 1, 1, ...] = (\sqrt{5} - 1)/2$. For critical circle maps with a cubic inflection point $\delta_1 = -2.833612\ldots$; a list of values of δ_p's for the shortest continued fraction blocks p is given in reference (Cvitanović, Gunaratne and Vinson 1990).

When the repeated block is not large, the rate of increase of denominators Q_r is not large, and (13) is a viable scheme for estimating δ's. However, for long repeating blocks, the rapid increase of Q_r's makes the periodic orbits hard to determine and better methods are required, such as the unstable manifold method employed in reference (Cvitanović, Gunaratne and Vinson 1990). This topic would take us beyond the space allotted here, so we merely record the golden-mean unstable manifold equation (Feigenbaum and Kenway 1983; Cvitanović 1985; Cvitanović, Jensen, Kadanoff and Procaccia 1985; Gunaratne 1986)

$$g_p(x) = \alpha g_{1+p/\delta}\left(\alpha g_{1+1/\delta+p/\delta^2}(x/\alpha^2)\right) \tag{14}$$

and leave the reader contemplating methods of solving such equations. We content ourselves here with stating what the extremal values of δ_p are.

For a given cycle length Q, the *narrowest* interval shrinks with a power law (Kaneko 1982, 1983a and 1983b; Jensen et al. 1983 and 1984; Bak et al. 1985; Feigenbaum, Kadanoff and Shenker 1982)

$$\Delta_{1/Q} \propto Q^{-3} \tag{15}$$

This leading behavior is derived by methods akin to those used in describing intermittency (Pomeau and Manneville 1980): $1/Q$ cycles accumulate toward the edge of $0/1$ mode-locked interval, and as the successive mode-locked intervals $1/Q$, $1/(Q-1)$ lie on a parabola, their differences are of order Q^{-3}. This should be compared to the subcritical circle maps in the number-theoretic limit (2), where the interval between $1/Q$ and $1/(Q-1)$ winding number value of the parameter Ω shrinks as $1/Q^2$. For the critical circle maps the $\ell_{1/Q}$ interval is *narrower* than in the k=0 case, because it is squeezed by the nearby broad $\Delta_{0/1}$ mode-locked interval.

For fixed Q the *widest* interval is bounded by $P/Q = F_{n-1}/F_n$, the n-th continued fraction approximant to the *golden mean*. The intuitive reason is that the golden mean winding sits as far as possible from any short cycle mode-locking. Herein lies the surprising importance of the golden mean number for dynamics; it corresponds to extremal scaling in physical problems characterized by winding numbers, such as the KAM tori of classical mechanics (Greene 1979; Shenker and Kadanoff 1982). The golden mean interval shrinks with a universal exponent

$$\Delta_{P/Q} \propto Q^{-2\mu_1} \tag{16}$$

where $P = F_{n-1}$, $Q = F_n$ and μ_1 is related to the universal Shenker number δ_1 (13) and the golden mean (9) by

$$\mu_1 = \frac{\ln|\delta_1|}{2\ln\rho} = 1.08218\ldots \tag{17}$$

The closeness of μ_1 to 1 indicates that the golden mean approximant mode-lockings barely feel the fact that the map is critical (in the k=0 limit this exponent is $\mu = 1$).

To summarize: for critical maps the spectrum of exponents arising from the circle maps renormalization theory is bounded from above by the harmonic scaling, and from below by the geometric golden-mean scaling:

$$3/2 > \mu_{m/n} \geq 1.08218\cdots. \tag{18}$$

7. Global Theory: Ergodic Averaging

So far we have discussed the results of the renormalization theory for isolated irrational winding numbers. Though the local theory has been tested experimentally (Stavans et al. 1985; Gwinn and Westervelt 1987), the golden-mean universality utilizes only a few of the available mode-locked intervals, and from the experimental point of view it would be preferable to test universal properties which are *global* in the sense of pertaining to a range of winding numbers. We first briefly review some of the attempts to derive such predictions using ideas from the ergodic number theory, and then turn to the predictions based on the thermodynamic formalism.

The ergodic number theory (Billingsley 1965; Khinchin 1964) is rich in (so far unfulfilled) promise for the mode-locking problem. For example, while the Gauss shift (6) invariant measure

$$\mu(x) = \frac{1}{\ln 2}\,\frac{1}{1+x} \tag{19}$$

was known already to Gauss, the corresponding invariant measure for the critical circle maps renormalization operator R has so far eluded description. It lies on a fractal set - computer sketches are given in references (Farmer and Satija 1985; Umberger Farmer and Satija 1986) - and a general picture of what the 'strange repeller' (in the space of limit functions for the renormalization operator (12) might look like is given by Lanford and Rand (Lanford 1987a and 1987b; Rand 1987 and 1988). In reference (Ostlund et al. 1983) the authors have advocated ergodic explorations of this attractor, by sequences of renormalizations R_{a_k} corresponding to the digits of the continued fraction expansion of a 'normal' winding number $W = [a_1, a_2, a_3, \ldots]$. A numerical implementation of this proposal (Farmer and Satija 1985; Umberger, Farmer and Satija 1986) by Monte Carlo generated strings $a_1, a_2, a_3, \ldots$ yields estimates of 'mean' scalings $\bar{\delta} = 15.5 \pm .5$ and $\bar{\alpha} = 1.8 \pm .1$. $\bar{\delta}^n$ is the estimate of the mean width of an 'average' mode-locked interval Δ_{P_n/Q_n}, where P_n/Q_n is the nth continued fraction approximation to a normal winding number $W = [a_1, a_2, a_3, \ldots]$. In this connection the following beautiful result by Khinchin, Kuzmin and Levy (Khinchin 1964) of the ergodic number theory is suggestive:

Theorem. *For almost all $W \in [0,1]$ the denominator Q_n of the n-th continued fraction approximant $W = P_n/Q_n + \epsilon_n$, $P_n/Q_n = [a_1, a_2, a_3, \ldots, a_n]$ converges asymptotically to*

$$\lim_{n\to\infty} \frac{1}{n} \ln Q_n = \frac{\pi^2}{12 \ln 2} \,. \tag{20}$$

In physics this theorem pops up in various guises; for example, $\pi^2/6\ln 2$ can be interpreted as the Kolmogorov entropy of 'mixmaster' cosmologies (see references in Csordás and Szépfalusy 1989). In the present context this theorem has been used (Umberger, Farmer and Satija 1986) to connect the ergodic estimate of $\bar{\delta}$ to $\hat{\delta}$ estimated (Jensen et al. 1983 and 1984; Bak et al. 1985) by averaging over all available mode-lockings up to given cycle length Q, but it is hard to tell what to make out of such results. The numerical convergence of ergodic averages is slow, if not outright hopeless, so we abandon henceforth the ergodic 'time' averages (here the 'time' is the length of a continued fraction) and turn instead to the 'thermodynamic' averages (averages over all 'configurations', here all mode lockings on a given level of a resolution hierarchy).

8. Global Theory: Thermodynamic Averaging

Consider the following average over mode-locking intervals (4):

$$\Omega(\tau) = \sum_{Q=1}^{\infty} \sum_{(P|Q)=1} \Delta_{P/Q}^{-\tau} . \tag{21}$$

The sum is over all irreducible rationals P/Q, $P < Q$, and $\Delta_{P/Q}$ is the width of the parameter interval for which the iterates of a critical circle map lock onto a cycle of length Q, with winding number P/Q.

The qualitative behavior of (21) is easy to pin down. For sufficiently negative τ, the sum is convergent; in particular, for $\tau = -1$, $\Omega(-1) = 1$, as for the critical circle maps the mode-lockings fill the entire Ω range (Swiatek 1988). However, as τ increases, the contributions of the narrow (large Q) mode-locked intervals $\Delta_{P/Q}$ get blown up to $1/\Delta_{P/Q}^{\tau}$, and at some critical value of τ the sum diverges. This occurs for $\tau < 0$, as $\Omega(0)$ equals the number of all rationals and is clearly divergent.

The sum (21) is infinite, but in practice the experimental or numerical mode-locked intervals are available only for small finite Q. Hence it is necessary to split up the sum into subsets $\mathcal{S}_n = \{i\}$ of rational winding numbers P_i/Q_i on the 'level' n, and present the set of mode-lockings hierarchically, with resolution increasing with the level:

$$\bar{Z}_n(\tau) = \sum_{i \in \mathcal{S}_n} \Delta_i^{-\tau}. \tag{22}$$

The original sum (21) can now be recovered as the $z = 1$ value of a 'generating' function $\Omega(z, \tau) = \sum_n z^n \bar{Z}_n(\tau)$. As z is anyway a formal parameter, and n is a rather arbitrary 'level' in some *ad hoc* partitioning of rational numbers, we bravely introduce a still more general, P/Q weighted generating function for (21):

$$\Omega(q, \tau) = \sum_{Q=1}^{\infty} \sum_{(P|Q)=1} e^{-\nu_{P/Q} q} Q^{2\mu_{P/Q} \tau} . \tag{23}$$

The sum (21) corresponds to $q = 0$. Exponents $\nu_{P/Q}$ will reflect the importance we assign to the P/Q mode-locking, i. e. the *measure* used in the averaging over all mode-lockings. Three choices of of the $\nu_{P/Q}$ hierarchy that we consider here correspond respectively to the Farey series partitioning (definition (1))

$$\Omega(q,\tau) = \sum_{Q=1}^{\infty} \Phi(Q)^{-q} \sum_{(P|Q)=1} Q^{2\mu_{P/Q}\tau} , \tag{24}$$

the continued fraction partitioning (definition (2))

$$\Omega(q,\tau) = \sum_{n=1}^{\infty} e^{-qn} \sum_{[a_1,\ldots,a_n]} Q^{2\mu_{[a_1,\ldots,a_n]}\tau} , \tag{25}$$

and the Farey tree partitioning (definition (3))

$$\Omega(q,\tau) = \sum_{k=n}^{\infty} 2^{-qn} \sum_{i=1}^{2^n} Q_i^{\tau\mu_i} , \quad Q_i/P_i \in T_n . \tag{26}$$

Other measures can be found in the literature, but the above three suffice for our purposes.

Sum (23) is an example of a 'thermodynamic' average. In the thermodynamic formalism a function $\tau(q)$ is defined by the requirement that the $n \to \infty$ limit of generalized sums

$$Z_n(\tau,q) = \sum_{i \in \mathcal{S}_n} \frac{p_i^q}{\ell_i^\tau} \tag{27}$$

is finite. For details on the thermodynamic formalism see the literature (Ruelle 1978; Grassberger 1983; Hentschel and Procaccia 1983; Benzi et al. 1984; Halsey et al. 1986; Feigenbaum 1987b and 1987c). Thermodynamic formalism was originally introduced to describe measures generated by strongly mixing ergodic systems, and for most practitioners p_i in (27) is the probability of finding the system in the partition i, given by the 'natural' measure. What we are using here in the Farey series and the Farey tree cases are the 'equipartition' measures $p_i = 1/N_n$, where N_n is the number of mode-locking intervals on the nth level of resolution. In the continued fraction partitioning this does not work, as N_n is infinite - in this case we assign all terms of equal continued fraction length equal weight. It is important to note that as the Cantor set under consideration is generated by scanning the parameter space, not by dynamical stretching and kneading, there is no 'natural' measure, and a variety of equally credible measures can be constructed (Jensen et al. 1983 and 1984; Bak et al. 1985; Artuso, Cvitanović and Kenny 1989; Cvitanović 1987a; Artuso 1988b and the above mentioned references on thermodynamical formalism). Each distinct hierarchical presentation of the irrational winding set (distinct partitioning of rationals on the unit interval) yields a *different* thermodynamics. As far as I can tell, no thermodynamic function $q(\tau)$ considered here (nor any of the $q(\tau)$ or $f(\alpha)$ functions studied in the literature in other contexts) has physical significance, but their qualitative properties are interesting; in particular, all versions of mode-locking thermodynamics studied so far exhibit phase transitions.

We summarize by succinctly stating what our problem is in a way suggestive to a number theorist, by changing the notation slightly and rephrasing (21) this way:

(4) Definition: The mode-locking problem. *Develop a theory of the following 'zeta' function:*

$$\hat{\zeta}(s) = \sum_{n=1}^{\infty} \sum_{(m|n)=1} n^{-2\mu_{m/n} s} , \tag{28}$$

where μ is defined as in (4).

For the shift map (2), $\mu_{m/n} = 1$, and this sum is a ratio of two Riemann zeta functions

$$\hat{\zeta}(s) = \frac{\zeta(2s-1)}{\zeta(2s)} .$$

For critical maps the spectrum of exponents arising from the circle maps renormalization theory is non-trivial; according to (18) it is bounded from above by the harmonic scaling, and from below by the geometric golden-mean scaling.

Our understanding of the $\hat{\zeta}(s)$ function for the critical circle maps is rudimentary – almost nothing that is the backbone of the theory of number-theoretic zeta functions has been accomplished here: no good integral representations of (28) are known, no functional equations (analogous to reflection formulas for the classical zeta functions) have been constructed, no Riemann-Siegel formulas, *etc.*. We summarize basically all that is known in the remainder of this lecture, and that is not much.

9. The Hausdorff Dimension of Irrational Windings

A finite cover of the set irrational windings at the 'n-th level of resolution' is obtained by deleting the parameter values corresponding to the mode-lockings in the subset $\mathcal{S}_n$; left behind is the set of complement *covering* intervals of widths

$$\ell_i = \Omega^{min}_{P_r/Q_r} - \Omega^{max}_{P_l/Q_l} . \tag{29}$$

Here $\Omega^{min}_{P_r/Q_r}$ ($\Omega^{max}_{P_l/Q_l}$) are respectively the lower (upper) edges of the mode-locking intervals Δ_{P_r/Q_r} (Δ_{P_l/Q_l}) bounding ℓ_i and i is a symbolic dynamics label, for example the entries of the continued fraction representation $P/Q = [a_1, a_2, ..., a_n]$ of one of the boundary mode-lockings, $i = a_1 a_2 \cdots a_n$. ℓ_i provide a finite cover for the irrational winding set, so one may consider the sum

$$Z_n(\tau) = \sum_{i \in \mathcal{S}_n} \ell_i^{-\tau} \tag{30}$$

The value of $-\tau$ for which the $n \to \infty$ limit of the sum (30) is finite is the *Hausdorff dimension* (see for example Falconer 1985) D_H of the irrational winding set. Strictly speaking, this is the Hausdorff dimension only if the choice of covering intervals ℓ_i is optimal; otherwise it provides an upper bound to D_H. As by construction the ℓ_i intervals cover the set of irrational winding with no slack, we expect that this limit yields the Hausdorff dimension. This is supported by all numerical evidence, but a proof that would satisfy mathematicians is lacking.

Jensen, Bak and Bohr (Jensen et al. 1983 and 1984; Bak et al. 1985) have provided numerical evidence that this Hausdorff dimension is approximately $D_H = .870\ldots$ and that it is universal. It is not at all clear whether this is the optimal global quantity to test - a careful investigation (Artuso, Cvitanović and Kenny 1989; Cvitanović 1987a) shows that D_H is surprisingly hard to pin down numerically. At least the Hausdorff dimension has the virtue of being independent of how one partitions mode-lockings and should thus be the same for the variety of thermodynamic averages in the literature. In contrast, the generalized dimensions introduced through the thermodynamical formalism differ from version to version.

10. A Bound on the Hausdorff Dimension

We start by giving an elementary argument that the Hausdorff dimension of irrational windings for critical circle maps is less than one. The argument depends on the reasonable, but so far unproven assumption that the golden mean scaling (17) is the extremal scaling.

In the crudest approximation, one can replace $\mu_{P/Q}$ in (28) by a 'mean' value $\hat{\mu}$; in that case the sum is given explicitly by a ratio of the Riemann ζ-functions:

$$\Omega(\tau) = \sum_{Q=1}^{\infty} \phi(Q)\, Q^{2\tau\hat{\mu}} = \frac{\zeta(-2\tau\hat{\mu}-1)}{\zeta(-2\tau\hat{\mu})} \tag{31}$$

As the sum diverges at $-\tau =$ Hausdorff dimension, the 'mean' scaling exponent $\hat{\mu}$ and D_H are related by the ζ function pole at $\zeta(1)$:

$$D_H\hat{\mu} = 1. \tag{32}$$

While this does not enable us to compute D_H, it does immediately establish that D_H for critical maps exists and is smaller than 1, as the μ bounds (18) yield

$$\frac{2}{3} < D_H < .9240\ldots \tag{33}$$

To obtain sharper estimates of D_H, we need to describe the distribution of $\mu_{P/Q}$ within the bounds (18). This we shall now attempt using several variants of the thermodynamic formalism.

11. The Hausdorff Dimension in Terms of Cycles

Estimating the $n \to \infty$ limit of (30) from finite numbers of covering intervals ℓ_i is a rather unilluminating chore. Fortunately, there exist considerably more elegant ways of extracting D_H. We have noted that in the case of the 'trivial' mode-locking problem (2), the covering intervals are generated by iterations of the Farey map (10) or the Gauss shift (11). The nth level sum (30) can be approximated by $\mathcal{L}^n$, where $\mathcal{L}(y,x) = \delta(x - f^{-1}(y))|f'(y)|^\tau$; this amounts to approximating each cover width ℓ_i by $|df^n/dx|$ evaluated on the ith interval. By nothing much deeper than use of the identity *log det* = *tr log* , the spectrum of $\mathcal{L}$ can be expressed (Ruelle 1978) in terms of stabilities of the prime (non-repeating) periodic orbits p of $f(x)$:

$$\begin{aligned}
\det(1 - z\mathcal{L}) &= \exp\left(-\sum_p \sum_{r=1}^{\infty} \frac{z^{rn_p}}{r} \frac{|\Lambda_p^r|^\tau}{1 - 1/\Lambda_p^r}\right) \\
&= \prod_p \prod_{k=0}^{\infty} \left(1 - z^{n_p} |\Lambda_p|^\tau / \Lambda_p^k\right) . \qquad (34)
\end{aligned}$$

In the 'trivial' Gauss shift (11) renormalization model, the Fredholm determinant and the dynamical zeta functions have been introduced and studied by Mayer (Mayer 1976) who has shown that the eigenvalues of the transfer operator are exponentially spaced, just as for the dynamical zeta functions (Ruelle 1976) for the 'Axiom A' hyperbolic systems.

The sum (30) is dominated by the leading eigenvalue of $\mathcal{L}$; the Hausdorff dimension condition $Z_n(-D_H) = O(1)$ means that $\tau = -D_H$ should be such that the leading eigenvalue is $z = 1$. The leading eigenvalue is determined by the $k = 0$ part of (34); putting all these pieces together, we obtain a pretty formula relating the Hausdorff dimension to the prime cycles of the map $f(x)$:

$$0 = \prod_p \left(1 - 1/|\Lambda_p|^{D_H}\right) . \qquad (35)$$

For the Gauss shift (11) the stabilities of periodic cycles are available analytically (Mayer 1976; Artuso, Aurell and Cvitanović 1990b), as roots of quadratic equations: For example, the x_a fixed points (quadratic irrationals with $x_a = [a, a, a \ldots]$ infinitely repeating continued fraction expansion) are given by

$$x_a = \frac{-a + \sqrt{a^2 + 4}}{2}, \qquad \Lambda_a = -\left(\frac{a + \sqrt{a^2 + 4}}{2}\right)^2 \qquad (36)$$

and the $x_{ab} = [a, b, a, b, a, b, \ldots]$ 2–cycles are given by

$$
\begin{aligned}
(37) \qquad x_{ab} &= \frac{-ab + \sqrt{(ab)^2 + 4ab}}{2b} \\
\Lambda_{ab} &= (x_{ab}x_{ba})^{-2} = \left(\frac{ab + 2 + \sqrt{ab(ab+4)}}{2}\right)^2
\end{aligned}
$$

We happen to know beforehand that $D_H = 1$ (the irrationals take the full measure on the unit interval; the continuous Gauss measure (19) is invariant under the Gauss shift (6); the Perron-Frobenius theorem), so is the infinite product (35) merely a very convoluted way to compute the number 1? Possibly so, but availability of this exact result provides a useful testing ground for trashing out the optimal methods for determining zeros of Fredholm determinants in presence of *non-hyperbolicities.* The Farey map (10) has one marginal stability fixed point $x_0 = 0$ which is excluded from the cycle expansion of (35), but its ghost haunts us as a nonhyperbolic 'intermittency' ripple in the cycle expansion. One has to sum (Artuso, Aurell and Cvitanović 1990b) infinities of cycles of nearly same stability

$$
(38) \qquad \prod_p (1 - |\Lambda_p|^{\tau}) = 1 - \sum_{a=1}^{\infty} |\Lambda_a|^{\tau} + (\text{curvatures})
$$

in order to attain the exponential convergence expected on the basis of the hyperbolicity (Mayer 1976) of this dynamical ζ function. We know from (36) that $|\Lambda_n| \propto n^2$, so the stability grows only as a power of the cycle length n, and these infinite sums pose a serious numerical headache for which we (as yet) know of no satisfactory cure. The sum (38) behaves essentially as the Riemann $\zeta(-2\tau)$, and the analytic number theory techniques might still rescue us.

Once the meaning of (35) has been grasped, the corresponding formula (Cvitanović, Gunaratne and Vinson 1990) for the *critical* circle maps follows immediately:

$$
(39) \qquad 0 = \prod_p (1 - 1/|\delta_p|^{D_H}) \ .
$$

This formula relates the dimension (introduced by Jensen et al.) of irrational windings to the universal Shenker parameter scaling ratios δ_p; its beauty lies in relating D_H to the universal scalings δ_p, thus rendering the universality of the Jensen et al. dimension manifest. As a practical formula for evaluating this dimension, (39) has so far yielded estimates of D_H of modest accuracy, but that can surely be improved. In particular, computations based on the infinite products (34) should be considerably more convergent (Christiansen et al. 1990; Cvitanović, Christiansen and Rugh 1990) but have not been carried out so far.

The derivation of (39) relies only on the following aspects of the 'hyperbolicity conjecture' of references (Cvitanović, Shraiman and Söderberg 1985; Lanford 1987a and 1987b; Rand 1987 and 1988; Kim and Ostlund 1989; Feigenbaum 1988b):

1) *limits* for Shenker δ's *exist* and are universal. This should follow from the renormalization theory developed in (Feigenbaum, Kadanoff and Shenker 1982; Ostlund et al. 1983; Mestel 1985), a general proof is still lacking.
2) δ_p grow *exponentially* with n_p, the length of the continued fraction block p.
3) δ_p for $p = a_1 a_2 \ldots n$ with a large continued fraction entry n grows as a *power* of n. According to (15), $\lim_{n\to\infty} \delta_p \propto n^3$. In the calculation of reference (Cvitanović, Gunaratne and Vinson 1990) the explicit values of the asymptotic exponents and prefactors were not used, only the assumption that the growth of δ_p with n is not slower than a power of n.

Explicit evaluation of the spectrum was first attempted in reference (Artuso, Aurell and Cvitanović 1990b)–prerequisite for attaining the exponential (or faster (Christiansen et al. 1990; Cvitanović, Christiansen and Rugh 1990)) convergence of the cycle expansions are effective methods for summation of *infinite* families of mode-lockings. At present, those are lacking - none of the tricks from the Riemann-zeta function theory (integral representations, saddle-point expansions, Poisson resummations, *etc.*) have worked for us - so we have been forced to rely on the rather treacherous logarithmic convergence acceleration algorithms (Levin 1973).

12. Farey Series and the Riemann Hypothesis

The *Farey series* thermodynamics (24) is obtained by deleting all mode-locked intervals $\Delta_{P'/Q'}$ of cycle lengths $1 \leq Q' \leq Q$. What remains are the irrational winding set covering intervals (29).

The thermodynamics of the Farey series in the number-theory limit (2) has been studied by Hall and others (Hall 1970; Kanemitsu et al. 1982 and 1984); their analytic results are instructive and are reviewed in references (Artuso, Cvitanović and Kenny 1989; Cvitanović 1987a).

The main result is that $q(\tau)$ consists of two straight sections

$$q(\tau) = \begin{cases} \tau/2 & \tau \leq -2 \\ 1+\tau & \tau \geq -2 \end{cases} . \tag{40}$$

and the Farey arc thermodynamics undergoes a first order phase transition at $\tau = -2$. What that means is that almost all covering intervals scale as Q^{-2} (the $q = 1+\tau$ phase); however, for $\tau \leq -2$, the thermodynamics average is dominated by a handful of fat intervals which scale as Q^{-1}. The number-theoretic investigations (Hall 1970; Kanemitsu et al. 1982 and 1984) also establish the rate of convergence as $Q \to \infty$; at the phase transition point it is very slow, logarithmic (Artuso, Cvitanović and Kenny 1989; Cvitanović 1987a). In practice, the Euler noise is such numerical nuisance that we skip here the discussion of the $q(\tau)$ convergence altogether.

For the *critical* circle maps the spectrum of scales is much richer. The $1/Q$ mode-locked intervals which lie on a parabolic devil staircase (Kaneko 1982, 1983a and 1983b; Jensen et al. 1983 and 1984; Bak et al. 1985; Cvitanović, Shraiman and Söderberg 1985) yield the broadest covering interval $\ell(1,Q) \simeq kQ^{-2}$, with the minimum scaling exponent $\mu_{min} = 1.$ and the narrowest covering interval $\ell(Q, Q-1) \approx kQ^{-3}$, with the exponent $\mu_{max} = 3/2$.

The Farey series thermodynamics is of a number theoretical interest, because the Farey series provide uniform coverings of the unit interval with rationals, and because they are closely related to the deepest problems in number theory, such as the Riemann hypothesis (Edwards 1974; Titchmarsh 1951). The distribution of the Farey series rationals across the unit interval is surprisingly uniform - indeed, so uniform that in the pre-computer days it has motivated a compilation of an entire handbook of Farey series (Neville 1950). A quantitative measure of the non-uniformity of the distribution of Farey rationals is given by displacements of Farey rationals for $P_i/Q_i \in \mathcal{F}_Q$ from uniform spacing:

$$\delta_i = \frac{i}{\Phi(Q)} - \frac{P_i}{Q_i}, \quad i = 1, 2, \cdots, \Phi(Q)$$

The Riemann hypothesis states that the zeros of the Riemann zeta function lie on the $s = 1/2 + i\tau$ line in the complex s plane, and would seem to have nothing to do with physicists' real mode-locking widths that we are interested in here. However, there is a real-line version of the Riemann hypothesis that lies very close to the mode-locking problem. According to the theorem of Franel and Landau (Franel and Landau 1924; Edwards 1974; Titchmarsh 1951), the Riemann hypothesis is equivalent to the statement that

$$\sum_{Q_i \leq Q} |\delta_i| = o(Q^{\frac{1}{2}+\epsilon})$$

for all ϵ as $Q \to \infty$. The mode-lockings $\Delta_{P/Q}$ contain the necessary information for constructing the partition of the unit interval into the ℓ_i covers, and therefore implicitly contain the δ_i information. The implications of this for the circle-map scaling theory have not been worked out, and is not known whether some conjecture about the thermodynamics of irrational windings is equivalent to (or harder than) the Riemann hypothesis, but the danger lurks.

13. Farey Tree Thermodynamics

The narrowest mode-locked interval (16) at the n-th level of the Farey tree partition sum (26) is the golden mean interval

$$\Delta_{F_{n-1}/F_n} \propto |\delta_1|^{-n}. \tag{41}$$

It shrinks exponentially, and for τ positive and large it dominates $q(\tau)$ and bounds $dq(\tau)/d\tau$:

$$q'_{max} = \frac{\ln|\delta_1|}{\ln 2} = 1.502642\ldots \tag{42}$$

However, for τ large and negative, $q(\tau)$ is dominated by the interval (15) which shrinks only harmonically, and $q(\tau)$ approaches 0 as

$$\frac{q(\tau)}{\tau} = \frac{3\ln n}{n\ln 2} \to 0. \tag{43}$$

So for finite n, $q_n(\tau)$ crosses the τ axis at $-\tau = D_n$, but in the $n \to \infty$ limit, the $q(\tau)$ function exhibits a phase transition; $q(\tau) = 0$ for $\tau < -D_H$, but is a non-trivial function of τ for $-D_H \leq \tau$. This non-analyticity is rather severe - to get a clearer picture, we illustrate it by a few number-theoretic models (the critical circle maps case is qualitatively the same).

An cute version of the 'trivial' Farey level thermodynamics is given by the 'Farey model' (Artuso, Cvitanović and Kenny 1989; Cvitanović 1987a), in which the intervals $\ell_{P/Q}$ are replaced by Q^{-2}:

$$Z_n(\tau) = \sum_{i=1}^{2^n} Q_i^{2\tau}. \tag{44}$$

Here Q_i is the denominator of the ith Farey rational P_i/Q_i. For example (see definition (3)),

$$Z_2(1/2) = 4 + 5 + 5 + 4.$$

Though it might seem to have been pulled out of a hat, the Farey model is as sensible description of the distribution of rationals as the periodic orbit expansion (34). By the 'annihilation' property of the Gauss shift (11), the n-th Farey level sum $Z_n(-1)$ can be written as the integral

$$Z_n(-1) = \int dx\delta(f^n(x)) = \sum 1/|f'_{a_1\ldots a_k}(0)|\,, \tag{45}$$

with the sum restricted to the Farey level $\sum a_1 + \ldots + a_k = n+2$. It is easily checked that $f'_{a_1\ldots a_k}(0) = (-1)^k Q^2_{[a_1,\ldots,a_k]}$, so the Farey model sum is a partition generated by the Gauss map preimages of $x = 0$, i. e. by rationals, rather than by the quadratic irrationals as in (34). The sums are generated by the same transfer operator, so the eigenvalue spectrum should be the same as for the periodic orbit expansion, but in this variant of the finite level sums we can can evaluate $q(\tau)$ *exactly* for $\tau = k/2$, k a non-negative integer. First one observes that $Z_n(0) = 2^n$. It is also easy to check (Williams and Browne 1947) that $Z_n(1/2) = \sum_i Q_i = 2\cdot 3^n$. More surprisingly, $Z_n(3/2) = \sum_i Q^3 = 54\cdot 7^{n-1}$. Such 'sum rules', listed in the table 1, are consequence of the fact that the denominators on a given level are Farey sums of denominators on preceding levels (Artuso, Cvitanović and Kenny 1989; these computations have been made in collaboration with A. D. Kennedy). Regrettably, we have not been able to extend this method to evaluating $q(-1/2)$, or to real τ.

A bound on D_H can be obtained by approximating (44) by

$$Z_n(\tau) = n^{2\tau} + 2^n \rho^{2n\tau}. \tag{46}$$

In this approximation we have replaced all $\ell_{P/Q}$, except the widest interval $\ell_{1/n}$, by the narrowest interval ℓ_{F_{n-1}/F_n} (see (16)). The crossover from the harmonic dominated to the golden mean dominated behavior occurs at the τ value for which the two terms in (46) contribute equally:

$$D_n = \hat{D} + O\left(\frac{\ln n}{n}\right), \quad \hat{D} = \frac{\ln 2}{2 \ln \rho} = .72\ldots \tag{47}$$

Table 1. Recursion relations for the Farey model partition sums (44) for $\tau = 1, 1/2, 1, \ldots, 7/2$; they relate the $2^{q(\tau)} = \lim_{n\to\infty} Z_{n+1}(\tau)/Z_n(\tau)$ to roots of polynomial equations.

τ	$2^{q(\tau)}$	$Z_n(\tau) =$
0	2	$2\,Z_{n-1}$
1/2	3	$3\,Z_{n-1}$
1	$(5+\sqrt{17})/2$	$5Z_{n-1} - 2Z_{n-2}$
3/2	7	$7\,Z_{n-1}$
2	$(11+\sqrt{113})/2$	$10Z_{n-1} + 9Z_{n-2} - 2Z_{n-3}$
5/2	$7+4\sqrt{6}$	$14Z_{n-1} + 47Z_{n-2}$
3	$26.20249\ldots$	$20Z_{n-1} + 161Z_{n-2} + 40Z_{n-3} - Z_{n-4}$
7/2	$41.0183\ldots$	$29Z_{n-1} + 485Z_{n-2} + 327Z_{n-3}$
$n/2$	ρ^n	$\rho=$ golden mean

For negative τ the sum (46)) is the lower bound on the sum (30), so $\hat{D}$ is a lower bound on D_H. The size of the level-dependent correction in (47) is ominous; the finite n estimates converge to the asymptotic value logarithmically. What this means is that the convergence is excruciatingly slow and cannot be overcome by any amount of brute computation.

14. Artuso Model

The Farey model (30) is difficult to control at the phase transition, but considerable insight into the nature of this non-analyticity can be gained by the following factorization approximation. Speaking very roughly, the stability $\Lambda \approx (-1)^n Q^2$ of a $P/Q = [a_1, \ldots, a_n]$ cycle gains a hyperbolic golden-mean factor $-\rho^2$ for each bounce in the central part of the Farey map (10), and a power-law factor for every a_k bounces in the neighbourhood of the marginal fixed point $x_0 = 0$. This leads to an estimate of Q in $P/Q = [a_1, \dots, a_n]$ as a product of the continued fraction entries (Artuso 1988a and 1988b)

$$Q \approx \rho^n a_1 a_2 \cdots a_n$$

In this approximation the cycle weights factorize, $\Lambda_{a_1 a_2 \dots a_n} = \Lambda_{a_1} \Lambda_{a_2} \cdots \Lambda_{a_n}$, and the curvature corrections in the cycle expansion (38) vanish *exactly*:

$$1/\zeta(q,\tau) = 1 - \sum_{a=1}^{\infty} (\rho a)^{2\tau} z^a, \quad z = 2^{-q}$$

The $q = q(\tau)$ condition $1/\zeta(q,\tau) = 0$ yields

$$\rho^{-2\tau} = \Phi(-2\tau, z) \tag{48}$$

where Φ is the Jonquière function (Fornberg and Kölbig 1975)

$$\Phi(s,x) = \sum_{n=1}^{\infty} \frac{x^n}{n^s} = \frac{1}{\Gamma(s)} \int_0^{\infty} dt \frac{t^{s-1} x}{e^t - x}$$

The sum (48) diverges for $z > 1$, so $q \geq 0$. The interesting aspect of this model, easy to check (Artuso 1988a and 1988b), is that the $q(\tau)$ curve goes to zero at $\tau = -D_H$, with all derivatives $d^n q / d\tau^n$ *continuous* at D_H, so the phase transition is of infinite order. We believe this to be the case also for the exact trivial and critical circle maps thermodynamics, but the matter is subtle and explored to more depth in references (Feigenbaum 1987a and 1988a).

There is one sobering lesson in this: the numerical convergence acceleration methods (Levin 1973) consistently yield finite gaps at the phase transition point; for example, they indicate that for the Farey model evaluated at $\tau = -D_H + \epsilon$, the first derivative converges to $dq/d\tau \to .64 \pm .03$. However, the phase transition is not of a first order, but logarithmic of infinite order (Cvitanović 1987a), and the failure of numerical and heuristic arguments serves as a warning of how delicate such phase transitions can be.

15. Summary and Conclusions

The fractal set discussed here, the set of all parameter values corresponding to irrational windings, has no 'natural' measure. We have discussed three distinct thermodynamic formulations: the *Farey series* (all mode-lockings with cycle lengths up to Q), the *Farey levels* (2^n mode-lockings on the binary Farey tree), and the *Gauss partitioning* (all mode-lockings with continued fraction expansion up to a given length). The thermodynamic functions are *different* for each distinct partitioning. The only point they have in common is the Hausdorff dimension, which does not depend on the choice of measure. What makes the description of the set of irrational windings considerably trickier than the usual 'Axiom A' strange sets is the fact that here the range of scales spans from the marginal (harmonic, power-law) scalings to the the hyperbolic (geometric, exponential) scalings, with a generic mode-locking being any mixture of harmonic and exponential scalings. One consequence is that all versions of the thermodynamic formalism that we have examined here exhibit phase transitions. For example, for the continued fraction partitioning choice of weights t_p, the cycle expansions of Artuso et al. (Artuso, Aurell and Cvitanović 1990a and 1990b) behave as hyperbolic averages only for sufficiently negative values of τ; hyperbolicity fails at the 'phase transition' (Artuso, Cvitanović and Kenny 1989; Cvitanović 1987a; Artuso 1988a and 1988b) value $\tau = -1/3$, due to the power law divergence of the harmonic tails $\delta_{...n} \approx n^3$.

The universality of the critical irrational winding Hausdorff dimension follows from the universality of quadratic irrational scalings. The formulas used are formally identical to those used for description of dynamical strange sets (Artuso, Aurell and Cvitanović 1990a), the deep difference being that here the cycles are not dynamical trajectories in the coordinate space, but renormalization group flows in the function spaces representing families of dynamical systems. The 'cycle eigenvalues' are in present context the universal quadratic irrational scaling numbers.

In the above investigations we were greatly helped by the availability of the number theory models: in the $k = 0$ limit of (1) the renormalization flow is given by the Gauss map (6), for which the universal scaling δ_p reduce to quadratic irrationals. In retrospect, even this 'trivial' case seems not so trivial; and for the critical circle maps we are a long way from having a satisfactory theory. Symptomatic of the situation is the fact that while for the period doubling repeller D_H is known to 25 significant digits (Cvitanović, Christiansen and Rugh 1990), here we can barely trust the first three digits.

The quasiperiodic route to chaos has been explored experimentally in systems ranging from convective hydrodynamic flows (Stavans et al. 1985) to semiconductor physics (Gwinn et al. 1987). Such experiments illustrate the high precision with which the experimentalists now test the theory of transitions to chaos. It is fascinating that not only that the number-theoretic aspects of dynamics can be measured with such precision in physical systems, but that these systems are studied by physicists for reasons other than merely testing the renormalization theory or number theory. But, in all fairness, chaos via circle-map criticality is not nature's preferred way of destroying invariant tori, and the critical circle map renormalization theory remains a theoretical physicist's toy.

Acknowledgements
These lectures are to large extent built on discussions and/or collaborations with R. Artuso, M.J. Feigenbaum, P. Grassberger, M.H. Jensen, L.P. Kadanoff, A.D. Kennedy, B. Kenny, O. Lanford, J. Myrheim, I. Procaccia, D. Rand, B. Shraiman, B. Söderberg and D. Sullivan. The author is grateful to P. Moussa for

reading the manuscript with TeXpert eye and thanks the Carlsberg Foundation for the support.

References

Arnold V. I. (1961) Izv. Akad. Nauk. SSSR Math. Ser. **25** (1961) 21
Arnold V. I. (1965) Am. Math. Soc. Trans. **46** (1965) 213
Arnold V. I. (1983) Geometrical Methods in the Theory of Ordinary Differential Equations (Springer, New York 1983)
Artuso R. (1988a) J. Phys. **A21** (1988) L923
Artuso R. (1988b) doctoral thesis, (University of Milano, 1988)
Artuso R., Aurell E. and Cvitanović P. (1990a) Nonlinearity **3** (1990) 325
Artuso R., Aurell E. and Cvitanović P. (1990b) Nonlinearity **3** (1990) 361
Artuso R., Cvitanović P. and Kenny B. G. (1989) Phys. Rev. **A39** (1989) 268
Bak P., Bohr T. and Jensen M. J., Physica Scripta **T9** (1985) 50, reprinted in (Cvitanović 1989)
Benzi R., Paladin G., Parisi G. and Vulpiani A. (1984) J. Phys. **A17** (1984) 3521
Billingsley P. (1965) Ergodic Theory and Information (Wiley, New York 1965)
Christiansen F., Paladin G. and Rugh H. H., (1990) Nordita preprint, submitted to Phys. Rev. Lett.
Csordás A. and Szépfalusy P. (1989) Phys. Rev. **A40** (1989) 2221, and references therein
Cvitanović P. (1985) in Instabilities and Dynamics of Lasers and Nonlinear Optical Systems, Boyd R. W., Narducci L. M. and Raymer M. G., editors. (Univ. of Cambridge Press, Cambridge, 1985)
Cvitanović P. (1987a) lectures in (Zweifel et al. 1987)
Cvitanović P. (1987b) in XV International Colloquium on Group Theoretical Methods in Physics, R. Gilmore editor. (World Scientific, Singapore 1987)
Cvitanović P. (1989) editor., Universality in Chaos, 2nd. edition, (Adam Hilger, Bristol 1989)
Cvitanović P., Christiansen F. and Rugh H. H. (1990) J. Phys. **A**, in press.
Cvitanović P., Gunaratne G. H. and Vinson M. (1990) Nonlinearity **3** (1990) 873
Cvitanović P., Jensen M. H., Kadanoff L. P. and Procaccia I. (1985) Phys. Rev. Lett. **55** (1985) 343
Cvitanović P. and Myrheim J. (1983) Phys. Lett. **A94** (1983) 329
Cvitanović P. and Myrheim J. (1989) Commun. Math. Phys. **121** (1989) 225
Cvitanović P., Shraiman B. and Söderberg B. (1985) Physica Scripta **32** (1985) 263
Edwards H. M. (1974) Riemann's Zeta Function (Academic, New York 1974)
Falconer K. M. (1985) The Geometry of Fractal Sets (Cambridge Univ. Press, Cambridge, 1985)
Farmer J. D. and Satija I. I. (1985) Phys. Rev. **A 31** (1985) 3520
Feigenbaum M. J. (1987a) lectures in (Zweifel et al. 1987)
Feigenbaum M. J. (1987b) J. Stat. Phys. **46** (1987) 919
Feigenbaum M. J. (1987c) J. Stat. Phys. **46** (1987) 925
Feigenbaum M. J. (1988a) J. Stat. Phys. **52** (1988) 527
Feigenbaum M. J. (1988b) Nonlinearity **1** (1988) 577
Feigenbaum M. J., Kadanoff L. P. and Shenker S. J. (1982) Physica **5D** (1982) 370
Feigenbaum M. J. and Kenway R. D. (1983), in Proceedings of the Scottish Universities Summer School (1983)
Fornberg B. and Kölbig K. S. (1975) Math. of Computation **29** (1975) 582
Franel J. and Landau E. (1924) Göttinger Nachr. 198 (1924)
Glass L., Guevara M. R., Shrier A. and Perez R. (1983) Physica **7D** (1983) 89, reprinted in (Cvitanović 1989)
Grassberger P. (1983) Phys. Lett. **97A** (1983) 227
Greene J. M. (1979) J. Math. Phys. **20** (1979) 1183
Gunaratne G. H. (1986), doctoral thesis, (Cornell University, 1986)
Gwinn E. G. and Westervelt R. M. (1987) Phys. Rev. Lett. **59** (1987) 157

Hall R. R. (1970) J. London Math. Soc., **2** (1970) 139
Halsey T. C., Jensen M. H., Kadanoff L. P., Procaccia I. and Shraiman B. I. (1986) Phys. Rev. **A33** (1986) 1141
Hardy G. H. and Wright H. M. (1938) 'Theory of Numbers' (Oxford Univ. Press, Oxford 1938)
Hentschel H. G. E. and Procaccia I. (1983) Physica **8D** (1983) 435
Herman M. (1979) Publ. IHES **49** (1979) 5
Jensen M. H., Bak P. and Bohr T. (1983) Phys. Rev. Lett. **50** (1983) 1637
Jensen M. H., Bak P. and Bohr T. (1984) Phys. Rev. **A30** (1984) 1960
Kaneko K. (1982) Prog. Theor. Phys. **68** (1982) 669
Kaneko K. (1983a) Prog. Theor. Phys. **69** (1983) 403
Kaneko K. (1983b) Prog. Theor. Phys. **69** (1983) 1427
Kanemitsu S., Sita Rama Chandra Rao R. and Siva Rama Sarma A. (1982) J. Math. Soc. Japan **34** (1982) 125
Kanemitsu S., Sita Rama Chandra Rao R. and Siva Rama Sarma A. (1984) Acta Arith. **44** (1984) 397
Khinchin A. Ya. (1964) Continued Fractions (Univ. of Chicago Press, Chicago, 1964)
Kim S. -H. and Ostlund S. (1989) Physica **39D** (1989) 365
Lanford O. E. (1985) Physica **14D** (1985) 403
Lanford O. E. (1987a), in: Proceedings 1986 IAMP Conference in Mathematical Physics, M. Mebkhout and R. Sénéor, editors., (World Scientific, Singapore 1987)
Lanford O. E. (1987b) lectures in (Zweifel et al. 1987)
Levin D. (1973) Inter. J. Computer Math. **B3** (1973) 371
MacKay R. S. (1982) doctoral thesis (Princeton University, 1982)
Mayer D. H. (1976) Bull. Soc. Math. France **104** (1976) 195
Mestel B. D. (1985) Ph. D. Thesis (Univ. of Warwick 1985).
Neville E. H. (1950) Roy. Soc. Mathematical Tables (Cambridge Univ. Press, Cambridge 1950)
Ostlund S., Rand D. A., Sethna J. and Siggia E. (1983) Physica **8D** (1983) 303
Pomeau Y. and Manneville P. (1980) Commun. Math. Phys. **74** (1980) 189
Rand D. A. (1987) Proc. R. Soc. London **A413** (1987) 45
Rand D. A. (1988) Nonlinearity **1** (1988) 78
Ruelle D. (1976) Inventiones Math. **34** (1976) 231
Ruelle D. (1978) Thermodynamic Formalism (Addison-Wesley, Reading 1978)
Shenker S. J. (1982) Physica **5D** (1982) 405
Shenker S. J. and Kadanoff L. P. (1982) J. Stat. Phys. **27** (1982) 631
Stavans J., Heslot F. and Libchaber A. (1985) Phys. Rev. Lett. **55** (1985) 569
Swiatek G. (1988) Commun. Math. Phys. **119** (1988) 109
Titchmarsh E. C. (1951) The Theory of Riemann Zeta Function (Oxford Univ. Press, Oxford 1951); chapter XIV.
Umberger D. K. , Farmer J. D. and Satija I. I. (1986) Phys. Lett. **A114** (1986) 341
Yoccoz J.-C. (1984) Ann. Scient. Ecol.. Norm. Sup. Paris **17** (1984) 333
Williams G. T. and Browne D. H. (1947) Amer. Math. Monthly **54** (1947) 534
Zweifel P., Gallavotti G. and Anile M., editors. (1987) Non-linear Evolution and Chaotic Phenomena (Plenum, New York 1987)

Chapter 14

An Introduction To Small Divisors Problems

by Jean-Christophe Yoccoz

1. Introduction

The problems linked with the so-called 'small divisors', i. e. the near resonances of frequencies in a quasiperiodic motion, have been known and studied since the last century. But it is only in the past fifty years, beginning with Siegel (Siegel 1942), that they have started to be overcome. To illustrate these problems, we have chosen, amongst others, two cases where they occur.

The first part of the text is devoted to a short survey of the dynamics of diffeomorphisms of the circle. Due to the works of Poincaré, Denjoy, Arnold, Moser and Herman, one can say that the theory is now fairly well understood, the only difficulty in this case being the small divisors in their simplest form.

On the opposite, the second part of the text is devoted to invariant tori in Hamiltonian dynamics, probably the part of small divisor theory which is most relevant to physics. The dynamics here are only very partially understood, and many important questions are still open. Nevertheless, the theorems of stability of quasiperiodic motions, known as KAM–theory after Kolmogorov, Arnold and Moser, are an essential step in the knowledge of these dynamics.

We have not tried to give a complete bibliography on the subject. The interested reader should consult the very good survey by Bost (Bost 1985), and the included references

2. Diffeomorphisms of the Circle

2.1 The Rotation Number

We refer in this Section to the work of Herman (Herman 1979).

2.1.1 Introduction. Let $\mathbb{T}^1 = \mathbb{R}/\mathbb{Z}$ be the circle (or 1-dimensional torus). We write $\mathrm{Diff}_+^r(\mathbb{T}^1)$ for the group of orientation preserving C^r-diffeomorphisms of $\mathbb{T}^1$. We may have $r = 0$, or $r \in [1, +\infty)$, or $r = \infty$ or $r = \omega$ (meaning real-analytic). We are interested in the dynamics under iteration of a diffeomorphism of the circle. But for practical reasons it is generally easier to work with a lift to $\mathbb{R}$ of such a diffeomorphism, i. e. to work in the group $\mathrm{D}^r(\mathbb{T}^1)$ formed by the C^r-diffeomorphisms of the real line such that $f(x) - x$ is $\mathbb{Z}$-periodic. We then denote by $\bar{f}$ the induced diffeomorphism of $\mathbb{T}^1$.

2.1.2 Examples. For $\alpha \in \mathbb{R}$ (resp. $\alpha \in \mathbb{T}^1$), we denote by R_α the translation $x \mapsto x + \alpha$ which belongs to $\mathrm{D}^\omega(\mathbb{T}^1)$ (resp. $\mathrm{Diff}_+^\omega(\mathbb{T}^1)$).

i) **Blaschke products** Let $n \geq 0$, $\lambda \in S^1 = \{z \in \mathbb{C}, |z| = 1\}$ and let $a_0, a_1, \dots, a_n, b_1, b_2, \dots, b_n \in \mathbf{D} = \{z \in \mathbb{C}, |z| < 1\}$. Then the formula

$$f(z) \;=\; \lambda \prod_{i=0}^{n} \left(\frac{z - a_i}{1 - \bar{a}_i z}\right) \prod_{j=1}^{n} \left(\frac{z - b_j}{1 - \bar{b}_j z}\right)^{-1}$$

defines a rational map which preserves S^1 and whose restriction to S^1 has degree 1. If it has no critical point on S^1 (which always occurs provided $|a_i|, |b_j|$ are small enough) it defines an element of $\mathrm{Diff}_+^\omega(\mathbb{T}^1)$).

ii) **The Arnold family** The formula

$$f_{a,t}(x) = x + t + a\sin(2\pi x) \;,\; |a| < \frac{1}{2\pi}, t \in \mathbb{R}$$

defines a 2-parameters family in $\mathrm{D}^\omega(\mathbb{T}^1)$, first studied by Arnold (Arnold 1965).

2.1.3 Definition of the Rotation Number. Let $f \in \mathrm{D}^0(\mathbb{T}^1)$. We give below several equivalent ways, after Poincaré, to define the rotation number of f, which is a real number denoted by $\rho(f)$.

(1) Proposition. *As n goes to ∞, the sequence $\frac{1}{n}(f^n(x) - x)$ converges uniformly in x to a constant limit $\rho(f)$.*

(2) Proposition. *Writing rational numbers p/q in irreducible form with $q \geq 1$, we define:*

$$D^+(f) \;=\; \left\{\frac{p}{q} \in \mathbb{Q},\ \forall x \in \mathbb{R},\ f^q(x) < x + p\right\}$$
$$D^-(f) \;=\; \left\{\frac{p}{q} \in \mathbb{Q},\ \forall x \in \mathbb{R},\ f^q(x) > x + p\right\} .$$

These two sets form a Dedekind section of the rational numbers defining $\rho(f)$.

(3) Proposition. *Let μ be a probability measure on $\mathbb{T}^1$ invariant under $\bar{f}$, then*

$$\rho(f) \;=\; \int_{\mathbb{T}^1} (f(x) - x)d\mu(x) .$$

(4) Proposition. *There exists a real number α and an non-decreasing map h : $\mathbb{R} \to \mathbb{R}$ satisfying, for all $x \in \mathbb{R}$:*

$$h(x+1) = h(x)+1 \ , \ h(f(x)) = h(x)+\alpha \ .$$

Moreover, one then has $\alpha = \rho(f)$.

2.1.4 Equivalence of the Definitions. We briefly indicate how the various above mentioned definitions are related. Write $f(x) = x + \phi(x)$. Then for $n \geq 1$, one has:

$$f^n(x) - x = \sum_{i=0}^{n-1} \phi \circ \bar{f}^i(x) \ .$$

Let μ be a probability measure on $\mathbb{T}^1$, invariant under $\bar{f}$ and ergodic. By Birkhoff's ergodic theorem, there exists $x_0 \in \mathbb{R}$ such that:

$$\lim_{n \to +\infty} \frac{1}{n}(f^n(x_0) - x_0) = \int_{\mathbb{T}^1} \phi(x) d\mu(x) \ . \tag{1}$$

On the other hand, as f^n is increasing, one has:

$$\max_{y \in \mathbb{R}}(f^n(y) - y) < \min_{y \in \mathbb{R}}(f^n(y) - y) + 1 \ . \tag{2}$$

Putting (1) and (2) together gives Propositions 1 and 3.

Defining h by $h(x) = \tilde{\mu}([0,x))$ for $x \geq 0$ and $h(x) = -\tilde{\mu}([x,0))$ for $x \leq 0$ (where $\tilde{\mu}$ is the lift of μ to $\mathbb{R}$), one has for all $x \in \mathbb{R}$:

$$\begin{aligned} h(x+1) - h(x) &= \tilde{\mu}([x, x+1)) = 1 \ , \\ h(f(x)) - h(x) &= \tilde{\mu}([x, f(x))) \\ &= \tilde{\mu}([0, f(0))) = h(f(0)) \ , \\ f^n(0) - 1 < h(f^n(0)) &= nh(f(0)) < f^n(0) + 1 \ . \end{aligned}$$

By Proposition 1, one has $h(f(0)) = \rho(f)$. This gives Proposition 4.

If $p/q \in D^+(f)$ and $p'/q' \in D^-(f)$, then for all $x \in \mathbb{R}$ one has:

$$x + p'q < f^{qq'}(x) < x + pq'$$

hence $p/q > p'/q'$. If p/q and p'/q' do not belong to $D^+(f) \cup D^-(f)$, there exist $x_0, x_1 \in [0,1)$ such that $f^q(x_0) = x_0 + p$ and $f^{q'}(x_1) = x_1 + p'$. Then, for $n \geq 1$:

$$-1 + np'q \leq f^{nqq'}(x_1) - 1 < f^{nqq'}(0) \leq f^{nqq'}(x_0) < 1 + npq' \ ;$$

this gives $p/q \geq p'/q'$. By symmetry, $p'/q' \geq p/q$, hence Proposition 2 holds.

2.1.5 Properties of the Rotation Number. The following properties hold:

i) For $\alpha \in \mathbb{R}$, one has $\rho(R_\alpha) = \alpha$.

ii) For $f \in \mathrm{D}^0(\mathbb{T}^1)$, $n \in \mathbb{Z}$, one has $\rho(f^n) = n\rho(f)$.

iii) If $f, g \in \mathrm{D}^0(\mathbb{T}^1)$ and $h : \mathbb{R} \to \mathbb{R}$ is an increasing map which satisfies $h \circ f = g \circ h$ and $h \circ R_1 = R_1 \circ h$, then $\rho(f) = \rho(g)$. In particular, the rotation number is a conjugacy invariant in the group $\mathrm{D}^0(\mathbb{T}^1)$, but is *not* a homomorphism into $\mathbb{R}$.

iv) For $f \in \mathrm{D}^0(\mathbb{T}^1)$, $p \in \mathbb{Z}$ one has $\rho(R_p \circ f) = p + \rho(f)$. This allows to define the rotation number (with values in $\mathbb{T}^1$) for $\bar{f} \in \mathrm{Diff}^0_+(\mathbb{T}^1)$.

v) Let p/q be a rational number ($q \geq 1$, $p \wedge q = 1$) and $f \in \mathrm{D}^0(\mathbb{T}^1)$. Proposition 2 says that one has $\rho(f) = p/q$ if and only if there exists $x_0 \in \mathbb{R}$ such that $f^q(x_0) = x_0 + p$. This gives for the induced homeomorphism $\bar{f}$ of $\mathbb{T}^1$ a periodic orbit of minimal period q cyclically ordered on $\mathbb{T}^1$ as an orbit of $R_{p/q}$.

vi) On the opposite, $\rho(f)$ is irrational if and only if $\bar{f}$ has no periodic orbit. In this case, one can show that the increasing map h of Proposition 4 is uniquely determined by its value at zero. This implies (see Section 2.1.4) that there is a *unique* $\bar{f}$-invariant probability measure on $\mathbb{T}^1$: in other terms, every orbit of $\bar{f}$ has the same statistical distribution on $\mathbb{T}^1$.

2.1.6 Denjoy's Alternative and Denjoy's Theorems. Let $f \in D^0(\mathbb{T}^1)$ have irrational rotation number α. Let $\bar{f}$ be the induced homeomorphism of $\mathbb{T}^1$ and let μ be the unique $\bar{f}$-invariant probability measure on $\mathbb{T}^1$. Two cases can occur:

i) Supp $\mu = \mathbb{T}^1$. Then the increasing map h of Proposition 4, normalized by the condition $h(0) = 0$, defines a homeomorphism in $D^0(\mathbb{T}^1)$ which conjugates f and R_α. The dynamical properties of f and R_α are the same fom the topological viewpoint. In particular, every orbit of $\bar{f}$ is dense in $\mathbb{T}^1$.

ii) Supp $\mu = K \neq \mathbb{T}^1$. One then speaks of f (or $\bar{f}$) as a *Denjoy counter-example.* The set K is a Cantor subset (compact, totally disconnected and without isolated points) of $\mathbb{T}^1$, invariant under $\bar{f}$, whose points are exactly the limit points of every orbit of $\bar{f}$.

One may ask whether Denjoy counter-examples exist at all. The following theorems of Denjoy (Herman 1979; Denjoy 1932; Denjoy 1946) answer this question.

(1) Theorem. *A C^2-diffeomorphism of $\mathbb{T}^1$ with no periodic orbits is topologically conjugated to an irrational rotation.*

(2) Theorem. *Let $\alpha \in \mathbb{R} - \mathbb{Q}$. There exist a Denjoy counter-example f with $\rho(f) = \alpha$ and $f \in D^{2-\epsilon}(\mathbb{T}^1)$ for all $\epsilon > 0$.*

2.1.7 Topological Stability Questions. Let $f \in D^2(\mathbb{T}^1)$ have no periodic orbit. Then by Denjoy's Theorem 1 above the topological dynamics of $\bar{f}$ are known, but this is not sufficient to answer further questions as:

i) Is the statistical distribution of $\bar{f}$-orbits in $\mathbb{T}^1$ given by a continuous density? by a C^∞- density? This amounts (see Section 2.1.4) to ask whether the topological conjugacy between f and R_α, where $\alpha = \rho(f)$, is C^1 or C^∞.

ii) Assuming f real-analytic, we are interested not only to the dynamics of $\bar{f}$ on $\mathbb{T}^1$ (or S^1), but also in a neighbourhood of it in $\mathbb{C}/\mathbb{Z}$ (or $\mathbb{C}$). For example, one wants to know whether the circle is *topologically stable* under $\bar{f}$ (meaning that for every neighbourhood U of $\mathbb{T}^1$ there exists a neighbourhood $V \subset U$ such that any $\bar{f}$-orbit starting in V stays in U). An easy argument shows that topological stability is actually equivalent to the analyticity of the conjugating function h.

In order to discuss the smoothness of the conjugacy to the rotation R_α, it is necessary to introduce some further arithmetical conditions on the rotation number α.

2.2 Diophantine Approximation for Irrational Numbers and Continued Fraction

2.2.1 Algebraic Formalism. An extended treatment of the subject can be found in textbooks (Lang 1966; Cassels 1957; Schmidt 1980).

For $x \in \mathbb{R}$, let $[x]$ be the integral part of x, and let

$$\{x\} = x - [x] \quad \text{and} \quad ||x|| = \inf_{p \in \mathbb{Z}} |x - p| \,.$$

Let $\alpha \in \mathbb{R} - \mathbb{Q}$. Define $a_0 = [\alpha]$, $\alpha_0 = \{\alpha\}$ and for $n \geq 1$:

$$a_n = \left[\alpha_{n-1}^{-1}\right] , \; \alpha_n = \left\{\alpha_{n-1}^{-1}\right\} .$$

One has $a_0 \in \mathbb{Z}$, $a_n \in \mathbb{N}^*$ for $n \geq 1$ and $\alpha_n \in (0,1)$ for $n \geq 0$. The continued fraction of α is:

$$\alpha = a_0 + \cfrac{1}{a_1 + \cfrac{1}{a_2 + \cfrac{1}{\ddots}}}$$

For $n \geq 0$, let $\beta_n = \prod_{i=0}^{n} \alpha_i$ and

$$\frac{p_n}{q_n} = a_0 + \cfrac{1}{a_1 + \cfrac{1}{a_2 + \cfrac{1}{\ddots + \cfrac{1}{a_n}}}}$$

Then, defining:

$$\beta_{-2} = \alpha \; , \; \beta_{-1} = 1 \; , \; q_{-2} = 1 \; , \; q_{-1} = 0 \; , \; p_{-2} = 0 \; , \; p_{-1} = 1 \; ,$$

the following algebraic formulas hold:

$$\begin{aligned}
\beta_n &= (-1)^n(q_n\alpha - p_n) && (n \geq -2)\\
q_n &= a_n q_{n-1} + q_{n-2} && (n \geq 0)\\
p_n &= a_n p_{n-1} + p_{n-2} && (n \geq 0)\\
\beta_{n-2} &= a_n \beta_{n-1} + \beta_n && (n \geq 0)
\end{aligned}$$

and

$$\begin{aligned}
p_{n+1}q_n - p_n q_{n+1} &= (-1)^n && (n \geq -2)\\
q_{n+1}\beta_n + q_n\beta_{n+1} &= 1 && (n \geq -2)\,.
\end{aligned}$$

Finally

$$\alpha = \frac{p_n + p_{n-1}\alpha_n}{q_n + q_{n-1}\alpha_n} \qquad (n \geq 0)\,.$$

The rational numbers $(p_n/q_n)_{n\geq 0}$ are called the *convergents* of α.

2.2.2 Approximation Properties. For $q \geq 1$, let

$$\sigma_q(\alpha) = \inf_{1\leq n\leq q} ||n\alpha||\,.$$

Then the convergents p_n/q_n of α are the best rational approximations to α in the sense that one has, for $n \geq 0$ and $q_n \leq q < q_{n+1}$:

$$\sigma_q(\alpha) = \beta_n = ||q_n\alpha||\,.$$

Moreover, by the penultimate formula in 2.2.1 one has:

$$\frac{1}{(a_{n+2}+2)q_n} \leq \frac{1}{q_{n+1}+q_n} < \beta_n < \frac{1}{q_{n+1}} \leq \frac{1}{a_{n+1}q_n}\,.$$

2.2.3 Diophantine Conditions. We introduce now various Diophantine conditions.

i) Let $\theta \geq 0$, $\rho > 0$. Define:

$$\begin{aligned}
\mathcal{C}(\theta,\rho) &= \{\alpha \in \mathbb{R} - \mathbb{Q},\ \forall q \geq 1\,,\ \sigma_q(\alpha) \geq \rho q^{-1-\theta}\}\\
\mathcal{C}(\theta) &= \bigcup_{\rho>0} \mathcal{C}(\theta,\rho)\\
\mathcal{C} &= \bigcup_{\theta\geq 0} \mathcal{C}(\theta)\,.
\end{aligned}$$

In terms of the convergents of α, one has:

$$\begin{aligned}
a_{n+1} &\leq \rho^{-1}q_n^{-\theta} - 2 &&\Longrightarrow && \alpha \in \mathcal{C}(\theta,\rho)\\
a_{n+1} &\leq \rho^{-1}q_n^{-\theta} &&\Longleftarrow && \alpha \in \mathcal{C}(\theta,\rho)\,,
\end{aligned}$$

and also:

$$\begin{aligned} \alpha \in \mathcal{C}(\theta) \Longleftrightarrow q_{n+1} &= O(q_n^{1+\theta}) \\ \Longleftrightarrow \alpha_{n+1}^{-1} &= O(\beta_n^{\theta}) , \\ \alpha \in \mathcal{C} \Longleftrightarrow \log(q_{n+1}) &= O(\log(q_n)) . \end{aligned}$$

ii) Clearly, one has $\mathcal{C}(\theta,\rho) \subset \mathcal{C}(\theta',\rho')$ for $\theta \leq \theta'$, $\rho \geq \rho'$. For $\theta \geq 0$, $\rho > 0$, the set $\mathcal{C}(\theta,\rho)$ is closed and has empty interior. The sets $\mathcal{C}(\theta)$ and $\mathcal{C}$ are countable unions of such sets, and this means that they are small in the topological sense (Baire category).

In the measure theoretical sense, the opposite is true: for $\theta > 0$, $\mathcal{C}(\theta)$ has full Lebesgue measure (but $\mathcal{C}(0)$ has Lebesgue measure zero).

iii) It is an easy exercise to show that the continued fraction of an irrational number α is preperiodic (meaning: $\exists k, n_0$ such that $a_{n+k} = a_n$ for $n \geq n_0$) if and only if α is algebraic of degree 2. Thus such a number belongs to $\mathcal{C}(0)$. A much harder theorem by Roth says that every irrational algebraic number belongs to $\mathcal{C}(\theta)$ for all $\theta > 0$, but it is not known for any such number (of degree ≥ 3) whether it belongs to $\mathcal{C}(0)$.

iv) The group $\mathrm{GL}(2,\mathbb{Z})$ acts on $\mathbb{R}$ by

$$Mx = \frac{ax+b}{cx+d} ,$$

where

$$M = \begin{pmatrix} a & b \\ c & d \end{pmatrix} \in \mathrm{GL}(2,\mathbb{Z}) ,$$

and this action is related to the continued fraction algorithm as follows. For $\alpha, \alpha' \in \mathbb{R} - \mathbb{Q}$, the two properties are equivalent:

$$\begin{aligned} &(1)\ \exists g \in \mathrm{GL}(2,\mathbb{Z}) \quad \text{such that} \quad g\alpha = \alpha' \\ &(2)\ \exists k \in \mathbb{Z},\ \exists n_0 \in \mathbb{N} \quad \text{such that} \quad a_n(\alpha) = a_{n+k}(\alpha') \quad \text{for all} \quad n \geq n_0 . \end{aligned}$$

This shows that for all $\theta \geq 0$, the set $\mathcal{C}(\theta)$ is invariant under the action of $\mathrm{GL}(2,\mathbb{Z})$.

v) The sets $\mathcal{C}(\theta)$ or $\mathcal{C}$ are appropriate for small divisors problems involving finite or infinite differentiability conditions. In the real-analytic case, different conditions are more appropriate.

Let $\alpha \in \mathbb{R} - \mathbb{Q}$. Define

$$\Delta_0(\alpha) = 10 ,$$

and

$$\Delta_{n+1}(\alpha) = \begin{cases} \alpha_n^{-1}(\Delta_n(\alpha) - \log(\alpha_n^{-1}) + 1) & \text{if } \Delta_n \geq \log(\alpha_n^{-1}) \\ \exp(\Delta_n(\alpha)) & \text{if } \Delta_n \leq \log(\alpha_n^{-1}) \end{cases}$$

and let

$$\begin{aligned} \mathcal{H}_0 &= \{\alpha \in \mathbb{R} - \mathbb{Q},\ \exists n_0,\ \forall\, n \geq n_0,\ \Delta_n \geq \log(\alpha_n^{-1})\}\ , \\ \mathcal{H} &= \{\alpha \in \mathbb{R} - \mathbb{Q},\ \forall\, g \in \mathrm{GL}(2,\mathbb{Z}),\ g \cdot \alpha \in \mathcal{H}_0,\}\ , \\ \mathcal{B} &= \{\alpha \in \mathbb{R} - \mathbb{Q},\ \exists\, g \in \mathrm{GL}(2,\mathbb{Z}),\ g \cdot \alpha \in \mathcal{H}_0,\}\ , \end{aligned}$$

It is easily seen that

$$\mathcal{C} \underset{\neq}{\subset} \mathcal{H} \underset{\neq}{\subset} \mathcal{H}_0 \underset{\neq}{\subset} \mathcal{B}\ ,$$

and

$$\mathcal{B} = \{\alpha \in \mathbb{R} - \mathbb{Q},\ \sum_n q_n^{-1} \log(q_{n+1}) < +\infty\}\ .$$

The set $\mathcal{B}$, named after Brjuno who first considered it (Brjuno 1971, 1972) is a countable union of closed sets with empty interior. The set $\mathcal{H}$ has a more complicated nature.

2.3 Smooth Conjugacy

2.3.1 The Linearized Equation. Let $f = R_\alpha + \Delta\phi$, $h = \mathrm{Id} + \Delta\Psi \in \mathrm{D}^0(\mathbb{T}^1)$. If we linearize at $f = R_\alpha$, $h = \mathrm{Id}$ (where Id is the identity map) the conjugacy equation $h \circ R_\alpha \circ h^{-1} = f$, we get

$$(*) \quad \Delta\psi \circ R_\alpha - \Delta\psi = \Delta\phi$$

Let m be the Lebesgue measure on $\mathbb{T}^1$ and:

$$C_0^r(\mathbb{T}^1) = \left\{\phi \in \mathrm{C}^r(\mathbb{T}^1),\ \int_{\mathbb{T}^1} \phi\, dm = 0\right\}\ .$$

For equation (*) to have a solution, one must have

$$\int_{\mathbb{T}^1} \Delta\phi\, dm = 0\ ,$$

which is the infinitesimal version of the condition $\rho(f) = \alpha$. One may ask for solution $\Delta\psi$ such that

$$\int_{\mathbb{T}^1} \Delta\psi\, dm = 0\ ;$$

this is just a normalizing condition on h. Besides this, equation (*) is trivially solved for Fourier coefficients:

$$\widehat{\Delta\psi}(n) = (\exp(2\pi i n\alpha) - 1)^{-1}\widehat{\Delta\phi}(n)\ ,\ n \neq 0$$

and we see that the relation between the smoothness properties of $\Delta\phi$ and $\Delta\psi$ depend on the *small divisors* $(\exp(2\pi i n\alpha) - 1)$. One has

$$4||n\alpha|| \leq |\exp(2\pi i n\alpha) - 1| \leq 2\pi||n\alpha||\ ,$$

and the Diophantine conditions of Section 2.2.3 thus appear in the discussion.

(1) Proposition. *If $\alpha \in \mathcal{C}$ and $\Delta\phi \in C_0^\infty(\mathbb{T}^1)$, then $\Delta\psi \in C_0^\infty(\mathbb{T}^1)$. If $\alpha \notin \mathcal{C}$, there exists $\Delta\phi \in C_0^\infty(\mathbb{T}^1)$ such that $\Delta\psi$ cannot even be a distribution.*

(2) Proposition. *Let $r, s > 0$, $r, s \notin \mathbb{N}$, $r - s = 1 + \theta \geq 1$. If $\alpha \in \mathcal{C}(\theta)$ and $\Delta\phi \in C_0^r(\mathbb{T}^1)$, then $\Delta\psi \in C_0^s(\mathbb{T}^1)$. If $\alpha \notin \mathcal{C}(\theta)$ there exists $\Delta\phi \in C_0^r(\mathbb{T}^1)$ such that $\Delta\psi \notin C_0^s(\mathbb{T}^1)$.*

(3) Proposition. *If $\lim_{n\to+\infty} q_n^{-1}\log(q_{n+1}) = 0$, and $\Delta\phi \in C_0^\omega(\mathbb{T}^1)$, then $\Delta\psi \in C_0^\omega(\mathbb{T}^1)$. If $\limsup_{n\to+\infty} q_n^{-1}\log(q_{n+1}) > 0$, there exists $\Delta\phi \in C_0^\omega(\mathbb{T}^1)$, such that $\Delta\psi$ cannot even be a distribution.*

Propositions 1 and 3 are easily proven. The proof of Proposition 2 requires standard techniques in harmonic analysis. In Proposition 2, in case r or s is an integer, the condition $r - s = 1 + \theta$ is replaced by $r - s > 1 + \theta$.

2.3.2 Local Theorems. The first theorem on smooth conjugacy for diffeomorphisms of the circle was proved by Arnold in 1961.

i) The analytic case

(1) Theorem. *Let $\alpha \in \mathcal{B}$, $\delta > 0$. There exists $\epsilon > 0$ such that, if $f \in D^\omega(\mathbb{T}^1)$ satisfies $\rho(f) = \alpha$ and extends holomorphically in the strip $\{z, |\mathrm{Im}(z)| < \delta\}$ with $|f(z) - z - \alpha| < \epsilon$ in this strip, then the conjugacy function h between f and R_α extends holomorphically in the strip $\{z, |\mathrm{Im}(z)| < \delta/2\}$.*

For details see the references (Arnold 1965; Herman 1979; Rüssmann 1972; Yoccoz 1989). Note that the Brjuno set $\mathcal{B}$ is *not* the one arising from the linearized equation (Proposition 3 above). The condition $\alpha \in \mathcal{B}$ is actually optimal.

(2) Theorem. *Let $\alpha \notin \mathcal{B}$, $\alpha \in \mathbb{R} - \mathbb{Q}$, and $a > 3$. Let $\lambda \in S^1$ be the unique value such that the Blaschke product*

$$f_{\lambda,a} = \lambda z^2 \frac{z+a}{1+az}$$

has rotation number α on S^1. Then $f_{\lambda,a}$ is not analytically conjugated to R_α.

See Yoccoz (Yoccoz 1989).

ii) The smooth case

(3) Theorem. *Let $\theta \geq 0$, $\alpha \in \mathcal{C}(\theta)$, $r_0 > 2 + \theta$. There exists a neighbourhood U of R_α in $D^{r_0}(\mathbb{T}^1)$ such that, if $r \geq r_0$ satisfies $r \notin \mathbb{N}, r - 1 - \theta \notin \mathbb{N}$ and $f \in D^r(\mathbb{T}^1) \cap U$, then the conjugacy function h between f and R_α belongs to $D^{r-1-\theta}(\mathbb{T}^1)$. In particular, if $f \in D^\infty(\mathbb{T}^1) \cap U$, then $h \in D^\infty(\mathbb{T}^1)$.*

For a proof see Moser (Moser 1966; Herman 1979). The arithmetical conditions above are again optimal.

(4) Theorem. *Let $r \geq s \geq 1$, $\theta = r - s - 1$, $\alpha \notin \mathcal{C}(\theta)$. Then there exists, arbitrarily near R_α in the C^r-topology, diffeomorphisms $f \in D^r(\mathbb{T}^1)$ with $\rho(f) = \alpha$, such that the conjugacy does not belong to $D^s(\mathbb{T}^1)$.*

See Herman (Herman 1979).

2.3.3 Global Theorems. We once more consider separately the smooth case and the analytic case

i) The smooth case

After Arnold and Moser proved around 1965 the local conjugacy theorems above, it was an important open question, formulated by Arnold, whether one could remove the hypothesis that f is near R_α and still have smooth conjugacy theorems. Herman answered this positively in 1976. For the following theorem, see the references (Herman 1979; Yoccoz 1984; Katznelson and Ornstein 1989a and 1989b; Khanin and Sinaï 1987).

(5) Theorem. *Let $r \geq s \geq 1$ and $\theta \geq 0$, such that $r - s - 1 > 0$. If $\alpha \in \mathcal{C}(\theta)$ and $f \in \mathrm{D}^r(\mathbb{T}^1)$ satisfies $\rho(f) = \alpha$, then the conjugacy function h belongs to $\mathrm{D}^s(\mathbb{T}^1)$. In particular if $\alpha \in \mathcal{C}$ and $f \in \mathrm{D}^\infty(\mathbb{T}^1)$ satisfies $\rho(f) = \alpha$ then $h \in \mathrm{D}^\infty(\mathbb{T}^1)$.*

Note the slight difference with the arithmetical condition of the local Theorem 3, which is known to be true even for $\theta = r - s - 1$ provided that r and s are not integers.

ii) The analytic case

(6) Theorem. *If $\alpha \in \mathcal{H}$ and $f \in \mathrm{D}^\omega(\mathbb{T}^1)$ satisfies $\rho(f) = \alpha$ then $h \in \mathrm{D}^\omega(\mathbb{T}^1)$. Conversely, if $\alpha \notin \mathcal{H}$ there exists $f \in \mathrm{D}^\omega(\mathbb{T}^1)$ such that $\rho(f) = \alpha$ and $h \notin \mathrm{D}^\omega(\mathbb{T}^1)$.*

See Yoccoz (Yoccoz 1989).

2.3.4 Some Remarks on the Proofs.

i) Functional analysis: Newton's method

A first natural approach to the conjugacy problem is to solve the conjugacy equation in an appropriate functional space.

The simplest method would be to present the conjugacy function as a fixed point of a contracting map. But no such proof is known at the moment, even in the simplest case, where the rotation number belongs to $\mathcal{C}(0)$.

In 1954, Kolmogorov proposed (Kolmogorov 1954) to replace in small divisors problems the Picard's iteration algorithm by Newton's algorithm. This was successfully carried out by Arnold in the analytic case and Moser in the smooth case (Arnold 1965; Arnold 1963a; Arnold 1963b; Moser 1966; Moser 1962; Moser 1969).

In the smooth case, one modifies slightly Newton's algorithm introducing smoothing operators to take care of the loss of differentiability at each step in the linearized equation. In the analytic case, the loss is controlled by changing the (complex) domains of definition at each step of the algorithm.

Later on, these methods gave rise to many versions of implicit function theorems in appropriate functional spaces (Moser 1961; Sergeraert 1972; Hörmander 1976; Lojasiewicz and Zehnder 1979; Zehnder 1976; Hamilton 1974 and 1982). Amongst those, the simplest and easiest to handle (but, as a consequence, not giving the more precise results) is the one introduced by Hamilton (Hamilton 1974 and 1982; Bost 1985; Herman 1980). Herman has given from Hamilton's implicit function theorem proofs of many small divisors theorems in the C^∞-case, including the local conjugacy theorem above and the KAM-theorems (Herman 1980; Bost 1985).

ii) Functional analysis: Schauder-Tichonoff's fixed point theorem

Herman has given (Herman 1983, 1985 and 1986) short and simple proofs of the local conjugacy theorem and the invariant curve theorem. They are based on Schauder-Tichonoff's fixed point theorem: any continuous map from a convex compact set in a locally convex topological vector space into itself has a fixed point. These proofs involve subtle manipulations of the functional equations and a fine study of the basic difference equation of Section 2.3.1.

iii) Global conjugacy

To prove his global conjugacy theorem, Herman had to develop non perturbative methods. In the smooth case, his proof starts from the following observation: a diffeomorphism f is C^r-conjugated to a rotation if and only if the iterates $f^n, n \geq 0$, form a bounded family in the C^r-topology. A variant of this is used by Katznelson and Ornstein (Katznelson and Ornstein 1989a and 1989b). The derivatives of the iterates are given by explicit formulas of the Faa-di-Bruno type. There is then a very delicate interplay between the geometry of the orbits of f and the size of the derivatives of the iterates. The crucial iterates in this process are the f^{q_n}, where the p_n/q_n are the convergents of the rotation number.

In the analytic case (Yoccoz 1987 and 1989) a different way is used to control the f^{q_n}, mimicking at the diffeomorphism level the $GL(2, \mathbb{Z})$ action on rotation number. The method also allows to construct the counterexamples. In the smooth case, counterexamples are constructed by Herman on the basis of a Baire category argument.

3. Invariant Tori

3.1 Diophantine Conditions

One of the most serious problems with small divisors involving more than two independent frequencies is the lack of a perfect analogue of the continued fraction algorithm. Nevertheless, one can still define Diophantine conditions as in Section 2.2.3.

For $\alpha = (\alpha_1, \alpha_2, \dots, \alpha_n)$, $\beta = (\beta_1, \beta_2, \dots, \beta_n)$ in $\mathbb{R}^n$, we let

$$|\alpha| = \sum_{i=1}^{n} |\alpha_i| \, , \quad < \alpha, \beta > = \sum_{i=1}^{n} \alpha_i \beta_i \, .$$

3.1.1 Discrete Time (Diffeomorphisms). Let $n \geq 1$. For $\alpha = (\alpha_1, \alpha_2, \dots, \alpha_n)$ in $\mathbb{R}^n$ and $q \geq 1$, define

$$\sigma_q(\alpha) = \inf_{k \in \mathbb{Z}^n, \; 1 \leq |k| \leq q} \| < k, \alpha > \| \, .$$

One has $\sigma_q(\alpha) \neq 0$ for all $q \geq 1$ if and only if $1, \alpha_1, \alpha_2, \dots, \alpha_n$ are linearly independent over $\mathbb{Q}$. For $\theta \geq 0$, $\rho > 0$, define:

$$\mathcal{C}(\theta, \rho) = \{\alpha \in \mathbb{R}^n, \; \forall q \geq 1 \, , \; \sigma_q(\alpha) \geq \rho q^{-n-\theta}\}$$

$$\mathcal{C}(\theta) = \bigcup_{\rho > 0} \mathcal{C}(\theta, \rho)$$

$$\mathcal{C} = \bigcup_{\theta \geq 0} \mathcal{C}(\theta) \, .$$

As for $n = 1$, one has $\mathcal{C}(\theta, \rho) \subset \mathcal{C}(\theta', \rho')$ for $\theta \leq \theta'$, $\rho \geq \rho'$ and $\mathcal{C}(\theta, \rho)$ is a closed totally disconnected subset of $\mathbb{R}^n$. Consequently, $\mathcal{C}(\theta)$ and $\mathcal{C}$ are small in the topological sense (Baire category). But $\mathcal{C}(\theta)$ has full Lebesgue measure for $\theta > 0$ ($\mathcal{C}(0)$ has Lebesgue measure 0).

3.1.2 Continuous Time (Flows). Let $n \geq 2$. For $\alpha = (\alpha_1, \alpha_2, \dots, \alpha_n)$ in $\mathbb{R}^n$ and $q \geq 1$, define

$$\sigma_q^*(\alpha) = \inf_{k \in \mathbb{Z}^n, \; 1 \leq |k| \leq q} | < k, \alpha > | \, .$$

One has $\sigma_q^*(\alpha) \neq 0$ for all $q \geq 1$ if and only if $\alpha_1, \alpha_2, \dots, \alpha_n$ are linearly independent over $\mathbb{Q}$. For $\theta \geq 0$, $\rho > 0$, define:

$$\mathcal{C}^*(\theta, \rho) = \{\alpha \in \mathbb{R}^n, \; \forall q \geq 1 \, , \; \sigma_q^*(\alpha) \geq \rho q^{1-n-\theta}\}$$

$$\mathcal{C}^*(\theta) = \bigcup_{\rho > 0} \mathcal{C}^*(\theta, \rho)$$

$$\mathcal{C}^* = \bigcup_{\theta \geq 0} \mathcal{C}(\theta) \, .$$

The set $\mathcal{C}^*(\theta)$ is the homogeneous analogue of the set $\mathcal{C}(\theta)$. Indeed one has:

$$(\alpha_1, \alpha_2, \ldots, \alpha_n) \in \mathcal{C}^*(\theta) \iff (\frac{\alpha_1}{\alpha_n}, \frac{\alpha_2}{\alpha_n}, \ldots, \frac{\alpha_{n-1}}{\alpha_n}) \in \mathcal{C}(\theta) .$$

One also has, for $t > 0$:

$$t\mathcal{C}^*(\theta, \rho) = \mathcal{C}^*(\theta, t\rho) .$$

The intersection of $\mathcal{C}^*(\theta, \rho)$ with the sphere $< \alpha, \alpha >= 1$ is a compact totally disconnected set. For $\theta > 0$, $\mathcal{C}^*(\theta)$ has full Lebesgue measure, but $\mathcal{C}^*(0)$ has Lebesgue measure 0.

3.2 Quasiperiodic Motions on Tori

3.2.1 Discrete Time (Diffeomorphisms). Let $n \geq 1$ and $\mathbb{T}^n = \mathbb{R}^n/\mathbb{Z}^n$ be the n-dimensional torus. Let $\mathrm{Diff}_0^r(\mathbb{T}^n)$ be the group of C^r-diffeomorphisms of $\mathbb{T}^n$ homotopic to the identity; the lifts to $\mathbb{R}^n$ of such diffeomorphisms form the group $\mathrm{D}^r(\mathbb{T}^n)$ of diffeomorphisms f of $\mathbb{R}^n$ such that $f - \mathrm{Id}$ is $\mathbb{Z}^n$-periodic.

For $\alpha \in \mathbb{R}^n$, (resp. $\alpha \in \mathbb{T}^n$), let R_α be the translation $x \mapsto x + \alpha$. For $n \geq 2$ there is unfortunately no rotation number for diffeomorphisms in $\mathrm{D}^r(\mathbb{T}^n)$: it is easy for instance to construct f in $\mathrm{D}^\omega(\mathbb{T}^2)$ such that $f(0,0) = (0,0)$ and $f(0,1/2) = (1,1/2)$. The best one can do is to define a rotation set: for $f \in \mathrm{D}^0(\mathbb{T}^n)$ let $\bar{f}$ be the induced homeomorphism of $\mathbb{T}^n$ and $\mathcal{M}(\bar{f})$ be the set of probability measures on $\mathbb{T}^n$ invariant under $\bar{f}$. Define:

$$R(f) = \left\{ \int_{\mathbb{T}^n} (f(x) - x) \, d\mu(x) \ , \ \mu \in \mathcal{M}(\bar{f}) \right\} .$$

This is a compact convex non empty subset of $\mathbb{R}^n$, and one can only speak of a rotation number when this set is reduced to one single point, for instance if $\bar{f}$ has only one invariant probability measure on $\mathbb{T}^n$.

It may be therefore rather surprising that a local smooth conjugacy theorem analogous to the case $n = 1$ still holds. The following theorem is due to Arnold and Moser (Arnold 1965; Moser 1966; Herman 1979).

(1) Theorem. *Let $\theta \geq 0$, $\alpha \in \mathcal{C}(\theta)$, $r_0 > n + 1 + \theta$. There exists a neighborhood U of R_α in $\mathrm{D}^{r_0}(\mathbb{T}^n)$ such that, if $r \geq r_0$ satisfies $r \notin \mathbb{N}$, $r - \theta \notin \mathbb{N}$ and $f \in \mathrm{D}^r(\mathbb{T}^n) \cap U$, then there exists a unique $\lambda \in \mathbb{R}^n$, near 0, and a unique $h \in \mathrm{D}^{r-n-\theta}(\mathbb{T}^n)$ near the identity such that*

$$f = R_\lambda \circ h \circ R_\alpha \circ h^{-1} \quad , \quad h(0) = 0 .$$

Moreover, the map $f \mapsto (\lambda, h)$ defined on $U \cap \mathrm{D}^\omega(\mathbb{T}^n)$ with values in $\mathbb{R}^n \times \mathrm{D}^\infty(\mathbb{T}^n, 0)$ (where $\mathrm{D}^\infty(\mathbb{T}^n, 0)$ is the subgroup of $\mathrm{D}^\infty(\mathbb{T}^n)$ formed by the diffeomorphism fixing 0) is a local C^∞-tame diffeomorphisms in the sense of Hamilton.

The concept of smooth tame maps, introduced by Hamilton, seems to be the natural one when one deals with Fréchet spaces such as spaces of C^∞-maps. For a definition see Hamilton and Bost (Hamilton 1974 and 1982; Bost 1985)

The main point of the theorem is the following: for diffeomorphisms f near R_α, to exhibit a quasiperiodic motion with frequency α is equivalent to have $\lambda = 0$ in the theorem, and this happens exactly (due to the last assertion in the theorem) on a submanifold of $\mathrm{D}^\infty(\mathbb{T}^n)$ of codimension n.

Another important fact is that the 'size' of the neighbourhood U of R_α in $\mathrm{D}^{r_0}(\mathbb{T}^n)$ with $\alpha \in \mathcal{C}(\theta, \rho)$ depends only on θ, ρ, and r_0.

When $\alpha \notin \mathcal{C}$, the set of diffeomorphisms which exhibit a quasiperiodic motion of frequency α, is very poorly understood, even in the neighbourhood of R_α.

3.2.2 Continuous Time (Flows). Let $\mathcal{X}^r(\mathbb{T}^n)$ be the vector space of C^r-vector fields on $\mathbb{T}^n$. For $\alpha \in \mathbb{R}^n$, we denote by X_α the constant vector field $\sum_{i=1}^n \alpha_i \frac{\partial}{\partial \theta_i}$, where $\theta_1, \theta_2, \ldots, \theta_n$ are the coordinates on $\mathbb{T}^n$.

For $h \in \mathrm{Diff}^1(\mathbb{T}^n)$ and $X \in \mathcal{X}^0(\mathbb{T}^n)$, let h_*X be the image under X under h:

$$h_*X(\theta) = \mathrm{D}h(h^{-1}(\theta))X(h^{-1}(\theta)) .$$

The local conjugacy for flows is then the following theorem (Arnold 1965; Moser 1966; Herman 1979).

(2) Theorem. *Let $\theta \geq 0$, $\alpha \in \mathcal{C}^*(\theta)$, $r_0 > n + \theta$. There exists a neighbourhood U of X_α in $\mathcal{X}^{r_0}(\mathbb{T}^n)$ such that if $r \geq r_0$ satisfies $r \notin \mathbb{N}$, $r - \theta \notin \mathbb{N}$, then for any $X \in \mathcal{X}^r(\mathbb{T}^n) \cap U$ there exists a unique $\lambda \in \mathbb{R}^n$ near 0 and a unique $h \in \mathrm{Diff}_0^{r+1-n-\theta}(\mathbb{T}^n, 0)$ near the identity such that:*

$$X = X_\lambda + h_*X_\alpha .$$

Moreover the map $X \mapsto (\lambda, h)$ defined on $\mathcal{X}^\infty(\mathbb{T}^n) \cap U$ with values in $\mathbb{R}^n \times \mathrm{Diff}_0^\infty(\mathbb{T}^n, 0)$ is a local C^∞-tame diffeomorphism.

3.3 Hamiltonian Dynamics

3.3.1 Lagrangian Tori. Let M be a smooth manifold of even dimension $2n$, and let ω be a symplectic form on M, i.e. a closed non-degenerate 2-form.

The fundamental example in our context will be $M = \mathbb{T}^n \times \mathbb{R}^n$ (or $M = \mathbb{T}^n \times V$, V open in $\mathbb{R}^n$), with $\omega = \sum_{i=1}^n dr_i \wedge d\theta_i$, where $r_1, r_2, \ldots, r_n$ are the coordinates in $\mathbb{R}^n$ and $\theta_1, \theta_2, \ldots, \theta_n$ are the coordinates in $\mathbb{T}^n$. One has then $\omega = d\lambda$, $\lambda = \sum_{i=1}^n r_i d\theta_i$.

A submanifold $\mathbb{T}$ of M is *Lagrangian* if its dimension is n and one has for all $x \in \mathbb{T}$:

$$\omega_x{}_{/\mathrm{T}_x\mathbb{T}} \equiv 0 .$$

We will be only interested in the case where $\mathbb{T}$ is diffeomorphic to the torus $\mathbb{T}^n$; we then speak of a Lagrangian torus.

The content of the following proposition is that in a neighbourhood of a Lagrangian torus, one can forget the ambient manifold M and work in $\mathbb{T}^n \times \mathbb{R}^n$.

(1) Proposition. *Let $\mathbb{T}$ be a C^∞ Lagrangian torus in M. There exists an open set V in $\mathbb{R}^n$ containing 0, an open set U in M containing $\mathbb{T}$, and a C^∞-diffeomorphism K from U onto $\mathbb{T}^n \times V$ such that*

$$\begin{aligned} K(\mathbb{T}) &= \mathbb{T}^n \times \{0\} . \\ K^*(\textstyle\sum dr_i \wedge d\theta_i) &= \omega . \end{aligned}$$

Moreover the restriction of K to $\mathbb{T}$ may be any given C^∞-diffeomorphism from $\mathbb{T}$ onto $\mathbb{T}^n \times \{0\}$.

3.3.2 Quasiperiodic Invariant Tori. Let H be a Hamiltonian on M, i.e. a C^∞-function on M. The associated Hamiltonian vector field X_H is defined by:

$$\forall x \in M,\ \forall v \in \mathrm{T}_x M \quad \omega(X_H(x), v) = \mathrm{D}_x H(v) .$$

An important property of Lagrangian submanifolds is the following proposition.

(2) Proposition. *Let $\mathbb{T}$ be a Lagrangian submanifold of M contained in an energy hypersurface of a Hamiltonian H. Then $\mathbb{T}$ is invariant under the flow of X_H.*

Indeed, for all $x \in \mathbb{T}$, $\mathrm{T}_x\mathbb{T}$ is its own ω_x-orthogonal and is contained in $\mathrm{Ker}(\mathrm{D}_x H)$, hence contains $X_H(x)$.

(3) Definition. *Let $H \in C^\infty(M)$ and $\alpha \in \mathbb{R}^n$. We say that a Lagrangian torus $\mathbb{T}$ is an α-quasiperiodic (Lagrangian) torus for the Hamiltonian vector field X_H if $\mathbb{T}$ is invariant under X_H and there exists a C^∞-diffeomorphism k from $\mathbb{T}$ onto $\mathbb{T}^n$ such that*

$$k_*\left(X_{H/\mathbb{T}}\right) = X_\alpha .$$

3.3.3 Complete Integrability, the Twist Condition. Let V be an open set in $\mathbb{R}^n$ and $H_0 \in C^\infty(V)$. Then H_0 may also be considered as a Hamiltonian on $\mathbb{T}^n \times V$, said to be *completely integrable*. The Hamiltonian vector field X_{H_0} is:

$$X_{H_0}(\theta, r) = \left(\frac{\partial H_0}{\partial r}(r), 0\right) .$$

Every (Lagrangian) torus $\{r = r_0\}$, $(r_0 \in V)$ is invariant under X_{H_0}, the restriction of X_{H_0} to it being the constant vector field $\sum_{i=1}^n \frac{\partial H_0}{\partial r_i}(r_0)\frac{\partial}{\partial \theta_i}$.

Defining $\alpha_i(r) = \frac{\partial H_0}{\partial r_i}(r)$, and $\alpha(r) = (\alpha_1(r), \alpha_2(r), \ldots, \alpha_n(r))$, the torus $\{r = r_0\}$ is an $\alpha(r_0)$-quasiperiodic torus for X_{H_0}. We say that H_0 satisfies the *twist condition* along $\{r = r_0\}$ if:

$$\det\left(\frac{\partial^2 H_0}{\partial r_i \partial r_j}(r_0)\right) \neq 0 .$$

In other terms, the map $r \mapsto \alpha(r)$ has to be a local diffeomorphism near $r = r_0$.

3.3.4 Normal Forms. Let H be a Hamiltonian on M, let $\alpha \in \mathbb{R}^n$, and let $\mathbb{T}$ be an α-quasiperiodic Lagrangian torus for X_H. By Proposition 1 in Section 3.3.1, there exists an open set V in $\mathbb{R}^n$, a neighbourhood U of $\mathbb{T}$ in M and a C^∞-diffeomorphism K from U onto $\mathbb{T}^n \times V$ such that:

$$\begin{aligned} K(\mathbb{T}) &= \mathbb{T}^n \times \{0\} , \\ K^*(\textstyle\sum dr_i \wedge d\theta_i) &= \omega , \\ (K_* \cdot X_H)/\mathbb{T}^n \times \{0\} &= (X_\alpha, 0) . \end{aligned}$$

(3) Proposition. *Assume moreover that $\alpha \in C^*$, and let $N \geq 1$. Then one can choose V, U, K as above and such that moreover one has:*

$$H \circ K^{-1}(\theta, r) = H_0(r) + \phi(\theta, r) ,$$

with $\phi(\theta, r) = o(||r||^N)$.

One says that the completely integrable Hamiltonian H_0 in $\mathbb{T}^n \times V$ is a normal form (of order N) for H in the neighbourhood of $\mathbb{T}$. Thus, near a quasiperiodic Lagrangian torus whose frequency belongs to C^*, a Hamiltonian may be approximated at any order by a completely integrable one.

In particular, taking $N = 2$, we say that H satisfies the *twist condition* along $\mathbb{T}$ if one has for the normal form H_0:

$$\det\left(\frac{\partial^2 H_0}{\partial r_i \partial r_j}(0)\right) \neq 0 .$$

Actually, even when the frequency α does not belong to C^*, it is still possible to define the twist condition along $\mathbb{T}$: taking U, V, K as in the beginning of this Section, write:

$$(K_* X_H)(\theta, r) = (\alpha + A(\theta) r, 0) + O(||r||^2)$$

where $A(\theta)$ is a symmetric $n \times n$ matrix depending on $\theta \in \mathbb{T}^n$. The *twist condition* for H along $\mathbb{T}$ is then:

$$\det\left(\int_{\mathbb{T}^n} A(\theta)\, dm(\theta)\right) \neq 0 .$$

3.4 KAM-Theorems

For the theorems of Kolmogorov Arnold and Moser (KAM-theorems) which are described below in an informal way, we refer to the following papers (Arnold 1963a and 1963b; Pöschel 1980 and 1982; Moser 1969; Herman 1980; Rüssmann 1970, 1972 and 1983).

3.4.1 Fixed Frequency. Let H be a Hamiltonian on a symplectic manifold (M, ω) of dimension $2n$. Let $\theta > 0$, $\rho > 0$, $\alpha \in \mathcal{C}^*(\theta, \rho)$. Assume that the vector field X_H exhibits an α-quasiperiodic Lagrangian torus $\mathbb{T}$, and that the twist condition is satisfied by H along $\mathbb{T}$.

Then there exists a neighbourhood W of α in $\mathbb{R}^n$, a neighbourhood U of $\mathbb{T}$ in M and a neighbourhood $\mathcal{U}$ of H in $\mathrm{C}^\infty(M)$, such that for any perturbation $\widetilde{H} \in \mathcal{U}$, and any frequency $\beta \in W \cap \mathcal{C}^*(\theta, \rho)$, the Hamiltonian vector field $X_{\widetilde{H}}$ has in U a β-quasiperiodic C^∞ Lagrangian torus $\mathrm{T}_\beta(\widetilde{H})$.

Moreover, for any given $\beta \in W \cap \mathcal{C}^*(\theta, \rho)$ the map $\widetilde{H} \mapsto \mathrm{T}_\beta(\widetilde{H})$ is smooth in $\mathcal{U}$ (actually, the map $(\beta, \widetilde{H}) \mapsto \mathrm{T}_\beta(\widetilde{H})$ is smooth in the sense of Whitney in $(W \cap \mathcal{C}^*(\theta, \rho)) \times U$.

Finally, for any $\widetilde{H} \in \mathcal{U}$, the union :

$$\bigcup_{\beta \in W \cap \mathcal{C}^*(\theta,\rho)} \mathrm{T}_\beta(\widetilde{H})$$

has *positive Lebesgue measure*, and even arbitrarily large relative Lebesgue measure in U (if one takes $\mathcal{U}$, U, W small enough).

3.4.2 Persistence of Tori. We want to give a very rough idea of the cause of the persistence of α-quasiperiodic Lagrangian tori ($\alpha \in \mathcal{C}^*$) under perturbations.

We only discuss the infinitesimal level, neglecting terms of order ≥ 2 in the perturbation.

Assume that a Hamiltonian H has an α-quasiperiodic Lagrangian torus $\mathbb{T}$, with $\alpha \in \mathcal{C}^*$, and that H satisfies the twist condition along $\mathbb{T}$. Then there are, in a neighbourhood of $\mathbb{T}$, symplectic coordinates $(\theta, r) \in \mathbb{T}^n \times V$ (V open in $\mathbb{R}^n$, $0 \in V$) such that the equation of $\mathbb{T}$ is $\{r = 0\}$ and that H may be written:

$$H(\theta, r) \;=\; \mu_0 + <\alpha, r> + <Ar, r> + \phi(\theta, r) \;,$$

with $\mu_0 \in \mathbb{R}$, A a symmetric *invertible* matrix, and $\phi(\theta, r) = O(||r||^3)$.

Let $\widetilde{H} = H + \Delta H$ a small perturbation of H, $\widetilde{\mathbb{T}}$ a small Lagrangian C^∞-perturbation of $\mathbb{T}$. Then there us a unique $\Delta\lambda \in \mathbb{R}^n$ and a unique $\Delta\psi \in \mathrm{C}_0^\infty(\mathbb{T}^n)$ such that:

$$\widetilde{\mathbb{T}} \;=\; \{(\theta, \Delta\lambda + d\Delta\psi(\theta)) \;, \theta \in \mathbb{T}^n\}$$

The condition on first-order terms for $\widetilde{\mathbb{T}}$ to be invariant under $X_{\widetilde{H}}$ (i. e. for $\widetilde{H}$ to be constant on $\widetilde{\mathbb{T}}$) is that there exists $\Delta\mu \in \mathbb{R}$ such that:

$$\Delta\mu \;= <\alpha, d\Delta\psi(\theta) + \Delta\lambda> + \Delta H(\theta, 0) \;. \tag{1}$$

Keeping only 0-th and first-order terms, the θ-component of $X_{\widetilde{H}}$ along $\widetilde{\mathbb{T}}$ is:

$$X_\alpha + A \cdot (\Delta\lambda + d\Delta\psi(\theta)) + \frac{\partial}{\partial r}\Delta H(\theta, 0)$$

and for this to be of the form $h_* X_\alpha$, $h = \mathrm{Id} + \Delta h$, $\Delta h \in (C_0^\infty(\mathbb{T}^n))^n$ one must have:

$$X_\alpha \cdot \Delta h \;=\; A \cdot (\Delta\lambda + d\Delta\psi(\theta)) + \frac{\partial}{\partial r}\Delta H(\theta, 0) \; . \tag{2}$$

Now, under the hypothesis $\alpha \in C^*$, the operator $L_\alpha \; : \; C_0^\infty(\mathbb{T}^n) \to C_0^\infty(\mathbb{T}^n)$

$$L_\alpha(\psi) \;=\; X_\alpha \cdot \psi$$

is invertible; this is easily proved considering Fourier coefficients. Hence (1) is solved by:

$$\begin{cases} \Delta\mu &= <\alpha, \Delta\lambda> + \int_{\mathbb{T}^n} \Delta H(\theta, 0) \; d\theta \\ \Delta\psi &= L_\alpha^{-1}(-\Delta H(\theta, 0) + \int_{\mathbb{T}^n} \Delta H(\theta, 0) \; d\theta) \; . \end{cases}$$

Then putting:

$$S(\theta) \;=\; \frac{\partial}{\partial r}\Delta H(\theta, 0) + A \cdot d\Delta\psi(\theta) \; ,$$

equation (2) is solved by:

$$\begin{cases} \Delta\lambda &= -A^{-1} \int_{\mathbb{T}^n} S(\theta) \; d\theta \\ \Delta h &= L_\alpha^{-1}(S + A \cdot \lambda) \; . \end{cases}$$

This shows that, at the infinitesimal level, $\widetilde{\mathbb{T}}$ and the normalized conjugacy h are uniquely determined by the perturbation ΔH; this gives at least some plausibility argument to the statements in Section 3.4.1.

3.4.3 Fixed Energy. The tori in the KAM-theorem of Section 3.4.1 are not necessarily in the same energy surface than was the initial torus $\mathbb{T}$. There is a modified version of the KAM-theorem if one looks for quasiperiodic Lagrangian tori in a fixed energy hypersurface , which requires two modifications.

i) The twist condition is modified as follows. Let H_0 be a normal form of order ≥ 2 for H near $\mathbb{T}$ (cf. Section 3.3.4) and $\alpha \; : \; V \to \mathbb{R}^n$ the frequency map of H_0 (cf. Section 3.3.3). We still assume that the frequency $\alpha(0)$ of $\mathbb{T}$ belongs to C^*. But instead of assuming that the derivative $\mathrm{D}\alpha(0)$ is a linear isomorphism of $\mathbb{R}^n$, we now assume that the restriction of $\mathrm{D}\alpha(0)$ to the hyperplane tangent to $\{H_0 = \text{constant}\}$ is a linear isomorphism onto a hyperplane of $\mathbb{R}^n$ not containing $\alpha(0)$. This is the modified twist condition for fixed energy.

ii) The conclusion of the theorem is modified as follows. Let $\widetilde{H} \in \mathcal{U}$ a perturbation of H and $\beta \in W \cap C^*(\theta, \rho)$ a frequency. Instead of a β-quasiperiodic torus in an unspecified energy surface, we obtain, in any given energy surface near the original one, a $t\beta$-quasiperiodic torus for some real number t near 1. The other conclusion of the theorem, mutatis mutandis, are the same.

3.4.4 Discrete Times. Recall that a diffeomorphism f of a symplectic manifold (M, ω) is *symplectic* if $f^*\omega = \omega$.

When ω is given as the differential of a 1-form λ (for instance: $M = \mathbb{T}^n \times V$, $\lambda = \sum_{i=1}^n r_i d\theta_i$), this means that the 1-form $f^*\lambda - \lambda$ has to be closed. We say that f is *exact symplectic* when this 1-form is exact.

Let f be a symplectic diffeomorphism of (M, ω). Assume that f leaves invariant some Lagrangian torus $\mathbb{T}$. Let $\alpha \in \mathbb{T}^n$. We say that $\mathbb{T}$ is an α-quasiperiodic torus for f if there is a diffeomorphism k from $\mathbb{T}$ onto $\mathbb{T}^n$ such that $k \circ f_{/\mathbb{T}} \circ k^{-1}$ is the translation R_α of $\mathbb{T}^n$.

Let $\mathbb{T}$ be an α-quasiperiodic torus for a symplectic diffeomorphism f. If we are only interested in the dynamics of f (and its perturbations) in a neighbourhood U of $\mathbb{T}$ we may assume that:

$$\begin{array}{rclcrcl} U & = & \mathbb{T}^n \times V & , & \omega & = & \sum dr_i \wedge d\theta_i \\ \mathbb{T} & = & \mathbb{T}^n \times \{0\} & , & f(\theta, 0) & = & (\theta + \alpha, 0) \, . \end{array}$$

The last condition implies that f is exact symplectic. Write the coordinates of f as $f(\theta, r) = (f_1(\theta, r), f_2(\theta, r))$, and let $A(\theta)$ be the symmetric matrix $\frac{\partial}{\partial r} f_1(\theta, 0)$. Then the twist condition for f along $\mathbb{T}$ is:

$$\det\left(\int_{\mathbb{T}^n} A(\theta)\, d\theta\right) \neq 0 \, .$$

The KAM-theorem for symplectic diffeomorphisms is now as follows. Let $\theta > 0$, $\rho > 0$, $\alpha \in \mathcal{C}(\theta, \rho)$. Let f be a symplectic C^∞-diffeomorphism of a symplectic manifold (M, ω). Let $\mathbb{T}$ be an α-quasiperiodic invariant Lagrangian torus for f. Assume that the *twist condition* is satisfied by f along $\mathbb{T}$.

Then there exists a neighbourhood W of $\alpha \in \mathbb{T}^n$, a neighbourhood U of $\mathbb{T}$ in M and a neighbourhood $\mathcal{U}$ of f in $\mathrm{Diff}^\infty(M)$ such that, for any symplectic perturbation $g \in \mathcal{U}$ exact near $\mathbb{T}$ and for any frequency $\beta \in W \cap \mathcal{C}(\theta, \rho)$, there exists in U a β-quasiperiodic Lagrangian torus $T_\beta(g)$ for g (exact near $\mathbb{T}$ means that g is exact in the coordinates $(\theta, r) \in \mathbb{T}^n \times V$).

The other conclusions of Section 3.4.1 on smoothness and measure are still true in this case.

As a final remark, we note that the three versions of the KAM-theorems presented above (fixed frequency, fixed energy, discrete time) may be proven along the same lines. But actually, it has been shown (R. Douady 1982) that any one of the three versions implies the other two.

3.4.5 Final Comment. The quasiperiodic invariant tori provided by the KAM-theorem are an essential step in our knowledge of Hamiltonian dynamics. But, although they describe the dynamics on a set of positive measure, this set is in general not open and one would like to know what happens 'between'the KAM-tori. This an extremely rich and very active field of study (Chenciner 1985; Herman 1989), with many important recent developments. But this would take us too far.

References

Arnold V. I. (1963a) Small denominators II: Proof of a theorem of A. N. Kolmogorov on the invariance of quasiperiodic motions under small perturbations of the Hamiltonian, Russian Math. Surveys **18**-5 (1963) 9-36

Arnold V. I. (1963b) Small denominators III: Small denominators and problems of stability of motion in classical and celestial mechanics, Russian Math. Surveys **18**-6 (1963) 85-193

Arnold V. I. (1965) Small denominators I: On the mapping of a circle onto itself, Trans. Amer. Math. Soc. Ser. 2, **46** (1965) 213-284

Bost J. B. (1985) Tores invariants des systèmes dynamiques hamiltoniens, Séminaire Bourbaki n^r. 639, Astérisque **133-134** (1985) 113-157

Brjuno A. D. (1971) Analytic form of differential equations, Trans. Moscow. Math. Soc. **25** (1971) 131-288

Brjuno A. D. (1972) Analytic form of differential equations, Trans. Moscow. Math. Soc. **26** (1972) 199-239

Cassels J.W. (1957) An introduction to Diophantine approximation, Cambridge tracts **45**, Cambridge University Press (1957)

Chenciner A. (1985) La dynamique au voisinage d'un point fixe elliptique conservatif: de Poincaré et Birkhoff à Aubry et Mather, Séminaire Bourbaki 622, Astérisque **121-122** (1985) 147-170

Denjoy A. (1932) Sur les courbes définies par les équations différentielles à la surface du tore, J. Math. Pures et Appl. **9** (1932) 333-375

Denjoy A. (1946) Les trajectoires à la surface du tore, C. R. Acad. Sci. Paris **223** (1946) 5-7

Douady R. (1982) Une démonstration directe de l'équivalence des théorèmes de tores invariants pour difféomorphismes et champs de vecteurs, C. R. Acad. Sci. Paris, **295** (1982) 201-204

Hamilton R. S. (1974) The inverse function theorem of Nash and Moser, preprint, Cornell University (1974)

Hamilton R. S. (1982) The inverse function theorem of Nash and Moser, Bull. Amer. Math. Soc. **7** (1982) 65-222

Herman M. R. (1979) Sur la conjugaison différentiable des difféomorphismes du cercle à des rotations, Publ. Math. I.H.E.S. **49** (1979) 5-234

Herman M. R. (1980) Démonstration du théorème des courbes translatées par les difféomorphismes de l'anneau; démonstration du théorème des tores invariants. Manuscript (1980)

Herman M. R. (1983) Sur les courbes invariantes par les difféomorphismes de l'anneau, vol. I, Astérisque **103-104** (1983)

Herman M. R. (1985) Simple proofs of local conjugacy theorems for diffeomorphisms of the circle with almost every rotation number, Bol. Soc. Bras. Mat. **16** (1985) 45-83

Herman M. R. (1986) Sur les courbes invariantes par les difféomorphismes de l'anneau, vol. II, Astérisque **144** (1986)

Herman M. R. (1989) Inégalités a priori pour des tores lagrangiens invariants par des difféomorphismes symplectiques, preprint (1989), Centre de Mathématiques de l'Ecole Polytechnique, Palaiseau, France

Hörmander L. (1976) The boundary problem of physical geodesy, Arch. Rat. Mech. Anal. **62** (1976) 1-52

Katznelson Y. and Ornstein D. (1989a) The differentiability of the conjugation of certain diffeomorphisms of the circle, Ergod. Th. and Dynam. Sys. **9** (1989) 643-680

Katznelson Y. and Ornstein D. (1989b) The absolute continuity of the conjugation of certain diffeomorphisms of the circle, Ergod. Th. and Dynam. Sys. **9** (1989) 681-690

Khanin K. M. and Sinaï Ya. (1987) A new proof of M. Herman's theorem, Commun. Math. Phys. **112** (1987) 89-101

Kolmogorov A. N. (1954) Théorie générale des systèmes dynamiques et mécanique classique, Proc. of the 1954 International Congress, North Holland, Amsterdam, 315-333

Lang S. (1966) Introduction to Diophantine approximation, Addison-Wesley, New York (1966)

Lojasiewicz S. and Zehnder E. (1979) An inverse function theorem in Fréchet spaces, Jour. of Funct. Anal. **33** (1979) 165-174

Moser J. (1961) A new technique for the construction of solutions of non-linear differential equations, Proc. Nat. Acad. Sci. U. S. A. **47** (1961) 1824-1831

Moser J. (1962) On invariant curves of area-preserving mappings of the annulus, Nachr. Akad. Wiss. Göttingen, Math. Phys. Kl. (1962) 1-20

Moser J. (1966) A rapidly convergent iteration method, part II, Ann. Scuola Norm. Sup. di Pisa, Ser. III **20** (1966) 499-535

Moser J. (1969) On the construction of almost periodic solutions for ordinary differential equations, Proc. Intern. Conf. on Funct. Anal. and Related Topics, Tokyo (1969) 60-67

Pöschel J. (1980) Über invariante Tori in differenzierbaren Hamiltonschen Systemen, Bonner Math. Schriften, **120** (1980)

Pöschel J. (1982) Integrability of Hamiltonian systems on Cantor sets, Comm. Pur. and Appl. Math. **35** (1982) 653-695

Rüssmann H. (1970) Kleine Nenner I: Über invariante Kurven differenzierbaren Abbildungen eines Kreisringes, Nachr. Akad. Wiss. Göttingen, Math. Phys. Kl. (1970) 67-105

Rüssmann H. (1972) Kleine Nenner II: Bemerkungen zur Newtonschen Methode, Nachr. Akad. Wiss. Göttingen, Math. Phys. Kl. (1972) 1-10

Rüssmann H. (1983) On the existence of invariant curves of twist mappings of an annulus, Lect. Notes in Math. **1007** (Springer 1983) 677-712

Schmidt W. M. (1980) Diophantine approximation, Springer Lecture Notes in Mathematics **785** (1980)

Sergeraert F. (1972) Un théorème de fonctions implicites sur certains espaces de Fréchet et quelques applications, Ann. Sc. de l'E. N. S., 4e série, **5** (1972) 599-660

Siegel C. L. (1942) Iteration of analytic functions, Ann. of Math. **43** (1942) 607-612

Yoccoz J. C. (1984) Conjugaison différentiable des difféomorphismes du cercle dont le nombre de rotation vérifie une condition diophantienne, Ann. Sc. de l'E. N. S., 4e série, **17** (1984) 333-359

Yoccoz J. C. (1987) Théorème de Siegel, polynômes quadratiques et nombres de Brjuno, preprint (1987)

Yoccoz J. C. (1989) Linéarisation analytique des difféomorphismes du cercle (1989, manuscript)

Zehnder E. (1976) Moser's implicit function theorem in the framework of analytic smoothing, Math. Annalen **219** (1976) 105-121

Index

Entries in bold type refer to chapter number

Printed by Books on Demand, Germany